Physics For Engineers

PHYSICS FOR ENGINEERS

SECOND EDITION

M.R. SRINIVASAN
Department of Physics
M.S. Ramaiah Institute of Technology
Bangalore, India

ANSHAN LTD
11a, Little Mount Sion
Tunbridge Wells, Kent
TN1 1YS

Co-Published in the U.K. by

ANSHAN LTD
11a, Little Mount Sion
Tunbridge Wells
Kent TN1 1YS

Tel.: +44 (0) 1892 557767
Fax: +44 (0) 1892 530358
E-mail: info@anshan.co.uk
Web site: www.anshan.co.uk

ISBN: 978 1848290 50 1

© 2011 by New Age International Publishers

All rights reserved. No part of this book may be reproduced in any form, by photostat, microfilm, xerography, or any other means, or incorporated into any information retrieval system, electronic or mechanical, without the written permission of the copyright owner.

British Library Cataloguing in Publication Data
A Catalogue record for this book is available from the British Library

Preface

The present day technological developments have been the result of joint efforts of physicists and engineers. Therefore, a proper and systematic study of physics is indispensable for engineering students to excel in their field. The present text is planned with the aim of presenting the principles of physics to engineering students.

Physics is a vast subject and the engineering applications are too numerous to be covered in a single book. To keep the size of the book within manageable proportion, topics have been chosen from acoustics, optics, modern physics and quantum mechanics. An introductory chapter on nanoscience and nanotechnology is also included to make the students aware of the present trends in technology. All the chapters in the book are complete by themselves. Experimental techniques and applications have been highlighted to make the students appreciate the interplay of physics and engineering. They are brief and introductory in nature. Solved examples, questions and problems are provided at the end of each chapter.

Listed below are the topics covered in the book. Students are expected to have a basic knowledge of differential and integral calculus.

- Vibrations and Resonance
- Acoustics of Buildings
- Ultrasonics
- Interference
- Diffraction
- Polarization of Light and Photoelasticity
- Lasers
- Holography
- Fiber Optics
- Modern Physics
- X-rays
- Basic Quantum Mechanics
- Quantum Computation
- Basics of Nanoscience and Nanotechnology

My wife Subhashree has enhanced the pleasure of writing this book by preparing the drawings. This text makes no claim to any originality. I have merely attempted to collect the pertinent material and present it in a form suitable and useful to students and teachers.

M.R. Srinivasan

Contents

Preface   *v*

1   VIBRATIONS AND RESONANCE   1

1.1   Introduction 1
1.2   Simple Harmonic Motion 2
1.3   Free Vibrations 4
1.4   Damped Vibrations 6
1.5   Steady State Response of an Oscillator Under the Action of a Periodic Force 14
1.6   Vibration Isolation 22
1.7   LCR Circuit Analysis—Analytical Approach 25
References 36
Solved Examples 36
Questions 41
Problems 41

2   ACOUSTICS OF BUILDINGS   44

2.1   Introduction 44
2.2   Properties of Sound 44
2.3   Room Acoustics 53
2.4   Measurement of Absorption Coefficient 61
2.5   Sound Absorbers 65
2.6   Design Considerations for Good Acoustics 69
References 85
Solved Examples 85
Questions 86
Problems 87

3   ULTRASONICS   91

3.1   Introduction 91
3.2   Properties of Ultrasonic Waves 92
3.3   Generation and Detection of Ultrasonic Waves 107
3.4   Measurement of Ultrasonic Velocity and Attenuation 120
3.5   Underwater Sonar 124
3.6   Ultrasonics in Industry 125

	3.7	Ultrasound in Medicine ... 141
	3.8	Ultrasonics in Electronics ... 146
	3.9	Ultrasonics in Optics ... 148
	3.10	Ultrasonics in Materials Science ... 152
	References ... 153
	Questions ... 153

4 INTERFERENCE   155

	4.1	Introduction ... 155
	4.2	Superposition of Waves ... 155
	4.3	Young's Double Slit Experiment ... 158
	4.4	Coherence ... 160
	4.5	Types of Interference ... 160
	4.6	Fresnel's Biprism ... 161
	4.7	Interference in Thin Films ... 163
	4.8	Colours of Thin Films ... 167
	4.9	Newton's Rings ... 167
	4.10	Michelson's Interferometer ... 169
	References ... 175
	Solved Examples ... 175
	Questions ... 184
	Problems ... 184

5 DIFFRACTION   186

	5.1	Introduction ... 186
	5.2	Fraunhofer Diffraction at a Single Slit ... 187
	5.3	Fraunhofer Diffraction due to Two Parallel Slits ... 191
	5.4	Fraunhofer Diffraction due to n Parallel Slits ... 193
	5.5	Diffraction-a Qualitative Description ... 196
	5.6	Diffraction Grating ... 199
	5.7	Grating Spectrum ... 200
	5.8	Fraunhofer Diffraction at a Circular Aperture ... 202
	5.9	Rayleigh's Criterion for Resolving Power ... 203
	5.10	Resolving Power of a Microscope ... 203
	5.11	Electron Microscope ... 205
	References ... 211
	Solved Examples ... 211
	Questions ... 215
	Problems ... 215

6 POLARIZATION OF LIGHT AND PHOTOELASTICITY   216

	6.1	Introduction ... 216
	6.2	Representation of Polarized and Unpolarized Light ... 216
	6.3	Production of Polarized Light ... 217

6.4	Circular and Elliptic Polarization	223
6.5	Calculation of the Phase Difference When a Linearly Polarized Light Passes Through a Double Refracting Crystal, the Optic Axis Being Normal to the Direction of Propagation	224
6.6	Plane and Circular Polariscope	225
6.7	Photoelasticity	227
6.8	Definitions of Stress and Strain at a Point	227
6.9	Stress-Optic Relations–Two Dimensional Case	230
6.10	Isochromatics and Isoclinics	231
6.11	Mathematical Analysis of Isoclinics and Isochromatics	233
6.12	Stress-Optic Law for Three Dimensions	234
6.13	Moire Fringes	235
	References	238
	Solved Examples	238
	Questions	239
	Problems	239

7 LASER — 240

7.1	Introduction	240
7.2	Characteristics of the Laser Light	240
7.3	Basic Concepts of Laser	247
7.4	Einstein Coefficients	250
7.5	Laser Amplifier and Laser Oscillator	252
7.6	Solid State Lasers	258
7.7	Semiconductor Lasers	262
7.8	Gas Lasers	274
7.9	Chemical Lasers	283
7.10	Liquid or Dye Lasers	284
7.11	X-Ray Lasers	286
7.12	Free Electron Lasers	287
7.13	Applications of Lasers	288
7.14	Materials Processing with Laser Beams	291
7.15	Laser Applications in Medicine	295
7.16	Laser Optical Disc (Compact Disc)	298
7.17	Laser Bar Code Scanner	301
7.18	Laser Printer	302
7.19	Laser Systems for Atmospheric Pollutant Detection	302
7.20	Laser Fusion	303
7.21	Laser Isotope Separation	304
7.22	Optical Communication	306
7.23	Applications in Pure Science	306
	References	306
	Solved Examples	307
	Questions	308
	Problems	309

8 HOLOGRAPHY — 310

- 8.1 Introduction — 310
- 8.2 Principle — 310
- 8.3 Mathematical Theory of Hologram — 311
- 8.4 Classification of Holograms — 313
- 8.5 Zone Plate Model of Transmission Holograms — 321
- 8.6 Applications of Holography — 323
- *References* — 327
- *Questions* — 327

9 FIBER OPTICS — 328

- 9.1 Introduction — 328
- 9.2 Basic Principles — 329
- 9.3 Optical Fibers and Cables — 330
- 9.4 Light Propagation in Fibers — 336
- 9.5 Signal Distortion or Dispersion in Optical Fibers — 340
- 9.6 Attenuation — 347
- 9.7 Basic Principles of Communication — 352
- 9.8 Multiplexing — 362
- 9.9 Asynchronous and Synchronous Transmission — 364
- 9.10 Types of Communication Systems — 364
- 9.11 Switching — 368
- 9.12 Network Devices — 371
- 9.13 OSI Communication Model — 371
- 9.14 Fiber Optic Communication System — 373
- 9.15 Advantages of Optical Fibers — 374
- 9.16 Coherent or Heterodyne Optical Communication — 392
- 9.17 Fiber Soliton Communication Systems — 394
- 9.18 Fiber Sensors — 396
- 9.19 Fiber Optics in Medicine and Industry — 407
- *References* — 408
- *Solved Examples* — 408
- *Questions* — 410
- *Problems* — 411

10 MATTER AND RADIATION–DUAL NATURE — 412

- 10.1 Introduction — 412
- 10.2 Black Body Radiation — 414
- 10.3 Models for the Black Body Radiation — 416
- 10.4 Pyrometry — 425
- 10.5 Particle Like Behaviour of Light — 440
- 10.6 Compton Effect — 446
- 10.7 Wavelike Behaviour of Particles — 450

10.8	Phase Velocity and Group Velocity of Waves	460
10.9	Heisenberg's Uncertainty Principle	466
10.10	Applications of Heisenberg's Uncertainty Principle	472
10.11	Significance of Uncertainty Principle	476

References 477
Solved Examples 477
Questions 490
Problems 493

11 X-RAYS   499

11.1	Introduction	499
11.2	Production of X-Rays	499
11.3	Origin of X-Ray Spectrum	502
11.4	Moseley's Law	504
11.5	Synchrotron Radiation as X-Ray Source	505
11.6	Interaction of X-Rays with Matter	506
11.7	Detection of X-Rays	508
11.8	X-Ray Radiography	512
11.9	Crystal Structure Determination Using X-Rays	519
11.10	Basics of X-Ray Scattering	530
11.11	X-Ray Techniques in Materials Science	538

References 544
Solved Examples 544
Questions 553
Problems 554

12 BASIC QUANTUM MECHANICS   557

12.1	Introduction	557
12.2	Wave Equation in Classical Mechanics	557
12.3	Schrödinger Wave Equation	558
12.4	Properties of Wave Functions	560
12.5	Operator Formalism of Schrödinger Equation	562
12.6	Free Particle	563
12.7	Particle Confinement in Potential Wells	564
12.8	Barrier Penetration and Tunneling	580
12.9	Applications of Tunneling	592
12.10	Electrons in a Periodic Potential of a Crystal-Band Theory of Crystalline Solids	605
12.11	Periodic Nature of Energies and the Brillouin Zone	612
12.12	Standing Waves at the Band Edges and the Origin of Band Gaps	612
12.13	Band Theory of Electrical Conduction in Solids	614
12.14	Analogue of Kronig-Penney Model	622

References 623
Solved Examples 623
Questions 625
Problems 626

13 QUANTUM INFORMATION AND QUANTUM COMPUTATION — 628

- 13.1 Introduction — 628
- 13.2 Basic Ideas — 630
- 13.3 Quantum Entanglement — 640
- 13.4 Computation-Basic Ideas — 645
- 13.5 Classical Computation — 650
- 13.6 Quantum Computation — 651
- 13.7 Design of Quantum Computers — 657
- 13.8 Disadvantages of Quantum Computers — 665
- 13.9 Quantum Information Theory — 667
- *References* — 678
- *Questions* — 678

14 BASICS OF NANOSCIENCE AND NANOTECHNOLOGY — 680

- 14.1 Introduction — 680
- 14.2 Scaling Laws in Miniaturisation — 680
- 14.3 Quantum Nature of the Nanoworld — 684
- 14.4 Size Dependent Properties — 685
- 14.5 Fabrication Processes—Top Down Approach — 688
- 14.6 Fabrication Process—Bottom UP Approach — 690
- 14.7 Chemical Synthesis of Nanoparticles — 698
- 14.8 Characterization of Nanomaterials — 708
- 14.9 Carbon Nanotubes — 713
- *References* — 721
- *Questions* — 721

INDEX — 723

1
Vibrations and Resonance

1.1 INTRODUCTION

Vibrations or oscillations constitute one of the most important fields of study in physics as well as engineering. The characteristic feature of vibration is its periodicity *i.e.*, there is a movement or displacement or a variation in the value of a physical quantity that repeats over and over again. In mechanical systems, the periodicity refers to displacement and force while in the electrical systems, it is related to current and voltage. At the microscopic level, atoms and molecules execute periodic vibration in the solid state. The propagation of light involves the vibrations of electric and magnetic fields while that of sound, the periodic motion of atoms and molecules. The complex and a variety of phenomena arising out of the interaction between matter and radiation can be understood in the conceptual framework of vibrations.

Vibratory motion of machines and mechanical structures is a source of noise. Consequently vibration analysis and control is an important discipline in engineering. The design of jet engines, rocket engines, wheeled vehicles which are made immune to shock during motion needs a thorough understanding of the phenomena of vibrations. Vibration isolators and dampers are routinely used in industry to minimise noise pollution. Vibratory motion is also put to use in many practical devices such as vibratory conveyors, hoppers, sieves, compactors, washing machines, dentists' drills etc. An understanding of the vibrations of strings, membranes, air columns in pipes is necessary for the design of musical instruments. The simulation of vibratory motion of earth's crust is often carried out to understand the phenomena of earthquake and the propagation of shock waves. The science of acoustics deals with the generation, transmission and reception of energy in the form of vibrational waves in matter. A study of the acoustics of building and noise control methods forms an integral part of architectural engineering.

Vibrations are classified into the following types:

Free vibrations: These refer to vibrations when the system is allowed to vibrate on its own. The system is not subjected to a periodic force.

Forced vibrations: If a system is subjected to a periodic force, the resulting vibrations are known as forced vibrations.

Undamped vibration: If no energy is lost or dissipated in friction during vibration, the vibration is termed undamped vibration.

Damped vibration: This refers to vibration in which there is loss of energy during vibration. Consequently a system which is set to vibrate will come to rest after in due course of time.

Linear/non-linear vibration: If the restoring force is proportional to displacement or when the frictional force is linearly proportional to velocity the vibrations are termed linear, otherwise they are non-linear.

Transient vibrations: When a system is subjected to a periodic force, it takes a finite though a short time, for the onset of steady state vibrations. The vibrations during this interim period is known as transient vibrations. We will begin with the study of Simple Harmonic Motion (SHM), which is the simplest of all periodic motion.

1.2 SIMPLE HARMONIC MOTION

We define Simple Harmonic Motion (SHM) as a motion in which the acceleration of the body (or force on the body) is directly proportional to its displacement from a fixed point and is always directed towards the fixed point. SHM possesses the following characteristics.

(*i*) The motion is periodic.

(*ii*) When displaced from the fixed point or the mean position, a restoring force acts on the particle tending to bring it to the mean position.

(*iii*) Restoring force on the particle is directly proportional to its displacement. The study of SHM is of practical interest. A vast variety of deformation of physical systems involving stretching, compression, bending or twisting (or combinations involving all of these) result in restoring forces proportional to displacement and hence, leads to SHM.

Now we will consider an example of a SHM. Let a particle A moves along the circumference of a circle with a constant speed $v(=r\omega)$ where r is the radius of the circle and 'ω' is its angular speed. Let the centre of the circle be O and a perpendicular AP be drawn from the particle on the diameter YY' of the circle (Fig. 1.1).

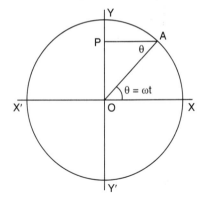

Fig. 1.1 Simple harmonic motion

Then as the particle moves along the circumference of the circle, the point P, the foot of the perpendicular vibrates along the diameter. Since the motion of A is uniform, the motion of P is periodic. At any instant, the distance OP from O is called the displacement. If the particle moves from X to A in time t, then

$$A\hat{O}X = P\hat{A}O = \omega t = \theta$$

i.e.,
$$OP = y = r \sin \omega t$$

$$\text{Velocity} = v = \frac{dy}{dt} = r\omega \cos \omega t \qquad (1.1)$$

$$= r\omega(1 - \sin^2 \omega t)^{1/2} = r\omega(1 - y^2/r^2)^{1/2}$$

$$\text{Acceleration} = \frac{d^2y}{dt^2} = -r\omega^2 \sin \omega t$$

$$= -\omega^2 y = -\omega^2 \times \text{displacement} \qquad (1.2)$$

Thus, acceleration is directly proportional to displacement and directed towards a fixed point. Hence, the above example corresponds to SHM.

It is instructive to learn how velocity and acceleration in a SHM vary with time. We notice that when the displacement is maximum ($+r$ or $-r$), the velocity = 0, because now the

VIBRATIONS AND RESONANCE

point P must change its direction. But when y is maximum ($+r$ or $-r$), the acceleration is also maximum ($-\omega^2 r$ or $+\omega^2 r$ respectively) and is directed opposite to the displacement. When $y = 0$, the velocity is maximum ($r\omega$ or $-r\omega$) and the acceleration is zero.

The time period T (the time required to complete one oscillation) is given by the following relation.

$$\text{Angular velocity} = \frac{\text{Angle described in one revolution}}{\text{Time taken for one revolution}}$$

i.e.,
$$\omega = \frac{2\pi}{T} \quad \text{or} \quad T = \frac{2\pi}{\omega}$$

Substituting the value of ω from eqn. (1.2), we have

$$T = \frac{2\pi}{\sqrt{\text{Acceleration} / \text{Displacement}}}$$

$$T = \frac{2\pi}{\sqrt{\text{Acceleration per unit displacement}}}$$

$$= 2\pi \sqrt{\frac{\text{Displacement}}{\text{Acceleration}}} \tag{1.3}$$

The frequency n is given by $1/T$.

The idea of phase is very important in SHM. Phase difference between two SHMs indicates how much the two motions are out of step with each other or by how much angle or how much time one is ahead of the other. In general, displacement is given by

$$y = r \sin(\omega t + \phi)$$

Clearly at $t = 0$, $y = r \sin \phi$. 'ϕ' is called the initial phase (Fig. 1.2.)

Now let us calculate the total energy associated with the particle executing SHM. When a body undergoes SHM, its total energy consists of potential energy and kinetic energy. The velocity and consequently kinetic energy is maximum at the mean position. Potential energy is zero at mean position and is maximum at the extreme position.

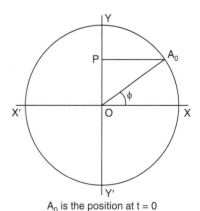

A_0 is the position at t = 0

Fig. 1.2 Simple harmonic motion

Let us calculate the potential energy.

Potential energy = Force × Distance

The work done in moving through dy is $m\omega^2 y \, dy$

∴
$$\text{P.E.} = \int m\omega^2 y \, dy$$

$$= \frac{m\omega^2 y^2}{2}$$

$$= \frac{m\omega^2 r^2 \sin^2 \omega t}{2}$$

$$\text{K.E.} = \frac{mv^2}{2} = \frac{mr^2 \omega^2 \cos^2 \omega t}{2}$$

$$\therefore \quad \text{T.E.} = \frac{mr^2\omega^2}{2}(\cos^2 \omega t + \sin^2 \omega t)$$

$$= \frac{mr^2\omega^2}{2} = \frac{mv^2}{2} \tag{1.4}$$

$$= \frac{1}{2} \times \text{mass} \times (\text{amplitude})^2 \times (\text{angular velocity})^2$$

1.3 FREE VIBRATIONS

Let us consider a body of mass m executing SHM. The equation of motion is of the form:

$$m\frac{d^2x}{dt^2} = -Kx \quad \text{or} \quad \frac{d^2x}{dt^2} = -\omega_n^2 x \quad \text{where} \quad \omega_n^2 = \frac{K}{m} \tag{1.5}$$

ω_n is the natural angular frequency of the simple harmonic oscillator. K is often called the spring constant. This ensures that the acceleration of the particle is always directed towards a fixed point on the line and proportional to the displacement from that point. For a solution of eqn. (1.5), consider

$$x = Ae^{st} \text{ so that } \frac{dx}{dt} = sAe^{st} \text{ and } \frac{d^2x}{dt^2} = s^2 Ae^{st} \tag{1.6}$$

From eqns. (1.5) and (1.6)

$$s^2 Ae^{st} + \omega_n^2 Ae^{st} = 0 \implies Ae^{st}(s^2 + \omega_n^2) = 0 \implies s = \pm i\omega_n$$

Thus eqn. (1.5) is satisfied by

$$x = e^{i\omega_n t} \text{ and } x = e^{-i\omega_n t}$$

A linear combination of these two also satisfies eqn. (1.5)

$$\therefore \quad x = A_1 e^{i\omega_n t} + A_2 e^{-i\omega_n t} \tag{1.7a}$$

where A_1 and A_2 are arbitrary constants. This solution may be written as:

$$x = A_1(\cos \omega_n t + i \sin \omega_n t) + A_2(\cos \omega_n t - i \sin \omega_n t)$$
$$= (A_1 + A_2)\cos \omega_n t + i(A_1 - A_2)\sin \omega_n t$$
$$= A \cos \omega_n t + B \sin \omega_n t$$
$$= C \sin(\omega_n t + \phi) = C \sin \omega_n t \cos \phi + C \cos \omega_n t \sin \phi \tag{1.7b}$$

Equating coefficients of $\sin \omega_n t$ and $\cos \omega_n t$,

$$A = C \sin \phi \text{ and } B = C \cos \phi$$

$$\therefore \quad C = \sqrt{A^2 + B^2} \text{ and } \phi = \tan^{-1}\left(\frac{A}{B}\right)$$

The velocity of the particle at any instant is given by

$$v = \frac{dx}{dt} = C\omega_n \cos(\omega_n t + \phi) \tag{1.8}$$

The values of C and ϕ depend upon the initial conditions.

Let $x = x_0$ and $\frac{dx}{dt} = v_0$ at $t = 0$

VIBRATIONS AND RESONANCE

Substituting in eqn. (1.7b) and its derivative, we get

$$x_0 = C \sin \phi \quad \Rightarrow \quad \sin \phi = \frac{x_0}{C} \tag{1.8a}$$

$$v_0 = \left(\frac{dx}{dt}\right)_{t=0} = C\omega_n \cos \phi \quad \Rightarrow \quad \cos \phi = \frac{v_0}{C\omega_n} \tag{1.8b}$$

Squaring and adding eqns. (1.8a) and (1.8b)

$$C = \left[x_0^2 + \frac{v_0^2}{\omega_n^2}\right]^{1/2} \tag{1.9a}$$

Dividing eqn. (1.8a) by eqn. (1.8b)

$$\phi = \tan^{-1} \frac{x_0 \omega_n}{v_0} \tag{1.9b}$$

Clearly C is the amplitude. The value of x repeats when t changes by $2\pi/\omega_n$ because

$$x = C \sin\left[\omega_n\left(t + \frac{2\pi}{\omega_n}\right) + \phi\right] = C \sin(\omega_n t + 2\pi + \phi) = C \sin(\omega_n t + \phi)$$

$$\therefore \quad \text{periodic time} = \frac{2\pi}{\omega_n} \quad \text{and frequency} = \frac{\omega_n}{2\pi}$$

The plot of displacement and velocity with time is shown in Fig. 1.3.

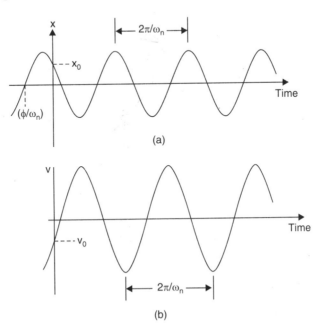

Fig. 1.3 (a) Displacement and (b) velocity in undamped vibration

It is possible to arrive at eqn. (1.5) by considering the principle of conservation of energy. In free vibrations without damping, the total energy is conserved, *i.e.*, $T + U$ = constant, where T and U are kinetic and potential energy respectively. They are given by:

$$T = \frac{1}{2} m \left(\frac{dx}{dt}\right)^2 \text{ and } U = \frac{1}{2} Kx^2 \qquad (1.10)$$

$$\therefore \qquad T + U = \text{constant or } \frac{d}{dt}(T + U) = 0 \qquad (1.11)$$

From eqns. (1.10) and (1.11)

$$\frac{d}{dt}\left[\frac{1}{2} m \left(\frac{dx}{dt}\right)^2 + \frac{1}{2} Kx^2\right] = 0 \quad \Rightarrow \quad \frac{1}{2} m \cdot 2 \cdot \frac{dx}{dt} \frac{d^2x}{dt^2} + \frac{1}{2} \cdot 2 \cdot Kx \frac{dx}{dt} = 0$$

$$\Rightarrow \qquad \frac{dx}{dt}\left[m \frac{d^2x}{dt^2} + Kx\right] = 0 \quad \text{or} \quad m\frac{d^2x}{dt^2} + Kx = 0 \quad \Rightarrow \quad \frac{d^2x}{dt^2} + \omega_n^2 x = 0$$

The above equation is identical with eqn. (1.5).

1.4 DAMPED VIBRATIONS

1.4.1 Introduction

In many practical systems, the vibrational energy is gradually converted to heat or sound. Due to the reduction in energy, the response, such as the displacement of the system gradually decreases. The mechanism by which the vibrational energy is gradually converted into heat or sound is known as damping. Although the energy loss due to damping may be small, consideration of damping becomes important for an accurate prediction of the vibrational response of the system. Damping is modelled as one or more of the following types:

Viscous Damping

When a mechanical system vibrates in a fluid such as air, gas, water or oil, the resistance offered by the fluid to the moving body causes energy to be dissipated. The amount of dissipated energy depends on many factors such as the size and shape of the vibrating body, the viscosity of the fluid, the frequency of vibration and the velocity of the vibrating body. In viscous damping, the damping force is proportional to the velocity of the vibrating body. Typical examples of viscous damping include

(a) fluid films between sliding surfaces
(b) fluid flow around a piston in a cylinder
(c) fluid flow through an orifice and
(d) the fluid flow around a journal in a bearing.

Coulomb or Dry Friction Damping

Here the damping force is constant in magnitude but opposite in direction to the motion of the vibrating body. It is caused by friction between rubbing surfaces that are either dry or have insufficient lubrication.

Material or Solid or Hysteretic Damping

Energy is absorbed or dissipated when a material is deformed under periodic stress. The effect is due to friction between the internal atomic planes, which slip or slide as the

VIBRATIONS AND RESONANCE

deformation takes place. When a body having material damping is subjected to a periodic stress, a periodic strain results which has a phase lag. The stress (σ)-strain (ε) curve shows a hysteresis loop as shown in Fig. 1.4b. The area of the loop denotes energy loss per unit volume of the body per cycle due to damping. This phenomenon is very much similar to ferromagnetic hysteresis where phase lag is observed between the magnetic induction (B) and the applied field (H). In a ferromagnetic material, the application of magnetic field leads to domain growth and rotation. This involves some sort of friction between the adjacent domains.

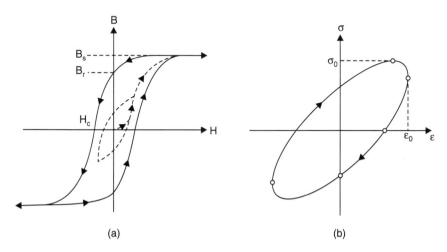

Fig. 1.4 Hysteretic damping (a) ferromagnetic hysteresis (b) elastic hysteresis

1.4.2 Free Vibration with Viscous Damping

Here the viscous force is proportional to velocity and acts opposite to the direction of velocity. Generally a damped oscillator is represented by a spring and a dashpot (Fig. 1.5). The equation of motion of a body subject to viscous drag is of the form

$$m\frac{d^2x}{dt^2} = -c\frac{dx}{dt} - Kx \quad \Rightarrow \quad \frac{d^2x}{dt^2} + \frac{c}{m}\frac{dx}{dt} + \frac{K}{m}x = 0 \tag{1.12}$$

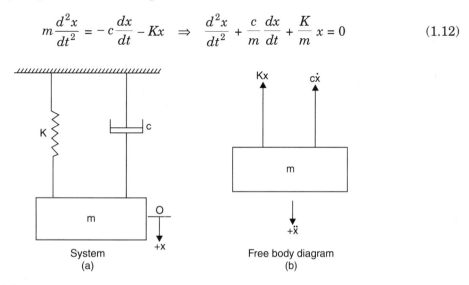

Fig. 1.5 One dimensional damped oscillator (a) system and (b) free body diagram

(c/m) denotes the frictional force per unit mass per unit velocity. For convenience, denote

$$\frac{K}{m} = \omega_n^2 \quad \text{and} \quad \frac{c}{m} = 2\xi\omega_n \tag{1.13}$$

ξ is a measure of damping. It is related to factors like quality factor (Q), logarithmic decrement (δ), relaxation time, loss coefficient and specific damping capacity, which are used to characterise a damped oscillator.

From eqns. (1.12) and (1.13) we have

$$\frac{d^2x}{dt^2} + 2\xi\omega_n \frac{dx}{dt} + \omega_n^2 x = 0 \tag{1.14}$$

Let $x = Ae^{st}$ be the solution of eqn. (1.14)

Then $$\frac{dx}{dt} = Ase^{st} \quad \text{and} \quad \frac{d^2x}{dt^2} = As^2 e^{st} \tag{1.15}$$

From eqns. (1.14) and (1.15)

$$Ae^{st}(s^2 + 2\xi\omega_n s + \omega_n^2) = 0 \quad \Rightarrow \quad s^2 + 2\xi\omega_n s + \omega_n^2 = 0$$

This is a quadratic equation in s. The solutions are given by

$$s = \omega_n \left[-\xi \pm \sqrt{\xi^2 - 1} \right]$$

The general solution to eqn. (1.12) is

$$x = A_1 e^{(-\xi + \sqrt{\xi^2 - 1})\omega_n t} + A_2 e^{(-\xi - \sqrt{\xi^2 - 1})\omega_n t} \tag{1.16}$$

Critical Damping Constant:

The critical damping constant (c_c) is defined as the value of the damping constant for which radical is eqn. (1.16) is zero. *i.e.*,

$$\xi = 1 \quad \Rightarrow \quad c_c = 2m\omega_n = 2\sqrt{mK} \quad \text{[from eqn. (1.13)]} \tag{1.17}$$

From eqns. (1.13) and (1.17)

$$\xi = \frac{c}{c_c} = \frac{\text{Damping constant}}{\text{Critical damping constant}} \quad \Rightarrow \quad c = \xi c_c = 2\xi m\omega_n = 2\xi\sqrt{mK} \tag{1.18}$$

The solution given by eqn. (1.16) can be analysed under the following three cases:

Case (i) Underdamped system

$$\xi < 1 \text{ or } c < c_c \text{ or } c < 2m\omega_n \text{ or } c < 2\sqrt{mK}$$

For this condition ($\xi^2 - 1$) is negative and the two roots can be expressed as

$$s_1 = (-\xi + i\sqrt{1-\xi^2})\omega_n \; ; \; s_2 = (-\xi - i\sqrt{1-\xi^2})\omega_n$$

Hence the solution eqn. (1.16) takes the form

$$x = A_1 e^{(-\xi + i\sqrt{1-\xi^2})\omega_n t} + A_2 e^{(-\xi - i\sqrt{1-\xi^2})\omega_n t} = e^{-\xi\omega_n t}\left[A_1 e^{i\sqrt{1-\xi^2}\omega_n t} + A_2 e^{-i\sqrt{1-\xi^2}\omega_n t}\right] \tag{1.19}$$

Following the procedure in simplifying as in the case of eqn. (1.7)

$$x(t) = Ce^{-\xi\omega_n t} \sin\left[\sqrt{1-\xi^2}\omega_n t + \phi\right] = Ce^{-\xi\omega_n t} \sin(\omega_d t + \phi) \tag{1.20}$$

where $\omega_d = \omega_n \sqrt{1-\xi^2}$ \hfill (1.21)

VIBRATIONS AND RESONANCE

C and ϕ can be evaluated by knowing the initial conditions and are given by eqn. (1.8). On comparing eqn. (1.20) with eqn. (1.7b) we find that eqn. (1.20) corresponds to a damped harmonic motion of angular frequency ω_d given by eqn. (1.21). The amplitude decreases exponentially with time given by:

$$Ce^{-\xi\omega_n t}$$

A sketch of displacement with time for various ξ is given in Fig. 1.6. Clearly $\omega_d < \omega_n$. The variation of ω_d with ξ is shown in Fig. 1.7.

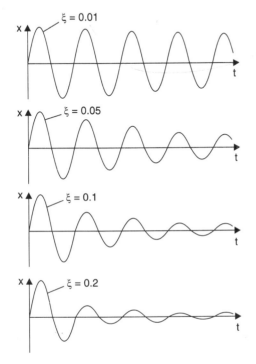

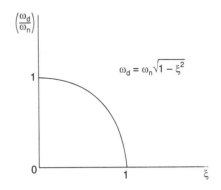

Fig. 1.7 Variation of (ω_d/ω_n) with ξ

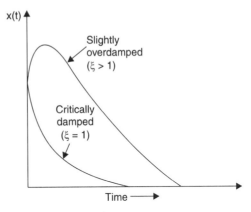

Fig. 1.6 Time variation of displacement as a function of damping ratio $\xi < 1$ (underdamping)

Case (ii) Critically Damped System

$$\xi = 1 \text{ or } c = c_c = 2m\omega_n = 2\sqrt{mK}$$

For this case the eqn. (1.16) takes the form

Fig. 1.8 Time variation of displacement with (a) $\xi = 1$ (critical damping) and (b) $\xi > 1$ (slightly greater than critical damping)

$$x = e^{-\xi\omega_n t}(A_1 + A_2)$$

Clearly the displacement decreases exponentially with time (Fig. 1.8). This solution does not provide any more information. On substituting this solution in eqn. (1.14) we get back the condition that $\xi = 1$. Hence let us assume that the expression within the radical sign is not equal to zero but a small quantity h. Later we will analyse the solution in the limit $h \to 0$.

i.e., assume $i\sqrt{\xi^2 - 1} = h$

Substituting the above expression in eqn. (1.19), we get

$$\therefore \quad x = A_1 e^{(-\xi+h)\omega_n t} + A_2 e^{(-\xi-h)\omega_n t} = e^{-\xi\omega_n t}\left[A_1 e^{h\omega_n t} + A_2 e^{-h\omega_n t}\right]$$

$$\approx e^{-\xi\omega_n t}\left[A_1(1 + h\omega_n t + \cdots) + A_2(1 - h\omega_n t + \cdots)\right]$$

$$\approx e^{-\xi\omega_n t}\left[(A_1 + A_2) + h\omega_n t(A_1 - A_2)\right] = e^{-\xi\omega_n t}\left[P + Qt\right] \quad (1.22)$$

Where $P = A_1 + A_2$ and $Q = h\omega_n (A_1 - A_2)$

It is clear from eqn. (1.22) that as t increases, the factor $P + Qt$ increases but the factor $e^{-\xi\omega_n t}$ decreases. Thus, the displacement x first increases due to the factor $(P + Qt)$ and then exponentially decreases to zero as it increases. The displacement is not periodic (Fig. 1.8).

Case (iii) Overdamped System

$$\xi > 1 \quad \text{or} \quad c > c_c \quad \text{or} \quad c > 2m\omega_n > 2\sqrt{mK}$$

For this condition $(\xi^2 - 1)$ is positive and the two roots are:

$$s_1 = (-\xi + \sqrt{\xi^2 - 1})\omega_n < 0$$

$$s_2 = (-\xi - \sqrt{\xi^2 - 1})\omega_n < 0$$

They are real. Hence x decays exponentially with time. Note that the exponent factor is greater in magnitude than in case (*ii*). Here the mass moves to equilibrium more slowly. Further, there is no oscillation in this case as well (Fig. 1.9). In critical damping the mass returns to rest in shortest possible time without overshooting. This property of critical damping is used in many practical applications. For example, large guns have dashpots with critical damping value so that they can return to their original position after recoil in the minimum time without vibrating. If the damping provided were more than critical value, some delay would occur before the next firing.

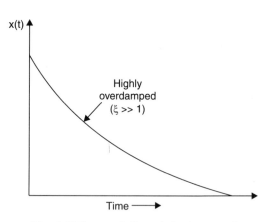

Fig. 1.9 Time variation of displacement for $\xi \gg 1$ (overdamping)

1.4.3 Energy of a Weakly Damped Oscillator

The amplitude of the oscillator is given by eqn. (1.20) to be $Ce^{-\xi\omega_n t}$

From eqn. (1.4) the energy of oscillator

$$E = \frac{1}{2} \times \text{mass} \times (\text{amplitude})^2 \times (\text{angular frequency})^2$$

$$= \frac{1}{2} m\omega_n^2 C^2 e^{-2\xi\omega_n t} = E_0 e^{-2\xi\omega_n t} \quad (1.23)$$

Thus the energy decays with time as shown in Fig. (1.10).

VIBRATIONS AND RESONANCE

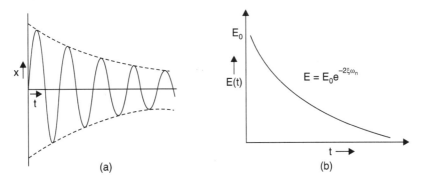

Fig. 1.10 Time variation of (a) amplitude and (b) energy of an underdamped oscillator

1.4.4 Logarithmic Decrement, Relaxation Time, Specific Damping Capacity, Loss Coefficient and Q-Factor

Several parameters such as logarithmic decrement, relaxation time, specific damping capacity, loss coefficient and quality factor are used to describe the characteristics of damped oscillator.

Logarithmic Decrement

The logarithmic decrement represents the rate at which the amplitude of a free damped vibration decreases. It is defined as the natural logarithm of the ratio of any two successive amplitudes. Let x_1 and x_2 be the two consecutive amplitudes. Let t_1 and t_2 denote the time corresponding to two consecutive amplitudes (Fig. 1.11).

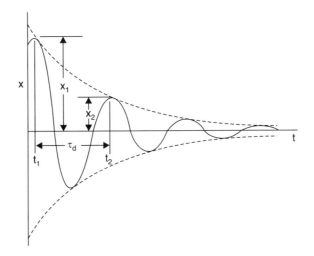

Fig. 1.11 Logarithmic decrement

$$\frac{x_1}{x_2} = \frac{Ce^{-\xi\omega_n t_1} \sin[\omega_d t_1 + \phi]}{Ce^{-\xi\omega_n t_2} \sin[\omega_d t_2 + \phi]} \quad \text{where } \omega_d \text{ is given by eqn. (1.21)} \quad (1.24)$$

But $t_2 = t_1 + \tau_d$ where $\tau_d = \dfrac{2\pi}{\omega_d}$ \hfill (1.25a)

$$\therefore \sin(\omega_d t_2 + \phi) = \sin\left[\omega_d\left(t_1 + \frac{2\pi}{\omega_d}\right) + \phi\right] = \sin(\omega_d t_1 + 2\pi + \phi) = \sin(\omega_d t_1 + \phi) \quad (1.25b)$$

From eqns. (1.24) and (1.25)

$$\frac{x_1}{x_2} = e^{\xi \omega_n \tau_d}$$

Taking natural logarithms and making use of eqn. (1.21)

$$\therefore \quad \ln\left(\frac{x_1}{x_2}\right) = \delta = \xi \omega_n \tau_d = \xi \omega_n \left(\frac{2\pi}{\omega_d}\right) = \frac{2\pi\xi}{\sqrt{1-\xi^2}} \approx 2\pi\xi \text{ for } \xi \ll 1 \quad (1.26)$$

Figure 1.12 shows the plot of logarithmic decrement as a function of ξ.

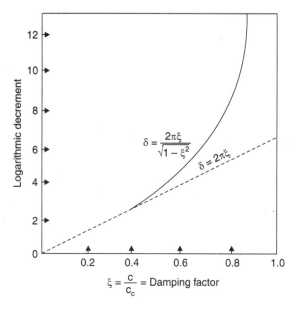

Fig. 1.12 Logarithmic decrement as a function of ξ

In general if x_1 and x_{m+1} denote the amplitudes corresponding to t_1 and $t_{m+1} = t_1 + m\tau_d$ where m is an integer, then

$$\frac{x_1}{x_{m+1}} = \frac{x_1}{x_2} \times \frac{x_2}{x_3} \times \frac{x_3}{x_4} \times \cdots \times \frac{x_m}{x_{m+1}} = \{e^{(\xi\omega_n\tau_d)}\}^m = e^{m\xi\omega_n\tau_d} = e^{m\delta}$$

Taking natural logarithms

$$\ln\left(\frac{x_1}{x_{m+1}}\right) = m\delta \quad \Rightarrow \quad \delta = \frac{1}{m} \ln\left(\frac{x_1}{x_{m+1}}\right) \quad (1.27)$$

To determine the number of cycles m elapsed for a 50% reduction in amplitude, we have from eqns. (1.26) and (1.27)

$$\delta \approx 2\pi\xi = \frac{1}{m} \ln 2 = \frac{0.693}{m} \quad \Rightarrow \quad m\xi = \frac{0.693}{2\pi} = 0.110$$

This is the eqn. of a rectangular hyperbola and is shown in Fig. 1.13.

VIBRATIONS AND RESONANCE

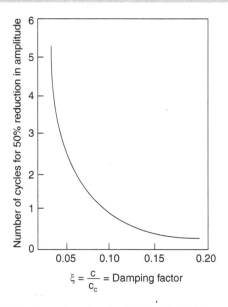

Fig. 1.13 Number of cycles for 50% reduction in amplitude

Relaxation Time

The amplitude is given by $x(t) = Ce^{-\xi\omega_n t}$

Relaxation time is defined as the time taken by the amplitude to decrease $(1/e)^{th}$ of its original value

i.e.,
$$\frac{x(\tau)}{x(o)} = \frac{1}{e} = \frac{Ce^{-\xi\omega_n \tau}}{C} = e^{-\xi\omega_n \tau} \Rightarrow \tau = \frac{1}{\xi\omega_n} \quad (1.28)$$

From eqns. (1.28) and (1.26)

$$\tau \Rightarrow \frac{1}{\xi\omega_n} = \frac{1}{\left(\frac{\delta}{2\pi}\right)\omega_n} = \left(\frac{2\pi}{\omega_n}\right)\frac{1}{\delta} = \frac{T}{\delta} \quad (1.29)$$

Relaxation time is inversely proportional to logarithmic decrement.

Natural time period = Relaxation time × Logarithmic decrement

Specific Damping Capacity, Loss Coefficient and Q-factor

The rate of decay of energy with time can be obtained by differentiating eqn. (1.23)

i.e.,
$$\frac{dE}{dt} = -E_0 \, 2\xi\omega_n e^{-2\xi\omega_n t} = -2\xi\omega_n E$$

∴ loss of energy in one cycle = $\left(\frac{dE}{dt}\right)$ × time period

$$= \Delta E = -2\xi\omega_n E \times \left(\frac{2\pi}{\omega_d}\right) \quad (1.30)$$

From eqn. (1.21) and (1.30)

$$\frac{\Delta E}{E} = \text{specific damping capacity} = \frac{4\pi\xi\omega_n}{\omega_d} = \frac{4\pi\xi}{\sqrt{1-\xi^2}} \approx 4\pi\xi$$

Quantity $\Delta E/E$ is called the specific damping capacity and is useful in comparing the damping capacity of materials. Another coefficient known as the loss coefficient is also used for comparing the damping capacity.

$$\text{Loss coefficient} = \frac{\text{Energy dissipated per radian}}{\text{Total energy}} = \frac{\left(\frac{\Delta E}{2\pi}\right)}{E} = \frac{1}{2\pi}\left(\frac{\Delta E}{E}\right) = 2\xi \approx \frac{\delta}{\pi} \quad (1.31)$$

$$Q\text{-factor} = Q = 2\pi \frac{\text{Energy stored}}{\text{Energy dissipated per cycle}} = 2\pi \left(\frac{E}{\Delta E}\right) = 2\pi \times \frac{1}{4\pi\xi} = \frac{1}{2\xi} = \frac{\pi}{\delta} \quad (1.32)$$

Note that Q-factor is the reciprocal of the loss coefficient.

1.5 STEADY STATE RESPONSE OF AN OSCILLATOR UNDER THE ACTION OF A PERIODIC FORCE

1.5.1 Introduction

In many situations the response of an oscillator to a periodic force has to be analysed both in the fields of physics and engineering. An important idea associated with the response of an oscillator to the periodic force is *resonance i.e.*, when the natural frequency of the oscillator is equal to the frequency of the periodic force. The differential equation which describes this phenomenon is used in the analysis of mechanical vibrations as well as a.c. circuits. The same mathematical formalism can be used to understand the response of electrons and ions to an electromagnetic field. In geophysics it is used to understand the phenomenon of tides which are the result of moon's periodic motion about the earth. Resonance is also encountered in many phenomena such as nuclear magnetic resonance, Mossbauer effect, infrared absorption dielectric dispersion and microwave absorption.

1.5.2 Undamped Oscillator

We consider an undamped system subjected to a periodic force of frequency ω

i.e., $F(t) = F_0 \sin \omega t$. The equation of motion is given by

$$m\frac{d^2x}{dt^2} = F_0 \sin \omega t - Kx \quad \Rightarrow \quad m\frac{d^2x}{dt^2} + Kx = F_0 \sin \omega t \quad (1.33)$$

Since the exciting force $F(t)$ is periodic, $x(t)$ is also periodic and has the same frequency

i.e., $$x(t) = A \sin \omega t \quad (1.34a)$$

$\therefore \quad \dfrac{dx}{dt} = A\omega \cos \omega t$ and $\dfrac{d^2x}{dt^2} = -A\omega^2 \sin \omega t \quad (1.34b)$

From eqns. (1.34a), (1.34b) and (1.33)

$$-mA\omega^2 \sin \omega t + KA \sin \omega t = F_0 \sin \omega t$$

$$\Rightarrow \quad A[K - m\omega^2]\sin \omega t = F_0 \sin \omega t$$

$$\Rightarrow \quad A = \frac{F_0}{(K - m\omega^2)} = \frac{F_0/K}{1 - \left(\dfrac{\omega}{\omega_n}\right)^2} = \frac{\delta_{st}}{1 - \left(\dfrac{\omega}{\omega_n}\right)^2} \quad (\because \omega_n^2 = K/m) \quad (1.35a)$$

VIBRATIONS AND RESONANCE

δ_{st} has the units of displacement and is given by (F_0/K). It denotes the deflection of mass under force F_0 and is sometimes referred to as static deflection because F_0 is constant. The quantity (A/δ_{st}) represents the ratio of the dynamic to static amplitude and is called magnification factor or amplitude ratio. It is given by eqn. (1.35b). Its variation with the frequency of the driving force is shown in Fig. 1.14

$$\frac{A}{\delta_{st}} = \frac{1}{1-\left(\dfrac{\omega}{\omega_n}\right)^2} \qquad (1.35b)$$

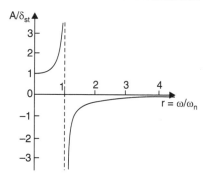

Fig. 1.14 Magnification factor for an undamped oscillator

It is instructive to consider the variation of the amplitude ratio for the following two cases:

Case (i) When $\omega < \omega_n$, the amplitude ratio (A/δ_{st}) is positive and the displacement is in phase with the force.

Case (ii) When $\omega > \omega_n$, the amplitude ratio (A/δ_{st}) is negative and the solution is expressed as

$$x = -A\sin\omega t \quad \text{and} \quad A = \frac{\delta_{st}}{\left(\dfrac{\omega}{\omega_n}\right)^2 - 1}$$

In this case there is a phase difference of π between the displacement and the force.

Case (iii) When $\omega = \omega_n$, the amplitude ratio (A/δ_{st}) becomes infinite and the phenomenon is known as *resonance*. But in practical situations, amplitude ratio does not become infinite since damping is invariably present.

1.5.3 Damped Oscillator

The response of a damped oscillator is of practical interest since most oscillatory motions in real life are damped. Some examples of damping are described below:

Resistive damping: In simple LCR circuit, the damping effect is produced by resistance. Energy is dissipated due to Joule heating and is referred to as *Resistive damping*.

Electromagnetic Damping: A galvanometer consists of a current carrying coil mounted on an axis in a magnetic field. The radial field produced by a properly shaped permanent magnet results in a deflection which is proportional to the current. The steady current in the coil gives rise to a torque which is proportional to the current. The coil comes to rest in a position when this electromagnetic torque is balanced by the elastic torque due to the stiffness of the suspension. A suspended coil is subjected to various mechanical damping processes. The viscosity of the atmosphere will produce a damping torque which is proportional to the angular velocity. There is also an electromagnetic source of damping. When the coil rotates towards its new equilibrium position, the magnetic field will induce a voltage proportional to the instantaneous angular speed. According to Lenz's law the induced voltage will reduce the current flowing in the coil by an amount proportional to the angular velocity and inversely proportional to the resistance of the circuit. Thus the electromagnetic torque will also be reduced.

Collision damping: Whenever an electron in a metal or in atmosphere is subjected to an electromagnetic field, the oscillatory motion of the electron is damped by collision with other electrons. This is often referred to as collision damping.

Radiation damping: An electron subjected to an oscillatory motion experiences a periodic acceleration. An electric charge in acceleration emits electromagnetic radiation and thus the electron loses energy. This is known as radiation damping.

We consider a damped oscillator subject to a periodic force of frequency ω. *i.e.*,
$F(t) = F_0 \sin \omega t$. The equation of motion is given by

$$m\frac{d^2x}{dt^2} = F_0 \sin \omega t - c\frac{dx}{dt} - Kx \;\Rightarrow\; m\frac{d^2x}{dt^2} + c\frac{dx}{dt} + Kx = F_0 \sin \omega t \tag{1.36}$$

Under steady state conditions $x(t)$ is also expected to be periodic. *i.e.*,

$$x(t) = A \sin(\omega t - \phi) \tag{1.37a}$$

so that
$$\frac{dx}{dt} = A\omega \cos(\omega t - \phi) \quad \text{and} \quad \frac{d^2x}{dt^2} = -A\omega^2 \sin(\omega t - \phi) \tag{1.37b}$$

From eqns. (1.36), (1.37a) and (1.37b)

$$m[-A\omega^2 \sin(\omega t - \phi)] + c[A\omega \cos(\omega t - \phi)] + KA \sin(\omega t - \phi)$$
$$= F_0 \sin \omega t = F_0 \sin[(\omega t - \phi) + \phi]$$
$$\Rightarrow\; A[(K - m\omega^2)] \sin(\omega t - \phi) + c\omega \cos(\omega t - \phi)]$$
$$= F_0 [\sin(\omega t - \phi) \cos \phi + \cos(\omega t - \phi) \sin \phi] \tag{1.38}$$

Equating coefficients of $\sin(\omega t - \phi)$ and $\cos(\omega t - \phi)$ in eqn. (1.38)

$$Ac\omega = F_0 \sin \phi \tag{1.39a}$$
$$A(K - m\omega^2) = F_0 \cos \phi \tag{1.39b}$$

Squaring and adding eqns. (1.39a) and (1.39b)

$$F_0^2 = A^2[(K - m\omega^2)^2 + c^2\omega^2] \;\Rightarrow\; A = \frac{F_0}{[(K - m\omega^2)^2 + c^2\omega^2]^{1/2}}$$

$$= \frac{F_0/K}{\left[\left(1 - \frac{m\omega^2}{K}\right)^2 + \frac{c^2\omega^2}{K^2}\right]^{1/2}} \tag{1.40a}$$

Dividing eqn. (1.39a) by eqn. (1.39b)

$$\tan \phi = \frac{c\omega}{(K - m\omega^2)} \tag{1.40b}$$

The A and ϕ for various values of Q is shown in Fig. 1.15.

It is convenient to recast eqn. (1.40a) in a simplified form by making the following substitutions:

$$\omega_n = \sqrt{\frac{K}{m}}\;;\; c = 2m\xi\omega_n\;;\; \delta_{st} = \frac{F_0}{K}\;;\; r = \frac{\omega}{\omega_n} \tag{1.41a}$$

VIBRATIONS AND RESONANCE

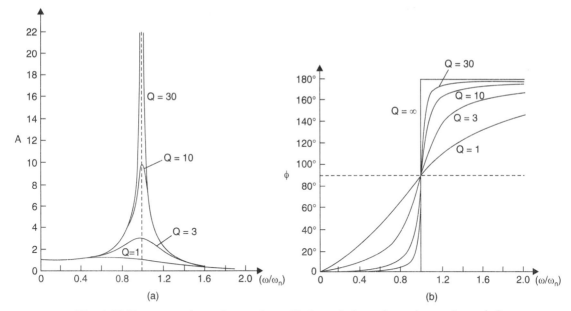

Fig. 1.15 Frequency dependence of amplitude and phase for various values of Q

$$\frac{A}{\delta_{st}} = \frac{1}{\sqrt{(1-r^2)^2 + (2r\xi)^2}} \tag{1.41b}$$

$$\phi = \tan^{-1}\left\{\frac{2r\xi}{1-r^2}\right\} = \tan^{-1}\left\{\frac{2\xi}{\frac{1}{r}-r}\right\} \tag{1.41c}$$

Equation (1.41) gives the amplitude and phase of the forced vibration. Depending on r, the following cases are possible.

Case (i) $r \ll 1$ i.e., when the driving frequency is very much less than the natural frequency.

In this case $(A/\delta_{st}) \sim 1$, since terms r^2, r^4 can be neglected. Note that r can lie only between 0 and 1. The amplitude of vibration is independent of frequency of external force. It depends only on the magnitude of the applied force F_0.

Further $\phi \sim \tan^{-1}(0) \sim 0$. i.e., the displacement and the force are in phase.

Case (ii) $r = 1$, when the frequency of the applied force is equal to the natural frequency of the oscillator.

In this case $A = \delta_{st}/2\xi = Q\delta_{st}$. i.e., the amplitude is proportional to the applied force and the quality factor.

Further $\phi = \tan^{-1}(\infty) = \pi/2$. i.e., the amplitude and the applied force are out of phase by 90°.

Case (iii) $r \gg 1$, when the frequency of the applied force is very much greater than the natural frequency.

Assuming that only r^4 is the dominant term, $A \sim \delta_{st}/r^2 \sim \delta_{st}\, \omega_n^2/\omega^2$. i.e., the amplitude decreases inversely as the square of the frequency of the applied force. It is also proportional to the magnitude of the force.

Further $\phi = \tan^{-1}(-0) = -\pi$. i.e., the displacement and force are out of phase by 180°.

1.5.4 Amplitude Resonance

Equation (1.41a) clearly indicates that amplitude is a function of frequency. It is interesting to find for what r, A will be maximum. Clearly A is maximum when the denominator is least i.e., when

$$\frac{d}{dr}\{(1-r^2)^2 + (2r\xi)^2\} = 0$$

$$\Rightarrow \quad 2(1-r^2)(-2r) + 8r\xi^2 = 0 \Rightarrow 4r[2\xi^2 - (1-r^2)] = 0 \text{ or } 1-r^2 = 2\xi^2 \text{ or } r = \sqrt{1-2\xi^2}$$

i.e., amplitude is maximum for angular frequency

$$\omega = \omega_n \sqrt{1-2\xi^2} \qquad (1.42a)$$

which is neither the natural frequency ω_n nor the frequency ω_d the damped oscillator.

1.5.5 Velocity Resonance

The solution to eqn. (1.36) is given by

$$x = A \sin(\omega t - \phi) \text{ where } A \text{ and } \phi \text{ are given by eqn. (1.41)}$$

$$v = \text{velocity} = \frac{dx}{dt} = A\omega \cos(\omega t - \phi) \Rightarrow v_{max} = A\omega$$

i.e.,

$$v_{max} = A\omega = \frac{\delta_{st} \cdot \omega}{\sqrt{(1-r^2)^2 + 4r^2\xi^2}} = \frac{\delta_{st} \cdot \omega_n \cdot r}{\sqrt{(1-r^2)^2 + 4r^2\xi^2}} = \frac{\delta_{st}\omega_n}{\sqrt{\left(\frac{1}{r} - r\right)^2 + 4\xi^2}} \qquad (1.42b)$$

The above expression is greatest when $r = 1$ i.e., $\omega = \omega_n$ and its value is $\dfrac{\delta_{st}\omega_n}{2\xi}$

Note that the amplitude resonance and velocity resonance occur at frequencies different from each other. For $\xi \ll 1$ (~ 0.05), both the resonances are at $\omega = \omega_n$.

1.5.6 Power Absorbed by a Driven Oscillator

Whenever an oscillator is driven by an external force, energy is absorbed by the oscillator. The energy absorbed by the oscillator is equal to the energy dissipated due to damping. The rate of energy absorption or power absorbed is a function of driving frequency. It is maximum at resonance i.e., when the frequency of the periodic force is equal to that of the natural frequency of the oscillator. The power absorbed by the oscillator is given by:

$$\text{Power} = \text{Viscous force} \times \text{Velocity}$$

From eqn. (1.36)

$$\text{Viscous force} = c\frac{dx}{dt} \quad \therefore \quad \text{power} = c\left(\frac{dx}{dt}\right)^2$$

From eqn. (1.37b)

$$x(t) = A \sin \omega t \quad \text{and} \quad \frac{dx}{dt} = A\omega \cos \omega t \text{ assuming } \phi = 0 \text{ at } t = 0$$

Power (ΔP) absorbed by the oscillator in one cycle is given by:

$$\Delta P = \int_0^{2\pi/\omega} c[A\omega \cos \omega t]^2 \, dt = cA^2\omega^2 \int_0^{2\pi/\omega} \cos^2 \omega t \, dt$$

$$= \frac{cA^2\omega^2}{2}\left[\int_0^{2\pi/\omega}\cos 2\omega t\,dt + \int_0^{2\pi/\omega} dt\right] = \frac{cA^2\omega^2}{2} \times \frac{2\pi}{\omega} = \pi c\omega A^2$$

From eqn. (1.37b)

Power = No. of cycles/sec × Average power per cycle

$$= \frac{\omega}{2\pi} \times \pi c\omega A^2 = \frac{1}{2} cA^2\omega^2 \tag{1.43}$$

The energy absorbed by the oscillator is equal to the energy dissipated due to damping. From eqns. (1.43) and (1.41a)

$$P = \frac{1}{2}cA^2\omega^2 = \frac{c}{2}\left[\frac{\delta_{st}^2\omega^2}{(1-r^2)^2 + (2\xi r)^2}\right]$$

$$= \frac{c}{2}\delta_{st}^2\omega_n^2 \frac{r^2}{(1-r^2)^2 + (2\xi r)^2} = \frac{c}{2}\delta_{st}^2\omega_n^2\left[\frac{1}{\left(\frac{1}{r}-r\right)^2 + 4\xi^2}\right] \tag{1.44}$$

A plot of P for various Q-factors is shown in Fig. 1.16.

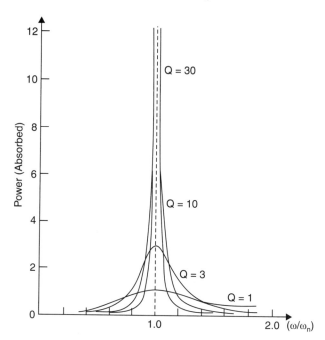

Fig. 1.16 Frequency dependence of mean power absorbed by an oscillator for various values of Q

Clearly this is maximum at $r = 1$ or $\omega = \omega_n$. The frequency of an undamped oscillator $(\omega_d) = \omega_n(1-\xi^2)^{1/2}$. Hence the power resonance occurs at a frequency different from the ω_d.

$$P_{max} = \frac{c}{2}\delta_{st}^2\omega_n^2 \frac{1}{4\xi^2} \tag{1.45}$$

From eqns. (1.44) and (1.45)

$$P = \frac{P_{max} 4\xi^2}{\left(\frac{1}{r} - r\right)^2 + 4\xi^2} \qquad (1.46)$$

1.5.7 Resonance, Quality Factor and Bandwidth

Let us find the value of r for which

$$P = \frac{P_{max}}{2}$$

$$\Rightarrow \qquad \frac{1}{2} = \frac{4\xi^2}{\left(\frac{1}{r} - r\right)^2 + 4\xi^2}$$

$$\Rightarrow \qquad \left(\frac{1}{r} - r\right)^2 = 4\xi^2 \quad \text{or} \quad \left(\frac{1}{r} - r\right) = \pm 2\xi \qquad 1 - r^2 = \pm 2r\xi \quad \text{or} \quad r^2 \pm 2r\xi - 1 = 0$$

Since $r > 0$

$$r_1 = \frac{\omega_1}{\omega_n} = \frac{\sqrt{4\xi^2 + 1} - 2\xi}{2} \quad \text{and} \quad r_2 = \frac{\omega_2}{\omega_n} = \frac{\sqrt{4\xi^2 + 1} + 2\xi}{2}$$

$$\Rightarrow \qquad 2(r_2 - r_1) = 4\xi \quad \text{or} \quad (\omega_2 - \omega_1) = (\Delta\omega) = 2\xi\omega_n \quad \Rightarrow \quad \frac{\omega_n}{\Delta\omega} = \frac{1}{2\xi} = Q \qquad (1.47)$$

The frequencies ω_1 and ω_2 corresponding to r_1 and r_2 are known as *half power points* (Fig. 1.17). $\omega_2 - \omega_1 = \Delta\omega$ is known as the *bandwidth*. It is possible to arrive at an expression for the bandwidth by considering the expression for the amplitude. For small values of damping ($\xi < 0.05$), from eqn. (1.41b), $\omega_d \approx \omega_n$.

$$\left(\frac{A}{\delta_{st}}\right)_{\omega = \omega_n} \approx \frac{1}{2\xi} \approx Q$$

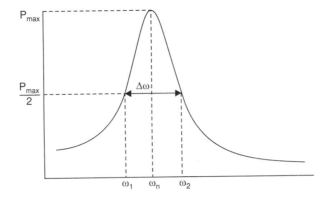

Fig. 1.17 Bandwidth and half power points

VIBRATIONS AND RESONANCE

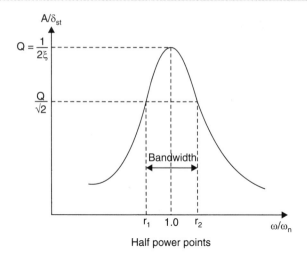

Fig. 1.18 Bandwidth and half power points

Thus Q is also equal to the amplitude ratio at resonance (Fig. 1.18). The points r_1 and r_2 where the amplification factor falls to $A/\sqrt{2}$ as called half-power points because the power absorbed by the damped oscillator responding harmonically at a given frequency, is proportional to the square of amplitude. The difference between the frequencies associated with the half power points r_1 and r_2 is called the bandwidth. To find the values of r_1 and r_2, from eqn. (1.41b), we have

$$\frac{Q}{\sqrt{2}} = \frac{1}{\sqrt{(1-r^2)^2 + (2\xi r)^2}} = \frac{1}{2\sqrt{2}\xi} \Rightarrow (1-r^2)^2 + (2\xi r)^2 = 8\xi^2$$

$$\Rightarrow \quad r^4 + r^2(4\xi^2 - 2) + (1 - 8\xi^2) = 0$$

This is a quadratic in r^2

$$\Rightarrow \quad r_1^2 = 1 - 2\xi^2 - 2\xi(1+\xi^2)^{1/2} \text{ and } r_2^2 = 1 - 2\xi^2 + 2\xi(1+\xi^2)^{1/2}$$

where $\omega = \omega_1$ at r_1 and $\omega = \omega_2$ at r_2

Making the approximation

$$(1+\xi^2)^{1/2} \approx \left(1 + \frac{1}{2}\xi^2\right), \text{ we get}$$

$$r_1^2 = \left(\frac{\omega_1}{\omega_n}\right)^2 = (1-2\xi^2) - 2\xi\left(1 + \frac{1}{2}\xi^2\right) \approx 1 - 2\xi$$

$$r_2^2 = \left(\frac{\omega_2}{\omega_n}\right)^2 = (1-2\xi^2) + 2\xi\left(1 + \frac{1}{2}\xi^2\right) \approx 1 + 2\xi$$

$$\omega_2^2 - \omega_1^2 = (\omega_2 + \omega_1)(\omega_2 - \omega_1) = (r_2^2 - r_1^2)\omega_n^2 \approx 4\xi\omega_n^2$$

But $\quad \omega_1 + \omega_2 = 2\omega_n$

$\therefore \quad \Delta\omega = \omega_2 - \omega_1 \approx 2\xi\omega_n$

$$\therefore \quad Q = \frac{1}{2\xi} \approx \frac{\omega_n}{\Delta\omega} \approx \frac{\omega_n}{\omega_2 - \omega_1} \quad (1.48)$$

Sharpness of Resonance

We have seen that the amplitude of forced vibration is maximum when the frequency of the applied force has the value $\omega = \omega_n(1 - 2\xi^2)^{1/2}$. If the frequency changes from this value, the amplitude falls. When the fall in amplitude for a small departure from the resonance condition is very large, the resonance is said to be sharp. On the other hand if the fall in amplitude is small, the resonance is termed as flat. Thus, the term sharpness of resonance means the rate of fall in amplitude with the change of forcing frequency on each side of the resonance frequency. Figure 1.15 shows the variation of amplitude with forcing frequency at different amounts of damping or Q-factor. Clearly smaller the damping, sharper the resonance and larger the damping flatter the resonance. As can be seen from the figure, larger the Q, smaller the bandwidth and sharper the resonance. In fact Q-factor is inversely proportional to the bandwidth as shown in eqns. (1.47) and (1.48).

1.6 VIBRATION ISOLATION

1.6.1 Introduction

Machines such as motors, fans and compressors produce a vibratory force at a particular or a range of frequencies. These are often a source of irritating noise that propagates through air and the ground. The vibratory force generated by such machines often leads to loosening of fasteners, excessive wear of bearings, formation of cracks and structural as well as mechanical failures. Electronic malfunctioning through fracture of solder joints and abrasion of insulation around conducting wires can also occur. Also there are many practical situations in which a delicate machinery has to be isolated from vibratory impacts that it is subjected to. All vehicles are provided with shock absorbers so that when they move on a rough and bumpy surface, the jerky motion is not communicated to the engine and the passengers. In all these cases one employs vibration isolation techniques, which reduce the undesirable effects of vibration. The design of a vibration isolation system is based on the theory of forced vibration. The vibration isolation system is said to be active or passive depending on whether the external power is required for the isolator to perform its function or not. A passive isolator consists of a resilient member (stiffner or spring) and an energy dissipator (dampner). Examples of passive isolators include metal springs, cork, felt, pneumatic springs and elastomer (rubber) springs. An active isolator is composed of a servomechanism with a sensor, signal processor and an actuator. A servomechanism essentially senses the vibratory motion and if it exceeds a preset value, activates an actuator for corrective action.

1.6.2 Basic Theory

Vibration isolation is used in two types of situations as shown in Figs. 1.19 and 1.20.

Case (i):

Consider for instance a heavy machine. To reduce the vibratory motion of a heavy machine that is communicated to the floor, it is mounted on a rigid base. In this case, efficiency of isolation is defined by force transmittability.

$$T = \frac{\text{Amplitude of the force transmitted to the base}}{\text{Amplitude of the force exerted by the machine}} = \frac{F_T}{F_0} = \sqrt{\frac{1 + 4\xi^2 r^2}{(1 - r^2)^2 + 4\xi^2 r^2}}$$

Case (ii):

This corresponds to the design of a shock absorber where a delicate machinery has to be insulated from the vibrations transmitted from the ground. Here one defines displacement transmittibility

VIBRATIONS AND RESONANCE

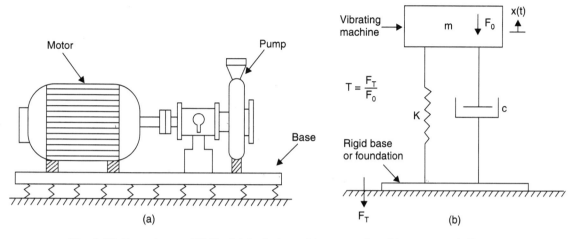

Fig. 1.19 Force transmittibility (a) typical machine mounting (b) conceptual diagram

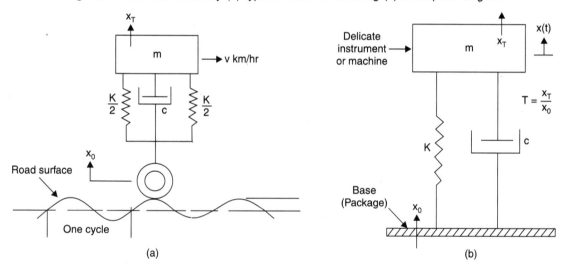

Fig. 1.20 Displacement transmittibility (a) Shock absorber (b) Conceptual diagram

$$T = \frac{\text{Amplitude of transmitted displacement to the machinery}}{\text{Amplitude of applied displacement}} = \frac{x_T}{x_0} = \frac{1}{\sqrt{(1-r^2)^2 + 4\xi^2 r^2}}$$

For a simple mass-spring system the force and displacement transmittibilities are equal and are given by the above equations. Figure 1.21 shows how T varies with frequencies for various amounts of damping. At frequencies below resonance $T = 1$, indicating that the mass and the base move in effect together as if rigidly connected. As the resonant frequency is approached, the transmittability increases greatly, indicating an amplification of the vibration being transmissied through the structure. The maximum transmittibility at resonance depends on the amount of damping. Above resonance, the transmittibility falls to 1 until at a frequency corresponding to $r = 1.414$. At higher frequencies T is less than 1, indicating that the vibrations are being alternated or isolated as they travel through the structure. In this region, well above resonance, it can be seen that the amount of damping affects the transmittibility. A lightly damped system has a lower T value at the same transmitted frequency ratio than a more heavily damped system.

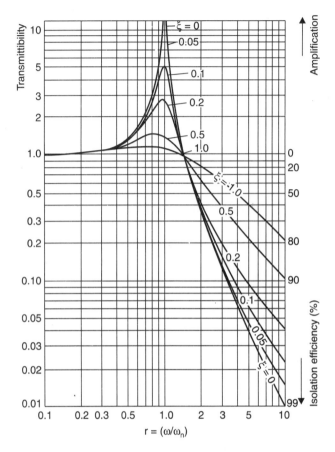

Fig. 1.21 Frequency dependence of transmittibility for various values of ξ

A machine such as a motor, fan or engine produces a vibratory force at a particular frequency ω, related to its rotational speed. The machine and the isolator form a mass-spring system with a resonating frequency ω_n, which is related to the mass of the machine and the stiffness of the isolator. The isolator must be selected such that ω_n is low enough to achieve a value of $r(=\omega/\omega_n)$ high enough to produce the required degree of isolation i.e., $r \gg 1$. The damping of the isolator should in theory be as small as possible to achieve the best reductions at a given frequency ratio. However, in practice, damping can be useful since in many machines the vibratory force passes through the resonance frequency of the system during run up or run down. The damping helps to limit the vibration amplitude of the machine as it passes through the resonance speed.

1.6.3 Vibration Absorbers

A machine or system will experience excessive vibration if it is acted upon by a force whose excitation frequency nearly coincides with the natural frequency of the machine or system. In such cases, the vibration of the machine or system can be reduced by using a vibration neutralizer or dynamic vibration absorber. This vibration absorber is another spring mass system. It is designed such that the natural frequency of the resulting system is away from the excitation frequency (Fig. 1.22).

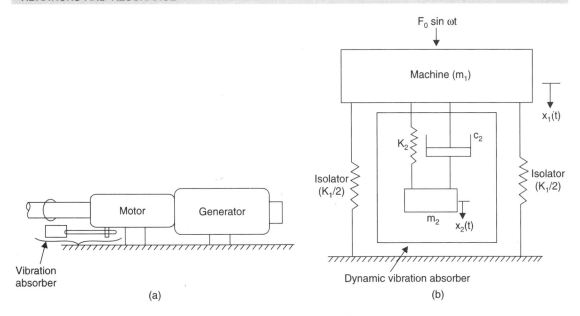

Fig. 1.22 Vibration absorber

1.6.4 Vibration Analysis and Control

Engineers routinely carry out vibration analysis of machines and structures. This is essential for optimal efficiency of operation. The measurement of the natural frequencies of a structure or a machine is useful in selecting the operation speeds of nearby machinery to avoid resonant conditions. This also helps in the design of vibration isolation systems. In addition it helps in understanding the operations of machines or structures under specific vibrational environment such as road surface conditions, fluctuating wind velocities, random variation of ocean waves and ground vibrations due to earthquake. Figure 1.23 shows the basic features of vibrational measurement. The motion of the vibrating body is converted into an electrical signal by a transducer or pick-up. Transducers measure displacement, velocity and acceleration. The output signal is usually amplified and the data presented on a display unit for visual inspection. The vibration measuring instrument is called a vibrometer.

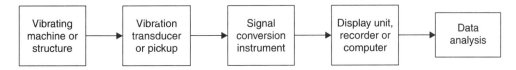

Fig. 1.23 Basic vibration measurement scheme

1.7 LCR CIRCUIT ANALYSIS—ANALYTICAL APPROACH

1.7.1 Introduction

The analytical approach that we have adopted in analysing the free, damped and forced vibrations can be adopted to analyse LCR circuits as well. This brings out the similarity between the mechanical and electrical oscillations.

1.7.2 Free Oscillation in an LC Circuit

Figure 1.24 shows a simple circuit consisting of a pure capacitor and a pure inductor. The capacitor is initially charged using a battery and then allowed to discharge across the inductor. At any time

The potential across the plate $V_C = \dfrac{q}{C}$

The potential across the inductor $V_L = -L\dfrac{di}{dt} = -L\dfrac{d^2q}{dt^2}$

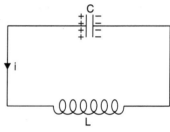

Fig. 1.24 LC circuit

The negative sign indicates that the voltage opposes the increase of current.
On applying Kirchhoff's second law

$$L\left(\dfrac{d^2q}{dt^2}\right) + \dfrac{q}{C} = 0 \text{ or } \dfrac{d^2q}{dt^2} + \dfrac{1}{LC}q = 0 \Rightarrow \dfrac{d^2q}{dt^2} = -\dfrac{1}{LC}q \quad (1.49)$$

Comparing eqns. (1.5) and (1.49) we get

$$\omega_n = \dfrac{1}{\sqrt{LC}} \text{ or } T = 2\pi\sqrt{LC} \quad (1.50)$$

Thus the magnitudes of charge on the plate varies periodically with a time

$$T = 2\pi\sqrt{LC}$$

A pure LC circuit is an electrical analogue of the undamped simple pendulum. Just as in the case of simple pendulum where the energy alternates between the potential and kinetic energy, here the energy is alternately stored in the capacitor as electric field and inductor as magnetic field.

1.7.3 Damped Oscillations in a Series LCR Circuit

Figure 1.25 shows a circuit consisting of a capacitor, inductor and a resistor. The capacitor is initially charged and then allowed to discharge across the inductor and the resistor. At any instant of time

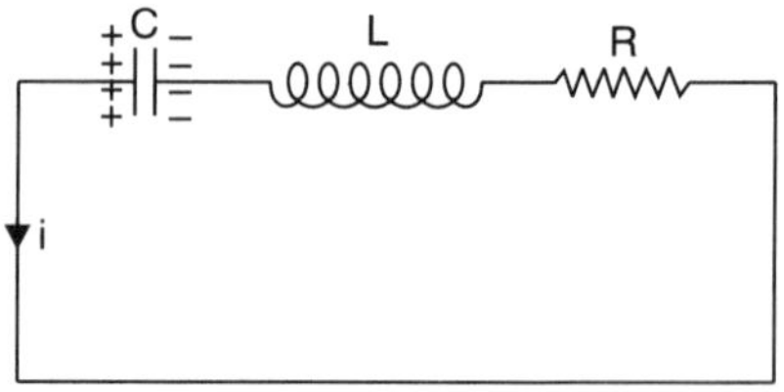

Voltage across the capacitor $= V_C = \dfrac{q}{C}$

Voltage across the inductor $= -L\dfrac{di}{dt} = -L\dfrac{d^2q}{dt^2}$

Fig. 1.25 LCR circuit

Voltage across the resistor $= iR = R\dfrac{dq}{dt}$

Applying Kirchhoff's second law

$$L\dfrac{d^2q}{dt^2} + R\cdot\dfrac{dq}{dt} + \dfrac{q}{C} = 0 \Rightarrow \dfrac{d^2q}{dt^2} + \dfrac{R}{L}\cdot\dfrac{dq}{dt} + \dfrac{q}{LC} = 0 \quad (1.51)$$

Comparing eqns. (1.51) and (1.14) we get

$$\omega_n^2 = \dfrac{1}{LC} \text{ and } 2\xi\omega_n = \dfrac{R}{L} \Rightarrow \xi = \dfrac{R}{2L\omega_n} = \dfrac{R}{2}\sqrt{\dfrac{C}{L}} \quad (1.52)$$

VIBRATIONS AND RESONANCE

Thus, the variation of q and $i(= dq/dt)$ is same as that of displacement and velocity with time and is shown in Fig 1.3. Note also that there is a phase lag of 90° between the charge oscillations and current oscillations. Here also one can differentiate between underdamped and overdamped cases.

Overdamping: This refers to the case when

$$\xi > 1 \Rightarrow \frac{R}{2}\sqrt{\frac{C}{L}} > 1 \Rightarrow R > 2\sqrt{\frac{L}{C}}$$

In this case charge on the plate decays with time. It is non-oscillatory and does not quickly decay with time.

Underdamping: This refers to the case when

$$\xi < 1 \Rightarrow \frac{R}{2}\sqrt{\frac{C}{L}} < 1 \Rightarrow R < 2\sqrt{\frac{L}{C}}$$

In this case charge on the plate oscillates with an angular frequency

$$\omega_d = \omega_n\sqrt{1-\xi^2} = \frac{1}{\sqrt{LC}}\sqrt{1-\frac{R^2 C}{4L}} = \sqrt{\frac{1}{LC}-\frac{R^2}{4L^2}} \qquad (1.53)$$

The energy stored in the capacitor decays with time as shown in Fig. 1.10

1.7.4 Forced Oscillations in a Series LCR Circuit

Figure 1.26 shows a series LCR circuit driven by an alternating applied voltage

$$V = V_0 \sin \omega t.$$

Once again applying Kirchhoff's second law

$$L\frac{d^2q}{dt^2} + R \cdot \frac{dq}{dt} + \frac{q}{C} = V_0 \sin \omega t$$

$$\Rightarrow \frac{d^2q}{dt^2} + \frac{R}{L}\cdot\frac{dq}{dt} + \frac{q}{LC} = \frac{V_0}{L}\sin \omega_0 t \qquad (1.54)$$

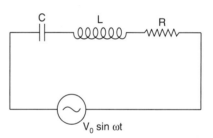

Fig. 1.26 Driven LCR circuit

Comparing eqn. (1.54) with eqn. (1.36) following analogy is possible

Table 1.1 Correspondence between physical quantities in mechanical and electrical oscillations

Displacement x	→	Charge q
Velocity dx/dt	→	Current dq/dt
Mass m	→	Inductance L
Damping coefficient c	→	Resistance R
Spring constant K	→	Reciprocal of capacitance $1/C$
Force amplitude F_o	→	Voltage amplitude V_0
Driving frequency ω	→	Oscillator ω
$\xi = \dfrac{c}{2m\omega_n}$	→	$\xi = \dfrac{R}{2}\sqrt{\dfrac{C}{L}}$

Comparing with eqn. (1.41a) we have

$$\delta_{st} = V_0 C; \quad \omega_n = \frac{1}{\sqrt{LC}}; \quad r = \frac{\omega}{\omega_n}; \quad \xi = \frac{R}{2}\sqrt{\frac{C}{L}} \tag{1.55a}$$

$$\therefore \quad \frac{q_0}{CV_0} = \frac{1}{\sqrt{(1-LC\omega^2)^2 + R^2C^2\omega^2}} \tag{1.55b}$$

$$\tan\phi = \frac{RC\omega}{1-LC\omega^2} = \left(\frac{R}{X_C - X_L}\right) \tag{1.55c}$$

where X_C and X_L are the impedances due to capacitor and inductor and are given by

$$X_C = \frac{1}{C\omega} \text{ and } X_L = L\omega$$

The variation of current with frequency can be obtained as follows:

$$q = q_0 \sin(\omega t - \phi)$$

$$i = \frac{dq}{dt} = q_0\omega \cos(\omega t - \phi) = i_0 \cos(\omega t - \phi) \tag{1.56}$$

From eqns. (1.55) and (1.56)

$$i_0 = \frac{V_0 C\omega}{\sqrt{(1-\omega^2 L_C)^2 + R^2 C^2 \omega^2}} = \frac{V_0}{\sqrt{\left(\frac{1}{C\omega} - L\omega\right)^2 + R^2}} = \frac{V_0}{\sqrt{(X_C - X_L)^2 + R^2}} \tag{1.57}$$

Power absorption:

From eqn. (1.45), (1.55) and table 1.1

$$P_{max} = \frac{C}{2}\cdot\delta_{st}^2\cdot\omega_n^2\cdot\frac{1}{4\xi^2} = \left(\frac{R}{2}\right)(CV_0)^2\cdot\left(\frac{1}{LC}\right)\left(\frac{1}{R}\sqrt{\frac{L}{C}}\right)^2 = \frac{V_0^2}{2R} = \frac{V_{rms}^2}{R} \tag{1.58}$$

Also from eqn. (1.46)

$$P = \frac{P_{max}\cdot 4\xi^2}{\left(\frac{1}{r}-r\right)^2 + 4\xi^2} = \frac{P_{max}}{\left(\frac{1}{2\xi r} - \frac{r}{2\xi}\right)^2 + 1} = \frac{P_{max}}{\left(\frac{1}{RC\omega} - \frac{L\omega}{R}\right)^2 + 1} = \frac{P_{max}}{\left(\frac{X_C - X_L}{R}\right)^2 + 1} \tag{1.59}$$

Q-factor

From eqns. (1.47) and (1.52)

$$Q = \frac{1}{2\xi} = \frac{\omega_n}{\Delta\omega} = \frac{1}{R}\sqrt{\frac{L}{C}} \tag{1.60}$$

1.7.5 Resonance in a Series LCR Circuit—Phasor Analysis

Consider and a.c. circuit consisting of resistance R, inductance L, and capacitance C connected in series as shown in Fig. 1.27.

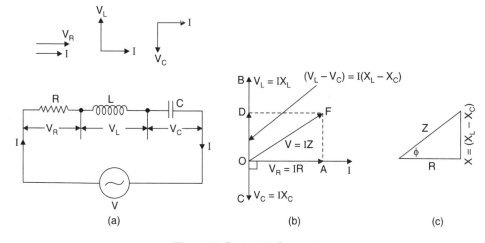

Fig. 1.27 Series RLC circuit
(a) circuit diagram (b) phasor diagram (c) impedence triangle

Let V be the r.m.s. value of applied voltage and I be the r.m.s. value of current.

∴ Voltage drop across R, $V_R = IR$ (in phase with I)

Voltage drop across L, $V_L = IX_L$ (leading I by 90°)

Voltage drop across C, $V_C = IX_C$ (lagging I by 90°)

The applied voltage is the vector sum of V_R, V_L and V_C

The phasor diagram is drawn as shown in Fig. 1.27b

$(OA = V_R = IR, OB = V_L = IX_L, OC = V_C = IX_C)$

Assuming that $V_L > V_C$. $OD = OB - OC$

$$= V_L - V_C = IX_L - IX_C = I(X_L - X_C)$$
$$= IX$$
$$OF = IZ$$

From triangle OAF

$$V = \sqrt{V_R^2 + (V_L - V_C)^2}$$
$$= \sqrt{(IR)^2 + (IX_L - IX_C)^2}$$
$$= I \times \sqrt{R^2 + (X_L - X_C)^2} = IZ$$

where $Z = \sqrt{R^2 + (X_L - X_C)^2}$ is the impedance of the circuit.

In the phasor diagram the current I lags behind the applied voltage by an angle ϕ such that

$$\tan \phi = \frac{AF}{OA} = \frac{I(X_L - X_C)}{IR} = \frac{(X_L - X_C)}{R} = \frac{X}{R}$$

Where $X = (X_L - X_C)$ is the net reactance of the circuit.

∴ Phase angle, $\phi = \tan^{-1}(X/R)$

Impedance triangle of the circuits is shown in Fig. 1.27c.

As in the previous cases, the power in the RLC series circuit can be obtained as
$$P = VI \cos \phi \, (= I^2 R)$$

Note that power factor $= \cos \phi = \dfrac{R}{\sqrt{R^2 + (X_L - X_C)^2}}$

Power is consumed only in the resistance (R) of the circuit.

If $V_C > V_L$, then the current will lead the applied voltage V, by an angle 'ϕ' such that $\phi = \tan^{-1}(X_C - X_L)/R$ and power factor $= \cos \phi = R/\sqrt{R^2 + (X_C - X_L)^2}$

If $V_L = V_C$ (i.e., $X_L = X_C$) then power factor $= \cos \phi =$ unity. In this situation current is maximum and corresponds to the series resonance condition.

The frequency of the resonance is given by
$$X_L = X_C$$
$$L\omega = \dfrac{1}{C\omega} \text{ i.e., } \omega^2 = \dfrac{1}{LC} \text{ i.e., } f_r^2 = \dfrac{1}{4\pi^2 LC}$$

or
$$f_r = \dfrac{1}{2\pi\sqrt{LC}}$$

Resonance Curve and Q-Factor

A plot of current versus frequency in the LCR circuits is known as resonance curve. The shape of such a curve for various values of R is shown in Fig. 1.28a. For smaller values of R, the current frequency curve is sharply peaked, but for large values of R the curve is flat. The variation of Z, power factor and I are separately shown in Fig. 1.28b.

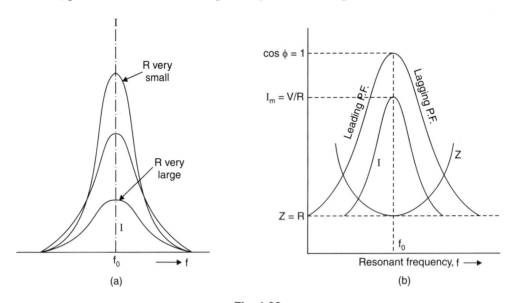

Fig. 1.28

VIBRATIONS AND RESONANCE

The ability of a reactive circuit to store energy is expressed in terms of the quality factor or Q-factor. It is a figure of merit which enables us to compare different coils. The Q-factor is defined as the ratio of the energy stored in the coil to the energy dissipated in the circuit across the resistance.

Thus
$$Q = 2\pi \frac{\text{Energy stored}}{\text{Energy dissipated per cycle}}$$

Thus larger the Q-factor, greater is the storing ability for a given dissipation.

In the case of an inductor the energy stored is $I^2 X_L t$. The energy dissipated is $I^2 R t$

$\therefore \quad Q_L = \dfrac{I^2 \omega L t}{I^2 R t} = \dfrac{\omega L}{R}$

$\quad = \dfrac{1}{R}\sqrt{\dfrac{L}{C}}$ at resonance $\qquad \left(\because \omega = \dfrac{1}{\sqrt{LC}}\right)$

In the case of a capacitor

$Q = \dfrac{I^2 (1/\omega C) t}{I^2 R t} = \dfrac{(1/\omega C)}{R} = \dfrac{1}{\omega C R}$

$\quad = \dfrac{1}{R}\sqrt{\dfrac{L}{C}}$ at resonance $\qquad \left(\because \omega = \dfrac{1}{\sqrt{LC}}\right)$

The Q-factor varies from 5 to 100 for inductive circuits and from about 1400 to 10000 for capacitive circuits. When a series circuit is in resonance the energy stored in the capacitor is equal to that stored in inductor. In one quarter cycle the inductor stores energy while in the next quarter cycle it is stored in the capacitor. The energy flows back and forth between inductance and capacitance. The only loss of energy is due to the energy loss in the resistance. Thus if the resistance is small, the energy oscillation continues for a long time even though there is no supply of energy from an external source.

In the case of an LCR circuit, Q-factor may also be defined as equal to the voltage magnification in the circuit at resonance. We have seen that at resonance, the current is maximum

i.e., $\qquad I_{max} = \dfrac{V}{R}$

Voltage across the coil $= I_{max} \cdot X_L$
Supply voltage $= V = I_{max} R$

$\therefore$ Voltage magnification $= \dfrac{I_{max} X_L}{I_{max} \cdot R} = \dfrac{X_L}{R} = \dfrac{\omega L}{R}$

$\therefore \qquad Q\text{-factor} = \dfrac{\omega L}{R} = \dfrac{2\pi f_r L}{R}$

But the resonant frequency $= f_r = \dfrac{1}{2\pi\sqrt{LC}}$

$\therefore \qquad Q = \dfrac{1}{R}\sqrt{\dfrac{L}{C}}$

Voltage across capacitor = $I_{max} \cdot X_C$

$\therefore$ Voltage magnification = $\dfrac{I_{max} X_C}{I_{max} R} = \dfrac{X_C}{R} = \dfrac{1}{\omega CR}$

$\therefore$ Q-factor = $\dfrac{1}{\omega CR} = \dfrac{1}{1/\sqrt{LC} \cdot CR} = \dfrac{1}{R}\sqrt{\dfrac{L}{C}}$

The Q-Factor of a Series LCR Circuit and Selectivity

At resonance the reactance is zero. Hence the current at resonance is $I_r = \dfrac{V}{R}$.

Consider two frequencies on either side of f_r where the reactance is equal to resistance (Fig. 1.29). Hence at these frequencies $Z = \sqrt{R^2 + R^2} = \sqrt{2} \cdot R$.

The current I at these frequencies is

$$I = \dfrac{V}{\sqrt{2} \cdot R}$$

Power dissipation at these frequencies

$$P_1 = \left(\dfrac{V}{\sqrt{2} \cdot R}\right)^2 R = \dfrac{V^2}{2R} = P_2$$

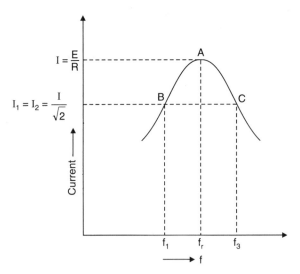

Fig. 1.29

The power dissipation at resonant frequency is

$$P_r = \left(\dfrac{V}{R}\right)^2 R = \dfrac{V^2}{R}$$

Thus the power dissipation at f_1 and f_2 is half that at resonance. Hence the points B and C are called half power points.

VIBRATIONS AND RESONANCE

At f_1 which is below f_r
$$X_L - X_C = -R$$

i.e.,
$$\omega_1 L - \frac{1}{\omega_1 C} = -R \quad \text{or} \quad \frac{1}{\omega_1 C} - \omega_1 L = R \tag{1.61}$$

the circuit is capacitive.

At f_2 which is above f_r
$$X_L - X_C = R$$

i.e.,
$$\omega_2 L - \frac{1}{\omega_2 C} = R \tag{1.62}$$

the circuit is inductive.

From (1.61) and (1.62), we get

$$(\omega_2 - \omega_1) L + \frac{1}{C}\left(\frac{\omega_2 - \omega_1}{\omega_1 \omega_2}\right) = 2R$$

or
$$(\omega_2 - \omega_1)\left[L + \frac{1}{\omega_1 \omega_2 C}\right] = 2R$$

Dividing by L

$$(\omega_2 - \omega_1)\left[1 + \frac{1}{\omega_1 \omega_2 LC}\right] = \frac{2R}{L} \tag{1.63}$$

As f_1 and f_2 are close together we can write to a first degree of approximation

$$\omega_1 L = \frac{1}{\omega_2 C}$$

Then $\omega_1 \omega_2 LC = 1$. Then (1.63) becomes

$$(\omega_2 - \omega_1)(1 + 1) = \frac{2R}{L}$$

$$(\omega_2 - \omega_1) = \frac{R}{L}$$

or
$$f_2 - f_1 = \frac{R}{2\pi L} \tag{1.64}$$

But
$$Q = \frac{\omega_r L}{R} = \frac{2\pi f_r L}{R}$$

or
$$f_r = \frac{QR}{2\pi L} \tag{1.65}$$

From (1.64) and (1.65)

$$\frac{f_2 - f_1}{f_r} = \frac{1}{Q} \tag{1.66}$$

This is identical to eqn (1.48).

$f_2 - f_1$ is called bandwidth B

$$B = f_2 - f_1 = f_r/Q \qquad (1.67)$$

If the bandwidth is small, the resonance is to be sharp.

Suppose the applied a.c. voltage has a number of frequency components. The LCR circuit will give maximum response to that component frequency that is equal or nearly equal to its resonant frequency. Thus the circuit exhibits selectivity. For this reason the circuit is called an acceptor circuit. Equation (1.67) shows that smaller the bandwidth greater the Q-factor. Hence a decrease in bandwidth results in better selectivity. The selectivity also increases as the ratio L/R is increased or the product CR is decreased.

1.7.6 Resonance in a Parallel LCR Circuit

We will consider the practical case of a coil in parallel with a capacitor as shown in Fig. 1.30. Such a circuit is said to be in electrical resonance when the reactive (or wattless) component of line current becomes zero. The frequency at which this happens is known as resonant frequency.

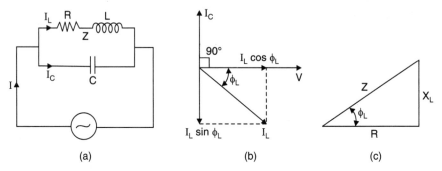

Fig. 1.30 RLC parallel circuit
(a) circuit diagram (b) phasor diagram (c) impedance triangle

The vector diagram for the circuit is shown in Fig. 1.30b.

Net reactive or wattless component $= I_C - I_L \sin \phi_L$

At resonance its value is zero.

$\therefore \qquad I_C - I_L \sin \phi_L = 0$ or $I_L \sin \phi = I_C$

Now $\qquad I_L = \dfrac{V}{Z}$; $\sin \phi_L = \dfrac{X_L}{Z}$; $I_C = \dfrac{V}{X_C}$

Hence the condition for resonance becomes

$$\frac{V}{Z} \frac{X_L}{Z} = \frac{V}{X_C} \quad \text{or} \quad X_L X_C = Z^2$$

Now $\qquad X_L = \omega L$ and $X_C = \dfrac{1}{\omega C}$

$$\frac{\omega L}{\omega C} = Z^2 \text{ or } \frac{L}{C} = Z^2$$

or $$\frac{L}{C} = R^2 + X_L^2$$

VIBRATIONS AND RESONANCE

or
$$\frac{L}{C} = R^2 + (2\pi f_r L)^2$$

or
$$(2\pi f_r L)^2 = \frac{L}{C} - R^2$$

$$2\pi f_r = \sqrt{\frac{1}{LC} - \frac{R^2}{L^2}} \quad \text{or} \quad f_r = \frac{1}{2\pi}\sqrt{\frac{1}{LC} - \frac{R^2}{L^2}}$$

This is the resonant frequency and is given in Hz if R is in ohms, L is in henrys and C is in farads. If R is negligible

$f_r = \dfrac{1}{2\pi \sqrt{LC}}$ which is same as for series resonance.

Because the wattless current is zero, the circuit current is minimum and is

$$I_{min} = I_L \cos \phi = \frac{V}{Z}\frac{R}{Z}$$

or
$$I_{min} = \frac{VR}{Z^2}$$

Putting the value $Z^2 = L/C$, we get

$$I_{min} = \frac{VR}{L/C} = \frac{V}{L/CR}$$

The denominator (L/CR) is known as the equivalent or dynamic impedance of the parallel circuit at resonance. It should be noted that this resistance is resistive only. Since current is minimum at resonance, L/CR must then represent the maximum impedance of the circuit.

Current at resonance is minimum. Hence such a circuit is sometimes known as rejector circuit because it rejects (or takes minimum current of) that frequency to which it resonates.

Q-Factor of a Parallel Circuit

It is defined as the ratio of the current through the coil or capacitor to the main current or as the current magnification at resonance.

Then Q-factor $= I_C/I$

Now
$$I_C = V/X_C = V/1/\omega C = V\omega C$$

$$I_{min} = \frac{V}{(L/CR)}$$

$$Q = \frac{I_c}{I_{min}} = \frac{V\omega C}{\dfrac{V}{(L/CR)}} = \frac{\omega L}{R} = \frac{2\pi f_r L}{R}$$

Now at resonant frequency when R is negligible,

$$f_r = \frac{1}{2\pi \sqrt{LC}}$$

$$Q = \frac{2\pi L}{R} \cdot \frac{1}{2\pi\sqrt{LC}} = \frac{1}{R}\sqrt{\frac{L}{C}}.$$

It should be noted that in series circuits, Q-factor gives the voltage magnification, whereas in parallel circuits, it gives the current magnification.

Note that Q is also given by I_L/I and that can also be shown to be equal to $\frac{1}{R}\sqrt{\frac{L}{C}}$.

REFERENCES

1. A.P. French, *Vibration and Waves*, Arnold-Heinemann India, New Delhi, 1973.
2. H.J. Pain, *The Physics of Vibrations and Waves*, John Wiley & Sons, New York, 2003.
3. S.S. Rao, *Mechanical Vibrations*, Pearson Education, New Delhi, 2004.
4. I.G. Main, *Vibrations and Waves in Physics*, Cambridge University Press, Cambridge, 1995.
5. W.H. Erickson and N.H.Bryant, *Electrical Engineering Theory and Practice*, John Wiley & Sons, New York, 1967.
6. Vincent Del Toro, *Principles of Electrical Engineering*, Prentice Hall of India Pvt. Ltd., New Delhi 1987.
7. H. Alex Romanowitz, *Introduction to Electric Circuits*, John Wiley & Sons, New York, 1971.
8. B.L. Theraja, *Fundamentals of Electric Engineering*, Niraja Construction and Development Pvt. Ltd., New Delhi, 1988.

SOLVED EXAMPLES

1. A massless spring, suspended from a rigid support, carries a flat disc of mass 100 g at its lower end, it is observed that the system oscillates with a frequency 10Hz and the amplitude of the damped oscillations reduces to half its undamped value in one minute. Calculate the resistive force constant and the relaxation time of the system.

   Solution: The amplitude of the damped oscillator at any instant t is given by
   $$A = Ce^{-\xi\omega_n t}$$
   since $\frac{A}{C} = \frac{1}{2}$ for $t = 1$ minute $= 60$s
   $$\frac{1}{2} = e^{-(\xi\omega_n)60} = e^{-60\xi\omega_n}$$
   $$\therefore \quad \xi\omega_n = \frac{\ln 2}{60} = 1.16 \times 10^{-2} \text{ rad/s}$$
   The resistive force constant $= c = 2m\xi\omega_n$
   $$= 2 \times (100 \times 10^{-3}) \times (1.16 \times 10^{-2})$$
   $$= 2.32 \times 10^{-3} \text{ newton-s/meter}$$
   Relaxation time $\tau = \dfrac{1}{\xi\omega_n} = \dfrac{1}{1.16 \times 10^{-2} \text{ red/sec}} = 86.96$ s.

2. A massless spring of spring constant 10 N/m is suspended from a rigid support and carries a mass of 0.1 kg at its lower end. The system is subjected to a resistive force $c(dx/dt)$.

VIBRATIONS AND RESONANCE

where c is the resistive force constant and dx/dt is the velocity. It is observed that the system performs damped oscillatory motion and its energy decays to $1/e$ of its initial value in 50 s.

(a) What is the value of resistive force constant c ?
(b) What is the natural angular frequency of the oscillator ?
(c) What is the damping ratio ξ and Q-factor ?
(d) What is the percentage change in frequency due to damping ?

Solution:

(a) $m = 0.1$ kg

The decay of the energy of the damped oscillator is given by

$$E(t) = E_0 e^{-2\xi\omega_n t}$$

where E_0 is the initial energy.

$$\frac{E(t)}{E_0} = \frac{1}{e} = e^{-1} = e^{-100\xi\omega_n}$$

$\Rightarrow \quad \xi\omega_n = 10^{-2}/s$

Hence the resistive force constant = c

$$= m \cdot 2\xi\omega_n = 0.1 \times 2 \times 10^{-2} = 2 \times 10^{-3} \text{ newton-s/meter} \tag{1}$$

(b) Since $K = 10$ N/m, the angular frequency ω in the absence of damping is

$$\omega_n = \sqrt{\frac{K}{m}} = \sqrt{\frac{10}{0.1}} = 10 \text{ rad/s} \tag{2}$$

(c) From (1) and (2)

$$\xi = \frac{1 \times 10^{-2}}{10} = 1 \times 10^{-3} \text{ and } Q = \frac{1}{2\xi} = \frac{1}{2 \times 10^{-3}} = 500$$

The angular frequency of damped oscillation is

$$\omega_d = \omega_n \sqrt{1-\xi^2}$$

$$= 10\sqrt{1-(1\times 10^{-3})^2}$$

$$\approx 10 \text{ radian/s} \tag{3}$$

(d) The fractional change in frequency is given by $\dfrac{\omega_n - \omega_d}{\omega_n}\left(1 - \dfrac{\omega_d}{\omega_n}\right)$

$$= \left(1 - \frac{\omega_d}{\omega_n}\right) = \{1 - (1-\xi^2)^{1/2}\} \approx \left\{1 - \left(1 - \frac{1}{2}\xi^2\right)\right\} \approx \frac{1}{2}\xi^2 \approx \frac{1}{2} \times (1\times 10^{-3})^2$$

$$\approx 0.5 \times 10^{-6}$$

$\therefore$ percentage change in frequency = $0.5 \times 10^{-6} \times 10^2 = 5 \times 10^{-5}$ \hfill (4)

3. An object of mass 0.1 kg is hung from a spring whose spring constant is 100 Nm^{-1}. A resistive force $c\ (dx/dt)$ acts on the object where dx/dt is the velocity in meters per second and $c = 1$ Nsm^{-1}. The object is subjected to a harmonic driving force of the form

$F_0 \cos \omega t$ where $F_0 = 2N$ and $\omega = 50$ radian/s. In the steady state what is the amplitude of the oscillations and the phase relative to applied force ?

Solution:

$m = 0.1$ kg ; $K = 100$ N/m ; $c = 1$ Nsm^{-1} ; $F_0 = 2N$; $\omega = 50$ rad/s

$$\omega_n = \sqrt{\frac{K}{m}} = \sqrt{\frac{100}{0.1}} = \sqrt{1000} = 31.6 \text{ rad/s}$$

$$r = \frac{\omega}{\omega_n} = \frac{50}{31.6} = 1.58 \; ; \; \delta_{st} = \frac{F_0}{k} = \frac{2}{100} = 2 \times 10^{-2} \text{ m}$$

$$\xi = \frac{c}{2m\omega_n} = \frac{1}{2 \times 0.1 \times 31.6} = 0.158$$

$$A = \frac{\delta_{st}}{\sqrt{(1-r^2)^2 + (2r\xi)^2}} = \frac{2 \times 10^{-2}}{\sqrt{(1-0.158^2) + (2 \times 1.58 \times 0.158)^2}} = 1.26 \times 10^{-2} \text{ m}$$

$$\tan \phi = \frac{2r\xi}{1-r^2} = \frac{2 \times 1.58 \times 0.158}{1 - 1.58^2} = -\frac{1}{3} \Rightarrow \phi = 161.7°$$

i.e., the oscillations lag behind the applied force by 161.7°.

4. A series circuit consists of a resistance of 15 ohms, an inductance of 0.08 henry and a condenser of capacity 30 microfarads. The applied voltage has a frequency of 500 radians/s. Does the current lead or lag the applied voltage and by what angle ?

Solution:

Here $\omega = 500$ radian/s, $L = 0.08$ H

$R = 15$ ohm, and $C = 30 \times 10^{-6}$ F

$L\omega = 0.08 \times 500 = 40$ ohm

$$\frac{1}{C\omega} = \frac{1}{30 \times 10^{-6} \times 500} = 66.7 \text{ ohm}$$

$$\tan \phi = \frac{L\omega - \frac{1}{C\omega}}{R} = \frac{40 - 66.7}{15} = -\frac{26.7}{15} = -1.78$$

$$\phi = -60.65°$$

The current leads the applied voltage by 60.65°.

5. A series circuit consists of a resistance, inductance and capacitance. The applied voltage and the current at any instant are given by

$$E = 141.4 \cos (3000t - 10°)$$
$$I = 5 \cos (3000t - 55°)$$

The inductance is 0.01 henry. Calculate the values of the resistance and capacitance.

Solution:

The E.M.F. is ahead of the current by $55° - 10° = 45°$

∴ $\phi = 45°$

$\tan \phi = \tan 45° = 1$

VIBRATIONS AND RESONANCE

Also
$$\tan \phi = \frac{\left[L\omega - \dfrac{1}{C\omega}\right]}{R} = 1$$

$$L\omega - \frac{1}{C\omega} = R$$

Also Impedance, $Z = \sqrt{R^2 + \left(L\omega - \dfrac{1}{C\omega}\right)^2}$

$$Z = \sqrt{R^2 + R^2} = \sqrt{2}\, R = 1.414\, R$$

But
$$Z = \frac{E_0}{I_0} = \frac{141.4}{5} = 28.28$$

∴ $1.414\, R = 28.28$

i.e., $R = 20$ ohm

$$L\omega - \frac{1}{C\omega} = 20$$

$\omega = 3000$ radian/s
$L = 0.01$ henry
$L\omega = 0.01 \times 3000 = 30$ ohm

∴ $30 - \dfrac{1}{C\omega} = 20$

i.e., $\dfrac{1}{C\omega} = 10$ ohm

$$C = \frac{1}{10\omega} = \frac{1}{10 \times 3000}$$

$C = 33.33 \times 10^{-6}$ F

$C = 33.33$ μF.

6. A series RLC circuit with a resistance of 50 Ω, a capacitance of 25 μF and an inductance of 0.15H is connected across 230-V, 50-Hz supply. Determine (i) impedance (ii) current (iii) power factor and (iv) power consumption of the circuit.

Solution:

$X_L = 2\pi f L = 2\pi \times 0.15 = 47.1$ Ω
$X_C = 1/2\pi f C = 10^6/\omega\pi \times 50 \times 25 = 127.3$ Ω

Net $X = X_L - X_C = 47.1 - 127.3 = 80.2$ Ω (capacitive)

(i) $Z = \sqrt{R^2 + X^2} = \sqrt{50^2 + 80.2^2} = 94.4$ Ω

(ii) $I = V/Z = 230/94.4 = 2.44$ A

(iii) p.f. $= \cos \phi = R/Z = 50/94.4 = 0.53$ (lead)

(iv) power consumed $= VI \cos \phi = 230 \times 2.44 \times 0.53 = 297$ W.

7. A coil of insulated wire of resistance of 8 Ω and inductance 0.03 H is connected to an a.c. supply at 240-V, 50-Hz. Calculate:
   (i) the current, p.f. and the power
   (ii) the value of capacitance which when connected in series with the above coil, causes no change in the values of the current and power taken from the supply.

   Solution:
   (i)
   $$X_L = 314 \times \times 0.03 = 9.42 \; \Omega$$
   $$Z = \sqrt{8^2 + 9.42^2} = 12.36 \; \Omega$$
   $$I = 240/12.36 = 19.4 \; A$$
   $$W = I^2 R = 19.42 \times 8 = 3011 \; W$$
   $$\text{p.f.} = R/Z = 8/12.36 = 0.65 \; (\text{lag})$$

   (ii) If the circuit is to draw the same current and at the same power factor, then the total reactance of the RLC circuit must be 9.42 Ω. This can be achieved by selecting a capacitor which should not only neutralize the inductive reactance of 9.42 Ω of the coil but must add a further capacitive reactance of 9.42 Ω. In other words, capacitor must have a reactance of 2 × 9.42 = 18.84 Ω.
   $$1/\omega C = 18.84 \quad \text{or} \quad C = 1/314 \times 18.84 = 169 \; \mu F$$

8. A resonant circuit consists of a 4 μF capacitor in parallel with an inductor of 0.25 H having a resistance of 50 Ω. Calculate the frequency of resonance.

   Solution:
   The resonance frequency is
   $$f_0 = \frac{1}{2\pi} \sqrt{\frac{1}{LC} - \frac{R^2}{L^2}}$$

   It is seen that f_0 is dependent on R. When R is large, f_0 is reduced. In a series circuit $f_0 = 1/2\pi \sqrt{LC}$. Obviously, f_0 does not depend on R.
   $$f_0 = \frac{1}{2\pi} \sqrt{\frac{10^4}{4 \times 0.25} - \frac{50^2}{0.25^2}} = 156 \; Hz.$$

9. A coil of resistance 30 Ω and inductance 20 mH is connected in parallel with a variable capacitor across a supply of 25 V and frequency 1000/π Hz. The capacitance of the capacitor is then varied until the current taken from the supply is a minimum (i.e., until the overall p.f. of the circuit is unity). For this condition, find
   (i) the capacitance of the circuit
   (ii) the value of the current.

   Solution:
   $$X_L = 2\pi f L = 2\pi \times (1000/\pi) \times 20 \times 10^{-3} = 40 \; \Omega$$
   $$Z = \sqrt{30^2 + 40^2} = 50 \; \Omega$$

   (i) At resonance $Z^2 = L/C$
   $$C = L/Z^2 = 20 \times 10^{-3}/50^2 = 8 \times 10^{-6} \; F = 8 \; \mu F.$$

$(ii)\ I_{min} = \dfrac{V}{L/CR}$

Dynamic impedance = $20 \times 10^{-3}/8 \times 10^{-6} \times 30 = 83.3\ \Omega$

$I_{min} = 25/80.3 = 0.3$ A.

QUESTIONS

1. Define (a) free vibrations (b) forced vibrations (c) transient vibrations (d) damped vibrations and (e) linear vibrations.
2. Show that the average energy of a weakly damped harmonic oscillator decays exponentially with time.
3. Establish the equation of motion of a damped oscillator subjected to a resistive force that is proportional to the first power of its velocity. If the damping is less than critical, show that the motion of the system is oscillatory with its amplitude decaying exponentially with time.
4. Discuss the theory of vibration isolator.
5. Discuss the theory of electrical oscillations in a series (a) LC and (b) LCR circuit by setting up the relevant differential equations.
6. Discuss the theory of forced electrical oscillations in a series LCR circuit by setting up the relevant differential equation.
7. Find the expression for current in the case of a series LCR circuit. Describe the frequency dependence of current, and hence discuss the concept of resonance.
8. Find the expression for current in the case of parallel LCR circuit. Describe the frequency dependence of current.
9. What are acceptor and rejector circuits ?
10. What do you understand by the term 'quality factor' and the 'relaxation time'?
11. What is Q-factor ? Obtain an expression for the same.

PROBLEMS

1. Prove that in simple harmonic motion the average potential energy equals the average kinetic energy when the average is taken with respect to time over one period of motion and that each average is equal to $(1/4)\ Ka^2$ where K is the spring constant and a is the amplitude. But when the average is taken with respect to position over one cycle, the average potential energy is equal to $(1/6)\ Ka^2$ and the average kinetic energy is equal to $(1/3)\ Ka^2$. Explain why the two results are different.
2. A simple pendulum consists of a rod of mass m and length l which is pivoted at the upper end and carries a mass M at the other end. Using energy consideration determines the frequency of the pendulum if (a) $m \ll M$ and (b) m is comparable with M.
3. A massless spring suspended from a rigid support carries a flat disc of mass 100g at its lower end. It is observed that the system oscillates with a frequency of 10Hz and the amplitude of the damped oscillations reduces to half its undamped value in one minute. Calculate the (a) the resistive force constant (b) relaxation time (c) the quality factor and (d) the spring constant.

(**Ans.** 0.0023 Nsm^{-1}, 86.58 s, 2720, 394.8 Nm^{-1})

4. The viscous force on sphere of radius a moving with a velocity v in a medium of coefficient of viscosity η is $6\pi\eta av$. Determine the effects of air viscosity on the amplitude and period of simple pendulum consisting of a aluminium bob of radius 0.5cm suspended by means of a 1m long thread. Take the density of aluminium to be 2.65 g/cc and η to be 1.78×10^{-4} gcm^{-1} s^{-1}.
5. According to classical electromagnetic theory, an electron orbiting around the nucleus can be thought of as executing simple harmonic oscillation. It is subjected to an acceleration which results in the emission of radiation. It radiates energy at the rate of $(ke^2\omega^4A^2/3c^3)$ watts, where $k = 9 \times 10^9$ Nm^2 c^{-2}, e is the charge on the electron, c is the velocity of light, A is the amplitude of

oscillation or the radius of the orbit and ω is the angular frequency. If the emitted radiation has a wavelength of 600 nm. Calculate the Q value of the oscillator and the radiation life time i.e., the time required for the energy to fall to e^{-1} of the original value. **(Ans. 5×10^7, 16.9 ns)**

6. For a mass-spring system, the spring constant = 10 N/m, m = 10 kg and resistive force constant = 8 Ns/m. Determine the motion of the mass when it is given a velocity of .068 m/s at t = 0.

7. An oscillator with small damping has mass 5 g and a force constant of 2 N/m. If the Q for the oscillator is 200 and the system oscillate in energy resonance with an applied force (a) what is the critical damping constant c_c (b) what is the frequency of the applied force (c) what frequency of the applied force will produce amplitude resonance ? (d) what is the amplitude of the displacement and velocity oscillations of the oscillator if the frequency of applied force is 90% of the natural frequency of the free oscillator ? Express the answer as fraction of the amplitude at energy resonance (e) what is the power delivered to the oscillator by the impressed force expressed as the fraction of the power at resonance (e) what is the bandwidth of the oscillator.

(Ans. 5×10^{-3} kg/s, 31.8Hz, ~ 31.8Hz, 0.0237, 0.000561 1rad/s)

8. A weakly damped harmonic oscillator is driven by a force $F_0 \cos \omega t$, whose amplitude F_0 is kept constant but its angular frequency is varied. It is experimentally observed that the amplitude of the steady state oscillations is 0.1 mm at very low ω and attains a maximum value of 10 cms when ω = 100 rad/s. Calculate (a) the Q value of the system (b) the time during which the energy of the oscillator falls to 1/e of its initial value and (c) half-width of the power resosnance.

(Ans. 1000, 10 s, 0.05 rad/s)

9. A vibrating system of natural frequency 500 Hz is forced to vibrate with a periodic force of amplitude 10^{-1} N/kg in the presence of damping coefficient of 10^{-5} N s/m. Calculate the maximum amplitude of the vibration of the system. **(Ans. 3.15×10^{-2} m)**

10. The energy of a piano string of frequency 256 Hz reduces to half its initial value in 2 s. What is the Q-factor of the string ? **(Ans. 4643)**

11. The quality factor of a sonometer wire of frequency 500 Hz is 5000. In what time will its energy reduce to 1/e of its value because of damping? **(Ans. 1.59s)**

12. An alternating EMF, $E = E_0 \sin \omega t$ is applied across a parallel combination of R, L and C as shown in Fig. 1.31. Calculate the current in each branch and express the total as a sine function.

$$\text{Ans. } I = E \sqrt{\left(\frac{1}{R}\right)^2 + \left(L\omega - \frac{1}{C\omega}\right)^2} \times \sin\left((\omega t + \tan^{-1}\left\{R\left(C\omega - \frac{1}{\omega}\right)\right\}\right)$$

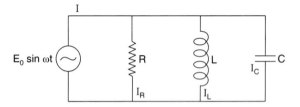

Fig. 1.31

13. A resistance of 10 ohms is connected in series with an inductance of 0.5 henry. What capacitance should be put in series with the combination to obtain the maximum current ? What will be the potential difference across the resistance, inductance and capacitance? The current is being supplied by 200 volts and 50 cycles per second mains. **(Ans. 200 V ; 3142 V ; 3142 V)**

14. A 60 cycles AC circuit has resistance of 2 ohms and inductance of 10 millihenries. What is the

VIBRATIONS AND RESONANCE

power factor ? What capacitance placed in the circuit will make the power factor unity ?

(**Ans.** 0.4687 ; 703 µF)

15. A series circuit consisting of a condenser of capacitative reactance 30 ohms, a noniductive resistance of 44 ohms and a coil of inductive reactance 90 ohms and resistance 36 ohms is connected to 200 V, 60 hertz line. Calculate (*i*) Impedance of the circuit (*ii*) Current in the circuit (*iii*) Potential difference across each component (*iv*) Power factor of the circuit (*v*) Power consumed.

(**Ans.** 1000 ohm ; 2 A, 194 V ; 60 V ; 320 Watt)

16. A circuit consists of a resistance of 20 Ω in series with an inductance of 95.6 mH and a capacitor of 318 µF. It is connected to a 500 V, 25 Hz supply. Calculate the current in the circuit and the power factor. (**Ans.** 24.3A ; 0.97 lead)

17. A circuit is made up of 10 Ω resistance, 12 µH inductance and 281.5 µF capacitance in series. The supply voltage is 10 V (constant). Calculate the value of the current when the supply frequency is (*i*) 50 Hz and (*ii*) 150 Hz. (**Ans.** (*i*) 8 A leading (*ii*) 8 A lagging)

18. An a.c. series circuit has a resistance of 10 Ω, an inductance of 0.2 H and a capacitance of 60 µF. Calculate (*i*) the resonant frequency (*ii*) the current (*iii*) the power at resonance given that applied voltage is 300 V. (**Ans.** (*i*) 46 Hz (*ii*) 20 A (*iii*) 4 kW)

19. A resistor and a capacitor are connected in series with a variable inductor. When the circuit is connected to a 240 V, 50 Hz supply, the maximum current given by varying the inductance is 0.5 A. At this current the voltage across the capacitor is 250 V. Calculate the value of : (*i*) the resistance (*ii*) the capacitance (*iii*) the inductance. Neglect the resistance of the inductor.

(**Ans.** (*i*) 480 Ω (*ii*) 6.36 µF (*iii*) 1.50 H)

20. A series circuit has the following characteristics : $R = 10\ \Omega$; $L = 100/\pi$ mH ; $C = 500/\pi$ µF. Find: (*i*) the current flowing when the applied voltage is 100 V at 50 Hz. (*ii*) the power factor of the circuit (*iii*) what value of the supply frequency would produce series resonance.

(**Ans.** (*i*) 7.07 A (*ii*) 0.707 lead (*iii*) 70.71 Hz)

21. An inductor of 0.5 H inductance and 90 Ω resistance is connected in parallel with a 20 µF capacitor. A voltage of 230 V at 50 Hz is maintained across the circuit. Determine the total power taken from the source. (**Ans.** 14.5 Ω)

22. A resistance of 50 Ω, an inductance of 0.15 H and a capacitance of 100 µF are connected in parallel across a 100 V, 50 Hz supply. Calculate (*i*) the current in each circuit (*ii*) the resultant current and (*iii*) the phase angle between the resultant current and supply voltage. Draw the phasor diagram. (**Ans.** (*i*) 2A in phases with voltage ; 212 A-lagging V by 90°

3.14 A-leading V by 90° (*ii*) 2.245A (*iii*) 12.8°)

23. A coil having a resistance of 8 Ω and inductance of 0.0181 H is connected in parallel with a capacitor having a capacitance of 398 µF and resistance of 5 Ω. If 100 V at 50 Hz are applied across the terminals of the above parallel circuit, calculate (*i*) the total current taken form the supply and (*ii*) its phase angle with respect the supply voltage. Draw a complete vector diagram for the circuit showing the 3 currents and the supply voltage. (**Ans.** (*i*) 13.97 A (*ii*) 12.8°)

24. A coil of inductance 31.8 mH and resistance of 10 Ω is connected in parallel with a capacitor across a 250 V 50 Hz supply. Determine the value of the capacitance if no reactive current is taken from the supply. (**Ans.** 159 µF)

25. A 100-V, 80 W lamp is to be operated on 230-V, 50-Hz a.c. supply. Calculate the inductance of the choke required to be connected in series with the lamp for this operation. The lamp can be taken as equivalent to a non-inductive resistance. If the p.f. of the lamp circuit is to be improved to unity, calculate the value of the capacitor which is to be connected across the circuit.

(**Ans.** $L = 0.823$ H ; 10 µF)

2
Acoustics of Buildings

2.1 INTRODUCTION

The subject of acoustics of buildings is of interest to speakers, musicians, theatre artists, listeners and architectural engineers. Musicians and speakers recognize the importance of good acoustics in communicating with the audience. In industry, commercial establishments, educational institutions and hospitals noise insulation procedures have to be strictly followed to create a conducive working environment. The acoustics of enclosed spaces had drawn the attention of many ever since people began to gather in large lecture halls, auditoria, concert halls and churches. However, a proper scientific basis for the acoustic design of buildings began with Sabine in 1900. Till then *room acoustics* was considered more of an art than as a subject of scientific study. The acoustical design of rooms as well as noise control, requires a thorough understanding of basic principles underlying the physics of sound waves and their propagation.

2.2 PROPERTIES OF SOUND

2.2.1 Frequency, Wavelength, Velocity and Intensity

Sound wave consists of pressure fluctuations traveling in the medium (Fig.2.1). In general, the sound waves due to speech or music will have complex waveforms. For sound wave of monotonic frequency propagating along the x-direction, the pressure fluctuations can be represented as:

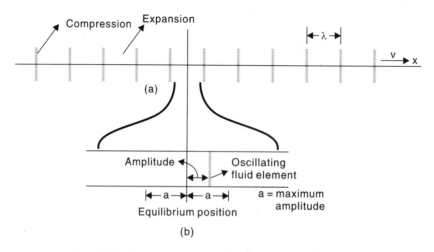

Fig. 2.1 (a) Longitudinal sound waves propagating in air
(b) Expanded view of the oscillating volume of air

$$\Delta p = \Delta p_{max} \sin(kx - \omega t) \qquad (2.1)$$

$+\Delta p$ and $-\Delta p$ correspond to compression and expansion of molecules constituting the medium. In fluids, sound wave is longitudinal *i.e.*, the movement of molecules of the medium is along the direction of propagation. In solids both longitudinal and transverse waves are possible (Fig. 2.2). The wavelength of sound waves corresponds to the distance between the consecutive pressure maxima. The frequency of sound waves corresponds to the number of times pressure fluctuations occur per second. Typical wavelengths of sound waves in the audible range (20Hz-20kHz) is given in Table 2.1

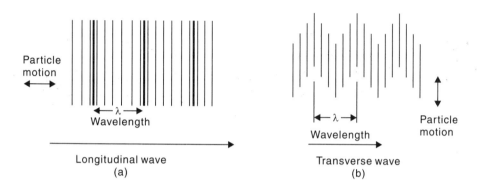

Fig. 2.2 (a) Longitudinal and (b) transverse waves

Table 2.1 Some examples of sound wavelengths in air in the audible range

Frequency	Wavelength	
	Sound waves in air ($c = 340 m/s$)	For light waves in vacuum ($c = 3 \times 10^8 m/s$)
20,000Hz or 20kHz	1.7 cm	1.5×10^4 m
16,000Hz or 16kHz	2.1 cm	1.88×10^4 m
10,000Hz or 10kHz	3.4 cm	3×10^3 m
5000Hz or 5kHz	6.8 cm	6×10^4 m
2000Hz or 2kHz	17 cm	1.5×10^5 m
1000Hz or 1kHz	34 cm	3×10^5 m
500Hz	68 cm	6×10^5 m
200Hz	1.7 m	1.5×10^6 m
100Hz	3.4 m	3×10^6 m
20Hz	17 m	1.5×10^7 m

The velocity of the wave is given by the product of wavelength and the frequency. The velocity of sound waves in the medium is governed by the elastic properties. The velocity of sound in air is given by

$$c = \sqrt{\frac{\gamma P}{\rho}} \qquad (2.2)$$

where c is the velocity, P is the pressure, γ is the ratio of specific heats (c_p/c_v) and ρ is the density. *In general the velocity of sound waves in a crystal is anisotropic and is governed by the various elastic constants of the medium.* The velocity of sound in various materials of interest is given in Table 2.2

At constant pressure, the volume of the gas is directly proportional to temperature. Thus the density of air decreases with temperature. *Hence the velocity of sound is directly proportional to square root of absolute temperature.* The variation with temperature can be represented by the formula.

c (m/s) = 331 + 0.6t where t is temperature in degree centigrade.

Sound wave is constituted by pressure fluctuations. The temperature tends to rise in those parts of the wave where the air is compressed and to fall where the air expands. The question is therefore whether appreciable thermal conduction can take place between these two sets of regions. The heat transfer has to occur in a time interval less than the reciprocal of the frequency of sound. It turns out that for ordinary acoustic wavelengths the pressure maxima and minima are so far apart that no appreciable conduction occurs. The behaviour can be described as *adiabatic*.

Table 2.2 Velocity of sound in different materials

Substance	Velocity of sound (m/s)
Fresh water	1480
Glass	5200
Concrete	3400
Steel	5000—5900
Wood	3000—4000
Carbon dioxide	259
Oxygen	316
Hydrogen	1284
Helium	965
Air	340

The sound wave is essentially a pressure wave arising out of the periodic displacements of the particles constituting the medium. Let a be the maximum displacement of the particle, ω be the angular frequency of the wave and ρ be the density of the medium. Let A be the area of cross-section, c be the velocity of sound and z (= $c\rho$) be the acoustic impedance, then

ACOUSTICS OF BUILDINGS

The maximum pressure $= ac\rho\omega = za\omega$ (2.3)

The power P carried by the sound wave is given by:

$$P = \frac{a^2\omega^2 \rho A c}{2} = \frac{zAa^2\omega^2}{2}$$ (2.4)

The power transferred across unit area, that is intensity I is therefore given by:

$$I = \frac{a^2\omega^2 \rho c}{2} = \frac{za^2\omega^2}{2}$$ (2.5)

In the case of longitudinal and shear waves the power is communicated parallel and perpendicular to the direction of propagation respectively. The intensity or the loudness of sound is expressed by either sound pressure level (SPL) or sound intensity level (SIL). The sound pressure level is defined as:

$$L = 10 \log \left(\frac{I}{I_0}\right) = 10 \log \left(\frac{p^2}{p_0^2}\right) = 20 \log \left(\frac{p}{p_0}\right)$$ (2.6)

Here p_0 refers to the threshold audible pressure level at 1000Hz and is equal to 20 micro pascals ($= 20 \times 10^{-6}$ pa). Pascal is the unit of pressure and is equal to one Newton/m². The threshold audible pressure is a function of frequency (Fig. 2.3). I_0 refers to the threshold intensity and is equal to one pico watts/m² ($= 10^{-12}$ watts/m²). *This logarithmic scale is chosen because the response of the ear is proportional to the logarithmic variations in intensity.*

In open space the intensity of sound obeys inverse square law *i.e.,*

$$I \propto (1/r^2)$$

This is often referred to as the *free field*. However in a closed room, in addition to the sound emitted from the source the intensity of sound undergoing multiple reflections at the walls, ceilings, floor and other objects present in the room has to be considered. This is referred to as the *diffuse field*.

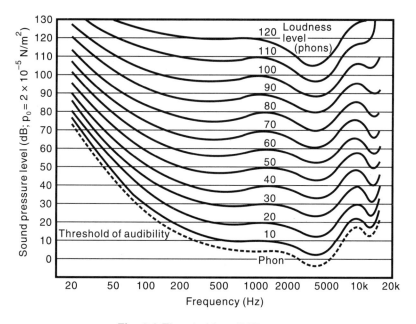

Fig. 2.3 Threshold audibility curve

2.2.2 Propagation of Sound

Reflection and Refraction of Sound

Reflection: The sound waves obey the laws of reflection. *i.e.*,

(a) the incident ray, the reflected ray and the normal at the point of incidence all lie in one plane.

(b) the angle of incidence is equal to the angle of reflection. These laws are useful in predicting the path of the sound waves when they encounter a planar or curved surface (Fig. 2.4). This is known as *specular reflection*. The reflection of sound waves at plane, convex, concave and corrugated surfaces is shown in Fig. 2.5. Sound gets focused due to reflection at a concave surface and diffused on undergoing reflection at convex surface. When the surface is corrugated, laws of reflection do not hold good. As a result sound is reflected in all directions.

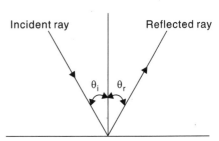

Fig. 2.4 Laws of reflection

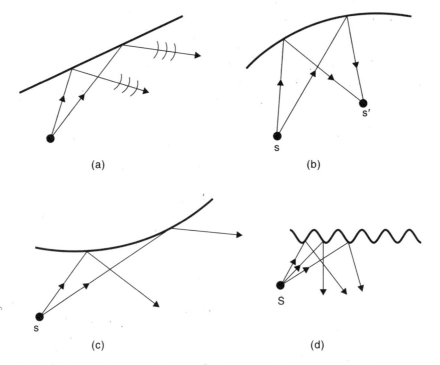

Fig. 2.5 Reflection of sound waves at (a) plane (b) convex (c) concave and (d) corrugated surface

Refraction: When sound waves travel from one medium to the other, their direction of propagation changes (Fig. 2.6). The laws governing the phenomenon of refraction are similar to those applicable to light waves. *i.e.*,

(a) the incident ray, the refracted ray and the normal at the point of incidence all lie in the same plane.

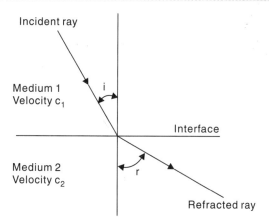

Fig. 2.6 Laws of refraction

(b) the sine of the angle of incidence bears a constant ratio to the sine of the angle of refraction which is equivalent to the ratio of the sound velocities in the media concerned, i.e.,

$$\frac{\sin i}{\sin r} = \text{constant} = \frac{c_1}{c_2} \tag{2.7}$$

where c_1 and c_2 are the velocities of ultrasonic waves in medium 1 and 2 and the sound wave is traveling from medium 1 and medium 2.

Acoustic Impedance: For plane harmonic waves, acoustic impedance (z) is ρc where ρ is the density of the medium and c is the velocity of sound waves. *The acoustic impedance is analogous to refractive index, i.e.,*

$$z = \rho c \tag{2.8}$$

The reflection coefficient (R) and the transmission coefficient (T) are given by:

$$R = \frac{I_r}{I_0} = \left(\frac{z_2 \cos i - z_1 \cos r}{z_2 \cos i + z_1 \cos r}\right)^2$$

and

$$T = \frac{I_t}{I_0} = \frac{4 z_1 z_2 \cos^2 r}{(z_2 \cos i + z_1 \cos r)^2} \tag{2.9}$$

Diffraction of Sound

The phenomenon of diffraction is observed when the wavelength is comparable to the size of the object scattering the sound waves. Typically the range of wavelength of audible sound waves lies in the region of 10^{-2}—20m. (Table 2.1). Hence diffraction effects can be expected for objects whose size lie in this range. Normally sound waves are reflected when they encounter an obstacle and a geometrical shadow is formed (Fig. 2.7). Diffraction refers to the bending of sound waves around obstacles. As a consequence perfect shadows which are expected to occur due to rectilinear propagation are not found. Figure 2.8 shows the diffraction of sound waves for various object sizes. Figure 2.9 shows the % reflection of sound waves versus the ratio of wavelength to the object size.

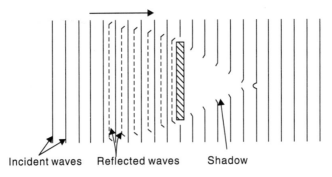

Fig. 2.7 Shadow caused by sound due to rectilinear propagation of sound

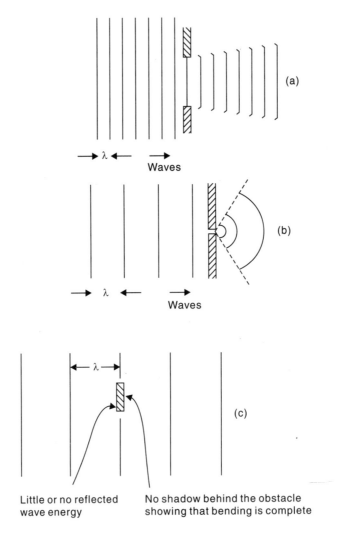

Fig. 2.8 Diffraction of sound waves around (a) big aperture and (b) small aperture (c) absence of shadows due to diffraction

ACOUSTICS OF BUILDINGS

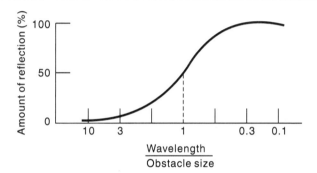

Fig. 2.9 % of sound reflection as a function of obstacle size

The diffraction of sound of a given frequency around a barrier depends on the Fresnel number N which is given by:

$$N = \frac{2}{\lambda}(d_1 + d_2 - d) \qquad (2.10)$$

where λ is the wavelength of sound and lengths d_1, d_2 and d are shown in Fig. 2.10. Due to diffraction there is attenuation of the sound beam intensity in the propagation direction. The attenuation A_d due to diffraction is then given as a function of the Fresnel number

$$A_d = 10 \log_{10}(20N) \qquad (2.11)$$

This is plotted in Fig. 2.11.

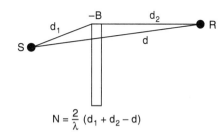

Fig. 2.10 Defining Fresnel number

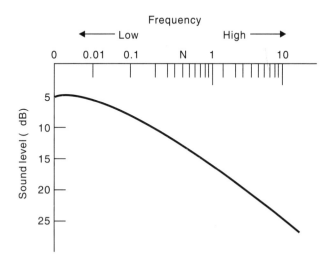

Fig. 2.11 Sound attenuation as a function of Fresnel number

Absorption of Sound

When incident on a surface, sound is partly reflected, partly transmitted and partly absorbed. The amount of energy absorbed depends on the nature of the material and the surface area. The absorption coefficient (a) is defined as:

$$a = \frac{\text{Energy absorbed by the surface}}{\text{Energy incident on the surface}} \quad (2.12)$$

An open window is an example of perfect absorber. The absorption coefficient of an open window is unity. *Absorption has units of area and is expressed as either metric Sabine (m^2) or English Sabine (ft^2).* The open window is taken to have a dimension of 1 m^2. Let a surface of area A m^2 absorb the same amount of energy as absorbed by 1 m^2 of an open window. Then the absorption coefficient $=1/A$. It is found that the normal person is equivalent of 0.4367 m^2 of an open window unit.

If $a_1, a_2, a_3, \ldots$ are the absorption coefficients of materials having surface areas $S_1, S_2, S_3,\ldots$ then the total absorption is given by:

$$A = \sum_i a_i S_i \quad (2.13)$$

The average absorption coefficient is given by:

$$a = \frac{\sum_i a_i S_i}{\sum_i S} \quad (2.14)$$

Sound Absorption Mechanisms

When sound waves travel through air a part of the energy is converted into random thermal energy. This leads to attenuation of sound waves. This is represented by the equation:

$$I(x) = I(o)\, e^{-\alpha x} \quad (2.15)$$

where $I(x)$ is the intensity at x and α is the attenuation coefficient.

Two important physical processes contribute to the absorption of sound waves. These arise due to viscosity and thermal conductivity of air.

The total attenuation coefficient α for a fluid may be written as

$$\alpha_{total} = \alpha_{vis} + \alpha_{th}$$

where α_{vis} and α_{th} are respectively the contributions arising from the two mechanisms mentioned above.

Viscous losses: Air may be considered to be made of different layers (Fig. 2.12). During the propagation of sound wave *there is relative motion between the adjacent layers, which gives rise to friction.* This arises on account of the finite viscosity of air. Viscous drag is also called the internal friction. A part of the acoustic energy is lost in overcoming this friction. The attenuation coefficient α_{vis} is given by:

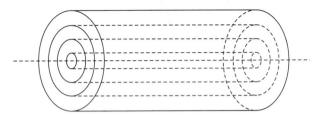

Fig. 2.12 Fluid can be thought of to be made of different layers. Relative motion between these layers gives rise to internal friction. This friction leads to sound attenuation

$$\alpha_{vis} = \left(\frac{\omega^2}{2\rho_0 c^2}\right)\left(\frac{4\eta}{3} + \eta_B\right) \quad (2.16)$$

where ω is the angular velocity, ρ is the density, c is the velocity, η is the shear viscosity and η_B is called the bulk viscosity. It is zero for monoatomic gases but can be finite in other fluids.

ACOUSTICS OF BUILDINGS

Thermal conduction losses: When a longitudinal wave passes through a medium there are regions of compression and rarefactions. *These fluctuations in pressure are described by an adiabatic process and not an isothermal process.* The regions where the molecules are compressed are at a slightly higher temperature while the regions where the molecules expand are at a slightly lower temperature (Fig.2.13). On account of the finite thermal conductivity of the medium, there is an irreversible flow of heat and consequently a loss of acoustic energy. The attenuation coefficient α_{th} is given by:

$$\alpha_{th} = \left(\frac{\omega^2}{2\rho c^3}\right)\frac{(\gamma - 1)k}{c_p} \qquad (2.17)$$

where ω is the angular velocity, γ is the ratio of specific heat at constant pressure to that at constant volume = (c_p/c_v) and k is the thermal conductivity, ρ is the density and c is the velocity of the sound wave.

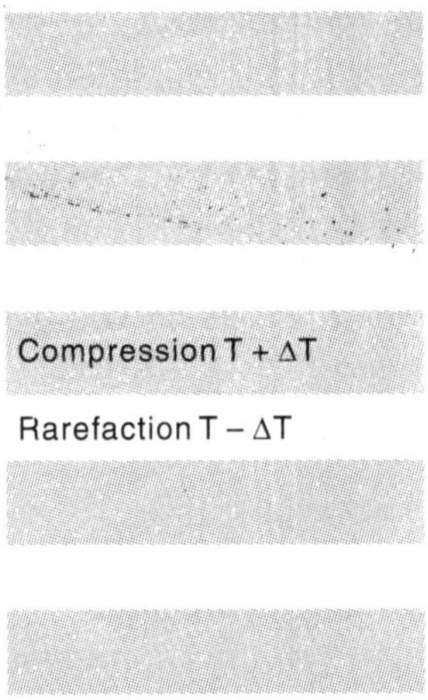

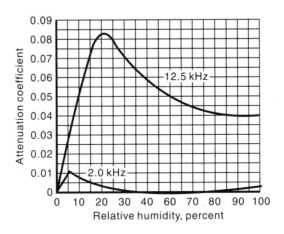

Fig. 2.13 Thermal conduction between regions of compressions and rarefactions lead to attenuation of sound

A further complication in sound absorption arises from the fact that molecules of oxygen and nitrogen are able to rotate and vibrate. The compressional heat energy, initially confined to the translational velocity of the molecules gradually becomes shared with the internal modes of the molecules. This accounts for the dependence of sound absorption on humidity. A plot of sound attenuation coefficient as a function of humidity is shown in Fig.2.14.

Fig. 2.14 Attenuation of sound as a function of relative humidity

2.3 ROOM ACOUSTICS

In an open space the sound intensity obeys inverse square law. The distribution of sound energy in a room depends on the room size, geometry and on the combined effects of reflection, diffraction and absorption. Sound is reflected from the walls, ceiling, floor and diffracted around objects in the room. Further, sound intensity in an enclosed room might build up at specific frequencies on account of room resonances. All these factors make the exact calculation of sound intensity difficult. However, far away from the source, the sound energy throughout the room may be considered to be uniform.

2.3.1 Direct, Early and Reverberation Sound

Consider the propagation of sound from the source to the receiver as shown in Fig.2.15. The sound will typically reach the listener (receiver) after $0.02s$ to $0.2s$ (20 m to 200 m) depending on the distance from the source to the listener. A short time later, the same sound will reach the listener from various reflecting surfaces, mainly the walls and the ceiling. In Fig. 2.16 these reflections are shown arriving with various time delays t_1, t_2, t_3, etc. The first group of reflections, reaching the listener within about 50ms of the direct sound, is often called the *early sound*. After the first group of reflections, the reflected sounds arrive from all directions. These reflections become smaller and closer together, merging into what is called the *reverberation sound*. If the source emits a continuous sound, the reverberant sound builds up until it reaches an equilibrium level. When the sound stops, the sound level decreases at a more or less constant rate until reaches inaudibility. Louder the sound the longer will this process take. *This gradual dying out is known as reverberation and it is the most important single factor in deciding whether a room has good acoustics or not.*

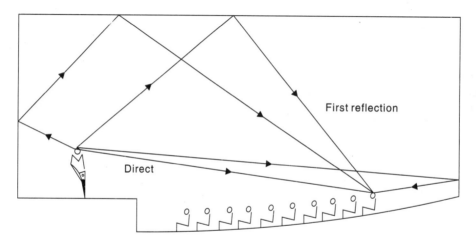

Fig. 2.15 Direct sound and reflections in the room

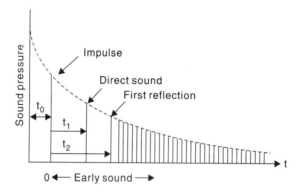

Fig. 2.16 Plot of time delays of direct and various reflections

The reverberation time is defined as the time taken for the sound intensity to fall to 10^{-6} of its intensity. This corresponds to a decrease in sound intensity by 60 dB. Unfortunately the two desirable characteristics, *loudness* and *clarity* are produced by opposite conditions of

ACOUSTICS OF BUILDINGS

reflecting surface. The better the reflecting power of walls and objects present in the room, the larger will be the intensity reached by the use of a given source. This is desired because with little efforts powerful sounds can be produced by a musician or a speaker. However, this will affect the *clarity* which is important for the audience. If the reverberation time is longer, in a piece of music, notes following one another in rapid succession will become confused while speech may be rendered unintelligible.

2.3.2 Reverberation Time—Sabine's Formula

Prior to about 1900 there was no scientific basis for the design of acoustically sound rooms. Sabine by conducting numerous experiments involving the measurement of reverberation time found that the reverberation time T was

(a) directly proportional to the volume of the room and

(b) inversely proportional to the sound absorption. i.e.,

$$T \propto \frac{V}{A} \text{ or } T = \frac{kV}{A} \text{ where } k = 0.162, \text{ when } V \text{ is in m}^3 \text{ and } A \text{ is in m}^2. \quad (2.18)$$

It is possible to arrive at Sabine's formula by the following analysis. Consider a room in which the source produces sound with a power of P. This sound energy will incident on the walls, the ceiling and the objects in the room. Part of the energy will be absorbed and part will be reflected. In steady state, it is assumed that there is a uniform distribution of energy of sound energy inside the room. *This is a reasonable assumption except in regions close to the proximity of the source.* Thus in steady state

Rate of sound energy produced by the source =

Rate of sound absorption in the room + Rate of growth of sound energy in the room.

(2.19)

Energy Density and Intensity of Sound

Let us consider a volume dV in the room and find the energy falling on a surface ΔS which is at a distance r from the volume. Let the radius vector r make an angle θ with the normal to the surface (Fig.2.17). Let the density of acoustic energy be E (*i.e.*, it is the sound energy per unit volume of the room). The energy present in the small volume dV is EdV. The amount of energy that will strike this area ΔS from this volume by direct transmission is

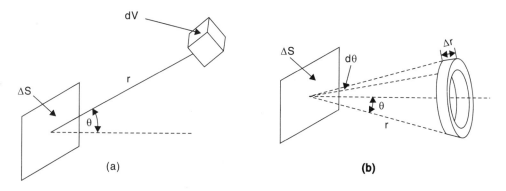

Fig. 2.17 (a) Surface and (b) volume elements used in deriving the relation between energy density and intensity

$$E \cdot dV \cdot \frac{\text{Projection of area } \Delta S \text{ on the sphere radius } r \text{ centered on } dV}{4\pi r^2}$$

$$= \frac{E \cdot dV \cdot \Delta S \cdot \cos\theta}{4\pi r^2}$$

Let dV be the part of the hemispherical shell of thickness Δr and radius r centered on ΔS. Then

$$dV = 2\pi r \cdot \sin\theta \cdot \Delta r \cdot d\theta$$

The acoustic energy ΔE contributed to ΔS by the entire shell is obtained by assuming that energy arrives from any direction with equal probability. Hence integrating over the entire hemisphere, we get

$$\Delta E = \frac{E \cdot \Delta S \cdot \Delta r}{2} \int_0^{\pi/2} \sin\theta \cdot \cos\theta \cdot d\theta = \frac{E \cdot \Delta S \cdot \Delta r}{4} \int_0^{\pi/2} \sin 2\theta \cdot d\theta = \frac{E \cdot \Delta S \cdot \Delta r}{4}$$

This energy arrives in a time interval $\Delta t = \Delta r/c$. Hence, the above expression can be rewritten as:

$$\Delta E = \frac{E \cdot \Delta S \cdot c \cdot \Delta t}{4}$$

$$\Rightarrow \quad \frac{\Delta E}{\Delta t} = \frac{E \cdot c \cdot \Delta S}{4} = \frac{E \cdot c}{4} \text{ (for unit area)} \qquad (2.20)$$

i.e., Rate of energy falling on the surface of unit area is equal to:

$$\frac{1}{4} \times (\text{Energy density} \times \text{Sound velocity})$$

If a is the absorption coefficient of the element ΔS, the energy absorbed per unit time is:

$$\frac{Eca\Delta S}{4}$$

Hence, the total rate of absorption at any time when the energy density is E is given by:

$$\frac{Ec \sum_i a_i \Delta S_i}{4} = \frac{EcA}{4} \qquad \left(\because A = \sum_i a_i \Delta S_i \right) \qquad (2.21)$$

From eqns. (2.19) and (2.21)

$$V \times \frac{dE}{dt} + \frac{E \cdot c \cdot A}{4} = P$$

$$\Rightarrow \quad \frac{dE}{dt} + \frac{E \cdot c \cdot A}{4V} = \frac{P}{V}$$

For convenience denote $\alpha = \frac{cA}{4V}$, so that the above equation can be written as:

$$\frac{dE}{dt} + \alpha E = \frac{4P\alpha}{cA}$$

Multiplying throughout by $e^{\alpha t}$:

$$\left[\frac{dE}{dt} + \alpha E \right] e^{\alpha t} = \frac{4P\alpha}{cA} e^{\alpha t} \quad \Rightarrow \quad \frac{d(Ee^{\alpha t})}{dt} = \frac{4P\alpha}{cA} e^{\alpha t}$$

Integrating we get: $Ee^{\alpha t} = \int \dfrac{4P\alpha}{cA} e^{\alpha t} = \dfrac{4P}{cA} e^{\alpha t} + K$ where K is the constant of integration.

(1) At $t = 0$, $E = 0$.

Hence $\quad K = -\dfrac{4P}{cA}.$

Substituting for K in the above expression and simplifying we get,

$\therefore \qquad E = \dfrac{4P}{cA}(1 - e^{-\alpha t})$ where $\alpha = \dfrac{cA}{4P}$

or $\qquad E = E_{max}(1 - e^{-\alpha t})$ where $E_{max} = \dfrac{4P}{cA}$ ...(2.22)

This expression shows the growth of E with time after the start of the sound. (Fig. 2.18a)

(2) If the source is cut off when E has reached the maximum, so that $P = 0$ and $E = E_{max}$ at $t = 0$ Then:

$$E = E_{max} e^{-\alpha t} \qquad ...(2.23)$$

This expression shows the decay of E with time after the source has been cut off. The decay of sound in a closed room is shown in Fig.2.18b. The case may be compared with the variation of potential difference across a capacitor during charging and discharging through a resistance.

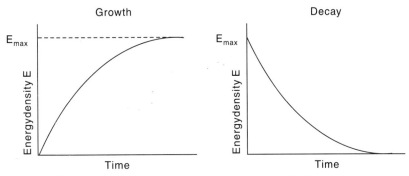

Fig. 2.18 Growth and decay of sound intensity in a room

The reverberation time is defined as the time taken for the intensity or the energy density to decay to one-millionth of its value before the cut off. Hence

$e^{-\alpha T} = 10^{-6} \quad \Rightarrow \quad \alpha T = 6 \log_e 10$

$\Rightarrow \qquad T = \dfrac{6 \log_e 10}{\alpha} = \dfrac{6 \log_e 10}{\left(\dfrac{cA}{4V}\right)} = \dfrac{0.162\, V}{A} \qquad (\because c = 340 \text{ m/s})$

where V is the volume in m^3 and A is the absorption in m^2.

Thus reverberation time is directly proportional to volume and inversely proportional to the absorption. Note that the reverberation time is the time taken for the sound intensity to decrease by 60 dB (Fig.2.19).

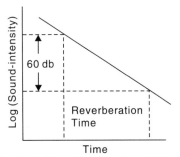

Fig. 2.19 Reverberation time is the time for 60 dB loss in sound intensity

It is important to note that reverberation time is affected by the relative humidity in the environment. A plot of attenuation coefficient versus relative humidity at various frequencies is shown in Fig.2.14. In large auditorium, air itself contributes a substantial amount of absorption of sound at high frequencies. The absorption of air depends on the temperature and relative humidity, and an additional term mV, proportional to the volume should be added to the absorption A. The constant m is given in the last two lines of Table 2.4. The reverberation time for large auditorium is:

$$T = 0.162 \left(\frac{V}{A + mV} \right)$$

2.3.3 Eyring's Formula

In the derivation of Sabine's formula it was assumed that the sound energy distribution was uniform throughout the room. This is achieved by a large number of reflections occurring at the walls, ceilings and other objects present in the room. *This large number of reflections can occur only if the absorption coefficients of the surfaces themselves are small.* Hence one does not expect the Sabine's formula to hold good when there are surfaces with large absorption coefficient. Sabine's formula is clearly not applicable in the limiting case when absorption coefficient is unity. For such surfaces the reflectivity is zero. The only energy present is that of the direct sound. Hence the reverberation time is zero. However, the Sabine's formula predicts a finite reverberation time. A new approach to the problem is therefore essential to the discussion of the growth and decay of sound in such dead rooms. It is found that the Sabine's formula gives incorrect results when the average absorption coefficient is greater than 0.2.

One such approach is due to Eyring. *Here the multiplicity of reflections from various surfaces is considered to be equivalent to a set of image sources* (Fig.2.20). *All these sources come into existence the instant the real source starts producing the sound.* The building up of acoustic energy at any point then consists of the summation of sound energy from the real source, from the first order or single reflection images whose strengths are $(1 - a)P$; from the second order or double reflection images whose strengths are $(1 - a)^2 P$ etc. All the image sources of appreciable strength have to be included in such a summation. When the true source in the room is stopped, all the images of this source are assumed to stop simultaneously. The decay of energy in the room therefore results from the successive losses of acoustic radiation, first from the source, then from first order images, the second order images etc.

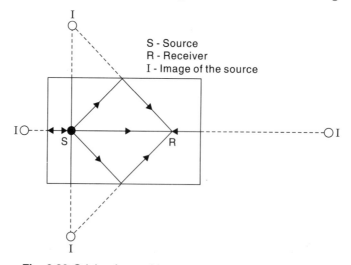

Fig. 2.20 Origin of sound image sources in Eyring's theory

On the basis of the assumption of the multiple sound sources which are the first, second, third,... nth order images of the original sound energy, it may be shown that the growth of acoustic energy density is given by:

$$E = \frac{4P}{-cS \ln(1-\bar{a})} \left[1 - e^{\frac{cS \ln(1-\bar{a})t}{4V}} \right] \quad (2.24)$$

This equation is similar to the corresponding eqn. (2.22) obtained by Sabine excepting that the total room absorption is now given by:

$$A = -S \ln(1-\bar{a})$$

where S is the total area of the interior surfaces of the room and $\bar{a}$ is an average sound absorption coefficient as defined by the equation

$$\bar{a} = \frac{\sum_i a_i S_i}{S}$$

Similarly, the equation for the decay of the sound is given by:

$$E = E_{max} \, e^{\frac{cS \ln(1-\bar{a})t}{4V}} \quad (2.25)$$

This corresponds to eqn. (2.23) of Sabine.

The reverberation time T is given by:

$$T = \frac{0.162V}{-S \ln(1-\bar{a})} \quad (2.26)$$

This formula for the reverberation time differs considerably from the Sabine's formula when a_i is large. When a_i is small we have

$$\ln(1-\bar{a}) = -\bar{a}$$

so that it reduces to Sabine's formula.

Also for the case of the dead room *i.e.*, $\bar{a} = 1$, the reverberation time is given by:

$$T = \frac{0.162V}{-S \ln(1-\bar{a})} = \frac{0.162V}{S \ln(1-\bar{a})^{-1}} = \frac{0.162V}{S \ln\left(\frac{1}{1-\bar{a}}\right)} = \frac{0.162V}{S \ln\left(\frac{1}{0}\right)} = \frac{0.162V}{S \ln(\infty)} = 0$$

which is the expected value.

2.3.4 Measurement of Reverberation Time

The reverberation time is of the order of seconds. It is also a function of frequency. Its measurement consists in measuring the sound pressure as a function of time. The source of sound is an octave or a third-octave band random noise from a white-noise generator. Sound of suitable amplitude is produced. It is amplified and connected to a pair of loud speakers as shown in Fig. 2.21a. The source of sound consists of different frequencies and the sound pressure is measured as a function of time. The desired frequency is chosen by switching the filters in the frequency analyzer. An alternative is to use a pistol (Fig.2.21b). This consists of a microphone connected to a frequency analyzer connected in turn to a level recorder. The sound level recorder will need to have a logarithmic potentiometer in the circuit to convert pressure measurement into dB. *The time required for the sound intensity to decrease by 60 dB gives the reverberation time. This is same as the time required for the sound pressure to decrease by 30 dB.* Modern microprocessor based equipment is able to provide graphs of the decay curve on a video screen and automatically calculate values of the reverberation time. In each case measurements are made in octave bandwidths, whose centre frequencies are 125Hz, 250Hz, 500Hz, 100Hz, 2000Hz, 4000Hz or in a third-octave bandwidths, whose centre frequencies are 100Hz, 125Hz, 160Hz, 200Hz, 250Hz, 315Hz, 400Hz, 500Hz, 630Hz, 800Hz, 1000Hz, 1250Hz, 1600Hz, 2000Hz, 2500Hz, 3150Hz, 4000Hz. A typical result is shown in Table 2.3.

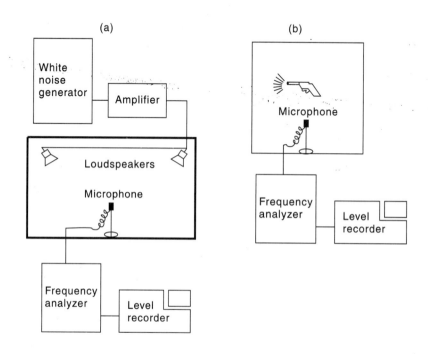

Fig. 2.21 Measurement of reverberation time using (a) continuous source and (b) pistol

Table 2.3 Typical reverberation times for a room

Third-octave bandwidth centre frequency(Hz)	Reverberation time (s)	Third-octave bandwidth centre frequency(Hz)	Reverberation time (s)
100	1.55	630	1.0
125	1.60	800	0.90
160	1.45	1000	1
200	1.30	1250	1.05
250	1.20	1600	1.05
315	1.05	2000	1.00
400	1.05	2500	0.95
500	1.00	3150	0.95
		4000	0.80

2.4 MEASUREMENT OF ABSORPTION COEFFICIENT

The Sabine's formula and Eyring's formula show that the reverberation time in a room is dependent on the volume and absorption in the room. It is essential that the absorption of the finished surfaces in a hall are known at the design stage so that an estimation of reverberation is possible. To calculate the total absorption in the room it is necessary to know the absorption coefficients of the various surfaces in the room. The absorption coefficient is defined as the fraction of the incident energy absorbed by the surface. The remaining $(1 - a)$ fraction of the incident energy is reflected assuming that the transmission coefficient is negligible. There are two main methods of measurement. One is based on measurement of reverberation time and the other based on measurement on the reflectivity of sound waves from the absorber.

2.4.1 Method Using a Reverberation Chamber

Method 1

In this method the reverberation time is measured with and without the material whose absorption coefficient has to be determined. The advantage of this method is that it allows for all angles of incidence. The minimum required volume of the reverberation chamber is about 200 m³ so that measurements down to 100Hz may be made. The lowest frequency should not be lower than about $125\,(180/V)^{1/3}$Hz to ensure a diffuse sound field, where V is the volume of the room. The room itself will have walls and ceiling all slightly out of parallel. There will also be some objects in the form of curved sheets of plywood or Perspex a few mm thick to obtain a uniform diffuse field. A long reverberation time is essential for the measurement. An audio frequency source of sound is used to measure the time of reverberation. Let T_1 be the reverberation time of the empty room. If $A = \sum a_i S_i$ denotes the absorption due to the walls of the room, the ceiling and the floor, then

$$T_1 = \frac{0.162V}{\sum_i a_i S_i} = \frac{0.162V}{A} \qquad (2.27)$$

Then a certain amount of absorbing material of area S and absorption coefficient a' is introduced into the room and the time of reverberation T_2 is again determined. Then

$$T_2 = \frac{0.162V}{A + a'S} \qquad (2.28)$$

Eliminating the unknown absorption A from equations (2.27) and (2.28) we get:

$$a' = \frac{0.162V}{S}\left[\frac{1}{T_1} - \frac{1}{T_2}\right] \qquad (2.29)$$

Method 2

In this method the reverberation times for two sources of emitting powers P and P' with and without the material are determined. P and P' need not be known absolutely. It is sufficient to know their ratio. Let the absorption of the empty room be A. The steady state energy densities of these sources will be

$$E_{max} = \frac{4P}{cA} \quad \text{and} \quad \frac{4P'}{cA} \quad \text{respectively.}$$

During the decay, let the sound energy density reach a fixed value E_0 in times T and T' respectively. Then

ACOUSTICS OF BUILDINGS

$$E_0 = \frac{4P}{cA} e^{-\alpha T} = \frac{4P'}{cA} e^{-\alpha T'} \text{ where } \alpha = \frac{cA}{4V}$$

$$\Rightarrow \quad \frac{P'}{P} = e^{\alpha(T-T')} \quad \text{or} \quad \alpha = \frac{\ln\left(\frac{P'}{P}\right)}{(T'-T)} = \frac{cA}{4V}$$

$$\Rightarrow \quad A = \frac{4V \ln\left(\frac{P'}{P}\right)}{c(T'-T)} \quad (2.30)$$

Then a certain amount of absorbing material of area S and absorption coefficient a' is introduced into the room and the absorption in the room is again determined. Let T_m and T'_m be the time taken for the energy density to reach a value E_0 with the source P and P' respectively. Then

$$A + a'S = \frac{4V \ln\left(\frac{P'}{P}\right)}{c(T'_m - T_m)} \quad (2.31)$$

From eqns. (2.30) and (2.31) we get:

$$a' = \frac{4V \ln\left(\frac{P'}{P}\right)}{cS} \left[\frac{1}{(T'_m - T_m)} - \frac{1}{(T'-T)} \right] \quad (2.32)$$

2.4.2 Impedance Tube Method

This method only measures the absorption coefficient at normal incidence. It is a useful indication of the sort of absorbent properties which a material may have. Its main use is in theoretical work, research work or in quality control for the production of acoustic absorbent material. Pure tones produced by an oscillator are used to excite the loudspeaker as shown in Fig.2.22 producing standing waves in the tube. Partial reflection will take place at the absorber surface resulting in a standing wave pattern as shown in Fig.2.23.

If the displacement at any time of the incident wave is represented by

$$d_1 = d_0 \sin(\omega t - kx), \quad (2.33)$$

Then the displacement of the reflected wave is given by

$$d_2 = f d_0 \sin(\omega t + kx) \text{ where } f d_0 \text{ is the reflected amplitude} \quad (2.34)$$

The resulting displacement at any point is given by

$$d = d_1 + d_2 = d_0 \sin(\omega t - kx) + f d_0 \sin(\omega t + kx)$$
$$= d_0 (1+f) \sin \omega t \cos kx + d_0 (1-f) \cos \omega t \sin kx \quad (2.35)$$

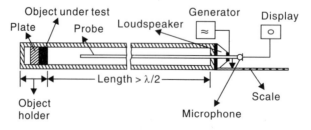

Fig. 2.22 Absorption coefficient measurement using impedance tube

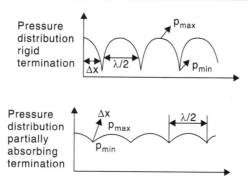

Fig. 2.23 Pressure amplitude of standing waves in the impedance tube

It can be seen that the maximum and minimum values will be $d_0(1+f)$ and $d_0(1-f)$ respectively separated by a distance of $\lambda/4$. The first term has maxima at $x = 0, \lambda/2, \lambda, 3\lambda/2, 2\lambda$, etc. while the second term has maxima at $\lambda/4, 3\lambda/4, 5\lambda/4, 7\lambda/4$, etc. If the maximum and minimum amplitudes are A_1 and A_2, then

$$\frac{A_1}{A_2} = \frac{d_0(1-f)}{d_0(1-f)} \Rightarrow f = \frac{A_1 - A_2}{A_1 + A_2} \qquad (2.36)$$

But the energy can be shown to be proportional to the square of the amplitude

$$\therefore \quad r = f^2 = \left(\frac{A_1 - A_2}{A_1 + A_2}\right)^2$$

where r is the fraction of reflected energy. $\qquad (2.37)$

The absorption coefficient, $\alpha = 1 - r$

$$\therefore \quad \alpha = 1 - \left(\frac{A_1 - A_2}{A_1 + A_2}\right)^2 = \frac{(A_1 + A_2)^2 - (A_1 - A_2)^2}{(A_1 + A_2)^2}$$

$$= \frac{4 A_1 A_2}{(A_1 + A_2)^2} = \frac{4}{\left(\dfrac{A_1}{A_2} + \dfrac{A_2}{A_1} + 2\right)} \qquad (2.38)$$

In the experiment, the ratio of maximum i.e., (A_1/A_2) is measured. Hence α can be calculated. It will normally be found that results from the standing wave tube, though reproducible, are less than those from the reverberation chamber. The size of the tube is also important. The maximum diameter of the sample should not be greater than about half of the wavelength under investigation. Thus for measurements in a tube of 100 mm diameter, the upper limiting frequency is about 1600Hz and for 6500Hz, the maximum diameter should be about 25 mm. It is possible to compare tube measurements of absorption coefficient with reverberation chamber measurements. But the values in the latter can vary depending upon the distribution of the absorbers in the room. Absorption coefficients of different materials are given in Table 2.4.

Table 2.4 Absorption coefficients of different materials

Material	Octave Band Centre Frequency (Hz)					
	125	250	500	1000	2000	4000
Brick, unglazed	0.05	0.03	0.03	0.04	0.05	0.07
Brick, unglazed, painted	0.01	0.01	0.02	0.02	0.02	0.03
Carpet on foam rubber	0.08	0.24	0.57	0.69	0.71	0.73
Carpet on concrete	0.02	0.06	0.14	0.37	0.60	0.65
Concrete block, coarse	0.36	0.44	0.31	0.29	0.39	0.25
Concrete block, painted	0.10	0.05	0.06	0.07	0.09	0.08
Floors, resilient flooring on concrete	0.02	0.03	0.03	0.03	0.03	0.02
Floors, hardwood	0.15	0.11	0.10	0.07	0.06	0.07
Glass, heavy plate	0.18	0.06	0.04	0.03	0.02	0.02
Glass, standard window	0.35	0.25	0.18	0.12	0.07	0.04
Gypsum, board, 0.5 inch thick	0.29	0.10	0.05	0.04	0.07	0.09
Panels, fiberglass, 1.5 inch thick	0.86	0.91	0.80	0.89	0.62	0.47
Panels, perforated metal, 4 inch thick	0.70	0.99	0.99	0.99	0.94	0.83
Panels, perforated metal with Fiberglass insulation, 2 inch thick	0.21	0.87	Values dependent on measuring conditions			
Panels, plywood, 3/8 inch thick	0.28	0.22	0.17	0.09	0.10	0.11
Panels, gypsum or lime, rough	0.02	0.03	0.04	0.05	0.04	0.03
Plaster, gypsum or lime, smooth	0.02	0.02	0.03	0.04	0.04	0.03
Polyurethane foam, 1 inch thick	0.16	0.25	0.45	0.84	0.97	0.87
Tile, ceiling, mineral fiber	0.18	0.45	0.81	0.97	0.93	0.83
Tile, marble or glazed	0.01	0.01	0.01	0.01	0.02	0.02
Wood, solid, 2 inch thick	0.01	0.05	0.05	0.04	0.04	0.04
Water surface	nil	nil	nil	0.003	0.007	0.02
One person	0.18	0.4	0.46	0.46	0.51	0.46
Air (20°C, 30% humidity, per m^3)	—	—	—	—	0.012	0.038
Air (20°C, 50% humidity, per m^3)	—	—	—	—	0.010	0.024

ACOUSTICS OF BUILDINGS

2.5 SOUND ABSORBERS

The sound absorbed by the material depends on the nature of the material. The absorption coefficient is a function of frequency. Sound absorbers may be divided into three main types

(a) Porous materials

(b) Membrane and panel absorbers and (c) Helmoltz resonators

Porous materials can be considered as broad band absorbers since they have some absorption at all frequencies. Membrane absorbers have far more critical absorption characteristics around the resonant frequency of the panel. Helmholtz resonators have very sharp critical absorption frequencies.

2.5.1 Porous Absorbers

These consist of such materials as fiberboard, mineral wools, insulation blankets, etc. All these have a network of interlocking pores. They act by converting sound energy into heat. Materials such as foamed plastics are far less effective as sound absorbers. Typical characteristics are shown in Fig.2.24. Sound absorption is far more efficient at high rather than at low frequencies. It may be slightly improved by increased thickness or mounting with an airspace behind (Fig.2.25). The displacement antinode, a point of maximum particle movement is $\lambda/4$ from the wall (Fig.2.26). This means that unless the absorber thickness is $> \lambda/4$ the absorption is not effective. Figure 2.27 shows absorption as a function of the absorber thickness.

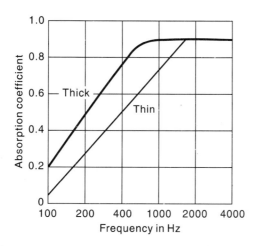

Fig. 2.24 Absorption characteristics of (a) thick porous absorbers and (b) thin porous absorbers

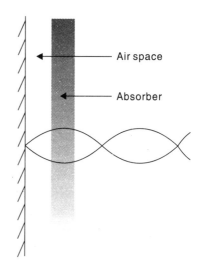

Fig. 2.25 Porous absorber with air space

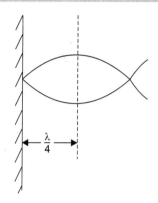

Fig. 2.26 Particle displacement at reflection

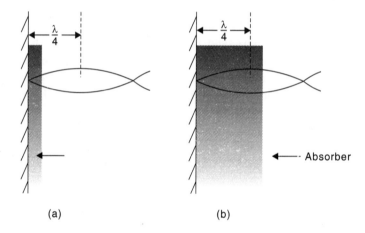

Fig. 2.27 Absorption as a function of absorber thickness
(a) minimum absorption (b) maximum absorption

Porous absorbers are available in three types: prefabricated tiles, plasters and spray on materials and acoustic blankets (glass wool). Some of the typical acoustical tiles are shown in Fig.2.28. The method of fixing can make a considerable difference to the efficiency of these materials, particularly at low frequencies. In general, it can be said that mounting the absorber away from the wall surface results in a marked increase in low frequency absorption. When lack of space on walls or ceilings prevents the addition of absorbers, they may be used in the form of space absorbers (Fig.2.29). These can be made from perforated sheets of steel, aluminum or hardboard in various shapes and sizes, such as cubes, prisms, spheres or cones and filled with glass wool or other suitable material. It is possible to make them with the underside reflecting while the top is absorbent. It is also possible to have absorption surfaces which can be covered to the extent needed as shown in Fig.2.29. This can be very helpful in preventing long delayed sound from a dome in a hall reaching the listeners and at the same time providing more reflection of sound to certain parts of the audience.

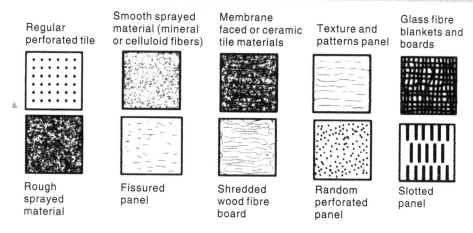

Fig. 2.28 Acoustic tiles used for sound absorption

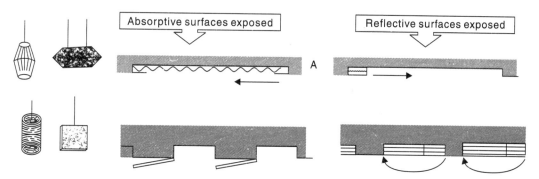

Fig. 2.29 Space absorbers with variable absorption

2.5.2 Membrane or Panel Absorbers

The panel may be an area of plywood backed with an energy absorbing material. Roofing felt and other similar bituminous substances are satisfactory. Behind the air space is an air space which provides the spring action against which the mass of the panel vibrates. These are useful because they can have good absorption characteristics in the low frequency range. The absorption is highly dependent upon frequency and is normally in the range of 50 to 500Hz (Fig. 2.30). The approximate resonant frequency f can be calculated from the formula

$$f = \frac{60}{\sqrt{md}}$$

where m is the mass of the panel in kg/m^2 and d is the depth of air space in m.

In practice, this is only an approximation as the method of fixing and the elastic constant of the individual panels can have a large effect. Panel absorbers can often be present fortuitously in the form of suspended ceilings or even closed double windows.

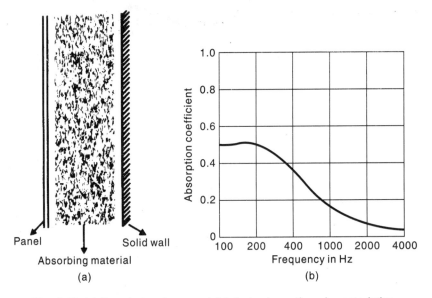

Fig. 2.30 (a) Panel absorbers and (b) their absorption characteristics

2.5.3 Helmholtz Resonators

These are containers with a small open neck and they absorb sound due to resonance of the air within the cavity. Porous material is often introduced into the neck to increase the efficiency of absorption. It can be shown that for a narrow-necked resonator of the type shown in Fig.2.31. The resonant frequency f is approximately

$$f = \frac{cr}{2\pi}\sqrt{\left[\frac{2\pi}{(2l + \pi r)V}\right]}$$

where c is the velocity of sound in air, r is the radius of the neck, l is the length of the neck and V is the volume of the cavity. If there is no neck the above formula reduces to

$$f = \frac{c}{2\pi}\sqrt{\frac{2r}{V}}$$

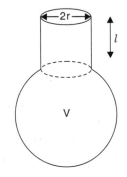

Fig. 2.31 Helmholtz resonator

Efficient absorption is only possible over a very narrow band, as shown in Fig.2.32. It is necessary to have many resonators tuned to slightly different frequencies if effective control of reverberation time is to be obtained. Typical Helmholtz resonators used in practice are shown in Fig.2.33. Cavity resonators are useful in controlling long reverberation times at isolated frequencies and are used in a number of concert halls.

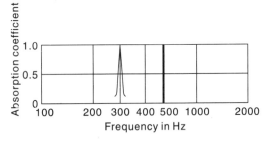

Fig. 2.32 Absorption characteristics of Helmholtz resonators

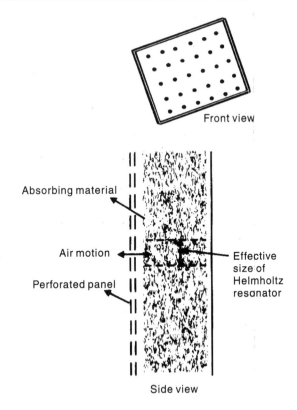

Fig. 2.33 Typical Helmholtz resonator used in practice

2.6 DESIGN CONSIDERATIONS FOR GOOD ACOUSTICS

The acoustic design of the building ought to be integrated with the overall plan of the building. The acoustic design depends on the nature of the building and the purpose for which it is built. The buildings could be broadly classified as lecture theatres, concert halls, general purpose auditoria, hospitals, educational institutions and commercial establishments. The acoustical design of a room consists of following steps:

(a) Defining the exact purpose for which the room is being put to use.

(b) Characterising the acoustic environment which consists in identifying the location of various sound sources both within the room and those which are external to the room.

(c) Optimising the acoustical design to enhance the desired sounds and eliminate or reduce the unwanted sound or noise.

(d) Choosing proper reverberation time, ensuring clarity and uniform distribution of sound, eliminating hot spots, dead spots, flutter echoes and background noise.

(e) Using sound reinforcement systems to enhance the acoustical quality of the room.

The design considerations for good room acoustics is summarised in Fig.2.34.

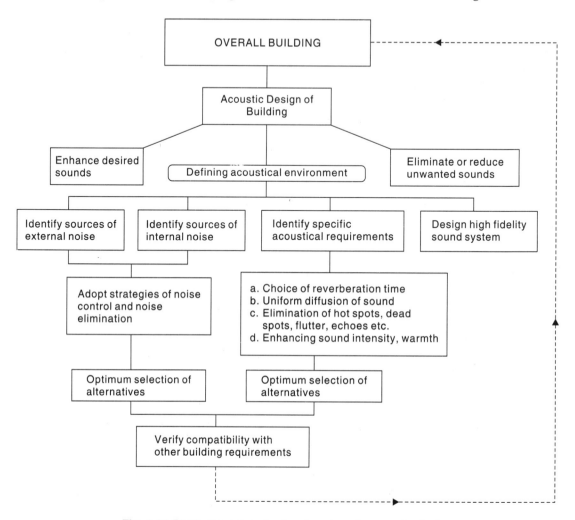

Fig. 2.34 Block diagram indicating the steps in acoustic design

2.6.1 Basic Characteristics of Good Acoustics

Choice of room shapes: There are basically three types of rooms which are in use. These correspond to rectangular, fan shaped and horse shoe all of which are illustrated in Fig.2.35. The choice is usually governed by specific requirements. Rectangular shaped rooms are usually preferred for small sized lecture halls. The fan shaped hall accommodates, through its spread, a larger audience within the close range from the stage. It features nonparallel walls that eliminate flutter echoes. Sound can also be reflected from the rear walls back to the stage, depending on balcony layout and the degree of sound absorptions. A disadvantage is the early time delay from echoes of side walls. Horseshoe shaped structures have been used as preferred design for opera houses and concert halls of modest hall capacity. This design provides for excellent lines of sight, short paths for direct sound and provides a sense of intimacy. In all types of auditoria, ceilings constitute design opportunities for transporting sound energy from the stage to distant listeners. The ceiling itself could be shaped to ensure uniform distribution

(Fig.2.36). Alternatively suitable reflectors could be suspended from the ceiling for this purpose (Fig.2.37).

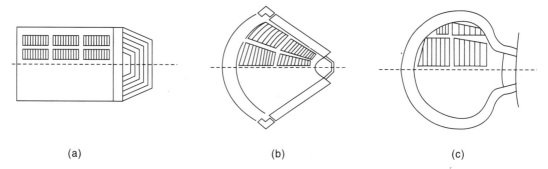

Fig. 2.35 (a) Rectangular (b) Fan shaped and (c) horse shoe shaped room

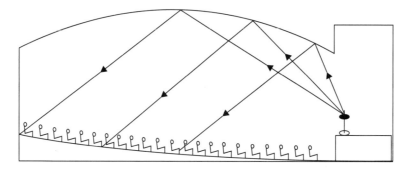

Fig. 2.36 Correctly shaped ceiling reflector can provide uniform distribution of sound energy

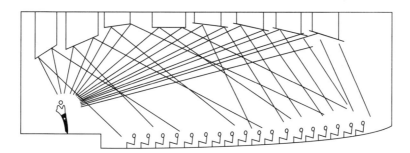

Fig. 2.37 Reflectors suspended from ceiling to achieve desired time delays and uniform distribution of sound in the room

Clarity: This ensures that the nature of the sound produced by the source is not distorted. This is possible if no strong echoes are heard after about 50ms after the arrival of the direct sound. Sound reflectors to reduce delay time are used for this purpose. Further the intensity of the early plus direct sound should be greater than the reverberant sound level at all locations.

Loudness of direct sound: The design must make sure that the sound produced by the source is heard without any feebleness and is of sufficient intensity throughout the auditorium. The auditorium should be designed such that no listener is seated too far from the sound source (since direct sound decreases with distance). If the hall is too large, sound amplification may be necessary. Ceiling should be a good sound reflector.

Diffusion or uniformity: Good spatial distribution of the sound is achieved by diffuse or irregular reflecting surfaces and by the avoidance of focused sound or sound shadows. An adequate intensity of sound must reach all parts of the room. The rear wall is usually made highly sound absorbing. The non-uniformity of the sound intensity in the room is usually indicated by the presence of more than single reverberation time as shown in Fig.2.38.

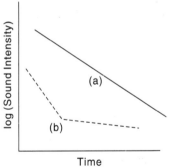

Fig. 2.38 Decay of sound intensity in (a) ideal room and (b) room which has non-uniform distribution of sound

Reverberation sound level: The intensity of the reverberant sound, which will be the same throughout the hall, depends on the power of the source and the reverberation time. Reverberation time should be optimum for the required use of the room. This is to ensure clarity for speech and fullness for music. The recommended reverberation times are given in Table 2.5. Air absorption should not be neglected.

Table 2.5 Recommended reverberation times (in seconds) for various hall dimensions

	$2.83 \times 10^3 \, (m^3)$	$4.25 \times 10^3 \, (m^3)$	$5.66 \times 10^3 \, (m^3)$	$8.9 \times 10^3 \, (m^3)$
Choir	1.0	1.0	1.2	1.25
Orchestra	0.8	0.85	0.9	1.00
Speech or Music	0.6	0.65	0.7	0.75

Balance of Direct and Reverberant Sound

Whenever a continuous source of sound is present in a room, two sound fields are produced. One, the *direct sound field* is the direct arrival from the source. The other, *reverberant sound field* is produced by reflections. The energy density E_d produced by the direct sound field of an omni directional source is given by:

$$E_d = \frac{P}{c}\left(\frac{1}{4\pi r^2}\right)$$

where P is the power of the source and r is the radial distance from the effective centre of the sound source. The maximum energy density of the reverberant field is given by:

$$E_{rev}^{max} = \frac{4P}{cA}$$

$$\therefore \quad \frac{E_{rev}^{max}}{E_d} = \frac{16\pi r^2}{A} = \left(\frac{r}{r_d}\right)^2 \quad \text{where } r_d = \frac{1}{4}\sqrt{\frac{A}{\pi}}$$

r_d is the distance at which the direct sound field has fallen to the same value as the reverberant field. This equation shows that for locations very close to the source ($r \ll r_d$), the shape or acoustic treatment of the room will have little influence on measured sound pressure levels. By contrast, at distances for which $r \gg r_d$, the sound pressure level will be reduced by $3 \, dB$ for each doubling of the total sound absorption A. Thus for example, a worker near a noisy machine will receive little benefit from increasing the total absorption of the room. However, the acoustic

ACOUSTICS OF BUILDINGS

exposure of other workers at some distance from the machine will be reduced by such treatment. Substituting for A in terms of the reverberation time from eqn. (2.15), we get

$$\frac{E_{rev}^{max}}{E_d} = \frac{16\pi r^2}{A} = \frac{16\pi r^2}{\left(\frac{0.162V}{T}\right)} = \frac{312 r^2 T}{V}$$

But, $r^2 = kV^{2/3}$ (where k is some constant)

$$\Rightarrow \quad \frac{E_{rev}^{max}}{E_d} = 312\, kV^{1/3}\, T = \frac{V^{1/3}\, T}{R'} \text{ where } R' \text{ is the room constant.}$$

Since the energy of the direct arrival falls off as the square of the distance from the source it is impossible to have a constant ratio throughout the room. However, for the distance where the maximum in reverberation sound is equal to the intensity of the direct sound, we get

$$TV^{1/3} = R'$$

This is an empirical relation which is useful. Approximate values of R' for various rooms are shown in Table 2.6.

Table 2.6 Approximate values of R' for rooms used for various purposes

Purpose	R'	Range of Volumes (m^3)
Concert hall	0.07	$10 \times 10^3 < V < 25 \times 10^3$
Opera house	0.06	$7 \times 10^3 < V < 20 \times 10^3$
Motion picture theatre	0.05	$V < 10 \times 10^3$
Auditorium	0.06	
Lecture hall	0.06	$V < 4 \times 10^3$
Conference hall	0.06	
Recording studio	0.04	$V < 1 \times 10^3$
Broadcasting studio	0.04	

Freedom from noise: The noise level should be sufficient less than about 50 dB at 50Hz and decrease to 10 dB at about 400Hz.

Variable acoustics: Many a time it is economically viable to have multipurpose halls. In multipurpose halls, there must be a provision for varying the absorption so that the reverberation time can be varied. This is achieved by absorbers whose area can be varied as shown in Fig.2.29.

Intimacy: A hall has acoustical intimacy when music sounds as if it were being played in a small hall. The time delay between the direct and first reflected sound should be less than 20ms in order for a hall to be intimate.

Liveness: This is a qualitative term used for describing the reverberation time. It is related primarily to the reverberation time for middle and high frequencies. The optimum reverberation time depends on size and function as shown in Fig.2.39. A hall with insufficient reverberation is termed dry.

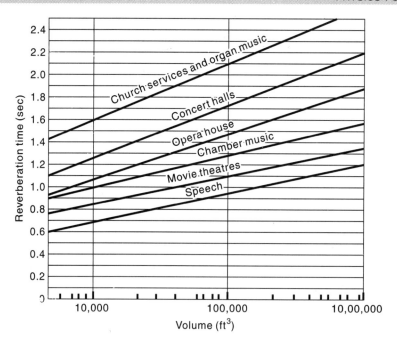

Fig. 2.39 Reverberation times for auditoria of varying volume and different purposes

Warmth: This is related to liveness and fullness of bass tone. Reverberation time at 250Hz and below, should be somewhat longer than at middle and high frequencies. Opposite of warmth is *brilliance*. In a *brilliant room* the reverberation time for higher frequencies is longer than at low frequencies. The reverberation times for ideal, warm and brilliant rooms is shown in Fig.2.40. This results in the persistence of the high frequency sound for longer time.

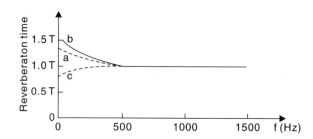

Fig. 2.40 Reverberation time for (a) normal (b) warm and (c) brilliant rooms

Balance and blend: Blend refers to the mixing of sound from all the instruments of the orchestra or ensemble over all points in audience. In a concert hall with a poor blend, a member of the audience in some specific location may hear one player louder than the other players. The simplest technique for achieving proper blend is to mix the sound from various instruments and voices before broadcasting. This need not be done through electronic equipment. By providing low ceiling and appropriate reflecting surfaces around the stage it is possible to achieve the proper blend. However, care should be taken to prevent the sound being preferentially reflected to a particular section of the audience from a particular part of the stage.

ACOUSTICS OF BUILDINGS

Ensemble: This refers to the ability of the members of an orchestra or an ensemble to hear each other. This is achieved by providing ample reflecting surfaces to the sides and above the stage.

2.6.2 Acoustical Defects and Their Remedies

Acoustical defects to be avoided include:

(a) Long delayed echoes (b) Flutter echoes
(c) Sound shadows (d) Sound distortion
(e) Sound concentration (f) Room resonances and
(g) Background noise.

Long delayed echoes: In large halls care must be taken to make certain that no strong reflections of sound are received by the audience after about 50 ms. Otherwise clarity of speech will be affected. Average speech is at the rate of about 15 to 20 syllables per second or roughly one syllable every 70 ms to 50 ms respectively. A member of the audience who hears strong echoes within this time will find it difficult to understand speech. These strong reflections can be prevented by covering the surfaces concerned with absorbent material or by making them into diffusing surfaces by means of convex shape. Quick reflections from the corners can be a problem but are easily overcome by the use of an acoustic plaster or some other absorbent material. The simple solution appears to be to cover as much of the surface as possible with an absorbent material. However, too much of absorption would lead to a very unpleasant effect and also make speech inaudible especially near the back of the hall. Sound amplification is then a must. Also, the reverberation time is also affected by the presence of the audience. The aim should be to use the minimum amount of absorbent material so that the hall of given floor area can accommodate maximum number of people.

Flutter echoes: Flutter echoes are a series of echoes that occur in rapid succession. They usually result from reflections between two parallel surfaces that are highly reflective (Fig.2.41). The remedy for this is to avoid parallel walls and parallel ceilings and floors.

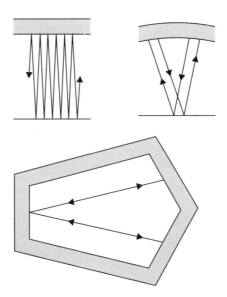

Fig. 2.41 Flutter echoes

Echelon effect: The echelon effect is the name given to musical tone produced by a set of steps as shown in Fig.2.42. If the spacing of the steps is d, then the distance traveled between two adjacent echoes will be $2d$. Suppose d is 34 cm, then the time interval between echoes will be (0.34 m × 2)/340 = 1/500 second. This corresponds to a musical tone of frequency 500Hz. This defect can be overcome by having steps at uneven intervals. Also, if a room has a flight of steps, with equal widths, this will act as a reflection grating. Hence this will lead to dispersion of sound and sound maximum occurs at different angles. This will in turn lead to non-uniform distribution of sound.

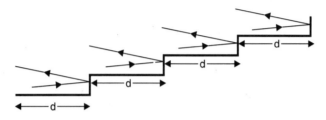

Fig. 2.42 Echelon effect

Sound focusing: Focusing of sound can be caused by reflection from large concave surfaces (Fig.2.43). Focusing prevents the uniform distribution of sound intensity throughout the room. Near the focus of the concave surface the sound intensity will be large.

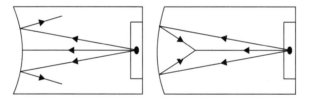

Fig. 2.43 Diffusing effect of convex and focusing effect of concave surface

Sound shadows: Balcony is used to increase seating capacity and to reduce the distance of the farthest row of seats. The audience occupying these seats is in direct line of sight to the performer and therefore receives the direct sound. However, under the balconies there may be insufficient early sound, since most of the reflections from the side walls and ceiling do not reach this area (Fig. 2.44). Balconies ought to be designed to avoid such regions of sound shadows.

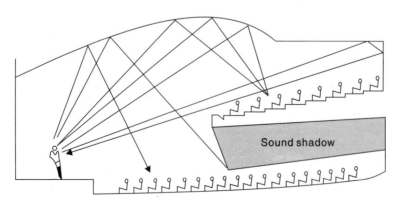

Fig. 2.44 Sound shadow

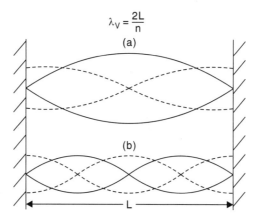

Fig. 2.45 Standing waves in one dimension (a) particle displacement and (b) pressure amplitude

Sound distortion due to room resonances: Any wave which is confined in a region gives rise to standing waves which can have only discrete wavelengths or frequencies. Figure 2.45 shows the particle displacement and pressure amplitudes for the first two modes. In any particular room the distribution of sound energy is not uniform on account of the presence of standing waves in the room. For a rectangular room of length l, width w and height h the resonant frequencies are given by:

$$f = \frac{c}{2}\sqrt{\left(\frac{p}{l}\right)^2 + \left(\frac{q}{w}\right)^2 + \left(\frac{r}{h}\right)^2}$$

where p, q and r are integers; c is the velocity of sound. This expression does not assume any absorption at the surface. In real life situations the absorption at the walls also has to be taken into account. Typical room resonances for a room are shown in Fig.2.46. Since only specific frequencies are amplified the presence of room resonances can affect the quality of music. Room resonances are generally indicated by the presence of spikes in the plot of decay of sound intensity with time (Fig.2.47). The effect of room resonances can be offset by tuning the sound amplifiers to have low gain at these frequencies.

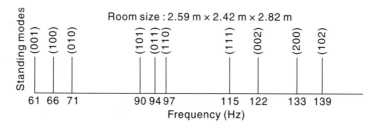

Fig. 2.46 Typical room resonances in a room

Selective absorption: In a room selective absorption of sound can occur due to room resonances and also due to the specific absorbers which have absorption in narrow range of frequencies. The frequency that is lost results in appreciable change in the quality of sound and music. This defect has to be overcome by using suitable sound reinforcement system.

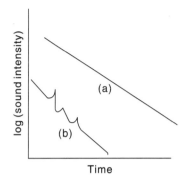

Fig. 2.47 Decay of sound intensity in (a) ideal and (b) room with resonances

Creep: Creep is the travel of sound around the perimeter of domes and other curved surfaces (Fig.2.48). This phenomenon is also responsible for whispering galleries in older structures with large domed roofs. Creep has to be avoided in designing the room since it leads to unwanted and delayed echoes.

Background Noise

Noise is unwanted sound. Sources of noise which mask the required sound must be reduced to a minimum. Background noise could be classified as (*a*) indoor noise and (*b*) outdoor noise. It is necessary to have sound insulation to protect the room from these noises. Permissible noise levels for various classes of rooms is shown in Table 2.7.

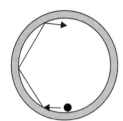

Fig. 2.48 Creep

Table 2.7 Permissible noise levels

Nature of Room	*Permissible Noise Level(dB)*
Concert hall	30—40
Lecture hall	35—45
Hospital ward	45
Large retail store	50—55
Open-plan office	45
Living room (urban house)	40
Recording and broadcasting studio	20—30

Noise could be further classified as (*a*) airborne noise and (*b*) structure borne noise. By air borne noise is meant sound waves that travel through air for considerable distance of propagation while structure borne noise refers to sound that travel through the glass of the windows, walls, ceilings etc. (Fig.2.49). Strategies for noise control are planned by considering the source, path, receiver diagram (Fig.2.50). It is always advisable to reduce the noise level at the source itself. The noise level can be reduced in the path by the use of partitions and absorbing surfaces. General guidelines for noise reduction are given further:

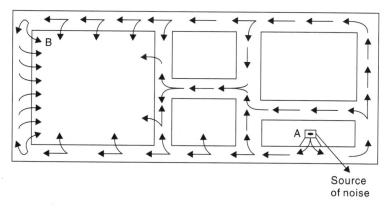

Fig. 2.49 Propagation of structural noise

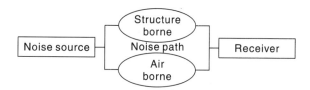

Fig. 2.50 General principles of noise reduction

(a) Structure borne sound essential arises because of the mounting of electric power generators or air-conditioners. All electrical, plumbing and other noise generating equipment have to located in the basement. Fans, refrigeration compressors, cooling towers, transformers, and emergency electricity generators have to be mounted on suitable vibration isolation mounting (Fig.2.51). Machine noise can also be reduced through the use of sound barriers (Fig.2.51).Plumbing lines and air-conditioning have to be suspended from the ceiling through vibration isolation springs (Fig.2.52). An important way of reducing noise due to movement of people is to have resilient floors (Fig.2.53). Carpeting floors also reduces the noise due to movement of people to a considerable extent. The ceiling has to be suspended from the roof by means of metal strips (Fig.2.54).

(b) Sound can be attenuated by using thick concrete walls. All walls (and ceilings, if necessary) should have as much mass per unit surface area as possible. The relationship between sound insulation effect (*i.e.,* the difference in sound pressure levels on the two sides of the wall) and the mass/area are given in the so-called mass law shown in Fig. 2.55. Doubling the mass/area increases the sound insulation by between 4—6 dB. This is assuming that the wall is large and there is no sound leakage round the sides. It is also better to cover the surface of the wall with wooden boards or other absorbing material (Fig. 2.56). In designing office rooms where speech privacy is important it is advisable to use single or double leaf partitions (Fig. 2.57). Windows must be properly sealed. All doors should be solid-core and gasketed around the entire perimeter. It is better to avoid sliding or roll-up doors.

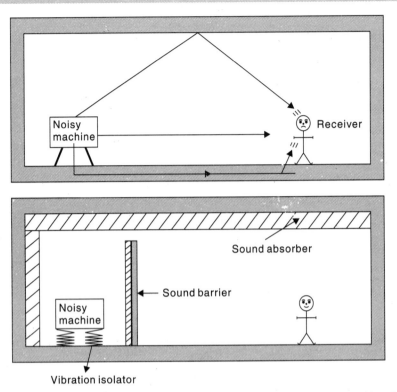

Fig. 2.51 Reducing machine noise through vibration isolators and sound barriers

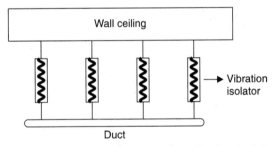

Fig. 2.52 Suspension of ducts using vibration isolators

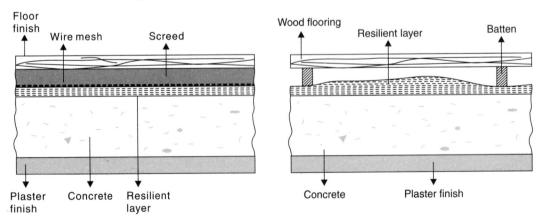

Fig. 2.53 Wooden or resilient floors

ACOUSTICS OF BUILDINGS

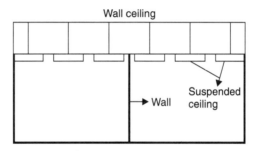

Fig. 2.54 Suspended ceilings

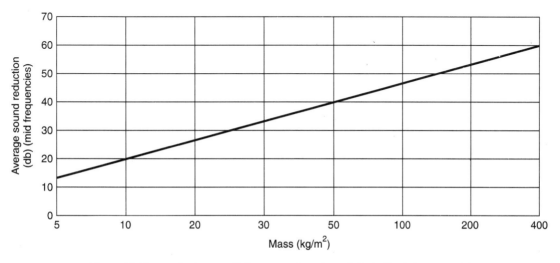

Fig. 2.55 Transmission coefficient as a function of (kg/m^2) of the material

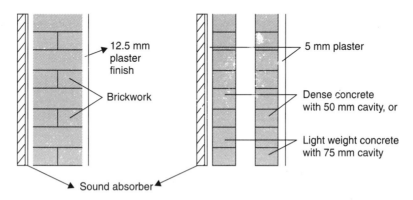

Fig. 2.56 Walls covered with absorbing material

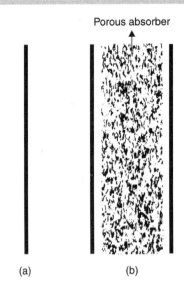

Fig. 2.57 (a) Single and (b) double leaf partition

Site considerations: The location of the room is of paramount importance. Providing sound insulation against vehicular traffic noise is inevitable and an important component of the acoustic design. In busy areas of the city where traffic noise is considerable, designing high acoustic quality rooms or auditoria is a challenge. Locating the rooms or auditoria in noise free environment is preferred. Natural landscaping can be used to reduce the traffic noise (Fig.2.58). Within the building corridors, closets, and quiet *buffer space* could be used to isolate the rooms.

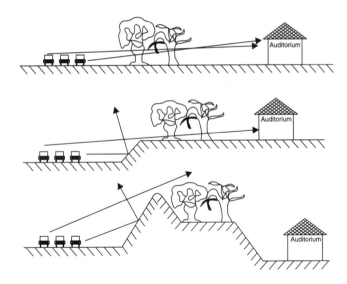

Fig. 2.58 Sound insulation from traffic noise

2.6.3 Sound Reinforcement Systems for Enhancing Acoustical Quality

One of the important parameters in the design of room acoustics is the choice of suitable sound system for sound amplification. *The ear is not equally sensitive to all ranges of frequency.* This can be overcome by the choice of a suitable sound system. A well designed sound reinforcing system should augment the natural transmission of sound from the source to the listener. It ought to be designed to provide adequate loudness and good distribution of sound. The loud speakers have to be suitable positioned to provide the best coverage over the listening area.

Most systems use a large single source or a distribution of small sources throughout the room. In most auditoriums, single-source systems are preferred because they preserve best the spatial pattern of the sound field. A single source generally consists of a cluster of loud speakers with directivity factors selected to give the best coverage of the audience. The preferred location of a single source is on the centerline of the room, near the front, over the speaker's head. The loudspeakers should be aimed toward listeners at the rear of the auditorium as shown in Fig. 2.59. Another arrangement, which provides satisfactory coverage in a long room with a low ceiling, is the distributed-speaker system shown in Fig.2.60. Each unit mounted in the ceiling covers 60° to 90°. If the room is long, it is important to have an electronic time delay for the rear speakers; otherwise, the direct sound arrives after the sound from the loudspeakers, so that it appears to be a distracting echo. Loudspeakers should not be placed along side walls of an auditorium, where their crossfire will cause the listener to hear sound from several loudspeakers at the same time. Another loudspeaker arrangement to be avoided is the all-too-common practice of putting one speaker on each side of the stage area or front wall as shown in Fig. 2.61. Listeners seated at points *A* or *B* hear sound from one of the speakers before the direct sound. If this arrangement of speakers is necessary, sound to both loudspeakers should be delayed electronically. Otherwise the single loudspeaker shown in Fig.2.60 is to be preferred. A sound system with loudspeakers at several positions in a large auditorium generally requires some type of time delay for best results.

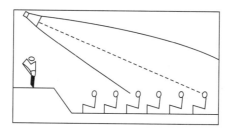

Fig. 2.59 Central loudspeaker in a large auditorium

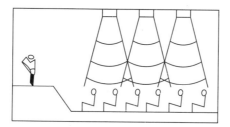

Fig. 2.60 Distributed loudspeaker system in low ceiling auditorium

Adjustable reverberation time: The fact that some halls are used for varied purposes makes the adjustment of reverberation time desirable. Maximum clarity of speech demands a short reverberation time. A solution that is becoming increasingly popular is the use of assisted resonance or the electronic enhancement of reverberation. One method used for reverberation enhancement is to place loudspeaker and a microphone in a *reverberation chamber*, a small room of 300 to 3000 m³ with highly reflecting walls, ceiling and floor. By amplifying the reverberant sound in this chamber and feeding it back into the main auditorium through loudspeakers, the reverberation time may be increased (Fig.2.62). The feed back of sound energy from loudspeaker to microphone by careful location of microphones out of the coverage pattern of the loudspeakers is absolutely essential. Feedback is the regeneration of a signal between the loudspeaker and microphone which is heard as *howling* or *screeching* (Fig.2.63).

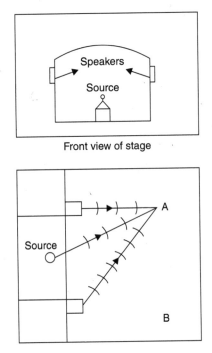

Fig. 2.61 Unsatisfactory arrangement of loudspeakers. Listeners on side A hear the sound from one speaker before the direct sound

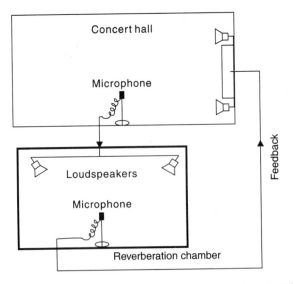

Fig. 2.62 Assisted resonance using a reverberation chamber

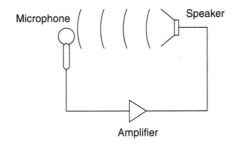

Fig. 2.63 Acoustic feedback from loudspeaker to microphone to amplifier can cause sound system to become unstable and go into oscillation

REFERENCES

1. A.H. Davis and G.W.C Kaye, *The Acoustics of Buildings*, G.Bell and Sons Ltd, London,1927.
2. L.E. Kinsler, A.R.Frey, A.B.Coppens and J.V.Sanders, *Fundamentals of Acoustics*, John Wiley and Sons, 2000.
3. M. David Egan, *Concepts in Architectural Acoustics*, McGraw-Hill Book Company, New York, 1972.
4. T.D. Rossing, *The Science of Sound*, Addison Wesley Publishing Co, New York,1990.
5. L.L. Doelle, *Environmental Acoustics*, McGraw-Hill Book Company, New York, 1975.

SOLVED EXAMPLES

1. A cinema hall has a volume of 7500 m³. It is required to have a reverberation of 1.2 sec. What should be the total absorption in the hall?

   Solution:

   The reverberation time $T = \dfrac{0.162V}{A}$

   where V is the volume in m³

   A is the absorption in m² sabine.

   $$T = \dfrac{0.162V}{A} \quad \text{or} \quad A = \dfrac{0.162V}{T}$$

   Here $T = 1.2$ s

   $V = 7500$ m³

   ∴ $A = \dfrac{0.162 \times 7500}{1.2} = 1012.5$ m² sabine.

2. Calculate the reverberation time of a hall of 1500 m³ having a seating capacity for 120 persons

   (*i*) when the hall is empty

   (*ii*) with full capacity of audience given the following data.

Surface	Area	Coefficient of absorption in O.W.U
Plastered wall	112 m²	0.03
Wooden floor	130 m²	0.06
Plastered ceiling	170 m²	0.04
Wooden doors	20 m²	0.06
Cushioned chairs	120 Nos.	0.50
Audience	120	0.4367

Solution:

Volume of the hall = 1500 m³

Absorption due to plastered wall	= 112 × 0.03 = 3.36 m² sabine
Absorption due to wooden floor	= 130 × 0.06 = 7.80 m² sabine
Absorption due to plastered ceiling	= 170 × 0.04 = 6.80 m² sabine
Absorption due to wooden doors	= 20 × 0.06 = 1.20 m² sabine
Absorption due to cushion chairs	= 120 × 0.50 = 60.00 m² sabine
Total absorption in hall	= 79.16 m² sabine

Case (i)

Reverberation time $T = \dfrac{0.162 \times 1500}{79.16} = 3.07$ s

Case (ii)

When the hall is with full capacity of 120 persons, then absorption due to 120 persons = 120 × 0.4367 = 52.404 m² sabine

Total absorption in this case = 131.56 m² sabine

$$T = \dfrac{0.162 \times 1500}{131.56} = 184 \text{ s}.$$

3. Calculate the reverberation time for a room 20 m × 15 m and 8 m high. The room contains 200 upholstered seats and half of them are occupied. The absorption coefficient of the wall, ceiling, floor and the seat are 0.09, 0.04 0.06 and 0.64 respectively. Calculate the reverberation time assuming the air absorption to be 0.012 per m³.

Solution:

Absorption due to walls	= A_1 = 2[(15 × 8) + (20 × 8)] × 0.09 = 50 m²
Absorption due to ceiling	= A_2 = (15 × 20) × 0.04 = 12 m²
Absorption due to floor	= A_3 = (15 × 20) × 0.60 = 180 m²
Absorption due to empty seats	= A_4 = 100 × 0.43 = 43 m²
Absorption due to occupied seats	= A_5 = 100 × 0.64 = 64 m²
Air absorption	= A_6 = 2400 × 0.012 = 29
Total absorption	= 50 + 12 + 180 + 43 + 64 + 29 = 349 m²
Volume of the room	= 20 × 15 × 8 = 2400 m³

$$T = 0.161 \left(\dfrac{2400}{349 + 29} \right) = 1.02 \text{ s}.$$

QUESTIONS

1. Explain the terms (a) reflection (b) refraction and (c) diffraction of sound.
2. Write a short note on the mechanisms responsible for sound absorption.
3. Define reverberation time.
4. Derive Sabine's formula and discuss its limitations.
5. Derive the relation between the sound energy density and intensity.
6. Write a short note on Eyring's formula.
7. Define absorption coefficient. Describe how it is measured using a reverberation chamber.
8. Explain how an impedance tube is used for measuring the absorption coefficient.

9. Write a short note on porous absorbers.
10. Write a short note on panel absorbers.
11. Write a short note on Helmholtz resonators.
12. Explain the design considerations involved for achieving rooms of good acoustical quality.
13. Define the terms (a) warmth (b) brilliance (c) clarity (d) blend and ensemble (e) liveness (f) intimacy.
14. Define the following acoustical defects and ways of remedying them
    (a) flutter echoes
    (b) echelon effect
    (c) non-uniform distribution of sound intensity
    (d) sound focusing
    (e) sound shadows
    (f) selective absorption
    (g) room resonances.
15. Discuss the general principles in reducing the background noise in a room.
16. Write a short note on sound reinforcement system as an aid to improve the acoustical quality of the room.

PROBLEMS

1. A hall has a volume of about 150000 m^3. Its reverberation time at 500 Hz is 11.7 s when empty, and 6.3 s when full. How many people would you expect to be present when it is full? (Assume that each person contributes 0.4 m^2 of absorption units).
2. The time of reverberation of an empty hall without and with 500 persons in the audience is 1.5 sec and 1.4 sec respectively. Find the time of reverberation with 1000 persons in the hall.
3. A lecture hall of volume 12×10^4 m^3 has a total absorption of 13200 m^2 of O.W.U. Entry of people into the hall raises the absorption by another 13200 m^2 of O.W.U. Find the change in reverberation time.
4. The volume of a room is 600 m^2. The wall area of the room is 220 m^2, the floor area is 120m^2 and the ceiling area is 120 m^2. The absorption coefficients for the walls, the ceiling the floor area are 0.03, 0.80 and 0.06 respectively. Calculate the reverberation time.
5. The time of reverberation of an empty hall is 1.5 secs. With 500 members of the audience present in the hall, the time of reverberation falls down to 1.4 secs. Find the number of persons present in the hall, if the time of reverberation falls down to 1.312 secs.
6. The time of reverberation of an empty hall without and with 500 persons in the audience is 1.5 sec and 1.4 sec respectively. Find the time of reverberation with 1000 persons in the hall.
7. A lecture hall of volume 12×10^4 m^3 has a total absorption of 13200 m^2 of O.W.U. Entry of people into the hall raises the absorption by another 13200 m^2 of O.W.U. Find the change in reverberation time.
8. The volume of a room is 600 m^3. The wall area of the room is 220 m^2, the floor area is 120 m^2 and the ceiling area is 120 m^2. The absorption coefficients for the walls, the ceiling the floor area are 0.03, 0.80 and 0.06 respectively. Calculate the reverberation time.
9. The time of reverberation of an empty hall is 1.5 secs. With 500 members of the audience present in the hall, the time of reverberation falls down to 1.4 secs. Find the number of persons present in the hall, if the time of reverberation falls down to 1.312 secs.
10. A college lecture theatre is to be built to hold 200 people and will be used mainly for speech.
    (a) Determine a suitable volume and reverberation time.
    (b) How many absorption units would be needed in the construction to achieve optimum conditions when the hall is about two-thirds full? (Assume that each person contributes 0.4m^2 of absorption units at 500 Hz.)
11. A concert hall has a volume of 5.7 m^3 per person and holds 1800 people. The reverberation time is 1.6s at mid frequencies. Assuming that Sabine's formula is accurate and that to achieve fullness of sound an orchestra requires one instrument for each 200 m^2 of absorption units, find the optimum size of orchestra.

12. The absorption coefficient of a certain material was measured in a reverberation chamber of volume 1300 m³ and the following average reverberation times were obtained:

Frequency (Hz)	R.T in s (Empty)	R.T in s (With 30 m² absorber)
125	16.8	10.2
250	20.1	10.4
500	18.5	9.4
1000	14.5	8.0
2000	9.1	6.1

The average absorption coefficient of all the surfaces is less than 0.2. Find the absorption coefficient of the material at the frequencies given.

13. Calculate the optimum and actual reverberation times at 500 Hz for a hall of volume 2500 m³ so as to be satisfactory for choral music using the following data. Allow 40 percent for shading of the floor.

Data:

Item	Area (or no.) (m²)	Absorption coefficient at 500 Hz
Wood block floor	160	0.05
Stage	80	0.30
Unoccupied seats	50	0.15/seat
Audience	200	0.4/person
Ceiling plaster	160	0.1
Canvas scenery	96	0.3
Perforated board	120	0.35
Glass	40	0.10
Plaster on brickwork	200	0.02

How many additional absorption units are required to make the actual reverberation time equal to the optimum?

14. (a) Calculate the optimum reverberation time for speech in a hall of volume 4000m³.
    (b) Calculate the actual reverberation time at 500 Hz, in a hall with the following surface finishes and seating conditions:

Item	Absorption coefficient at 500 Hz
750 m² brick walls	0.02
540 m² plaster on solid backing	0.02
65 m² glass windows	0.10
70 m² curtain	0.40
130 m² acoustic board	0.70
300 m² wood block floor	0.05

(allow 40 percent for shading).

In addition there are 500 occupied seats each contributing 0.4 m² units of absorption. The volume of the hall is 4000 m³.

15. The reverberation time was measured in a lecture room of volume 150 m³ and was found to be:

Octave band centre frequency (Hz)	R.T. in s
125	1.0
250	1.1
500	0.95
1000	1.0
2000	0.9
4000	0.8

Calculate the amount of extra absorption needed for each octave band.

16. A room 16 m long, 10 m wide and 5 m high which was previously used as a laboratory is to be converted to use as lecture room for 200 people. The original wall and floor surfaces are hard plaster and concrete whose average absorption coefficient is 0.05. Acoustical tiles of absorption coefficient 0.75 are available for wall or ceiling finishes. What is the desirable reverberation time for the new use of the room? Calculate the area of the tile to be applied to achieve this. Absorption of seated audience (per person) is 0.4 m² units.

17. (a) A hall 60 m long, 25 m wide and 8 m high has seating for 1200, and generally hard surfaces whose average absorption coefficient is 0.05. Calculate the reverberation time with a two-thirds capacity audience for the frequency at which this data applies. The audience has absorption of 0.4 m² units per person, but effectively reduces floor absorption by 40 per cent. The empty seats have absorption of 0.28 m² units.

(b) What reduction in noise level would occur if the ceiling was then covered with acoustic tiles whose absorption coefficient is 0.5.

18. A lecture room 16 m long, 12.5 m wide and 5 m high has a reverberation time of 0.7s. Calculate the average absorption coefficient of the surfaces using Eyring's formula.

19. A theatre of dimensions 30 m long, 35 m wide and 10 m high has a seating capacity of 1000 persons on wooden chairs. The area of floor is 90% of the total and consists of timber boards. The remaining 10% is stage. The walls and ceiling are concrete. The entire ceiling, rear wall and side walls are to be treated with the same material. What values of absorption coefficient are required

for this material to achieve a reverberation time of 1s at each frequency, in the theatre with a capacity audience?

Material	Absorption coefficient		
	125Hz	*500Hz*	*2kHz*
Timber boards	0.15	0.10	0.10
Concrete	0.02	0.02	0.05
Stage	0.4	Nil	Nil
Audience per person	0.15	0.40	0.45

20. A room 8m in length, 4m in width, 2.8 m high contains four walls faced with gypsum boards. The only exceptions to the wall area are a glass window 1 m × 0.5 m and a plywood-paneled door 2.2 m × 0.5 m. In addition the door has a gap underneath 1.5 cm high. The ceiling is mineral fiber tile and the floor is hardwood. Given the absorption coefficient of mineral fiber tile, floor, glass, plywood panel door to be 0.05, 0.015, 0.18 and 0.17 respectively, predict the reverberation time T.

3
Ultrasonics

3.1 INTRODUCTION

The term *ultrasonics* or *ultrasound* is applied to mechanical vibrational waves whose frequencies are higher than the audibility limit, *i.e.*, 20 kHz. The upper limit, though not strictly agreed upon, could be 10^{12} Hz. The term *supersonics* is applied to speeds greater than that of sound in air (~340 m/s) and ought not to be confused with *ultrasonics*. Typical frequencies and wavelengths of ultrasonic waves are given in Table 3.1. The *physics* of ultrasonic waves is the same as that of *sound waves*. However, on account of the development of modern and sophisticated ultrasonic generators and detectors, ultrasound has found immense applications in industry, medicine, electronic communication, materials science and oceanography.

The technology of ultrasound got an impetus during the First World War. Using piezoelectric quartz discs underwater *Sonar* was developed for the detection of enemy submarines. Later, the use of ultrasound for non-destructive testing of materials in industry was recognized. The techniques of ultrasonic welding, ultrasonic soldering, ultrasonic cavitation, ultrasonic cleaning and ultrasonic emulsification were employed in many industries with obvious advantages. The technique of generating *surface acoustic waves* (SAW) has lead to novel techniques in electronic communication. SAW devices have been used to design delay lines, band pass filters, matched filters, resonators and sensors. They have also led to accurate devices for *frequency control* and *calibration*. The development of techniques for imaging or the visual display of ultrasound has enhanced its applications in medicine. Echocardiography (for monitoring the function of heart), echoencephalography (for detecting brain tumors), lithotripsy (for removing kidney stones), ultrasonic diathermy (for physiotherapy) have become ubiquitous tools in medicine these days.

The diffraction of light in a medium carrying ultrasonic waves has led to *acousto-optic devices* which are extremely useful in optical communication. *Acoustic microscope* and *acoustic holography* have become useful tools for an engineer. *Acoustic emission* and *photoacoustic spectroscopy* have proved useful in the study of engineering materials. A study of ultrasonic attenuation in a medium can provide information about the molecular processes occurring at the microscopic level. The propagation of high intensity ultrasound in liquids leads to formation of bubbles and subsequent emission of light. This phenomenon of *sonoluminescence* has attracted the attention of many researchers in recent years. Ultrasound is also used by many animals as a means of communication and this has engaged the attention of many biologists (Fig. 3.1).

Table 3.1 Some examples of ultrasonic wavelengths

Frequency	Wavelength		
	(for c = 1000 m/s)	(for c = 3000 m/s)	For electromagnetic radiation
20 kHz = 2×10^4 Hz	5 cm	15 cm	1.5×10^4 m
100 kHz = 10^5 Hz	1 cm	3 cm	3×10^3 m
1 MHz = 10^6 Hz	1 mm	3 mm	300 m
50 MHz = 5×10^7 Hz	20 μm	60 μm	6 m
1 GHz = 10^9 Hz	1 μm	3 μm	30 cm

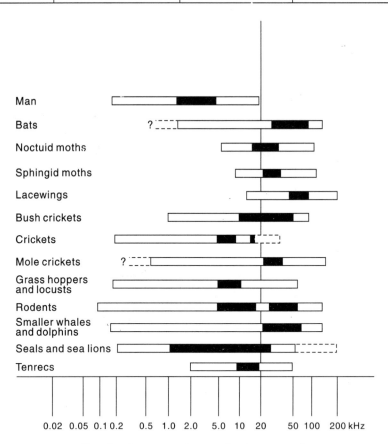

Fig. 3.1 The frequency range of hearing in man and approximate ranges for some groups of animals

3.2 PROPERTIES OF ULTRASONIC WAVES

Ultrasonic waves propagate through the material as stress or strain waves. The speed of propagation of these waves through the material medium depends upon the elastic properties. Like light, ultrasonic waves obey laws of reflection, refraction and diffraction. *But unlike light they cannot travel through vacuum.*

3.2.1 Types of Ultrasonic Waves

Based on particle displacement of the media, ultrasonic waves are classified as *longitudinal waves, transverse waves, Rayleigh waves* and *Lamb waves*.

Longitudinal waves are also known as compressional waves. These waves travel through the material media as a series of alternate compression and rarefaction, in which particles vibrate back and forth in the direction of wave propagation (Fig. 3.2a).

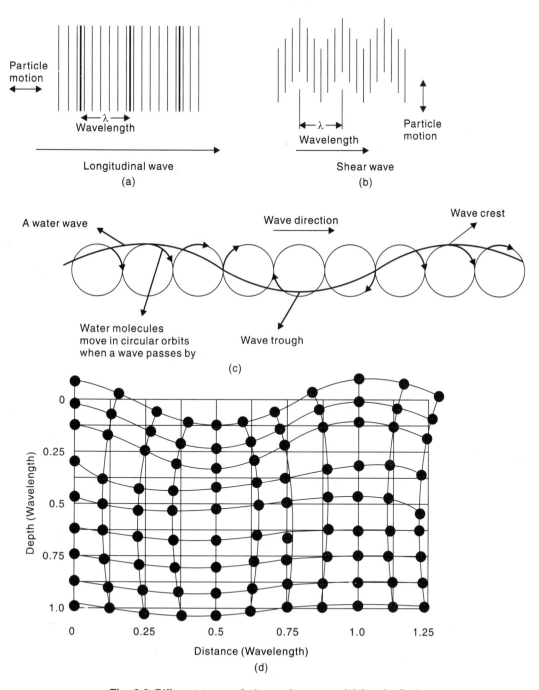

Fig. 3.2 Different types of ultrasonic waves: (a) longitudinal, (b) transverse (c), surface or Rayleigh waves in ocean and (d) surface waves in a solid

In *transverse* or *shear waves* particles of the medium vibrate up and down in a plane perpendicular to the direction of motion (Fig. 3.2b).

Surface waves or *Rayleigh waves* are neither longitudinal nor transverse. Waves on the ocean are surface waves. In these waves water molecules undergo circular motion as shown in Fig. 3.2c. These waves travel along the flat or curved surface of thick solids without influencing the bulk of the medium below the surface. The depth to which these waves propagate below the surface with effective intensity is of the order of wavelength only. Practically all its energy is attenuated at this depth. During the propagation of surface waves the particles follow elliptical orbits (Fig. 3.2d).

Lamb waves or *flexural* waves are also known as *plate waves* and are produced in a thin metal whose thickness is comparable to wavelength.

In solids ultrasonic waves can propagate as both longitudinal and transverse waves. In a crystal, these waves can be purely longitudinal or purely transverse in some specific directions. In any arbitrary direction waves possess a hybrid character. Since fluids do not support shear force, ultrasonic waves in liquids and gases are necessarily longitudinal.

3.2.2 Reflection and Refraction of Ultrasonic Waves

Reflection: The ultrasonic waves obey the laws of reflection. *i.e.,*

(a) the incident ray, the reflected ray and the normal at the point of incidence all lie in one plane.

(b) the angle of incidence is equal to the angle of reflection. These laws are useful in predicting the path of the ultrasonic waves when they encounter a planar or curved surface (Fig. 3.3). This is known as *specular reflection*.

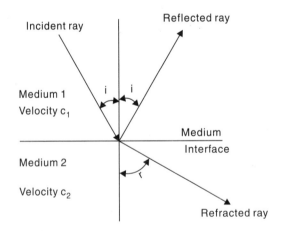

Fig. 3.3 Reflection and refraction of ultrasonic waves at a plane surface

Refraction: When ultrasonic waves travel from one medium to the other, their direction of propagation changes (Fig. 3.3). The laws governing the phenomenon of refraction are similar to those applicable to light waves, *i.e.,*

(a) the incident ray, the refracted ray and the normal at the point of incidence all lie in the same plane.

(b) the sine of the angle of incidence bears a constant ratio to the sine of the angle of refraction which is equivalent to the ratio of the sound velocities in the media concerned, *i.e.,*

$$\frac{\sin i}{\sin r} = \text{constant} = \frac{c_1}{c_2}$$

where c_1 and c_2 are the velocities of ultrasonic waves in mediums 1 and 2 as the ultrasonic wave is traveling from medium 1 to medium 2.

Acoustic Impedance: For plane harmonic waves, acoustic impedance (z) is ρc where ρ is the density of the medium and c is the velocity of ultrasonic waves. *The acoustic impedance is analogous to refractive index, i.e.,*

$$z = \rho c$$

Tables 3.2 and 3.3 give the velocity and acoustic impedance of ultrasound in some gases, liquids and solids.

Table 3.2 Velocities and acoustic impedances for some gases and liquids

Gases and liquids	Velocity(m/s)	Acoustic impedance ($kgm^{-2}s^{-1}$)
Air	331.46	431
Cabon dioxide	259	512
Helium	971.9	173
Hydrogen	1286	116
Neon	434	391
Nitrogen	337	421
Distilled water	1482.3	1.48×10^6
Acetic acid	1173	1.23×10^6
Acetone	1190	9.37×10^5
Carbon tetrachloride	940	1.94×10^6
Ethanol	1161	9.17×10^5
Glycerol	1860	2.34×10^6
Mercury	1454	1.97×10^7

Table 3.3 Velocities and acoustic impedances for some solids

Solids	Velocity for rod waves (m/s)	Velocity for compressional waves (m/s)	Velocity for shear waves (m/s)	Acoustic impedance for compressional waves ($kgm^{-2}s^{-1}$)
Aluminum	5102	6374	3111	1.7×10^7
Beryllium	?	12890	8880	2.3×10^7
Brass	3451	4372	2100	3.7×10^7
Crown glass	5342	5660	3420	1.4×10^7
Lead	1188	2160	700	2.4×10^7
Perspex	2177	2700	1330	3.2×10^6
Sand stone	2820	2920	1840	4.7×10^7
Soft iron	5189	5957	3224	4.7×10^7
Zinc	3826	4187	2421	3.0×10^7

Reflection and Transmission Coefficient

When ultrasonic waves are incident on a surface they are reflected and transmitted, usually with a change in amplitude (Fig. 3.4a). If z_1 and z_2 are the acoustic impedances of the media then

$$\text{Transmission Coefficient} = \frac{\text{Amplitude of the transmitted wave}}{\text{Amplitude of the incident wave}} = \frac{2z_1}{z_1 + z_2}$$

$$\text{Reflection Coefficient} = \frac{\text{Amplitude of the reflected wave}}{\text{Amplitude of the incident wave}} = \frac{z_2 - z_1}{z_2 + z_1}$$

Figure 3.4b shows a plot of the reflection and transmission coefficients as a function of (z_1/z_2). The reflection coefficient is zero when $z_1 = z_2$. When the acoustic impedances of the two media are identical, the wave is transmitted with 100% efficiency. This is similar to the concept of *impedance matching* in electrical circuits. When an ultrasonic transducer is bonded to a medium to send or receive ultrasound, *acoustic impedance matching* between the transducer material and the medium is necessary. As in the case of light waves, the reflection and transmission coefficients are also functions of the incident angle.

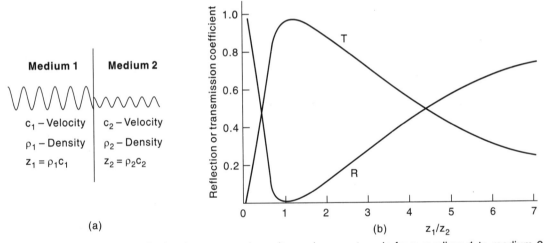

Fig. 3.4 (a) Change of amplitude of pressure when ultrasonic wave travels from medium 1 to medium 2 (b) Variation of reflection and transmission coefficients as a function of z_1/z_2

Mode Conversion

When ultrasonic wave strikes an interface between two materials having different acoustic impedances at oblique incidence, some of its energy is converted into modes of vibration other than the incident mode. This is shown in Fig. 3.5.

For any angle of incidence other than the normal, every longitudinal wave has reflected and refracted components of both longitudinal and shear waves. As the angle of incidence increases, the angle of refraction for longitudinal waves reaches 90°. The angle of incidence corresponding to 90° angle of refraction is called the *first critical angle*. For greater angle of incidence the longitudinal wave is totally internally reflected in medium 1 and no longitudinal mode exists in medium 2. If the angle of incidence is increased further the angle of refraction for shear wave mode reaches 90°. This incident angle is called the *second critical angle*. If the angle of incidence is further increased, total internal reflection for both longitudinal and shear modes occurs. Similar situation holds good for shear wave incident at the interface, when the angle of incidence is less than the first critical angle. Besides these, it is found that Rayleigh and Lamb waves are also generated.

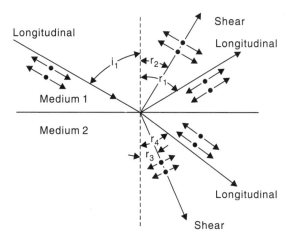

Fig. 3.5 Mode conversion—shear and longitudinal vibrations produced for oblique incidence

3.2.3 Doppler Effect

Doppler effect is a change in the observed frequency when there is a relative motion between the source and the detector (Fig. 3.6). When the source is moving, there is a change in the wavelength of the sound, but the velocity of the ultrasound remains the same. When the detector is moving, the velocity of the sound changes but the wavelength of the ultrasound remains the same. The observed frequency f_D is given in terms of the source frequency f_S by the formula

$$f_D = f_S \frac{c \pm v_D}{c} \quad \begin{pmatrix} \text{Detector moving; Source stationary; +ve sign when the detector} \\ \text{is moving towards the source and } -ve \text{ sign when the detector} \\ \text{is moving away} \end{pmatrix}$$

$$f_D = f_S \frac{c}{c \pm v_S} \quad \begin{pmatrix} \text{Detector stationary; Source moving; +ve sign when the source} \\ \text{is moving away from the detector and } -ve \text{ sign when the source} \\ \text{is moving towards the detector} \end{pmatrix}$$

Both the formulae can be combined to give

$$f_D = f_S \frac{c \pm v_D}{c \pm v_S}$$

where v_D and v_S are the speeds of the detector and source relative to the medium, c is the speed of the sound in the medium. The upper sign on v_S (or v_D) is used when the source (or detector) moves toward the detector (or source), while the lower sign is used when it moves away.

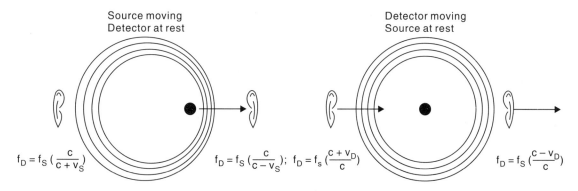

Fig. 3.6 Doppler effect (a) moving source and stationary detector
(b) moving detector and stationary source

3.2.4 Diffraction

The phenomenon of diffraction is observed when the wavelength is comparable to the size of the object scattering the ultrasonic waves. Typically the range of wavelength of ultrasonic waves lies in the region of 10^{-2}—10^{-6} m (Table 3.1).

Diffraction of Light by Sound Waves

The refractive index of an optical medium is altered by the presence of ultrasound waves. When light passes through such a medium it gets diffracted. This phenomenon is known as *acousto-optic effect* (Fig. 3.7). The sound wave consists of pressure variations and this modifies the refractive index in a periodic manner (Fig. 3.8). There is both spatial and temporal variations in refractive index. *However, since optical frequencies are very much larger than acoustic frequencies, the temporal variations in refractive index can be neglected.* Hence the medium can be treated as static with a periodic variation in refractive index. This period is equal to the wavelength of the sound wave travelling through the medium. A medium in which there is periodic variation of refractive index is often referred to as *phase grating. A plane wave front passing through such a phase grating gets corrugated and gives rise to diffraction.* The diffraction is termed as Bragg diffraction or Raman-Nath diffraction depending on whether the sound beam is wide or narrow. It is conventional to define a parameter Q given by $Q = \dfrac{2\pi \lambda_{optic} L}{\lambda^2_{acoustic}}$

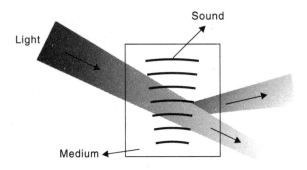

Fig. 3.7 Diffraction of light waves by sound: Acousto-optic effect

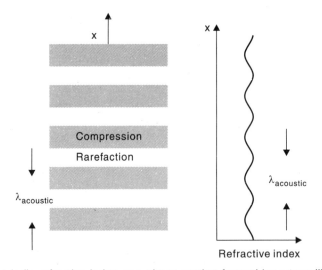

Fig. 3.8 Spatially periodic refractive index or a phase grating formed by a travelling ultrasonic wave

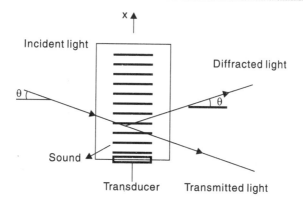

Fig. 3.9 Bragg diffraction

where λ_{optic} is the wavelength of light, $\lambda_{acoustic}$ is the wavelength of the ultrasonic wave and L is the interaction length. The interaction length L is the distance measured along the direction of propagation, over which the light and acoustic waves overlap. For $Q < 1$, Raman-Nath diffraction occurs and for $Q > 10$ Bragg diffraction takes place. Acousto-optic effect has been observed in both liquids as well as solids.

In the case of Bragg diffraction, a parallel beam of light is reflected from the set of planes representing the refractive index variations caused by the ultrasound. It satisfies the Bragg's condition observed in X-ray diffraction (Fig. 3.9).

$$\sin \theta_B = \frac{\lambda_{optic}}{2\lambda_{acoustic}}$$

where λ_{optic} is the wavelength of light, $\lambda_{acoustic}$ is the wavelength of the ultrasonic wave and θ_B is the Bragg angle. Further, since the light wave is diffracted by a travelling wave, the frequency of the diffracted light wave ω_1 is upshifted with respect to the incident light ω_0, by the acoustic frequency ω_a due to Doppler effect, i.e.,

$$\omega_1 = \omega_0 + \omega_a$$

In the case of Raman-Nath diffraction, light is incident normal to the grating. In this case the ultrasound beam is narrow. It can be shown that light gets diffracted at angles (Fig. 3.10).

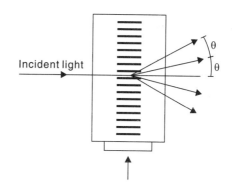

$$\sin \theta_n \approx \theta_n = \frac{n\lambda_{optic}}{\lambda_{acoustic}}$$

where $n = 0, \pm 1, \pm 2, \pm 3$ and the frequency of the diffracted light of order n is given by :

$$\omega_n = \omega_0 + n\omega_a$$

The intensity of the diffracted light in Bragg diffraction and the intensity of light in Raman-

Fig. 3.10 Raman-Nath diffraction.

Nath diffraction for the order n = ± 1, are to a good approximation proportional to the ultrasound intensity. This feature of acousto-optic effect is useful in optical communication. This phenomenon could also be used to determine the velocity of ultrasonic waves in the medium.

3.2.5 Pressure, Intensity and Power in an Ultrasonic Wave

The ultrasonic wave is essentially a pressure wave arising out of the periodic displacements of the particles constituting the medium.

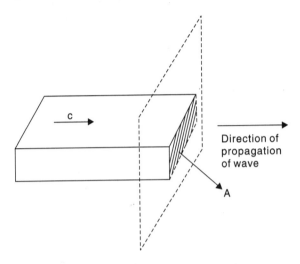

Fig. 3.11 Power carried by an ultrasonic wave

The maximum pressure is $= ac\rho\omega = za\omega$

The power P carried by the ultrasonic wave is given by (Fig. 3.11):

$$P = \frac{a^2\omega^2 \rho A c}{2} = \frac{zAa^2\omega^2}{2}$$

where a is the maximum displacement of the particle, ω is the angular frequency of the wave, ρ is the density of the medium, A is the area of cross-section, c is the velocity of the wave and z is the acoustic impedance. The power transferred across unit area, that is intensity I, is therefore given by:

$$I = \frac{a^2\omega^2 \rho c}{2} = \frac{za^2\omega^2}{2}$$

In the case of longitudinal and shear waves the power is communicated parallel and perpendicular to the direction of propagation respectively.

3.2.6 Attenuation of Ultrasonic Waves

The attenuation refers to the decrease in the intensity or power of the ultrasonic waves as it traverses through the medium. Consider the propagation of ultrasonic waves along the x-direction (Fig. 3.12). The attenuation will result from

(a) the absorption of energy by the medium between x_1 and x_2.
(b) the deflection of energy from the path of the beam by reflection, refraction, diffraction and scattering. These losses depend on the geometry of the system as well as the physical properties of the medium or media.

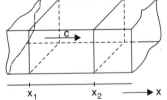

Fig. 3.12 Ultrasonic attenuation

Absorption involves the conversion of ultrasonic waves into other forms of energy. It will depend on the nature of the medium between x_1 and x_2 and can therefore furnish information about the physical properties of the medium. Reflection and refraction will occur at boundaries between regions with different acoustic impedances. Diffraction will occur at barriers that are interposed in the path of the beam. Scattering losses are characteristic of the structure of the material. In polycrystals it will also depend on the grain size distribution.

Let δI be the decrease in intensity I on traveling a distance δx in the medium. δI will be proportional to I as well as δx. Therefore,

$$-\delta I = \alpha_I I \, \delta x$$

$$\frac{\delta I}{\delta x} = -\alpha_I I \text{ or in the limit of } \delta x \to 0$$

$$\frac{dI}{dx} = -\alpha_I I \quad \Rightarrow \quad \frac{dI}{I} = -\alpha_I \, dx$$

Integrating the above expression

$$\ln(I) = -\alpha_I x \quad \Rightarrow \quad I = I_0 \, e^{-\alpha_I x}$$

Alternately one can consider the excess pressure p such that

$$p = p_0 e^{-\alpha_p x}$$

Since the intensity of the ultrasonic wave is proportional to the square of the pressure

$$e^{-\alpha_I x} = \frac{I}{I_0} = \frac{p^2}{p_0^2} = e^{-2\alpha_p x} \quad \Rightarrow \quad \alpha_I = 2\alpha_p$$

Attenuation is measured in terms of napiers (Np) or of decibels (dB).
The power level is defined as:

$$\text{Power level} = \log_e\left(\frac{p_0}{p}\right) \text{Np} = 10 \log_{10}\left(\frac{I_0}{I}\right) \text{dB}$$

But $\left(\dfrac{I}{I_0}\right) = \left(\dfrac{p^2}{p_0^2}\right)$ and $\log_{10} x = 0.4343 \log_e x$, so that

$$10 \log_{10}\left(\frac{I_0}{I}\right) \text{dB} = 20 \log_{10}\left(\frac{p_0}{p}\right) \text{dB} = 20 \times 0.4343 \log\left(\frac{p_0}{p}\right) = 8.686 \log_e\left(\frac{p_0}{p}\right) \text{dB}$$

$$\therefore \quad 1 \text{ Np} = 8.686 \text{ dB}$$

Ultrasonic Attenuation Mechanisms

When ultrasonic waves travel through a medium, a part or a considerable portion of the energy is converted into random thermal energy. The sources of this dissipation may be divided into two categories:

(a) those intrinsic to the medium and

(b) those associated with the boundaries of the medium.

Several physical processes may contribute to the absorption of ultrasonic waves as they pass through the medium. Those which are actually important may be different for different materials. In general, acoustic losses in the medium may be further divided into three basic types:

(a) viscous losses

(b) thermal conduction losses and

(c) thermal relaxation losses due to various molecular processes.

It is convenient to discuss ultrasonic absorption in fluids (liquids and gases) and solids separately.

Absorption in Fluids

The total attenuation coefficient α for a fluid may be written as

$$\alpha_{total} = \alpha_{vis} + \alpha_{th} + \alpha_{relax}$$

where α_{vis}, α_{th} and α_{relax} are respectively the contributions arising from the three mechanisms mentioned above.

Viscous losses: This arises on account of the finite viscosity in fluids. Viscous drag is also called the internal friction. Consider a fluid moving in a tube (Fig. 3.13). The fluid may be considered to be made of different layers. The layer in contact with the pipe will have zero velocity while the layer at the axis of the cylinder will be moving with the greatest velocity. The layer moving with the slower velocity will act as a drag on the one moving with a greater velocity. To overcome this friction, a constant shear force or pressure is needed to maintain the flow. *We can say that there is diffusion of momentum by molecular collisions between different layers.* Thus viscosity is a measure of internal frictional force between different layers of the fluid. *During the propagation of ultrasonic wave, there is relative motion between the adjacent layers of the medium, which gives rise to friction.* A part of the acoustic energy is lost in overcoming this friction. The attenuation coefficient α_{vis} is given by:

$$\alpha_{vis} = \left(\frac{\omega^2}{2\rho_0 c^2}\right)\left(\frac{4\eta}{3} + \eta_B\right)$$

where ω is the angular velocity, ρ is the density, c is the velocity, η is the shear viscosity and η_B is called the bulk viscosity. It is zero for monoatomic gases but can be finite in other fluids.

Thermal conduction losses: When a longitudinal wave passes through a medium there are regions where the molecules are compressed and there are regions where they expand. *These fluctuations in pressure are described by an adiabatic process and not an isothermal process.* The regions where the molecules are compressed are at a slightly higher temperature while the regions where the molecules expand are at a slightly lower temperature (Fig. 3.14). On account of the finite thermal conductivity of the medium, there is an irreversible flow of heat and consequently a loss of acoustic energy. The attenuation coefficient α_{th} is given by:

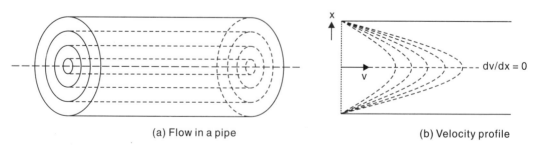

(a) Flow in a pipe   (b) Velocity profile

Fig. 3.13 Viscosity in fluids

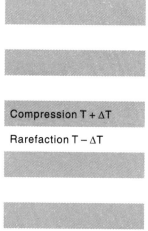

Fig. 3.14 Thermal conduction loss

$$\alpha_{th} = \left(\frac{\omega^2}{2\rho c^3}\right)\frac{(\gamma-1)\kappa}{c_p}$$

where ω is the angular velocity, γ is the ratio of specific heat at constant pressure to that at constant volume = (c_p/c_v) and κ is the thermal conductivity, ρ is the density and c is the velocity of the ultrasonic wave. Table 3.4 gives the acoustic absorption in some selected fluids.

Table 3.4 Acoustic absorption in fluids

	\multicolumn{4}{c}{$\alpha/f^2 (Np \cdot s^2/m)$}			
	Shear viscosity	Thermal conductivity	Classical	Observed
Gases	\multicolumn{4}{c}{*Multiply all the values by 10^{-11}*}			
Argon	1.08	0.77	1.85	1.87
Helium	0.31	0.22	0.53	0.54
Oxygen	1.14	0.47	1.61	1.92
Nitrogen	0.96	0.39	1.35	1.64
Air (dry)	0.99	0.38	1.37	α/f peaks at 40 kHz
Carbon dioxide	1.09	0.31	1.40	α/f peaks at 30 kHz
Liquids	\multicolumn{4}{c}{*Multiply all the values by 10^{-15}*}			
Glycerine	3000.0	—	3000.0	3000.0
Mercury	—	6.0	6.0	5.0
Acetone	6.5	0.5	7.0	30.0
Water	8.1	—	8.1	25
Sea water	8.1	—	8.1	α/f peaks at 1.2 kHz and 1.36 kHz

All data at 20°C and atmospheric pressure.

Thermal relaxation losses: This occurs in polyatomic molecules. In a monoatomic gas all the thermal energy of the molecules will be in the form of the translational kinetic energy of the molecules. For a polyatomic gas the thermal energy will be partly in the form of translational energy, rotational energy and the vibrational energy (Fig. 3.15). The vibrational energy is made of both kinetic energy and potential energy as in the case of a pendulum. Under the influence of an ultrasonic wave there could be inter-conversion between these three forms of energy. This process has a time scale of its own and is known as thermal relaxation time. This process leads to acoustic losses. The thermal relaxation time can be calculated by taking into account the internal structure of the molecules and the interactions between them that lead to internal vibrations, rotations.

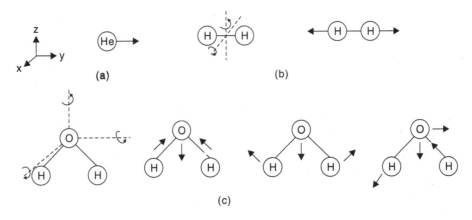

Fig. 3.15 (a) Translation (b) & (c) Rotational and vibrational modes in diatomic and triatomic molecule

Structural relaxation losses: In some materials there could be losses in acoustic energy due to additional mechanisms. This is illustrated by considering the example of cyclohexane. Cyclohexane can exist in different isomeric forms as shown in Fig. 3.16.

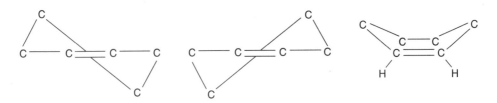

Fig. 3.16 Half-chair form and half-boat form of cyclohexane

Cyclohexane (C_6H_{10}) is a ring compound with one double bond, four of the six carbon atoms (and two of the hydrogen atoms) are coplanar. A cyclohexane molecule may exist in two isomeric forms, a half-chair form and a half-boat form. There are actually two equivalent half-chair forms. The half-chair forms are more stable than the half-boat forms by about 11 kJ/mole. Thermal relaxation between isomeric forms, as a mechanism for ultrasonic absorption in such a liquid, involves individual molecules undergoing transitions from the form with lower energy to the form with higher energy, at the expense of the energy of the ultrasonic beam.

The ultrasonic attenuation in air for various humidities and that of sea water as well as freshwater is shown in Figs. 3.17 and 3.18.

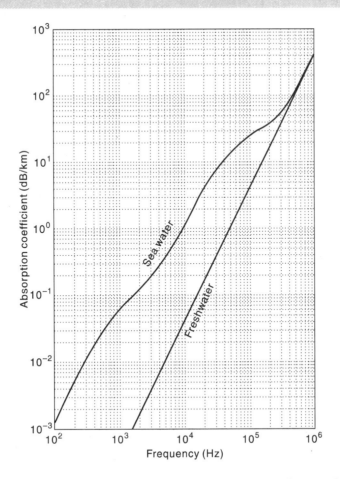

Fig. 3.17 Frequency dependence of ultrasonic attenuation in ordinary and sea water

Ultrasonic Attenuation in Solids

In the case of solids ultrasonic attenuation is due to the following:

(a) Thermoelastic effect: This is similar to the attenuation due to thermal conduction which has been described for fluids.

(b) Scattering by phonons: In solids atoms vibrate about their equilibrium positions. The motion of these atoms is described in terms of traveling waves. The energy of these traveling waves is quantized and is known as phonon. Phonons are classified as optical and acoustic depending on the nature of atomic displacements. The acoustic energy is scattered by phonons leading to acoustic losses.

(c) Dislocation damping: In a crystalline solid there are dislocations which can move under the action of stress. An ultrasonic wave passing through the crystal can lead to motion of dislocations which in turn results in acoustic loss.

(d) Magnetoelastic interactions: In magnetic materials there exists magnetic domains which can move under the action of stress. This leads to acoustic losses.

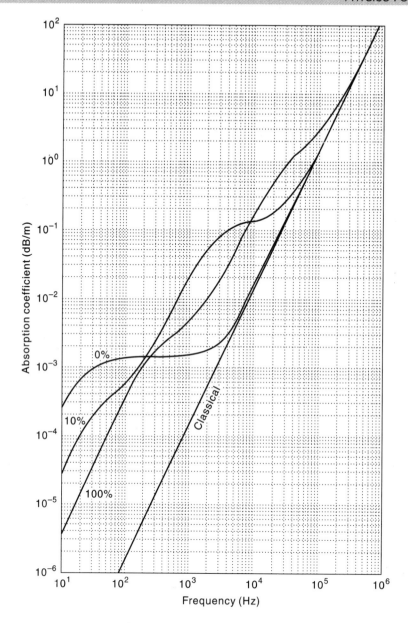

Fig. 3.18 Frequency dependence of ultrasonic attenuation in air for different humidities

(e) Electron-phonon interactions: In metals, the ultrasonic wave can get attenuated due to electron-phonon interactions.

(f) Scattering by inhomogeneities in the solids: This is the principal cause of loss of acoustic energy. In a polycrystalline material, the grain size may be comparable to the ultrasonic wavelength which can lead to scattering losses. The ultrasonic velocity is different in different directions within the grain. They undergo loss when they travel across the grain boundaries. This scattering is found to be significant for inhomogeneities larger than about one-hundredth of the ultrasonic wavelength. The attenuation increases as the third power of the grain size and the fourth power of frequency as long as the wavelength remains greater than the grain size (Rayleigh scattering).

3.3 GENERATION AND DETECTION OF ULTRASONIC WAVES

There are a number of ways of generating ultrasonic waves. The method to be employed depends upon the needed power output and the desired frequency range.

3.3.1 Mechanical Methods

Galton's Whistle

It is a miniature organ pipe in the form of a whistle to generate ultrasonic waves of low frequency up to 10,000 Hz. Figure 3.19 shows the schematic of Galton's whistle. It consists of a cylinder terminated by the end surface of a piston which can be adjusted in position to provide resonance at the required frequency *i.e.*, for which the length of the cavity is one quarter-wavelength. Air flows through an annular slit at high speed and strikes the rim of the tube where vortices appear and produce *edge* tones. The frequency of the edge-tones depends on the velocity of the fluid which can be adjusted until the cavity resonates. For air, at a frequency of 20 kHz, a fundamental resonance takes place for a cavity length of approximately 3 mm. It is difficult to excite pure tone resonances of frequencies of much higher than this. It is possible to estimate the upper limit of the frequencies that can be obtained with ultrasonic whistles. A whistle can be regarded as a resonant cavity of length l, which is open at one end and closed at the other (Fig. 3.20). The wavelengths of the first few harmonics are given by:

$$l + \delta = \frac{\lambda}{4}, \frac{3\lambda}{4}, \frac{5\lambda}{4}, \ldots\ etc.$$

where δ is the end correction of the tube. The corresponding frequencies are given by:

$$\nu = \frac{c}{4(l+\delta)}, \frac{3c}{4(l+\delta)}, \frac{5c}{4(l+\delta)}, \ldots\ etc.$$

If we take $l + \delta$ to be of the order of 2 to 3 mm and $c = 330$ m/s, the upper limit on ν is about 30–40 kHz.

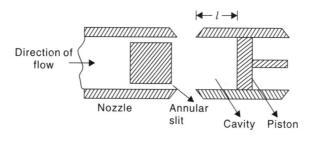

Fig. 3.19 Galton's whistle

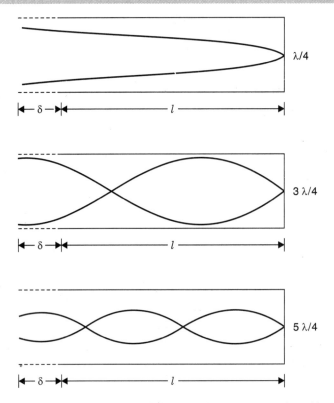

Fig. 3.20 Standing waves in a whistle which is treated as an open tube with one end closed

Hartmann Generator

This is similar in design to the Galton's whistle except that the annular slit is replaced by a conical nozzle (Fig. 3.21). Air is forced through the nozzle which emerges at a supersonic velocity to produce shock waves, which cause the cavity to be excited at a high intensity. Resonance is achieved by adjusting the velocity of air. This device is much more powerful than Galton's whistle and at a frequency of 20 kHz, it is possible in air to produce up to 50 W of acoustical energy.

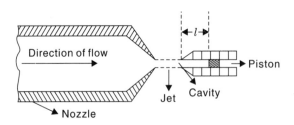

Fig. 3.21 Hartmann generator

Siren

Sirens may also be used as generators of sound or ultrasound. The jet of air is interrupted every time the blank part of the disc between adjacent holes passes in front of the nozzle (Fig. 3.22), so that the frequency is given by

$$\nu = Nn$$

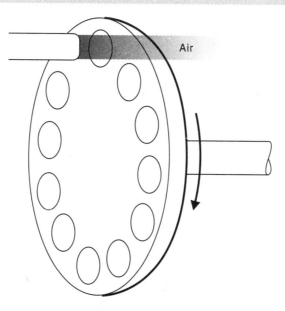

Fig. 3.22 Siren

where N is the number of holes in the disc and n is the number of rotations of the disc per second. Although mechanical generators of ultrasound are limited to rather low frequencies from modern standards, there are some applications for which they are suitable.

3.3.2 Piezoelectric Transducers

Piezoelectric materials develop electric charges on the surface when subjected to mechanical stress (Fig. 3.23). Thus the phenomenon of piezoelectricity refers to the origin of electric polarization under the influence of a mechanical stress. The converse piezoelectric effect refers to the origin of the mechanical strain or the change in the dimensions of the material when subjected to an electric field (Fig. 3.24). Thus by applying an alternating electric field it is possible to induce periodic strain in the material and generate ultrasonic waves. Piezoelectricity is not restricted to crystals alone. Many of the polymers exhibit piezoelectricity. Collagen which is constitutive protein of the bone is piezoelectric. Some of the important piezoelectric materials used for ultrasound generation are quartz (SiO_2), lead zirconate titanate ($PbZr_xTi_{1-x}O_3$; PZT) and polyvinyledene fluoride. The crystal structures of quartz and PZT are shown in Fig. 3.25.

Symmetry considerations restrict the existence of piezoelectricity to crystals which do not possess a centre of symmetry. Thus there are only 20 piezoelectric classes of crystals. *All ferroelectric crystals are necessarily piezoelectric while the converse is not true.* Thus quartz is not ferroelectric but exhibits piezoelectricity. Lead zirconate titanate (PZT) is ferroelectric and consequently piezoelectric. In general, physical properties of crystals are anisotropic. An acoustic wave propagating through a crystal in an arbitrary direction (which does not coincide with rotational axes of symmetry or lie in the planes of symmetry of the crystal) is in a combination of longitudinal and shear modes. This type of wave is called either *quasilongitudinal* or *quasishear*, depending on which component is predominant. In crystals there are certain directions which are called *pure mode directions*. In these directions wave can propagate in either pure shear or pure longitudinal mode. In most acoustic devices using quartz, *pure mode directions* are chosen as propagating directions. This facilitates the desired wave excitation and maximizes the electrical to mechanical conversion efficiency. Figure 3.26 shows the different vibratory modes of the quartz crystal. Quartz cut in different directions generates ultrasound

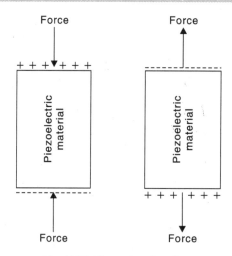

Fig. 3.23 Piezoelectric effect

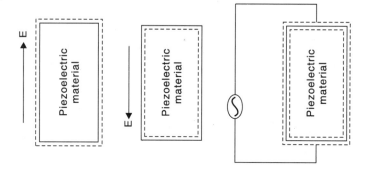

Fig. 3.24 Converse piezoelectric effect

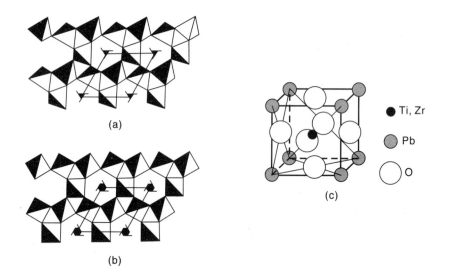

Fig. 3.25 Crystal structures of (a) low temperature α-quartz (b) high temperature β-quartz and (c) PZT (lead zirconate titanate)

ULTRASONICS

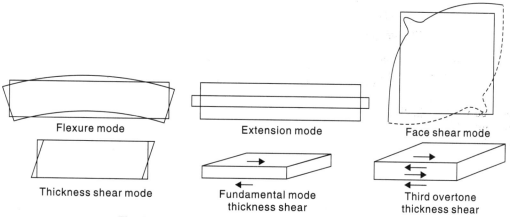

Fig. 3.26 Different vibratory modes of quartz crystal

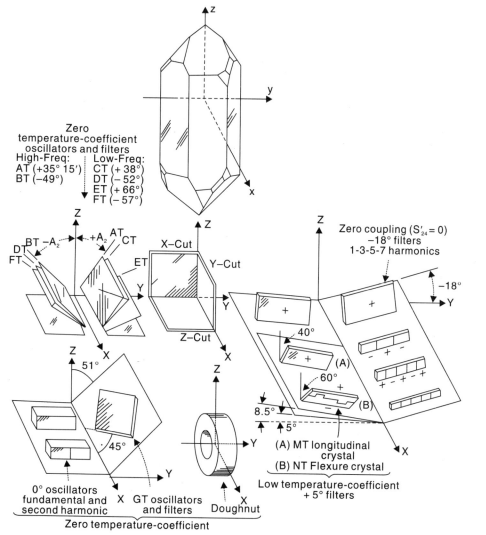

Fig. 3.27 Different quartz cuts used for ultrasonic transducer

of different ranges of frequencies. The value of the frequency depends on the actual dimensions of the material as well. Figure 3.27 and Table 3.5 summarises the important characteristics of the quartz transducers. *PZT is used in the ceramic form.* Typical configurations of the PZT transducer are shown in Fig. 3.28. Figure 3.29 shows the construction of a typical transducer. The electrodes are evaporated or sputtered onto each of the transducer, and the backing material is chosen to give the required frequency characteristics. The backing material is usually designed to have the same acoustic impedance as the transducer. A typical material is an epoxy resin loaded with tungsten powder and rubber powder. If the ultrasound is transmitted as pulses, it is usual to use the same transducer as both transmitter and receiver. If the ultrasound is to be transmitted continuously, separate transmitting and receiving crystals are mounted side by side on backing blocks separated by an acoustic insulation. Ultrasonic imaging is carried out using arrays of transducers.

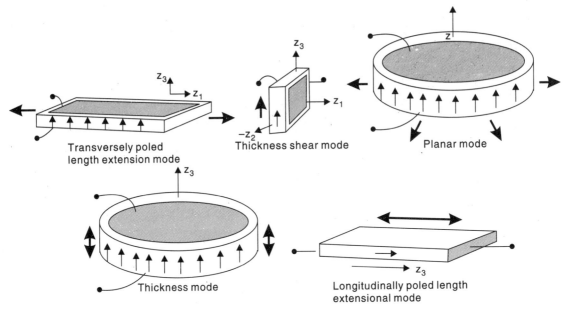

Fig. 3.28 Different vibratory modes of PZT ceramic transducer. Arrows within the material indicates the poling direction

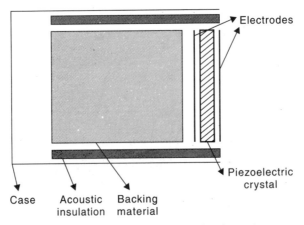

Fig. 3.29 A single piezoelectric element

ULTRASONICS

Table 3.5 Designations for quartz transducers, their vibration modes and frequency ranges

Element Designation	Reference	Mode of Vibration	Frequency Range
A	AT	Thickness shear	0.5—100 MHz
B	BT	Thickness shear	5—15 MHz
C	CT	Plate shear	300—1000 kHz
D	DT	Plate shear	200—500 kHz
E	+ 5° X-cut	Longitudinal	60—300 kHz
F	− 18° X-cut	Longitudinal	60—300 kHz
G	GT	Longitudinal	100—556 kHz
H	+ 5° X-cut	Length-width flexure	10—100 kHz
J	+ 5° X-cut 2 plates	Duplex length-thickness flexure	1.2—10 kHz
M	MT	Longitudinal	10—100 kHz
N	NT	Length-width flexure	10—100 kHz

3.3.3 Magnetostrictive Transducers

Magnetostriction is a phenomenon analogous to piezoelectric effect. A solid material, subjected to a magnetic field changes its dimensions (Fig. 3.30). The strain induced by the applied magnetic field is in the direction of the magnetic field. In the converse effect, called *piezomagnetism,* a magnetic field is generated when the material is subjected to a mechanical strain. There is a quadratic relationship between the magnetostrictive strain and the magnetic flux (Fig. 3.31)

i.e., $\frac{\Delta l}{l} = d_m B_m^2$ where B_m is the flux density inside the material, Δl is the change in length

and l is the original length in the absence of the magnetic field. The quadratic relation is valid at low flux densities. *As a consequence of this quadratic relationship the direction of the magnetostrictive strain is independent of the direction of the field.* Hence, if an alternating magnetic field is applied, then the magnetostrictive strain resembles a rectified sine wave.

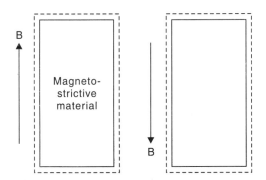

Fig. 3.30 Magnetostrictive effect

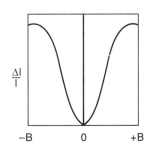

Fig. 3.31 Strain as a function of the magnetic field with bias

In order to obtain the desired strain relationship out of a magnetostrictive device it is essential to bias the material magnetically. Thus in the presence of the biasing magnetic field B_{bias}, the above equation becomes

$$\frac{\Delta l}{l} = d_m (B_m \sin \omega t + B_{bias})^2$$

$$= d_m(B_m^2 \sin^2 \omega t + B_{bias}^2 + 2B_{bias} B_m \sin \omega t)$$

$$= d_m \left[\frac{B_m^2}{2}(1 - \cos 2\omega t) + B_{bias}^2 + 2B_{bias} B_m \sin \omega t \right]$$

$$= d_m \left[\frac{B_m^2}{2} + B_{bias}^2 + 2B_{bias} B_m \sin \omega t - \frac{B_m^2}{2} \cos 2\omega t \right]$$

If $B_{bias} > B_m$ then $\frac{\Delta l}{l} \approx d_m [B_{bias}^2 + 2B_{bias} B_m \sin \omega t]$

Thus applying a biasing magnetic field ensures that the magnetostrictive strain follows the sinusoidal variation of the applied magnetic field at the same frequency. Magnetic biasing is achieved either by a small permanent magnet included in the magnetic circuit or passing a d.c. biasing current through an appropriate coil or winding.

Magnetostrictive Materials

Materials which exhibit magnetostriction include

(a) ferromagnetic metals such as nickel, iron and cobalt;

(b) ferromagnetic alloys such as permalloy, an alloy with iron and nickel with additives and

(c) ceramic materials called ferrites.

Developments over the last decade have led to highly magnetostrictive materials such as TbF_2. With some materials (e.g., nickel) the magnetostrictive effect is negative and with others such as permalloy it is positive. The magnitude of the magnetostrictive effect depends on the processing conditions of the material. Nickel is often used for low power ultrasonic applications. Permalloy has the advantage of a high Curie point but the magnetomechanical coupling coefficient is low. Ferrites have poor mechanical properties and are not often used. The magnetostrictive effect decreases as the magnetic Curie point is approached and disappears completely above it.

The magnitude of the magnetostrictive strain is given by:

$$\frac{\Delta l}{l} = \frac{3}{2} \lambda_s \left(\cos^2 \theta - \frac{1}{3} \right)$$

where λ_s is the saturation magnetization coefficient and θ is the angle between the magnetic field and measurement direction. λ_s for various materials is given in Table 3.6.

Table 3.6 Saturation magnetization coefficients

Material	$10^6 \lambda_s(T)$	Material	$10^6 \lambda_s(T)$
Fe	9	$SmFe_2$	1560
Ni	35	$TbFe_2$	1753
Co	60	Fe_3O_4	40
60%Co–40%Fe	68	Metallic glass	40
60%Ni–40%Fe	25	Terefenol($Tb_{0.27}Dy_{0.73}Fe_{1.95}$)	1600
$TbCO_3$	65	$Tb_{0.6}Dy_{0.4}$	6000

Basic Design

The simplest design of a magnetostrictive transducer is a nickel rod supported at its nodal point and provided with suitable electrical wirings (Fig. 3.32). For resonance the rod should be half a wavelength long. Nickel rod of about 2 cm long gives a frequency of about 140 kHz. Magnetically this design is inefficient because there is an open magnetic circuit. In power producing transducers the design is greatly improved magnetically by using a window geometry or a ring geometry. The window design generates ultrasound from its ends and may be supported rigidly at its nodes or a flexible mounting at one end. The ring design generates ultrasound radially and concentrates it at the centre of the ring. The resonant frequency of these devices depends on the detailed design, construction and load. In high power transducers there is a great deal of energy dissipated in the magnetostrictive material as a result of hysteresis and eddy currents. Magnetostrictive transducers are particularly useful when it is required to generate high power, but low frequency ultrasound. This is the case for ultrasonic machining, drilling, cleaning etc. For these applications frequencies of 20—50 kHz are typical. A potential advantage of magnetostrictive transducers is their possible use at temperatures which cannot be readily tolerated by piezoelectric transducers.

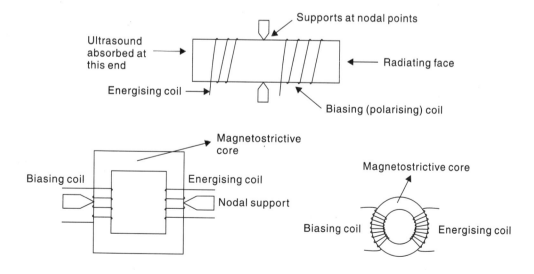

Fig. 3.32 Typical configurations of magnetostrictive transducers

3.3.4 Electromagnetic Acoustic Transducers (EMATs)

Electromagnetic acoustic transducer (EMAT) has become an useful way of generating ultrasonic waves and is used in applications involving electronic communications. They are particularly valuable for two features:

(a) they are good at producing special kinds of acoustic waves such as beams at angle, and surface acoustic waves.

(b) they can be designed in a way which does not require contact between the transducer and the material in which the ultrasound is being generated.

EMAT makes use of the Lorentz force. A current carrying conductor in a magnetic field experiences a mechanical force (Fig. 3.33). Consequently the electrons inside the conductor experience a force and in a magnetic field these electrons have helical trajectories. These electrons collide with the ions embedded in the crystal lattice and impart momentum to them. This will generate ultrasonic waves in the crystal lattice. In applying the Lorentz force mechanism to the transducer design, the simplest approach would be to make an acoustic bond between the dielectric and a current carrying conductor. This is done by having a meander sheet made of a good conductor. This is firmly bonded between the several pieces of the dielectric in which one would like to excite acoustic waves (Fig. 3.34). Another geometry is to have meander line bonded to the surface of the dielectric (Fig. 3.35). Thick film techniques of metal deposition followed by masking and etching, as for printed circuits can be used with all the attendant advantages. By arranging the spacing between the meanders to be $\lambda/2$, an ultrasonic surface wave of wavelength λ is generated. With a pancake meander line and a magnetic field parallel to the surface compressional bulk waves can be generated (Fig. 3.36). If the applied magnetic field is parallel to the surface shear waves are generated (Fig. 3.37). Shear waves can also be generated by having a periodic alternating magnetic field perpendicular to the surface (Fig. 3.38). The intensity of the ultrasound can be controlled by the current and the magnetic field. If the dielectric is piezoelectric then the waves are also generated due to piezoelectric effect (Fig. 3.39).

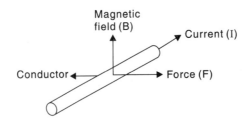

Fig. 3.33 Lorentz force

Non-contact EMAT:

In non-contact EMAT the Lorentz force is not transmitted by mechanical means. Instead the force is generated directly in the material itself. This is achieved by having a metal substrate. When the meander lines are above the metal, eddy currents are induced due to electromagnetic induction (Fig. 3.40). Eddy currents follow paths parallel to the parent current but in the opposite direction. Typical geometries for the generation of compressive and shear waves are shown in Figs. 3.41 and 3.42. By making a thin substrate, surface waves and Lamb waves can be generated (Fig. 3.43). If the metal is magnetic, magnetostrictive effect also has to be taken into account. The choice of the electrode geometry gives plenty of scope for the generation of a wide choice of acoustic waves.

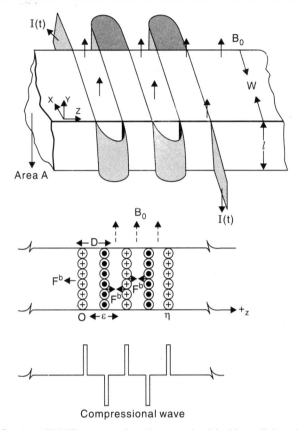

Fig. 3.34 Contact EMAT— meander sheet embedded in a dielectric; Forces and the waves generated are shown

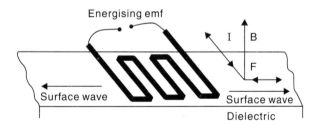

Fig. 3.35 Contact EMAT—meander line deposited on a dielectric surface

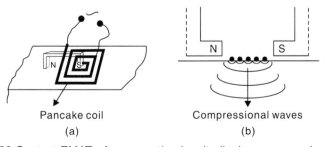

Fig. 3.36 Contact EMAT—for generating longitudinal or compressive waves

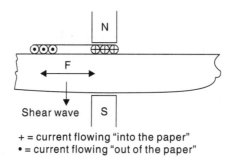

+ = current flowing "into the paper"
• = current flowing "out of the paper"

Fig. 3.37 Contact EMAT for generating shear waves

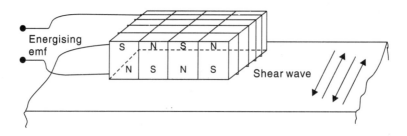

Fig. 3.38 Contact EMAT with periodic and alternating magnetic field for generating shear waves

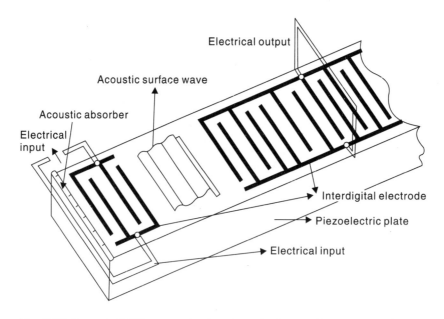

Fig. 3.39 Contact EMAT—meander line deposited on a piezoelectric substrate

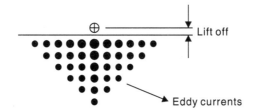

Fig. 3.40 Origin of eddy currents

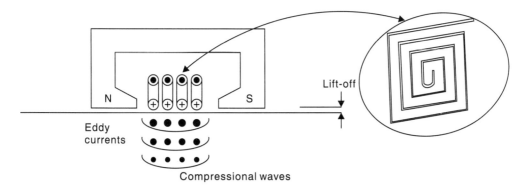

Fig. 3.41 Non-contact EMAT for generating compressional waves

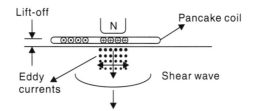

Fig. 3.42 Non-contact EMAT for generating shear waves

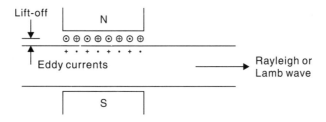

Fig. 3.43 Non-contact EMAT for generating Rayleigh or Lamb wave

3.3.5 Ultrasound Production by Laser Pulsing

When laser pulse (~ nano to picosecond duration) strikes a solid or a liquid surface, depending on the energy, two phenomena can occur which can generate ultrasound. If the energy is low (up to about 0.5 MWcm^{-2}) some or all the laser energy is converted to heat near the surface without causing a phase change. The rise in temperature causes rapid expansion of the heated region. This thermoelastic process will produce stresses and strains in the solid and they will in turn generate an ultrasonic wave front progressing through the solid. If, however, the energy is much higher the heating of the surface will cause a phase change and rapid evaporation occurs. There may even be the production of high temperature plasma. This process is called laser ablation. This ablation process will compress the medium on account of reactive effect. Molecules or atoms evaporating at a rapid rate out of the surface will impart a momentum to the medium according to Newton's third law. A major disadvantage of the method is that the laser may cause unacceptable damage to the material. This is overcome by coating the surface of a metal with a thin film of water or other easily vaporized liquid or solid. The ablation of this coating will no longer cause much damage to the material in which the ultrasound has to be generated. Typical beams for thermoelastic and ablation processes are shown in Fig. 3.44. The major advantage of this method is the generation of ultrasound in a remote medium.

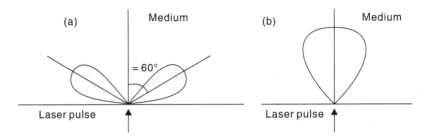

Fig. 3.44 Laser generation of ultrasound. Ultrasonic beam profile (a) thermoelastic process and (b) ablation process

3.4 MEASUREMENT OF ULTRASONIC VELOCITY AND ATTENUATION

The velocity of ultrasonic waves is given by $c = \nu \times \lambda$ where ν is the frequency and λ is the wavelength. The frequency is determined by the transducer. Hence from a measurement of λ the velocity of the waves can be calculated. The second method is based on the echo. Here one measures the time of transit of an ultrasonic pulse and the distance traveled by the pulse in the medium. The distance traveled divided by the time of travel gives the velocity.

The available experimental methods for determining these parameters can be grouped under the following headings:

(*a*) Standing wave methods

(*b*) Pulse-echo methods

(*c*) Other methods.

Standing Wave Methods

At audible frequencies the apparatus used for measurements would be Kundt's tube (Fig. 3.45). In Kundt's tube some convenient powder is sprinkled in the tube and when standing waves are excited the powder is agitated into little heaps at the nodes. By measuring the distance ($\lambda/2$) between the successive nodes, the wavelength and hence the velocity can be determined. At audiofrequencies the wavelengths involved are such that the standing waves have a very suitable value of λ for this experiment. Thus, for example, in air $\lambda \approx 66$ mm at a frequency of 5 kHz. However, at higher frequencies the method is not convenient since the wavelength becomes very small.

ULTRASONICS

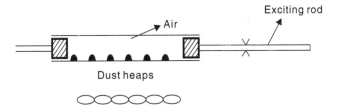

Fig. 3.45 Kundt's tube and the standing waves

It is possible to form stationary ultrasonic waves in much the same way that stationary acoustic waves are formed in Kundt's tube. The stationary ultrasonic waves will be formed by having a transducer, which is excited electrically as the source of ultrasound and by having a reflector to reflect the ultrasound. To form stationary waves the distance between the transducer and the reflector will have to be an integral multiple of $\lambda/2$. In the frequency range of ultrasound the transducer or the reflector ought to be capable of being moved to an accuracy of a fraction of 1 mm. The transducer itself acts as the detector. When the transducer is at the node, the output voltage will be minimum. The voltage will be maximum when the transducer is at an antinode. A plot of the voltage output versus the distance between the transducer and the reflector is shown in Fig. 3.46. The distance between two consecutive maxima gives $\lambda/2$. Knowing the frequency of the transducer the velocity can be calculated. This method is convenient for fluids.

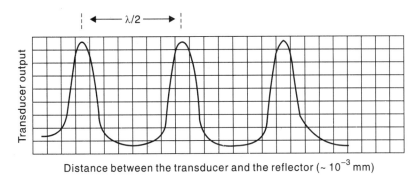

Fig. 3.46 Transducer output as a function of the distance between the transducer and reflector

A slight variation of this method is possible for solids. Here one makes use of a variable frequency transducer. The transducer output is measured as a function of frequency. Each time the number of loops of the standing wave changes by one, the transducer output becomes minimum. If d is the length of the solid, then

$$\lambda_n = \frac{2d}{n} \quad \text{or} \quad \nu_n = \frac{c}{\lambda_n} = \frac{cn}{2d}$$

$$\therefore \quad \Delta\nu = \nu_{n+1} - \nu_n = \frac{c}{2d}$$

A plot of transducer output versus the frequency of the transducer is shown in Fig. 3.47. Knowing the length of the solid and change in frequency of the transducer to produce voltage maxima, the velocity can be calculated.

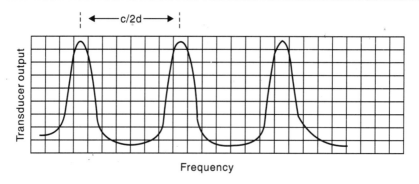

Fig. 3.47 Transducer output as a function of the frequency

Pulse Echo Method

This method is suitable for solids, liquids as well as gases. In the pulse method the ultrasonic velocity is computed by measuring the distance traveled by the pulse and the time it takes for traversing this distance. The experimental set up is shown in Fig. 3.48. If t is the time taken for observing the first echo, then the velocity is given by $2d/t$, where d is the length of the medium. This method allows the measurement of attenuation as well. For measurement of attenuation the multiple echo pattern is obtained *i.e.*, the pulses after first, second, third reflections are obtained. Thus the first echo will be obtained after the ultrasonic pulse has traveled a distance of $2d$, the second after a distance of $4d$, the third after a distance of $6d$ and so on. Thus there will be a sequence of echoes received by the detector at decreasing amplitudes (Fig. 3.49). Hence from measurements of the heights of the successive peaks the value of attenuation coefficient can be got. Thus if h_n and h_{n+1} are the pulse heights corresponding to two successive reflections then the attenuation α is given by:

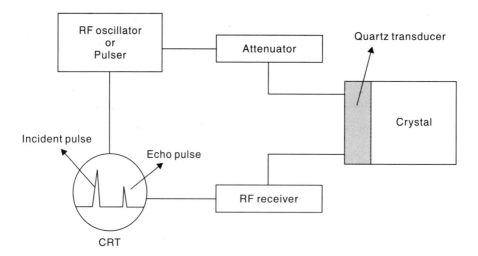

Fig. 3.48 Pulse echo method

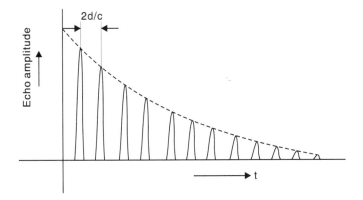

Fig. 3.49 Experimental determination of ultrasonic attenuation by pulse height analysis

$$h_n = h_0 e^{-\alpha x} \quad \text{and} \quad h_{n+1} = h_0 e^{-\alpha(x+2d)}$$

$$\therefore \quad \frac{h_n}{h_{n+1}} = e^{2d\alpha} \Rightarrow \alpha = \frac{1}{2d} \ln\left(\frac{h_n}{h_{n+1}}\right)$$

Other Methods

In the case of liquids the ultrasonic waves propagating through it gives rise to a spatial periodic variation in the refractive index. Such a medium acts like a *phase grating* which gives rise to diffraction of light. Hence by measuring the angle of diffraction the wavelength of the ultrasonic wave and hence its velocity can be measured. A typical experimental set up is shown in Fig. 3.50. A narrow parallel beam of light is passed through the optically transparent solid or liquid in which ultrasonic waves are made to propagate using a transducer. On activating the transducer, there is a shift in the direction of the incident beam (from I_1 to I_2). By measuring this shift, the angle of diffraction is calculated. Assuming the condition for Raman-Nath diffraction, the wavelength is calculated using the formula:

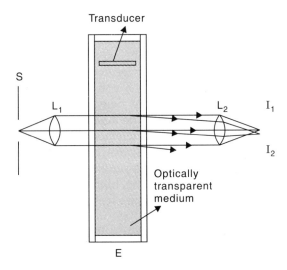

Fig. 3.50 Experimental set up for observing Raman-Nath diffraction

$$\lambda_{acoustic} = \frac{\lambda_{optic}}{\theta} = \frac{\lambda_0}{\mu \times \theta}$$

where μ is the refractive index of the medium, λ_0 is the wavelength of light in vacuum and θ is the angle of diffraction for first order. Velocity of the ultrasonic wave is got by the product of the wavelength and its frequency.

Ultrasonic Velocity and Elastic Constants

A measurement of ultrasonic velocity permits the calculation of elastic constants. In the case of fluids the velocity is related to the elastic constant by the relation

$$c = \sqrt{\frac{K}{\rho}} \text{ where } K \text{ is the bulk modulus and } \rho \text{ is the density}$$

In the case of crystalline solids, in general, the velocity is anisotropic and the number of elastic constants is also large and depends on the crystal symmetry. Hence the determination of elastic constants involves measuring the velocity along different directions and finally computing the elastic constants. For the case of cubic crystals, the velocities of the longitudinal waves propagating along [100] and [110] and that of the shear wave propagating along [100] are given by:

$$v_l^{[100]} = \sqrt{\frac{c_{11}}{\rho}}, \; v_t^{[100]} = \sqrt{\frac{c_{44}}{\rho}}, \; v_l^{[110]} = \sqrt{\frac{(c_{11} + c_{12} + c_{44})}{2\rho}}$$

where ρ is the density and c_{11}, c_{12} and c_{44} are the elastic constants for the cubic system.

3.5 UNDERWATER SONAR

Sound waves are absorbed and scattered in water to a much less degree than electromagnetic waves. Because of this property sound waves have proved to be particularly useful in detecting undersea distant objects by means of *sonar* (acronym of Sound Navigation and Ranging). *Passive sonar* is strictly a *listening* device that detects sound radiation emitted (sometimes unintentionally) by a target. In *active sonar* the process involves sending out a sound pulse and listening for a returning echo (Fig. 3.51). Thus, for example, if d is the depth of the ocean and t is the time taken for the return echo, then the depth of the ocean is given by $ct = 2d$ where c is the velocity of the ultrasound. Though the principle of echo is simple to comprehend, ranging or locating the distance of the target is a non-trivial task. This is on account of the variation of the sound velocity with temperature, depth of the sea and the degree of salinity. The loudness of the echo hinges principally on the amount of energy absorption in the water and the degree of reflection from an intercepting surface. It also depends on the distance of the target surface from the emitter. Also there are stray echoes due to sea bottom, sea surface and other inhomogeneities in the water. Modern echo-ranging sonar consists of an elaborate array of equipment to send out signals in the form of long, high power pulses in designated directions vertically and horizontally. Novel signal processing techniques present the echoed data to the observer. Much of the initial developments of ultrasonics resulted because of their utility in detecting enemy submarines during the First World War. In general, sonar is used to locate any underwater object of interest.

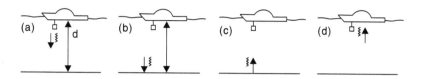

Fig. 3.51 Principle of SONAR

Sonar is also made use of in the following fields:

(a) Marine fishing: Sonar enables to locate the regions of high density fish population and makes fishing more efficient (Fig. 3.52).

(b) Marine biology: Sonar is used to study fish migration and behaviour of whales, dolphins and porpoises by tracking the sounds made by them.

(c) Oceanography: Sonar is employed to map ocean floor and determine ocean depth. It is also used to locate undersea oil wells.

(d) Communications: It is used to transmit communications and telemetric data.

3.6 ULTRASONICS IN INDUSTRY

Ultrasound is primarily used for non-destructive testing of materials. However, many industrial processes such as welding, cleaning, emulsification or homogenization etc. make use of ultrasound.

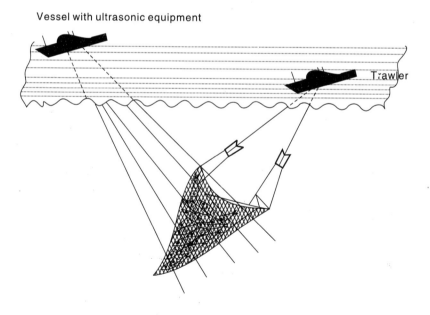

Fig. 3.52 Artist's impression of marine fishing

3.6.1 Nondestructive Testing

Nondestructive testing (NDT) is employed to examine a material, to detect imperfections, determine its properties or assess its quality without causing a physical or chemical damage. In recent years ultrasound imaging has enhanced the scope of NDT. Manual observations of ultrasonic waveforms are still the most widely used industrial ultrasonic test method. This has been primarily due to the low cost of the instrumentation in comparison to the cost of an imaging system. Scanned imaging systems have steadily expanded in industrial use, especially where the highest probability of detection for flaws is required.

ULTRASONIC TESTING SYSTEMS

There are three basic ultrasonic test systems that are commonly used in industries. They are pulse echo system, transmission system and resonance system.

Pulse Echo System

This system is most commonly employed. Here short pulses of ultrasonic waves are transmitted into the material being tested. These pulses get reflected from discontinuities on their path or from any boundary of the material on which they strike. The received echoes are then displayed on a cathode ray tube screen (Fig. 3.53). The CRT screen furnishes specific data as to the relative size of a discontinuity in terms of signal amplitude. The location of the discontinuity with respect to the scanning surface can be obtained by proper calibration of the CRT time base scale. Sometimes a single transducer is used both as a transmitter and a receiver. Otherwise two transducers are separately used for this purpose.

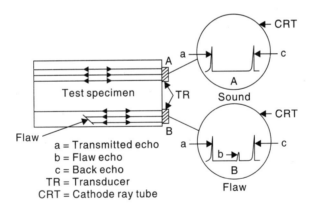

a = Transmitted echo
b = Flaw echo
c = Back echo
TR = Transducer
CRT = Cathode ray tube

Fig. 3.53 Pulse echo system

Through Transmission System

In this method two transducers are required; one is used as a transmitter and the other as a receiver. Short ultrasonic pulses are transmitted into the material. The receiver is aligned with the transmitter on the opposite side to pick up the sound waves which pass through the material. The quality of the material is indicated by the energy lost by the sound beam as it traverses the material (Fig. 3.54). A marked reduction in received energy amplitude indicates a discontinuity.

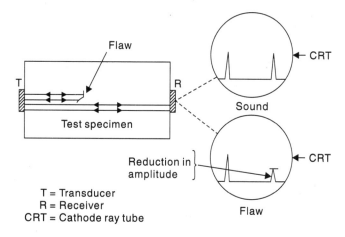

Fig. 3.54 Through transmission system

Resonance System

This system makes use of the resonance phenomenon to detect flaws. Here continuous longitudinal waves are transmitted into the material. The frequency of the waves is varied till the standing waves are set up within the specimen causing the specimen to vibrate or resonate at greater amplitude (Fig. 3.55). Resonance is then sensed by a device and presented on the cathode ray tube. A change in resonant frequency which cannot be traced to a change in material thickness is usually an indication of a discontinuity.

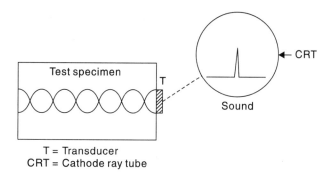

Fig. 3.55 Resonance system

ULTRASONIC TESTING METHODS

Basically there are two methods of ultrasonic testing. They are
  (i) Contact method and
  (ii) Immersion method.

Contact method: In this method the transducer is brought in contact with the test specimen through the use of a suitable couplant like water, oil, grease etc. The ultrasonic wave reflected from the flaw or the discontinuity is displayed on the CRT. There are two methods — one that uses straight beam and the other which uses angle beam. In the case of a straight beam, the ultrasonic beam is transmitted into the test specimen at angle of 90° to the test surface by placing a normal beam probe in a plane parallel to the test surface. In the case of angle beam, the ultrasonic beam is transmitted at an angle to the test surface.

Immersion method: In this method, the test specimen and the special type leak-proof transducer are immersed in a liquid, usually water. The liquid acts as a couplant in the transfer of sound energy from the transducer to the test specimen. This method provides testing flexibility since transducer can be moved under water to introduce the sound beam at any desired angle. The transducer does not touch the specimen and therefore is not subjected to wear. In the immersion method we can have pulse-echo (straight as well as angle beams) and through the transmission systems as shown in Figs. 3.56–3.58.

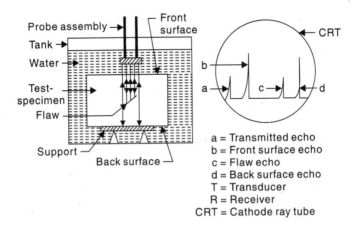

Fig. 3.56 Immersion pulse echo method

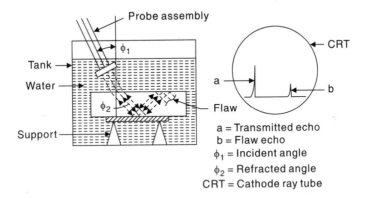

Fig. 3.57 Angle beaming using longitudinal waves

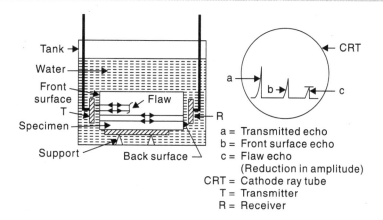

Fig. 3.58 Immersion through transmission system

Surface wave techniques: Surface waves are also employed for the detection of flaws near the surface such as cracks, tool marks etc.

3.6.2 Ultrasonic Imaging

Ultrasonic imaging is employed both in industry as well as in medicine. It is most useful for presenting the data of NDT. This is more so in medicine where the clinical data has to be made easily comprehensible both to the patient and the doctor who do not have any background in engineering. There are mainly three or four types of techniques or scans which are used for the construction of an ultrasound image of a system. These are designated as *A*-scan, *B*-scan, *C*-scan and *M*-scan. *M*-scan or the motion scan is used exclusively in medicine.

Ultrasonic Transducer Arrays for Scanning

A typical ultrasound imaging system is shown in Fig. 3.59*a*. Essentially one gathers data regarding the intensity of echo pulses from flaws within the component. The location of these flaws within the component has also to be determined and this is obtained by measuring the echo time. The component to be examined is immersed in water. The ultrasonic beam generated by a transducer then probes the whole volume of the component. The echo of the ultrasound is received by a transducer and the data is processed using a minicomputer. Results are displayed on CRT and also stored in the computer memory for further analysis of the data. Figure 3.59*b* shows the probe employed for scanning the entire volume. The movement of the transducer in three dimensions is controlled by a scan controller which in turn is linked to the computer. A single element ultrasound transducer tends to radiate a rather narrow beam or receive signals over a narrow spatial range. In order to cover a wider volume and also to avoid the mechanical movement of the transducer, *scanning* is done by employing an *array of detectors*. These detectors emit more powerful signals than is possible with a single element. Different types of transducer arrays are shown in Fig. 3.60.

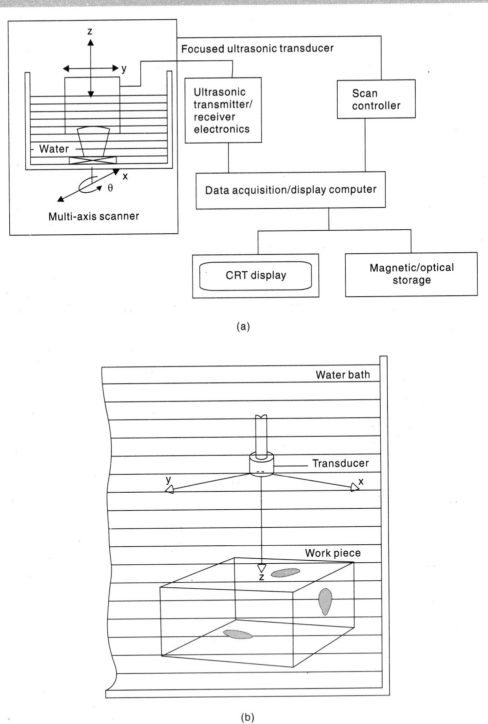

Fig. 3.59 (a) Ultrasonic imaging system (b) Coordinate system for defining various scans

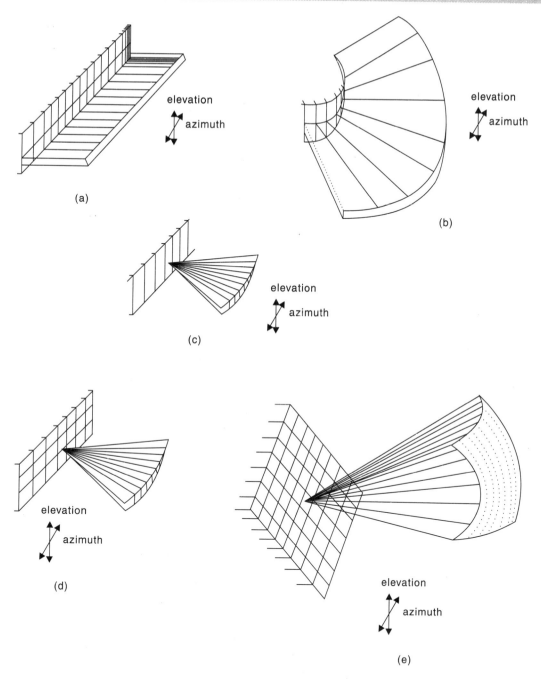

Fig. 3.60 Various configurations of the array elements and the corresponding region scanned by the acoustic beam (a) sequential linear array scanning a rectangular region (b) curvilinear array scanning a sectored region (c) linear phased array sweeping a sectored region (d) 1.5D array scanning a sectored region and (e) 2D array sweeping a pyramidal region

In Fig. 3.60 a, the scanning lines are perpendicular to the transducer face. The beam is directed straight out. The disadvantage is that the field of view is limited to the rectangular region directly facing the transducer. Linear-array transducers also require a large footprint to obtain a sufficient wide field of view.

Another type of array configuration is that of the *curvilinear array*. Because of its convex shape, the curvilinear array scans a wider field of view than does a linear array configuration.

In the linear phased array (Fig. 3.60c) each element is used to emit and receive each line of data. The scanner steers the beam through a sector-shaped region in the azimuth plane. The 1.5D array is structurally similar to a 2D array but operates as 1D. The 1.5D array consists of elements along both the azimuth and elevation directions. However, number of elements in the elevation is less. This improves image quality. Steering is not possible in the elevation direction.

In the 2D phased array a large number of elements are employed both in the azimuth and elevation. This permits focusing and steering of the acoustic beam along both the dimensions. A 2D array can scan a pyramidal volume in real time to yield a volumetric image.

In a linear or 2D phased array the acoustic beam is steered by applying time delay pulses to various individual elements as shown in Fig. 3.61. Thus the mechanical movement of the transducer is avoided.

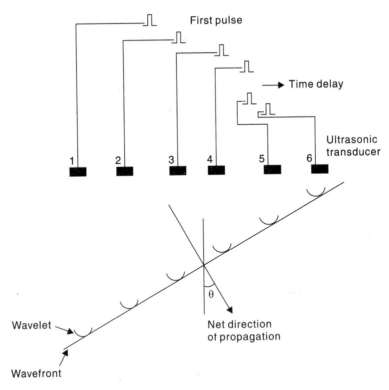

Fig. 3.61 Linear phase array for steering the acoustic beam

A-scan

In *A*-scan the amplitudes of the received echoes are plotted on the vertical axis of the display. The horizontal axis would indicate the time taken by the echo pulse to reach the transducer. In this scan the transducer is stationary. If the velocity of the ultrasonic wave in the material is known, the horizontal axis can be calibrated in terms of the distance traveled by the echo. Thus A-scan can be converted electronically into a thickness gauge. Diagrammatic representations of such displays are shown in Fig. 3.62a.

B-scan

Consider a work piece in which the ultrasonic wave propagates along the z-axis (Fig. 3.59b). In the B-scan, the horizontal axis of the display indicates the distance z in the work piece. The distance along the vertical axis of the scan is proportional to the distance y in the work piece normal to the propagation direction z. This dimension is achieved by moving the transducer laterally or sweeping it in an arc or using a phased array of transducer elements. The display is in grey scale with brightness proportional to the amplitude of any reflection occurring at the y-z coordinate of the work piece (Fig. 3.62b).

C-scan

This type of scan is carried out when the actual depth information of the defects in the work piece is not important. This scan gives only the distribution of defects perpendicular to a particular direction. In the C-scan the transducer is swept rapidly back and forth in the x-direction while being stepped in the y-direction after each sweep. The repeatition rate of the pulse is rapid compared with the x sweep rate. Thus in the sweep sequence the complete volume of the specimen is scanned. The method is most often used in a tank of liquid. The x-y display is a grey scale with amplitude proportional to the reflection amplitudes within the specimen in the liquid tank (Fig. 3.62c).

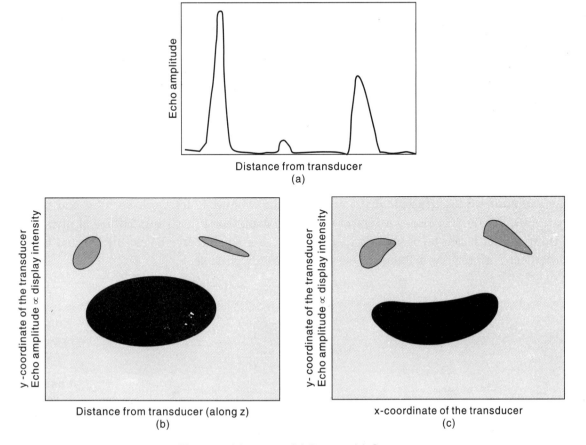

Fig. 3.62 (a) A-scan, (b) B-scan, (c) C-scan

3.6.3 Ultrasonic Welding

Ultrasonic welding of metals can be performed at room temperatures without the need for any special surface preparation. Also there is practically no deformation of the work pieces. Its application has been limited to the welding of thin metal sheets and foils. It is also ideal for plastic materials. The welding of plastics takes place under the action of longitudinal vibrations as opposed to shear vibrations for the welding of metals. The molecular process probably involves the diffusion of one material into the other when compressed. Figure 3.63 shows a typical arrangement using a magnetostrictive transducer working at 25 kHz. The two pieces to be welded are held in good mechanical contact by the application of forces as shown. When high intensity ultrasound is passed through them, the surfaces are subjected to high compressive forces at elevated temperatures generated by the wave itself during its propagation. The technique is found to be suitable for spot and seam welding.

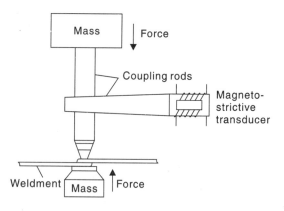

Fig. 3.63 Ultrasonic welding

3.6.4 Ultrasonic Drilling

Ultrasound has also been employed for drilling holes in metal sheets. The ultrasonic drill bit is coupled mechanically to a transducer in which longitudinal ultrasonic oscillations are generated. Thus the drill bit is made to vibrate longitudinally. In an ultrasonic drill the action is reciprocal, like that of a pneumatic road drill or a hand saw. This is in contrast to the rotary action of hand or electric drill. The cutting is performed by an abrasive slurry which is circulated continuously between the drill bit and material which is being drilled. The drill bit is therefore not itself cutting the material directly but is thrusting the abrasive violently into contact with the material. *Consequently, the drill bit need not be made of a special hard material and can be made of mild steel.* Ultrasonic drills are particularly suitable for work on hard and brittle materials. Figure 3.64 shows a typical ultrasonic drill. One interesting feature of ultrasonic drilling is that it can be used to drill square or other non-circular holes in hard and brittle materials. With regular drilling only circular holes are possible.

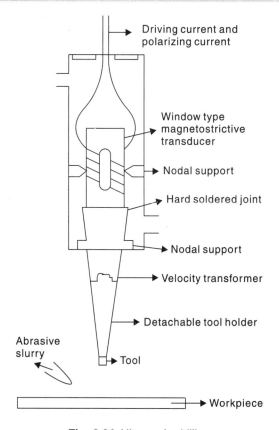

Fig. 3.64 Ultrasonic drilling

3.6.5 Ultrasonic Cleaning

The process ultrasonic cleaning is based on the phenomena of cavitation.

Ultrasonic Cavitation

Let a bubble of radius r exists in a liquid through which ultrasound is passing. The bubble may contain either the vapour of the liquid itself or some gas which has been dissolved in the liquid. The bubble will be subjected to the changes in pressure associated with the ultrasound (Fig. 3.65). The bubble will contract and expand as the pressure variations occur. At the extreme of the rarefaction half of the cycle the bubble has its maximum radius r_{max} and at the extreme of the compression half of the cycle it has its minimum radius r_{min}. If the excess pressure amplitude of the ultrasound is sufficiently high, and the initial radius of the bubble is less than a certain critical radius r_0, the bubble will suddenly collapse during the compressional half of the cycle with the sudden release of a comparatively large amount of energy. *This collapse and the associated release of energy which occurs is known as cavitation.* The critical radius is given by:

$$\omega^2 r_0^2 = \left(\frac{3\gamma}{\rho}\right)\left(p_o + \frac{2T_s}{r_o}\right)$$

where $\gamma = \dfrac{c_p}{c_v}$; p_0 is the hydrostatic pressure in the absence of the ultrasound, ρ is the density of the liquid, T_s is the surface tension and ω is the angular frequency of the ultrasound.

The pressure in a bubble just before it finally collapses may be very large. Thus, when the bubble finally collapses an extremely powerful shock wave is produced and it is the energy in this shock wave which is responsible for many of the effects arising from cavitation. For example a piece of metal placed in a liquid in which cavitation is occurring may become seriously pitted or eroded.

Following are the important experimental observations:

(a) The amount of energy released in cavitation depends on the ratio of r_{max}/r_0 and hence the intensity of the ultrasound.

(b) The energy released will be greater for large values of the surface tension of the bubble and for smaller values of the vapour pressure. Hence, a bubble with large surface tension will involve large cavitational energies. The cavitational energy can be increased by adding alcohol, say upto 10%.

(c) Although the presence of bubbles produced by the release of a dissolved gas facilitates the onset of cavitation, it is also possible for cavitation to occur in gas-free liquids if the excess pressure amplitude of the ultrasound exceeds the hydrostatic pressure in the liquid.

(d) For a given liquid there is a certain minimum ultrasonic intensity needed to produce cavitation. This threshold intensity for the onset of cavitation varies with frequency.

(e) If energy is released during cavitation in the visible region, the phenomenon is known as *sonoluminescence*.

Cavitation plays an important role in several applications of ultrasound. Ultrasonic cleaning is achieved by immersing an article in a suitable cleaning fluid and passing ultrasound into the fluid. The cleaning is found to be efficient if both agitation and cavitation are involved. The shock waves produced during cavitation will reach any solid surface that is in the liquid and will scour or scrub that surface. The probability of scratching or otherwise damaging the specimen that would occur in conventional scrubbing is less. Also through ultrasonic cleaning it is possible to extend the scrubbing action to all the *nooks and corners* that would be inaccessible with mechanical scrubbing. Figure 3.66 shows the set up normally employed for ultrasonic cleaning. Most cleaning applications are executed in the frequency range 20—40 kHz where cavitation effects occur more strongly. Either piezoelectric or magnetostrictive transducers are used. The work piece to be cleaned is immersed in a tank containing a liquid selected on the basis of its susceptibility to cavitation, its detergent properties, ability to degrease and so on. Trichloroethylene and cyclohexane are among the more satisfactory fluids used for ultrasonic cleaning.

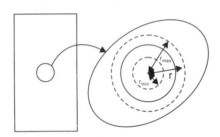

Fig. 3.65 Bubble in the pressure field of ultrasound

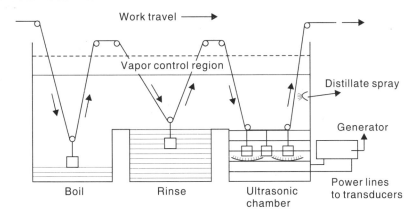

Fig. 3.66 Ultrasonic cleaning

Ultrasound cleaning lends itself to continuous processing in which a series of work pieces can be transported on a conveyor belt through a series of processes in separate tanks as shown in Fig. 3.66. Ultrasonic cleaning supersedes other usual methods of cleaning, particularly when these methods are ineffective and liable to cause damage. Applications include

(a) removal of lapping paste from lenses without scratching after grinding,

(b) flush out grease and machining particulates from small crevices in engine components,

(c) removal of blood and other organic material from surgical instruments, after use and so on.

Very delicate parts that can be damaged by cavitation are cleaned by wave agitation at much higher frequencies from 100 kHz to 1 MHz.

3.6.6 Ultrasonic Emulsifier or Homogeniser

An emulsion consists of two liquids which are normally immiscible but which have been coerced into forming a mixture. This is done by dispersing very small particles of one liquid as a suspension in the second liquid. Emulsification can occur as a direct consequence of agitation. An example of this is the mixing together of two liquids having widely differing densities, *e.g.*, water and mercury. However, the emulsion so formed is not very stable and the liquids separate after a very short time. Some emulsions are stable and can be stored on the shelf for many years without separating. Some emulsions like milk occur naturally.

An important application of ultrasonically induced cavitation is the emulsification of two immiscible liquids such as oil and water. Oil is injected into the water, which is then excited ultrasonically to produce cavitation. As a result, very fine droplets of oil are forced with high energies into the body of the water and the minute particles of oil rapidly disperse to form a highly stable emulsion. The name homogenization is often applied to this process. The production of emulsions by ultrasonically induced cavitation has extensive applications in food, pharmaceutical and cosmetics industry. This technique is found to be more effective than the method of mechanical agitation or stirring. Ultrasonic techniques are advantageous because there is less mechanical wear of the equipment and no unwanted air bubbles are introduced. There is no need to raise the temperature, an advantage where heating might otherwise affect the quality of the product. Further no time is wasted in waiting for the emulsion to cool down.

In the food industry, the method is used in the preparation of dairy products, sauces, gravies, mayonnaises, salad creams and synthetic creams. In the frozen-food industry, the sauces prepared by ultrasonic homogenization will withstand repeated freezing and thawing because of the high stability of emulsions. This technique is also applied in the preparation of pharmaceutical products, including antibiotic preparations. Manufacture of cosmetics, paints and polishes using ultrasonic cavitation is routinely done.

Figure 3.67 illustrates the method of preparing an oil-water emulsion with this device. First an emulsion is formed ultrasonically as the oil is introduced into the water and then the

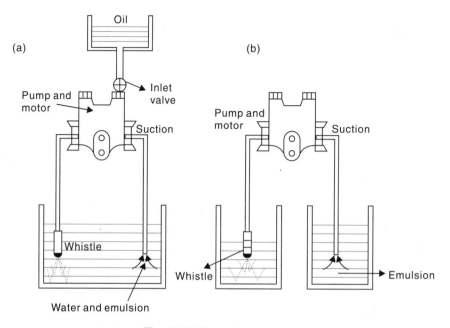

Fig. 3.67 Ultrasonic emulsifier

mixture is refined by circulation. Finally the oil supply is disconnected and the emulsion is forced through the resonant wedge whistle into the second container to produce an even fine suspension. The diameters of the suspended particles may be as small as 2 μm. The important component of the emulsifier is the resonant wedge whistle which is used for generating the ultrasonic waves with liquids. Figure 3.68a shows the diagram of a jet-edge generator which is used for generating the ultrasound. Ultrasound vibrations may be set up either in the fluid in the region XY or in the edge itself. In the liquid whistle both types of vibrations are important. One can design a liquid whistle so that these vibrations occur at a common frequency. In the air whistle it would only be the former type of vibrations which would be important and the whistle would be designed with a suitable resonant cavity surrounding this region. For a straight edge it was shown by Rayleigh that the frequency of sound or ultrasound generated in the fluid is given by:

$$\nu \cong \frac{nv}{b}\left(A + \frac{B}{R}\right)$$

where n (= 1, 2 or 3) characterises a particular stage in the operation of the jet, A and B are constants, v is the velocity of the fluid through the jet and R is the Reynold's number for the fluid.

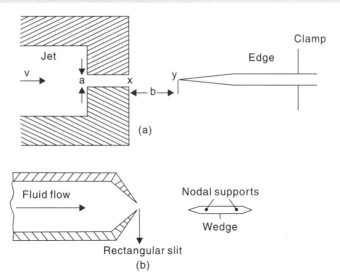

Fig. 3.68 (a) Jet-edge system and (b) liquid whistle used in emulsifier

However, in a liquid whistle the mechanical resonant vibrations of the edge itself are extremely important. The edge consists of a steel blade, with a beveled edge facing the jet. The plate is mounted in such a way that one particular normal mode of flexural vibrations will be excited. The plate may be mounted as a cantilever or at its half-wave nodal points. For a plate of thickness a, length l, density ρ and Young's modulus E, the frequency ν of the fundamental flexural vibrations is given by

$$\nu = \frac{na}{2\sqrt{12l^2}} \sqrt{\frac{E}{\rho}} \times (0.597)^2$$

for cantilever mounting shown in Fig. 3.68a and

$$\nu = \frac{na}{2\sqrt{12l^2}} \sqrt{\frac{E}{\rho}} \times (2.5)^2$$

for mounting shown in Fig. 3.68b

In ultrasonic homogenizer the two liquids which are to be homogenized are forced together through the jet and the ultrasound is therefore either generated directly in the liquids as the edge tones or is very quickly transferred to the liquids from the vibrations of the blade.

3.6.7 Acoustic Emission

Acoustic emission (AE) involves stress waves internally generated during dynamic processes in a material. The dynamic processes may be the result of an externally applied stress or the result of some other unstable situation. Fracture, plastic deformation, crack initiation and growth, inter particle motion in soils and rocks, first order phase transitions involving strain changes lead to acoustic emission. The dynamic nature of acoustic emission makes it a highly potential technique for monitoring the integrity of critical structures and components in various industries related to nuclear and fossil fuel power plants, aerospace, chemical, petrochemical, transportation, manufacturing, fabrication etc. Acoustic emission testing (AET) as a technique for monitoring and evaluating structural integrity is superior to other techniques because of its capability for

(*a*) continuous monitoring

(*b*) inspection of complete volume of the component

(c) issue of advance warning and

(d) location of any crack initiation and propagation and system leaks.

AE might occur over a range of frequencies depending on the phenomenon. Figure 3.69 shows a detailed experimental set up for detecting and analyzing acoustic emission. The stress waves are detected using a piezoelectric transducer. The resulting charge or current pulse is amplified and displayed using a conventional counter or an oscilloscope. Figure 3.70 shows a typical AE signal and the various parameters used for interpretation. Ringdown counts is the number of times the signal crosses a threshold level set for eliminating background noise. This could be used independently or as the cumulative counts with respect to time, load or any other parameter. Count rate is another parameter commonly used.

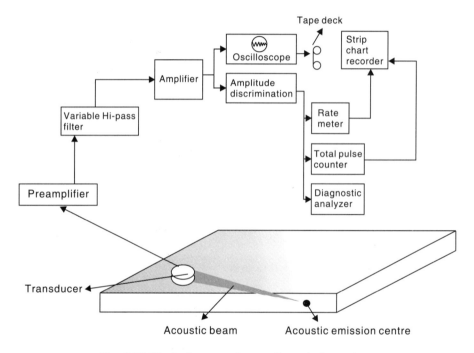

Fig. 3.69 Block diagram of acoustic emission set-up

The most common ways in which AE signals are processed are the following:

(a) Counting: Ringdown counts, rates, events

(b) Energy analysis: To arrive at the energy involved in these emissions

(c) Amplitude analysis: To characterize the events

(d) Frequency analysis: Fourier analyse the various frequency components of AE and identify the processes leading to acoustic emission.

Depending on the nature of energy release two types of AE are observed. These are (a) continuous type and (b) burst type. Continuous emission is characterized by low amplitude emissions. The amplitude varies with AE activity. In metals and alloys, this type of emission occurs during plastic deformation by dislocation movement, diffusion controlled phase transformations and fluid leakage. Burst emissions are characterized by short duration (~ μs) and high amplitude peaks to discrete release of strain energy. This type of emission occurs during diffusionless phase transformations, crack initiation and propagation, stress corrosion cracking etc. High intense AE precedes earthquakes since it involves large movements of rocks and soils.

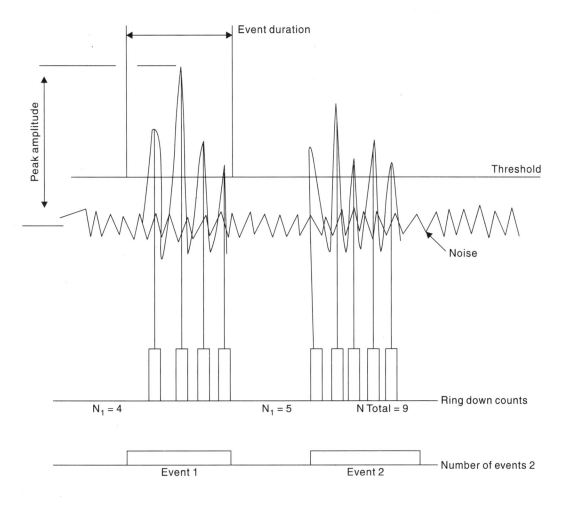

Fig. 3.70 Acoustic emission signal parameter

Acoustic emission can detect growing of small sized flaws which are inaccessible to NDT. Hence it is emerging as a powerful tool for NDT of plant components such as pressure vessels, pipes, welds etc. A wide variety of materials have been studied using AE. These include metals, ceramics, wood, rocks and composites. Some of the processes studied include phase transitions, welding, and magnetic processes. AE is a statistical process. An important feature of AE is its irreversibility. AE is used for statistical quality control of industrial components.

3.7 ULTRASOUND IN MEDICINE

The ultrasound waves have been widely used for examining the shape and movement of organs within the body. Normally ultrasound is transmitted in short bursts at a repetition rate of 1 to 12 kHz. The images of ultrasonic waves that are reflected from the boundaries of the organs are obtained. The ultrasonic waves reflected from moving objects such as red blood cells exhibit a Doppler shift which is used to estimate the speed of blood flow. This technique is also used to monitor the motion of heart valves. Ultrasonic imaging can provide valuable information regarding the size, location, displacement or velocity of a given structure without the necessity of surgery or the use of potentially harmful radiation. Tumors and other regions of an organ that differ in density from surrounding tissues can be detected. It is normal to employ A and B scans. In addition, M-scan is also employed.

M-scan or Motion scan display: M-scan is used to study motion as that of the heart and the heart valves. M-scan combines certain features of the A-scan and B-scan. The transducer is held stationary as in the A-scan and the echoes appear as dots as in the B-scan, *i.e.*, the intensity of the dot is proportional to the amplitude of the echo. However, the vertical axis provides the time variation of the reflecting interface, *i.e.*, the vertical axis displays the periodic displacement of the reflecting surface (Fig. 3.71).

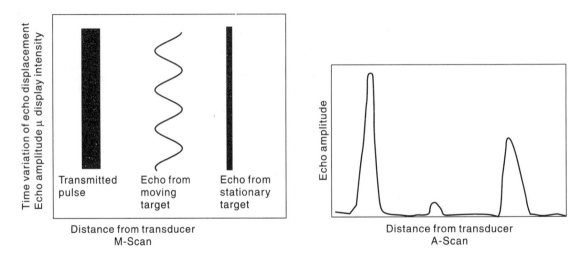

Fig. 3.71 M-scan and the corresponding A-scan

3.7.1 Echoencephalography

This pertains to the method of detecting tumors in the brain. Pulses of ultrasound are sent into a thin region of the skull slightly above the ear and echoes from the different structures within the head are displayed using A-scan (Fig. 3.72). The leftmost reflection corresponds to the skull wall nearest to the transducer. The rightmost feature is the reflection for the far side of the patient's skull. The features between the skull wall reflections depend upon transducer placement and certain other factors. The middle line feature is caused by reflections from the lateral ventricles, the third ventricle and the septum pellucidum between the lateral ventricles. Ordinarily, the septum lies within ± 2 mm of the centre line of the skull and its reflection will be located exactly midway between the spikes representing the third ventricle as shown in Fig. 3.72. A tumor on one side of the brain tends to shift the midline structure. A shift of more than ±3 mm from the correct position is often considered to be pathological and may indicate the presence of tumor. The usual procedure is to compare the echoes from the left side of the head to those from the right side.

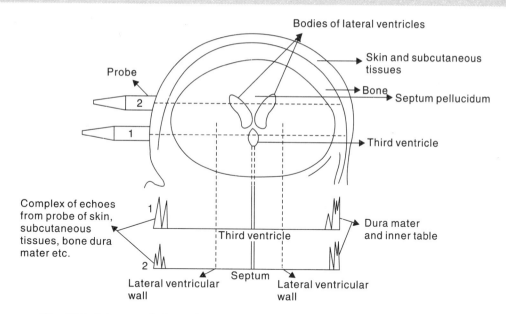

Fig. 3.72 Echoencephalograph transducer placement and the patterns generated

3.7.2 Ultrasound in Ophthalmology

Applications of A-scans in ophthalmology can be divided into two areas: one is concerned with obtaining information for use in the diagnosis of eye diseases; the second involves biometry or measurements of the distances in the eye. Ultrasound diagnostic techniques are supplementary to the generally practiced ophthalmological examinations. They can provide information about the deeper regions of the eye and are especially useful when the cornea or lens is opaque. Tumors, foreign bodies and detachment of the retina are some of the problems that can be diagnosed with ultrasound. Figure 3.73 shows typical A-scan of the eye. Without ultrasound ophthalmologists can look into the living eye up to the optic nerve. But measurements of the eye have been largely confined to the exterior segment. With ultrasound it is possible to measure distances in the eye such as lens thickness, depth from cornea to the lens, the distance to the retina and the thickness of the vitreous humor.

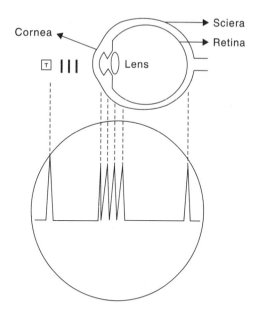

Fig. 3.73 A-scan of the eye

3.7.3 Doppler Effect in Ultrasound to Measure Motion

The Doppler effect can be used to measure the speed of moving objects or fluids within the body, such as that of blood or a heart valve. When continuous ultrasound beam is *received* by some red blood cells in an artery the blood cells *hear* frequency which differs from that of the source. If the blood cells are moving towards the source they *hear* a frequency slightly greater than that of the source and if they are moving away from the source they *hear* a frequency slightly less than that of the source. The blood sends back the echoes to the transducer with different frequencies. In the first case the transducer *hears* the sound from a source moving towards it and hence the frequency is shifted to still higher frequency. In the second case, the transducer *hears* the sound from a source moving away from it and hence the frequency is downshifted further. Hence in effect there is a *double* Doppler shift in frequency and is given by:

$$f_D = 2f_0 \frac{v_B}{c} \cos \theta$$

where f_0 is the frequency of the source, f_D is the frequency of the sound at the detector, v_B is the velocity of the red blood cells and θ is the angle between the direction of the propagation of the sound and the direction of motion of the red blood cells. Thus knowing f_D, f_0, c and θ, v_B can be calculated. f_D is positive or negative depending on whether the blood is moving towards the detector or away from it.

The shift in frequency is extracted from the returning signal by multiplying the signal by the emitted signal. On multiplying we have

$$\cos 2\pi f_0 t \times \cos 2\pi (f_0 + f_D)t$$

which can be rewritten as

$$[\cos 2\pi(2f_0 + f_D)t + \cos 2\pi f_D t]/2$$

The first term is at a very high frequency and can be filtered out using a low-pass filter to reveal the shift frequency. However, this does not indicate whether the second term is positive or negative (If the blood is moving towards the transmitter the term will be positive and if it is moving away it will be negative). Hence returning signal is multiplied by the sine of the transmitted signal to give

$$\sin 2\pi f_0 t \times \cos 2\pi (f_0 + f_D)t$$

which can be rewritten as

$$[\sin 2\pi (2f_0 + f_D)t + \sin 2\pi f_D t]/2$$

Hence by measuring both $\cos 2\pi f_D t$ and $\sin 2\pi f_D t$, both the magnitude and sign of v_D can be determined.

The Doppler effect is also used to detect the motion of the fetal heart, umbilical cord and placenta in order to establish fetal life during 12 to 20 week period of gestation when radiological and clinical signs are unreliable. The most common use of the Doppler effect in obstetrics is in locating the point of entry of the umbilical cord (artery) into the placenta. This information is very useful if there is bleeding due to a misplaced placenta.

Following are the techniques which are used to study the Doppler effect:

Continuous Doppler: Here a continuous ultrasonic signal is transmitted while returning echoes are picked up by a separate moving transducer (Fig. 3.74). Frequency shifts due to moving interfaces are detected and recorded. The average velocity of the targets is usually determined as a function of time. This mode always requires two transducer crystals, one for transmission and the other for receiving. This is in contrast to pulsed mode where a single transducer can be used for

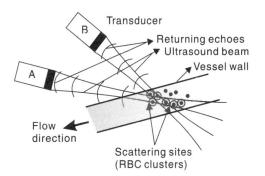

Fig. 3.74 Doppler effect to measure the speed of blood

transmission and reception. Continuous Doppler ultrasound is used in blood flow measurements and in certain other applications in which the average velocity is measured without regard to the distance of the source.

Pulsed Doppler: As in pulsed ultrasound, short bursts of ultrasonic energy are transmitted and the returning echoes are received. However, in this mode frequency shifts due to movement of reflected interfaces as well as the distance of the reflecting interfaces can be measured.

Range-gated pulsed Doppler: This mode is a refinement of pulsed-Doppler ultrasound, in which a gating circuit permits measurement of the velocity of targets at a specific distance from the transducer. The velocity of these targets can be measured as a function of time. With range-gated pulsed Doppler ultrasound, the velocity of blood can be measured not only as a function of time, but also as a function of the distance from the vessel wall.

3.7.4 Therapeutic Uses of Ultrasound

Dentistry: A 25 kHz ultrasound scrubber combined with a water jet is used to remove plaque from teeth. Dental and medical tools are usually subjected to ultrasound cleaning.

Shockwave lithotripsy has completely changed the treatment of kidney stones. Kidney stones are calcified particles that tend to block the urinary tract. In this type of treatment, the patient is immersed in water to equalize as much as possible the acoustic impedances between the transducer and the patient's body. A focused high pressure ultrasonic pulse is directed through the water and into the patient's torso to break the stones into small pieces. The pulverized material can pass out of the body unhindered. Lithotripsy causes very little damage to the kidney tissue.

Ultrasonic diathermy is helpful in the treatment of joint disease and joint stiffness. Ultrasound is an effective deep heater of bones and joints. The treatment is similar to hot fomentation.

Acoustic surgery: A promising procedure for therapeutic ultrasound is the laser-guided ablative acoustic surgery in which sound supersedes the scalpel in destroying benign or malignant tissues. The ultrasound focused by a specially shaped set of transducers converges inside the body to create a region of intense heat that can destroy tumor cells. The spot of destruction is so small that a boundary of only six cells lies between the destroyed tissue and completely unharmed tissue. This precision is beyond any current method of surgical incision.

Acoustic hemostatis: Ultrasound can also be used to stop internal bleeding through an effect called *acoustic hemostatis*. With sufficient power, ultrasonic pulses can elevate the body temperature at selected sites from 37°C to between 70°C and 90°C in an extremely short time, less than one second. This causes the tissue to undergo a series of phase transitions. The protein based bodily fluids and blood coagulate as the result of proteins undergoing cross-linking. This process is similar to what happens when egg is boiled in water.

Targeted Drug Delivery: A newer method of targeted drug delivery is that of *sonophoresis*, which uses sound waves instead of needles to inject drugs such as insulin and interferon through the skin. The high frequency ultrasound opens tiny holes in cell membranes. This renders the cells temporarily permeable in localized regions thus allowing better penetration of the drug into the blood vessels below the skin. This results in greater effectiveness of the drug, lessens the dosage requirements and toxicity, and allows for more precise localization of drug delivery. An early application of this technique has been to dissolve life-threatening blood clots by injecting thrombolytic drugs. Other long term application is administering insulin through the skin for treating diabetes. It is hoped that ultrasound can penetrate blood-brain barrier, which insulates the brain from foreign substances and also prevents many drugs from reaching diseased tissues there, so that the effects of chemotherapy can be enhanced.

3.8 ULTRASONICS IN ELECTRONICS

3.8.1 Delay Line

The speed of sound or the ultrasound is very much lower than the speed of electromagnetic radiation. This leads to an interesting use for ultrasonic waves in radar, television and in digital computers. This is to introduce time delay in electromagnetic waves. This delay could be achieved by sending the electromagnetic wave on a round trip in air or vacuum. But to achieve a delay of, say, 1 ms would require a distance of $3 \times 10^8 \times 10^{-3}$ m = 300 km, which is highly inconvenient. However, with the device shown in Fig. 3.75 one can achieve a delay of 1 ms with quite a short path length. In such a *delay line*, an electromagnetic wave is converted by a *transducer* into an ultrasonic wave which is launched at A into the solid medium AB. The ultrasonic wave travels from A to B and then at B it is converted back into an electromagnetic wave by a second transducer. If the ultrasonic wave travels in the solid at, say, 3000 m/s, the path length needed for a 1 ms delay is only 3000×10^{-3} = 3 mm. Multiple reflections instead of a single transit enable the length of the line needed for a given delay to be reduced still further (Fig. 3.76).

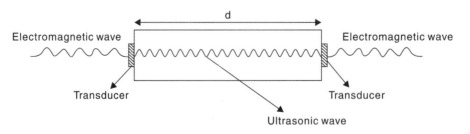

Fig. 3.75 Delay line

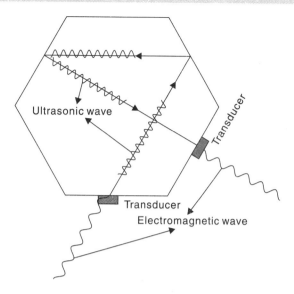

Fig. 3.76 Polygonal delay line

3.8.2 Frequency Control

Modern electronics depends on frequency control devices which provide *precise time and frequency*. A vibrating quartz crystal, *i.e.*, a quartz resonator, is the *heart* of nearly all frequency control devices. Quartz clocks provide accurate time and quartz oscillators are the sources of precise frequency. Precise measurement of time is important not only for the functioning of a computer but also for determining the frequencies of radio and TV transmissions, radar system, communication systems and navigation systems, surveillance, missile guidance etc. A quartz resonator has a high Q factor (Q factor is the ratio of energy stored in an oscillator to the energy dissipated per cycle). Its resonant frequency does not vary with temperature and consequently has high frequency stability. Quartz is also chemically stable.

A simple LC tank circuit behaves like an oscillator with an angular frequency given by the relation $\omega^2 LC = 1$. However to sustain oscillations a part of the output of the circuit is used to provide input as shown in Fig. 3.77. In ordinary oscillators the frequency stability is not good. If a quartz crystal is used as a dielectric between the capacitor plates, then the oscillator can be made to vibrate at the resonant frequency characteristic of the quartz crystal. Such an oscillator possesses high frequency stability. Typical circuits using quartz are shown in Fig. 3.78.

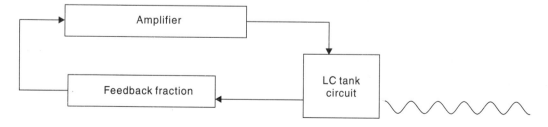

Fig. 3.77 Basic principle of sustaining electrical oscillations

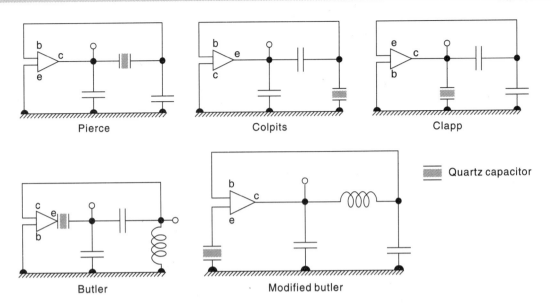

Fig. 3.78 Different circuits used for quartz oscillator

SAW Devices

Surface acoustic wave devices have become an important component of modern electronics. The greatest advantage of the SAW devices is its compatibility with thin film technology and the fabrication techniques of microelectronics. It can be designed for specific application and can generate the desired type of acoustic waves. *Non-contact* EMATs possess unique advantages in some specific applications. They are used for transmission and receival of microwave frequencies. They are used as delay lines, resonators, filters, band pass filters, matched filters etc.

3.9 ULTRASONICS IN OPTICS

3.9.1 Acoustic Microscope

Ultrasonic waves obey the laws of geometrical optics. Hence an acoustic microscope can be constructed along the same principles as that of an optical microscope. However, since the eye cannot form an image out of ultrasonic waves, visualisation of the object has to be achieved by optical means. Figure 3.79 shows a simple acoustic microscope. This is used to visually observe the flaws inside a material. The object whose acoustic image has to be obtained is immersed in a tank of water. A transducer at the bottom acts as a source of ultrasound. An image is formed using a converging lens. Usually acoustic lenses are made of metals or suitable plastics. The velocity of ultrasound in solids is higher than that in liquids. Hence solid concave lenses are convergent and solid convex lenses are divergent, unlike in the case of light. The acoustic image is formed as a ripple pattern at the surface of the liquid. It is viewed with light incident obliquely on the surface. Alternatively, the surface can be scanned using a scanner and the image can be obtained on a TV screen.

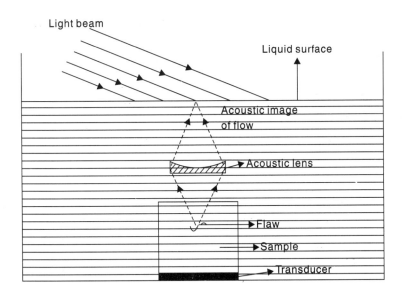

Fig. 3.79 Acoustic microscope

Scanning Acoustic Microscope

In scanning acoustic microscopy, images are formed of the surface or interior regions of materials by mechanically scanning using piezoelectric transducers that produce focused acoustic beams. The image data are acquired by scanning line by line. The resolution of the image is controlled by the diameter of the focused acoustic beam and by the size and spacing of the pixels that display the assembled data. A single focused transducer may be used to transmit and receive reflected signal. Figure 3.80 shows the schematic diagram of a scanning acoustic microscope in pulse transmission and pulse echo modes.

3.9.2 Acoustic Holography

Holography is a technique of obtaining *three dimensional image* of an object. A *three dimensional image* of an object is a replica of the object and exhibits *parallax* when viewed from different directions. An ordinary photograph is a two dimensional recording of a three dimensional scene. It does not exhibit *parallax*. The photographic emulsion is sensitive only to intensity (*i.e.*, square of the amplitude) of the light scattered by the object and not to the phase of the light waves. This loss of phase information in collecting the light scattered by the object results in a two dimensional representation of the image. *Holography is a technique of recording both the amplitude and phase of the light waves scattered by the object.* This is done in two steps: (*a*) formation of hologram and (*b*) reconstruction of hologram, as shown in Fig. 3.81. In the first step, light scattered by the object and the reference beam are made to interfere on a photographic plate. The photographic plate is developed in the usual way to get a negative. This negative will have an interference pattern characteristic of the object. This interference is possible because *coherent laser beams* are used for the purpose. *The interference pattern contains both the amplitude as well as the phase information of the waves scattered by the object.* Hologram bears no resemblance to the object. When the hologram is illuminated by the reference beam two images, one real and one virtual, are formed. These images are three dimensional in character and possess the property of *parallax* which is the stamp mark of three dimensionality.

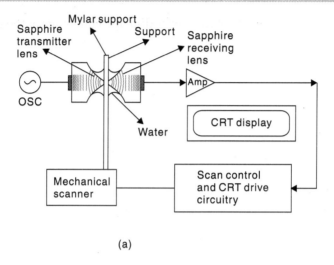

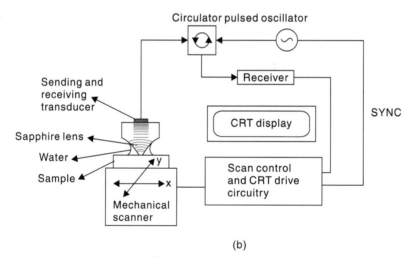

Fig. 3.80 (a) Scanning acoustic microscope (b) Scanning pulse-echo acoustic microscope

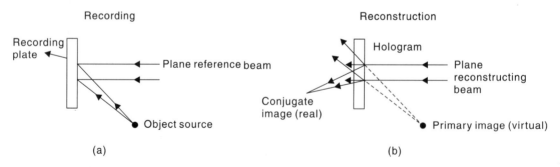

Fig. 3.81 Basic principle of holography (a) recording of hologram (b) reconstruction of 3D image

In optical holography both the reference beam and the beam used for reconstruction are laser beams.

Acoustic holography is used to obtain three dimensional image of the underwater objects. It is identical to optical holography except that the formation of the hologram is done using ultrasound while the reconstruction is done using a laser beam. Instead of a photographic plate, hologram formation takes place on the surface of water itself. Figure 3.82 shows the principle of acoustic holography. The ultrasound produced by the first source and scattered by the object reaches the surface of water. The reference ultrasound wave is obtained by generating a separate ultrasonic wave through a second synchronous transducer. On a still water surface, these two beams form an interference pattern or hologram on the surface of water. This surface may be photographed and then the photograph used for reconstruction using a laser beam. Alternatively, reconstruction may be done in real time by coherent light reflected off the surface as shown. The advantage of acoustic holography is that on reconstruction a *magnified image* is formed. *Magnification factor is equal to the ratio of the wavelength of the ultrasound to that of the wavelength of the laser beam.*

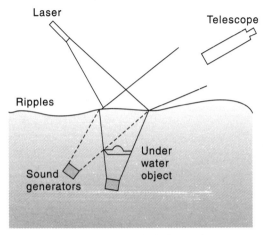

Fig. 3.82 Acoustic holography

3.9.3 Acousto-Optic Modulator

Light is diffracted by the ultrasonic waves in a medium. This is because of the phase grating formed due to variations in refractive index. The intensity of the diffracted light in Bragg diffraction and the intensity of light in Raman-Nath diffraction for the order n = ± 1, are to a good approximation proportional to the ultrasound intensity. This feature of acousto-optic effect is useful in optical communication. A simple acousto-optic modulator for modulating the signal is shown in Fig. 3.83. The signal in the form of ultrasound passes through the medium. The

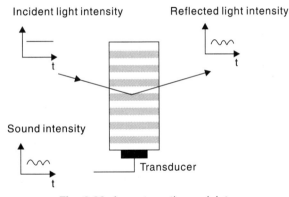

Fig. 3.83 Acousto-optic modulator

reflected light from the medium has the intensity of light characteristic of the sound wave in the medium.

3.10 ULTRASONICS IN MATERIALS SCIENCE

3.10.1 Photoacoustic Spectroscopy

Short pulses of light typically from a laser are focused and scanned over the surface of the object. The light pulses are formed by chopping or pulsing the light source. Acoustic pulses result from the rapid localized heating of the object surface by each of the light pulses. These acoustic pulses undergo attenuation which results from the various processes mentioned in previous sections. The acoustic pulses are in turn monitored by a piezoelectric receiver. An analysis of these pulses can provide information about the various molecular processes occurring within the material. A block diagram of the photoacoustic spectroscopy is shown in Fig. 3.84. The photoacoustic effect in materials can be used to design acoustic microscope as well. Here the ultrasound that is generated by the laser beam is viewed as in the case of an acoustic microscope (Fig. 3.85)

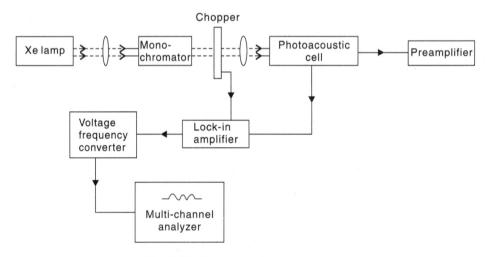

Fig. 3.84 Photoacoustic spectroscopy

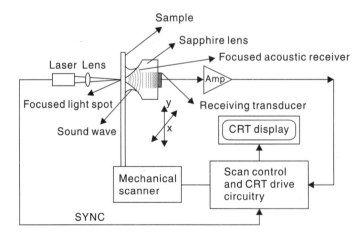

Fig. 3.85 Photoacoustic microscope

3.10.2 Molecular Processes in Materials

Ultrasonic attenuation across the phase transition temperature in the case of magnetic, ferroelectric and superconducting transitions shows anomalous behaviour (Fig. 3.86). A comparison of the observed attenuation with the theoretical models of phase transitions has led to a better understanding of mechanisms of phase transitions.

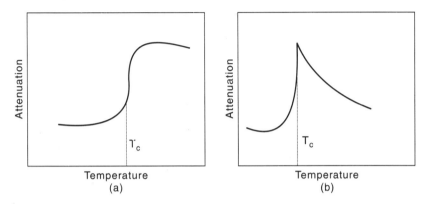

Fig. 3.86 Typical behaviour of ultrasonic attenuation across (a) ferromagnetic and (b) ferroelectric transitions.

REFERENCES

1. L.E. Kinsler, A.R. Frey, A.B. Coppens and J.V. Sanders *Fundamentals of Acoustics*, John Wiley & Sons, Inc. New York, 2000.
2. P. Crackmell *Ultrasonics*, Wykehm Publications (London) Ltd. London, 1980.
3. Crocker *Encyclopedia of Acoustics*, Vols. 1—4, John Wiley 1987.
4. B. Carlin *Ultrasonics*, McGraw-Hill Book Company, New York, 1960.
5. B. Raj, T. Jayakumar and M. Thavasimuthu *Practical Nondestructive Testing*, Narosa Publishing House, New Delhi, 1997.
6. J. Prasad, T. Rangachari and B.N.S. Murthy *Theory and Practice of Ultrasonic Testing*, NDT Centre, HAL, Bangalore.
7. J.J. Carr and J.M. Brown *Introduction to Biomedical Equipment*, Pearson Education Asia, New Delhi, 2001.

QUESTIONS

1. Describe the various types of ultrasonic waves.
2. Discuss the various properties of ultrasonic waves.
3. Discuss Doppler effect in the case of ultrasonic waves and describe the applications where this principle is in use.
4. Write a short note on acousto-optic effect.
5. Distinguish between Bragg diffraction and Raman-Nath diffraction of ultrasonic waves.
6. Explain how acousto-optic effect can be employed to determine the velocity of ultrasound.
7. Discuss the functioning of an acousto-optic modulator.
8. Define ultrasonic attenuation. Discuss the various mechanisms contributing to ultrasonic attenuation in (a) solids and (b) fluids.
9. Discuss the mechanical methods of generating ultrasonic waves.

10. Define piezoelectricity and magnetostriction.
11. Describe the construction of transducers based on piezoelectricity and magnetostriction.
12. Describe the construction EMAT (Electromagnetic Acoustic Transducer).
13. Write a short note on production of ultrasound using laser.
14. Describe the methods of determining ultrasonic velocity and attenuation.
15. Write a short note on (a) ultrasonic cleaning (b) ultrasonic emulsifier (c) ultrasonic drilling and (d) ultrasonic welding.
16. Discuss the use of ultrasound in the nondestructive testing of engineering materials.
17. Write a short note on ultrasonic cavitation.
18. Write a short note on acoustic emission and its utility in engineering.
19. Discuss the use of ultrasound in medicine.
20. Describe the principle of ultrasonic imaging with special reference to A-scan, B-scan, C-scan and M-scan.
21. Discuss the principle on which acoustic holography is based. Compare it with optical holography.
22. Define photoacoustic effect. Write a short note on photoelastic spectroscopy.
23. Write a short note on acoustic microscope.
24. Write a short note on photoacoustic microscope.
25. Write a short note on SAW devices.
26. Discuss the use of ultrasound in the field of electronics by giving specific examples.

4

Interference

4.1 INTRODUCTION

In this chapter we will discuss the phenomena associated with the interference of light waves. At any point where two or more wave trains cross one another they are said to interfere. In studying the effects of interference we are interested to know the physical effects of superimposing two or more wave trains.

It is found that the resultant amplitude and consequently, the intensity of light gets modified when two light beams interfere. This modification of intensity obtained by the superposition of two or more beams of light is called interference. In order to find out resultant amplitude, when two waves interfere, we make use of the principle of superposition. The truth of the principle of superposition is based on the fact that after the waves have passed out of the region of crossing, they appear to have been entirely uninfluenced by the other set of waves. Amplitude, frequency and all other characteristics of each wave are just as if they had crossed an undisturbed space. *The principle of superposition states that the resultant displacement at any point and at any instant may be found by adding the instantaneous displacements that would be produced at the point by the individual wave trains if each were present alone.* In the case of light wave, by displacement we mean the magnitude of electric field or magnetic field intensity.

4.2 SUPERPOSITION OF WAVES

4.2.1 Superposition of Waves of Equal Phase and Frequency

Let us assume that two sinusoidal waves of the same frequency are travelling together in a medium. The waves have the same phase, without any phase angle difference between them. Then the crest of one wave falls exactly on the crest of the other wave and so do the troughs. The resultant amplitude is got by adding the amplitudes of each wave point by point. The resultant amplitude is the sum of the individual amplitudes (Fig. 4.1).

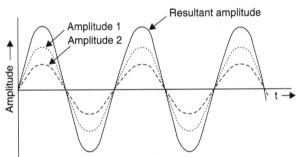

Fig. 4.1 Superposition of waves of equal phase and frequency

$$A = A_1 + A_2 + \ldots\ldots$$

The resultant intensity is the square of the sum of the amplitudes

$$I = (A_1 + A_2 + A_3 + \ldots\ldots)^2 \qquad (4.1)$$

4.2.2 Superposition of Waves of Constant Phase Difference

Let us consider two waves that have the same frequency but have a certain constant phase angle difference between them. The two waves have a certain differential phase angle ϕ. In this case the crest of one wave does not exactly coincide with the crest of the other wave (Fig. 4.2). The resultant amplitude and intensity can be obtained by trigonometry.

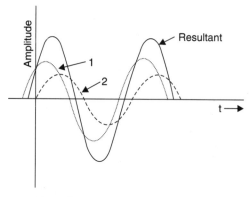

Fig. 4.2 Superposition of two sine waves of constant phase difference

The two waves having the same frequency ($\omega = 2\pi f$) and a constant phase difference (ϕ) can be represented by the equations

$$Y_1 = a \sin \omega t$$
$$Y_2 = b \sin (\omega t + \phi) \qquad (4.2)$$

where ϕ is the constant phase difference, a, b are the amplitudes and ω is the angular frequency of the waves. The resultant amplitude Y is given by

$$\begin{aligned}
Y &= Y_1 + Y_2 \\
&= a \sin \omega t + b \sin (\omega t + \phi) \\
&= a \sin \omega t + b(\sin \omega t \cos \phi + \cos \omega t \sin \phi) \\
&= a \sin \omega t + b \sin \omega t \cos \phi + b \cos \omega t \sin \phi \\
&= (a + b \cos \phi) \sin \omega t + b \cos \omega t \sin \phi
\end{aligned} \qquad (4.3)$$

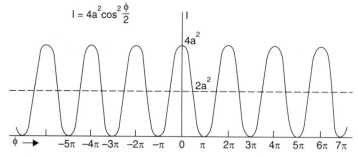

Fig. 4.3 Intensity distribution for the interference fringes from two waves of same frequency and amplitude

INTERFERENCE

If R is the amplitude of the resultant wave and θ is the phase angle then

$$Y = R \sin(\omega t + \theta)$$
$$= R \{\sin \omega t \cos \theta + \cos \omega t \sin \theta\}$$
$$= R \cos \theta \sin \omega t + R \sin \theta \cos \omega t \quad (4.4)$$

Comparing Eqns. 4.3 and 4.4

$$R \cos \theta = a + b \cos \phi$$
$$R \sin \theta = b \sin \phi$$
$$\Rightarrow \quad R^2 = a^2 + b^2 + 2ab \cos \phi$$

$$\theta = \tan^{-1} \frac{b \sin \phi}{a + b \cos \phi} \quad (4.5)$$

Clearly, R is maximum when $\phi = 2n\pi$
and is minimum when $\phi = (2n + 1)\pi$, where $n = 0, 1, 2, 3, \ldots$

When ϕ is an even multiple of π we say that waves are in phase and when ϕ is an odd multiple of π, the waves are out of phase.

When the amplitude of waves are equal to a say, then

$$I = 2a^2 (1 + \cos \phi) = 4a^2 \cos^2 \phi/2 \quad (4.6)$$

A plot of I versus ϕ is shown in Fig. 4.3. Clearly, this reveals that the light distribution from the superposition of waves will consist of alternatively bright and dark bands called interference fringes. Such fringes can be observed visually if projected on a screen or recorded photo-electrically. In the above discussion we have not considered travelling waves (*i.e.*, waves in which displacement is also a function of distance). If λ is the wavelength, then the change of phase that occurs over a distance λ is 2π. Thus, if the difference in phase between two waves arriving at a point is 2π, then difference in the path travelled by these waves is λ. Let the phase difference of two waves arriving at a point be δ and the corresponding path difference be x. For a path difference of λ, the phase difference = 2π. Therefore, for a path difference of x,

$$\text{Phase difference} = \delta = \frac{2\pi}{\lambda} \cdot x = \frac{2\pi}{\lambda} \cdot \text{path difference}$$

and Path difference $= x = \frac{\lambda}{2\pi}$ phase difference

4.2.3 Superposition of Waves of Different Frequencies

So far we have assumed that the waves have the same frequency. But light is never truly monochromatic. Many light sources emit quasimonochromatic light *i.e.*, light emitted will be predominantly of one frequency but will still contain other ranges of frequencies. When waves of different freqencies are superimposed, the result is more complicated.

4.2.4 Superposition of Waves of Random Phase Differences

When waves having random phase differences between them superimpose, no discernible interference pattern is produced. The resultant intensity is got by adding the square of the individual amplitudes,

i.e.,
$$I = \sum_{i=1}^{N} A_i^2 = A_1^2 + A_2^2 + A_3^2 + \ldots \ldots \quad (4.7)$$

4.3 YOUNG'S DOUBLE SLIT EXPERIMENT

We have seen in the previous section that two waves with a constant phase difference will produce an interference pattern. Let us see how it can be realized in practice. Let us use two conventional light sources (like two sodium lamps) illuminating two pin holes (Fig. 4.4). Then we will find that no interference pattern is observed on the screen. This can be understood from the following reasoning. In a conventional light source, light comes from a large number of independent atoms each atom emitting light for about 10^{-9} second i.e., light emitted by an atom, is essentially a pulse lasting for only 10^{-9} second. Even if the atoms were emitting under similar conditions, waves from different atoms would differ in their initial phases. Consequently light coming out from the holes S_1 and S_2 will have a fixed

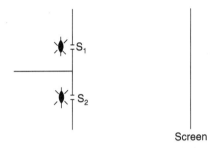

Fig. 4.4 If two sodium lamps illuminate two pin holes S_1 and S_2 no interference pattern is observed on the screen

phase relationship for a period of about 10^{-9} sec. Hence, the interference pattern will keep on changing every billionth of a second. The human eye can notice intensity changes which last at least for a tenth of a second and hence we will observe a uniform intensity over the screen. However, if we have a camera whose time of shutter can be made less than 10^{-9} sec., then the film will record an interference pattern. *We can summarize the above argument by noting that light beams from two independent sources do not have a fixed phase relationship over a prolonged time period and hence, do not produce any stationary interference pattern.*

Thomas Young in 1802 devised an ingenious but simple method to lock the phase relationship between two sources. The trick lies in the division of a wavefront into two. These two split wavefronts act as if they emanated from two sources having a fixed phase relationship and therefore, when these two waves were allowed to interfere, a stationary interference pattern was produced. In the actual experiment a light source illuminated a tiny pin hole S (Fig. 4.5).

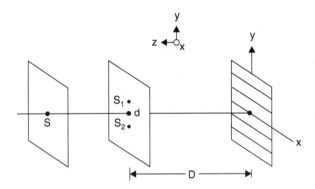

Fig. 4.5 Young's arrangement to produce interference pattern

Light diverging from this pin hole fell on a barrier containing two rectangular apertures S_1 and S_2 which were very close to each other and were located equidistant from S. Spherical waves travelling from S_1 and S_2 were coherent and on the screen beautiful interference fringes (Fig. 4.6) could be obtained. In the centre screen, where the light waves from two slits have travelled through equal distances and where the path difference is zero, we have zeroth-order maximum. But maxima will also occur whenever the path difference is one wavelength λ or an integral multiple of wavelength $n\lambda$. The integer n is called the order of interference.

INTERFERENCE

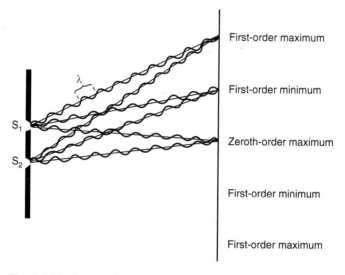

Fig. 4.6 Maxima and minima in Young's double slit experiment

When the path difference is a multiple of $(n + 1/2)\lambda$, we observe a dark fringe.

In order to calculate the position of the maxima, we proceed as follows. Let d be the distance between the slits and D be the distance of the screen from the slits.

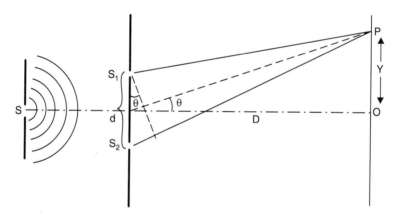

Fig. 4.7 Path difference in Young's double slit experiment

Let P be the position of the maximum (Fig. 4.7). Then the path difference between the two waves reaching P is

$$d \sin \theta = n\lambda \quad \text{or} \quad \sin \theta = \frac{n\lambda}{d} \quad (n = 1, 2, 3, \ldots\ldots)$$

where λ is the wavelength of light used and θ is the angle as shown in Fig. 4.7. If Y is the distance of point P from O, the centre of the screen, then we have

$$Y = D \tan \theta$$

For small angles of θ, $\quad Y = D \tan \theta \approx D \sin \theta$

$$Y = \frac{Dn\lambda}{d} \quad \text{or} \quad \lambda = \frac{dY}{Dn} \qquad (4.8)$$

Clearly, fringe width = $Y_{n+1} - Y_n = \beta = \dfrac{D\lambda}{d}$ \hfill (4.9)

Hence, by measuring the distance between slits, the distance to the screen and the distance from the central fringe to some fringe on either side, the wavelength of light producing the interference pattern may be determined.

4.4 COHERENCE

An important concept associated with the idea of interference is coherence. Coherence means that two or more electromagnetic waves are in a fixed and predictable phase relationship to each other. In general the phase between two electromagnetic waves can vary from point to point (in space) or change from instant to instant (in time). There are thus two independent concepts of coherence namely temporal coherence and spatial coherence.

Temporal Coherence: This type of coherence refers to the correlation between the field at a point and the field at the same point at a later time *i.e.*, the relation between $E(x, y, z, t_1)$ and $E(x, y, z, t_2)$. If the phase difference between the two fields is constant during the period normally covered by observations, the wave is said to have temporal coherence. If the phase difference changes many times and in an irregular way during the shortest period of observation, the wave is said to be non-coherent.

Spatial Coherence: The waves at different points in space are said to be space coherent if they preserve a constant phase difference over any time t. This is possible even when two beams are individually time incoherent, as long as any phase change in one of the beams is accompanied by a simultaneous equal phase change in the other beam (this is what happens in Young's double slit experiment). With the ordinary light sources, this is possible only if the two beams have been produced in the same part of the source.

Time coherene is a characteristic of a single beam of light whereas space coherence concerns the relationship between two separate beams of light. Interference is a manifestation of coherence.

Light waves come in the form of wave trains because light is produced during deexcitation of electrons in atoms. These wave trains are of finite length. Each wave train contains only a limited number of waves. The length of the wave train Δs is called the *coherence length*. It is the product of the number of waves N contained in wave train and their wavelength λ *i.e.*, $\Delta s = N\lambda$. Since velocity is defined as the distance travelled per unit of time, it takes a wave train of length Δs, a certain length of time Δt, to pass a given point

$$\Delta t = \Delta s/c$$

where c is the velocity of light. The length of time Δt is called the *coherence time*. The degree of temporal coherence can be measured using a Michelson's interferometer.

It is clear from the above discussion that the important condition for observing interference is that the two sources should be coherent. The observations of interference are facilitated by reducing the separation between the sources of light producing interference. Further, in the Young's double slit experiment the distance between two sources and the screen should be large. The contrast between the bright and dark fringes is improved by making equal the amplitudes of the light sources producing interference. Further, the sources must be narrow and monochromatic. The concept of coherence is discussed in greater detail in the chapter on lasers.

4.5 TYPES OF INTERFERENCE

The phenomenon of interference is divided into two classes depending on the mode of production of interference. These are (*a*) interference produced by the division of wavefront and

(b) interference produced by the division of amplitude. In the first case the incident wavefront is divided into two parts by making use of the phenomenon of reflection, refraction or diffraction. The two parts of the wavefront travel unequal distances and reunite to produce interference fringes. Young's double slit experiment and Fresnel's biprism are classic examples for this. In Young's double slit experiment one uses two narrow slits to isolate beams from separate portions of the primary wavefront. In Fresnel's biprism the phenomenon of refraction is made use of. In the second case the amplitude of the incident light is divided into two parts either by parallel reflection or refraction. These light waves with divided amplitude reinforce after travelling different distances and produce interference. Newton's rings and Michelson's interferometer are examples for this type. We will now discuss Fresnel's biprism, interference in thin films and Michelson's interferometer in detail.

4.6 FRESNEL'S BIPRISM

Critics of the Thomas Young's experiment argued that the observed dark and bright fringes were probably due to some complicated modification of the light by the edges of the slits and not due to true interference. In order to answer the critics of Young's experiment Fresnel thought of a new experiment for demonstrating the interference of light. He made use of a biprism for this purpose.

A schematic diagram of the biprism experiment is shown in Fig. 4.8. Biprism is actually a simple prism, the base angles of which are extremely small (1/2°). The base of the prism is shown in figure and the prism is assumed to stand perpendicular to the plane of the paper. S represents the slit which is also placed perpendicular to the plane of paper. Light from slit S gets refracted by the prism and produces two virtual images S_1 and S_2. These images act as coherent sources and produce interference fringes on the right of the prism. The fringes can be viewed through an eye piece.

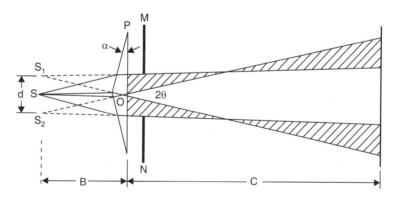

Fig. 4.8 Diagram of the Fresnel's biprism experiment

The theory of the biprism experiment is the same as that of Thomas Young's double slit experiment. Of course the distance between two sources d has to be determined and this is done in the following manner.

To determine the distance between the virtual sources of light in the biprism experiment, one makes use of what is called the displacement method. A lens with a focal length less than one-fourth of the distance between the biprism and eye piece is mounted between the biprism and eye piece as shown in Fig. 4.9. The lens is adjusted in two positions L_1 and L_2 till sharp images of S_1 and S_2 are obtained in the field of view of the eye piece. The distances d_1 and d_2 between the real images in two cases are measured.

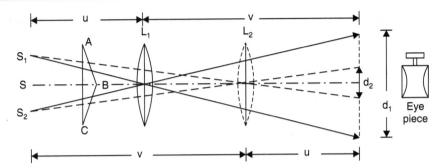

Fig. 4.9 Fresnel's biprism arrangement showing the position of lens

Then we have

$$\frac{d_1}{d} = \frac{v}{u} \text{ and } \frac{d_2}{d} = \frac{u}{v} \text{ where } u \text{ and } v \text{ are the object and image distances.}$$

i.e., $\quad d_1 d_2 = d^2 \quad \text{or} \quad d = \sqrt{d_1 d_2} \quad$ (4.10)

The second method of finding d is to measure accurately the refracting angle α. As the angle is small the deviation produced is $\theta = (\mu - 1)\alpha$. Therefore the total angle $S_1 O S_2$ is $2\theta = 2(\mu - 1)\alpha$. If the distance between the prism and the slit is B then

$$d = 2(\mu - 1)\alpha B$$

4.6.1 Fringes with White Light Using the Biprism

When white light is employed, the centre of the fringe at C is white while the fringes on both sides of C are coloured because the fringe width depends on wavelength. In the biprism the two coherent virtual sources are produced by refraction and the distance between two sources depends on the refractive index which in turn depends on the wavelength of the light used. Therefore, for blue light the distance between two virtual sources is different from that for red. The distance Y_n of the nth fringe from the centre as given by eqn.(4.8) is

$$Y_n = \frac{n\lambda D}{d} \quad \text{where } d = 2(\mu - 1)\alpha B$$

or $\quad Y_n = \dfrac{n\lambda D}{2(\mu - 1)\alpha B} \quad$ (4.11)

Therefore for blue and red rays, the nth fringe will be

$$Y_{nb} = \frac{n\lambda_b D}{2(\mu_b - 1)\alpha B}$$

$$Y_{nr} = \frac{n\lambda_r D}{2(\mu_r - 1)\alpha B}$$

4.6.2 Determination of the Thickness of a Thin Sheet of Transparent Material Using the Fresnel's Biprism

The biprism experiment can be used to determine the thickness of a given thin sheet of transparent material such as glass or mica. Suppose S_1 and S_2 are the two virtual coherent sources, the point C is equidistant from S_1 and S_2. When a transparent plate G of thickness t and refractive index μ is introduced in the path of one of the beams, the fringe which was originally at C shifts to P (Fig. 4.10). The time taken by the wave from S_2 to P in air is the same as the time taken by the wave from S_1 to P partly through air and partly through the plate. Suppose c is the velocity of light in air and v is the velocity in the medium

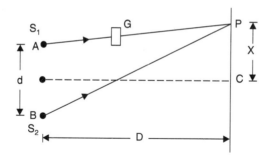

Fig. 4.10 Thickness of a plate using Fresnel's Biprism

$$\frac{S_2P}{c} = \frac{S_1P - t}{c} + \frac{t}{v}$$

$$S_2P = S_1P - t + \frac{c}{v}t, \text{ But } \frac{c}{v} = \mu$$

$$S_2P - S_1P = \mu t - t = (\mu - 1)t \quad (4.12)$$

If P is the point originally occupied by the nth fringe, then the path difference

$$S_2P - S_1P = n\lambda$$

i.e., $$(\mu - 1)t = n\lambda \quad (4.13)$$

The distance X through which the fringe is shifted $= \dfrac{n\lambda D}{d} = n\beta$

where $\dfrac{\lambda D}{d} = \beta$, the fringe width

$$\therefore \quad X = \frac{n\lambda D}{d} \quad \text{or} \quad n\lambda = \frac{Xd}{D} \quad \text{or} \quad (\mu - 1)t = \frac{Xd}{D} \quad (4.14)$$

Therefore, knowing X the distance through which the central fringe is shifted, D, d and μ, the thickness of the transparent plate can be calculated. Generally X is determined in the following way. If a monochromatic source of light is used the fringes will be similar and it will be difficult to locate the position to which the central fringe shifts after the introduction of the transparent plate. Therefore, white light is used. The fringes will be coloured but the central fringe will be white. When the cross wires are at the central white fringe, without the transparent plate in the path, the reading is noted. When the transparent plate is introduced, the position to which the central white fringe shifts is observed. The difference between the two positions gives the value of the shift which is equal to X.

The thickness of the plate can also be determined using equation (4.13). X is determined as described earlier. Now with the monochromatic light the micrometer eye piece is moved through the same distance X and the number of fringes that cross the field of view is observed. Suppose n fringes cross the field of view, then from the relation

$$(\mu - 1)t = n\lambda$$

the value of t can be calculated. Also if t is known μ can be calculated.

4.7 INTERFERENCE IN THIN FILMS

The colours of thin films, soap bubbles and oil slicks can be explained as due to the phenomena of interference. In all these examples, the formation of interference pattern is by the division of amplitude. For example, if a plane wave falls on a thin film then the wave reflected from the upper surface interferes with the wave reflected from the lower surface. Such studies have many practical applications as provided by the example of production of non-reflecting coatings.

4.7.1 Interference in Plane Parallel Films due to Reflected Light

Let us consider a plane parallel film as shown in the Fig. 4.11. Let light be incident at A. Part of the light is reflected toward B and the other part is refracted into the film towards C. This second part is reflected at C and emerges at D, and is parallel to the first part. At normal incidence, the path difference between rays 1 and 2 is twice the optical thickness of the film.

$$\Gamma = 2\mu d$$

At oblique incidence the path difference is given by

$$\Gamma = \mu(AC + CD) - AB = \frac{2\mu d}{\cos r} - AB$$

$$= \frac{2\mu d}{\cos r} - 2\mu d \tan r \sin r \quad [\because AB = AD \sin i = 2AE . \sin i = 2d \tan r . \sin i = 2d \tan r . \mu \sin r]$$

i.e.,
$$\Gamma = 2\mu d \left\{ \frac{1}{\cos r} - \tan r \sin r \right\} = 2\mu d \left\{ \frac{1 - \sin^2 r}{\cos r} \right\} = 2\mu d . \cos r$$

where μ is the refractive index of the medium between the surfaces. Since for air $\mu = 1$, the path difference between rays 1 and 2 is given by

$$\Gamma = 2d \cos r$$

While calculating the path difference, the phase change that might occur during reflection has to be taken into account. Whenever light is reflected from an interface beyond which the medium has lower index of refraction, the reflected wave undergoes no phase change. When the medium beyond the interface has a higher refractive index there is phase change of π. The transmitted waves do not experience any phase change.

Hence, the condition for maxima for the air film to appear bright is

$$2\mu d \cos r + \frac{\lambda}{2} = n\lambda$$

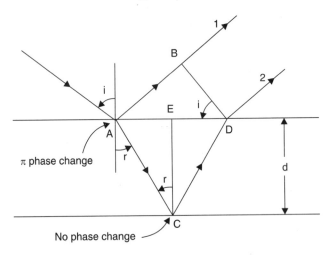

Fig. 4.11 Interference in plane parallel films (Reflection geometry)

INTERFERENCE

or
$$2\mu d \cos r = n\lambda - \frac{\lambda}{2}$$
$$= (2n - 1)\frac{\lambda}{2} \quad \text{where } n = 1, 2, 3, \ldots$$

The film will appear dark in the reflected light when
$$2\mu d \cos r + \frac{\lambda}{2} = (2n + 1)\frac{\lambda}{2}$$

or
$$2\mu d \cos r = n\lambda \quad \text{where } n = 0, 1, 2, 3, \ldots$$

4.7.2 Interference in Plane Parallel Films due to Transmitted Light

Figure 4.12 illustrates the geometry for observing interference in plane parallel films due to transmitted light. We have two transmitted rays CT and EU which are derived from the same point source and hence, are in a position to interfere. The effective path difference between these two rays is given by

$$\Gamma = \mu(CD + DE) - CP$$

But
$$\mu = \frac{\sin i}{\sin r} = \frac{CP/CE}{QE/CE} = \frac{CP}{QE} \Rightarrow CP = \mu(QE)$$

or
$$\Gamma = \mu(CD + DQ + QE) - \mu(QE)$$
$$= \mu(CD + DQ) = \mu(ID + DQ) = \mu(QI)$$
$$= 2\mu d \cos r$$

In this case it should be noted that, phase change occurs when the rays are refracted unlike in the case of reflection. Hence, the condition for maxima is $2\mu d \cos r = n\lambda$ and the condition for minima is $2\mu d \cos r = (2n - 1)\frac{\lambda}{2}$.

Thus, the conditions of maxima and minima in transmitted light are just reverse of the condition for reflected light.

4.7.3 Interference in Wedge Shaped Film

Let us consider two plane surfaces GH and G_1H_1 inclined at an angle α and enclosing a wedge shaped film (Fig. 4.13). The thickness of the film increases from G to H as shown in the figure. Let μ be the refractive index of the material of the film. When this film is illuminated there is

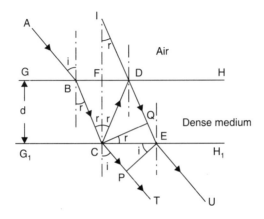

Fig. 4.12 Interference in plane parallel films (Transmission geometry)

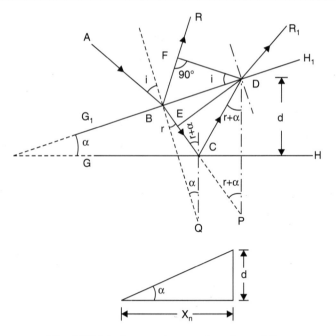

Fig. 4.13 Interference in a wedge shaped film

interference between two systems of rays, one reflected from the front surface and the other obtained by internal reflection at the back surface.

The path difference Γ is given by

$$\Gamma = \mu(BC + CD) - BF$$
$$\Gamma = \mu(BE + EC + CD) - \mu BE$$

$$\left[\because \sin i = \frac{BF}{BD}; \sin r = \frac{BE}{BD}; \mu = \frac{\sin i}{\sin r} \Rightarrow \mu = \frac{BF}{BE}\right]$$

$$\Gamma = \mu(EC + CD) = \mu(EC + CP) = \mu EP = 2\mu d \cos(r + \alpha)$$

Due to reflection an additional phase difference of $\lambda/2$ is introduced.

Hence, $\Gamma = 2\mu d \cos(r + \alpha) + \lambda/2$

For constructive interference

$$2\mu d \cos(r + \alpha) + \lambda/2 = n\lambda$$

or
$$2\mu d \cos(r + \alpha) = (2n - 1)\lambda/2 \quad \text{where } n = 1, 2, 3, \ldots$$

For destructive interference

$$\therefore \quad 2\mu d \cos(r + \alpha) + \frac{\lambda}{2} = (2n + 1)\frac{\lambda}{2}$$

or
$$2\mu d \cos(r + \alpha) = n\lambda \quad \text{where } n = 0, 1, 2, 3, \ldots$$

Spacing between two consecutive bright bands is obtained as follows.

For nth maxima

$$2\mu d \cos(r + \alpha) = (2n - 1)\frac{\lambda}{2}$$

Let this band be obtained at a distance X_n from thin edge as shown in Fig. (4.13). For near normal incidence, $r = 0$. Assuming, $\mu = 1$,

INTERFERENCE

From the figure, $d = X_n \tan \alpha$

$\therefore \quad 2X_n \tan \alpha \cos \alpha = (2n - 1) \dfrac{\lambda}{2}$

$2X_n \sin \alpha = (2n - 1) \dfrac{\lambda}{2}$

For $(n + 1)$th maxima

$2X_{n+1} \sin \alpha = (2n + 1) \dfrac{\lambda}{2}$

$\therefore \quad 2(X_{n+1} - X_n) \sin \alpha = \lambda$

or fringe spacing, $\beta = X_{n+1} - X_n = \dfrac{\lambda}{2 \sin \alpha} = \dfrac{\lambda}{2\alpha}$

where α is small and measured in radians.

4.8 COLOURS OF THIN FILMS

The discussion of the interference due to a parallel film and at a wedge should now enable us to understand as to why films appear coloured. To summarize, the incident light is split up by reflection at the top and bottom of the film. The split rays are in a position to interfere and interference of these rays is responsible for colours. Since the interference condition is a function of thickness of the film, the wavelength and the angle of refraction, different colours are observed at different positions of the eye. The colours for which the condition of maxima will be satisfied will be seen and others will be absent. It should be noted here that the conditions for maxima and minima in transmitted light are opposite to that of reflected light. Hence, the colours that are absent in reflected light will be present in transmitted light. The colours observed in transmitted and reflected light are complimentary.

4.9 NEWTON'S RINGS

When a plano-convex lens with its convex surface is placed on a plane glass plate, an air film of gradually increasing thickness is formed between the two. If monochromatic light is allowed to fall normally and viewed as shown in the Fig. 4.14 then alternate dark and bright circular fringes are observed. The fringes are circular because the air film has a circular symmetry. Newton's rings are formed because of the interference between the waves reflected from the top and bottom surfaces of the air film formed between the plates as shown in the Fig. 4.15.

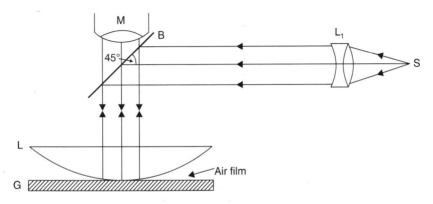

Fig. 4.14 Experimental set up for viewing Newton's rings

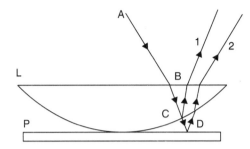

Fig. 4.15 Interference in Newton's rings setup

The path difference Γ between these rays (*i.e.*, rays 1 and 2) is

$$2\mu d \cos r + \frac{\lambda}{2}$$

i.e., Since $r \approx 0$, $\mu = 1$, $\Gamma = 2d + \frac{\lambda}{2}$

At the point of contact $d = 0$, the path difference is $\frac{\lambda}{2}$. Hence, the central spot is dark.

The condition for bright fringe is

$$2d + \frac{\lambda}{2} = n\lambda \quad \text{or} \quad 2d = \frac{(2n-1)\lambda}{2}, \text{ where } n = 1, 2, 3, \ldots$$

and the condition for dark fringe is

$$2d + \frac{\lambda}{2} = (2n+1)\frac{\lambda}{2} \quad \text{or} \quad 2d = n\lambda, \text{ where } n = 0, 1, 2, 3, \ldots$$

Now let us calculate the diameters of these fringes. Let LOL' be the lens placed on the glass plate AB (Fig. 4.16). The curved surface LOL' is part of the spherical surface with the centre at C. Let R be the radius of curvature and r be the radius of Newton's ring corresponding to constant film thickness d.

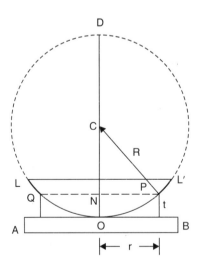

Fig. 4.16 Calculation of diameter of Newton's ring

INTERFERENCE

From the property of the circle.

i.e., $\quad NP \times NQ = NO \times ND$

i.e., $\quad r \times r = d(2R - d) = 2Rd - d^2 \approx 2Rd$

i.e., $\quad r^2 = 2Rd \quad$ or $\quad d = r^2/2R$

Thus, for a bright fringe

$$\frac{2r^2}{2R} = \frac{(2n-1)\lambda}{2} \quad \text{or} \quad r^2 = \frac{(2n-1)\lambda R}{2}$$

Replacing r by $D/2$ where D is the diameter we get

$$D_n = \sqrt{2\lambda R}\sqrt{2n-1}$$

Similarly, for a dark fringe

$$\frac{2r^2}{2R} = n\lambda \quad \text{or} \quad r^2 = n\lambda R$$

$$D_n^2 = 4n\lambda R$$

$$D_n = 2\sqrt{n\lambda R}$$

Thus, the diameters of the rings are proportional to the square roots of the natural numbers.

By measuring the diameter of the Newton's rings, it is possible to calculate the wavelength of light as follows. We have for the diameter of the nth dark fringe.

$$D_n^2 = 4n\lambda R$$

Similarly diameter for the $(n+p)$th dark fringe

$$D_{n+p}^2 = 4(n+p)\lambda R$$

$\therefore \quad D_{n+p}^2 - D_n^2 = 4pR\lambda$

or $\quad \lambda = \dfrac{D_{n+p}^2 - D_n^2}{4pR}$

λ can be calculated using this formula.

Newton's rings set up could also be used to determine the refractive index of a liquid. First the experiment is performed when there is air film between the lens and the glass plate. The diameters of the nth and $(n+p)$th fringes are determined. Then we have

$$D_{n+p}^2 - D_n^2 = 4p\lambda R$$

Now the liquid whose refractive index is to be determined is poured into the container without disturbing the entire arrangement. Again the diameter of the nth and $(n+p)$th dark fringes are determined. Again we have

$$D'^2_{n+p} - D'^2_n = \frac{4p\lambda R}{\mu}$$

from the above equations

$$\mu = \frac{D_{n+p}^2 - D_n^2}{D'^2_{n+p} - D'^2_n}.$$

4.10 MICHELSON'S INTERFEROMETER

An interferometer is an instrument for measuring small changes in length and is based on the principle of interference. The Michelson's interferometer is shown in Fig. 4.17. Light from an extended source falls on beam splitter A. The beam splitter A has a reflective coating on its rear surface. The beam splitter divides the light into two beams of nearly equal intensity, one transmitted towards mirror C and the other reflected towards D. The two mirrors, C and D, return the light to A. There they recombine and proceed towards E, where interference is observed. One of the mirrors is so mounted that it can be moved along the axis. Since the reflection at A occurs at the rear surface as shown, the light reflected at D will pass through A three times while the light reflection at C will pass through only once. For this reason a compensating plate B of the same thickness and inclination as A, is inserted into the A-C path. If we look into the instrument from E, we see mirror, D and in addition we see a virtual image, C′ of mirror C. Depending on the position of the mirrors, image C′ may be in front of or behind or exactly coincident with mirror D.

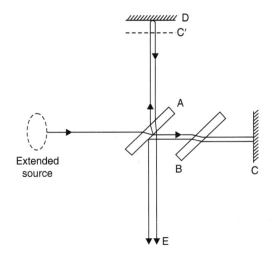

Fig. 4.17 Michelson's interferometer

4.10.1 Theory Behind the Formation of Different Kinds of Fringes

(*i*) If the two mirrors are absolutely perpendicular to each other and have the same axial distance from the rear face of A, then image C′ is coincident with mirror D.

At this coincidence position, the two paths are of equal length. But actually this is not the case because a π phase change occurs on external reflection (air-to-glass) only. No phase change occurs on internal (glass-to-air) reflection and none occurs on transmission or refraction. From the figure it is clear that only the light that comes from C and goes to E is reflected (air-to-glass) and therefore, undergoes the π phase change. This means that at the coincidence position the path difference will be minimum: the centre of the field will be dark.

(*ii*) Now let us move one of the mirrors. If the mirror is moved through a distance of quarter of a wavelength, $d = \lambda/4$, the path length changes by $\lambda/2$ (because light passes through the distance twice). Now the two beams get out of phase by π and hence the phase change compensates. Therefore, we will have a maximum. Moving the mirror by another $\lambda/4$ gives another minimum, another $\lambda/4$ another maximum and so on.

(*iii*) Let us assume that we look obliquely into the interferometer and that our line of sight makes and angle α with the axis. (Figs. 4.18 and 4.19). The two planes D and C′ are at a

INTERFERENCE

distance d apart and the two virtual images I and I' are separated by $2d$. But for oblique incidence the path difference between the two lines of sight becomes less and we have

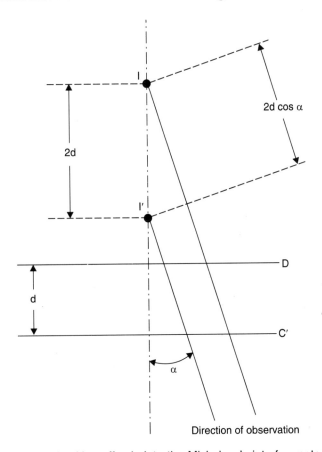

Fig. 4.18 Looking off-axis into the Michelson's interferometer

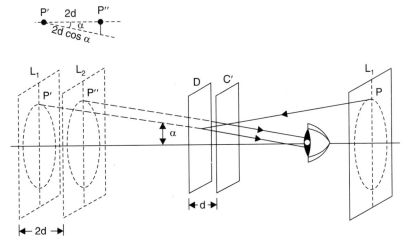

Fig. 4.19 Formation of circular fringes in Michelson's interferometer

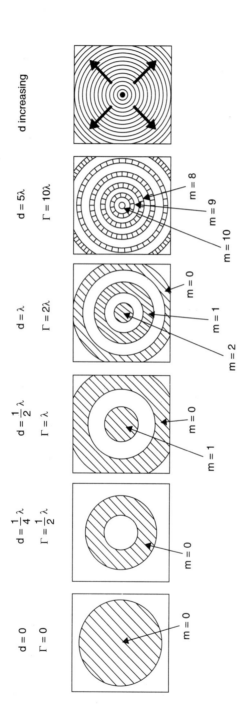

Fig. 4.20 Appearance of fringes in Michelson's interferometer

$$2d \cos \alpha + \frac{\lambda}{2} = n\lambda \text{ for bright fringe}$$

Furthermore, for a given d and λ, α is constant. Therefore, fringes are of rotational symmetry. They appear in the form of cirlces concentric around the axis. They are fringes of equal inclination.

(iv) Let us move the mirrors further away from each other. Consequently, as d becomes larger, a given ring which has a certain order n increases in size because the product $2d \cos \alpha$ must remain constant. The rings therefore expand and new rings appear in the centre, one ring appearing each time one mirror is moved by one-half of a wavelength (Fig. 4.20). Also note that as d increases, the rings in the periphery disappear slower as compared to the appearance of new rings at the centre. Then the field of view becomes more crowded with thinner rings. Conversely, as d is made smaller, the rings appear to contract and to disappear in the centre.

(v) When the mirrors are tilted and are not exactly perpendicular to each other, the fringes will appear as shown in Fig. 4.21.

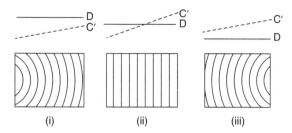

Fig. 4.21 Formation of fringes with inclined mirrors in Michelson interferometer

4.10.2 Applications of Michelson's Interferometer

Michelson's interferometer can be used for the following purposes:

(i) Determination of wavelength of monochromatic light

For this purpose the Michelson's interferometer is set for circular fringes with central bright spot. Then we have the relation

$$2d + \frac{\lambda}{2} = n\lambda \text{ assuming } \cos \alpha = 1$$

If now mirror D is moved $\lambda/2$ away from C, then an additional path difference of λ will be introduced and hence $(n + 1)$th bright spot appears at the centre of the field. Thus, each time mirror D moves through a distance $\lambda/2$, next bright spot appears at the centre of the field. Let n be the number of fringes that cross the centre of the field when the mirror D is moved from initial position X_1 to a final position X_2, then

$$\frac{n\lambda}{2} = X_2 - X_1 \qquad \text{or} \qquad \lambda = \frac{2(X_2 - X_1)}{n}$$

Then difference $(X_2 - X_1)$ can be measured with a micrometer screw and n is actually counted. Thus the evaluation of λ is possible.

(ii) Determination of difference in wavelength

Michelson's interferometer is set for circular fringes. Let the source have two wavelengths λ_1 and λ_2 ($\lambda_1 > \lambda_2$) which are very close to each other. The two wavelengths form their separate fringe patterns, but because of the minute change in wavelength, the two patterns overlap. As the mirror D is moved slowly the two patterns separate slowly and when the path difference is such that dark fringe of λ_1, falls on the bright fringe of λ_2, the result is maximum indistinctness. Now the mirror D is moved say through a distance X so that the next indistinct position is reached. In this position if nth fringe of λ_2 appear at the centre, then $(n + 1)$th fringe of λ_2 should appear at the centre of the field of view,

Hence,
$$X = \frac{n\lambda_1}{2} = (n+1)\frac{\lambda_2}{2}$$

or
$$n = \frac{2X}{\lambda_1} \quad \text{and} \quad (n+1) = \frac{2X}{\lambda_2}$$

Subtracting
$$1 = \frac{2X}{\lambda_2} - \frac{2X}{\lambda_1} = 2X\left\{\frac{\lambda_1 - \lambda_2}{\lambda_1\lambda_2}\right\}$$

or
$$\lambda_1 - \lambda_2 = \frac{\lambda_1\lambda_2}{2X} = \frac{\lambda_{\text{mean}}^2}{2X}$$

Thus, by measuring the distance X moved by mirror D, and the distance between two consecutive indistinctiveness, the difference between two wavelengths can be determined.

(iii) Determination of thickness of a thin plate

Michelson's interferometer is set for parallel fringes. White light is employed. The cross wire is set on the central fringe. Now the thin plate whose thickness has to be measured is introduced in one of the interfering beams. The introduction of the plate of thickness d and refractive index μ increases the path by $2(\mu - 1)d$. Thus, a shift in the fringe system occurs. Now the mirror D is adjusted backward or forward till the central fringe coincides with the cross wire. The distance X moved by mirror D is measured using micrometer screw. Hence, we have

$$2X = 2(\mu - 1)d \quad \text{or} \quad d = \frac{X}{\mu - 1}$$

Using this equation, d can be calculated. (Note that $2X = n\lambda$ where n is the number of fringes that cross the field of view on introducing the thin plate in the path of the beam).

(iv) Standardization of meter

An important measurement made with the Michelson's interferometer was the standardization of meter. A problem with the standard meter was that measurements with it could only be repeated to a few parts in 10^7. Michelson was the first to show that an improvement by an order of magnitude was possible with interferometric measurements using the red cadmium line. After an extensive search for a suitable spectral line, the standard meter was finally abandoned in 1980 and meter was redefined interms of the wavelength of the orange line from a ^{86}Kr discharge lamp. It was defined as the length which was equal to 1,850,763.73 wavelengths of the orange light of ^{86}Kr (λ = 6057.80211 Å). However when frequency stabilised lasers became available, comparisons of their wavelengths with ^{86}Kr standard showed that the accuracy of such measurements was limited to a few parts in 10^9 by uncertainties associated with ^{86}Kr standard. This led to a renewed search for a better definition of the meter. The primary standard of the time is the ^{133}Cs clock. Since the laser frequencies can be compared with the ^{133}Cs clock with in an accuracy of a few parts in 10^{11}, the meter was redefined in 1983 as follows. The meter is the length of the path travelled by light in vacuum during a time interval of 1/2997792458 of a second.

INTERFERENCE

REFERENCES

1. F.A. Jenkins and H.E. White, *Fundamentals of Optics*, McGraw-Hill Book Company, New York, 1985.
2. J.R. Meyer-Arendt, *Introduction to Classical and Modern Optics*, Prentice Hall Pvt. Ltd., New York, 1984.
3. A. Ghatak, *Optics*, Tata McGraw-Hill Publishing Co. Ltd., New Delhi, 1977.
4. R.K. Gaur & S.L. Gupta, *Engineering Physics*, Dhanpat Rai and Sons, 1987.
5. N. Subrahmanyan and Brijlal, *A Text of Optics*, Niraj Prakashan, 1968.

SOLVED EXAMPLES

1. Two narrow and parallel slits 0.08 cm apart are illuminated by light of frequency 8×10^{11} kHz. It is desired to have a fringe width of 6×10^{-4} m. Where should the screen be placed from the slits?

   Solution:

   $d = 0.08$ cm $= 0.08 \times 10^{-2}$ m, $\beta = 6 \times 10^{-4}$ m

   frequency $\nu = 8 \times 10^{11}$ kHz

   i.e., $\lambda = \dfrac{c}{\nu} = \dfrac{3 \times 10^8}{8 \times 10^{11} \times 10^3}$ m, $D = ?$

   From $\beta = \dfrac{\lambda D}{d}$ we have $D = \dfrac{\beta d}{\lambda}$

   $\therefore$ $D = \dfrac{6 \times 10^{-4} \times 0.08 \times 10^{-2} \times 8 \times 10^{14}}{3 \times 10^8} = 1.28$ m.

2. In Young's double slit experiment, a source of light of wavelength 4200 Å is used to obtain interference fringes of width 0.64×10^{-2} m. What should be the wavelength of the light source to obtain fringes 0.46×10^{-2} m wide, if the distance between screen and the slits is reduced to half the initial value?

   Solution:

   In the first case $\lambda = 4200$ Å $= 4200 \times 10^{-10}$ m

   $\beta = 0.64 \times 10^{-2}$ m

   $\therefore$ $0.64 \times 10^{-2} = \dfrac{4200 \times 10^{-10} \times D}{d}$ ...(i)

   In the second case $\beta = 0.46 \times 10^{-2}$ m, $\lambda = ?$

   $0.46 \times 10^{-2} = \dfrac{\lambda \times D/2}{d} = \dfrac{\lambda D}{2d}$ ...(ii)

   Dividing equation (i) by (ii)

   $\dfrac{0.64 \times 10^{-2}}{0.46 \times 10^{-2}} = \dfrac{4200 \times 10^{-10} \times D}{d} \times \dfrac{2d}{\lambda D}$

   $\therefore$ $\lambda = \dfrac{4200 \times 10^{-10} \times 2 \times 0.46}{0.64} = 6037.5$ Å

3. Fresnel's biprism having angle of 1° and a refractive index 1.5 forms fringes on a screen placed 0.8 m from biprism. If the distance between the source and the biprism is 0.02 m, find fringe width when wavelength of light used is 6900 Å.

Solution:

Here $D = 0.8 + 0.02 = 0.82$ m, $B = 0.02$ m

$\lambda = 6900 \times 10^{-10}$ m

$\mu = 1.5, \alpha = 1° = \dfrac{\pi}{180} \times 1 \text{ radian}$

We have $d = 2(\mu - 1)\alpha B = 2(1.5 - 1) \times \dfrac{\pi}{180} \times 1 \times 0.02$

$= \dfrac{2 \times 0.5 \times \pi \times 1 \times 0.02}{180}$

$\beta = \text{Fringe width} = \dfrac{\lambda D}{d} = \dfrac{6900 \times 10^{-10} \times 0.82 \times 180}{2 \times 0.5 \times \pi \times 1 \times 0.02}$ m

$= 1.621 \times 10^{-3}$ m.

4. Fringes are formed by a Fresnel's biprism in the focal plane of a reading microscope which is 100 cm from the slit. A lens inserted between the biprism and the microscope gives two images of the slit in two positions. In one case the two images of the slit are 4.05 mm and in the other, they are 2.90 mm apart. If sodium light ($\lambda = 5893$ Å) is used, find the distance between the interference fringes.

Solution:

Here $D = 100$ cm $= 1$ m

$\lambda = 5893 \times 10^{-10}$ m

$d_1 = 4.05 \times 10^{-3}$ m

$d_2 = 9.90 \times 10^{-3}$ m

∴ $d = \sqrt{d_1 d_2} = \sqrt{4.05 \times 2.90} \times 10^{-3}$ m $= 3.427 \times 10^{-3}$

$\beta = \dfrac{\lambda D}{d} = \dfrac{5893 \times 10^{-10} \times 1}{3.427 \times 10^{-3}} = 1.72 \times 10^{-10}$ m.

5. In a biprism experiment, bandwidth 0.0195 cm are observed at 100 cm from the slit. On introducing a convex lens 30 cm from the slit, the two images of the slit are seen 0.7 cm apart, at 100 cm distance from the slit. Calculate the wavelength of light used.

Solution:

We have

$\dfrac{O}{I} = \dfrac{\text{Size of the object}}{\text{Size of the image}} = \dfrac{u}{v}$

Here $O = ?$ $I = 0.7$ cm $= 0.7 \times 10^{-2}$ m

$u = 30$ cm $= 0.3$ m

$v = 70$ cm $= 0.7$ m

∴ $O = \dfrac{Iu}{v} = \dfrac{0.7 \times 10^{-2} \times 0.3}{0.7} = 3 \times 10^{-3}$ m

∴ $d = 3 \times 10^{-3}$ m

INTERFERENCE 177

Here $\beta = 0.0195 \times 10^{-2}$ m, $D = 100$ cm $= 1$ m

$$\lambda = \frac{\beta d}{D} = \frac{0.0195 \times 10^{-2} \times 3 \times 10^{-3}}{1} = 5850 \times 10^{-10} \text{ m}.$$

6. In Young's double slit experiment, the distance between the slits is 1 mm. The distance between the slit and the screen is 1 meter. The wavelength used in 5893 Å. Compare the intensity at a point distance 1 mm from the centre to that at its centre. Also find the minimum distance from the centre of a point where the intensity is half of that at the centre.

Solution:

Path difference at a point on the screen distance y from the central point

$$= \frac{Y \cdot d}{D}$$

Here $Y = 1$ mm $= 1 \times 10^{-3}$ m

$D = 1$m

$d = 1$ mm $= 1 \times 10^{-3}$ m

∴ Path difference $= \dfrac{1 \times 10^{-3} \times 1 \times 10^{-3}}{1} = 1 \times 10^{-6}$ m $= \Delta$

Phase difference $= \dfrac{2\pi}{\lambda} \Delta = \dfrac{10^{-6} \times 2 \times \pi}{5893 \times 10^{-10}} = 3.394\ \pi$ radian

∴ Ratio of intensity with the central maximum

$= \cos^2 \delta/2 = \cos^2 (1.697\pi) = 0.3372$

When the intensity is half of the maximum, if δ is the phase difference, we have

$$\cos^2 \delta/2 = 0.5 \quad \text{or} \quad \delta/2 = 45° \quad \text{or} \quad \delta = 90° = \pi/2$$

Path difference $= \Delta = \delta \dfrac{\lambda}{2\pi} = \dfrac{\pi}{2} \times \dfrac{\lambda}{2\pi} = \dfrac{\lambda}{4}$

Distance of the point on the screen from the centre $= Y = \Delta \cdot \dfrac{D}{d}$

$$= \frac{\lambda}{4} \times \frac{1}{1 \times 10^{-3}} = \frac{5893 \times 10^{-10}}{4 \times 10^{-3}} = 1.473 \times 10^{-4} \text{ m}.$$

7. In an experiment with a biprism a convex lens is kept in-between the biprism and the eye piece. In two different positions of the lens the distance between the images obtained in the eye piece are 0.42 mm and 1.21 mm. In one position of the eye piece the band width is 0.4 mm; when the eye piece is moved away 60 cm the bandwidth increases to 0.5 mm. Calculate the wavelength of the source used.

Solution:

Distance d between the sources $= \sqrt{d_1 d_2}$

Here $d_1 = 0.42$ mm $= 0.42 \times 10^{-3}$ m

$d_2 = 1.21$ mm $= 1.21 \times 10^{-3}$ m

$d = \sqrt{d_1 d_2} = \sqrt{0.42 \times 1.21} \times 10^{-3}$ m $= 0.7128 \times 10^{-3}$ m

Bandwidth $\beta_1 = \dfrac{D_1 \lambda}{d}$; $\beta_2 = \dfrac{D_2 \lambda}{d}$

$$\beta_2 - \beta_1 = (D_2 - D_1)\dfrac{\lambda}{d} \quad \text{or} \quad \lambda = \dfrac{(\beta_2 - \beta_1)d}{(D_2 - D_1)}$$

Here $D_2 - D_1 = 60$ cm $= 0.60$ m, $\beta_2 - \beta_1 = 0.9$ mm $= 0.9 \times 10^{-3}$ m

$\therefore \quad \lambda = \dfrac{0.7128 \times 10^{-3} \times 0.9 \times 10^{-3}}{0.60} = 5940 \times 10^{-10}$ m.

8. In a biprism experiment, the bandwidth obtained is 0.3 mm. A mica sheet of thickness 5 μm and refractive index 1.45 is kept in the path of one of the interfering beams. Calculate the shift of the central band of the pattern. Given wavelength of light used is 5860 Å. When another mica sheet is kept in the path of one of the beams the central bright band is shifted to the position occupied by 5th dark band earlier. Find the thickness of the mica sheet.

Solution:

Shift X of the central band is given by

$$X = \dfrac{(\mu - 1)tD}{d} = \dfrac{(\mu - 1)t\beta}{\lambda}$$

Here $\mu = 1.45$, $t = 5 \times 10^{-6}$ m, $\beta = 0.3 \times 10^{-3}$ m

$\lambda = 5860 \times 10^{-10}$ m

$$X = \dfrac{0.45 \times 5 \times 10^{-6} \times 0.3 \times 10^{-3}}{5860 \times 10^{-10}} = 1.52 \times 10^{-3} \text{ m}$$

In the next set up we have $\dfrac{X_0}{\beta} = 4.5$

$\therefore$ Thickness $t = \dfrac{X_0 \times \lambda}{\beta(\mu - 1)} = \dfrac{4.5 \times 5880 \times 10^{-10}}{0.45}$

$= 5.860 \times 10^{-6}$ m.

9. The refractive index of a biprism is 1.5. With a monochromatic source of light, the band width obtained is 0.2 mm. The whole set up is immersed in a liquid of refractive index 1.3. What will be the bandwidth now?

Solution:

The distance between the two sources
$$d = 2(\mu - 1)\alpha B$$

On immersing in the liquid of refractive index μ_0

$$d' = 2\left(\dfrac{\mu - \mu_0}{\mu_0}\right)\alpha B$$

$\therefore \quad d' = d\dfrac{(\mu - \mu_0)}{(\mu - 1)\mu_0}$

Bandwidth $\beta = \dfrac{D\lambda}{d}$. On immersing in the liquid

INTERFERENCE

$$\beta' = \frac{D}{d'}\lambda' = \frac{D}{d}\frac{\lambda(\mu-1)\mu_0}{\mu_0(\mu-\mu_0)} = \frac{\beta(\mu-1)}{(\mu-\mu_0)} \qquad \left(\because \lambda' = \frac{\lambda}{\mu_0}\right)$$

∴ Now bandwidth $\beta' = \dfrac{0.2 \times 10^{-3} \times 0.5}{0.2} = 5 \times 10^{-4}$ m.

10. In a biprism experiment the cross wire of the eye piece is set at the position of the 8th dark band when the source used has a wavelength of 6360 Å. When the source is changed the cross wire is found to be in the position (i) 8th bright band (ii) 9th dark band. Calculate the wavelengths of the source in each case.

Solution:

The distance between the centre and the cross wire

$$X = 7.5\beta_1 = 8\beta_2 = 8.5\beta_2$$

i.e., $\quad 7.5\dfrac{D}{d}\lambda_1 = 8\dfrac{D}{d}\lambda_2 = 8.5\dfrac{D}{d}\lambda_2$

∴ Wavelengths $\quad \lambda_2 = \dfrac{7.5}{8}\lambda_1 = \dfrac{15}{16} \times 6360 = 5962.5$ Å

$$\lambda_2 \frac{7.5}{8.5}\lambda_1 = \frac{15}{17} \times 6360 = 5611.8 \text{ Å}.$$

11. In a double slit experiment, fringes are produced using light of wavelength 4800 Å. One slit is covered by a thin plate of glass of refractive index 1.4 and the other slit by another plate of glass of the same thickness but of refractive index 1.7. On doing so the central bright fringe shifts to the position originally occupied by the fifth bright fringe from the centre. Find the thickness of the glass plate.

Solution:

We have $\quad n\lambda = (\mu - \mu')t$

Here $\quad n = 5$

$\quad \mu - \mu' = 0.3$

$\quad \lambda = 4800 \times 10^{-10}$ m

∴ $\quad t = \dfrac{5 \times 4800 \times 10^{-10}}{0.3} = 8.0 \times 10^{-8}$ m.

12. A Fresnel's biprism arrangement is set with sodium light ($\lambda = 5893$ Å) and in the field of view of the eye piece 62 fringes are observed. How many fringes will we get in the same field of view if we replace the source by mercury lamp using green filter ($\lambda = 5461$ Å).

Solution:

Let the length of the field of view be l and the number of fringes be n.

Then $\quad \beta = \dfrac{\lambda D}{d} = \dfrac{l}{n}$

Clearly $\quad n\lambda = \dfrac{ld}{D} =$ Constant

Now $\quad n_1\lambda_1 = n_2\lambda_2$

Here $\quad n_1 = 62, \lambda_1 = 5893$ Å, $\lambda_2 = 5461$ Å, $n_2 = ?$

$$62 \times 5893 = n_2 \times 5461 \quad \text{or} \quad n_2 = \frac{62 \times 5893}{5461} = 67.$$

13. A drop of oil of volume 0.2 cc is dropped on a surface of tank of water of area 1 m². The film spreads uniformly over the whole surface. White light which is incident normally is observed through a spectrometer. The spectrum is seen to contain one dark band whose centre has a wavelength 5.5×10^{-5} cm in air. Find the refractive index of oil.

Solution:

The thickness of the film = $d = \dfrac{0.2}{100 \times 100} = 2 \times 10^{-5}$ cm

The film appears dark by reflected light for a wavelength λ given by the relation

$$2\mu d \cos r = n\lambda$$

For normal incidence $r = 0$, $\cos r = 1$

Further $n = 1$ and $\lambda = 5.5 \times 10^{-5}$ cm

$$\mu = \frac{n\lambda}{2t \cos r} = \frac{1 \times 5.5 \times 10^{-5}}{2 \times 2 \times 10^{-5} \times 1} = 1.375.$$

14. A soap film 5×10^{-5} cm thick is viewed at an angle of 35° to the normal. Find the wavelengths of light in the visible spectrum which will be absent from the reflected light ($\mu = 1.33$).

Solution:

Let i be the angle of incidence and r the angle of refraction.

Then $\quad \mu = \dfrac{\sin i}{\sin r}; \quad 1.33 = \dfrac{\sin 35°}{\sin r}$

$\Rightarrow \quad r = 25.55° \quad \Rightarrow \quad \cos r = 0.90$

Applying the relation, $2\mu d \cos r = n\lambda$

where $\quad d = 5 \times 10^{-5}$ cm

(i) For the first order $n = 1$

$$\lambda_1 = 2 \times 1.33 \times 5 \times 10^{-5} \times 0.90 = 12.0 \times 10^{-5} \text{ cm}$$

which lies in the infrared (invisible) region.

(ii) For the second order $n = 2$

$$2\lambda_2 = 2 \times 1.33 \times 5 \times 10^{-5} \times 0.90 = 6.0 \times 10^{-5} \text{ cm}$$

which lies in the visible region.

(iii) Similarly, taking $n = 3$, $\lambda_3 = 4.0 \times 10^{-5}$ cm which also lies in the visible region.

(iv) If $n = 4$, $\lambda_4 = 3.0 \times 10^{-5}$ cm

which lies in the ultraviolet (invisible region). Hence, absent wavelengths in the reflected light are 6.0×10^{-5} and 4.0×10^{-5} cm.

15. Two glass plates enclose a wedge shaped air film, touching at one edge and separated by a wire of 0.05 mm diameter at a distance 15 cm from that edge. Calculate the fringe width. Monochromatic light of $\lambda = 6000$ Å from a broad source falls normally on the film.

INTERFERENCE

Solution:

Fringe width $\beta = \dfrac{\lambda}{2\alpha}$

Clearly $\alpha = \dfrac{0.05 \text{ mm}}{15 \text{ cm}} = \dfrac{0.005}{15}$ radian

$\beta = \dfrac{\lambda}{2\alpha} = \dfrac{6000 \times 10^{-9} \times 15}{2 \times 0.005} = 0.09$ cm.

16. An air wedge of angle 0.01 radian is illuminated by monochromatic light of 6000 Å falling normally on it. At what distance from the edge of the wedge, will the 10th fringe be observed by reflected light.

Solution:

Here $\alpha = 0.01$ radian, $n = 10$

$\lambda = 6000 \times 10^{-10}$ m

$2d = \dfrac{(2n-1)\lambda}{2}$

where d is the thickness of wedge.

But $\alpha = \dfrac{d}{x}$

$d = \alpha x$

$\therefore \quad 2\alpha x = \dfrac{(2n-1)\lambda}{2}$

Here $n = 10$

$x = \dfrac{(2n-1)\lambda}{4\alpha} = \dfrac{19 \times 6000 \times 10^{-10}}{4 \times 0.01} = 2.85 \times 10^{-4}$ m.

17. A thin equiconvex lens of focal length 4 meters and refractive index 1.50 rests on and is in contact with an optical flat. Using light of wavelength 5460 Å, Newton's rings are viewed normally by reflection. What is the diameter of the 5th bright ring?

Solution:

The diameter of the nth bright ring is given by

$D_n = \sqrt{2(2n-1)\lambda R}$

Here $n = 5$, $\lambda = 5460 \times 10^{-6}$ cm

$f = 400$ cm, $\mu = 1.50$

We have

$\dfrac{1}{f} = (\mu - 1)\left(\dfrac{1}{R_1} + \dfrac{1}{R_2}\right)$

Here $R_1 = R_2 = R$

$\therefore \quad \dfrac{1}{f} = (\mu - 1)\dfrac{2}{R}$

i.e., $\dfrac{1}{400} = (1.50 - 1)\dfrac{2}{R} \quad \Rightarrow \quad R = 400$ cm

$\therefore \quad D_n = \sqrt{2 \times (2 \times 5 - 1) \times 5460 \times 10^{-6} \times 400} = 0.627$ cm.

18. In Newton's ring experiment, the diameters of the 4th and 12th dark rings are 0.400 cm and 0.700 cm respectively. Find the diameter of the 20th dark ring.

Solution:

We have $\quad D_{n+p}^2 - D_n^2 = 4p\lambda R$

Here $\quad (n+p) = 12, n = 4, p = 12 - 4 = 8$

$\therefore \quad D_{12}^2 - D_4^2 = 4 \times 3 \times \lambda R \qquad (i)$

$D_{20}^2 - D_4^2 = 4 \times 16 \times \lambda R \qquad (ii)$

Dividing (ii) by (i)

$$\frac{D_{20}^2 - D_4^2}{D_{12}^2 - D_4^2} = \frac{4 \times 16 \times \lambda R}{4 \times 8 \times \lambda R} = 2$$

$$\frac{D_{20}^2 - (0.4)^2}{(0.7)^2 - (0.4)^2} = 2 \quad \Rightarrow \quad D_{20} = 0.906 \text{ cm}.$$

19. In a Newton's ring experiment the diameter of the 10th ring changes from 1.40 to 1.27 cm when a liquid is introduced between the lens and the plate. Calculate the refractive index of the liquid.

Solution:

When the liquid is used the diameter of the 10th ring is given by

$$(D'_{10})^2 = \frac{4 \times 10 \times \lambda R}{\mu} \qquad (i)$$

For air medium

$$(D_{10})^2 = 4 \times 10 \times \lambda R \qquad (ii)$$

Dividing (i) by (ii)

$$\mu = \frac{D_{10}^2}{D'^2_{10}} = 1.215.$$

20. In a Newton's ring experiment the diameter of the 5th dark ring was 0.3 cm and the diameter of the 25th ring was 0.8 cm. If the radius of curvature of the plano-convex lens is 100 cm, find the wavelength of the light used.

Solution:

$$\lambda = \frac{D_{n+p}^2 - D_n^2}{4pR}$$

Here $\quad D_{25} = 0.8 \text{ cm}, D_5 = 0.3 \text{ cm}$

$p = 25 - 5 = 20 \quad$ and $\quad R = 100 \text{ cm}$

$\therefore \quad \lambda = \dfrac{(0.8)^2 - (0.3)^2}{4 \times 20 \times 100} = 4.87 \times 10^{-5}$ cm.

21. In a Michelson's interferometer 200 fringes cross the field of view when the movable mirror is displaced through 0.05896 mm. Calculate the wavelength of the monochromatic light used.

INTERFERENCE

Solution:

In Michelson's interferometer
$$\frac{n\lambda}{2} = X_2 - X_1$$

Here $n = 200$, $X_2 - X_1 = 0.05896$ mm
$$= 0.05896 \times 10^{-3} \text{ m}.$$

$\therefore \quad \lambda = \dfrac{2 \times 0.05896 \times 10^{-3}}{200} = 5896 \times 10^{-10}$ m.

22. In an experiment with Michelson's interferometer, scale readings for a pair of maximum indistinctness were found to be 0.6939 mm and 0.9884 mm. If the mean wavelength of the two components of the D line is 5893 Å, deduce the difference between wavelengths.

Solution:

In a Michelson's interferometer we have
$$\lambda_1 - \lambda_2 = \frac{\lambda_{\text{ave}}^2}{2X}$$

Here $\lambda_{\text{ave}} = 5893 \times 10^{-10}$ m
$$X = 0.9884 - 0.6939 \text{ mm}$$
$$= 0.2945 \text{ mm} = 0.2945 \times 10^{-3} \text{ m}$$

$\therefore \quad \lambda_1 - \lambda_2 = \dfrac{(5893 \times 10^{-10})^2}{2 \times 0.2945 \times 10^{-3}} = 5.895 \times 10^{-10}$ m.

23. When a thin plate of glass of refractive index 1.5 is interposed in the path of one of the interfering beams of Michelson's interferometer, a shift of 30 fringes of sodium light is observed across the field of view. If the thickness of the plate is 0.018 mm, calculate the wavelength of light used.

Solution:

Here $\mu = 1.5$, $n = 30$, $d = 0.018$ mm, $\lambda = ?$

Now we have
$$2(\mu - 1)d = n\lambda$$
$$\lambda = \frac{2(\mu - 1)d}{n} = \frac{2 \times 0.5 \times 0.018}{30} = 6000 \times 10^{-10} \text{ m}.$$

24. An air cell of thickness 3 cm is introduced in one path of a Michelson's interferometer and fringes are obtained with mercury green light ($\lambda = 5.46 \times 10^{-5}$ cm). Calculate the number of fringes that pass the field of view as the pressure in the cell is changed from 76 cm to 16.4 cm. (μ of air at 76 cm pressure is 1.000293 and ($\mu - 1$) is proportional to pressure).

Solution:

($\mu - 1$) at 16.4 cm pressure
$$= 0.000293 \times \frac{16.4}{76} = 0.000063$$

$\therefore$ $\delta\mu$ in changing the pressure from 76 to 16.4 cm = 0.000230

Change in one way path $= t \cdot \delta\mu$
$$= 3 \times 0.000230$$
$$= 0.000690 \text{ cm}$$

No. of fringes passing = $\dfrac{2 \times 0.000690}{5.46 \times 10^{-5}} = 25.3$

Note: The converse of this method is used to measure the refractive index of gases.

QUESTIONS

1. What is interference of light waves? What are the conditions necessary for obtaining interference fringes?
2. Two independent non-coherent sources of light cannot produce an interference pattern. Why?
3. Define spatial and temporal coherence.
4. Describe Young's double slit experiment and obtain an expression for fringe width.
5. Explain how interference fringes are obtained using a biprism.
6. Describe in detail how the wavelength of monochromatic light can be found using Fresnel's biprism.
7. How do you determine the thickness of a thin sheet of transparent material using Fresnel's biprism?
8. Write a note on colours of thin films.
9. Show that colours exhibited by reflected and transmitted systems are complementary.
10. Find an expression for the width of the fringes obtained in the case of air wedge. How would you use the result to find the wavelength of a given monochromatic radiation?
11. What are Newton's rings? How are they formed? Why are they circular?
12. Explain why the centre of Newton's rings is dark in the reflected system.
13. Describe how you would use Newton's rings to determine the wavelength of a monochromatic radiation and derive the relevant formula.
14. Obtain an expression for the radius of the nth dark ring in the case of Newton's rings.
15. Show that the radii of Newton's rings are in the ratio of the square roots of the natural numbers.
16. Describe the construction and action of Michelson's interferometer.
17. How would you use Michelson's interferometer to find the wavelength of a given monochromatic radiation?
18. A given light consists of two close wavelengths. How would you determine the difference between these wavelengths using Michelson's interferometer?
19. With Michelson's interferometer under what circumstances will you get (a) circular fringes (b) straight line fringes? Explain.

PROBLEMS

1. The distance between the slit and the biprism and between the biprism and the screen are 50 cm each. The angle of the prism is 179° and its refractive index is 1.5. If the distance between successive fringes is 0.0135 cm, calculate the wavelength of light used. (**Ans.** 5893 Å)
2. Interference fringes are formed on a screen which is at a distance of 0.8 m. It is found that the fourth bright fringe is situated at a distance of 0.00108 m from the central fringe. Calculate the distance between the two coherent sources, (Given $\lambda = 5896$ Å). (**Ans.** 1.75×10^{-19} m)
3. In a biprism experiment, the eye piece is placed at a distance of 0.8 m from the source. The distance between the virtual sources is found to be 0.0005 m. If the fringe width is 0.000943 m, calculate the wavelength of the source. (**Ans.** 5893.8×10^{-10} m)
4. In an experiment with Fresnel's biprism sodium light is used and interference bands 0.0235×10^{-2} m in width, are observed at a distance of 1 m from the slit. A convex lens is then put between the observer and the biprism, so as to give an image at a distance 1 m from the slit. The distance between the images on the screen is found to be 0.75×10^{-2} m when the lens is 0.25 m from the slit. Calculate the wavelength of light used. (**Ans.** 5875×10^{-10} m)
5. A biprism of angle 1° and refractive index 1.5 forms interference fringes on a screen 1 m from the biprism. The distance between the slit and the biprism is 0.1 m. Calculate the fringe width when the slit is illuminated with a light of wavelength 6000×10^{-10} m. (**Ans.** 3.438×10^{-4} m)

INTERFERENCE

6. A parallel beam of light ($\lambda = 5890 \times 10^{-10}$ m) is incident on a thin glass plate ($\mu = 1.5$) such that the angle of refraction into the plate is 60°. Calculate the smallest thickness of plate which would appear dark by reflection. **(Ans. 3.926×10^{-7} m)**

7. White light falls normally on a film of soapy water whose thickness is 5×10^{-5} cm and $\mu = 1.33$. Which wavelength in the visible region will be reflected most strongly? **(Ans. 5320×10^{-10} m)**

8. White light is incident on a soap film at an angle of $\sin^{-1} 4/5$ and the reflected light on examination by a spectroscope shows dark bands. Two consecutive bands correspond to wavelength 6.1×10^{-5} and 6.0×10^{-5} cm. If $\mu = 4/3$, calculate its thickness. **(Ans. 1.7×10^{-5} m)**

9. In an experiment with Fresnel's biprism fringes for light of wavelength 5×10^{-5} cm are observed 0.2 mm apart at a distance of 175 cm from the prism. The prism is made of glass of refractive index 1.50 and is at a distance of 25 cm from the illuminated slit. Calculate the angle at the vertex of the biprism. **(Ans. 179° 42′)**

10. If the angle of the air wedge is 0.25° and the wavelengths of sodium lines are 5896 and 5890 Å, find the distance from the apex at which the maximum due to two wavelengths first coincide when observed in reflected light. **(Ans. 6.63 cm)**

11. A monochromatic light of wavelength 5893×10^{-10} m falls normally on an air wedge. If the length of the wedge is 0.05 m, calculate the distance at which the 12th dark and 12th bright fringes will form the line of contact of the glass plates forming the wedge. (Given the thickness of the specimen = 154×10^{-6} m). **(Ans. 9.61×10^{-4} m, 9.21×10^{-4} m)**

12. A square piece of cellophane film with refractive index 1.5 has a wedge shaped section so that its thickness at two opposite sides is t_1 and t_2. If with a light of $\lambda = 6000$ Å, the number of fringes appearing in the film is 10, calculate the difference $t_2 - t_1$. **(Ans. 2×10^{-4} cm)**

13. A Newton's ring arrangement is used with a source emitting two wavelengths $\lambda_1 = 6 \times 10^{-5}$ m and $\lambda_2 = 4.5 \times 10^{-5}$ m. It is found that nth dark ring due to λ_1 coincides with $(n + 1)$th dark ring for λ_2. If radius of curvature of the lens is 90 cm find the diameter of the nth dark ring. **(Ans. 0.254 cm)**

14. Light containing two wavelengths λ_1 and λ_2 falls normally on a planoconvex lens of radius of curvature R resting on a glass plate. If the nth dark ring due to λ_1, coincides with $(n = 1)$th dark ring due to λ_2, prove that the radius of the nth dark ring of λ_1 is $\sqrt{\dfrac{\lambda_1 \lambda_2 R}{(\lambda_1 - \lambda_2)}}$.

15. Newton's rings formed by sodium light between a flat glass plate and a convex lens are viewed normally. What will be the order of the dark ring which will have double the diameter of 40th ring? **(Ans. 160)**

16. The movable mirror of Michelson's interferometer is moved through a distance of 0.2603 mm. Find the number of fringes shifted across the cross wire of eye piece of the telescope if wavelength of 5206 Å is used. **(Ans. 100)**

17. Calculate the distance between two successive positions of a movable mirror of a Michelson's interferometer giving best fringes in the case of sodium light having lines of wavelengths 5890 Å and 5896 Å. **(Ans. 0.02894 cm)**

18. A thin glass of refractive index 1.5 is introduced normally in the path of one of the interfering beams of Michelson's interferometer which is illuminated with light of wavelength 4800 Å. This causes 500 fringes to sweep across the field. Determine the thickness of the glass. **(Ans. 0.024 cm)**

19. A shift of 100 circular fringes is observed when the movable mirror of the Michelson's interferometer is shifted by 0.0295 mm. Calculate the wavelength of light. **(Ans. 5900 Å)**

20. On introducing a thin glass of refractive index 1.5 and thickness 0.00345 cm in the path of one of the interfering beams S of Michelson's interferometer, a certain number of fringes passes across the field of view. If the wavelength of light used is 6900 Å, calculate the number of fringes that pass across the field of view. **(Ans. 50)**

5
Diffraction

5.1 INTRODUCTION

Let us consider a plane wave incident on a long narrow slit of width 'b' (Fig. 5.1). According to geometrical optics one expects the region AB in the screen SS' to be illuminated and the remaining portion (known as the geometrical shadow) to be absolutely dark. However, if observations are made carefully one finds that if the width of the slit is not very large compared to wavelength, then the light intensity in the region AB is not uniform and there is also some intensity inside the geometrical shadow. Further if the width of the slit is made smaller, larger amount of energy reaches the

Fig. 5.1 If a plane wave is incident on an aperture then according to geometrical optics a sharp shadow will be cast in the region AB of the screen

geometrical shadow. *This spreading out of wave when it passes through a narrow opening is known as the phenomenon of diffraction.* The intensity distribution on the screen is known as the diffraction pattern. In this chapter we will investigate quantitatively the diffraction of light based on wave theory of light. The diffraction phenomena are usually divided into two classes, (a) Fresnel diffraction (b) Fraunhofer diffraction.

In the Fresnel class of diffraction the source of light and the screen are in general at a finite distance from the diffracting aperture. In the Fraunhofer class of diffraction, the source and the screen are at infinite distance from the aperture. This is easily achieved by placing the source on the focal plane of the convex lens and placing the screen on the focal plane of another convex lens. The two lenses effectively move the source and the screen infinity because the first lens makes the light beam parallel and the second lens makes the screen receive a parallel beam of light. The two types of diffraction are shown in Fig. 5.2.

Fresnel diffraction is more general and it includes Fraunhofer diffraction as a special case. But Fraunhofer diffraction is much simpler and in this chapter we will discuss only the Fraunhofer diffraction.

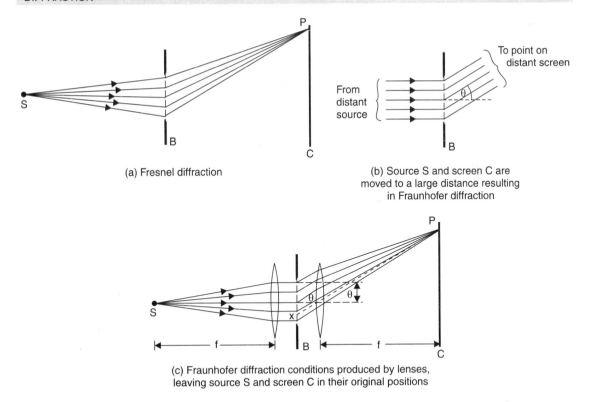

Fig. 5.2

The correct interpretation of diffraction phenomena was provided by Fresnel. According to him, the diffraction phenomenon is due to mutual interference of secondary wavelets originating from various points of the wavefront which are not blocked off by the obstacle. Thus by considering the interference of secondary wavelets one can calculate the diffraction pattern. It is important to understand the distinction between the phenomenon of interference and diffraction. In the phenomenon of interference, interaction takes place between two separate wavefronts originating from two coherent sources while in the phenomenon of diffraction interaction takes place between the secondary wavelets originating from different points of the same wavefront. In the interference pattern the regions of minimum intensity are usually almost perfectly dark while this is not the case in the diffraction pattern. Further in the interference pattern all the maxima are of same intensity but in diffraction pattern they are of varying intensity. The fringe widths in a diffraction pattern are never equal while in the case of interference the fringe widths could be equal in some cases.

5.2 FRAUNHOFER DIFFRACTION AT A SINGLE SLIT

A slit is a rectangular aperture whose length is large compared to its breadth. In Fig. 5.3 AB is a slit of width a and its length is perpendicular to the plane of the paper. Source point S and lens L_1 render the wavefronts parallel which reach the slit. Lens L_2 produces an image S' which should be a point image of S according to geometrical optics. However, diffraction occurs and the explanation for it lies in the interference of the Huygen's secondary wavelets. These wavelets can be thought of as being sent out from every point on the wavefront at the instant

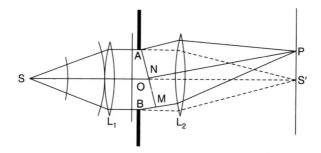

Fig. 5.3 Diffraction at a slit

when it occupies the plane of the slit. Let us consider secondary wavelets from AB moving in a general direction like OP inclined at an angle θ from the direct direction OS'. The addition of the secondary wavelets reaching P from different parts of the slit AB can be done by integration. Let us do this calculation with reference to Fig. 5.4.

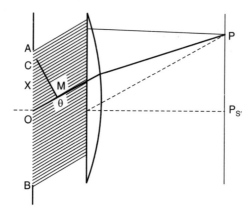

Fig. 5.4 Diffraction at a slit

Let the disturbance caused at P by the wavelet from unit width of the slit at O be

$$y_0 = A \cos \omega t$$

Then the wavelet from width dx at C when it reaches P has the amplitude Adx and phase $\omega t + \dfrac{2\pi}{\lambda} x \sin \theta$. Calling this disturbance dy we have

$$dy = Adx \cos \left\{ \omega t + \frac{2\pi x \sin \theta}{\lambda} \right\}$$

For the total disturbance at the point of observation at an angle θ, we get

$$y = \int_{-a/2}^{+a/2} dy = \int_{-a/2}^{+a/2} A \cos \left\{ \omega t + \frac{2\pi x \sin \theta}{\lambda} \right\} dx$$

$$= A \cos \omega t \int_{-a/2}^{+a/2} \cos \frac{2\pi x \sin \theta}{\lambda} dx - A \sin \omega t \int_{-a/2}^{+a/2} \sin \frac{2\pi x \sin \theta}{\lambda} dx$$

DIFFRACTION

$$= \left\{ A \frac{\sin\left[\frac{\pi a \sin\theta}{\lambda}\right]}{\frac{\pi a \sin\theta}{\lambda}} \right\} \cos\omega t \quad (5.1)$$

$$= A_\theta \cos\omega t$$

The quantity in brackets gives the amplitude A_θ of the resultant disturbance. For $\theta = 0$ it becomes Aa which we call A_0. Thus the result is

$$y = Aa \left\{ \frac{\sin\left[\frac{\pi a \sin\theta}{\lambda}\right]}{\frac{\pi a \sin\theta}{\lambda}} \right\} \cos\omega t$$

$$y = A_0 \left\{ \frac{\sin\left[\frac{\pi a \sin\theta}{\lambda}\right]}{\frac{\pi a \sin\theta}{\lambda}} \right\} \cos\omega t$$

where $A_0 = Aa$ is the amplitude for $\theta = 0$

Let $\quad \frac{\pi a \sin\theta}{\lambda} = \beta$, then

$$y = A_\theta \cdot \cos\omega t = A_0 \cdot \frac{\sin\beta}{\beta} \cdot \cos\omega t \quad (5.2)$$

The intensity distribution is given by

$$I = I_0 \frac{\sin^2\beta}{\beta^2} \quad (5.3)$$

where I_0 represents the intensity at $\theta = 0$.

5.2.1 The Positions of Maxima and Minima

The variation of intensity with β is shown in Fig. 5.5. It is obvious that intensity is zero when $\beta = m\pi, m \neq 0$, when $\beta = 0$, $\frac{\sin\beta}{\beta} = 1$ and $I = I_0$ which corresponds to the maximum of intensity. Substituting the value for β one obtains $\frac{\pi a \sin\theta}{\lambda} = m\pi$.

or $\quad a \sin\theta = m\lambda; \; m = \pm 1, \pm 2, \pm 3, \pm 4, \ldots\ldots$(minima) $\quad (5.4)$

as the condition for minima.

The first minimum occurs at $\theta = \pm \sin^{-1}\left(\frac{\lambda}{a}\right)$; the second minimum at $\theta = \pm \sin^{-1}\left(\frac{2\lambda}{a}\right)$.

Since $\sin\theta$ cannot exceed unity, the maximum value of m is the integer which is less than $\frac{a}{\lambda}$.

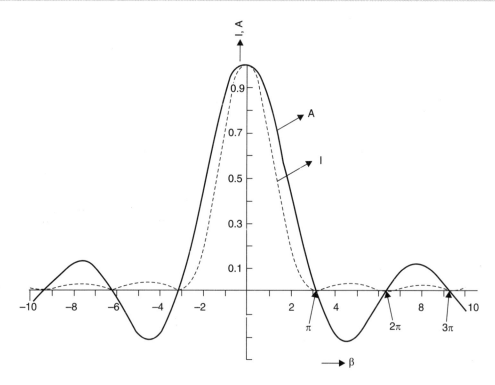

Fig. 5.5 Amplitude distribution (solid line) and intensity distribution (dashed line) in Fraunhofer diffraction by a single slit

In order to determine the positions of maxima we differentiate equation (5.3) with respect to β and set it equal to zero. Thus

or
$$\frac{dI}{d\beta} = I_0 \left\{ \frac{2\sin\beta \cdot \cos\beta}{\beta^2} - \frac{2\sin^2\beta}{\beta^3} \right\} = I_0 \frac{\sin 2\beta}{\beta^3} \{\beta - \tan\beta\} = 0 \qquad (5.5)$$

The condition $\beta = 0$ corresponds to maxima. The conditions for other maxima are roots of the following transcendental equation

$$\tan\beta = \beta \text{ (maxima)}$$

The root $\beta = 0$, corresponds to central maxima. The other roots can be found by determining the points of intersection of the curve $y = \beta$ and $y = \tan\beta$ (Fig. 5.6). The intersection occurs at $\beta = 1.43\pi$, $\beta = 2.46\pi$ etc., and are known as the first maximum, the second maximum etc.

Since $\left\{ \dfrac{\sin(1.43\pi)}{1.43\pi} \right\}^2$ is about 0.0496, the intensity of the first maximum is about 4.96% of the central maximum. Similarly the intensity of the second and third maxima are about 1.68% and 0.83% of the central maxima respectively.

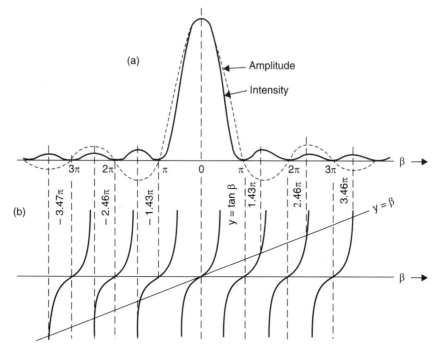

Fig. 5.6 Amplitude and intensity contours for Fraunhofer diffraction of a single slit showing positions of maxima and minima

5.3 FRAUNHOFER DIFFRACTION DUE TO TWO PARALLEL SLITS

In Fig. 5.7, let A_1B_1, A_2B_2 be the two slits of width a and separated by distance e. Lengths of the slits are perpendicular to the plane of the paper. Consider a monochromatic light of wavelength λ falling normally on the slits from the left. All the wavelets proceeding in a direction θ from the normal are collected by a lens at a given point in its focal plane. The problem is to find the resultant intensity and its variation with θ.

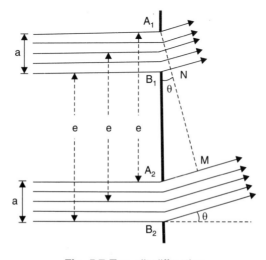

Fig. 5.7 Two-slit diffraction

We have here interference of two waves of amplitude A_θ each, having a phase difference $\Delta = \dfrac{2\pi}{\lambda} \cdot e \sin \theta$. The resultant amplitude R is therefore given by

$$R_\theta^2 = A_\theta^2 + A_\theta^2 + 2A_\theta A_\theta \cos \Delta = 2A_\theta^2 \{1 + \cos \Delta\} = 4A_\theta^2 \cos^2 \Delta/2 = 4I_\theta \cos^2 \dfrac{\Delta}{2}$$

I_θ is the result of diffraction at each slit and it is given by equation 5.3. Substituting for I_θ we get for intensity

$$I_\theta = 4I_0 \dfrac{\sin^2 \beta}{\beta^2} \cos^2 \dfrac{\Delta}{2} \qquad (5.6)$$

Equation (5.5) tells us that intensity is zero whenever
$$\beta = \pi, 2\pi, 3\pi, \ldots$$
or when
$$\Delta = \pi, 3\pi, 5\pi, \ldots$$

The corresponding angles of diffraction are given by
$$a \sin \theta = m\lambda; \ (m = 1, 2, 3, \ldots)$$
$$e \sin \theta = \left(n + \dfrac{1}{2}\right)\lambda; \ (n = 0, 1, 2, 3, \ldots)$$

The interference maxima occurs when
$$\Delta = 0, 2\pi, 4\pi, \ldots$$
or when
$$e \sin \theta = 0, \lambda, 2\lambda, 3\lambda, \ldots$$

The interference maxima as a function of β remains the same as in the case of single slit diffraction.

The double-slit intensity distribution is shown in Fig. 5.8. Figure 5.8a denotes the interference term. Figure 5.8b denotes the diffraction term while Fig. 5.8c represents the resultant distribution.

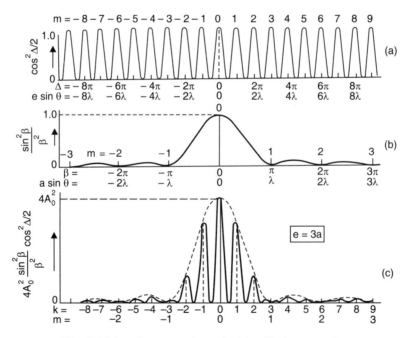

Fig. 5.8 Intensity curves for a double slit where $e = 3a$

DIFFRACTION

5.4 FRAUNHOFER DIFFRACTION DUE TO N PARALLEL SLITS

In Fig. 5.9 a number (N) of equidistant and equally wide slits are represented, the extreme members being A and B. The slit lengths are perpendicular to the paper. Widths are all equal to a and the separation between slits is equal to e. The arrangement is illuminated by a monochromatic beam of light of wavelength λ falling normally from the left. At angle θ on the right we now have N waves, each of amplitude $A_\theta = A_0 \dfrac{\sin \beta}{\beta}$ and having successive phase differences $\Delta = \dfrac{2\pi}{\lambda} e \sin \theta$. The resultant disturbance y is given by

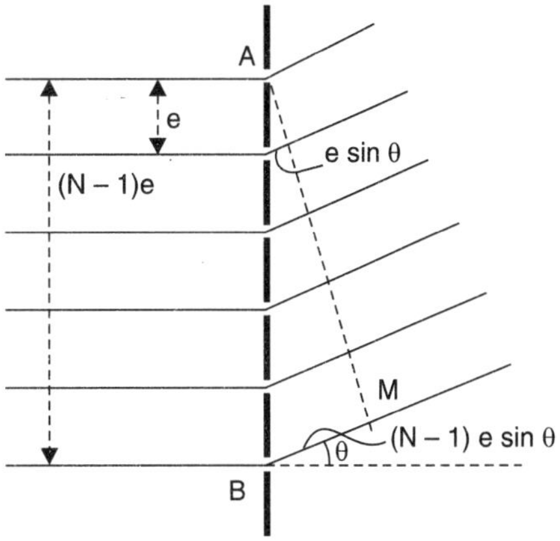

$$y = A_\theta \{\cos \omega t + \cos(\omega t + \Delta) + \cos(\omega t + 2\Delta) + \ldots\ldots N \text{ terms}\}$$

$$= \text{Real part of } A_\theta\, e^{i\omega t} \{1 + e^{i\Delta} + e^{2i\Delta} + \ldots\ldots N \text{ terms}\}$$

$$= A_\theta\, e^{i\omega t} \left\{ \frac{1 - e^{iN\Delta}}{1 - e^{i\Delta}} \right\} \tag{5.7}$$

Fig. 5.9 N-parallel slits

To find the intensity, we multiply the amplitude with its complex conjugate

$$A^2 = I = A_\theta^2 \left\{ \frac{(1 - e^{iN\Delta})(1 - e^{-iN\Delta})}{(1 - e^{i\Delta})(1 - e^{-i\Delta})} \right\}$$

$$= A_\theta^2 \left\{ \frac{1 - \cos N\Delta}{1 - \cos \Delta} \right\} = A_\theta^2 \frac{\sin^2 \dfrac{N}{2}\Delta}{\sin^2 \dfrac{\Delta}{2}}$$

$$= A_\theta^2 \frac{\sin^2 \left[\dfrac{N\pi e \sin \theta}{\lambda} \right]}{\sin^2 \left[\dfrac{\pi e \sin \theta}{\lambda} \right]}$$

$$I = I_0 \frac{\sin^2 \beta}{\beta^2} \frac{\sin^2 \left[\dfrac{N\pi e \sin \theta}{\lambda} \right]}{\sin^2 \left[\dfrac{\pi e \sin \theta}{\lambda} \right]}$$

$$I = I_0 \frac{\sin^2 \beta}{\beta^2} \frac{\sin^2 N\alpha}{\sin^2 \alpha} \quad \text{where } \alpha = \frac{\pi e \sin \theta}{\lambda} \tag{5.8}$$

As can be seen, the intensity distribution is a product of two terms. Ths first term $\dfrac{\sin^2 \beta}{\beta^2}$ represents the diffraction produced by single slit and the second term $\dfrac{\sin^2 N\alpha}{\sin^2 \alpha}$ represents the interference patterns produced by N equally spaced slits. For $N = 1$, the equation (5.8) reduces to the single slit diffraction pattern and for $N = 2$, the double slit diffraction pattern. Now let us analyse equation (5.8) in more detail.

5.4.1 Principal Maxima

These intensities as given by equation (5.8) would be maximum when $\sin \alpha = 0$ or when $\alpha = \pm k\pi$ where $k = 0, 1, 2, 3, \ldots$. Please note that when $\alpha = k\pi$, $\dfrac{\sin N\alpha}{\sin \alpha}$ becomes indeterminant. Hence it has to be evaluated by applying L' Hospital's rule *i.e.*,

$$\lim_{\alpha \to \pm k\pi} \frac{\sin N\alpha}{\sin \alpha} = \lim_{\alpha \to \pm k\pi} \frac{\frac{d}{d\alpha}(\sin N\alpha)}{\frac{d}{d\alpha}(\sin \alpha)} = \lim_{\alpha \to \pm k\pi} \frac{N \cos N\alpha}{\cos \alpha} = \pm N$$

Hence,
$$\lim_{\alpha \to \pm k\pi} \left\{ \frac{\sin N\alpha}{\sin \alpha} \right\}^2 = N^2$$

Thus the resultant intensity when $\alpha = \pm k\pi$ is $I_0 \dfrac{\sin^2 \beta}{\beta^2} N^2$. The maxima are more intense and are called as principal maxima.

Note that when

$$\alpha = \pm k\pi, \quad \frac{\pi e \sin \theta}{\lambda} = \pm k\pi \quad \text{or} \quad e \sin \theta = \pm k\lambda \qquad (5.9)$$

This is known as the grating equation. $k = 0$, corresponds to the zero order maximum for $k = 1, 2, 3$ etc. We obtain first, second and third order maxima respectively. The $\pm$ sign shows that the principal maxima lie on either side of zero order maximum.

5.4.2 Minima and Secondary Maxima

From equation (5.8) we see that I is zero when $\beta = n\pi$ or $\dfrac{\pi a \sin \theta}{\lambda} = n\pi$ or

$$a \sin \theta = n\lambda \quad \text{where } n = 1, 2, 3, \ldots \qquad (5.10)$$

I is also 0 when

$$N\alpha = \pm r\pi \quad \text{where } r \approx 0, N, 2N, 3N, \ldots \qquad (5.11)$$

Equation (5.10) gives the minima corresponding to single slit diffraction pattern. The angles of diffraction corresponding to (5.11) are $e \sin \theta = \dfrac{\lambda}{N}$,

$$\frac{2\lambda}{N} \ldots \frac{(N-1)\lambda}{N}, \frac{(N+1)\lambda}{N}, \frac{(N+2)\lambda}{N}, \ldots \frac{(2N-1)\lambda}{N}, \frac{(2N+1)\lambda}{N}, \frac{(2N+2)\lambda}{N} \ldots$$

(Please note that when $r = 0, N, 2N, 3N$ etc., it obeys the condition for principal maxima.) Thus between two principal maxima we have at least $(N-1)$ minima. Between two consecutive minima, the intensity has to have a maxima. These maxima are known as secondary maxima. There are $(N-2)$ such secondary maxima. To find out the position of these secondary maxima we differentiate equation (5.8) with respect to α and equate it to zero.

$$\frac{dI}{d\alpha} = \frac{I_0 \sin^2 \beta}{\beta^2} \frac{2 \sin N\alpha}{\sin \alpha} \times \left\{ \frac{N \cos N\alpha \cdot \sin \alpha - \sin N\alpha \cdot \cos \alpha}{\sin^2 \alpha} \right\}$$

or $\qquad N \cos N\alpha \cdot \sin \alpha - \sin N\alpha \cdot \cos \alpha = 0$

or $\qquad N \tan \alpha = \tan N\alpha \qquad$ (5.12)

The roots of this equation other than those for which $\alpha = \pm k\pi$ (which correspond to principal maxima) give the position of secondary maxima. To find out the value of $\dfrac{\sin^2 N\alpha}{\sin^2 \alpha}$ from equation $N \tan \alpha = \tan N\alpha$ we make use of the triangle as shown in Fig. 5.10.

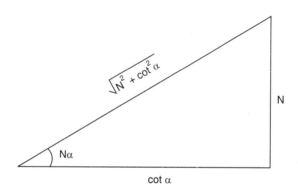

Fig. 5.10

$$\sin N\alpha = \frac{N}{\sqrt{N^2 + \cot^2 \alpha}}$$

$\therefore \qquad \dfrac{\sin^2 N\alpha}{\sin^2 \alpha} = \dfrac{N^2}{(N^2 + \cot^2 \alpha)\sin^2 \alpha} = \dfrac{N^2}{N^2 \sin^2 \alpha + \cos^2 \alpha}$

$$= \frac{N^2}{1 + (N^2 - 1)\sin^2 \alpha} \qquad (5.13)$$

$\therefore \qquad \dfrac{\text{Intensity of secondary maxima}}{\text{Intensity of principal maxima}} = \dfrac{1}{1 + (N^2 - 1)\sin^2 \alpha}$

As N increases the intensity of the secondary maximum relative to principal maxima decreases and becomes negligible when N becomes large.

We may mention one relevant point here. A particular principal maximum may be absent if it corresponds to the angle which also determines the minimum of the single slit diffraction. This will happen when

$$e \sin \theta = k\lambda \qquad (5.14)$$

and $\qquad a \sin \theta = m\lambda \qquad (5.15)$

are satisfied simultaneously and is usually referred to as absence of order. It should be noted that even when equation (5.15) does not hold exactly (*i.e.*, $a \sin \theta$ is close to an integral multiple of λ). The intensity of the corresponding principal maxima will be weak.

The diffraction pattern due to six slits is shown in Fig. 5.11.

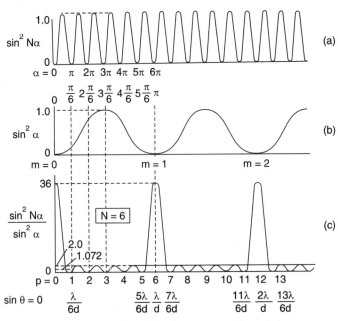

Fig. 5.11 Fraunhofer diffraction by a grating of six very narrow slits and details of the intensity pattern

5.4.3 Angular Width of the Principal Maxima
Our interest is in the minima nearest to nth order maximum for which
$$r = nN \pm 1.$$
If this occurs at an angle $\theta \pm \delta\theta$,
$$e \sin(\theta + \delta\theta) = \left\{\frac{nN \pm 1}{N}\right\}\lambda = n\lambda \pm \frac{\lambda}{N}$$
Taking difference with $e \sin\theta = n\lambda$ we get
$$e \cos\theta \cdot \delta\theta = \frac{\lambda}{N} \quad \text{or} \quad |\delta\theta| = \frac{\lambda}{Ne \cos\theta} \tag{5.16}$$
This is called the angular width of the nth order maximum. Note that it varies in proportion $\frac{1}{N}$; the interference maxima become sharper as the number of slits is increased.

5.5 DIFFRACTION–A QUALITATIVE DESCRIPTION
5.5.1 Diffraction at a Single Slit
In section 5.2 we discussed diffraction at a single slit and arrived at the diffraction pattern through rigorous mathematical analysis. The same results can be understood in qualitative way. In Fig. 5.12b, AB is slit of width a. Consider the slit to be split into $2n$ (where $n = 1, 2, 3, \ldots$) equal sections of width $a/2n$. Let these sections be AC, CD, DE, EF etc. Consider some direction θ to the normal and suppose that λ is the wavelength of the light being used. If θ is such that $AN = \lambda/2$, the wave from A will be completely out of phase with that from C. Thus the wave from A and C will destroy each other when they come together at P on the screen. The presence of the lens does not affect the phase relationship between the waves from A and C. Similarly light from each point between A and C will be destroyed by that form the corresponding

point between C and D. The same is true of every other pair of sections, such as DE and EF. Therefore there is no diffracted light in those directions of θ which are such that

$$AN = \lambda/2$$

i.e.,
$$AC \sin A\hat{C}N = \lambda/2$$

From simple geometry, $A\hat{C}N = \theta$ and therefore
$$AC \sin \theta = \lambda/2$$

i.e,
$$a/2n \sin \theta = \lambda/2$$

i.e., the angular positions, θ, of the minima are given by $a \sin \theta = n\lambda$ ($n = 1, 2, 3$)

In Fig. 5.12a, all the waves arriving at O, at any one time, will have left the various points on the wavefront at AB at the same time. Since all these waves are in phase with each other when they leave AB, they are still in phase when they reach O. Thus at O the radiation from each point on AB enhances that from every other and therefore there is brightness at O.

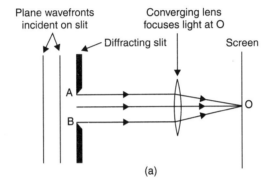

(a)

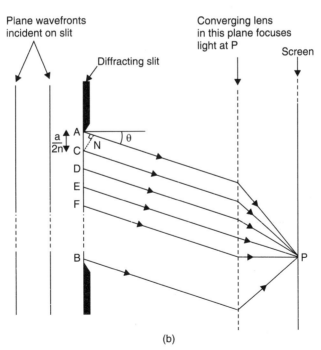

(b)

Fig. 5.12 Diffraction at a single slit

The positions of the maxima are difficult to determine and has to be obtained only through a mathematical analysis, which we discussed in section 5.2.

5.5.2 Diffraction due to Multiple Slits

In Fig. 5.13a, XY is the plane of the multiple slits and MN is the screen. The slits are all parallel to one another and perpendicular to the plane of the paper. Here AB is a slit and BC is the opaque portion. The width of each slit is a and the distance between any two consecutive slits is b.

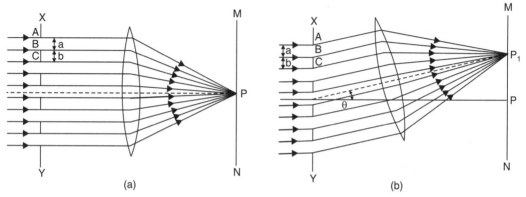

Fig. 5.13 Diffraction due to multiple slits

Let a plane wavefront be incident on the surface of the slits. Then all the secondary waves travelling in the same direction as that of the incident light will come to focus at the point P on the screen. The screen is placed at the focal plane of the collecting lens. The point P where all the secondary waves reinforce one another corresponds to the position of the central bright maximum.

Now consider the secondary wave travelling in a direction inclined at an angle θ with the direction of the incident light (Fig. 5.13b). The collecting lens is also turned such that the axis of the lens is parallel to the direction of the secondary waves. These secondary waves come to focus at a point P_1 on the screen. The intensity at P_1 will depend on the path difference between the secondary wave originating from the corresponding points A and C of the two neighbouring slits. In Fig. 5.13a, AB = a and BC = b. The path difference between the secondary waves starting from A and C is equal to AC sin θ.

But AC = AB + BC = a + b

Therefore path difference = (a + b) sin θ

The point P_1 will be of maximum intensity if this path difference is equal to an integral multiple of λ.

In general, $(a + b) \sin \theta_n = n\lambda$

where θ_n is the direction of the nth principal maxima.

In this equation (a + b) is called the grating element.

Now, let the angle of diffraction be increased by a small amount $d\theta_n$ such that the path difference between the secondary waves from the points A and C increases by λ/N where N is the total number of slits. Then the path difference between the secondary waves from the extreme points on the multiple slit surface will be (λ/N) N = λ. Assuming that the whole wavefront to be divided into two halves, the path difference between the corresponding points of the two halves will be $\lambda/2$ and all the secondary waves cancel one another. Thus $(\theta_n + d\theta_n)$ will give the direction of the first secondary minimum after the nth primary maximum. Similarly if the path difference between the secondary waves from the points A and C is $2\lambda/N$, $3\lambda/N$ etc., for gradual increasing values of $d\theta_n$, these angles correspond to the directions of 2nd, 3rd etc., secondary minima after the nth primary maximum. If the value of the path difference between the secondary waves from the extreme points of the multiple slit surface is $(2\lambda/N) N = 2\lambda$ and considering the wavefront to be divided into four portions, the concept of 2nd secondary minimum can be understood. The number of secondary minima in-between the two primary maxima is $N - 1$ and the number of secondary maxima is $N - 2$. The positions of the secondary maxima has to be obtained only by rigorous mathematical analysis which was described in section 5.4.

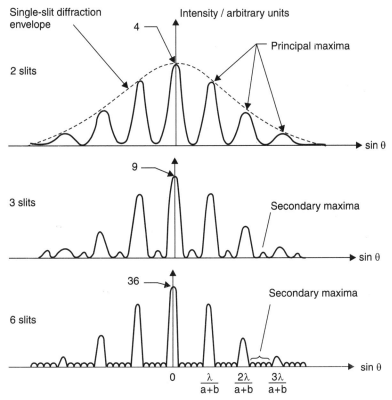

Fig. 5.14 Effect on the diffraction pattern of varying N while keeping a and $a + b$ fixed

The effect on the diffraction pattern of varying N while keeping a and b constant is shown in Fig. 5.14.

5.6 DIFFRACTION GRATING

In section 5.4 we discussed the diffraction pattern produced by a system of parallel equidistant slits. An arrangement which essentially consists of a large number of equidistant slits is known as diffraction grating. The corresponding diffraction pattern is known as diffraction spectrum. Since the exact positions of the principal maxima in the diffraction pattern depend on the wavelength, (equation 5.9), the principal maxima corresponding to different spectral lines (associated with the source) will correspond to different angles of diffraction. Thus the grating spectrum provides us with an easily obtainable experimental set up for the determination of wavelengths. Gratings have proved to be a powerful instrument for the study of the spectra.

From equation (5.16), we see that for narrow principal maxima (*i.e.*, sharper spectral lines) a large value of N is required. A good quality grating therefore requires a large number of slits (typically about 30,000 per inch). This is achieved by ruling grooves with a diamond point on an optically transparent material. The grooves act as opaque spaces. After each groove is ruled, the machine lifts the diamond points and moves the sheet forward for the ruling of the next groove. Since the distance between two consecutive grooves is extremely small, the movement of the optically transparent sheet is obtained with the help of the rotation of a screw which drives the carriage carrying it. Further, one of the important requirements of a good quality grating is that the line should be as equally spaced as possible, consequently the pitch of the screw must be constant. In fact, it was not until the manufacture of a nearly perfect screw (which was achieved by Rowland in 1882) that the problem of construction of gratings was successfully solved. Rowland's arrangement gave 14, 438 lines per inch corresponding to $e = 2.54/14, 438 = 1.693 \times 10^{-4}$ cm.

Commercial gratings are produced by taking the cast of an actual grating on a transparent film like that of cellulose acetate. A solution of cellulose acetate of appropriate strength is poured on the ruled surface and allowed to dry to form a strong thin film, detachable from the parent grating. These impressions of a grating are preserved by mounting the film between two glass sheets. Now-a-days gratings are also produced holographically, by recording the interference pattern between two plane or spherical waves. In contrast to ruled gratings, the holographic gratings have a much larger number of lines/cm.

5.7 GRATING SPECTRUM

While discussing the diffraction due to N slits we have shown that the positions for the principal maximum is

$$e \sin \theta = k\lambda \qquad (5.17)$$

This relation which is also called the grating equation, can be used to study the dependence of the angle of diffraction θ on wavelength λ. The zeroth order principal maximum occurs at $\theta = 0$, irrespective of the wavelength. Thus if we use the white light, then central maximum will also be white. However for k not equal to 0, the angles of diffraction are different for different wavelengths and hence various spectral components appear at different positions. Thus by measuring the angles of diffraction for various colours and knowing the value of k one can determine the values of wavelengths. It may be mentioned here that the intensity is maximum for the zeroth order maximum and it falls off as the value of k increases.

If we differentiate eqn. (5.17), we get

$$\frac{\Delta\theta}{\Delta\lambda} = \frac{k}{e \cos \theta} \qquad (5.18)$$

From this relation we can conclude the following.

(*i*) Assuming θ to be small (*i.e.*, $\cos \theta \approx 1$), we see that angle $\Delta\theta$ is directly proportional to order k of the spectrum for a given $\Delta\lambda$. For a given k, $\Delta\theta/\Delta\lambda$ is constant. Such a spectrum is known as a normal spectrum. In this, difference in the angle for two spectral lines is directly proportional to difference in wavelength. For large θ, it can be shown that the dispersion is greater at the red end of the spectrum.

(*ii*) Equation (5.18) tells us that $\Delta\theta$ is inversely proportional to e the grating element. Smaller the grating element greater will be the angular dispersion.

DIFFRACTION

5.7.1 Resolving Power of a Grating

The resolving power of a grating refers to the power of distinguishing two nearby spectral lines defined by the following equation

$$R = \frac{\lambda}{\Delta \lambda}$$

where $\Delta \lambda$ is the separation of two wavelengths which a grating can resolve. Obviously the smaller the value of $\Delta \lambda$, the larger the resolving power.

According to Rayleigh's criterion, if the principal maximum corresponding to wavelength $\lambda + d\lambda$ falls on the first minimum of wavelength λ, then the wavelengths are said to be resolved (see section 5.9). If this common diffraction angle is represented by θ and if we are looking at the mth order spectrum, then the two wavelengths λ and $\lambda + d\lambda$ will be just resolved if the following two equations are simultaneously satisfied.

$$e \sin \theta = m(\lambda + d\lambda)$$

$$e \sin \theta = m\lambda + \frac{\lambda}{N}$$

$$\therefore \quad R = \frac{\lambda}{\Delta \lambda} = mN$$

which implies that the resolving power depends on the total number of lines in the grating (only those lines which are exposed to the beam have to be taken into account). Further, resolving power is proportional to the order of the spectrum.

5.7.2 Oblique Incidence

Till now we have assumed that only plane waves were incident on the grating. In experiment it is difficult to achieve normal incidence to a great precision. The wavelength measurement is therefore done using minimum deviation method employing oblique incidence.

If the angle of incidence is i, angle of diffraction θ, then the path difference of the diffracted rays from two corresponding points in adjacent slits will be $e(\sin \theta + \sin i)$.

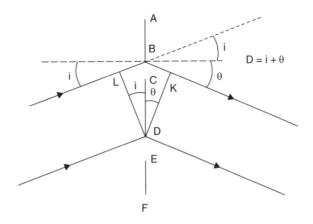

Fig. 5.15 Diffraction grating-oblique incidence

The condition for principal maxima is therefore

$$e(\sin \theta + \sin i) = k\lambda$$

Differentiating w.r.t. 'i'

$$e\left\{\cos\theta\frac{d\theta}{di} + \cos i\right\} = 0$$

i.e.,
$$e\{\cos\theta d\theta + \cos i\, di\} = 0$$

Now $D = i + \theta$

For minimum deviation
$$\delta D = \delta i + \delta\theta = 0$$

i.e., $$di = -d\theta$$

Substituting we have, $e\{\cos\theta(-di) + \cos i\, di\} = 0$

$\cos\theta\, di = \cos i\, di \quad \Rightarrow \quad i = \theta$

$\therefore \quad e\{\sin i + \sin i\} = k\lambda$

$2e \sin i = k\lambda = 2e \sin D/2$

Hence at the position of minimum deviation the grating condition becomes

$$2e \sin \frac{D}{2} = k\lambda \tag{5.19}$$

The minimum deviation position can be obtained in a manner similar to that used in the case of a prism. Since the adjustments are relatively simpler, this provides a more accurate method for the determination of λ.

5.8 FRAUNHOFER DIFFRACTION AT A CIRCULAR APERTURE

The diffraction pattern formed by plane waves from a point source passing through a circular aperture is of practical interest because most lenses and stops are round. The problem was first solved by Airy in 1835. The resultant diffraction pattern again consists of a series of maxima and minima which take the form of a concentric ring. The bright central maximum is called Airy's disk. The pattern looks similar to that formed behind a slit but the dimension are different as shown in the Fig. 5.16.

The mathematical analysis of diffraction due to a circular aperture is much more difficult than diffraction due to a slit. As before aperture is divided into a series of narrow strips of equal width. But since these strips are not of equal length, the amplitudes are unequal too. The resultant amplitude is found by integration. Table 5.1 shows the comparison between the diffraction maxima and minima intensities for a single slit and a circular aperture.

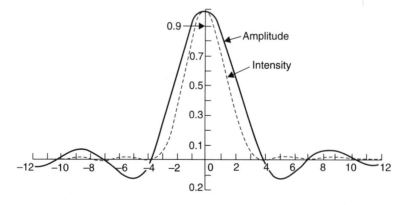

Fig. 5.16 Amplitude and intensity distributions in the diffraction pattern behind a circular aperture. Compare it with Fig. 5.5.

DIFFRACTION

Table 5.1

Ring	Circular Aperture		Single Slit	
	m	I_{max}	m	I_{max}
Central maximum	0	1	0	1
First dark	1.220		1.000	
Second bright	1.635	0.01750	1.430	0.0472
Second dark	2.233		2.000	
Third bright	2.679	0.00416	2.459	0.0165
Third dark	3.238		3.000	
Fourth bright	3.699	0.00160	3.471	0.0083
Fourth dark	4.241		4.000	
Fifth bright	4.710	0.00078	4.477	0.0050
Fifth dark	5.243		5.000	

5.9 RAYLEIGH'S CRITERION FOR RESOLVING POWER

By resolving power of an optical instrument we mean its ability to produce separate images of objects very close together. Using the laws of geometrical optics, one designs a microscope or a telescope. However in the final analysis it is the diffraction pattern which sets a limit on the

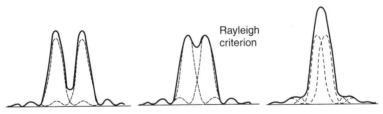

Fig. 5.17 Rayleigh's criterion note the separation of the maxima in the left hand plot and close overlap on the right

resolving power. For example, consider two nearby sources. Only when the diffraction patterns of these two sources are separate will they appear separate. When the central maxima fuse, the two sources appear as one. When the central maximum of one source coincides with the first minimum of the other, the resolution is marginal, a condition called Rayleigh's criterion (Fig. 5.17).

Therefore from Table 5.1, the minimum angle of resolution provided by a lens of diameter D at a wavelength λ is

$$\theta_{min} = 1.22 \frac{\lambda}{D} \tag{5.20}$$

5.10 RESOLVING POWER OF A MICROSCOPE

In the case of a microscope the object is held very close to the objective and the object subtends an angle say 2α as shown in the Fig. 5.18. Then we would like to know the smallest distance between A and B which will produce images A' and B' that are just resolved.

In the figure, MN is the aperture of objective of the microscope. A and B are two object points at a distance d. A' and B' are the corresponding Fraunhofer's diffraction patterns. A' is the position of central maximum of A and B' that of B. A' and B' are surrounded by alternate dark and bright diffraction bands. The two angles are said to be resolved when the central maximum of B' also corresponds to that first minimum of the image of A'.

The path difference between the extreme rays from B and reaching A' is given by

$$(BN + NA') - (BM + MA') = BN - BM$$

To evaluate $BN - BM$, consider Fig. 5.18b

$$BN - BM = (BC + CN) - (DM - DB)$$

But $\quad CN = AN = AM = DM$

∴ Path difference = $BC + DB$

But $\quad BC = AB \sin \alpha = d \sin \alpha$

$\quad DB = AB \sin \alpha = d \sin \alpha$

∴ Path difference = $2d \sin \alpha$

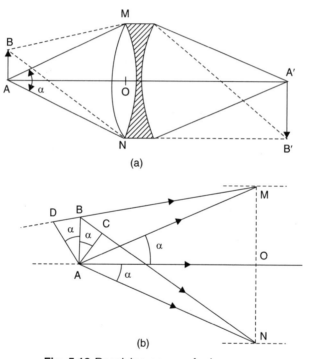

Fig. 5.18 Resolving power of microscope

If this path difference $2d \sin \alpha = 1.22\lambda$, then A' corresponds to the first minimum of the image B' and the two images appear just resolved

$$2d \sin \alpha = 1.22\lambda$$

$$d = \frac{1.22\lambda}{2 \sin \alpha}$$

The above equation is derived on the condition that A and B are self luminous objects and as such the light given out by each has no constant phase relative to that from other. Actually the objects used in microscopes are not self luminous but are illuminated with light from the condenser. In this case it is impossible to have the light scattered by two points on the object entirely independent in phase. This greatly complicates the problems since the resolving

DIFFRACTION

power is found to depend somewhat on the mode of illumination of the object. Abbe investigated the problem in detail and concluded that a good working rule for calculating the resolving power was given by the above equation omitting factor 1.22. If the space between the object and the lens is filled with a liquid of refractive index μ, the effective wavelength becomes λ/μ, so that

$$\therefore \quad d = \frac{\lambda}{2\mu \sin \alpha} \quad (5.21)$$

The product $\mu \sin \alpha$ is called NA (Numerical Aperture). In practice the largest value of NA obtainable is about 1.6. With white light of effective wavelength 5.6×10^{-5} cm, $d = 1.8 \times 10^{-5}$ cm. This then is the diameter of the smallest particle that we can see or photograph using light waves. It is clear from equation that decrease of λ will increase the resolving power. This is achieved using electrons instead of white light.

5.11 ELECTRON MICROSCOPE

The human eye is a wonderful optical instrument enabling us to see things around us. However we would like to see things which unaided eyes cannot see. To see the far off planets and stars, man has invented the telescope. And to see the smallest things he has invented the microscope. Before studying the electron microscope, let us study the eye in a little detail. The normal eye has a very wide field of vision and great range in distance. However beyond a certain range of distance eye is unable to see. It has been found that if the angle subtended at the eye by the object becomes less than about 1.4 minutes (~1/50°) the object ceases to be distinguishable. Thus whether an object is visible to the eye or not depends both on its size and distance (see Table 5.2).

From this it follows that whether any particular object is visible to the eye or not depends both on its size and its distance from the eye. Now an object may be so small that it is invisible at a distance of 25 cm from the eye. This is true if it subtends an angle at the eye less than 1.4 minutes. Since the eye cannot focus anything which is closer to it than 10 inches, we must conclude that the unaided eye cannot see any object less than .01 cm diameter. It also cannot make out any detail of any marking on a large visible object if the linear dimensions of the object is less than .01 cm.

An electron microscope is a device for obtaining pictures of high resolution employing a beam of electrons. There are 3 principal kinds of electron microscopes. (a) transmission (b) scanning and (c) emission. In the first two types free electrons are discharged from an electronic gun to act upon the atomic nuclei of the specimen. Whereas in the field emission type, the specimen itself is a source of radiation.

Table 5.2

Distance of the object from the eye	Linear dimensions of the object visible
16 km	6 m
1.6 km	.76 m
91.5 m	2.5 cm
3 m	.125 cm
25 cm	.01 cm

Two events in the 1920's brought about the development of electron microscope. One was the realisation from de Broglie's theory (1924) that particles have wave properties. According to de Broglie a particle having a momentum p is associated with a wavelength $\lambda = h/p$. Thus electrons with high energy can have very short wavelengths (0.05 Å). The other event was the demonstration by Busch in 1926–27, that a suitably shaped magnetic field could be used as a lens in electron microscope. The production, propagation and focusing of electrons properly belongs to the subject of electron optics. The deflection and focusing of electron beam is achieved through the use of electric and magnetic fields produced by suitable arrangement of electrodes and magnets. Such an arrangement of electrodes and magnets is referred to as electron lenses. The image formation by electron lenses can be treated in the same way as is done for the light rays in geometrical optics.

The deflection of electrons in electric field is best understood by considering the refraction of electrons at the equipotential surface. Suppose an electron with a velocity of v_0 approaches an equipotential surface specified by constant potential V_0 and V_1 on either side of the surface (Fig. 5.19). The angles of incidence i and refraction r are defined as in optics with respect to the perpendicular at the point of incidence O. The energy of the electron is given by $eV_0 = \frac{1}{2} m v_0^2$

or $v_0 \propto V_0^{1/2}$.

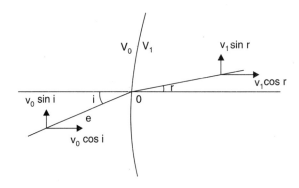

Fig. 5.19 Refraction of an electron beam at a potential discontinuity ($V_1 > V_0$)

In crossing the surface the tangential velocity of the electron will be unchanged, there being no force in that direction. The normal component will however change from $v_0 \cos i$ to $v_1 \cos r$. Then we have

$$v_0 \sin i = v_1 \sin r$$

or
$$\frac{\sin i}{\sin r} = \frac{v_1}{v_0} = \sqrt{\frac{V_1}{V_0}} \qquad (5.22)$$

This is very much similar to the Snell's law of refraction in optics. In practical electron optics one has to determine the configuration of equipotential surfaces of a given field and plot step by step the path of electrons through it. Some of the typical electron lenses employed in electron microscope for focusing electrons using electric field are shown in Fig. 5.20.

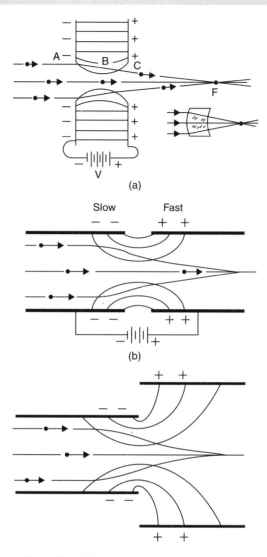

Fig. 5.20 (a) Double-aperture electron lens
(b) Symmetrical electron lens
(c) Asymmetrical electron lens

Magnetic focusing: The electron beam can also be focused using a magnetic field as the following discussion shows. Let an electron enter the magnetic field with a velocity v at an angle θ with the magnetic field (Fig. 5.21). The direction of the force acting on the electron is given by $\vec{v} \times \vec{H}$. Hence, it traces out a circle, the radius of which is given by

$$\frac{mv^2 \sin^2 \theta}{r} = Hev \sin \theta$$

$$v \sin \theta = \frac{He\,r}{m}$$

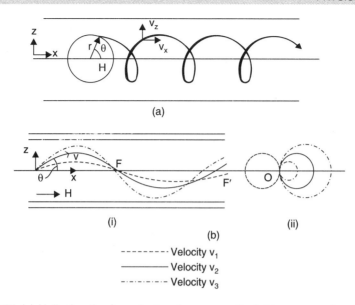

Velocity v_1
Velocity v_2
Velocity v_3

Fig. 5.21 (a) Helical path of an electron in a magnetic field. *e.g.,* a solenoid lens
(b) Electron paths in a solenoidal magnetic lens
(i) Along the field (ii) Normal to the field

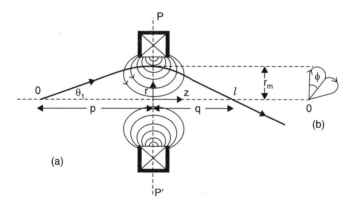

Fig. 5.22 Trajectory in a short magnetic lens
(a) In a rotating meridional plane
(b) In projection on transverse plane

The time taken by the electron to describe the circle of radius r is

$$t = \frac{2\pi r}{v \sin \theta} = \frac{2\pi rm}{He\, r} = \frac{2\pi m}{He} = \frac{2\pi}{H(e/m)} \qquad (5.23)$$

The electron possessing a component of velocity parallel to $\vec{H}$, describes a helical path. The pitch of the helix is given by

$$l = v \cos \theta \times t = v \cos \theta \, \frac{2\pi}{H(e/m)} = \frac{2\pi m v \cos \theta}{He}$$

From equation (5.23), it is clear that the time t is independent of θ. Therefore all the electrons will arrive at the same point of focus for a given θ. Thus a divergent beam of electron is brought to focus at the same point. The focal length can be varied by changing the strength of the magnetic field.

It can be similarly shown that the short magnetic lens has also the focusing action and this is the one which is usually employed in electron microscope. In this case both the object and the image lie outside the field (Fig. 5.22).

Construction: Transmission electron microscope is arranged very much like an ordinary microscope (Figs. 5.23 and 5.24). The electrons originate in an electronic gun. The gun consists of three components; the filament, shield and anode (Fig. 5.25). The filament is a thermoionic cathode made of tungsten. When a current is passed through, it emits electrons. By the use of apertures and lens systems these electrons are channeled. The shield lies immediately in front of the filament with a small aperture and relatively small potential difference is applied to it. The anode is an aperture disc and is held at a constant positive potential with respect to the cathode. The emitted electrons are thus accelerated.

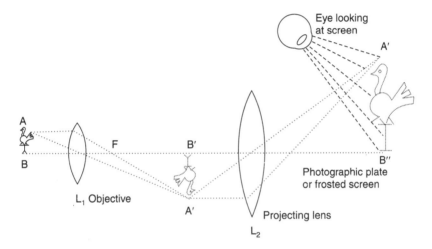

Fig. 5.23 A compound microscope

The basic function of the condenser lens is to focus the electron beam from the gun on to the specimen to permit illumination and recording of the image. The different parts of the specimen absorb electrons differently and an image is thus formed. The objective lens forms an initial enlarged image of the illuminated specimen in a plane that is suitable for further enlargement in projector lens. The projector lens, as the name indicates, is used to project the final magnified image on to the screen or photographic emulsion. The electron image is converted into a visible light image by means of a fluorescent screen and photographs. Since the electron beam would be scattered by collision with air molecules, the interior of the electron microscope has to be evacuated using a diffusion pump and fore pump.

Image formation in the scanning electron microscope differs from that of conventional TEM. In the SEM the image is formed on a cathode ray tube, synchronized with the electron probe as it scans the surface of the object. When electron beam hits the surface, secondary electrons are emitted which are collected and amplified. The advantage of SEM is that even a thick specimen can be examined.

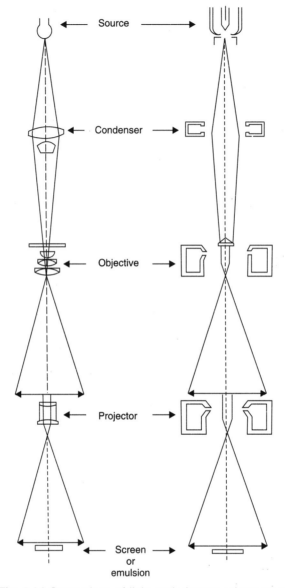

Fig. 5.24 Comparison of light and electron microscopes

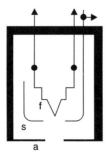

Fig. 5.25 Electron gun. The three components of the elec the filament 'f' the shield 's' and anode 'a'.

DIFFRACTION

The electron microscope differs from the light microscope in the following ways:

(i) The focal length formed by electric and magnetic fields can be varied by varying the fields whereas focal length of lenses in a light microscope is fixed.

(ii) In an electron microscope the objective and projective lens need not be changed when different magnifications are needed unlike in a light microscope.

(iii) The resolution obtainable with electron microscope is ~2 Å while that with a light microscope is ~2000 Å.

Electron microscopes have been extensively used in biology and material science for routine analysis.

REFERENCES

1. Benjamin N. Siegel, *Modern Developments in Electron Microscopy*, Academic Press, New York, 1964.
2. V.K. Zworykin, G.A. Morton, E.G. Ramberg, J. Hillier and A.W. Vance, *Electron Optics and Electron Microscope,* John Wiley & Sons, New York, 1945.
3. V.E. Cosslett, *Introduction to Electron Optics,* Oxford at the Clarendon Press, 1948.
4. S.Wischnitzer, *Introduction to Electron Microscopy,* Pergamon Press, New York, 1970.
5. E.F. Burton and W.H. Kohl, *The Electron Microscope,* Reinhold Publishing Corporation, New York, 1942.
6. M.A. Hayat, *Principles and Techniques of Scanning Electron Microscopy,* Van-Nostrand Reinhold Company, New York, 1978.
7. O.C. Wells, *Scanning Electron Microscopy*, McGraw-Hill Book Company, New York, 1974.
8. T.G. Rochow and E.G. Rochow, *An Introduction to Microscopy by Means of Light,* Electrons; X-rays or ultrasound, Plenum Press, New York, 1978.

SOLVED EXAMPLES

1. In a plane transmission grating the angle of diffraction for the second order principal maximum for the wavelength 5×10^{-5} cm is $30°$. Calculate the number of lines in one centimeter of the grating surface.

    Solution:

    We have $e \sin \theta = k\lambda$

    Here $k = 2, \lambda = 5 \times 10^{-5}$ cm, $\theta = 30°$

    $$e = \frac{k\lambda}{\sin \theta} = \frac{2 \times 5 \times 10^{-5}}{0.5} = 10^{-3} \text{ cm}$$

    $\therefore$ No. of lines per centimeter $= \dfrac{1}{e} = 10^3$.

2. Light of wavelength 5000 Å is incident normally on a plane transmission grating. Find the difference in the angles of deviation in the first and third order spectra. The number of lines per centimeter on the grating surface is 6000.

    Solution:

    Here $\lambda = 5000$ Å, $e = \dfrac{1}{6000}$

    For Ist order $e \sin \theta_1 = 1 \times \lambda$

$$\sin\theta_1 = \frac{\lambda}{e} = 5\times 10^{-5}\times 6000 = 0.30$$

∴ $\theta_1 = 17.5°$

For the third order,

$$e\sin\theta_3 = 3\lambda$$

$$\sin\theta_3 = \frac{3\lambda}{e} = 3\times 5\times 10^{-5}\times 6000 = 0.90$$

∴ $\theta_3 = 64.2°$

$\theta_2 - \theta_1 = 64.2° - 17.5° = 46.7°$.

3. Calculate the minimum number of lines per centimeter in a 2.5 cm wide grating which will just resolve the sodium lines (5890 Å and 5896 Å) in the second order spectrum.

   Solution:

   Let the total number of lines required on the grating be N

   Then $\quad \dfrac{\lambda}{d\lambda} = kN$

   Here $\quad \lambda = 5890\times 10^{-8}$ cm

   $\quad d\lambda = 6\times 10^{-8}$ cm

   $\quad k = 2$

   $\quad N = ?$

   ∴ $\quad N = \dfrac{\lambda}{k\cdot d\lambda} = \dfrac{5890\times 10^{-8}}{2\times 6\times 10^{-8}} = 491$

   Width of the grating = 2.5 cm

   No. of lines per cm = $\dfrac{491}{2.5} = 196.4$.

4. Examine if two spectral lines of wavelengths 5890 Å and 5896 Å, can be clearly resolved in the (*i*) first order and (*ii*) second order by a diffraction grating 2 cm wide and having 425 lines/cm.

   Solution:

   Total no. of lines on the grating = $2\times 425 = 850$

   (*i*) For the first order

   $$\frac{\lambda}{d\lambda} = kN,\ \text{where}\ k = 1$$

   $$N = \frac{\lambda}{d\lambda} = \frac{5890\times 10^{-8}}{6\times 10^{-8}} = 982\ \text{lines}.$$

   As the total number of lines required for the just resolution in the first order is 982 and the total number of lines on the grating is 850, the lines will not be resolved.

   (*ii*) For the second order

   $$\frac{\lambda}{d\lambda} = kN,\ \text{where}\ k = 2$$

   $$N = \frac{\lambda}{2d\lambda} = \frac{5890\times 10^{-8}}{2\times 6\times 10^{-8}} = 491\ \text{lines}.$$

DIFFRACTION

As the total number of lines required is 491 and the given grating has a total of 850 lines, the lines will appear resolved in the second order.

5. A plane grating has 15,000 lines per inch. Find the angle of separation of the 5048 Å and 5016 Å lines of helium in the second order spectrum.

   Solution:

   Here
   $$\lambda_1 = 5016 \text{ Å}$$
   $$\lambda_2 = 5048 \text{ Å}$$
   $$k = 2$$
   $$e = \frac{2.54}{15,000} \text{ cm.}$$

   Let θ_1 and θ_2 be the angle of diffraction for the second order for wavelengths λ_1 and λ_2 respectively.

   ∴
   $$e \sin \theta_1 = 2\lambda_1; \ e \sin \theta_1 = 2\lambda_2$$
   $$\sin \theta_1 = \frac{2\lambda_1}{e} = \frac{2 \times 5016 \times 10^{-8} \times 15,000}{2.54} = 0.5924$$
   $$\theta_1 = 36°20'$$
   $$\sin \theta_2 = \frac{2\lambda_2}{e} = \frac{2 \times 5048 \times 10^{-8} \times 15,000}{2.54} = 0.5962$$
   $$\theta_2 = 36°36'$$

   ∴ Angle of separation $\theta_2 - \theta_1 = 16'$.

6. A diffraction grating which has 4000 lines per cm is used at normal incidence. Calculate the dispersive power of the grating in the third order spectrum in the wavelength region 5000 Å.

   Solution:

   $$\frac{d\theta}{d\lambda} = \frac{k}{e \cos \theta}; \ e \sin \theta = k\lambda$$

   Here
   $$k = 3 \ e = \frac{1}{4000} \text{ cm}$$
   $$\lambda = 5000 \text{ Å}$$
   $$\sin \theta = \frac{k\lambda}{e} = 3 \times 5000 \times 10^{-8} \times 4000$$
   $$\sin \theta = 0.6$$
   $$\cos \theta = \sqrt{1 - \sin^2 \theta} = 0.8$$

   ∴
   $$\frac{d\theta}{d\lambda} = \frac{3 \times 4000}{0.8} = 15,000.$$

7. What is the highest order spectrum which may be seen with monochromatic light of wavelength 6000 Å, by means of a diffraction grating with 5000 lines/cm?

   Solution:

   Here $e \sin \theta = k\lambda$

   Max. possible value of $\sin \theta = 1$

∴ $e = \dfrac{1}{5000}$ cm, $\lambda = 6000 \times 10^{-8}$ cm

$$\dfrac{1}{5000} = k \times 6000 \times 10^{-8} \quad \text{or} \quad k = 3.33.$$

The highest order of the spectrum that can be seen is 3.

8. In the second order spectrum of a plane diffraction grating, a certain spectral line appears at an angle 10°, while another line of wavelength 5×10^{-9} cm greater appears at an angle 3° greater. Find the wavelength of the lines and the maximum grating width required to resolve them.

Solution:

We have $\quad e \sin\theta = k\lambda \quad$ (i)

Differentiating,

$\quad e \cos\theta\, d\theta = k\, d\lambda \quad$ (ii)

From (1) and (2)

$$\dfrac{\sin\theta}{\cos\theta\, d\theta} = \dfrac{\lambda}{d\lambda} \quad \text{or} \quad \lambda = \dfrac{\sin\theta\, d\lambda}{\cos\theta\, d\theta}$$

Here $\quad \theta = 10°, d\theta = 3°, d\lambda = 5 \times 10^{-9}$ cm

∴ $\quad \lambda = \dfrac{\sin 10° \times 5 \times 10^{-9}}{\cos 10° \times \dfrac{3}{60\times 60} \times \dfrac{\pi}{180}} = 6063 \times 10^{-9}$ cm

Now $\quad \lambda + d\lambda = 6063 \times 10^{-9} + 0.5 \times 10^{-9}$ cm

$\quad\quad\quad\quad = 6063.5 \times 10^{-9}$ cm

Again $\quad \dfrac{\lambda}{d\lambda} = kN \quad \text{or} \quad \dfrac{6063}{0.5} = 2N$

∴ $\quad N = \dfrac{6063}{0.5 \times 2} = 6063$

Hence minimum grating width required

$$Ne = \dfrac{Nk\lambda}{\sin\theta} = \dfrac{6063 \times 2 \times 6063 \times 10^{-9}}{\sin 10°} = 4.2 \text{ cm}.$$

9. Light of $\lambda = 5000$ Å falls on a grating normally. Two adjacent principal maxima occur at $\sin\theta = 0.2$ and $\sin\theta = 0.3$ respectively. Calculate the grating element. If the width of the grating surface is 2.5 cm, calculate its resolving power in the second order.

Solution:

The position of the principal maxima of order k is given by

$$e \sin\theta = k\lambda$$

The two adjacent orders say of k and $k+1$ occur at $\sin\theta = 0.2$ and $\sin\theta = 0.3$.

Thus

$\quad e \times 0.2 = k\lambda$

$\quad e \times 0.3 = (k+1)\lambda$

Subtracting $\quad e \times 0.1 = \lambda = 5000$ Å

$\quad\quad\quad\quad e = 5 \times 10^{-4}$ cm

Now number or rulings

$$N = \frac{2.5}{e} = \frac{2.5}{5 \times 10^{-4}} = 5000$$

Resolving power = $2 \times 5000 = 10,000$.

QUESTIONS

1. What is the difference between interference and diffraction bands?
2. Explain Fresnel and Fraunhofer diffraction.
3. Explain how a plane diffraction grating produces the spectrum of a given light.
4. Give the theory of a plane diffraction grating for normal incidence.
5. Give the theory of a plane diffraction grating for oblique incidence.
6. Show that for a plane diffraction grating the incident and the diffracted beams are equally inclined to the grating normal for mimimum deviation.
7. Describe the method of determining the wavelength of sodium light using a plane diffraction grating. Derive the necessary formula.
8. Derive an expression for the dispersive power of a plane diffraction grating.
9. What do you understand by the resolving power of a grating? Derive an expression for it.
10. Explain Rayleigh criterion for resolution.
11. Describe the construction and working of an electron microscope.

PROBLEMS

1. Light in incident normally on a grating 0.5 cm wide with 2500 lines. Find the angles of diffraction for the principal maxima of the two sodium lines in the first order spectrum (λ_1 = 5896 Å, λ_2 = 5890 Å). Are the two lines resolved? **(Ans. 17.1°, 17.2°. The lines are resolved)**
2. Light of wavelength 6×10^{-7} m is incident normally on a grating having 6000 lines per cm. Calculate the number of orders of the spectra that can be seen. **(Ans. 2)**
3. A parallel beam of monochromatic light is incident normally on a plane diffraction grating having 6000 lines per cm. Calculate the angular separation between the two lines of wavelengths 5890 Å in the second order. **(Ans. 18′)**
4. Calculate the least width of a plane diffraction grating having 500 lines/cm which will just resolve in the second order the sodium lines of wavelengths 5890 Å and 5896 Å. **(Ans. 0.982 cm)**
5. When light of wavelength 5690 Å falls normally on a grating, the deviation is found to be 11° in the first order spectrum. Calculate the number of lines per metre length of the grating.

    (Ans. 3.35×10^5 lines/meter)
6. A grating of width 2″ is ruled with 15,000 lines per inch. Find the smallest wavelength separation that can be resolved in second order at a mean wavelength of 5000 Å. **(Ans. 0.082 Å)**
7. A plane transmission grating has 40,000 lines in all with grating element 12.5×10^{-8} cm. Calculate the maximum resolving power for which it can be used in the range of wavelength 5000 Å.

    (Ans. 80,000)
8. A diffraction grating used at normal incidence gives a line (5400 Å) in a certain order superimposed on the violet line (4050 Å) of the next higher order. If the angle of diffraction is 30°, how many lines per cm are there in the grating. **(Ans. 3086)**
9. The separation of sodium lines (mean λ = 5893 Å) in the second order spectrum of a transmission grating containing 5000 lines per cm is 2.5 minutes for normal incidence. What is the difference in wavelength of the two yellow lines. **(Ans. 6 Å)**

6
Polarization of Light and Photoelasticity

6.1 INTRODUCTION

The phenomena of interference and diffraction which we discussed in the previous chapters tell you that light is a form of wave motion. To be precise it is a form of wave motion involving electric and magnetic fields (Fig. 6.1). In general there are two forms of wave motion (a) longitudinal and (b) transverse. In the case of longitudinal waves the vibrations are always parallel to the direction of propagation. In the plane at right angles to the direction of propagation there is no motion. In the case of transverse wave motion the vibrations are always perpendicular to the direction of propagation. Thus in the case of transverse wave motion two directions have to be specified; one, the direction of propagation and the other that of vibration.

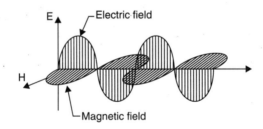

Fig. 6.1 Light is an electromagnetic wave

Maxwell's electromagnetic theory predicts the light waves to be transverse. This is supported by numerous experiments as well. In this chapter we consider some optic phenomena that depend not merely on the fact that light is a wave motion, but that these waves are transverse too. These phenomena are named polarization effects. They can be observed only with transverse waves.

6.2 REPRESENTATION OF POLARIZED AND UNPOLARIZED LIGHT

We say that light is plane polarized if the electric field vector is confined to a single plane (Fig. 6.2a). In the case of unpolarized light the orientation of the electric field is not fixed and any orientation is equally probable (Fig. 6.2b). In both these figures the direction of propagation is normal to the plane of the paper. In the case of unpolarized light the electric field will be vibrating in all directions normal to the direction of propagation. However for convenience we can regard half the number of waves to be vibrating in vertical direction and the other half to be vibrating in horizontal direction. Such a thing is feasible because waves not vibrating in either of these two directions can be resolved into two components, one component vibrating in a vertical direction and the other vibrating in a horizontal in a vertical direction. Figure 6.3

POLARIZATION OF LIGHT AND PHOTOELASTICITY

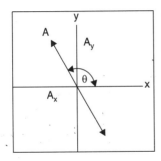

Fig. 6.2 (a) Vibrations in a plane polarized light viewed normal to the plane of paper

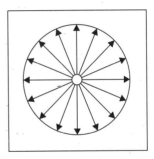

Fig. 6.2 (b) Vibrations in unpolarized light viewed normal to the plane of paper

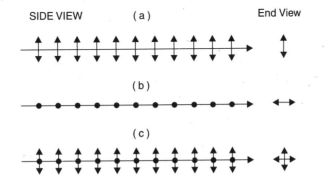

Fig. 6.3 Pictorial representation of the side and end views of (a) vertically polarized light (b) horizontally polarized light (c) unpolarized

shows a pictorial representation of the side and end views of plane polarized and ordinary unpolarized.

The methods by which polarized waves are produced are classified under the following headings:

(*i*) Polarization by reflection

(*ii*) Polarization by double refraction

(*iii*) Polarization by selective absorption and

(*iv*) Polarization by scattering

6.3 PRODUCTION OF POLARIZED LIGHT

6.3.1 Polarization by Reflection

When natural light strikes a reflecting surface, there is found to be a preferential reflection for those waves in which the electric vector is vibrating perpendicular to the plane of incidence (the plane of incidence is the plane containing the incident ray and the normal to the surface). However at normal incidence all directions of polarization are reflected equally. At one particular angle of incidence, known as the 'polarizing angle' (Brewster angle) only the light for which the electric field vector is perpendicular to the plane of incidence is reflected (Fig. 6.4). When incident at the polarizing angle, none of the components parallel to the plane of incidence are reflected *i.e.,* they are 100% transmitted in the refracted beam. Of the components perpendicular to the plane of incidence about 15% are reflected if the reflecting surface is a transparent material like glass (the fraction reflected depends upon the refractive index of the reflecting material). Hence the reflected light is weak and completely linearly polarized. The refracted light is a mixture of the parallel components all of which are refracted and the remaining 85% of the perpendicular component. It is therefore strong, but only partially polarized. To increase the intensity of reflected light a pile of thin glass plates is often used (Fig. 6.5).

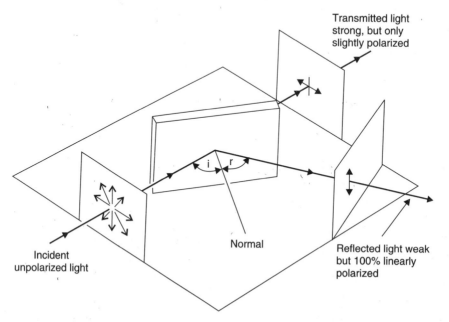

Fig. 6.4 When light is incident at the polarizing angle, the reflected light is linearly polarized

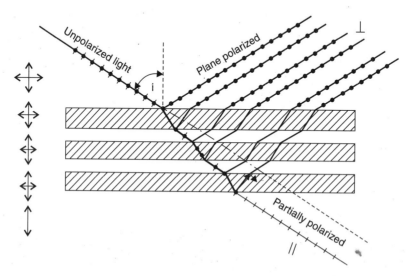

Fig. 6.5 Polarization of light by a pile of glass plates

The polarization by reflection is also shown in Fig. 6.6. It was Sir David Brewster, a Scottish physicist, who first discovered that at the polarizing angle the reflected and refracted rays are 90° apart. This is known as the Brewster's Law.

POLARIZATION OF LIGHT AND PHOTOELASTICITY

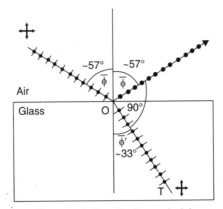

Fig. 6.6 Brewster's law for the polarizing angle

Because these two rays make 90° with each other, the angle of incidence 'i' and the angle of refraction r are complements of each other and $\sin r$ in the Snell's Law ($\sin i / \sin r = \mu$) can be replaced by $\cos i$ giving

$$\frac{\sin i}{\cos i} = \tan i = \mu.$$

This formula is useful in calculating the polarizing angle. For example with water, $\mu = 1.33$, $i = 53°$ whereas for glass $\mu = 1.52$, $i = 57°$. In general if light is travelling from medium 1 to medium 2 of refractive indices μ_1 and μ_2 respectively then the polarizing angle 'ϕ' is given by

$$\tan \phi = \frac{\mu_2}{\mu_1}$$

6.3.2 Polarization by Double Refraction

When a beam of light is passed through a transparent crystal like calcite ($CaCO_3$) or quartz (SiO_2) it is split into two beams (Fig. 6.7). Substances having this property are called doubly refracting or birefringent. If experiments are carried out for various angles of incidence, one of the beam is found to obey the Snell's Law and is called the O-ray (ordinary ray). The other beam does not obey Snell's Law and is not even found to be in the plane of incidence. This is called the extraordinary ray or E-ray. It is found that

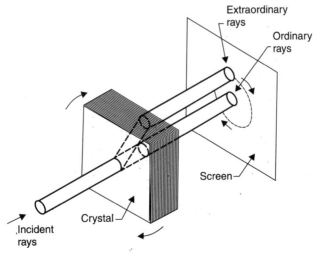

Fig. 6.7 A beam of light is split into two by a doubly refracting crystal

(i) *O*-ray travels in the crystal with the same speed in all the directions. In other words the crystal is characterized by a single value of refractive index for the *O*-ray.

(ii) *E*-ray travels in the crystal with a speed that varies with direction. In other words refractive index for the *E*-ray varies with the direction and is in fact described by an ellipsoid.

(iii) *Birefringence:* The difference between the refractive index for *O* and *E*-rays is called birefringence.

(iv) In the case of calcite and quartz there is one direction in which there is no double refraction. This direction is called the optic axis. In a class of crystals called biaxial crystals there are two directions in which there is no double refraction.

(v) *O*-ray and *E*-ray are polarized at right angles to each other and the actual polarization direction depends on the direction of propagation. The vibration direction of *O*-ray is perpendicular to the plane defined by the direction of propagation and the optic axis. The vibration direction of *E*-ray is in the plane defined by direction of propagation and the optic axis.

(vi) Even though *O*-ray and *E*-ray are derived from the same beam, they will not exhibit interference fringes when they are made to interfere. This crucial experimental observation enabled Thomas Young to conjecture that light waves were transverse in character.

Since *O*- and *E*-rays are polarized perpendicular to each other, if some means can be found to separate them, then this could be a way of producing polarized light from unpolarized one. This is in fact achieved using a Nicol prism. The Nicol prism is an optical device made from a calcite crystal (Fig. 6.8). This is used in many optical instruments for producing and analyzing polarized light. Nicol prism is made by cutting a calcite crystal along a diagonal and cementing it back together again with a special cement called Canada balsam. Canada balsam is a transparent substance. It is optically more dense than calcite for the *E*-ray and less dense for the *O*-ray (for sodium light $\mu_0 = 1.65836$, $\mu_{Canada\ balsam} = 1.55$, $\mu_e = 1.48641$). There exists therefore a critical angle of refraction for *O*-ray but not for the *E*-ray. After both rays are refracted at the first crystal surface, the *O*-ray is totally reflected by the first Canada balsam surface as shown in Fig. 6.9, while the *E*-ray passes on through to emerge parallel to the incident light. Thus starting with ordinary unpolarized light, a Nicol prism transmits only the plane polarized light.

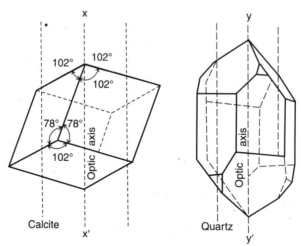

Fig. 6.8 Crystal form of calcite and quartz

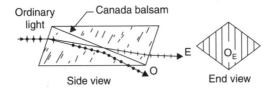

Fig. 6.9 Cross-section and end view of a Nicol prism showing the elimination of O-ray by total reflection

Two Nicols lined up one behind the other as in Fig. 6.10a is often used in optical microscopes for studying optical properties of crystals. The first Nicol which is used to produce the plane polarized light is called the polarizer and the second Nicol which is used to test the light is called the analyzer. In the parallel position light from the polarizer passes on through the analyzer. Upon rotating the analyzer through 90° as in Fig. 6.10b, no light is transmitted. In this case E-ray in the second Nicol is totally reflected. Thus when Nicols are crossed no light is transmitted.

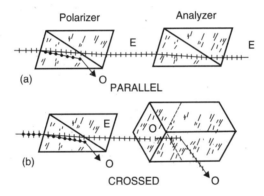

Fig. 6.10 Two Nicol prism mounted as polarizer and analyzer
(a) Parallel Nicols (b) Crossed Nicols

6.3.3 Polarization by Selective Absorption

When ordinary light enters a crystal of tourmaline, double refraction takes place as in calcite, but with a difference. The O-ray is entirely absorbed in the crystal while the E-ray passes through. This phenomenon is called selective absorption because the crystal absorbs light waves vibrating in one plane and not those vibrating in the other. Tourmaline crystals are therefore like Nicol prisms for they take in unpolarized light and transmit only the plane polarized light. When two such crystals are lined up parallel, the plane polarized light from the first crystal passes through the second although with less intensity (Fig. 6.11a). If the crystals are turned 90° with respect to each other i.e., in the crossed position the light is completely absorbed and none passes through (Fig. 6.11b).

Tourmaline crystals are not used in optical instruments because they are yellow in colour and do not transmit white light. A more satisfactory substance for this purpose is a newly manufactured material called polaroid. This material is made in the form of thin films. They are made from small needle shaped crystals of an organic compund iodosulphate of quinine. Lined up parallel to each other and embedded in a nitrocellulose mastic these crystals act like tourmaline by absorbing one component of polarization and transmitting the other. As in the case of tourmaline, in crossed position no light passes through two films whereas in parallel position light vibrating in the plane indicated by parallel lines is transmitted (Fig. 6.12).

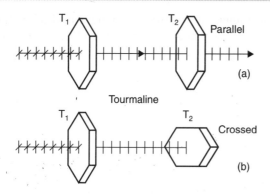
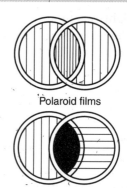

Fig. 6.11 (a) Tourmaline crystal plates in parallel
(b) Crossed tourmaline crystal plates

Fig. 6.12 Polaroid films in parallel and cross positions

6.3.4 Polarization by Scattering

Sky is blue. Sunrise and Sunset are red. Sky light is found to be predominantly linearly polarized. It turns out that one and the same phenomena is responsible for all the three effects viz., scattering of light by air molecules.

In Fig. 6.13, sunlight comes from the left along the z-axis and passes over an observer looking vertically upward along y-axis. Consider an air molecule at O. The electric field due to the beam of sunlight sets the electrical charges of the molecule in vibration. Since light is a transverse wave the direction of the electric field lies in the xy plane and the motion of charges takes place in this plane. There is no electric field along z and hence there is no vibration in

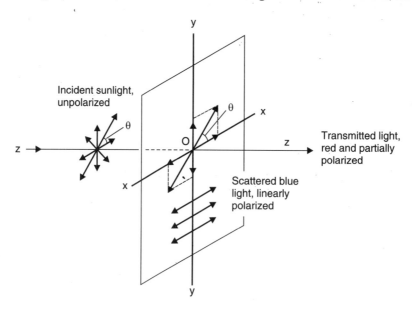

Fig. 6.13 Scattered light is linearly polarized

this direction. Let an arbitrary component of the incident light vibrating at an angle θ with the x-axis set the electric charge in motion as indicated by the heavy line through O. In the usual way we can resolve this vibration into two, one along x-axis and the other along y-axis. Then we have molecular antennas along x and y. The antenna along y-axis does not send any light to the observer directly in line with it. It does of course send out light in other directions. The only light reaching the observer comes from molecular x-antenna along x and as in the case with the waves from any antenna the light is linearly polarized with the electric field parallel to the antenna. The vectors along the y-axis below the point O show the direction of vibrations of light reaching the observer.

This process described above is called scattering. Both theory and experiment indicate that shorter waves are scattered more than the longer ones. In fact scattering power is proportional to λ^{-4} where λ is the wavelength of light. Thus short waves of violet and blue are scattered more than the red. That is why sky appears blue. Towards evening, when sunlight travels a large distance through the atmosphere to reach a point over the observer a large portion of blue light in the sunlight is removed from it by scattering. White light without blue is light yellowish and reddish in hue. This is the reason for yellowish-red sun sets. From this explanation it follows that if the earth had no atmosphere we would receive no sky light and the sky would appear black even during the day. This is true on the moon, as it does not have any atmosphere. If moon had atmosphere it would have appeared bluish.

6.4 CIRCULAR AND ELLIPTIC POLARIZATION

Linearly polarized light represents a special and relatively simple type of polarization. When the ordinary and extraordinary rays in a doubly refracting crystal are separated, each ray is linearly polarized, but with the directions of vibration at right angles. When, however, the crystal is cut with its faces parallel to the optic axis, so that light, incident normally on one of the faces, traverses the crystal in a direction perpendicular to the optic axis, as shown in Fig. 6.14, the ordinary and extraordinary rays are not separated. They traverse the same path, but with different speeds. Upon emerging from the second face of the crystal, the ordinary and the extraordinary rays are out of phase with each other and give rise to either eliptically polarized, circularly polarized or linearly polarized, depending on sample thickness.

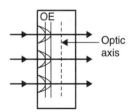

Since in the ordinary ray the direction of vibration is perpendicular to that in the extraordinary ray, we have to consider the following question. What sort of vibration results from the combination of two simple harmonic vibrations at right angles to each other and differing in phase? It can then be seen that two simple harmonic motions at right angles to each other never produce destructive interference, no matter what the phase difference.

Fig. 6.14 Light incident on a doubly refracting crystal perpendicular to the optic axis

(i) When the phase difference is 0, 2π or any even multiple of π, the result is a linear vibration at 45°, to both original vibrations.

(ii) When the difference is π, 3π or any odd multiple of π, the result is also a linear vibration, but at right angles to those corresponding to even multiple of π.

(iii) When the phase difference is $\pi/2$, $3\pi/2$ or any odd multiple of $\pi/2$, the resulting vibration is a circle.

(iv) For all the other phase differences, the resulting vibration is an ellipse.

The above results are summarized in Fig. 6.15.

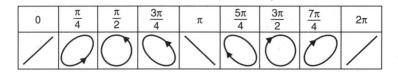

Fig. 6.15 Vibrations which result from the combination of a horizontal and vertical simple harmonic motion of the same frequency and same amplitude, for various values of phase difference

The phase difference between E-and O-rays of the crystal depends on the following:

(i) Frequency of light.
(ii) Indices of refraction of the crystal for E-and O-rays.
(iii) Thickness of the crystal.

If a given crystal has such a thickness as to give rise to a phase difference of $\pi/2$ for a given frequency, then according to Fig. 6.15 a circular vibration results and the light emerging from this crystal is found to be circularly polarized. The crystal itself is called a quarter wave plate. If the crystal has such a thickness so as to give rise to a phase difference of π for a given frequency, then according to the Fig. 6.15 a linear vibration results. A crystal plate of this sort is called a half-wave plate for a given frequency of light.

6.5 CALCULATION OF THE PHASE DIFFERENCE WHEN A LINEARLY POLARIZED LIGHT PASSES THROUGH A DOUBLE REFRACTING CRYSTAL, THE OPTIC AXIS BEING NORMAL TO THE DIRECTION OF PROPAGATION

Consider a crystalline plate of thickness d. Let a plane polarized light be incident normally as shown in the Fig. 6.16. Let the polarizer makes an angle θ with respect to the optic axis. Two refracted beams travel through the medium with velocities v_1 and v_2. Both these are plane polarized. In this derivation we have assumed the optic axis to be in the crystal plane.

The incident ray propagating along z-axis can be represented by the equation,

$A \cos \dfrac{2\pi}{\lambda} (z - ct + \varepsilon)$, where A is the amplitude and λ is the wavelength, c is the velocity and ε is the arbitrary phase factor.

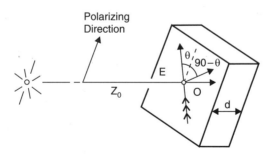

Fig. 6.16 Passage of polarized light through a crystal

At the interface, the amplitude gets resolved into $A \cos \theta$ and $A \sin \theta$. Both these components are polarized perpendicular to each other. The two vibrating components will travel with different velocities inside the crystal and when they emerge there will be a certain amount of phase difference between them. Our object is to determine this phase difference. The equation for the plane wave can be written as

$$A \cos \frac{2\pi}{\lambda}(z - ct + \varepsilon) = A \cos \frac{2\pi c}{\lambda}\left(\frac{z}{c} - t + \varepsilon'\right)$$

$$= A \cos 2\pi f \left(\frac{z}{c} - t + \varepsilon'\right) \tag{6.1}$$

where f is the frequency and ε' is some constant.

Just inside the front face, each vibratory component will have a phase, say ϕ. The thickness d of the crystal is traversed by the two components with different velocities v_1 and v_2. Hence just at the exit, the phases of the two components are given by

$$\phi + \frac{d}{v_1} f 2\pi \quad \text{and} \quad \phi + \frac{d}{v_2} f 2\pi \tag{6.2}$$

In the above expressions $\frac{d}{v_1}$ and $\frac{d}{v_2}$ are the times taken by the two waves to travel a distances d. The quantities $\frac{d}{v_1} f$ and $\frac{d}{v_2} f$ gives the number of cycles or oscillations during this travel and $\frac{d}{v_1} f 2\pi$ and $\frac{d}{v_2} f 2\pi$ express the accumulated phase in radians. Hence eqn. (6.2) gives the phases just at the exit. A light wave not passing through the crystal (known as the unimpeded ray) will have a phase equal to $\phi + \frac{d}{c} f 2\pi$.

The relative phase difference between two vibrating components is therefore given by

$$\left(\phi + 2\pi d \frac{f}{v_1}\right) - \left(\phi + 2\pi d \frac{f}{v_2}\right) = 2\pi df \left(\frac{1}{v_1} - \frac{1}{v_2}\right)$$

$$= 2\pi d \frac{c}{\lambda}\left(\frac{1}{v_1} - \frac{1}{v_2}\right) = \frac{2\pi d}{\lambda}(\mu_1 - \mu_2) \tag{6.3}$$

where μ_1 and μ_2 are the absolute refractive indices of the medium for the two rays.

The number of wavelengths of the relative path difference is $\frac{d}{\lambda}(\mu_1 - \mu_2)$.

6.6 PLANE AND CIRCULAR POLARISCOPE

A polarizer and an analyzer kept crossed with respect to each other constitutes a plane polariscope. Such a set up is often used in experiments related to photoelasticity. Polariscope is essentially an instrument for stress or strain analysis. If the polarizer and analyser are crossed, light cannot pass through the combination. If, however, a piece of glass which is under stress is placed between the polarizer and the analyzer, a pattern of light intensity variations can be seen (Fig. 6.17). Glass becomes doubly refracting when it is under stress and the resultant ray produced from the E-ray and the O-ray, is no longer plane-polarized at right angles to the analyzer. The light intensity at any point in the field of view depends on the thickness of the glass and the amount of strain or stress in the corresponding region of the specimen. Many plastic materials also become doubly refracting when under stress. By making plastic models of bridges etc., and examining them in the polariscope, engineers are able to detect the potential weak points.

Sometimes a crossed circular polariscope is also employed. In addition to polarizer (P) and analyzer (A) it has two quarter wave plates Q_P and Q_A as shown in Fig. 6.18. It is interesting to consider the effect of introducing a crystal plate in a crossed circular polariscope. Following cases can be considered.

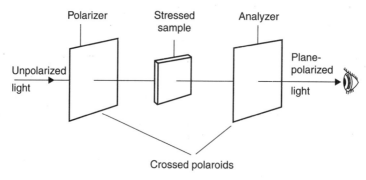

Fig. 6.17 A plane polariscope

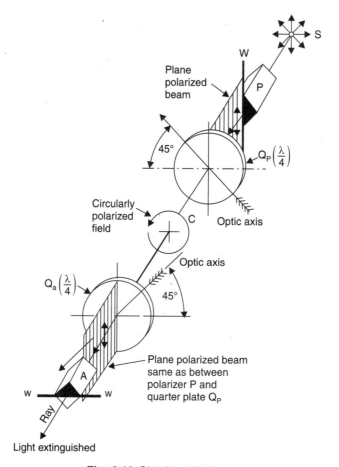

Fig. 6.18 Circular polariscope

Case (i) Thickness of plate causes a difference of one wavelength. Then if the polariscope is set for extinction, the screen will remain dark after the interposition of such an additional plate. If on the other hand the polariscope is set to give a bright background that brightness will remain essentially unaltered.

Case (ii) The plate has thickness which causes a phase difference of half wavelength. In this case the insertion of a half plate into a polariscope of any setting alters the intensity of illumination of the screen from darkness to maximum brightness or from maximum brightness to darkness.

Case (iii) For a crystal of arbitrary thickness there is a change in the intensity of illumination.

The explanations for the above settings are summarized in Fig. 6.19.

6.7 PHOTOELASTICITY

The subject of photoelasticity deals with the artificial birefringence developed in a solid under the application of a mechanical stress. Today it is the most powerful and indispensable tool in solving intricate problems in structural engineering. This phenomenon was first discovered by Sir David Brewster in 1815. In this investigation, he placed a strip of glass between two crossed Nicols. When the strip of glass was stretched, the field of view brightened up, there by showing that artificial birefringence was induced in the glass strip. The subject of photoelasticity remained as academic interest for nearly hundred years. It became of practical value when Coker and Filon (1902) applied it to measuring stresses in machine parts.

An engineer is interested in strength of materials. He is concerned with the causes and methods of prevention of failure of machine parts. Hence he is interested in stress analysis. The great majority of machine failures are caused by faulty design, a design which ignores or neglects exact stress distribution, especially the presence of stress concentrations. These stress concentrations provide starting point for the propagation of cracks which wreck the machine or structure. It is desirable to experimentally determine the stress distribution and it is here that the science of photoelasticity comes in. The subject of stress analysis using photoelasticity has opened a new chapter and has shed much light on the causes of failure and led to the rational improvement of design. The science of photoelasticity enables one to visualize a stress pattern, in many cases which are beyond the reach of calculation.

6.8 DEFINITIONS OF STRESS AND STRAIN AT A POINT

6.8.1 Strain

When a body which is not allowed to move is subjected to force it gets deformed. This deformation is called strain. The nature and extent of strain vary from point to point of the body. To define the strain at P consider a small rectangular parallelepiped in the unstrained body of which P is the centre (Fig. 6.20). The semi axis PA, PB, PC are taken as parallel to the axis of coordinates and their length are small quantities which are made to vanish in the limit. When the body is strained, such a right six face becomes to the first order approximation an oblique parallelepiped of which semi axis are $P'A', P'B', P'C'$, where A', B', C', P' are the new positions of P, A, B, C. The strain at P is defined when we know the size and shape of the oblique parallelepiped. This is entirely determined when the length of the new semi axis are $P'A', P'B', P'C'$ and their mutual inclinations are known. These are related according to the relations given below.

i.e.,
$$P'A' = PA\,(1 + \varepsilon_{xx}), \quad \cos B'\hat{P}'C' = \gamma_{yz}$$
$$P'B' = PB\,(1 + \varepsilon_{yy}), \quad \cos C'\hat{P}'A' = \gamma_{zx}$$
$$P'C' = PC\,(1 + \varepsilon_{zz}), \quad \cos A'\hat{P}'B' = \gamma_{xy}$$

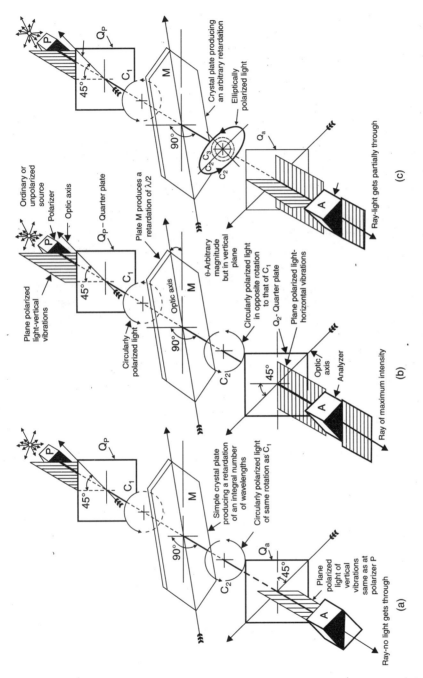

Fig. 6.19 Optical transformations in a standard polariscope produced by crystal plates of different thickness. Retardation due to birefringence are (a) An integral number of wavelengths (b) An odd number of half wavelengths (c) An arbitrary retardation

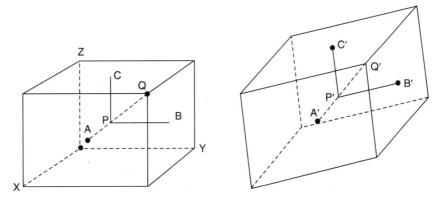

Fig. 6.20 A deformed body

Then the quantities $\varepsilon_{xx}, \varepsilon_{yy}, \varepsilon_{zz}, \gamma_{yz}, \gamma_{zx}, \gamma_{xy}$ all of which are of zero dimensions and therefore mere numbers determine completely the three edges $P'A'$, $P'B'$, $P'C'$ and therefore the strain. They are spoken as the six component of strain at P. The quantities $\varepsilon_{xx}, \varepsilon_{yy}, \varepsilon_{zz}$ are termed stretches to the axis of the coordinates. The quantities $\gamma_{yz}, \gamma_{zx}, \gamma_{xy}$ are often spoken as shear strains.

6.8.2 Stress

When a body is strained, every part of it exerts certain forces upon the surrounding parts. These in turn exert equal and opposite forces upon it. Thus across any small area in the material there is a pair of equal and opposite forces representing the action and reaction of the parts of the material on opposite sides of the small interface. The limiting value of such forces per unit area, when the dimensions of the interface are made indefinitely small, is called the stress across the surface.

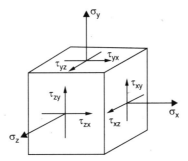

Fig. 6.21 Rectangular stress components

In general, the stress across any small interface has two components, one normal (σ) and one tangential (τ) to the surface. The latter may be again resolved into two components, along given rectangular directions in the plane of the surface. The tangential and shear stresses acting on a cube is shown in Fig. 6.21. In general τ_{pq} represents a stress parallel to q across an interface perpendicular to p. τ's represent the shearing stresses whereas the σ's represent the normal stresses.

It is necessary at this stage to define what are known as principal stresses. Normally, any stress across any small interface can be decomposed into normal and shear stresses. However there exists specific orientations of the stress for which the accompanying shear stresses are equal to zero. These are called principal stresses. In fact for a three dimensional stress, there are three principal stresses acting along three directions for which shear stresses are zero. The three principal directions are orthogonal to each other. For the case of a two dimensional stress, there are two mutually perpendicular directions in which there are two principal stresses.

There are six components of stress at a point. They are the σ_{xx}, σ_{yy}, σ_{zz} and the three shears τ_{yz}, τ_{zx} and τ_{xy}. It can be shown $\tau_{zy} = \tau_{yz}$, $\tau_{zx} = \tau_{xz}$, $\tau_{xy} = \tau_{yx}$.

6.8.3 Plane State of Stress

If in a given state of stress there exists a coordinate system $Oxyz$ such that for this system

$$\sigma_z = 0, \tau_{xz} = 0 \quad \text{and} \quad \tau_{yz} = 0$$

then this state is said to have a plane state of stress parallel to xy plane. This state is also known as a two dimensional state of stress. In the plane state of stress, stresses are confined to a plane. From the mechanics of solids, it can be shown that principal stresses σ_1 and σ_2 and their orientations with θ and $\theta + \pi/2$ with respect to the Ox is given by

$$\sigma_1, \sigma_2 = \frac{\sigma_x + \sigma_y}{2} \pm \sqrt{\left[\frac{\sigma_x - \sigma_y}{2}\right]^2 + \tau_{xy}^2}$$

$$\tan 2\theta = \frac{2\tau_{xy}}{\sigma_x - \sigma_y}$$

$$\sigma_x - \sigma_y = (\sigma_1 - \sigma_2) \cos 2\theta \qquad (6.4)$$

$$\tau_{xy} = \frac{1}{2}(\sigma_1 - \sigma_2) \sin 2\theta.$$

6.9 STRESS-OPTIC RELATIONS–TWO DIMENSIONAL CASE

Let us consider a model of uniform thickness made of glass, epoxy or some transparent high polymer material. Let the model be loaded such that it is in a plane state of stress (Fig. 6.22). Then the state of stress can be characterized by σ_x, σ_y and τ_{xy} or by the principal stresses σ_1, σ_2 and their orientation with respect to a set of axes. Let μ_0 be the refractive index of the material when it is in a free stare (unstressed). When the model is put in a stress, experiments show that the

(i) Model become doubly refractive.

(ii) The directions of polarization of the light at the point P coincide with the direction of principal stress axis at that point.

(iii) If μ_1 and μ_2 are the refractive indices for vibrations corresponding to these two directions, then

$$\mu_1 - \mu_0 = c_1 \sigma_1 - c_2 \sigma_2$$
$$\mu_2 - \mu_0 = c_1 \sigma_2 - c_2 \sigma_1 \qquad (6.5)$$

c_1 is called the direct stress optic coefficient and c_2 the transverse stress optic coefficient. Since the stresses vary uniformly, σ_1, σ_2 and θ are continuously distributed functions over the model in the xy plane. The directions of the polarizing axes as well as the values of μ_1 and μ_2 vary uniformly over the xy-face of the model.

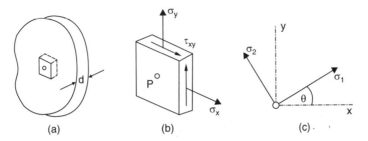

Fig. 6.22 Plane state of stress

If a plane polarized light is incident normally at any point P of the model, the incident light gets resolved along σ_1 and σ_2 and these two vibrating components travel through the thickness of the model with different velocities. The velocities of propagation are governed by equation (6.5). When they emerge, there will be a certain amount of relative phase difference between these two components. The phase difference is given by

$$\varepsilon = \frac{2\pi d}{\lambda}(\mu_1 - \mu_2) = \frac{2\pi d}{\lambda}(c_1 + c_2)(\sigma_1 - \sigma_2) \quad \text{(From Eqns. 6.3 and 6.5)}$$

if $c_1 + c_2$ is set equal to c, the relative retardation is given by

$$\varepsilon = \frac{2\pi d}{\lambda} c(\sigma_1 - \sigma_2) \tag{6.6}$$

The number of wavelengths of relative path difference is given by

$$N = \frac{\varepsilon}{2\pi} = \frac{d}{\lambda} c(\sigma_1 - \sigma_2) \tag{6.7}$$

Equations 6.6 and 6.7 are called stress-optic relations or stress optic law. These equations relate the state of stress at a point to the optical behaviour of the model. In practice one evaluates the values of $(\sigma_1 - \sigma_2)$ from the observed values of ε or N. Then

$$(\sigma_1 - \sigma_2) = \frac{N\lambda}{dc} = \frac{N}{d} F \tag{6.8}$$

F is called the material fringe value. If $d = 1$ cm and $N = 1$ wavelength, then F gives the values of $(\sigma_1 - \sigma_2)$. This produces a relative phase difference of 2π radian on a model of unit thickness. This is a property of the model material and wavelength of light used. The quantity $F/d = f$ is called the model fringe value. If θ is the orientation of the σ_1 axis with reference to x-axis as shown in Fig. 6.22 then from equation (6.4).

$$(\sigma_x - \sigma_y) = \frac{NF}{d} \cos 2\theta \tag{6.9}$$

$$\tau_{xy} = \frac{NF}{2d} \sin 2\theta$$

6.10 ISOCHROMATICS AND ISOCLINICS

Let us consider an arrangement shown in Fig. 6.23a. S is the source of monochromatic light, P is a polarizer, M is the model under a plane state of stress. A called the analyzer is a second polarizing element kept at 90° to the polarizer and B is the screen. We shall assume that through a suitable optical arrangement the image of the model is projected on the screen. We shall trace the passage of a typical ray of light through the various optical elements in the assembly.

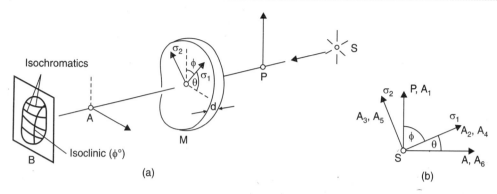

Fig. 6.23 Planar photoelastic model in polariscope

The arrangement shown in Fig. 6.23 (a) is known as a plane polariscope. The polarizer and the analyzer are always kept crossed. But their combined orientation can be arbitrary. Now we can make a few important observations based on the knowledge of crystal optics. Where the model is stressed, it behaves like a double refracting crystal and at the point where the ray passes, the polarizing axes coincide with the principal stress axes, σ_1, σ_2 at that point.

In general the polarizer makes an angle ϕ with the σ_1 axis. If ϕ happens to be zero or $\pi/2$, i.e., direction of the polarizer coincides with either σ_1, (or σ_2) then a plane polarized light incident on the model at that point will emerge as a plane polarized light. Since the analyzer is kept crossed with respect to the polarizer, no light comes out of A. Consequently, at all those points of the model where the directions of the principal stresses happen to coincide with the particular orientation of the polarizer-analyzer combinations, the light coming out of the analyzer is zero. If the polarizer-analyzer combination happens to coincide with the directions of σ_1, σ_2 stresses at one point of the model, then in general, there will be a locus of points in the model along which this condition is satisfied. This is so because in general, the stresses are distributed in a continuous manner in the model.

The locus of points where the directions of principal stresses coincide with a particular orientation of the polarizer-analyzer combination is known as the isoclinic (meaning same inclination). For example, if the polarizing element is kept vertical and the analyzer kept horizontal, then on the screen a dark band will be seen which is the locus of points where σ_1 and σ_2 directions happen to be vertical and horizontal. If one measures angles from the vertical reference axis, this isoclinic will be called the 0° isoclinic. If now the polarizer is turned through 30° and the analyzer is also rotated through an equal amount (so that the analyzer is always kept crossed with respect to the polarizer) then the previously observed 0° isoclinic vanishes and a new dark band is observed on the screen. This is the 30° isoclinic and it represents the locus of points in the model where the principal stress axes are oriented at 30° and 30° + $\pi/2$ with respect to the vertical.

Let us now consider another situation. Suppose at a particular point of the model, the values of σ_1, and σ_2 are such as to cause a relative phase difference of $2m\pi$ where m is an integer. Recall that if the relative phase difference is $2m\pi$, the model behaves like a full wave plate at that particular point. Therefore at all those points of the model where the values of σ_1 and σ_2 are such as to cause a relative phase difference of $2m\pi$ (m = 0, 1, 2, ...) the intensity of light will be zero (if the polarizer and analyzer are crossed.) On the screen a series of dark bands or fringes corresponding to these locus of points are observed. These dark bands or fringes are called isochromatics. An isochromatic is a locus of points where the values of σ_1, σ_2 are such as to cause a relative phase difference of $2m\pi$ (where m = 0, 1, 2, 3, ...) when the

POLARIZATION OF LIGHT AND PHOTOELASTICITY   233

background is dark. The locus of points, where the values of $\sigma_1 - \sigma_2$ are such as to cause zero radian of phase difference, is called the zero-order fringe. The locus of points, where the values of σ_1, σ_2 are such as to cause 2π radian phase difference (equivalent to a relative path difference of λ) is known as first order fringe. Similarly, one can observe second order fringe, third order fringe and so on. Typical isoclinic and isochromatics for a stressed circular disc is shown in Fig. 6.24.

6.11 MATHEMATICAL ANALYSIS OF ISOCLINICS AND ISOCHROMATICS

Let the light coming from the polarizer be $A_1 = a \cos \omega t$.

On entering the model, the light vector gets resolved along principal stress axes (Fig. 6.23b). Thus

$$A_2 = a \cos \phi \cos \omega t$$
$$A_3 = a \sin \phi \cos \omega t \qquad (6.10)$$

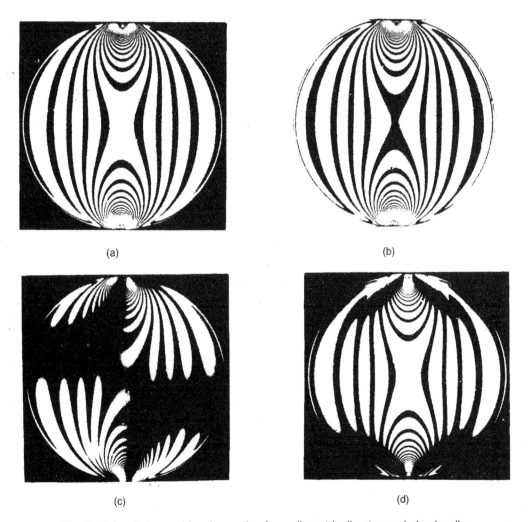

Fig. 6.24 Isoclinics and Isochromatics for a diametrically stressed circular disc
(a) Isochromatics-in dark field set up (b) Isochromatics-in bright field view
(c) 15°- Isoclinic (d) 45°- Isoclinic

In traversing the thickness d of the model the two components acquire a relative phase difference ε. Hence on leaving the model the vibrating components are

$$A_4 = a \cos \phi \cos (\omega t + \varepsilon)$$
$$A_5 = a \sin \phi \cos \omega t \qquad (6.11)$$

Generally speaking A_4 and A_5 will both acquire certain amounts of absolute phase differences. However, the final result in the process of analysis depends essentially on the relative phase difference. On entering the analyzer only the components along A are allowed to emerge.

$$A_6 = A_4 \sin \phi - A_5 \cos \phi$$
$$= a \cos \phi \sin \phi \cos (\omega t + \varepsilon) - a \cos \phi \sin \phi \cos \omega t$$
$$= \frac{a}{2} \sin 2\phi [\cos (\omega t + \varepsilon) - \cos \omega t]$$
$$= \frac{a}{2} \sin 2\phi [\cos \omega t (\cos \varepsilon - 1) - \sin \omega t \sin \varepsilon]$$
$$= \frac{a}{2} \sin 2\phi \left[-2 \cos \omega t \sin^2 \frac{\varepsilon}{2} - 2 \sin \omega t \sin \frac{\varepsilon}{2} \cos \frac{\varepsilon}{2} \right]$$
$$= -a \sin 2\phi \sin \frac{\varepsilon}{2} \left(\cos \omega t \sin \frac{\varepsilon}{2} + \sin \omega t \cos \frac{\varepsilon}{2} \right)$$
$$= -a \sin 2\phi \sin \frac{\varepsilon}{2} \sin \left(\omega t + \frac{\varepsilon}{2} \right)$$
$$= -b \sin \left(\omega t + \frac{\varepsilon}{2} \right) \qquad (6.12)$$

where $b = a \sin 2\phi \sin \frac{\varepsilon}{2}$ is the amplitude of the light emerging from the analyzer. A measure of the intensity of the light is given by the square of the amplitude. The intensity of the emerging light is therefore

$$I = a^2 \sin^2 2\phi \sin^2 \varepsilon/2 \qquad (6.13)$$

The intensity of light is zero under two conditions :

(i) when $\phi = 0$ or $\pi/2$; or/and

(ii) when $\varepsilon = 2m\pi$ ($m = 0, 1, 2, 3, ...$)

Condition (i) tells us that light extinction occurs at a point where the direction of the principal stresses coincide with the direction of the polarizer and the analyzer. According to the condition (ii) the intensity is zero when the relative phase difference is equal to $2m\pi$. The locus of points where the direction of principal stresses has a common orientation with reference to a given axis is called an isoclinic. The locus of points satisfying condition (ii) is called an isochromatic (or equal retardation points).

6.12 STRESS-OPTIC LAW FOR THREE DIMENSIONS

So far we have discussed stress-optic law for the case of two dimensions. In general the state of stress at a point is characterized by σ_x, σ_y, σ_z, τ_{xy}, τ_{yz}, τ_{zx} or equivalently by the three principal stresses σ_1, σ_2, σ_3. In solids which exhibit birefringence under stress, Maxwell found that the principal axes of the refractive index ellipsoid coincide with the principal stress axes at the point considered. If μ_0 is the refractive index of the unstressed medium and if μ_1, μ_2, μ_3 are the principal indices of refraction at a point (corresponding to the three semi axes of the ellipsoid)

POLARIZATION OF LIGHT AND PHOTOELASTICITY

when the model is under stress, then the relationship between $\mu_0, \mu_1, \mu_2, \mu_3$ and the stresses $\sigma_1, \sigma_2, \sigma_3$ are expressed by Maxwell's stress-optic relations as

$$\mu_1 - \mu_0 = c_1 \sigma_1 - c_2 (\sigma_2 + \sigma_3)$$
$$\mu_2 - \mu_0 = c_1 \sigma_2 - c_2 (\sigma_3 + \sigma_1) \quad (6.14)$$
$$\mu_3 - \mu_0 = c_1 \sigma_3 - c_2 (\sigma_1 + \sigma_2)$$

where c_1 is called the direct stresses optic coefficient and c_2 the transverse optic coefficient.

Taking the difference between the above expressions we have

$$\mu_1 - \mu_2 = (c_1 + c_2)(\sigma_1 - \sigma_2) = 2c\tau_3$$
$$\mu_2 - \mu_3 = (c_1 + c_2)(\sigma_2 - \sigma_3) = 2c\tau_1 \quad (6.15)$$
$$\mu_3 - \mu_1 = (c_1 + c_2)(\sigma_3 - \sigma_1) = 2c\tau_2$$

where τ_1, τ_2, τ_3 are the principal shear stress at the point under consideration. Three dimensional photoelasticity is more complex than the two dimensional one. The frozen-stress method is well suited for three dimensional studies. Certain optically sensitive materials like bakelite when annealed in a stress condition retain the deformation and birefringent characteristics of the initially stressed state when the load is removed. A three-dimensional model may therefore be cut into slices that may then be analyzed individually on somewhat the same principles as the planer models.

6.13 MOIRE FRINGES

The word 'Moire' derives it origin from a silk fabric which when superimposed on itself exhibits patterns of light and dark bands. The Moire fringes are produced by the superposition of two sets of grating under certain circumstances. The gratings could be linear, circular or elliptical.

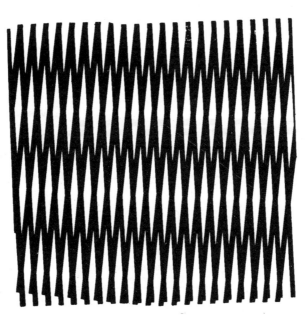

Fig. 6.25 Moire fringes when two linear gratings are tilted with respect to one another

Beautiful Moire patterns could be generated using such gratings. The Moire pattern which results by tilting one line grating with respect to another is shown in Fig. 6.25. The idea of utilizing such Moire patterns for analyzing strain was thought of as early as 1859, when a method of testing lenses and optical systems by using low frequency line gratings was introduced by Foucault. In 1874 British Physicist Lord Rayleigh suggested that Moire patterns could be used to test the perfection of ruled diffraction grating. By placing the plastic repilca over the original grating one can immediately see any periodic errors made by the ruling engine or any distortions resulting from the production of the replica.

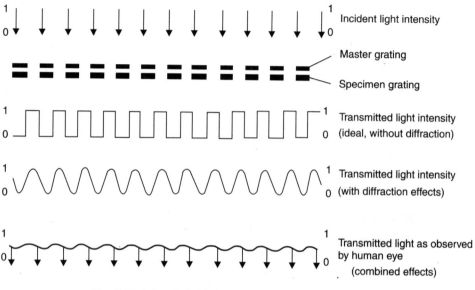

Fig. 6.26 Intensity of light transmitted through identical, superposed and aligned specimen and master gratings

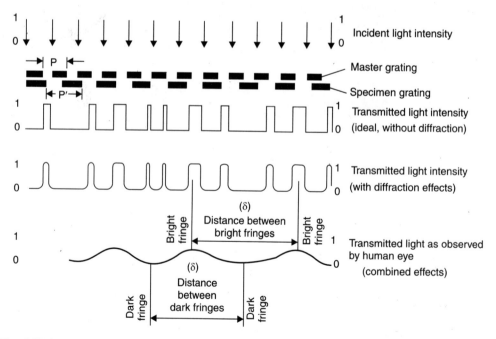

Fig. 6.27 Intensity of light transmitted through linearly deformed specimen and master gratings

The basic principle for analyzing strain using Moire fringes is as follows. Of the two gratings used for fringe formation, one grating is either bonded to or etched or printed on the specimen being analyzed and is termed 'Model' grating or 'Specimen grating'. The other grating is known as the 'Master grating' or 'Reference grating'. The specimen grating undergoes deformation depending on the state of the strain on the surface and is accompanied by a change of spacing (*i.e.*, pitch) between the lines of grating. The master grating which is not strained does not undergo any change in the spacing between the lines of grating. Light passing through the strained specimen grating and the unstrained master grating interfere to form Moire fringes. Figures 6.26 and 6.27 show the intensity of the light patterns when the specimen and master gratings are perfectly aligned or misaligned. The Moire pattern when the specimen grating undergoes a shear strain is indicated in Fig. 6.28. It can be shown that the Moire fringe is the locus of points having the same component of displacement in the principal direction of the master grating. Such a locus is called Isothetic. The experimental set up for obtaining Moire fringes is shown in Fig. 6.29. The exact determination of strain from isothetics is beyond the scope of this book.

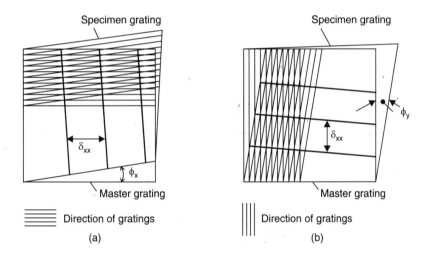

Fig. 6.28 Moire patterns for a pure shear strain

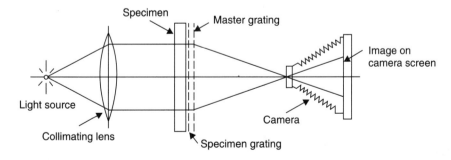

Fig. 6.29 Experimental setup for Moire method of strain analysis

REFERENCES

1. F.A. Jenkins and H.E. White, *Fundamentals of Optics,* McGraw-Hill Book Company, New Delhi, 1985.
2. E.G. Coker and L.N.G. Filon (Revised by H.T. Jessop), *A Treatise on Photoelasticity,* Cambridge University Press, 1957.
3. M.M. Frocht, *Photoelasticity,* Vol. I and II, John Wiley and Sons, New York, 1941.
4. L.S. Srinath M.R. Raghvan, K. Lingaiah, G. Gargesha, B. Pant and K. Ramachandra, *Experimental Stress Analysis,* Tata McGraw-Hill Publishing Company Limited, New Delhi, 1984.
5. L.S. Srinath, *Scattered Light Photoelasticity,* Tata McGraw Hill Publishing Company Limited, 1983.
6. A.J. Durelli and V.J. Parks, *Moire Analysis of Strain,* Prentice Hall Inc., New Jersey, 1970.
7. A.J. Dureli, *Applied Stress Analysis,* Prentice Hall Inc., New Jersey, 1967.

SOLVED EXAMPLES

1. The refractive index of glass is 1.5. Calculate the polarising angle.

   Solution:

   If the polarising angle is i_p, then tan i_p = 1.5. From Tan tables we get polarising angle $i_p = 56°18'$.

2. Calculate the thickness of a quarter wave plate for light of wavelength 6000Å ($\mu_0 = 1.554$; $\mu_e = 1.544$).

   Solution: The number of wavelengths of relative path difference

   $$= \frac{d(\mu_0 - \mu_e)}{\lambda} = N$$

   Here $N = 1/4$

   $$\therefore \quad d = \frac{\lambda}{4(\mu_0 - \mu_e)} = \frac{6000 \times 10^{-10}}{4(1.554 - 1.544)} = 1.5 \times 10^{-5} \text{ m.}$$

3. A quarter wave plate is 12.5×10^{-6} thick. Calculate the wavelength for which it acts as a quarter wave plate. The difference in the principal refractive indices is 0.01.

   Solution:

   Here $\quad d = \dfrac{\lambda}{4(\mu_0 - \mu_e)}$

   We are given $\quad d = 12.5 \times 10^{-6}$ m

   $\mu_0 - \mu_e = 0.01$

   $\therefore \quad \lambda = 4 \times 0.01 \times 12.5 \times 10^{-6} = 5 \times 10^{-7}$ m.

4. Plane polarized light passes through a quartz plate with its axis parallel to the face. Calculate the thickness of the plate so that the emergent light may be plane polarized. For quartz $\mu_e = 1.553$, $\mu_0 = 1.542$, $\lambda = 5.5 \times 10^{-5}$ cm.

   Solution:

   The minimum thickness is that of a half wave plate

   $$d = \frac{\lambda}{2(\mu_0 - \mu_e)} = \frac{5.5 \times 10^{-5}}{2(1.553 - 1.542)} = 2.5 \times 10^{-5} \text{ m.}$$

POLARIZATION OF LIGHT AND PHOTOELASTICITY

QUESTIONS

1. What is polarized light? What is the difference between polarized and unpolarized light?
2. Explain Brewster's Law.
3. Describe the construction and action of a nicol prism.
4. Describe how elliptically and circularly polarized light are obtained.
5. What are quarter and half wave plates?
6. What are the different methods of producing polarized light?
7. Define the phenomenon of photoelasticity.
8. What are plane and circular polariscope?
9. What is stress-optic law?
10. Describe isoclinics and isochromatics.
11. Write a short note on Moire fringes.

PROBLEMS

1. Find the polarizing angle for a glass of refractive index 1.732. **(Ans. 60°)**
2. Calculate the thickness of quarter wave plate of quartz for light of wavelength 5893×10^{-10} m given $\mu_e = 1.553$ and $\mu_0 = 1.544$. **(Ans. 1.673×10^{-5})**
3. The refractive index of calcite is 1.648 for ordinary ray and 1.486 for extraordinary ray. A plane parallel phase of calcite is cut of thickness 0.01 mm. For what wavelengths in the visible region will this plate behave as (a) ouarter wave plate (b) a half wave plate.
   (Ans. (a) 720, 589.9, 498.5 and 432 nm (b) 648 and 462.9 nm)
4. A beam of plane polarized light is converted into circularly polarized light by passing it through a crystal slice of thickness 3×10^{-5} m. Calculate the difference in the refractive indices of the two rays inside the crystal assuming the above thickness to be the minimum value required to produce the observed effect. Wavelength of light = 600 nm. **(Ans. 5×10^{-3})**

7

Laser

7.1 INTRODUCTION

Laser is an acronym for light amplification by stimulated emission of radiation. In a laser, *stimulated emission* is used to amplify light waves. Lasers are essentially highly directional, highly intense, highly monochromatic and highly coherent optical sources. On account of these properties lasers have become a useful tool in industry, medicine and basic sciences. Spontaneous emission and stimulated emission constitute two modes of emission of radiation by atoms. Stimulated emission was postulated by Einstein as early as in 1917. However it was only in 1960 that Maiman built a Ruby laser on this principle. Since then the development of lasers has been extremely rapid and laser action has been demonstrated with gases, liquids, solids, semiconductors, free electrons and plasma.

7.2 CHARACTERISTICS OF THE LASER LIGHT

The most outstanding characteristics of the laser beam are its high degree of directionality, intensity, monochromaticity and coherence.

7.2.1 Directionality

The laser beam is highly directional. As the laser beam travels through space it diverges. This is because of the diffraction due to the finite length of the wavefront. The angle of divergence of a parallel beam is directly proportional to the wavelength and inversely proportional to the beam diameter. Figure 7.1 shows the divergence of a laser beam with circular area of cross-section. At the output aperture of the laser, the beam diameter is d. Its beam divergence angle is θ, usually expressed in mill radian. In traversing a distance l, the beam diverges to a circle of diameter d'. If θ is a small angle, the diameter d' of the beam at a distance l from the output aperture is given by :

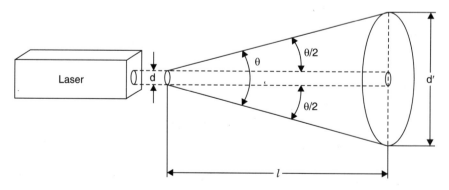

Fig. 7.1 Directionality of the laser beam

$$d' \simeq l\theta + d$$

The beam divergence angle varies considerably from one family of lasers to another. Continuous wave lasers generally have the smallest values of beam divergence (1 m rad or less). Solid state and organic dye lasers usually have larger beam divergence angles (5 – 20 m rad). Semiconductor lasers have a relatively wide cone of radiation as much as 30° or 0.524 radian. On account of high directionality of the laser beam it is possible to direct a laser beam to a far off object and obtain the reflection of the same. The distance between the earth and the moon has been accurately measured by directing the laser beam to the moon and obtaining the laser reflection from a mirror placed on the moon's surface by the astronauts. The time taken for the laser pulse to travel back and forth multiplied by the velocity of light gives the distance.

7.2.2 Intensity

The laser beam is highly intense as compared to ordinary sources of light. That is why it can be used for such operations as welding of metals which involve high temperatures. According to geometrical optics light travels along straight lines. A point source of light can be focused into a point image. However, light has a wave nature and this plays a major role in the propagation and focusing of laser beams. Figure 7.2 shows the focusing of a laser beam by means of simple lens of radius r and focal length f. The beam has full width angular divergence θ. In geometrical optics, it is shown that the distance d_1, of the focused beam is approximately equal to the focal length of the lens f, multiplied by the divergent angle θ i.e., $d_1 \sim f\theta$. To compute the amount of power per unit area concentrated on the spot using a laser as the light source, the area A_{spot} of the focused beam :

$$A_{spot} = \pi \frac{d_1^2}{4} = \pi \frac{(f\theta)^2}{4} = 0.785 \, (f\theta)^2$$

The power density at the focused spot is given by :

$$I = \frac{P}{A_{spot}} = \frac{P}{0.785 \, (f\theta)^2}$$

There is a limit to which a laser beam can be focused. This is the *diffraction limited beam spot size*. This can be shown to be equal to $\sim 5\lambda^2$, where λ is the wavelength of the light used.

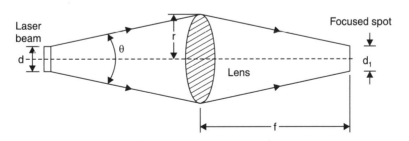

Fig. 7.2 Laser beam focusing

7.2.3 Monochromaticity

An optical source of light does not emit light of single wavelength. The emission actually occurs over a range of wavelengths. The spectrum of light emitted by a source can be represented by a mathematical function $g(\nu)$, which is called the *line shape function*. The function $g(\nu) \, d\nu$

denotes the probability that a given transition between the two energy levels result in the emission of a photon whose frequency lies between ν and $\nu + d\nu$. The line shape function $g(\nu)$ is normalised in such a way that

$$\int g(\nu)\, d\nu = 1$$

The form of the line shape function $g(\nu)$ is usually a Lorentzian or a Gaussian.

The Lorentzian line shape function may be written as:

$$g(\nu)_L = \frac{\Delta \nu}{2\pi}\left[\frac{1}{(\nu - \nu_0)^2 + (\Delta\nu/2)^2}\right] \quad \text{and} \quad g(\nu_0)_L = \frac{2}{\pi \Delta\nu} = \frac{2\tau}{\pi}$$

where $\Delta\nu$ is the line width (i.e., the separation between two points on the frequency curve where the function falls to half its peak value, the peak value itself occurring at a frequency ν_0.
τ is called the lifetime of the energy level.

The Gaussian line shape function is given by :

$$g(\nu)_G = \frac{2}{\Delta\nu}\left(\frac{\ln 2}{\pi}\right)^{1/2} \exp\left[-\ln 2\left(\frac{\nu - \nu_0}{\Delta\nu/2}\right)^2\right] \quad \text{and} \quad g(\nu_0)_L = \frac{2}{\Delta\nu}\left(\frac{\ln 2}{\pi}\right)^{1/2} \approx \frac{1}{\Delta\nu} \approx \tau$$

The Lorentzian and Gaussian line shape functions are shown in Fig. 7.3.

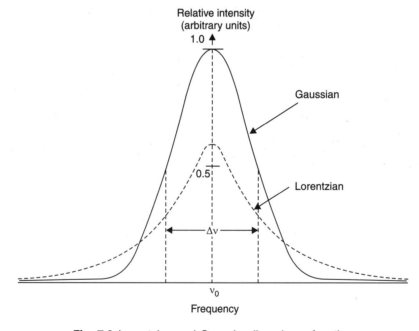

Fig. 7.3 Lorentzian and Gaussian line shape function

The important mechanisms, which contribute to the line width, are :

(a) Natural broadening (b) Doppler broadening and (c) Collision broadening.

Natural Broadening: This essentially arises on account of the Heisenberg's uncertainty principle. There is always an uncertainty in energy associated with each energy level. As a consequence a photon emitted during the transition between the two energy levels has a range of frequencies and not a single value. The natural broadening is shown in Fig. 7.4.

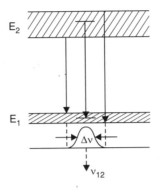

Fig. 7.4 Natural line width of a spectral line transition

Doppler broadening: Doppler broadening arises because the atoms are in motion. The Doppler effect refers to the apparent change in the frequency of the source whenever there is a relative motion between the source and the observer. The frequency as measured by the observer increases if the source and the observer approach one another and decreases as they recede from one another. This effect is observed in a collection of atoms emitting at an optical frequency v_{12} due to the transition between energy levels E_1 and E_2. The observed frequency v'_{12} is given by

$$v'_{12} = v_{12}\left(1 \pm \left(\frac{v}{c}\right)\right)$$

where v is the component of velocity of the atom along the direction of observation (Fig. 7.5b). If the atom is moving towards the observer the frequency of the photon will be higher while if the atom moves away from him the frequency of the photon will be lower. Since the atoms are in random motion, an observer would measure a range of frequencies depending on the magnitude and direction of v. Hence as far as the observer is concerned the collection of atoms will be emitting frequencies over a range of frequencies, resulting in the broadening of the spectral line (Fig. 7.5c).

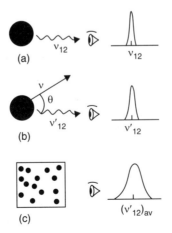

Fig. 7.5 Doppler broadening of a spectral line (a) Atom at rest. The observed frequency is v_{12}.
(b) Doppler effect (c) Broadening

The kinetic energy of gas atoms is proportional to temperature. Hence the velocity is directly proportional to the square root of temperature. Thus the half width of the line shape function is proportional to the square root of the temperature. Doppler broadening is the predominant mechanism in most of the gas lasers emitting in the visible region.

Collision Broadening: If an atom undergoes collision while emitting a photon, the phase of the wave train associated with the photon is suddenly altered. This in effect shortens the length of the emitted wave train (Fig. 7.6). *According the Fourier theorem, smaller the length of the wave train greater is the spread in the frequency domain.* Hence in a collection of atoms where collision between them is a natural occurrence, there is a spread in the frequency of emitted photons. This is called collision broadening. In the case of gas atoms, at high temperature and high pressure, the atoms will undergo collisions more frequently. Thus line broadening will be more at higher temperatures and higher pressures. Collision broadening also occurs in the case of doped insulator lasers. In these lasers, the electrons of the active medium may be thought of as undergoing collisions with phonons *i.e.*, quantised lattice vibrations.

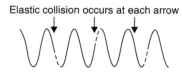

Fig. 7.6 Collision broadening

Line broadening mechanisms can be classified as *homogeneous* and *inhomogeneous*. If all the atoms of the collection have the same transition centre frequency and identical line shape, then the broadening is termed *homogeneous*. Homogeneous broadening is represented by a Lorentzian line shape. In some situations, each atom has a slightly different centre frequency or line shape for the same transition. The observed line shape is then the average of the individual atoms. This is called *inhomogeneous broadening* and is represented by the Gaussian line shape.

Because of all these line broadening mechanisms, the light emitted by any source is strictly not monochromatic. Instead there is always a spread in the frequency of emittted photons. However, in the case of lasers, the line widths are much smaller compared to the ordinary sources and therefore are more coherent. Typical line widths of various sources of light are given in Table 7.1.

7.2.4 Coherence

Coherence is related to phenomenon of interference. *Interference is observed only with coherent sources.* The laser beam is temporally and spatially coherent to an extraordinary degree. It is possible to observe interference effects from two independent laser beams. Temporal coherence is referred to as longitudinal coherence while the spatial coherence is called lateral coherence.

Temporal coherence. Light is emitted whenever an atom makes a transition between two energy levels E_1 and E_2. According to Heisenberg's uncertainty principle, each level E_1 and E_2 has a width associated with it. Consequently, the light which is emitted is not strictly monochromatic. It is always characterised by a line width (Fig. 7.4). *An alternate way of describing this is to say that real light sources emit wave trains of finite length. (This follows from Fourier theorem since energy or frequency and time are conjugate variables).* In such a wave train, the phase difference $\Delta\phi$ between two fixed points x_1 and x_2, spaced any distance along the wavefront is time independent (Fig. 7.7). Hence temporal coherence is often called longitudinal coherence. In contrast, in spatial coherence, the phase difference for any two fixed points in a plane normal to the ray direction remains time-independent.

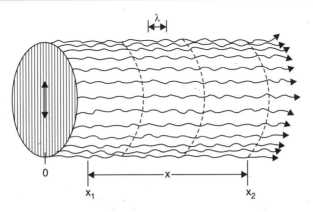

Fig. 7.7 Longitudinal or temporal coherence

The coherence length, $l_c = c \times \tau_c$, where c is the velocity of light and τ_c is called the coherent time. $\tau_c = 1/\Delta v$ (Δv being the line width)

$$l_c = c \times \tau_c = \frac{c}{\Delta v} = \frac{\lambda^2}{\Delta \lambda}$$

$$\left(\because c = v\lambda \text{ and } v\Delta\lambda + \lambda\Delta v = 0 \Rightarrow \Delta v = -v\frac{\Delta\lambda}{\lambda} = -c\frac{\Delta\lambda}{\lambda^2} \right)$$

Temporal coherence can also be understood well, by considering the Young's double slit experiment (Fig. 7.8). In the Young's double slit experiment, if the source emits only one wavelength λ_1, then the fringes are formed everywhere on the screen. The positions of maxima and minima are given by

$$Y_n = \frac{Dn\lambda_1}{d} \text{ (for intensity maxima)}$$

and
$$Y_n = \frac{D\lambda_1}{d}\left(n - \frac{1}{2}\right) \text{ (for intensity minima); } \beta \text{ (fringe width)} = \frac{D\lambda_1}{d}$$

Here Y_n denotes the distance from the centre of the screen, d is the distance between the slits, D is the distance of the screen from the plane of the two slits and n is an integer which denotes the order of the interference maxima or minima.

Assume that the source emits another wavelength λ_2 along with λ_1. At the center of the screen in Fig. 7.8, $Y = 0$ and the maxima of both wavelengths coincide exactly. As we move away from the center, since the fringe spacings of λ_1 (which is equal to $\lambda_1 D/d$) and of λ_2 (equal to $\lambda_2 D/d$) are unequal, at some value of Y, the maxima of λ_1 and the minima of λ_2 will start to overlap. In such a region fringes are not observed. For this to happen at a value of Y, the optical path difference (OPD) of λ_1 for intensity maximum, should be equal to the OPD of λ_2 for intensity minima. Thus

$$n\lambda_1 = \left(n - \frac{1}{2}\right)\lambda_2 \quad \text{or} \quad n = \frac{\lambda_2}{2(\lambda_2 - \lambda_1)} = \frac{\lambda_2}{2\Delta\lambda}$$

$$\therefore \quad \text{OPD} = n\lambda_1 = \frac{\lambda_1\lambda_2}{2\Delta\lambda} \approx \frac{\lambda^2}{2\Delta\lambda}$$

where $\Delta\lambda = \lambda_1 - \lambda_2$ and we have assumed $\lambda_1\lambda_2 \cong \lambda^2$, where λ is the average wavelength.

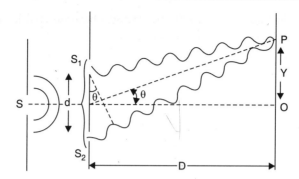

Fig. 7.8 Temporal coherence and Young's double slit experiment

The quantity $\lambda^2/2\Delta\lambda$ represents the maximum optical path difference so as to obtain an observable interference pattern in the presence of two wavelengths spaced by $\Delta\lambda$. If the source emits a continuous spectrum of width $\Delta\lambda$, then one can show that for good contrast the path difference should be less than $\lambda^2/\Delta\lambda$. This quantity is referred to as the coherent length. The coherent time τ_c is defined as

$$\tau_c = \frac{l_c}{c} = \frac{1}{\Delta\nu}$$

Thus, more monochromatic a wave is, the larger is the coherent length and coherent time. The coherent length, coherent time and line widths for a number of sources are shown in Table 7.1.

Table 7.1 Line width of a number of light sources together with their coherence times and coherence lengths in free space

Source	$\Delta\nu(Hz)$	$\tau_c = 1/\Delta\nu$	$l_c = c\tau_c$
Filtered Sunlight ($\lambda = 0.4$–0.8 mm)	3.75×10^{14}	2.67 fs	800 nm
LED ($\lambda_0 = 1$ µm)	1.5×10^{13}	67 fs	20 µm
Low pressure sodium lamp			
He-Ne Laser (Multi mode) ($\lambda_0 = 633$ nm)	5×10^{11}	2 ps	600 µm
	1.5×10^9	0.67 ns	20 cm
He-Ne laser (Single mode) ($\lambda_0 = 633$ nm)	1×10^6	1 µs	300 m

Spatial Coherence. The waves at different points in space are said to be space coherent, if they preserve a constant phase difference over any time. *This is possible even when two beams are individually time incoherent, as long as any phase change in one of the beams is accompanied by a simultaneous equal phase change in the other beam.* With the ordinary light sources, this is possible only if the two beams have been produced in the same part of the source. In Young's double slit experiment, the two slits S_1 and S_2 are illuminated by an extended source S. It can be shown that fringes are washed out (*i.e.*, not observed), if illumination emanates from a source of angular diameter $\theta_s > \lambda/2a$ (Fig. 7.9). '$2a$' is the distance between the slits. If the distance $2a$ is smaller than λ/θ_s, the fringes become visible. To characterise spatial coherence, one defines a coherent distance $\rho = \lambda/\theta_s$.

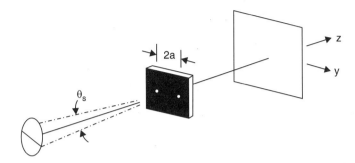

Fig. 7.9 Spatial coherence and Young's double slit experiment

7.3 BASIC CONCEPTS OF LASER

Some of the basic concepts associated with the laser are absorption, spontaneous emission, stimulated emission, population inversion and pumping.

7.3.1 Absorption

Energy levels of atoms or molecules in a material are classified as electronic, vibrational and rotational levels. The separation between rotational and vibrational energy levels corresponds to photon energies in the infrared and microwave region. The separation between electronic energy levels corresponds to photon energies in the visible and ultraviolet region. According to quantum mechanics all these levels are quantized. Consider an atom in the lower energy state E_1. A photon of frequency ν can excite the atom to higher energy state E_2 provided, $h\nu = E_2 - E_1$. This is called absorption (Fig. 7.10a).

7.3.2 Spontaneous Emission

In the excited state E_2 the atom stays for a finite time before returning to the energy state E_1. The duration of the stay is called the *lifetime* of the level. On undergoing transition to E_1, energy difference $(E_2 - E_1)$ must be released by the atom. When this energy is delivered in the form of an electromagnetic wave or photon, the process is called spontaneous emission (Fig. 7.10b). This is termed as *radiative transition*. The frequency ν of the radiated wave is then given by the expression, $h\nu = E_2 - E_1$. *The photon which is emitted by an atom is in random direction.* Hence in an ensemble of atoms, photons which are emitted are in random directions.

The transition to the state E_2 can also occur in *a non-radiative* way for some levels. In this case the energy difference $E_2 - E_1$ is delivered in some form other than electromagnetic radiation (*e.g.*, it may go into kinetic energy of the surrounding molecules). An example of this type is the transition between energy levels across the band gap in indirect band gap semiconductors. In this case the energy is used to excite the lattice vibrations.

7.3.3 Stimulated Emission

A photon of energy $h\nu = E_2 - E_1$ can interact with an atom in the excited state E_2 and force it to undergo a transition to E_2 (Fig. 7.10c). In this case the energy difference $E_2 - E_1$ is emitted in the form of an electromagnetic wave in the same direction as that of the incident photon. Further, this is in phase with the incident photon and has the same polarization as that of the incident photon. Hence this leads to amplification of the incident photon. This is the phenomenon of stimulated emission. Unlike spontaneous emission which is random in direction, stimulated emission is directional and also more intense. The concept of stimulated emission was first put forward by A. Einstein in 1917.

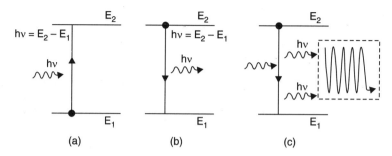

Fig. 7.10 (a) Absorption (b) Spontaneous emission and (c) Stimulated emission

7.3.4 Transition Probabilities

According to quantum mechanics, the transitions between energy levels due to absorption, spontaneous emission and stimulated emission are probabilistic in nature. It is possible to work out these probabilities by the using the methods of quantum mechanics. If the probability of transition is zero, then the transition is termed as forbidden transition.

7.3.5 Population Inversion

When in thermal equilibrium at a temperature T, the number of atoms or molecules is distributed among various energy levels according the Maxwell-Boltzmann's distribution function given by :

$$N_i = N_0 e^{-E_1/k_B T} \qquad (7.1)$$

where N_i is the number of atoms having the energy E_i, N_0 is the number of atoms in the ground state. k_B is the Boltzmann's constant and T is the temperature of the system in Kelvin. Thus the population is maximum in the ground state and decreases exponentially as one goes to higher energy states (Fig. 7.11a). If N_1 and N_2 are the populations in the lower state E_1 and higher state E_2, we have

$$\frac{N_2}{N_1} = e^{(E_1 - E_2)/k_B T} = e^{-h\nu/k_B T} < 1 \qquad (7.2)$$

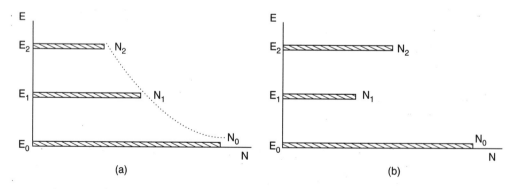

Fig. 7.11 (a) Maxwell Boltzmann distribution (b) Population inversion

Thus it follows that $N_2 < N_1$. Hence when atoms and radiation are in thermal equilibrium, the number of atoms undergoing spontaneous emission is greater than the number of atoms undergoing stimulated emission.

For laser action to take place, it is absolutely necessary that the number of atoms undergoing stimulated emission predominate over the number of atoms undergoing spontaneous emission. This is possible only if $N_2 > N_1$ (i.e., the upper levels are more populated than the lower levels). This situation in which $N_2 > N_1$ is called population inversion (Fig. 7.11b).

7.3.6 Pumping

The population inversion cannot be achieved thermally. It is achieved usually by exciting the medium with suitable form of energy. This process is called pumping. It is impossible to achieve population inversion between any two energy levels without making use of the other energy levels. Population inversion is usually achieved using a three level or four level schemes which are shown in Fig. 7.12. A knowledge of the various energy levels of the system and their lifetimes is necessary for the choosing the energy levels that are involved in laser action.

Three level scheme: In this scheme three energy levels are involved (Fig. 7.12a). One of the levels, E_2 is a metastable state *i.e.*, it has a long life time. Energy level E_3 has a short life time. Hence the atoms undergo rapid decay to level E_2. Further the transition probability from level E_1 to E_3 is very high. Thus population inversion is achieved between E_2 and E_1.

It is also possible that E_3 is a metastable state and E_2 has a short life time (Fig. 7.12b). In such a case laser action is achieved between E_3 and E_2. Here also transition probability from E_1 to E_3 is high.

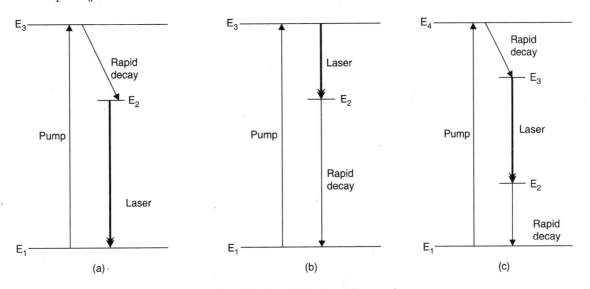

Fig. 7.12 Three level and Four level laser schemes

Four level scheme: In this scheme four energy levels are involved (Fig. 7.12c). The transition probability from level E_1 to E_4 is high. Both E_4 and E_2 have short life times compared to E_3. Hence population inversion is achieved between E_2 and E_3.

There are several ways of pumping a laser and producing population inversion necessary for stimulated emission to occur. Most commonly used methods are the following.

Optical pumping: In optical pumping, a light source is used to supply luminous energy. Most often this energy comes in the form of short flashes of light. This method was first used by Maiman in his ruby laser and is still widely employed today in solid state lasers. The laser material is simply placed inside a helical xenon flash lamp of the type used in photography.

Electric discharge: In this method the atoms are excited by collisions with fast electrons in an electric discharge. This method is preferred in gaseous ion lasers, of which the argon ion laser is a good example. The electric field typically several kV/m, causes electrons emitted by the cathode to be accelerated towards the anode. Some of these electrons will collide with the atoms of the active medium, ionize the medium and raise it to the higher level. This produces the required population inversions.

Inelastic-atom-atom collisions: This method is used in gas lasers consisting of two species of atoms. Pumping by electrical discharge raises one type of atoms to their excited states. These atoms collide inelastically with another type of atoms. It is these latter atoms that provide the population inversion needed for the laser emission. An example for this is the helium neon laser.

Direct conversion: In these lasers which are based on semiconductor p-n junctions, electrons and holes are made to recombine across the depletion region by applying a forward bias to the semiconductor diode. Electrons and holes recombine to emit radiation. Thus direct conversion of electrical energy into radiation occurs in light emitting diodes (LED's) and semiconductor lasers.

Chemical reactions: In a chemical laser the energy comes from a chemical reaction without any need for other energy source. Hydrogen can react with fluorine to produce hydrogen fluoride according to the reaction

$$H_2 + F_2 \rightarrow 2HF + heat$$

This reaction generates enough heat to pump a CO_2 laser.

7.4 EINSTEIN COEFFICIENTS

Einstein showed that the parameters describing absorption, spontaneous emission and stimulated emission are related. Consider an enclosure, which contains a large collection of atoms in equilibrium with radiation at a temperature T (Fig. 7.13). In such a system all the three processes *viz.*, spontaneous emission, stimulated emission and absorption will be taking place simultaneously. *Since the system is in thermal equilibrium the rate of upward transitions (from E_1 to E_2) must equal the rate of the downward transition processes (from E_2 to E_1).*

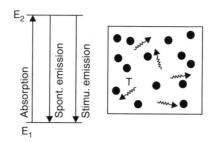

Fig. 7.13 Atoms and photons being in thermal equilibrium

If there are N_1 atoms per unit volume with energy E_1 in the ensemble, then the upward transition or absorption rate will be proportional to both N_1 and to the number of photons available at the correct frequency. The energy density at frequency, γ is given by $U_v = Nh\nu$

where N is the number of photons per unit volume having frequency between ν and $\nu + d\nu$. Therefore the upward transition rate is $N_1 U_\nu B_{12}$, where B_{12} is a constant.

$$\left(\frac{dN_1}{dt}\right)_{\text{absorption}} = B_{12} U_\nu N_1 \tag{7.3}$$

Similarly if there are N_2 atoms per unit volume with energy E_2 in the ensemble then the induced transition rate (i.e., stimulated transition rate) from E_2 to E_1 is $N_2 U_\nu B_{21}$, where again B_{21} is a constant.

$$\left(\frac{dN_2}{dt}\right)_{\text{simulated emission}} = B_{21} U_\nu N_2 \tag{7.4}$$

The spontaneous transition rate from E_2 to E_1 is simply $N_2 A_{21}$ and is independent of the energy density.

$$\left(\frac{dN_2}{dt}\right)_{\text{spontaneous emission}} = A_{21} N_2 \tag{7.5}$$

At equilibrium the number of atoms undergoing absorption per second must be equal to the number of atoms undergoing emission per second. Hence from eqns. (7.3)—(7.5), we get

$$N_1 U_\nu B_{12} = N_2 U_\nu B_{21} + N_2 A_{21} \tag{7.6}$$

Thus from equations (7.6) and (7.2)

$$U_\nu = \frac{N_2 A_{21}}{N_1 B_{12} - N_2 B_{21}} = \frac{A_{21}}{B_{12}\left(\frac{N_1}{N_2}\right) - B_{21}} = \frac{A_{21}}{B_{12} e^{h\nu/k_B T} - B_{21}} \tag{7.7}$$

This must be of the same form as that of the Planck's formula for the energy density:

$$U_\nu = \frac{8\pi h \nu^3}{c^3} \frac{1}{e^{h\nu/k_B T} - 1} \tag{7.8}$$

Comparing equations (7.7) and (7.8), we get

$$B_{12} = B_{21} \tag{7.9a}$$

$$\frac{A_{21}}{B_{21}} = \frac{A_{21}}{B_{12}} = \frac{8\pi h \nu^3}{c^3} \tag{7.9b}$$

Equations (7.9a) and (7.9b) are referred to as the *Einstein relations*. The second relation enables us to evaluate the ratio of the rate of spontaneous emission to that of stimulated emission for a given pair of energy levels. This ratio is given by

$$\frac{\text{Rate of spontaneous emission}}{\text{Rate of stimulated emission}} = \frac{A_{21} N_2}{B_{21} U_\nu N_2} = \frac{A_{21}}{B_{21} U_\nu} = e^{h\nu/k_B T} - 1 \approx e^{h\nu/k_B T} = \frac{N_1}{N_2}$$

The above discussion indicates that the process of stimulated emission competes with the process of spontaneous emission. *To amplify a beam of light by stimulated emission N_2 should be greater than N_1.* This is known as population inversion. *Population inversion is a necessary condition for laser action.*

Physical significance of Einstein coefficients

Equation (7.5) can be written as:

$$\frac{dN_2}{dt} = -A_{21} N_2$$

(–ve sign indicates that the population in energy level E_2 decreases with time)

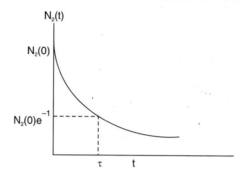

Fig. 7.14 The exponential decay of population due to spontaneous emission

$$\frac{dN_2}{N_2} = -A_{21}dt$$

Integrating the above expression, we get

$$\ln N_2 = -A_{21}t \implies N_2(t) = N_2(0)e^{-A_{21}t}$$

This equation indicates that the population in the energy level E_2 decreases exponentially with time (Fig. 7.14). Let the population N_2 decrease by a factor of e at $t = \tau$, where τ denotes the life time of the level E_2.

$$\frac{N_2(\tau)}{N_2(0)} = \frac{1}{e} = e^{-1} = e^{-A_{21}\tau} \implies A_{21} = \frac{1}{\tau}$$

Thus A_{21} denotes the reciprocal of the life time of the energy level.

Rewriting equation (7.5)

$$A_{21} = \frac{(dN_2/dt)}{N_2}$$

Thus A_{21} denotes the probability of an atom undergoing spontaneous emission. Similarly we have :

$$B_{12} = \frac{(dN_1/dt)}{N_1} \times \frac{1}{U_v} \quad \text{and} \quad B_{21} = \frac{(dN_2/dt)}{N_2} \times \frac{1}{U_v}$$

Thus B_{21} denotes the probability of an atom undergoing stimulated emission per unit energy density and Thus B_{12} denotes the probability of an atom undergoing absorption per unit energy density.

7.5 LASER AMPLIFIER AND LASER OSCILLATOR

7.5.1 Basic Principle

Laser is a coherent optical amplifier. It increases the amplitude of an optical field while maintaining its phase. Coherent amplification results on account of stimulated emission. When laser is used as an amplifier, an external source (called the pump) excites the active medium. If the population inversion is achieved between any two energy levels, then the laser acts like a coherent amplifier. The frequency of the amplified optical field corresponds to the difference between the two energy levels, in which the population inversion exits. The block diagram of a laser amplifier is shown in Fig. 7.15. Such amplifiers are used in fiber optic communication systems.

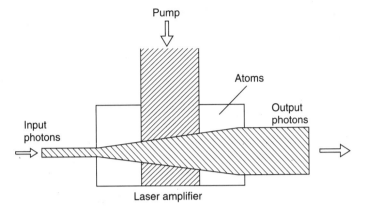

Fig. 7.15 Laser amplifier

An electronic amplifier is converted into an oscillator by having a positive feed back. The same principle is used in the laser. When laser is used as an oscillator, a part of its output is fed back into its input with matching phase. Feedback is achieved by placing the active medium in an optical resonator consisting of highly reflecting mirrors. These mirrors reflect light back and forth. This enhances the interaction between the atoms of the active medium and the radiation. Two conditions must be satisfied for laser oscillation to occur.

(a) The amplifier gain must be greater than the loss in the feedback system so that net gain is incurred in a round trip through the feedback loop.

(b) The total phase shift in a simple round trip must be a multiple of 2π so that the feedback input phase matches the phase of the original input.

Similarity between laser oscillator and *RF* oscillator is shown in Fig. 7.16.

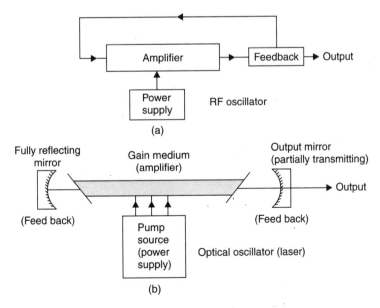

Fig. 7.16 (a) *RF* oscillator and (b) Laser oscillator

7.5.2 Construction and Components of a Laser

A laser requires three components for operation.

(a) *Active medium*: This refers to the medium in which the laser action takes place. The energy levels of the atoms or molecules which are involved in laser action are identified. Accurate information about the energy levels and their life times helps in identifying the levels between which population inversion can be achieved. Laser action is observed in a variety of media such as solids, liquids, gases, plasma and free electrons.

(b) *Pump*: This refers to the energy source which will raise the atoms and molecules of the active medium to higher energy states.

(c) *Optical cavity*: Optical cavity essentially consists of two mirrors facing each other. The active medium is enclosed by this cavity. One of the mirrors is 100% reflecting. The other mirror is partially transparent to let some of the radiation pass through. The optical cavity is made use of to make stimulated emission possible in more number of atoms in the active medium. This naturally increases the intensity of the laser beam. These components are shown in Fig. 7.17. In solid state lasers, the solid is usually in the shape of a rod whose ends are polished to act as mirrors. In semiconductor lasers also no external mirrors are employed. In gas and liquid lasers external mirrors are used.

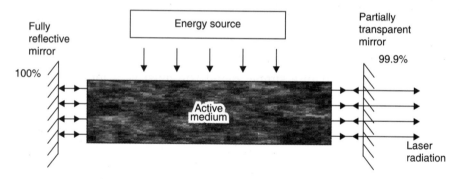

Fig. 7.17 Block diagram of a laser

(d) *Laser modes:* The light waves which are reflected back and forth between the mirrors give rise to *standing waves*. Only certain wavelengths or frequencies are allowed to be amplified by the optical cavity. This is similar to standing waves on a string (Fig. 7.18). If L is the length of the cavity and μ is the refractive index of the active medium, then the allowed wavelengths and frequencies are given by:

$$n\left(\frac{\lambda}{2\mu}\right) = L \quad \text{or} \quad \lambda_n = \frac{2\mu L}{n}$$

or

$$f_n = \frac{nc}{2\mu L}$$

where n is an integer and λ refers to the wavelength in vacuum. Note that in the active medium the wavelength is given by (λ/μ).

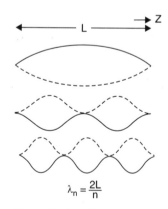

Fig. 7.18 Standing waves in a stretched string

These allowed frequencies correspond to *longitudinal modes*. The number of allowed modes is generally very large. *It is possible to make only one mode operative and such lasers are known as single mode lasers.*

The modes of oscillation which describe the functional dependence of the electromagnetic field on x and y coordinates are the transverse modes. The transverse mode TEM_{lmn}, the subscript l, m and n specify the number of times the electric or magnetic field crosses the x, y and z-axes. Thus the subscript indicates the number of modes in the standing wave pattern along each coordinate axis. The subscript n is often dropped because the number of longitudinal modes is very large. The notation is then abbreviated as TEM_{lm}. Several patterns of TEM modes are shown in Fig. 7.19.

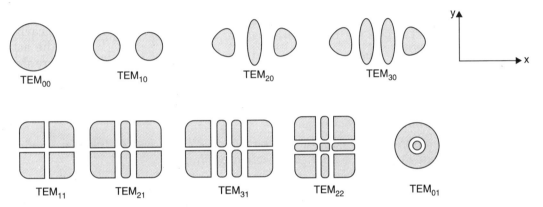

Fig. 7.19 Optical cavity and laser modes (appearance of laser beam cross-section)

7.5.3 Types of Lasers

Depending on the nature of the active material lasers are classified into the following categories :

Solid-state (rod-type) lasers: Lasers of this type are ruby (Cr^{3+} : Al_2O_3), neodymium-doped glass, neodymium-doped yttrium-aluminium-garnet (Nd-YAG), alexandrite (Cr^{3+} : $BeAl_2O_4$), titanium doped sapphire (Ti : Al_2O_3) and a number of rare earth ion doped materials. Pumping is achieved by means of an optical flash lamp. The active medium is in the form of a rod with highly polished ends. No external mirrors are employed.

Semiconductor or injection lasers: In these lasers, laser action is obtained by the recombination of electrons and holes in a semiconductor diode. These lasers are usually made of compound semiconductors such gallium arsenide, gallium phosphide, indium arsenide and indium phosphide and their alloys. They are extremely attractive on account of their miniature size, low power consumption and the ease with which laser action over a wide range of wavelengths can be obtained.

Colour centre lasers: In these solid state lasers the active medium is an alkali halide containing interstitial or substitutional defects. The characteristic feature of these lasers is their tunability in the infrared region.

Liquid or dye lasers: In these lasers optically clear solutions of various organic dyes form the active medium. The dyes are pumped by another laser, such as Nd-YAG, Nd-glass, nitrogen or argon ion laser. The dyes used are oxazine, rhodamine 6, coumarin and a great many others. Dye lasers are generally weak in light intensity but are important because of the tunability of laser wavelength.

Gas lasers: In these lasers the active medium is a gas subjected to an electric field by means of an electric discharge. Pumping is by excitation of atoms by electrons and in some cases by energy transfer between ions and atoms. Typical lasers of this kind are neutral gas lasers (helium-neon), ion gas laser (Ar^+, Kr^+, Xe^+), molecular gas laser (CO_2, N_2). Highly polished mirrors are provided at each end of the discharge tube to form the optical cavity.

Excimer lasers: These are also gas lasers. An inert gas combined with a halogen (*e.g.*, fluorine or bromine) form the active medium. On account of their toxic and corrosive nature, they are classified separate from other gas lasers. Laser action is possible on account of the formation of short lived dimer molecule. Examples of excimers are ArF, XeF, KrF, KrCl, XeBr, etc.

Metal vapour lasers: These also come under the category of gas lasers. In these lasers the active medium is an ionized metal vapour, which is mixed with one of the inert gases, such as helium or argon. Examples are He-Cd laser, copper-vapour laser, and gold vapour laser.

Free electron lasers: These are based on the movement of fast electrons through an alternating magnetic field.

X-ray lasers: These are sources which produce laser action in the X-ray region of the electromagnetic spectrum. The active medium is a plasma generated by impinging a high intensity laser pulse on a target material.

Fiber laser amplifier: Laser action can be achieved in an optical fiber for amplifying optical signal. Erbium doped glass fiber is the active medium.

Based on the function, lasers are classified as tunable or continuous or pulsed.

Tunable lasers generate coherent radiation over a range of wavelengths. This is possible because the lower levels happens to be vibrational levels which are very closely spaced. The desired output wavelength is selected by turning a grating in the optical cavity as shown in Fig. 7.20.

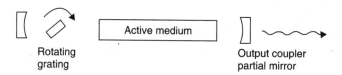

Fig. 7.20 Tunable laser

Continuous lasers generate uninterrupted coherent radiation.

Pulsed lasers generate pulses of coherent radiation whose time duration can be as short as nano to picoseconds (*i.e.*, $10^{-9} - 10^{-12}$ s). The laser pulses are generated by a process known as *Q-switching*. The Q-factor of an oscillator is defined as the ratio of the energy stored to the energy dissipated per cycle. In *Q*-switching the *Q*-value of the optical cavity is changed from a low value to a very high value in a very short time of the order of nanoseconds. When the active medium is being pumped the *Q*-value is kept at a low value so that maximum population inversion takes place. The *Q*-value is changed to a high value when a giant laser pulse is generated. This is shown in Fig. 7.21.

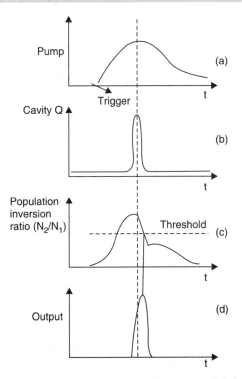

Fig. 7.21 Time variation of (a) pumping output (b) Q-value (c) population inversion ratio and (d) laser pulse

Q-switching can be done by mechanical means using a rotating prism or by electrical means which is based on electro-optic effect. In the rotating prism method a highly polished and reflecting prism is placed in the optical cavity and is rotated at a great angular velocity (Fig. 7.22). Note that only when the prism surface is parallel to the mirror of the cavity Q-value is high.

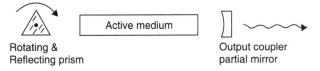

Fig. 7.22 Rotating mirror method

Some crystals become birefringent or doubly refracting on applying an electric field. This is known as electro-optic effect. From naturally birefringent crystals quarter wave plates and half wave plates are made through a proper choice of the thickness of the crystal. Electro-optic crystals can be made to behave like a quarter wave plate or a half wave plate by applying a suitable voltage. For Q-switching an electro-optic crystal is placed in the optical cavity (Fig. 7.23). Suitable voltage is applied to the crystal to produce a quarter wave plate. This converts linear polarized light incident on it to a circularly polarized light. The laser mirror reflects this light and in so doing reverses its sense of rotation. Thus on passing through the electro-optic crystal during its return trip, it emerges as a plane polarized light but it is at 90° to the original direction of polarization. Hence the light is not transmitted by the polarizer and the optical cavity is switched off. When the voltage is reduced to zero, optical cavity is restored to its normal state and the Q-value is very high. The changes in voltage are synchronized with pumping and can be achieved in less than 10 ns.

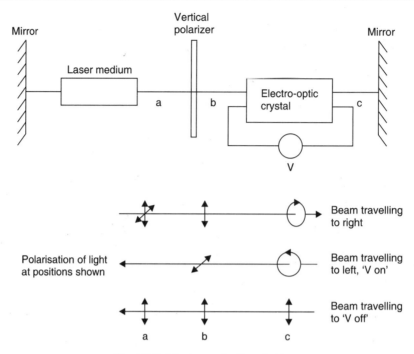

Fig. 7.23 Electro-optic Q-switching

7.6 SOLID STATE LASERS

7.6.1 Ruby Laser

Ruby laser is a classic example of a three-level solid state laser. Ruby is synthetic aluminium oxide, (Al_2O_3) with 0.05 percent weight of chromium oxide Cr_2O_3 added to it. The addition of Cr^{3+} induces a pinkish colour suggesting that there are strong absorption bands in the visible region. Laser action takes place between the energy levels of the Cr^{3+} ions. Al_2O_3 is an insulator and the introduction of Cr^{3+} results in additional energy levels within the band gap. The ruby crystal is cut into a cylindrical rod several centimeters long and several millimeters in diameter. The ends of the ruby are polished flat, to act as the cavity mirrors. Pumping is by light from a xenon flash tube. A typical ruby laser is shown in Fig. 7.24. Elliptical reflectors are some times used for pumping (Fig. 7.25).

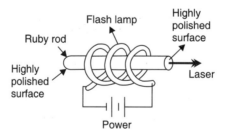

Fig. 7.24 Ruby laser

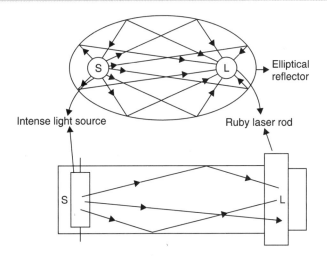

Fig. 7.25 Optical pumping using an elliptical mirror

Chromium doped ruby has three energy levels E_1, E_2 and E_3 (Fig. 7.26). The upper most band, E_3 is fairly wide and hence can accept a wide range of wavelengths. It has a short life time. The excited Cr^{3+} ions rapidly relax and drop to a next lower state E_2. This transition is non-radiative. The E_2 state is metastable. It has a life time of about 10^{-3} sec, several orders of magnitude longer than that of E_3. Consequently, the Cr^{3+} ions remain longer in E_2 before they drop down to the ground state E_1. The $E_2 \to E_1$ transition is radiative. It produces the spontaneous incoherent red fluorescence typical of ruby, with a peak near 6940 Å. But as the pumping energy is increased beyond the threshold, population inversion occurs in E_2 with respect to E_1. The output becomes coherent and the system lases with a sharp peak at 6943 Å. The Ruby laser is a pulsed one.

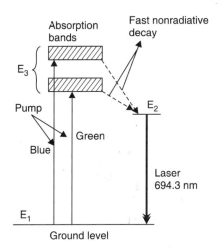

Fig. 7.26 Energy level diagram of Cr^{3+} in Ruby laser

7.6.2 Neodymium Laser

Another solid state laser in the category of narrow line width lasers is the rare-earth Nd^{3+} ion doped in several different host materials. Neodymium is perhaps the most useful laser material developed to date and is most commonly doped in either YAG (Yttrium Aluminium Garnet) or glass matrix. The energy level diagram of Nd^{3+} doped in YAG is shown in Fig. 7.27. The relevant laser levels are slightly lowered for doping in glass as compared to YAG and are shown in Fig. 7.28. The excitation bands occur in the blue and the green region. Neodymium ions doped in YAG crystal are limited to a maximum concentration of 1 – 1.5% whereas it can be doped to much higher concentration in glass. These lasers can be pumped by flash lamp as in the case of ruby laser. They can also be pumped using a GaAs laser diode and this has led to much more compact design of the Neodymium lasers.

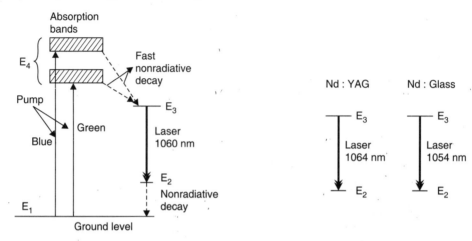

Fig. 7.27 Energy level of Nd^{3+} in YAG   **Fig. 7.28** Energy level diagram of Nd^{3+} in YAG and glass

Neodymium laser can also be used to generate laser in the visible region. This is done by second harmonic generation. On passing the infrared laser beam through crystals like KTP (Potassium titanyl phosphate), or KDP (Potassium dihydrogen posphate) it is possible to obtain the second harmonic which lies in the visible region (Fig. 7.29).

Fig. 7.29 Second harmonic generation using Nd-YAG

7.6.3 Alexandrite Laser

This is similar to ruby laser except that the host lattice is $BeAl_2O_4$ (Chrysoberyl). In this laser the emission wavelength can be tuned ranging from 700 to 818 nm. The pumping wavelength band ranges from 380 to 630 nm with peaks at 410 and 590 nm. The energy levels of the Cr^{3+} in chrysoberyl are shown in Fig. 7.30. The terminal laser level is a vibrational level and hence tuning is possible.

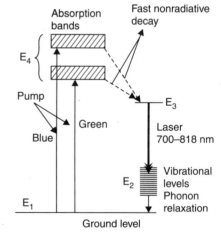

Fig. 7.30 Energy levels of Cr^{3+} in $BeAl_2O_4$ (Alexandrite laser)

7.6.4 Titanium-Sapphire Laser

The titanium doped sapphire (Ti : Al_2O_3) is probably the most well known material for use as a tunable solid state laser. The energy levels of Ti^{3+} ions in the Al_2O_3 host lattice are shown in Fig. 7.31. It has a broad emission ranging from 660 to 1180 nm. The pump band ranges from under 400 nm to just beyond 600 nm. Flash lamp pumping is not found to be effective. Hence it is pumped by Argon ion laser and frequency doubled Nd : YAG laser. The experimental set up is shown in Fig. 7.32. The three mirror optical cavity is used for optical pumping.

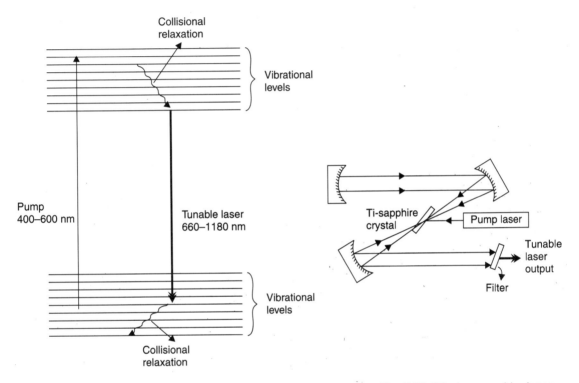

Fig. 7.31 Energy levels of titanium in sapphire   **Fig. 7.32** Titanium sapphire laser

7.6.5 Erbium–Glass Laser Amplifier

Laser action is achieved in the energy levels of Er^+ ion in the glass matrix. The energy levels are indicated in Fig. 7.33. Erbium has strong absorption bands at ~980 nm and 1480 nm. Since the fiber optic communication systems operate at 1550 nm, it is convenient for amplifying over a range of wavelengths. Erbium based fiber amplifiers are used in fiber optic communication systems (Fig. 7.34) of over a range of wavelengths.

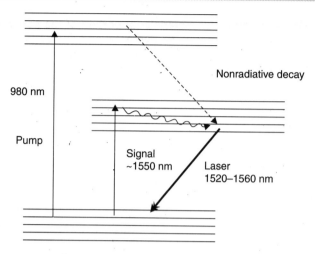

Fig. 7.33 Energy level diagram of Er^{3+}

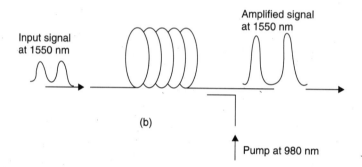

Fig. 7.34 Erbium doped fiber amplifier

7.7 SEMICONDUCTOR LASERS

7.7.1 Physics of Semiconductors

Semiconductors are a class of materials having electrical conductivities intermediate between those of metals and insulators. In these materials changes in dopant concentration and temperature result in large variation in electrical conductivity. They also exhibit interesting optical properties. They are technologically very important on account of their electrical and optical properties as well as the ease with which miniature devices can be made from them.

Semiconductor materials are found in group IV and neighbouring groups of periodic table (Table 7.2). Silicon and germanium which belong to the group IV are called elemental semiconductors. Compound semiconductors which are made from elements of group III and group V are referred to as III–V compounds. Similarly II–VI semiconductor compounds refer to compounds made from elements of group II and group VI. Silicon is used in a majority of semiconductor devices. Group III–V semiconductor compounds are used in the making of lasers and light emitting devices. The advantages of semiconductor lasers are their high efficiency and the ease with which they can be integrated with fiber optic communication systems. Group II–VI semiconductor compounds are used for light detectors.

Table 7.2 Common semiconductor materials (a) portion of the periodic table where semiconductors occur and (b) elemental and compound semiconductors

(a)	II	III	IV	V	VI
		B	C		
		Al	Si	P	S
	Zn	Ga	Ge	As	Se
	Cd	In	Sn	Sb	Te

(b)	Elemental	IV Compounds	III–V Compounds	II–VI Compounds
	Si, Ge	SiC	AlP, AlAs, AlSb, GaP, GaAs, GaSb, InP, InAs, InSb	ZnS, ZnSe, ZnTe, CdS, CdSe, CdTe

Intrinsic Semiconductors: An isolated Si or Ge atom has four valence electrons. Si and Ge crystallize in the diamond cubic lattice. Each atom has a coordination number of 4. Thus each atom has tetravalent bonding around them. This is represented in Fig. 7.35a. At a finite temperature some of the bonds can be broken and there are free electrons in the lattice. A missing bond is referred to as hole. This is shown in Fig. 7.35b.

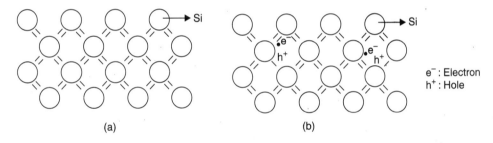

Fig. 7.35 Intrinsic semiconductor (a) at absolute zero (b) at a finite temperature

In the band theory of solids, the semiconductor has a filled valence band and an empty conduction band at 0K. The valence band and conduction band are separated by a gap often referred to as *band gap*. The band gap is ~1–3 eV. At finite temperature > 0K, the electrons get excited to the conduction band. An empty energy state in the valence band is known as *hole*. Thus at a finite temperature electron-hole pairs are formed. In an intrinsic semiconductor (*i.e.*, without doping) the number of conduction electrons is equal to number of holes. The electron or the hole concentration is given by :

$$n_e = n_h = 2\left(\frac{2\pi k_B T}{h^2}\right)^{3/2} (m_e\, m_h)^{3/2} \exp\left(-\frac{E_g}{2k_B T}\right)$$

Here E_g is the energy gap and m_e and m_h refer to the mass of the electron and hole respectively. The electrons and holes are distributed among energy levels according the Fermi-Dirac distribution (Fig. 7.36) :

$$f_e(E) = \frac{1}{e^{(E-E_F)/k_B T} + 1} \quad \text{and} \quad f_h(E) = 1 - f_e(E)$$

where E_F is known as the Fermi energy and the probability of occupancy of this level is equal to half. In intrinsic semiconductor the Fermi level lies exactly at the middle of the energy gap. *Fermi level also plays the role of a potential as will be discussed later.*

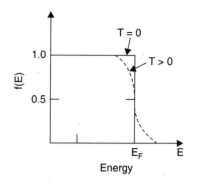

Fig. 7.36 Fermi-Dirac distribution

It is conventional to plot E versus k plots for electrons and holes (Fig. 7.37a). Electrons and holes can be treated as particles with an effective mass m^* or waves with the de Broglie wavelength λ.

$$E = \frac{p^2}{2m^*} = \frac{1}{2m^*}\left(\frac{h}{\lambda}\right)^2 = \frac{h^2}{8\pi^2 m^*}k^2$$

$$\left(\because \lambda = \frac{h}{p} \text{ and } k = \frac{2\pi}{\lambda}\right)$$

Thus $E_{vs}\ k$ is a parabola. The curvature of this curve is related to the effective mass of the electrons and holes at the particular value of k. The effective mass is given by

$$\frac{1}{m^*} = \frac{4\pi^2}{h^2}\left(\frac{d^2E}{dk^2}\right)$$

The population of electrons and holes is shown in Fig. 7.37b.

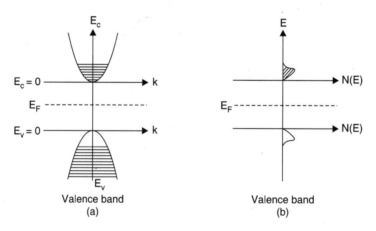

Fig. 7.37 Band diagram of intrinsic semiconductor (a) energy levels (b) population

Extrinsic semiconductors: The electrical and optical properties change on doping. On doping with a trivalent atom the four fold coordination is no longer satisfied and a hole is created on account of the missing bond. This is known as *p*-type semiconductor and there are excess of holes. A pentavalent dopant atom has an additional bond or an electron which is loosely bound to the parent atom. Hence there are excess of electrons. This is known as *n*-type semiconductors (Fig. 7.38). *Thus in an extrinsic*

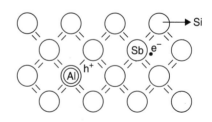

Fig. 7.38 Extrinsic semiconductor with p-type and n-type dopants

semiconductor the electron concentration is not equal to the hole concentration. The electron and hole concentration are given by :

$$n_e = 2\left(\frac{2\pi m_e k_B T}{h^2}\right)^{3/2} \exp\left[(E_F - E_c)/k_B T\right]$$

$$n_h = 2\left(\frac{2\pi m_h k_B T}{h^2}\right)^{3/2} \exp\left[(E_v - E_F)/k_B T\right]$$

Also, $n_e n_h = n_i^2$, where n_i denotes the concentration in intrinsic semiconductor at the same temperature.

The effect of dopants is to introduce additional energy levels in the band gap. When pentavalent and trivalent dopants are introduced, additional energy levels called the donor levels and acceptor levels are created in the forbidden energy gap. This is shown in Fig.7.39. The Fermi level for *n*-type is closer to the conduction band while for the *p-type* it is closer to the valence band. The distribution of electrons and holes in *p*-and *n*-type semiconductor are shown in Fig. 7.40.

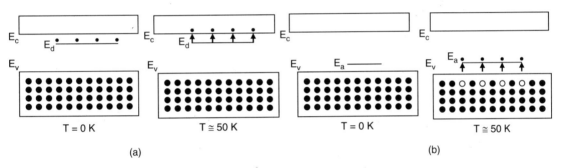

Fig. 7.39 Additional energy levels in the forbidden gap in extrinsic semiconductor
(a) n-type at T = 0 and (b) p-type at T = 0 and 50K

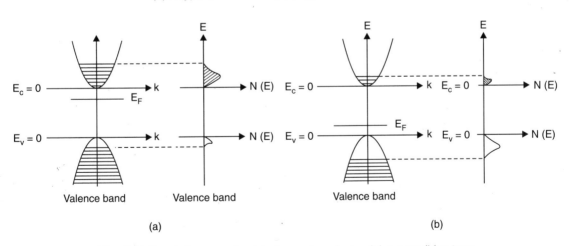

Fig. 7.40 Band diagram of extrinsic semiconductor (a) n-type (b) p-type

Direct and Indirect band gap semiconductors: In a direct band gap semiconductor the minimum of the valence band and the maximum of the conduction band occur at the same value of k i.e., $k = 0$ (Fig. 7.41). Consequently in a direct band gap semiconductor, when electrons and holes recombine all the energy is radiated in the form of a photon. The emission wavelength of the common semiconductors which are used for making LEDs and lasers are given in Table 7.3. In an indirect band semiconductor the minimum of the conduction band and the maximum of the valence do not occur at the same value of k. Hence whenever electron hole pairs are formed or recombine there is a change in the momentum ($= hk/2\pi$) (Fig. 7.41). This momentum is given to the lattice atoms. Thus electron hole recombination in an indirect band gap semiconductor gives rise to non-radiative transition and leads to the heating of the lattice.

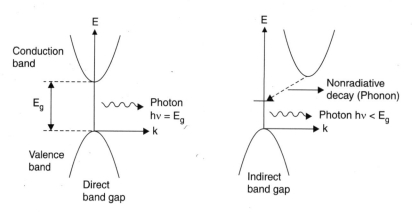

Fig. 7.41 Band diagram of (a) direct band gap and (b) indirect band gap semiconductor

Table 7.3 Emission wavelengths of materials used for LEDs and laser diodes

Material	$\lambda = hc/E_g$ (nm)	Material	$\lambda = hc/E_g$ (nm)
In AsSbP/InAs	4200	$Al_{0.11}Ga_{0.89}$ As : Si	830
InAs	3800	$Al_{0.4}Ga_{0.6}$ As : Si	650
GaInAsP/GaSb	2000	$GaAs_{0.6}P_{0.4}$	660
GaSb	1800	$GaAs_{0.4}P_{0.6}$	620
$Ga_x In_{1-x} As_{1-y} P_y$	1100–1600	$GaAs_{0.15}P_{0.85}$	590
$Ga_{0.47}In_{0.53}$ As	1550	$(Al_x Ga_{1-x})_{0.5} In_{0.5}$ P	655
$Ga_{0.27}In_{0.73}As_{0.63}P_{0.37}$	1300	GaP	690
GaAs : Er, InP : Er	1540	GaP : N	550–570
Si : C	1300	$Ga_x In_{1-x}$ N	340, 430, 590
GaAs : Yb, InP : Yb	1000	SiC	400–460
$Al_x Ga_{1-x}$ As : Si	650–940	BN	260, 310, 490
GaAs : Si	940		

7.7.2 p-n Junction

The production of light by applying electric fields to solid materials has been known for a considerable time, long before the invention of the laser. The phenomenon is called *electro-luminescence*. In electro-luminescence the emitted light is incoherent. Light emitting diode (LED) is based on electroluminescence in semiconductors. At higher current densities the light emission becomes coherent and this is known as laser diode.

A semiconductor laser is made by forming a junction between p and n materials in the same host lattice so as to form a *p-n* junction. In the *n-region* of the *p-n* junction, the concentration of the free electrons is higher than that of free electrons. Therefore, when a *p-n* junction is formed, some electrons from the *n*-region diffuse into the *p*-region. Since hole is nothing but the vacancy of an electron, an electron diffusing from the *n*-region simply fills the vacancy *i.e.*, it completes the chemical bond. This process is called electron hole recombination. However, this diffusion cannot go on indefinitely since the immobile positive and negative ions on either side of the junction oppose the incoming holes and electrons respectively. Thus at the junction a thin region which is depleted of free charge carriers (electrons and holes) is formed. This region which consists of only immobile positive and negative ions is called the *depletion region* (Fig. 7.42).

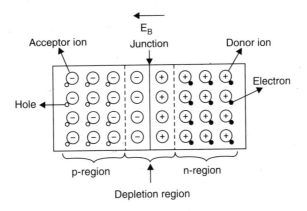

Fig. 7.42 Formation of depletion region

As a result of the accumulation of positive and negative immobile charges in the depletion region, an electric field E_b is established in the region. It is this field which opposes the further diffusion of electrons and holes across the junction. Thus equilibrium is reached. Due to the existence of the potential barrier V_B across the junction, an electron from the *n-region* can cross the junction only if energy $> eV_B$ is supplied to it from outside. An equal amount of energy is required to move a hole from the *p*-region to cross over to the *n-region* across the barrier. The value of the barrier potential for germanium is about 0.3V and for silicon about 0.7 V. The width of the depletion region is $\sim 10^{-6}$ m. Assuming that $V_B = 0.5$V and depletion width 10^{-6} m, the barrier field is 5×10^5 V/m which is fairly high.

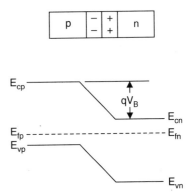

Fig. 7.43 Band diagram of p-n junction in equilibrium

Fermi level also plays the role of a potential which directs the motion of electrons. When a *p-n* junction is formed the electron flow continues till the Fermi energy levels in both the *p* and *n* sections of the junctions are equalized. The band diagram for a *p-n* junction in equilibrium is shown in Fig. 7.43.

In a semiconductor laser the *p-n* junction is operated in the forward bias (Fig. 7.44*a*). The doping is extremely heavy. The basic mechanism responsible for light emission from the *p-n* junction is the recombination of electrons and holes. As the forward bias is increased, more and more photons are emitted and the light instantly becomes stronger. The band diagram for a *p-n* junction in the forward bias is shown in Fig. 7.44*c*. The applied voltage, is opposite in direction to the barrier potential. Consequently there is a threshold for the flow of current as shown in the *V-I* characteristics of the *p-n* junction in the forward bias (Fig. 7.44*b*). Initially the light emitted is incoherent. LED's operate in this region. After the current density across the junction exceeds a particular value, light emitted is coherent (Fig. 7.45). The frequency of the laser beam is E_g/h where E_g is the band gap. This type of laser action was first achieved in GaAs. Subsequently InP, InAs, InSb, PbTe, PbSe, PbS have proved to be useful materials for laser action. Figure 7.46 shows the laser wavelength emitted for various semiconductors.

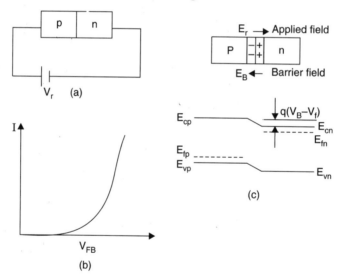

Fig. 7.44 Forward bias (a) circuit (b) characteristic and (c) band diagram

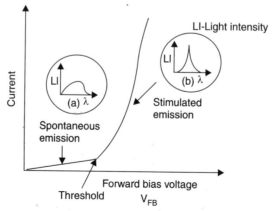

Fig. 7.45 Current density for (a) LED and (b) laser diode

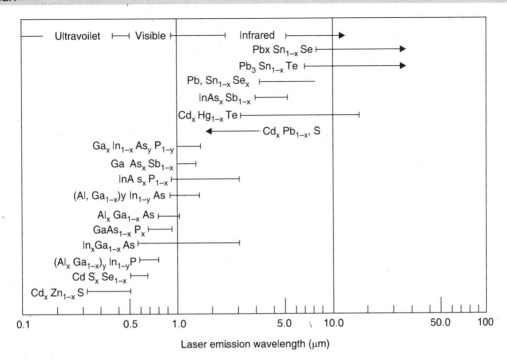

Fig. 7.46 Laser wavelengths from different semiconductors

The lasing action is enhanced by having highly polished surfaces which act like mirrors as in the case of a laser (Fig. 7.47). The advantage with laser diodes is the narrow line width of the source. Further the active region forms a resonant cavity that resembles the Fabry-Perot optical cavity used for gas lasers. If l is the length of the cavity then the allowed frequencies are given by

$$f_m = \frac{mc}{2ln}$$

where $m = 1, 2, 3,...$ etc., denotes the various modes, n is the refractive index and c is the velocity of light. The gain is however maximum at the central frequency.

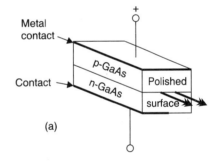

Fig. 7.47 Typical construction of a semiconductor laser

7.7.3 Typical Semiconductor Laser Structures

Homojunctions and Heterojunction Lasers

If the p-n junctions involve only one p-type material and one n-type material they are known as *homojunctions*. However, threshold current density required for laser action in such junctions is high. It is highly desirable to have smaller area of junction so that the current densities are high. For many applications it is desirable to confine the laser generation within as small a volume as possible. This is achieved by having *heterojuctions and quantum well lasers*.

Simple heterojunction: This consists of an upper p-type layer of AlGaAs followed by a p layer of Ga As and then a substrate of n-type GaAs. In this arrangement the p layer of GaAs has an active region that is only 0.1–0.2 µm thick (Fig. 7.48a). This is the only region where the current can flow owing to the increased energy of the conduction band for the AlGaAs which serves as a barrier. Consequently recombination of charge carriers occurs in the thin *p* layer. This is shown in Fig. 7.48*b*. Because the charge carriers in heterojunction lasers are confined to a much smaller region than in homojunction lasers, the heat dissipation is much lower. In addition the change in refractive index at the interface between the *p* type GaAs and *p* type AlGaAs can provide guiding effect for the laser beam (7.48*c*).

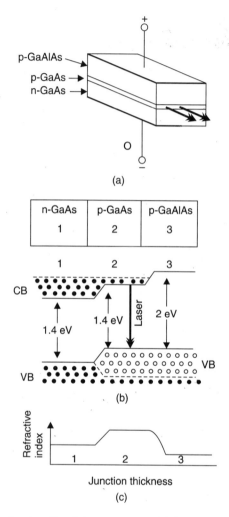

Fig. 7.48 Simple heterojunction (a) structure and (b) band diagram (c) RI profile

Double heterojunction: Double heterojunction structures provide even more control over the size of the active region. It also provides additional variations in refractive index that allow for confinement of light when the semiconductor is used as a laser. Figure 7.49*a* shows a simple double heterostructure laser composed of various doping concentrations of GaAs and AlGaAs. In addition to the control of layer thickness during fabrication, the side walls of the active region are narrowed via lithographic techniques. This confines the laser to a narrow region and thus reduces the threshold current. Figure 7.49*b* shows the band structure. The refractive index variation also provides the guiding effect.

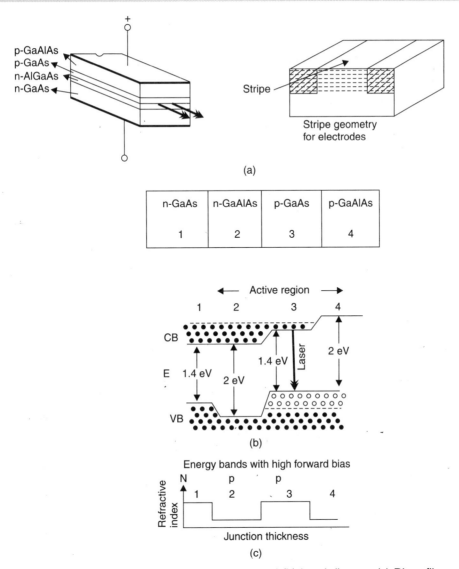

Fig. 7.49 Double heterojunction (a) structure and (b) band diagram (c) RI profile

Multi quantum well laser diodes: It is a great advantage to obtain laser action at low current densities. This can be achieved by making the active region as thin as possible. At such small length (~ nm) quantum effects begin to play a dramatic role. The band-gap energy of the active layer is smaller than the surrounding layers. Also the density of states for a two dimensional well is different from that of a bulk material. This enables one to construct efficient semiconductor lasers. For a bulk material the density of states is proportional to the $\sqrt{E}$. For a quantum well in two dimensions, the density of states is constant (Fig. 7.50). This favours the lasing action. There are two representative examples of the active region structures of AlGaAs quantum well lasers. Lasers comprising multiple wells are termed *multiple quantum well* (MQW) lasers. MQW lasers employ a parallel stack of single quantum wells. A typical structure of MQW laser is shown in Fig. 7.51. Figure 7.52 shows the active region of a GRIN (graded index) SCH (separate confinement heterostructure)-SQW (single quantum well) laser. In the latter structure, the injected carriers are confined in the SQW while the laser light is guided by the GRIN waveguide.

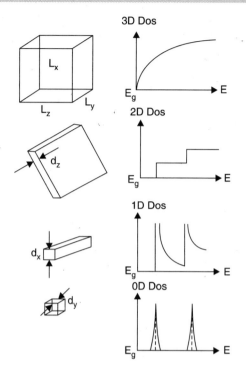

Fig. 7.50 Density of states in zero, one, two and three dimensional quantum wells

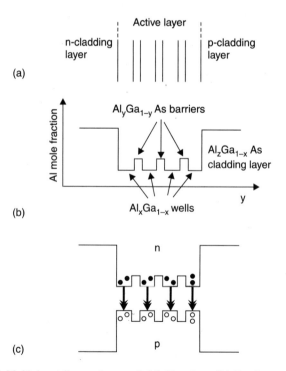

Fig. 7.51 Multiple well quantum well (a) Structure (b) Barriers and wells (c) Electron-hole recombination

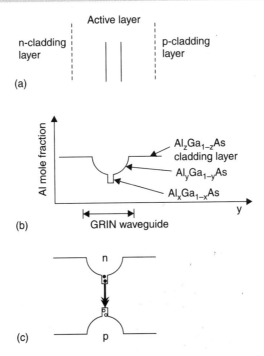

Fig. 7.52 GRIN SCH-SQW (a) Structure (b) Formation of a single well (c) Electron-hole recombination

Distributed Feed Back Laser and Distributed Bragg Reflector: The spectral width of light emitted by a laser diode can be reduced further by making only one of the modes active and suppressing all the other modes. This is achieved by replacing the cleaved sufaces usually used as mirrors with frequency selective reflectors such as gratings parallel to the junction plane. The grating is a periodic structure that reflects light only when the grating period equals an integral times half the wavelength. These are called *distributed Bragg reflectors* (Fig. 7.53a). In another approach the grating itself is placed directly adjacent to the active layer. This is known as the distributed feed back laser (Fig. 7.53b).

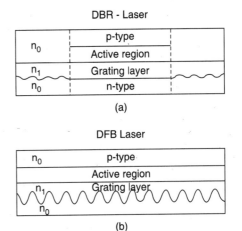

Fig. 7.53 (a) Distributed Bragg reflector laser (b) Distributed feed back laser

Vertical Cavity Surface Emitting Lasers: Conventional semiconductor lasers utilizing cleaved ended faces as a means of optical feed back are not fit for monolithic integration. For monolithic integration, it is desirable to have laser output normal to the wafer surface. Figure 7.54 is the basic configuration of a surface emitting laser incorporating two mirrors parallel to the surface to form a vertical cavity typically 5 – 10 μm long. It is crucially important to increase the reflectivity of mirrors since the reflection loss otherwise becomes very high. Bragg reflectors composed of multilayered semiconductors or dielectrics are exploited to provide reflectivities exceeding 95%. Carrier injection is performed through a pair of electrodes in the upper and lower surfaces to excite a cylindrical active region typically of 10 μm diameter, and 1 to 3 μm thick. The surface emitting laser has several advantages over the conventional edge emitting laser. These include the capability of being integrated on a wafer to form a two dimensional densely packed laser array, the attainability of a narrow circular output beam and single mode laser (Fig. 7.54b).

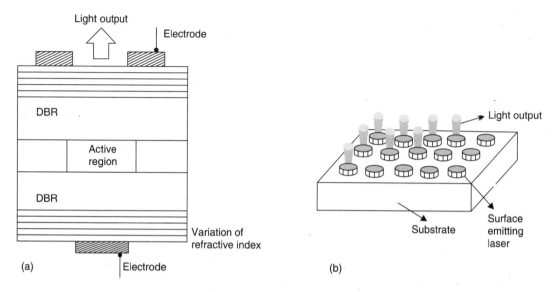

Fig. 7.54 (a) Vertical cavity surface emitting laser (b) An array

7.8 GAS LASERS

7.8.1 Introduction

In the case of solid state laser such as ruby, the pumping is done using a flash lamp. Such a technique is efficient if the lasing system has broad absorption bands. In gases, atoms are characterized by sharp energy levels as compared to those in solids. Hence an electric discharge is used to pump the atoms in a gas laser. At ordinary pressures most gases are insulators. However, they become conductors under certain conditions. External agencies such as ultraviolet light, X-rays, electric sparks, radioactive substances render gases conducting. The nature of electrical discharge also depends on the gas pressure, gas composition, applied potential and the electrode geometry. In a discharge tube the electrons and negative ions are attracted to anode while positive ions move towards the cathode. The positive ions that migrate and strike the cathode can cause secondary electron emission from the cathode.

Thus in a discharge tube atoms, ions and electrons undergo frequent collisions with each other. Collisions could be elastic or inelastic. In elastic collisions the kinetic energy of the particles involved are conserved. While in inelastic collisions only the total energy is conserved. However, the momentum is conserved in both elastic as well as inelastic collisions. From these conservation requirements it turns out that electrons can transfer only a small amount of their kinetic energy to ions by elastic collision and that ions can transfer nearly all of their kinetic energy to neutral gas atoms. Because of their mass, ions do not acquire much velocity either from the applied field or from the impact of electrons. The electrons, on the other hand, acquire and maintain their velocities easily and thus can contribute to further ionization and excitation of neutral atoms by inelastic collision processes. *In a gas discharge energy transfer between excited atoms and neutral atoms takes place.* This mechanism is often used to achieve population inversion.

7.8.2 Physics of gas discharge

The experimental arrangement for determining the *I-V* characteristics of a discharge tube is shown in Fig. 7.55. It consists of a glass tube in which the gas is filled upto a desired pressure. The anode and the cathode of the discharge tube are connected to the power supply. The current and the voltage across the tube are measured. The *I-V* characteristics are shown in Fig. 7.56. The various regions of the characteristic curve are described below :

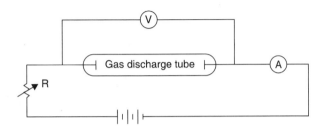

Fig. 7.55 Discharge tube

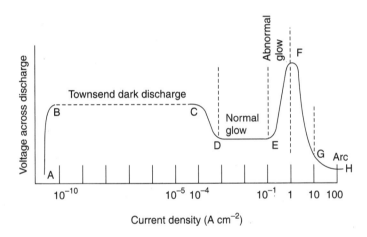

Fig. 7.56 I-V characteristics of the discharge tube

Region A-B: Upto about 10^{-11} amp, the current in the discharge tube is not sustained. The external agency required for sustaining the discharge may be provided in many ways. A hot cathode which emits electrons is normally used. X-rays or ultraviolet light can be made to pass through the tube to produce ionization of the gas and thus sustain the discharge.

Region B-C: This region corresponds to the Townsend discharge after the physicist who discovered it. The electrons have insufficient energies to produce secondary ionization. The voltage is practically constant. There is no emission of light from the discharge tube. In the Townsend discharge there is no appreciable space charge between anode and cathode and the electric field is more or less uniform across the tube.

Region C-D: More number of electron-ion pairs are produced due to high electron energies. Consequently voltage across the tube falls.

Region D-E: This is referred to as the glow region. The discharge emits light depending on the nature of the gas. The discharge is now self-sustaining due to secondary ionization. Electrons are liberated by the bombardment of ions at the cathode. The glow region is shown in Fig. 7.57a. There is also accumulation of ions and electrons within the tube as shown in Fig. 7.57b. The voltage drop across the tube is shown in Fig. 7.57c. The discharge is characterized by a number of distinct luminous and non-luminous or dark regions. Secondary electrons produced by ion bombardment accumulate in the Aston dark space and some of them recombine with the incoming positive ions to form the cathode glow at the outer edge of this dark space. Other electrons from the Aston dark space move through the cathode glow without being captured and are accelerated in the Crookes dark space by the positive ion space charge. Sufficient energy is gained in the Crookes dark space to cause inelastic collisions and excitation of the gas to produce negative glow. The anode edge of the negative glow depends on the point at which the electron energy is too low to produce further excitation. The electrons are further accelerated in the Faraday dark space and produce the positive column that extends almost to the anode. In the positive column the net space charge is zero. The positive column is a neutral plasma. The number density of electrons in the positive column is also determined by the loss of electrons at the walls of the discharge tube. The successive acceleration and deceleration of the electrons can cause positive column to break up into bright and dark segments called striations (Fig. 7.58).

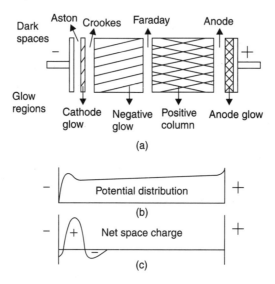

Fig. 7.57 (a) Glow region (b) Accumulation of charges and (c) Voltage drop across the tube

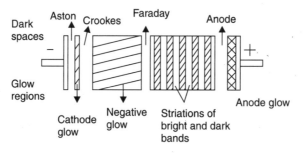

Fig. 7.58 Striations

Region E-F: As the current of the normal glow discharge is increased, the current density at the cathode does not change. Instead more and more of the cathode area is involved in the discharge. However, when the whole cathode surface has been covered in the operation of the glow discharge, the sustaining voltage must now be increased to provide additional current flow.

Region F-G: This corresponds to the transition to the arc operation. It might occur abruptly and may not be obviously observed in practice.

Region G-H: This region is called the arc region. It is characterized by high current flow and low voltage. The energy deposited at the cathode by the positive ions is great enough that the thermal emission of electrons begin to occur.

7.8.3 Helium Neon Laser

The first gas laser to be operated successfully was the He-Ne laser. It consists of a long and narrow discharge tube (diameter ~2 to 8 mm and length 10 to 100 cm) which is filled with helium and neon with typical partial pressures of 1 torr and 0.1 torr. *The actual lasing atoms are the neon atoms and helium is used for selective pumping of the upper laser level of neon.* The pressure of helium is nearly ten times that of neon. The electrodes are connected to a high voltage source of few kV d.c. The mirrors are placed outside the tube as shown in Fig. 7.59a. The end windows of the tube are set at Brewster's angle. When a ray of light is incident at the Brewster's angle, the reflected light is polarized perpendicular to the plane of reflection. Thus only the component oscillating parallel to the plane of incidence becomes dominant and will sustain laser emission. The light that emerges from the laser is linearly polarized.

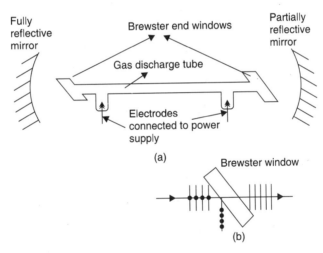

Fig. 7.59 (a) Helium Neon laser (b) Plane polarized light using Brewster window

Figure 7.60 shows the energy levels of helium and neon. The levels E_4 and E_6 of neon atoms have almost the same energy as F_2 and F_3. When an electrical discharge is passed through the gas, the electrons which are accelerated down the tube collide with helium and neon atoms and excite them to higher energy levels. The helium atoms tend to accumulate at levels F_2 and F_3 due to their long life times 10^{-4} and 5×10^{-6} s respectively. When the excited helium atoms collide with the neon atoms in the ground state, neon atoms are excited to E_4 and E_6. *This is referred to as energy transfer due to resonant collisions.* Since the pressure of helium is ten times that of neon, the levels E_4 and E_6 of neon are selectively populated as compared to other levels of neon.

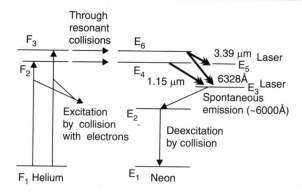

Fig. 7.60 Energy levels of the helium and neon atoms

Transition between E_6 and E_3 produces the very popular 6328 line of the He-Ne laser, the other two important wavelengths from the He-Ne laser are 1.15 μm and 3.39 μm which correspond to the $E_4 \to E_3$ and $E_6 \to E_5$ transitions. It is interesting to observe that both 3.39 μm and 6328 Å transitions share the same laser level. Typical power from the laser is $\sim mW$.

7.8.4 The Argon Ion Laser

Laser action has been observed in a number of inert gas ions. In an argon ion laser one uses the energy levels of an ionized argon atom. Figure 7.61 shows the typical construction of the argon

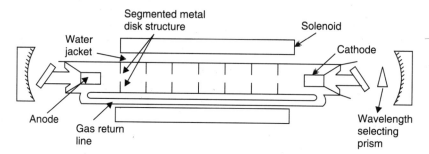

Fig. 7.61 Argon ion laser

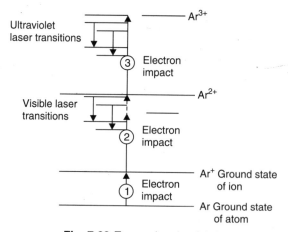

Fig. 7.62 Energy levels of Ar+

ion laser. The discharge design of the argon laser is much more complicated than that of the He-Ne laser. This is because higher energy is required to pump the ionic levels and the need to dissipate the heat energy. The current density is increased by concentrating the discharge with a magnetic field applied along the axis of the tube This has the added advantage of preventing the collisions of the ions with the walls of the tube. The tube is made of refractory material such as graphite or beryllium oxide. Most ion lasers are water cooled and often include a series of metal disks to act as heat exchangers.

Table 7.4 Laser wavelengths in nm of argon and krypton laser

	Visible	*Ultraviolet*
Ar	528.7, 514.5, 501.7, 496.5, 488, 476.5, 472.7, 465.8, 457.9, 454.5	385.1, 351.1, 363.8, 333.6, 335.8, 300.3, 305.5, 275.4
Kr	743.6	219, 225, 350.7

7.8.5 Krypton Ion Laser

Krypton ion laser is similar to argon ion laser. Table 7.4 shows the energy levels of krypton ion and the laser transitions.

7.8.6 Carbon Dioxide Laser

In CO_2 laser the transitions occur between different vibrational states of the carbon dioxide molecule. Hence the laser operates in the infrared region. Figure 7.63 shows the three independent modes of vibrations of CO_2 molecule. For a given electronic configuration in a molecule, there are a number of vibrational levels. Further, for each vibrational level there are a number of rotational levels. The energy levels for the electronic configuration of the ground state are shown in Fig. 7.64. The addition of nitrogen

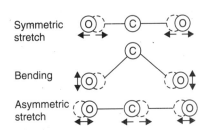

Fig. 7.63 Modes of vibration of CO_2 molecule

gas in the discharge enables the energy transfer to CO_2 molecules. This is very much similar to the relative transfer of excitation energy from helium to neon atoms in the He-Ne laser. Operating at high pressure in a longitudinal discharge is extremely difficult. At atmospheric pressure the breakdown voltage is $1.2 \, kVmm^{-1}$. Hence even for a 1 metre laser the voltage that has to be applied is quite high. To overcome this problem, the discharge is struck transversely across the tube so that the discharge path length is about a centimeter. In a transverse discharge the two electrodes are placed parallel to each other over the length of the discharge, separated by a few centimeters are more and a high voltage is applied across the electrodes. For obtaining higher energy output per unit volume of the gas, transverse excited atmospheric pressure (TEA) configuration is used. This is shown in Fig. 7.65. Prior to the application of the high voltage, a form of pre-ionization is used to ionize the space between the electrodes uniformly and there by fill it with electrons. Gas flow is achieved between the electrodes as shown.

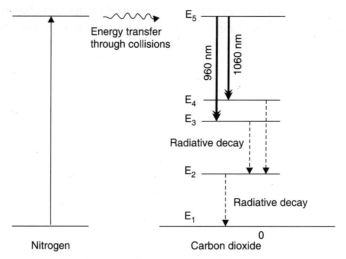

Fig. 7.64 Energy levels of CO_2 laser

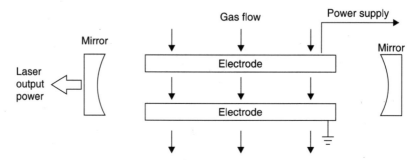

Fig. 7.65 Transverse Electric Atmospheric (TEA) CO_2 laser

The second approach that has been employed is to apply the principles of thermodynamics to obtain population inversion. This forms the basis for *gas dynamic laser*. A nitrogen-carbon dioxide mixture is heated and compressed. Then it is allowed to expand into a low pressure region (Fig. 7.66). During heating and compression the population of the energy states reaches the Boltzmann's distribution appropriate to higher temperature. At high temperatures most of the energy is stored in the vibrational modes of the nitrogen molecule. At lower temperatures, after expansion into the low pressure region, resonant collisions of the nitrogen molecules with carbon dioxide molecules, creates population inversion between the energy levels of the carbon dioxide molecule. Gas dynamic lasers have the disadvantage of large size and high noise levels generated during gas expansion.

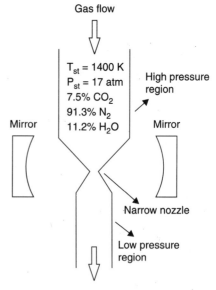

Fig. 7.66 Gas dynamic CO_2 laser

The CO_2 laser is highly efficient. This is attributed to the fact that the vibrational and rotational states require little energy for excitations and a good part of this energy is transferred to the laser beam. Whereas it requires some 20 eV of energy to excite a helium atom to its metastable state, only 0.3 eV is required to excite a CO_2 molecule to one of its lower vibrational and rotational levels.

7.8.7 Nitrogen Laser

The nitrogen laser produces a laser output at 337.1nm. The energy levels involved in the laser transition are shown in Fig. 7.67. The laser transition involves a change in both electronic and vibrational energy levels. It operates at pressures ranging from 20 torr to atmospheric pressure in a sealed chamber at repetition rates of up to 200 Hz. It can also be operated in the TEA configuration. Nitrogen laser is used primarily for pumping a dye laser.

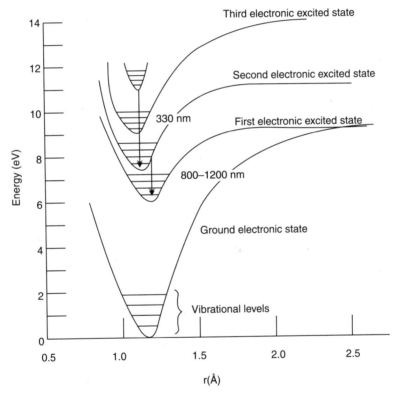

Fig. 7.67 Energy levels of N_2 laser

7.8.8 Helium-Cadmium Laser

The helium-cadmium laser is similar in operation to that of helium-neon laser. The laser tube contains a mixture of helium gas and vapour of cadmium through which a d.c. discharge is struck. It has a small reservoir near the anode to contain cadmium. The reservoir is heated to ~423 K to produce cadmium vapour. A typical set up of the helium cadmium laser is shown in the Fig. 7.68. The laser transitions occur between the electronic levels of Cd^+ (Fig. 7.69). The laser radiation has wavelengths 416 nm (blue) and 325 nm (ultraviolet). Pumping of Cd^+ is achieved through the Penning ionization process:

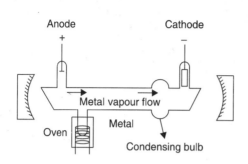
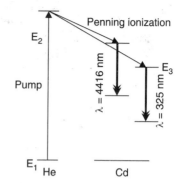

Fig. 7.68 Helium cadmium laser

Fig. 7.69 Energy levels of helium and cadmium

$$He^* + Cd \rightarrow He + Cd^+ + e^-$$

This is a non-resonant collision unlike in the case of helium-neon laser.

7.8.9 Copper and Gold Vapour Lasers

Copper and gold vapour lasers have become important on account of their high gain. An alumina discharge tube with a buffer gas of helium or neon is employed. The optimum pressure for these lasers is of the order of 1 torr of metal vapour and 40–50 torr of the buffer gas. The metal vapour requires a laser bore temperature of ~ 1500°C for copper vapour laser and ~1650°C for the gold vapour laser. Metal ions are excited by striking an electric discharge. The excitation process produces a pulsed laser output ranging from 10 to 50 ns at repetition rate of 100 kHz. The energy level diagram for the copper and gold vapour lasers is shown in Fig. 7.70.

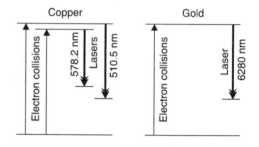

Fig. 7.70 Energy levels of Copper and gold

7.8.10 Excimer Lasers

Excimer lasers are pulsed lasers that utilise electronic transitions within short lived molecules. Such lasers are most often composed of the combination of rare-gas such as argon, krypton or xenon and halogens such as fluorine, chlorine, bromine or iodine. The word *excimer* is a combination of the phrase *exci*ted di*mer*. The excimer molecule exists only as an excited molecule because the ground state is extremely short lived owing to the repulsive force between the two atoms of the molecule. Thus in xenon-fluoride, the ground state (A) is repulsive, and (X) is loosely attractive. The states B and C correspond to stable electronic states with very small lifetimes of the order of ns. Thus the molecules do not normally exist in nature (Fig. 7.71). However, molecules can be formed in their excited state by various special excitation techniques. After the excited states are formed, the populations rapidly decay to the repulsive ground state in a time typically ranging from 1 – 5 ns and the

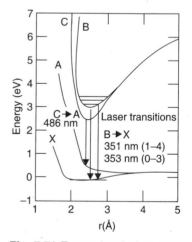

Fig. 7.71 Energy level of an excimer molecule (XeF)

atoms subsequently fly apart. Thus the molecules exist only for a duration corresponding to the lifetime of the excited state. The excimer laser is produced by forming the excited state species of the excimer molecule. Since the ground state of these molecules is not stable, it is not possible to produce the upper laser level by direct pumping from such a ground state. Hence indirect pumping is necessary. Thus the electric discharge produces excited Ar and individual fluorine atoms. Many of the fluorine atoms gather an electron to become F⁻ negative ion. The excited excimer molecule is then produced by a collision between the two species via the following reactions :

$$Ar^+ + F^- \rightarrow ArF^*$$
$$Ar^* + F^* \rightarrow ArF^*$$

Due to the corrosive nature of the halogen species, the entire laser structure is made of stainless steel with polyvinyl and Teflon components. The discharge is transverse and a continuous gas flow is maintained.

The most common excimer molecules used for lasers are xenon fluoride (XeF), xenon chloride (XeCl), krypton fluoride (KrF), argon fluoride (ArF), mercury bromide (HgBr), molecular xenon (Xe_2) and molecular argon (Ar_2).

7.9 CHEMICAL LASERS

In these lasers the laser levels are the vibrational and rotational levels of the molecules. The pumping energy is obtained from a chemical reaction. Most chemical lasers operate in the near to middle infrared portion of the electromagnetic spectrum. The most well known of chemical lasers are those operating on vibrational transitions of the hydrogen fluoride and deuterium fluoride.

The pumping mechanism involved in the hydrogen fluoride laser results from the chemical reactions :

$$H_2 + e \rightarrow 2H + e$$
$$H + F_2 \rightarrow HF(v^{**}) + F + 4.24 \text{ eV (excitation possible upto } v = 6)$$
$$F + H_2 \rightarrow HF(v^*) + H + 1.38 \text{ eV (excitation possible upto } v = 3)$$

Each of these reactions liberates energy which can emerge in part as vibrational energy of the hydrogen molecule. Sufficient energy is liberated by reaction to excite the HF to vibrational levels up to $v = 6$. Laser action is possible on vibrational transitions $6 \rightarrow 5, 5 \rightarrow 4, 4 \rightarrow 3, 3 \rightarrow 2, 2 \rightarrow 1$ and $1 \rightarrow 0$ as shown in Fig. 7.72. The wavelength ranges from 2.41 μm to 3.38 μm. Since the decay can occur to various vibrational levels, a range of operating wavelengths is possible. The HF laser emits in the wavelength range of 2.6 – 3.3 μm. By using deuterium instead of hydrogen, the laser wavelength range is shifted to 3.5 – 4.2 μm. Other chemical lasers include the following:

CO laser: The CO chemical laser uses the following reaction sequence :

$$CS_2 + O \rightarrow CS + SO_2$$
$$CS + O \rightarrow CO(v^*) + S + 3.6 \text{ eV}$$

The CO laser operates in the region 4.9 – 5.8 μm

HCl laser: The HCl laser uses the reaction scheme

$$H + Cl_2 \rightarrow HCl(v^{**}) + Cl$$
$$Cl + H_2 \rightarrow HCl(v^*) + H$$

HCl laser operates in the region 3.57 – 4.11 μm.

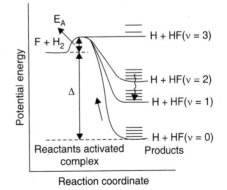

Fig. 7.72 Energy level-Chemical laser

HBr laser: The HBr laser uses the reaction scheme

$$H + Br_2 \rightarrow HBr\ (v^{**}) + Br$$
$$Br + H_2 \rightarrow HBr\ (v^{*}) + H$$

HBr laser operates in the region 4.0 – 4.7 μm.

7.10 LIQUID OR DYE LASERS

Organic dye molecules are the active media in these lasers. Organic dye molecules are very complex molecules that exhibit fluorescence. Fluorescence refers to the emission of light at longer wavelengths on being irradiated with visible or ultraviolet light. Table 7.5 gives the wavelengths of the fluorescent radiation for various dyes.

Table 7.5 Fluorescent wavelength of some dyes

Dyes	Wavelength of fluorescent radiation (nm)
Polymethane dyes	700–1500
Xanthene dyes	500–700
Coummarin dyes	400–500
Scintillator dyes	320–400

When they are used as laser media, dye molecules are dissolved in liquid solvents such as water, alcohol and ethylene glycol. In such solutions the concentration of dye molecules is of the order of one part in ten thousand. Hence the dye molecules are somewhat isolated from each other and are surrounded only by solvent molecules.

Laser dyes have strong absorption in the ultraviolet and visible spectral regions, as shown for a typical dye molecule Rhodamine *B* in Fig. 7.73. If they are irradiated or pumped with light at these wavelengths, then the dye molecules radiate very efficiently at somewhat longer wavelengths than the pump wavelength as shown in the emission spectrum. Because of their efficient optical absorption, dye lasers utilise optical pumping either using a flash lamp or other lasers.

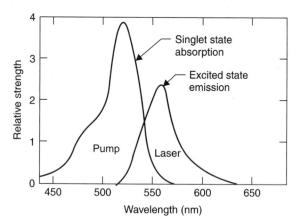

Fig. 7.73 Fluorescence of Rhodamine

The energy levels associated with organic dyes suspended in solvents are shown in Fig. 7.74. There are two manifolds (sets of energy levels) within which excitation and decay can occur. These are referred to as the singlet (S) and triplet (T) levels. In the singlet state of the molecule, the spin of the excited electron is antiparallel to the spin of the remaining molecule. In the triplet state, the spins are parallel. As in the case of atoms and simple molecules, transitions within singlet states or triplet states are much more likely to occur than transitions between a single and a triplet state. When these latter transitions do occur, they are referred to as intersystem crossings. As in the case of simpler molecules, the various electronic energy levels S_0, S_1, T_1 and T_2 are associated with the vibrational and rotational energy levels.

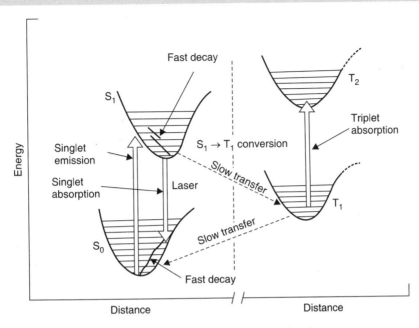

Fig. 7.74 Energy levels of dye molecule

When excitation occurs in dyes from the ground state S_0 to the excited state S_1, all of the vibrational states of S_1 can be populated if the excitation or pump energy occurs over a broad spectrum. However, decay of the rotational and vibrational levels of S_1 occurs very rapidly (in $10^{-12} - 10^{-13}$ s) to the lowest lying excited level of S_1 via collision decay. The population temporarily accumulates at this lowest level of S_1. The decay to S_0, which is predominantly radiative, occurs over a much longer time of the order of 1 – 5 ns. If one considers radiative transitions between all the many possible combinations of allowed level pairs between the lowest level of S_1 and the full manifold of S_0, the resulting emission is a continuous spectrum of radiation as shown in the Fig. 7.73.

The schematic diagram of a tunable continuous dye laser is shown in Fig. 7.75. The tuning can be done using a prism or a diffraction grating, which acts as dispersing element.

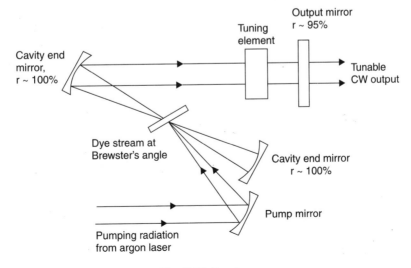

Fig. 7.75 Dye laser

7.11 X-RAY LASERS

Laser action has been achieved with X-rays as well. However, achieving X-ray laser action is more difficult compared to laser action in the visible and infrared region. Firstly, it is not so easy to achieve population inversion in the X-ray region. Also, it is difficult to fabricate high quality mirrors in the X-ray region because the refractive index does not vary appreciably from material to material. The active medium for X-ray lasers is ionized plasma. Short, intense laser pulses are used for creating the ionized plasma. Normally a nanosecond laser prepulse is used to illuminate a solid target to create the preplasma and a picosecond laser pulse is used to heat the plasma to lasing conditions. A typical experimental set up is shown in Fig. 7.76. The plasma is radially confined by the use of magnetic field.

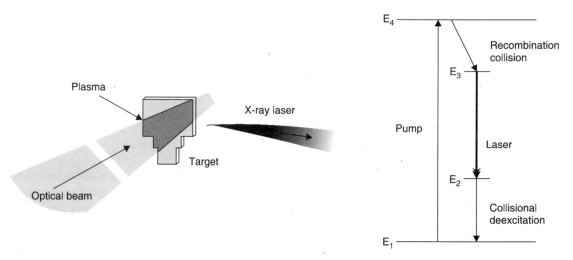

Fig. 7.76 X-ray laser   **Fig. 7.77** Energy levels of a typical X-ray laser

The population inversion between two levels E_3 and E_2 is achieved by recombination collisions or collisional deexcitation. This is done by cooling the plasma rapidly. This is achieved by adiabatic expansion of the plasma. The other way is by collisional excitation. This relays on the fast decay of the lower level E_2. The higher level is populated by one of the mechanisms mentioned above. Figure 7.77 shows the typical energy level diagram. Table 7.6 summarises the X-ray wavelengths achieved in various target materials.

Table 7.6 Wavelengths of some of the prominent X-ray lasers

Target material (atomic number)	Number of electrons stripped	Ion stage (atomic number)	X-ray wavelength (nm)
Ar (18)	8	Ne (10)	46.9
Xe (54)	8	Pd (46)	41.8
Kr (36)	8	Ni (28)	32.8
Ti (22)	12	Ne (10)	32.6
Ti (22)	12	Ne (10)	30.1

(Contd...)

Ge (32)	22	Ne (10)	19.6
Mo (42)	14	Ni (28)	18.9
Pd (46)	18	Ni (28)	14.7
Ag (47)	19	Ni (28)	13.9
Li (3)	2	H (1)	13.5
Cd (48)	20	Ni (28)	13.2
Ti (50)	22	Ni (28)	11.9
Xe (54)	26	Ni (28)	10.0
Sa (62)	34	Ni (28)	7.3

7.12 FREE ELECTRON LASERS

In conventional lasers population inversion is achieved between two discrete bound energy levels of the active medium. However, this is not the case in free electron lasers. In these lasers free electrons are the active medium and there are no discrete energy levels. According to classical electromagnetic theory, an oscillating electron radiates in a typical dipole radiation pattern that is maximum in a direction perpendicular to the oscillation. For this situation, the electrons can be thought of as making transitions between continuous energy states rather than between discrete states. These transitions are produced by projecting electrons at very high velocities through an alternating magnetic field structure. The experimental arrangement is shown in Fig. 7.78. In such an alternating magnetic field, periodic Lorentz force acting on the electron leads to an oscillatory motion. The frequency of oscillating electrons and hence the wavelength of the radiation emitted by them, depends on the kinetic energy of electrons as well as the period of the alternating magnetic field structure. The relativistic speed of the electrons causes the oscillating frequency to shift from the low frequency produced in the electron rest frame to a very high frequency observed in the laboratory frame. Laser mirrors are placed at opposite ends of the magnetic field structure and normal to the direction of electrons as in a traditional laser. This reflects back a portion of the radiated energy so that a standing wave pattern is formed. This standing wave pattern of the electric field further contributes to the oscillation of electrons and stimulates additional radiation in the desired direction. Thus a highly intense optical beam is produced within and beyond the laser cavity. The electrons are prevented from impacting and damaging the mirrors by turning them away with the help of bending magnets. Free electron lasers have produced stimulated emission over a wide range of wavelengths ranging from 248 nm to 8 nm. The quality of the laser output relies heavily on the quality of the electron beam. A spread in the energy of the electrons leads to laser line broadening. Pulsed electron beams lead to pulsed lasers and continuous electron beams lead to continuous laser output.

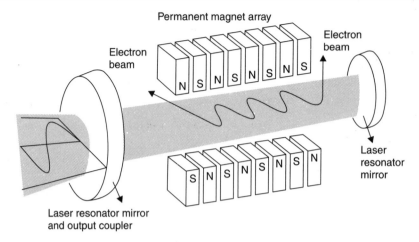

Fig. 7.78 Free electron laser

7.13 APPLICATIONS OF LASERS

7.13.1 Introduction

Lasers are used in a variety of applications. These are based on the four important properties of the laser beam *viz.*, high directionality, high intensity, high monochromaticity and high coherence. The high directionality of the laser beam makes it useful for ranging *i.e.*, to measure distance of an object and the speed with which it moves. High intensity of laser beam makes it convenient for material processing and surgery. High intense laser beams have also been used to generate high temperatures and high pressures at which nuclear fusion reactions occur. The high monochromaticity of laser beam has been used to separate isotopes. Holography, the technique of three dimensional photography is based on the high coherent nature of the laser beam. Many of the interferometric techniques in optics use laser sources on account of their coherence.

7.13.2 Measurement of Distance and Motion

Lasers can be utilised for measuring distance and motion of objects. Here the high directional properties of the laser beam are made use of. Lasers can be utilised to measure very large as well as minute distances. They can also be utilised to determine the velocity of an object. The techniques for measuring the distance falls under the following categories :

Time of flight or the pulse echo method. This is based on the principle of echo. The system consists of a pulsed laser beam and a telescope to collect the reflected light from an object whose distance has to be measured. A photodetector is used for detecting the reflected light (Fig. 7.79). By measuring the round trip time of transit for a short pulse reflected from an object, the distance of the object from the point of observation can be calculated. This technique of often referred to as *optical radar* or LIDAR (Light detection and Ranging). Accuracy of the ~ 5 m in 10 km has been achieved.

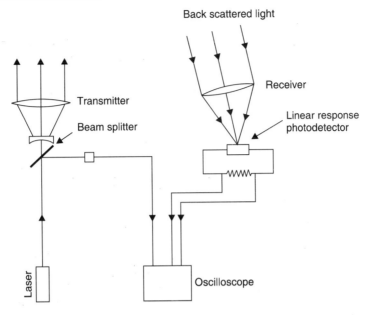

Fig. 7.79 LIDAR

Distance Estimation by Measurement of the Phase Difference. In this technique, the difference in the phase of the transmitted and the reflected laser beam from the target is measured. The laser beam is amplitude modulated by changing the laser beam power sinusoidally. This can be conveniently accomplished by the sinusoidal variation of current in a diode laser. The modulation wavelength $\lambda_m = c/f_m$, where c is the velocity of light and f_m is the

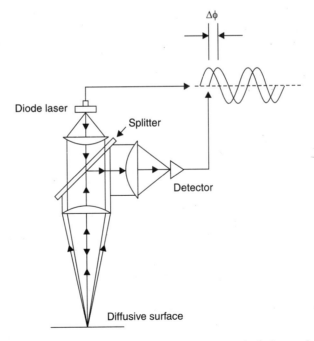

Fig. 7.80 Distance measurement by measurement of phase change

modulation frequency. The modulated beam is focused on to the target surface using a converging lens. The same lens can be used to focus the light onto the detector with the help of a beam splitter as shown in the Fig. 7.80. The phase of the laser diode current is compared with the phase of the detector output to determine the phase difference. The phase difference between the two signals depends on the time required for the laser beam to make a round trip from the transmitter to the target and back. If ϕ is the phase difference, then

$$\phi = \left(\frac{2\pi}{\lambda_m}\right) \times (2nL)$$

where λ_m is the modulation wavelength, n is the refractive of the medium in which the laser beam propagates and L is the total distance traversed by the laser beam. Thus by measuring the phase difference the distance of the target can be calculated.

Interferometric techniques for measuring small displacements. Small distances can be measured by interferometry. The arrangement consists of the Michelson interferometer (Fig. 7.81). The laser beam is split into two beams 1 and 2 by a beam splitter. Beam1 is reflected to a stationary retroreflector, which reflects the beam back to the beam splitter. Beam 2 is transmitted through the beam splitter to a movable retroreflector, which also reflects the beam to the beam splitter. At the beam splitter, a portion of beam1 is transmitted and a portion of beam 2 is reflected, thus superimposing the two beams. The superposed beams are incident of the detector. If the optical path lengths traveled by the two beams are identical, the two beams will interfere to give maximum brightness at the detector. By translating the movable

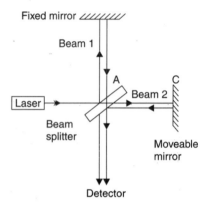

Fig. 7.81 Michelson interferometer

retroreflector a distance of $\lambda/4$ directly away or towards the beam splitter, the optical wavelenghts of the beam 2 is increased or decreased by $\lambda/2$. Now the beams are 180° out of phase and the light intensity at the detector is minimum. Thus the laser light intensity at the detector can be correlated with the small displacement of the movable retroreflector. Since the change from a constructive to destructive interference or maximum to minimum intensity of laser light corresponds to a change of distance of $\lambda/4$, one can measure minute displacements traversed by the surface to a great accuracy.

7.13.3 Laser Doppler Velocimetry

Laser could be utilised to determine the velocity of the object. When light is reflected by a moving object, the frequency of the reflected light is different from that of the incident light. This is due to the Doppler effect. The frequency of the reflected light will increase if the object is moving towards the observer and will decrease if the object is moving away from the observer. The change in frequency or the Doppler shift depends on the velocity of the object. Thus if ν represents the incident light frequency and V is the velocity of the object moving at an angle θ with respect to the incident light beam, then the Doppler shift $\Delta \nu$ is given by the formula :

$$\frac{\Delta \nu}{\nu} = \pm \left(\frac{2V}{c}\right) \cos \theta$$

(+ve sign if the object is moving towards the observer and
−ve sign if the object is moving away from the observer)

where c is the velocity of light in free space. Thus by measuring the Doppler shift, one can estimate the velocity of the object. The basic arrangement for velocity measurement essentially is similar to that of Michelson interferometer. A laser beam is split by a beam splitter. One of the components is reflected from a fixed mirror and the other component from the moving object. The two beams are then combined and made to interfere as shown in the Fig. 7.82. Because of the Doppler shift, the two beams reaching the detector have slightly differing frequencies and hence one observes the phenomenon of beats. The beat frequency can be measured and hence the velocity of the object can be determined. Laser Doppler velocimetry has been used to measure the velocity of fluid flows and is employed in chemical industries, where monitoring of fluid flows is necessary. In medicine, it has been utilised to determine the velocity of blood flows, which is used as a diagnostic tool for locating the blocks in the arteries.

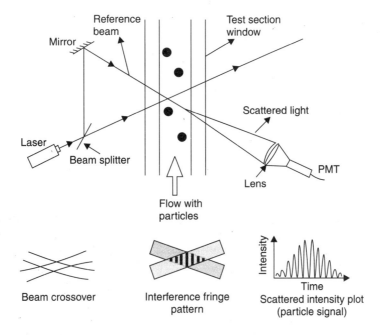

Fig. 7.82 Laser Doppler velocimetry

7.14 MATERIALS PROCESSING WITH LASER BEAMS

Lasers have been used for various processes such as welding, drilling, cutting, scribing, thermal treatment and annealing. The following are some of the important requirements :

(a) the laser beam must be of sufficient energy to transmit heat to the material to be processed

(b) processing time must be optimized for a particular operation

(c) proper choice of the laser wavelength must be made depending on the optical absorption characteristics of the material

(d) optimum choice of the beam focus and the laser pulse shape is necessary

(e) in some cases initial preparation of the material surface may be necessary

The laser beam is usually a few millimeters in diameter. For material processing applications, the laser beam has to be focused using lens or a combination of lenses. If λ is the wavelength of light, a is the radius of the beam and f is the focal length of the lens, the incoming beam will get focused into a region of radius b given by

$$b \approx \lambda f/a$$

If P is the power of the incoming beam, then the intensity I obtained at the focused region is given by

$$I \approx \frac{P}{\pi b^2} = \frac{Pa^2}{\pi \lambda^2 f^2}$$

Thus on focusing a $1W$ laser beam (with $\lambda = 1.06$ μm and having a beam radius of 1 cm) by a lens of focal length 2 cm, the intensity at the focused spot is given by $\approx 8 \times 10^8$ W/cm². Because of such large intensities, enormous heat is generated which can melt metals.

Lasers have been employed to modify the surface properties of a material. Laser surface modification (LSM) refers to any process in which the heat produced by the interaction of the laser beam and the surface of the material is used to bring about a desirable alteration in material properties. It is used to increase wear resistance, corrosion resistance or strength. The techniques usually employed are laser chemical vapour deposition (LCVD), cladding, alloying, surface hardening and surface melting. The set-up used for LSM is relatively simple. The beam is directed perpendicular or nearly perpendicular to the surface to be treated. The spot size of the laser beam is usually large and the beam or the surface is moved to facilitate the scanning of the entire surface.

In conventional or thermally activated chemical vapour deposition (CVD) reactive gases are passed over a heated surface. The thermal energy derived from the surface drives the reactions, depositing a film. CVD is commonly used in semiconductor processing to deposit thin films such as metals, dielectrics and semiconductors. Dopants which are controlled impurities, are added to these films in order to obtain the desired electrical properties. Laser chemical vapour deposition (LCVD) is a process whereby the laser is used to heat a substrate so that a vapourous material can be deposited on it. It is one of the several techniques used in thin film technology. The substrate is heated for better adhesion. Laser is an inappropriate heat source for depositing on large areas the substrate. It is ideal and appropriate for localized heating and for complex substrate geometries which are encountered in integrated circuits. In some cases laser could also be used to dissociate molecules or excite atoms in the gas phase or on a surface to form a thin film, without actually heating the substrate.

Cladding and Alloying: Cladding and alloying are similar processes. In both these techniques an appropriate powder is laid down on the surface and is melted by the heat from the laser beam. During cooling the melt forms a chemical bond with the base metal surface. This process is also known as hard facing. In alloying, the base metal is melted to a substantial depth. The alloying material is mixed into the base metal to produce a new alloy. These two processes are shown in Fig. 7.83. Cladding and alloying may be used for improving wear resistance, corrosion resistance and impact strength.

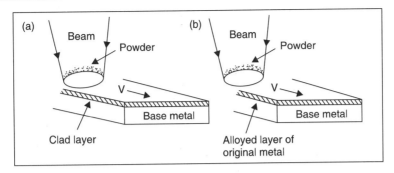

Fig. 7.83 Cladding/Alloying

Surface hardening: It is well known that many materials such as cast iron and steel become hard after heat treatment. The laser is an ideal tool for selective and localized heating. Surface hardening with the laser improves wear resistance, impact strength and fatigue strength. The structural changes that occur on the surface due to rapid heating and cooling introduce residual stresses. These residual compressive stresses are responsible for the improved properties.

Soldering and welding: Lasers are used for both welding and soldering. In soldering, the laser acts as a heat source to reflow the solder which has previously been placed on the metallic part or parts. Laser soldering is preferred because the heat input can be localised and minimal heating is required. Hot plate or hydrogen flame techniques heat either the entire substrate or a large portion of it. This can damage sensitive electronic components.

Two regimes of operation occur in laser welding depending on the incident power and power density. These are referred to as *thermal conduction welding* and *keyhole welding*. In thermal conduction welding, the laser welding is predominantly controlled by thermal conduction. This means that the power is absorbed at the surface of the metal and diffuses into the metal. Melting occurs for a depth determined by the thermal characteristics of the material and the welding parameters. Obviously, there must be sufficient heat to raise the material's temperature above the melting point. In thermal conduction welding a great deal of energy is lost due to reflection. Although the reflectance of metals decreases as the temperature increases, molten metal is still highly reflective.

In key-hole welding the welding process is made more effective. The vapour pressure of the heated metal overcomes the surface tension of the molten pool and opens up a cavity into the metal which allows the laser beam to penetrate deeper (Fig. 7.84). The molten metal flows around the key hole and fills the hole, after the laser beam has passed over. Typical laser weld configurations are shown in the figure. Usually the welding operation is carried out in an atmosphere of flowing inert gas such as helium, argon and nitrogen (Fig. 7.85). This ensures the protection of the focusing optics and control of weld plasma. The plasma is produced by the heating of the work piece which emits small particles, electrons and ionised atoms. The advantages of laser welding are (*a*) flexibility (*b*) no need of a filler material (*c*) weld in most atmospheres (*d*) weld many dissimilar metals (*e*) deep penetration (*f*) high speed (*g*) no special joint preparation (*h*) minimal part distortion and (*i*) small heat affected zone.

Hole drilling: There are two methods of piercing holes with lasers. They are percussive drilling and gas assist process. In percussive drilling, the energy from the laser pulse causes material to vapourise rapidly enough so that the molten and the solid material is expelled form the hole due to the rapid high pressure build-up. In the gas assist process air or an inert gas is blown into the hole so that the molten material and solid particles are expelled. Holes ranging from 0.025 mm in diameter upto 1.5 mm in diameter are routinely drilled by lasers up to depths of 25 mm.

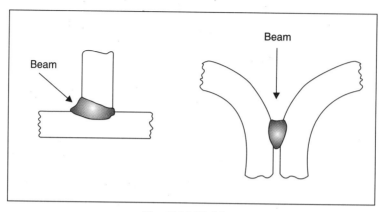

Fig. 7.84 Welding

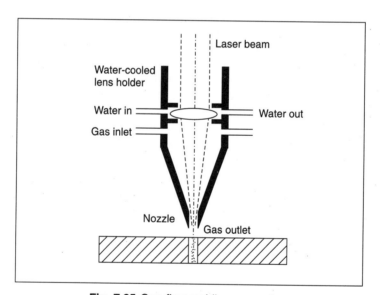

Fig. 7.85 Gas flow welding apparatus

Cutting: Cutting of metals with lasers is a gas assisted process. The purpose of using the gas is to blow away the molten material and prevent it from damaging the focusing optics. Since the molten material is blown away, the amount of vapourisation required is minimised. This improves both the cutting rate and the quality of the cut. Laser cutting is also employed in garment industry.

Marking: One of the major uses of lasers is in marking (sometimes referred to as engraving). CO_2 lasers are used to produce entire patterns, bar codes, serial numbers, logos etc. with one or two laser pulses.

Figure 7.86 shows the typical energies employed for various operations involved in materials processing.

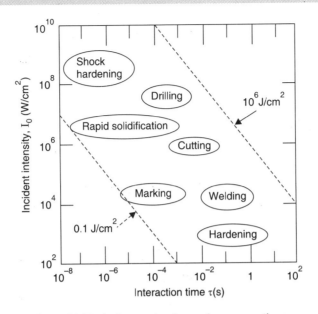

Fig. 7.86 Typical energies for various operations

7.15 LASER APPLICATIONS IN MEDICINE

Lasers have been used in medicine both for therapeutics as well as for diagnosis. When a human tissue is exposed to laser radiation its temperature raises. The extent of damage depends on the time for which the tissue is at elevated temperatures (Fig. 7.87). The nature of laser-tissue interaction process may be divided into several regions, determined primarily by the intensity of the laser beam and its interaction time with the tissue. Lasers are used for tissue

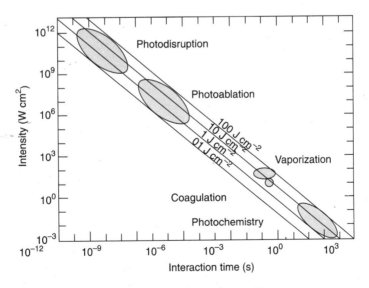

Fig. 7.87 Laser tissue interaction

cutting and removal at relatively high beam intensities with exposure times of milliseconds to seconds. This results in rapid deposition of heat and subsequent vapourisation or decomposition. These are processes that are characteristic of thermal or thermoacoustic ablation. For nanosecond pulses at relatively high photon energies, photons can directly break specific chemical bonds resulting in strong absorption and particularly clean cut. Lasers have been extensively used in ophthalmology, oncology, dermatology, cardiology, gynecology, dentistry and acupuncture. In many cases the laser is precisely targeted with the help of an optical fiber. The advantages of using laser include

   (a) greater precision in targeting the diseased tissue
   (b) less bleeding and swelling
   (c) greater control over the input energy by means of varying the laser pulse duration as well as the number of pulses per second
   (d) less pain and dispensing with general anesthesia.

Ophthalmology: The eye is roughly spherical in shape and consists of an outer transparent wall called cornea (Fig. 7.88). This is followed by iris which controls the amount of light entering the eye and the eye lens. The space between the cornea and the lens is filled with aqueous humor. The rear portion of the eye consists of retina. Light falling on the eye is focused onto the retina. The photosensitive pigment contained in the retinal cells convert the light energy to electrical pulses. These are carried by the optic nerve to the brain for processing. Some times the retina many get detached from the underlying tissue causing blurred vision or blindness. Lasers are primarily used for treating the retinal detachment. It is used for photocoagulation of the retinal blood cells *i.e.*, the blood vessels are heated to the point where the blood coagulates and forms a viscous mass. Photocoagulation is useful for repairing retinal tears and holes that develop prior to retinal detachment.

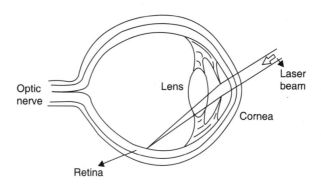

Fig. 7.88 Laser eye surgery

Lasers can be used for correction of focusing defects of the eyes. In the method referred to as LASIK (Laser In-Situ Keratomileules), the cornea of the eye can be crafted to adjust its curvature. This leads to proper focusing of light on to the retina. This treatment is suitable for people who use glasses with high lens powers.

Oncology: Laser is an ideal tool for burning away the cancerous tissues. The treatment has the advantage that the laser beam can be focused to a very small region. This helps in retaining the healthy tissues which are otherwise destroyed along with cancerous cells. Some cancerous cells selectively absorb a phototoxic dye. These dyes become toxic when exposed to light and hence destroys the cancer cells on exposure to laser. This is known as photodynamic therapy.

Dermatology: Lasers can be used to remove tattoos and skin blemishes. It is also used to remove port-wine birth marks, which appear as a result of concentration of blood vessels in the patch of the skin.

Cardiology: Cholesterol clots and plaques in coronary arteries obstruct the smooth flow of blood and cause malfunctioning of the heart. Lasers have been used to vapourise cholesterol clots and plaques which are deposited in the coronary arteries. The presence of clots and plaques in coronary arteries alters the speed of blood flow. Hence a measurement of the speed of blood could provide an early diagnosis for taking remedial measures. Laser Doppler velocimetry is employed to measure the speed of blood flow.

Gynecology: Blocks in the fallopian tubes is one of the major reasons for the infertility in women. These blocks can be removed with laser. Cysts and tumors in uterus have been cauterized using laser.

Urology: Lasers have been successfully used to remove kidney stones from the bladder.

Dentistry: Lasers have been used to drill out the decayed portion of the tooth or sterilize the decayed tooth cavity. It has been used to fuse the enamel when there is a slight porosity or cracks. Laser brushing of teeth is found to alter the surface texture of the teeth and make it resistant to decay. In dental restorative operations laser is used for welding gold and other dental alloys with various dental bridge work elements.

Acupuncture: Pain due to disorders of many internal organs of the body travels away from the source of disorder and is experienced at sites remote from the ailing organs. This is because the pain manifests itself due to the irritation of a nerve trunk, its root or its terminal ends. There are a number of areas in the human skin where pain from the organ appears and this is known as *referred pain* or *reflex pain.* About 700 sites on the skin surface are identified which corresponds to various inner organs. Acupuncture is traditionally performed with sharp pointed needles. Depending on the diagnosis of the medical experts, these needles are inserted into the relevant areas of the skin to overcome pain and cure the disease. In recent years acupuncture has been tried with laser pulses instead of sharp pointed needles. Table 7.7 summarises the use of various lasers for treating different ailments

Table 7.7 Use of lasers in medicine

Laser type	*Spectral region (nm)*	*Body tissue*	*Treatment*
He-Ne	Visible (632.8)	Dermal	Photradiation, wrinkle removal, acupuncture
Argon	Visible (450)	Oncological, cholesterol clot, fallopian tube, gastrointestinal tract	Photocoagulation, skin blemishes, skin therapy, incision, ulcer
Argon	Ultraviolet (350)	Nasal mucous membranes, dermal tissue	Ultraviolet therapy, germicidal agent, radiotherapy
Nd-YAG	Infrared (1064)	Stomach, liver, lungs, heart, kidneys, pancreas, brain, skin	Incision, control of hemophilia, cancer
CO_2	Infrared (1060)	Stomach, liver, pancreas, skin, profuse bleeding, syphilitic tissue, herpes sores, melanoma, vocal cords, tonsils	Removal of tissue, sprained joint, cancerous tissue, fallopian tube

(Contd...)

Table 7.7 (Contd.)

Laser	Spectral type	Body tissue region (nm)	Treatment
Er-YLF	Infrared (1228)	Eye tissue	Ophthalmology
Ruby	Visible (694.3)	Retinopathy, melanoma, skin blemishes	Eye disorders, cancerous tissue
Copper vapour	Visible (510.5)	Herpes sores	Excision

7.16 LASER OPTICAL DISC (COMPACT DISC)

Lasers are used to read out data from the optical disc, also known as the compact disc. A major advantage of the optical disc is its contact-less read-out. In a compact disc, the information is stored in digital form. The sequence of 1s and 0s are recorded in the form of pits or depressions along a spiral track on a plastic material with a metallic coating (Fig. 7.89). The usual coding is such that any transition from pit to land (flat area) or land to pit is read as 1. The duration in the pit or land is read as zero. Typically the radial distance between the adjacent tracks is 1.6 µm. The length of the pit is about 0.6 µm.

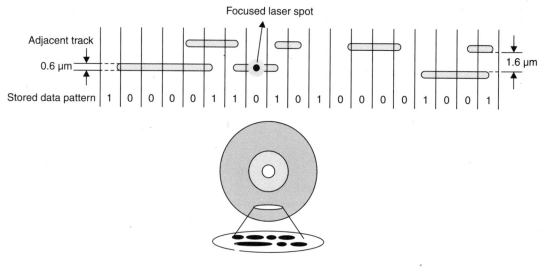

Fig. 7.89 Optical disc

In compact disc systems the laser beam is focused into a tiny spot (beam diameter ~1 µm), which reads the information (Fig. 7.90). The information layer of the optical disc is inside a transparent plastic layer so that the defects (scratches and dust) only occur on the outer surface where the beam of light is out of focus and has diameter of the order of a mm. Thus scratches and dust do not affect the reading quality. Further, since it is a contact-less read-out, the disc has a very long life compared to the tapes where the information is read-out through actual physical contact. Thus an optical disc can be viewed as a microscope which scans the pits on the rapidly rotating disc.

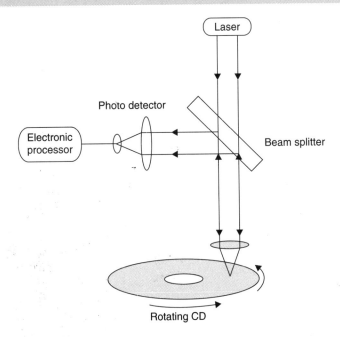

Fig. 7.90 Optical readout

When the disc is scanned with the focused laser beam, the diffraction of the beam on striking the depression modulates the light intensity reflected back into the optical pick-up unit. The optical pick-up head will detect a modulation of the reflected light with varying frequency and varying pulse width. The modulation of the reflected light by the pick-up is transformed into a current modulation. The photocurrent is amplified and decoded to obtain information as standard input signals for video and audio equipment. Digital signal processing techniques are employed.

The information layer of the optical disc is a reflective metallic layer, inside a plastic disc. The audio/video information is present in the disc as a relief pattern imprinted on the plastic material, covered with a reflective layer. The basic principle involved in the making of the disc is shown in Fig. 7.91. A glass disc is coated with a thin layer of photo resist. The recorder uses a laser beam whose intensity is modulated by the signal on a master tape, to write a pattern in the photo resist. After photographic development, the real pit structure is obtained. For the replication process, the master plate is silvered and coated with nickel. A so-called *metal father* is obtained by separating the metal from the glass. This *father* is the negative of the pit structure required on the disc. This is directly used as a mould for replication.

The maximum information storing capacity of an optical disc is determined by the size of the focused spot of a laser beam. Smaller spot sizes can be achieved by smaller wavelengths and smaller focal length lenses. Since the CD is covered by a protective layer, the focusing needs to be done through the protective layer. A CD uses a 1.2 mm clear substrate and the data is recorded on the recordable layer through the clear substrate. The substrate also acts as a protective layer for the data. The reading wavelength is typically 780 nm. In contrast in a DVD two clear substrates each of 0.6 mm thickness are bonded together and data is recorded on both sides of each substrate. The reading wavelength in DVDs is typically 650 nm (Fig. 7.92). Thus DVDs can store much more data than CDs. Recent developments of blue lasers emitting a wavelength of 405 nm have triggered development of DVDs with much higher capacities. Table 7.8 provides a comparison of CDs, DVDs and blue ray DVDs.

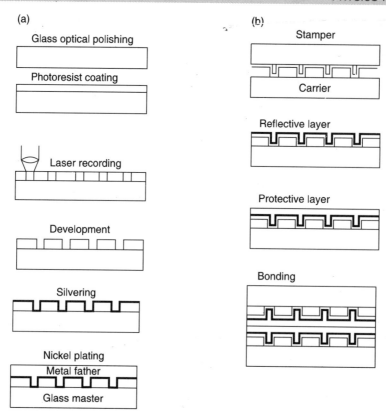

Fig. 7.91 Production of a disc

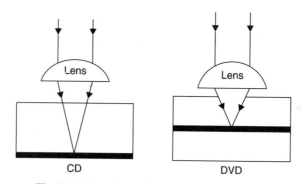

Fig. 7.92 Focusing of laser in a CD and DVD

Table 7.8 Comparison of CD, DVD and Blue Ray DVD

Parameter	CD	DVD	Blue ray DVD
Laser wavelength (nm)	780	650	405
Track to track spacing (nm)	1600	740	320
Spot size (nm)	1600	1100	480
User capacity (Gb)	0.68	9	50

7.17 LASER BAR CODE SCANNER

The Universal Product Code (UPC) and European Article Numbering (EAN) systems are widely used for identifying any product, its manufacturer, its price and other specifications. A special form of the EAN code is the International Standard Book Numbering (ISBN) system for identifying books. A bar code consists of a series of strips of dark and white bands (Fig. 7.93). These white and dark bands are of different widths separated from each other by specific distances that contain all the information about the product. A laser is used to scan the bar code with the help of a rotating mirror (Fig. 7.94). Typical scanning speeds are about 200 m/s. Such high speeds enable the product to be scanned even when in motion in a delivery line. When the laser beam is incident on the bar code, the amount of light scattered depends on whether the strip is black or white. Since the bars are separated by variable distance, light intensity varies with time and is recorded by the photodetector. The signal is fed into an amplifier and later to a decoder which displays the information on a TV monitor and also sends it to product inventory system.

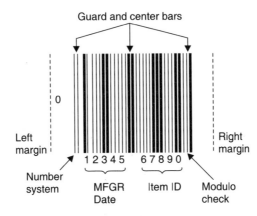

Fig. 7.93 Bar code

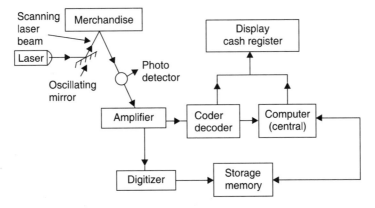

Fig. 7.94 Laser bar code scanner

7.18 LASER PRINTER

A laser printer is essentially a xerographic device (Fig. 7.95). The photoconductor in a xerographic copier allows light to control the distribution of electric charge. An optical image of the original document is projected on to the charged region of the photoconductor. Wherever light falls, the photoconductor conducts electricity and the charge on that region escapes. The result is a charge image on the surface of the photoconductor. Tiny black toner particles, oppositely charged from the unilluminated portions of the photoconductor are brought near the charge image. These toner particles stick to the charged portions of the photoconductor, forming a visible image of the document. In a laser printer, a laser beam is used to write a charge image directly onto the photoconductor drum. This rotating drum is charged by a coratron and then a spinning mirror scans laser light rapidly across the surface. Wherever laser light falls on the drum charge flows through the photoconductor. A computer in the printer turns the laser on and off as it systematically constructs the charge image one dot at a time. The printer then sticks the toner particles to the charge image and transfers it to the paper. Most laser printers use special electrostatic tricks to reverse the charge image, so that the toner particles stick only to the portions of the drum that were exposed to the laser light. This reversal makes it easier for the printer to produce white backgrounds for its prints.

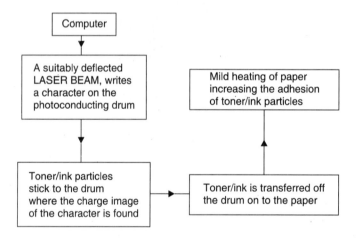

Fig. 7.95 Laser printer

A colour laser printer works like a single colour model, except that the process is repeated four times and a different toner colour is used for each pass. The four colours used are cyan, magenta, yellow and black. Laser printers have resolution ranging from 300 dots per inch (dpi) to 1800 dpi.

7.19 LASER SYSTEMS FOR ATMOSPHERIC POLLUTANT DETECTION

The atmospheric pollutants such as CO_2, SO_2, CH_4, N_2O, C_2H_6 etc. can be detected using laser. This system consists of a laser beam transmitter which traverses a certain distance (~km) in the atmosphere and then gets reflected from a mirror. The attenuation in the intensity of the reflected beam is measure of the concentration of the pollutants (Fig. 7.96). If I_0 is the initial intensity of the beam and $I(x)$ is the intensity after travelling a distance x, then

$$I(x) = I_0 e^{-\alpha x}$$

where α is the attenuation coefficient. α depends on the wavelength as well as the nature of the pollutant. The laser absorption is generally high for specific frequencies depending on the vibrational frequencies of these molecules. The laser beam is tuned to different frequencies characteristic of the different pollutant molecules to measure the concentration of different pollutants.

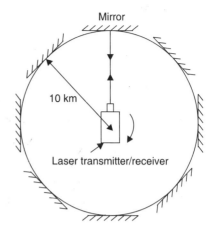

Fig. 7.96 Atmospheric pollutant detection system

7.20 LASER FUSION

The enormous energy released from the sun and the stars is due to thermonuclear fusion reactions. Attempts are being made to generate fusion energy even in the laboratory. Fusion reactions occur only at very high temperature ($\sim 10^8$ K) and very high pressures ($\sim 10^{12}$ atm). This is necessary to overcome the coulomb repulsion between the nuclei. With the availability of intense laser pulses, a new idea of fusion system has emerged. The idea is essentially compressing and heating the thermonuclear material with the help of lasers. The thermonuclear material is usually in the form of a pellet and is irradiated by a large number of laser beams. Laser fusion facilities exist in USA, UK, France and Japan. CO_2 or Nd : glass laser has been put into use. Typical nuclear reactions employed are :

$$_1H^2 + _1H^2 \rightarrow _2He^3 + _0n^1 + 3.23 \text{ MeV}$$
$$_1H^2 + _1H^2 \rightarrow _1H^3 + _1H^1 + 4.03 \text{ MeV}$$
$$_1H^2 + _1H^3 \rightarrow _2He^4 + _0n^1 + 17.6 \text{ MeV}$$
$$_1H^2 + _2He^3 \rightarrow _2He^4 + _1H^1 + 18.3 \text{ MeV}$$

The sustainability of a nuclear fusion is characterised in terms of the Lawson criterion. This states that at a temperature of 5 keV (i.e., $k_B T = 5$ keV; 1 eV = 11,600 K), the product of the ion number density (n_i) and containment time 'τ' must satisfy the relation $n_i \tau > 10^{20}$ m^{-3}s. In a magnetically confined reactor, confinement times are typically in the range of milliseconds to seconds.

The principle of laser fusion is that of inertial confinement. The idea is to heat a small pellet of nuclear fuel so fast that the mass of the fuel pellet prevents it from dispersing, because of its own inertia. The Lawson criterion is thereby satisfied with a large ion density and smaller confinement time. For a fuel pellet of radius a the natural inertial confinement time is

$$\tau_i \sim a/v_s$$

where v_s is the velocity of sound in the fuel at elevated temperature. At a temperature of 10 keV, $v_s \sim 10^6$ m/s, so for a fuel pellet of radius 1 mm, the inertial confinement time is only 1 ns.

The fuel pellet is irradiated simultaneously by a large number of high energy short pulse laser beams to heat the fuel pellet in such a short time (Fig. 7.97). The irradiation should be as spherically symmetric as possible. Typically Nd-glass lasers operating at about 1.05 µm are used. The irradiation of the fuel pellet vapourises its outer surface to form plasma. The plasma allows the laser beam to penetrate a certain distance, to the critical-density surface, where substantial energy is absorbed. The hot plasma conducts thermal energy inwards to the solid boundary of the fuel pellet. The rapid heating of the solid boundary ablates the material, which explodes away from the rest of the pellet like a rocket engine. The equal and opposite reaction force generated according to Newton's third law, exerts an enormous pressure, ($\sim 10^{12}$ atm) and compresses the remainder of the fuel pellet inwards. The core of the pellet is compressed to as much as 100 times the solid density. The culmination of the compression and the heating process is that perhaps 10% of the original fuel pellet is sufficiently heated as well as compressed. Thus conditions for fusion reaction is realized.

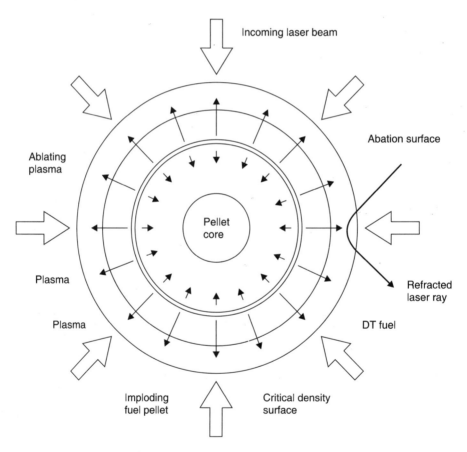

Fig. 7.97 Laser induced nuclear fusion

7.21 LASER ISOTOPE SEPARATION

The different isotopes of an element are chemically indistinguishable. Hence chemical means cannot be used for isotope separation. Lasers are employed for separating the various isotopes of an element. In this technique slight differences in the energy levels of the atoms of the isotopes due to differences in nuclear mass is made use of. The high monochromaticity of the laser beam makes this practically feasible. Thus laser light of a certain wavelength may be absorbed by one isotope while the other isotope of the element may not absorb it. Since light emitted from a laser is highly monochromatic selective excitation of isotope of one kind is possible. Later on, various techniques are employed for separating the excited isotope.

This technique has enormous use for the large scale enrichment of uranium (separation of U^{235} from U^{238}) for use in nuclear power reactors. There are other applications for pure isotopes in medicine. U^{235} is fissile and is used in nuclear fission. The natural uranium contains only 0.7% of U^{235}. The methods utilised for separating them using lasers are discussed here.

The use of lasers for isotope separation is based on the slight difference in the energy levels of isotopes. The energy levels of U^{235} and U^{238} slightly differ and this is exploited for isotope separation (Fig. 7.98). Natural uranium atoms emerge from an oven at 2600 K. At this temperature 27% of uranium atoms are in the level E_1. U^{235} atoms are excited from this level E_1 (620 cm^{-1}) by a xenon laser with wavelength $\lambda_1 = 378.1$ nm to an intermediate level of energy E_2. The wavelength of the transition from E_1' to E_2' in U^{238} is slightly longer. Hence these atoms are not excited. Excited U^{235} atoms that reach level E_2 are ionized with krypton laser operating at 350.7 nm or 246 nm. By this two step photo dissociation process U^{235+} ions are selectively produced. The ions are collected by deflecting them to a collector with a strong electric field.

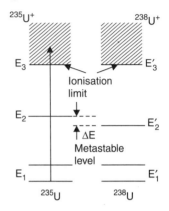

Fig. 7.98 Laser isotope separation

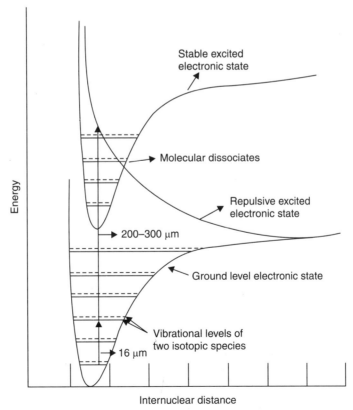

Fig. 7.99 Laser isotope separation using vibrational levels of UF_6

Another scheme has also been employed. Here differences in the vibrational levels of $U^{235}F_6$ and $U^{238}F_6$ is made use of (Fig. 7.99). A laser radiation that is resonance with vibration-vibration absorption in the ground electronic state of UF_6 selectively excites $^{235}UF_6$. This vibrationally excited molecule is then dissociated with a second energetic photon to produce a $^{235}UF_5$ fragment and a flourine atom. The first selective excitation in UF_6 is provided with a 16 μm photon and the second with a 200–300 nm photon from an excimer laser. Initially UF_6 is cooled by expansion through a supersonic nozzle. This ensures that the absorption lines are sharp and the higher vibrational levels are not populated thermally.

7.22 OPTICAL COMMUNICATION

In the field of communication lasers offer two unusual advantages. The first of these pertains to bandwidth. It is known that the rate at which information can be transmitted is proportional to bandwidth of the information carrier. Bandwidth is proportional to the carrier frequency. The step from microwaves to the optical region expands the available bandwidth by a factor of 10,000 or more. Therefore, if proper methods of modulation and demodulation are found for the laser light, an extremely potent information carrier could be achieved. The second consideration in connection with information transmission is the ability to aim it in proper direction. A point to point communication requires the concentration of the information carrier into a narrow beam. The high directionality of the laser beam is amply suited for this.

A typical optical communication system is shown in Fig. 7.100. Modulation is achieved in a number of ways such as amplitude modulation, frequency modulation or pulse code modulation. The earth's atmosphere makes transmission difficult. It could be overcome by using optical fibers (see chapter 8).

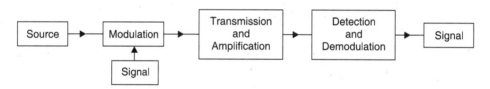

Fig. 7.100 Optical communication system

One aspect of communication is the determination of the position of a distant object, in short, ranging. The first application of laser actually demonstrated was in this field.

7.23 APPLICATIONS IN PURE SCIENCE

Some of the laser applications are in the field of pure science. Lasers offer a wonderful opportunity to investigate the basic laws of interaction of atoms and molecules with electromagnetic waves of high intensity. Many a new optical phenomenon has been observed with lasers, which otherwise would not have been possible. This forms the subject of *nonlinear* optics. In recent years using laser beams it has been possible to cool atoms to very low temperatures.

REFERENCES

1. B.A. Lengyel, *Introduction to Laser Physics,* John Wiley & Sons, NY, 1967.
2. M.J. Beesley, *Lasers and Their Applications,* Tylor and Francis Ltd., London, 1976.
3. O. Svelto, *Principles of Lasers,* Plenum Press, NY, 1976.
4. A.K. Ghatak and K. Thyagarajan, *Optical Electronics,* Cambridge Press, NY, 1989.

LASER

5. H.M. Muncheryan, *Laser and Optoelectronic Engineering,* Hemesphere Publishing Corporation, NY 1990.
6. W.T. Silfast, *Laser Fundamentals,* Cambridge Press, NY, 1996.
7. C.C. Davis, *Lasers and Electro-optics—Fundamentals and Engineering,* NY, 1996.

SOLVED EXAMPLES

1. Calculate the electric field of a laser beam of wavelength 630 nm, power 1 mW and cross section 3 mm^2.

    Solution:

    Assuming a beam of uniform intensity

    $$I = \frac{10^{-3}}{3 \times 10^{-6}} = \frac{1}{3} \times 10^3 \text{ W/m}^2 = \frac{n}{2c\mu_0} E_0^2$$

    For air $n \cong 1$ and using $c = 3 \times 10^8$ m/s, $\mu_0 = 4 \times 10^{-7}$ SI units, we have the corresponding electric field as

    $$E_0 \cong 501 \text{ V/m}.$$

2. Calculate the electric field of a bulb emitting 10W of optical power at a distance of 10 m from the source.

    Solution:

    Since the emission is uniform along all directions, the corresponding intensity at a distance of 10 m is

    $$I = \frac{100W}{4\pi(10)^2} = 7.95 \times 10^{-2} \text{ W/m}^2$$

    The corresponding electric field is given by

    $$E_0 = \left(\frac{2c\mu_0 I}{n}\right)^{1/2} \cong 2.4 \text{ V/m}.$$

3. Calculate the electric field intensity at a point when the laser beam of power 1mW is focused to a spot of radius 6 mm.

    Solution:

    The resulting intensity is given by :

    $$I = \frac{1 \times 10^{-3}}{\pi \times (6 \times 10^{-6})^2} = 8.8 \times 10^6 \text{ W/m}^2$$

    The resulting electric field is given by:

    $$E = \sqrt{\frac{2c\mu_0 I}{n}} = \sqrt{\frac{2 \times 3 \times 10^8 \times 4 \times 10^{-7} \times 8.8 \times 10^6}{1.0}}$$
    $$= 8.1 \times 10^4 \text{ volt/m}$$

4. What is the ratio of populations of the two energy levels corresponding to the lasing wavelength of 694.3 nm in Ruby laser?

Solution:

We have
$$\frac{N_1}{N_2} = e^{\Delta E/k_B T} = \exp\left(\frac{hc}{\lambda k_B T}\right)$$

$$= \exp\left(\frac{6.63 \times 10^{-34} \times 3 \times 10^8}{694.3 \times 10^{-9} \times 1.38 \times 10^{-23} \times 300}\right) = e^{69.2} = 1.13 \times 10^{30}.$$

5. The ratio of population of two energy levels at 300 K is 10^{-30}. Find the wavelength of the radiation emitted.

   Solution:

   We have the relation
   $$\frac{N_2}{N_1} = \exp\left(-\frac{hc}{\lambda k_B T}\right)$$

   ∴
   $$10^{-30} = \exp\left(-\frac{6.63 \times 10^{-34} \times 3 \times 10^8}{\lambda \times 1.38 \times 10^{-23} \times 300}\right)$$

   Taking natural logarithms on both sides
   $$\lambda = \frac{6.63 \times 10^{-34} \times 3 \times 10^8}{\lambda \times 1.38 \times 10^{-23} \times 300} = 48.6 \ \mu m.$$

6. Calculate the ratio of the stimulated emission to spontaneous emission for laser wavelength 694.3 nm.

   Solution:

   We have the relation
   $$R = \frac{B_{21} U_v N_2}{A_{21} N_2} = \frac{1}{e^{h\nu/k_B T} - 1}$$

   Here
   $$\frac{h\nu}{k_B T} = \frac{hc}{\lambda k_B T} = \frac{6.63 \times 10^{-34} \times 3 \times 10^8}{694.3 \times 10^{-9} \times 1.38 \times 10^{-23} \times 300} = 68.9$$

   ∴
   $$R = \frac{1}{e^{68.9} - 1} \approx e^{-68.9} = 4.98 \times 10^{-14}.$$

7. Calculate the number of photons emitted by ruby laser of output power 1 W. The lasing frequency of the Ruby laser is 694.3 nm.

   Solution:

   Let n be the number of photons emitted per second. Then
   $$nh\nu = P \Rightarrow \frac{P}{h\nu} = \frac{P\lambda}{hc} = \frac{1 \times 694.3 \times 10^{-9}}{6.63 \times 10^{-34} \times 3 \times 10^8} = 3.49 \times 10^{18}.$$

QUESTIONS

1. What are the characteristics of a laser beam ?
2. Describe the (a) spontaneous emission (b) stimulated emission (c) absorption and (d) population inversion.
3. Describe the various methods of pumping lasers with suitable examples.
4. Define Einstein's coefficients and obtain the relation between them.

5. Distinguish between continuous, pulsed and tunable lasers.
6. What are laser modes ?
7. Discuss the advantages of Q-switching in laser action.
8. Describe the laser action in the following lasers with suitable energy level diagram : (a) Ruby (b) Nd-Glass (c) Nd-YAG (d) Alexandrite (e) Ti-Saphire
9. Write a short note on semiconductor lasers.
10. Write a short note on quantum well lasers.
11. Discuss the laser action in a mixture of He-Ne.
12. Describe the laser action inert gas ion lasers with suitable energy level diagrams.
13. Discuss the laser action in N_2 and CO_2.
14. What are excimer lasers ? Explain with relevant energy level diagram.
15. Write a short note on chemical lasers.
16. Discuss the working of dye lasers. Mention their advantages.
17. What are X-ray lasers ? Give examples.
18. Discuss the operation of a free electron laser.
19. Discuss the industrial applications of laser.
20. Explain the use of laser in material processing.
21. Write a short note on LIDAR.
22. Discuss the technique of laser Doppler velocimetry.
23. Explain how lasers are used in separating isotopes.
24. Write a short note on laser induced fusion.
25. Discuss the medicinal applications of laser.

PROBLEMS

1. What is the ratio of population for energy levels corresponding to 1060 nm at 300K ?
2. Calculate the ratio between Einstein coefficients for a lasing wavelength of 200 nm.
3. The ratio of population of two energy states is e^{-1}. Determine the temperature if the lasing wavelength is 600 nm.
4. A laser pulse duration is 12 ps and energy is 1 W. Calculate the power and the number of photons in each pulse if the laser wavelength is 1060 nm.
5. What is the spatial interaction distance of a pulse of duration 10 ns ?

8
Holography

8.1 INTRODUCTION

The advent of lasers has made the art of holography possible. Holography can be thought of as a new approach to the problem of generating images of objects. It enables to generate a three dimensional image of an object. An ordinary photograph represents a two dimensional recording of a three dimensional scene. The light scattered by the object is focused on to a photographic film. The emulsion on the photographic plate is sensitive only to the intensity variations. In this process, the phase information carried by the electromagnetic wave scattered from different points on the object is lost. Since only the intensity pattern is recorded, the three dimensional character of the object is lost. (A spatial sense of depth and parallax accompanies a three dimensional object. Parallax refers to the apparent shift in the position of the object with respect to far off background when viewed from different angles). One way of overcoming this problem is to take a photograph of the object from two slightly different view angles and then superimpose the two transparencies. *This results in an image having a sense of depth but is still devoid of parallax.* This is referred to as stereograph.

Interferometry is a technique of recording phase difference between two waves. In the Young's double slit experiment, from the interference pattern it is possible to deduce the phase difference between two waves. At the central maximum, the two waves from the two slits have zero phase difference. For every intensity maximum encountered on either side of the central maximum there is a phase difference of 2π. However, interference is possible only with coherent sources. Holography is based on interferometric technique with lasers. High temporal and spatial coherence of the laser makes such a technique practically feasible. The inherent limitation in conventional photography is overcome in holography.

8.2 PRINCIPLE

In the year 1948 Dennis Gabor conceived of an entirely new idea of recording the phase and the amplitude of the electromagnetic wave emanating from the object. Figure 8.1a shows the typical experimental arrangement used for recording a hologram. A laser beam is split into two beams by a beam splitter. One beam is scattered by the object and is called the object wave. The other beam is called the reference wave. The object wave and the reference wave are made to interfere. The interference fringes, characteristic of the object are formed. This interference pattern has in it not only the amplitude distribution but also the phase of the electromagnetic wave scattered by the object. Since the intensity pattern has both the amplitude and phase recorded in it, Gaber called the recording a hologram (Holos in Greek means "whole"). The hologram has little resemblance to the object. It has in it a coded form of the object wave.

HOLOGRAPHY

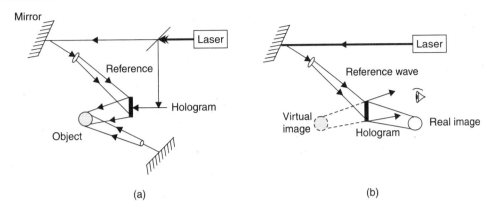

Fig. 8.1 Holography-principle (a) formation of hologram and (b) reconstruction

The technique by which one reproduces the image is called reconstruction. In the reconstruction process, the hologram is illuminated by a wave called the reconstruction wave. This reconstruction wave is identical in wavelength to the reference wave used for recording the hologram. When the hologram is illuminated by the reconstruction wave, two waves are produced (Fig. 8.1b). One wave appears to diverge from the object and provides the virtual image of the object. The second wave converges to form an image which is real and thus can be recorded on a screen or photographed. The image possesses a sense of depth and parallax just as the object. The spatial perspective will change as the viewer moves his head. As the observer moves his eyes to different positions, the rays of light entering each pupil come through different sections of the fringe pattern on the hologram. As a result he sees the object in different perspective. If he finds an object hidden behind the other, he can move his head and look around to view the hidden object. *This is best expressed by saying that a single hologram is equivalent to a thousand photographs.*

Following conditions are to be satisfied for good recording of the hologram :

(*a*) The laser beam should have spatial coherence over the extent of the object and temporal coherence over its depth.

(*b*) The experimental configuration ought to be mounted on a vibration isolation table. No part of the apparatus should move by an amount that may change the optical path by more than $\lambda/8$ during exposure.

(*c*) The polarization of the laser beam is kept perpendicular to the horizontal table and reflections are confined to the horizontal plane. This ensures that the plane of polarization does not rotate.

(*d*) Photoemulsion must consist of very fine grains for obtaining high resolving power.

Although the principle of holography was laid down in 1948, it was not until the lasers arrived in 1960, that holography attained practical importance. Holography relies on the coherence properties of the laser light. The technique involves the observation and recording of interference fringes, which is easily possible with coherent beams since they have large coherence length compared to ordinary sources.

8.3 MATHEMATICAL THEORY OF HOLOGRAM

Let the hologram be perpendicular to z-axis during the recording. The object wave and the reference waves interfere on the hologram to give the interference pattern characteristic of the object.

Let
$$E_0(x, y) = A_0(x, y)e^{-i\phi_0(x, y)} \tag{8.1}$$
represent the field distribution of the object wave falling on the recording plate. A_0 and ϕ_0 are the amplitude and phase distribution of the object wave. Similarly let
$$E_r(x, y) = A_r(x, y)e^{-i\phi_r(x, y)} \tag{8.2}$$
represent the field distribution of the reference wave on the plane of the recording medium. A_r and ϕ_r represent the corresponding amplitude and phase distribution of the reference beam. In equations (8.1) and (8.2), it is assumed that the plane of the recording medium is at $z = 0$. The object wave and the reference wave interfere on the recording medium. The resultant intensity distribution is given by

$$I(x, y) = |E_0(x, y) + E_r(x, y)|^2 = A_0^2 + A_r^2 + A_r A_0 e^{-i(\phi_0 - \phi_r)} + A_r A_0 e^{i(\phi_0 - \phi_r)}$$

$$I(x, y) = A_0^2 + A_r^2 + 2A_r A_0 \cos(\phi_0 - \phi_r) \tag{8.3}$$

Equation (8.3) indicates that the recorded intensity pattern has information about the phase of the object wave E_0.

The photographic plate which has recorded the above intensity variation is developed to obtain the hologram of the object. The amplitude transmittance (T) of the hologram i.e., the ratio of the transmitted amplitude to the incident field amplitude on the hologram is a function of $I(x, y)$. By a suitable developing process, the amplitude transmittance is linearly related to $I(x, y)$.

i.e., $\qquad T(x, y) = I(x, y)$, where the constant of proportionality is unity. $\qquad$ (8.4)

In order to reconstruct the object, the hologram is illuminated with a reconstruction wave as shown in Fig. 8.1b. The reconstruction wave is in most cases identical to the reference wave. It is assumed that the reconstruction wave has a field distribution on the plane of the hologram (at $z = 0$) given by

$$E_c(x, y) = A_r(x, y)e^{-i\phi_r(x, y)} \tag{8.5}$$

Hence using Eqns. (8.3), (8.4) and (8.5), the field distribution of the transmitted wave on the plane of the hologram will be

$$E_t(x, y) = E_c(x, y) T(x, y)$$
$$= A_r(A_0^2 + A_r^2)e^{-i\phi_r(x, y)} + A_r^2 A_0 e^{-i\phi_0(x, y)} + A_r^2 A_0 e^{i[\phi_0(x, y) - 2\phi_r(x, y)]} \tag{8.6}$$

Consider the three terms on the RHS of Equation (8.6).

The first term represents the reconstruction wave itself but whose amplitude is modulated due to the term $(A_0^2 + A_r^2)$.

The second term is identical within a constant multiple of A_r^2 *to the object wave, which was present on the plane of the hologram when the hologram was recorded. If one views the object wave one would see a virtual image of the object exactly at the location where the object was placed during recording. In addition, since the original object wave itself has been reproduced, viewing the virtual image would have all the effects of depth and parallax.*

The last term is a conjugate of the object wave (due to $e^{i\phi}$ *term) and is also phase modulated by a phase variation* $e^{-2i\phi}$. *The conjugate wave would result in the reconstruction of a real image of the object. If the original wave is a diverging wave, the conjugate wave will be converging. The effect of this modulation of the phase is to rotate the direction of propagation of the conjugate wave (Fig. 8.1). It must be noted that in order to view the image of the object without overlap of the other waves, the three waves in Eqn. (8.6) must be travelling along different directions.*

8.4 CLASSIFICATION OF HOLOGRAMS

The angle between the reference and the object beam affects the properties of the hologram. In what follows various types of holograms will be discussed.

In-line or Gabor Hologram: This was the arrangement first thought of by Gabor. Here the reference beam as well as the object or the signal beam is located along the axis normal to the photographic plate used for making the hologram (Fig. 8.2a). During reconstruction, the observer focusing on the real image is also forced to view the virtual image and the coherent beam used for reconstruction, as shown in Fig. 8.2b. This is the serious limitation of this arrangement.

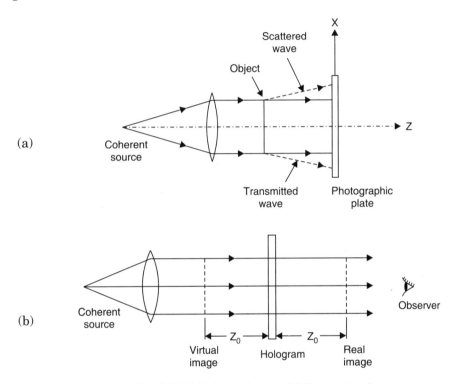

Fig. 8.2 (a) Gabor hologram (b) Reconstruction

Off-axis or Leith-Upatnieks Hologram: This arrangement for recording and reconstruction was mooted by Leith and Upatnieks to overcome the limitations of the Gabor hologram. Here the reference beam and object beam make an angle with respect to each other. They interfere on the photographic plate to form the hologram. During the reconstruction, one finds that the real and virtual images are angularly well separated as shown in the Fig. 8.1b. Thus the inherent limitations in the Gabor hologram are overcome.

Fourier Hologram: It can be shown that the optical field distribution produced at the back focal plane of an aberration free converging lens is the two dimensional Fourier transform of the optical field distribution in the front focal plane of the lens. Thus a lens can be thought of a device which produces the Fourier transform of the optical field corresponding to an object. A Fourier hologram is formed by the interference of the reference beam and the Fourier transform of the object wave. The experimental arrangement corresponding to the recording and reconstruction of the Fourier hologram is shown in Fig. 8.3. *A Fourier hologram has an interesting property that the reconstructed image is stationary even when the hologram is translated in its own plane.*

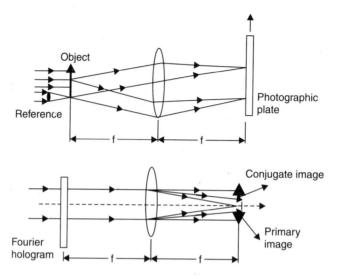

Fig. 8.3 Fourier hologram

Lensless Fourier Hologram: A hologram with the same properties as Fourier hologram can be obtained even without a lens to produce the Fourier transform of the object wave, provided the reference wave is produced by a point source in the plane of the object. Figure 8.4 shows the experimental arrangement for the same.

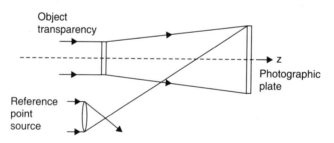

Fig. 8.4 Lensless Fourier hologram

Amplitude and Phase Holograms: A hologram recorded on a photographic plate and processed normally is equivalent to a grating with a spatially varying transmittance. Such holograms are known as amplitude holograms. During the reconstruction of the object wave, the diffraction efficiency of the amplitude hologram is found to be small. To overcome this problem, one makes a phase hologram by bleaching the photographic plate. To appreciate the difference between the amplitude and phase holograms it is necessary to understand the chemical processes involved in developing the photographic plate.

An unexposed photographic plate or film generally consists of a very large number of tiny silver halide (often AgBr) grains suspended in a gelatin support. The gelatin is attached to a firm base consisting of acetate or mylar for films and glass for plates. The soft emulsion has a thin layer of protective coating on its exposed surface (Fig. 8.5). Light incident on the emulsion initiates a complex physical process that is outlined below :

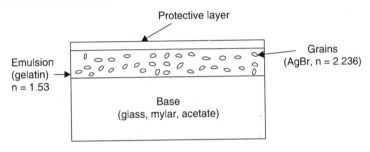

Fig. 8.5 Photographic plate

(a) A photon incident on a silver halide grain may or may not be absorbed by that grain. If it is absorbed, an electron-hole pair is released within the grain.

(b) The resulting electron is in the conduction band. It is mobile within the silver halide grain and gets trapped at a dislocation within the grain.

(c) The trapped electron electrostatically attracts a silver ion. Such ions are mobile even before exposure by light, due to thermal agitation. The electron and the silver ion combine to form a single atom of metallic silver at the dislocation site. The lifetime of their combination is rather short, of the order of a few seconds. If within the lifetime of this first silver atom, a second silver atom is formed by the same process at the same site, a more stable two-atom unit is formed with a life time of at least several days. The speck of silver formed as above is referred to as a developmental speck and the collection of development specks present in an exposed emulsion is called the latent image.

(d) The exposed photographic transparency is immersed in a chemical bath containing the developer. The developer acts on the developed silver specks. For such grains, the developer causes the entire grain to be reduced to metallic silver.

(e) At this point, the processed emulsion consists of two types of grains, those that have been turned to silver and those that did not absorb enough light to form development speck. The latter grains are still silver halide and without further processing will eventually turn to metallic silver themselves simply through thermal processes. Thus in order to assure the stability of the latent image, it is necessary to remove the undeveloped silver halide grains. This is done by a process called fixing. For fixing emulsion transparency, the emulsion is immersed in a chemical bath containing a fixer, which removes the remaining silver halide crystals from the emulsion leaving only the stable metallic silver. The processes involved in exposure, development and fixing are shown in the Fig. 8.6.

Exposure: The energy incident per unit area on a photographic emulsion during the exposure process is called the Exposure (E). Assuming the photographic plate or the film to be the x-y plane, we can write

$$E(x, y) = I(x, y) \times t$$

where $I(x, y)$ is the intensity and t is the exposure time.

Intensity transmittance: The ratio of intensity transmitted by a developed transparency to the incident intensity on the transparency is called intensity transmittance. It is averaged over a region that is large compared with a single grain but small compared with the finest structure in the original exposure pattern.

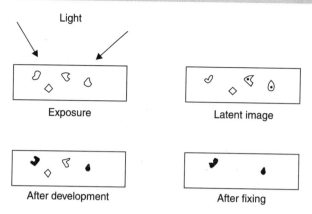

Fig. 8.6 Photographic process

$$\tau(x, y) = \left\{ \frac{\text{intensity of transmitted light at } (x, y)}{\text{intensity of incident at } (x, y)} \right\}_{\text{local average}}$$

Photographic density: It is experimentally found that the logarithm of the reciprocal of the intensity transmittance $\tau(x, y)$ is proportional to the silver mass per unit area $D(x, y)$ on the developed photographic plate. *i.e.*,

$$D = \log_{10}(1/\tau) \quad \text{or} \quad \tau = 10^{-D}$$

The plot of D vs τ is shown in Fig. 8.7. The linear region of the graph is made use of in making of the hologram.

Thus the conventional photographic emulsion modulates light primarily through the absorption caused by metallic silver present in the transparency. As a consequence, the diffraction efficiency is poor. To overcome this limitation, one makes a phase hologram. The phase hologram is produced by bleaching the photographic emulsion *i.e.*, the amplitude hologram. The bleaching process removes the metallic silver from the emulsion and leaves in its place an emulsion thickness variation or a refractive index variation within the emulsion (Fig. 8.8). The chemical processes that lead to these two different phenomenon are in general different. A thickness variation results when so-called tanning bleach is used while the refractive index modulation occurs when a non-tanning bleaching used.

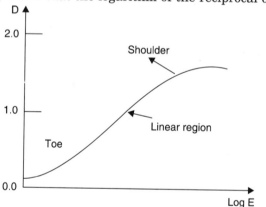

Fig. 8.7 Photographic density versus exposure time

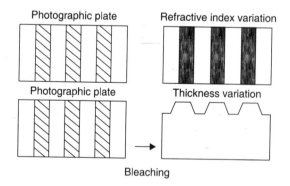

Fig. 8.8 Phase hologram (a) thickness variation (b) refractive index variation

Thick and thin holograms: Consider the simple case when the object wave and the reference wave are plane waves, with wave vectors k_o and k_r as shown in Fig. 8.9a. The recording medium extends between $z = 0$ and $z = D$ and thus has a thickness Δ. The recording medium is in the x-y plane. Clearly the interference pattern is function of x, y and z. It is given by

$$I(x, y, z) = [I_r^{1/2} \cdot \exp(-ik_r \cdot r) + I_0^{1/2} \cdot \exp(-ik_0 \cdot r)]^2$$
$$= I_r + I_0 + 2(I_r \cdot I_0)^{1/2} \cos k_g \cdot r$$

where
$$k_g = (k_0 - k_r)$$

This is a sinusoidal pattern with a period $\Lambda = 2\pi/|k_g|$ and with the surfaces of constant intensity normal to the vector k_g. For example if the reference wave points in the z-direction and the object wave makes an angle θ, with the z-axis, as shown in Fig. 8.9b, then

$|k_g| = 2k \sin(\theta/2)$ where $k = 2\pi/\lambda$, λ being the wavelength of the light employed.

Then the period Λ of the interference pattern is given by

$$\Lambda = \frac{\lambda}{2n \sin(\theta/2)}$$

where n is the refractive index of the recording medium. If the thickness of the recording material is small compared to Λ, then it is called a *thin hologram* (i.e., $\Delta <<< \Lambda$). Otherwise it is called a *thick hologram* (i.e., $\Delta >>> \Lambda$). Thick hologram is also known as the *volume hologram*.

Transmission Holograms: The distinction between transmitting and reflecting holograms is due to the angle between the object and the reference beam. In the case of a transmitting hologram, both the reference and the object wave are introduced from the same side of the recording medium as shown in Fig. 8.10a. The interference pattern produced by the two wave fronts contain horizontal strips of maxima of spacing given by

$$\Lambda = \frac{\lambda}{2n \sin \alpha}$$

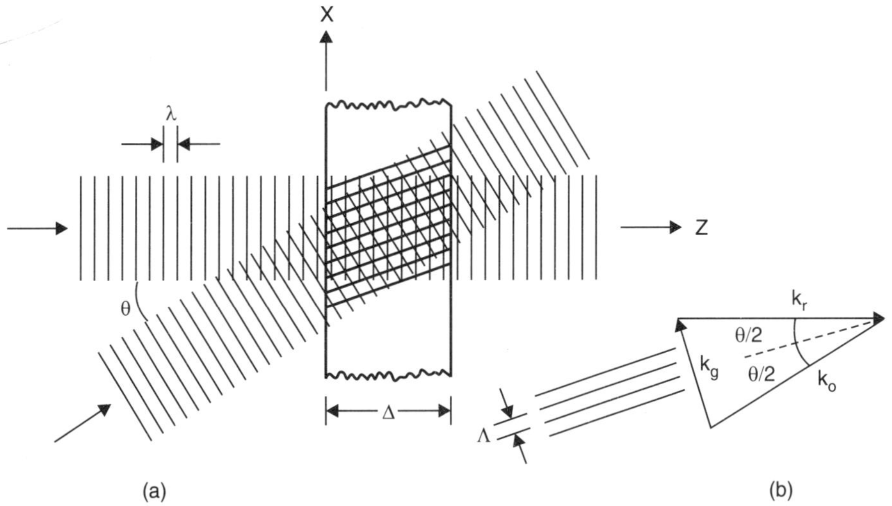

Fig. 8.9 Thick and thin hologram

where n is the index of refraction of the recording medium and α is the angle shown in the figure. The fringes recorded in the medium are shown in Fig. 8.10b. In the reconstruction process, the fringes act like reflective layers to the incident reference wave. The situation is very similar to that occurring in X-ray diffraction from crystals. There is interference between reflected light from different layers. At the angle given by the well known Bragg condition, there is reinforcement and a bright image is reconstructed. *In the case of transmitting hologram these reflecting layers are perpendicular to the plane of the hologram.* The incident reconstruction wave is reflected toward the other side of the hologram so that the reconstructed wave appears as the wave transmitted through the hologram.

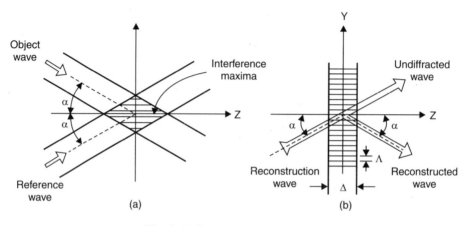

Fig. 8.10 Transmission hologram

Reflection Hologram: In the case of reflecting hologram, the object beam is introduced almost 180° with respect to the reference beam. For the simple case of plane wavefronts, the interference pattern contains vertical strips at a spacing

$$\Lambda = \frac{\lambda}{2n \cos \alpha}$$

where α is the angle as shown in Fig. 8.11. Here the reflecting layers formed in the recording medium are parallel to the surface of the hologram. In the reconstruction process, the reflecting layers reflect the incident wave back and reinforce at the angle given by the Bragg condition. The reconstructed wave is on the same side of the hologram and thus appears as a wave reflected from the hologram.

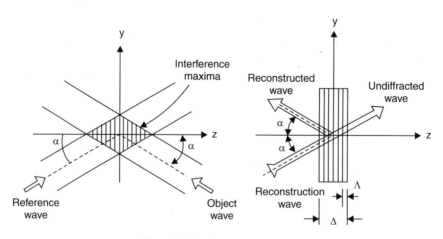

Fig. 8.11 Reflection hologram

HOLOGRAPHY

Reflection holograms are useful since laser is not necessary for reconstruction. The hologram when viewed in white light (sunlight is an excellent source), the appropriate wavelength is selected to produce the reflected image, since the Bragg condition has to be satisfied. Moreover if the hologram is produced by illumination with lasers which produce the three additive primary colours (red, green and blue), the resulting hologram will be seen in full colour when viewed with white light.

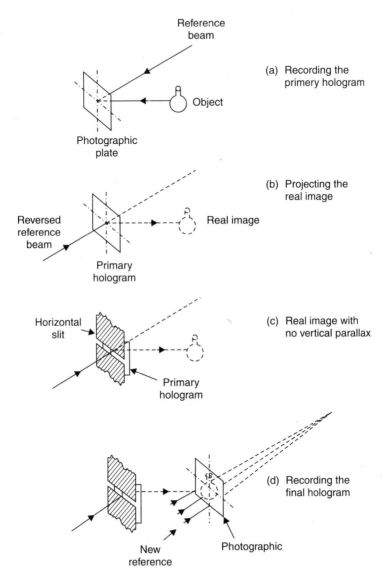

Fig. 8.12 Rainbow hologram

Rainbow Hologram: A major advance in the technology of hologram for displays is the *rainbow hologram*. This is a new type of transmission hologram capable of reconstructing a bright, sharp and monochromatic image when illuminated by white light. In this technique, a part of the information content of the hologram is deliberately sacrificed to gain these advantages. What is given up is parallax in the vertical plane. There is a considerable gain in the brightness of the reconstructed image. Following steps are involved in making a rainbow hologram (Fig. 8.12). First, a conventional transmission hologram of the object is made. The hologram is then illuminated by the conjugate of the original reference beam (*i.e.*, an identical beam propagating in a direction opposite to that of the reference beam). This generates a diffracted wave, which is the real image of the object. Vertical parallax is then eliminated by placing a horizontal slit over the primary hologram. A second hologram is now recorded and that is used for reconstructing the object.

360° Hologram: One of the most impressive holographic images is formed by a 360° circular film. Here the photographic emulsion is mounted on a cylindrical surface surrounding the object. A divergent laser beam from the top is used for illuminating the object as well as the emulsion (Fig. 8.13). A convex mirror at the bottom provides for the illumination of the object, which does not face the beam directly. When the cylindrical hologram is illuminated, the virtual image is observed in the centre of the cylinder and can be viewed from all the sides.

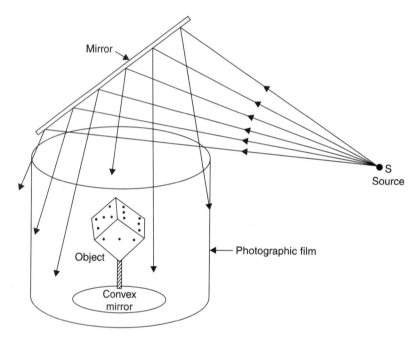

Fig. 8.13 360° hologram

Copying Holograms: Many of the commercial products bear a holographic logo and trademark as a mark of the identity and the authenticity of the manufacturer. This requires the bulk production of holograms. Copies of the original 'master' hologram can be made using an arrangement as shown in Fig. 8.14. Light reflected or a transmitted from the master hologram and a direct beam is made to interfere on a copy plate. This results in a copy hologram. Excellent copies can be made this way. An embossing method that resembles the one used in stamping out plastic copies of a master disc in the recording industry has also been used. A phase hologram where the thickness of the emulsion varies can be used for such a purpose. This 'master' hologram has a surface relief structure that can be metallised. The metallised hologram is used for impressing the pattern onto thin sheets of plastic.

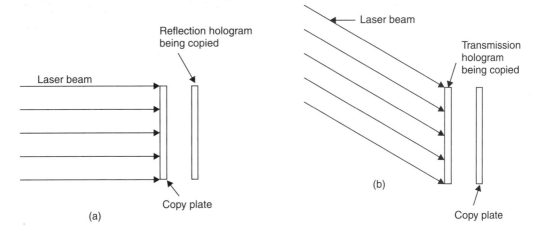

Fig. 8.14 Copy hologram

8.5 ZONE PLATE MODEL OF TRANSMISSION HOLOGRAMS

A simple way to visualise the image formation by a transmission hologram is to take the analogy of image formation by Fresnel zone plate. A zone plate consists of annular circular zones more or less of equal area, such that light is blocked from every alternate zone. A zone plate is shown in Fig. 8.15. To make such a zone plate one draws a concentric circles whose radii are proportional to the square root of the natural numbers *viz.*, $\sqrt{1K}$, $\sqrt{2K}$, $\sqrt{3K}$, $\sqrt{4K}$, $\sqrt{5K}$, etc., and alternate regions are blackened. It can be shown that such a zone plate acts like a converging spherical lens with multiple focal lengths given by:

K^2/λ, $K^2/3\lambda$, $K^2/5\lambda$, $K^2/7\lambda$, This is shown in Fig. 8.16.

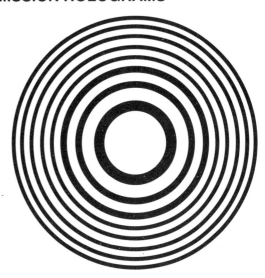

Fig. 8.15 Fresnel zone plate

Consider the Gabor hologram made with a point source as shown in the Fig. 8.17. The interference pattern resembles a zone plate but with its transmitivity varying smoothly and not abruptly as in the case of a Fresnel zone plate. The nature of the transmitivity is shown in Fig. 8.18a. This is called a *sinusoidal zone plate*. It is also called a Gabor zone plate. The analysis of diffraction of light from such a plate can be made and image formation can be understood. It can be shown that such a plate leads to a real and a virtual image (Fig. 8.18b).

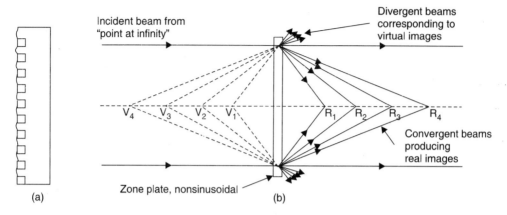

Fig. 8.16 (a) Fresnel zone plate and (b) Image formation using Fresnel zone plate

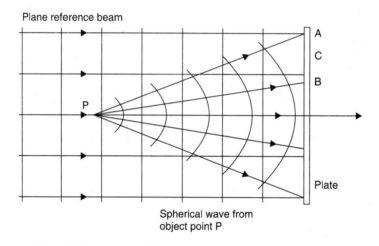

Fig. 8.17 Formation of Gabor zone plate with a point source

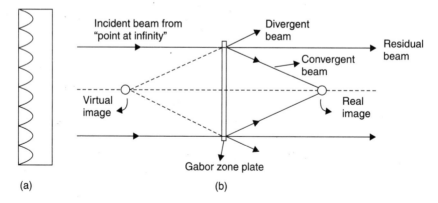

Fig. 8.18 (a) Gabor zone plate and (b) Image formation with Gabor zone plate

8.6 APPLICATIONS OF HOLOGRAPHY
8.6.1 Holographic Interferometry

One of the most important applications in holography has been in the field of non destructive testing and vibrational analysis. Determination of vibrational frequencies, amplitudes and mode patterns of machines and other structures are of considerable importance in mechanical and civil engineering. Three distinct interferometric techniques have been developed.

In *double exposure holographic interferometry*, a hologram of the undisturbed object is formed on a photographic plate. Now the object is disturbed (let us say due to stress or vibration) and the photographic plate is subjected to another exposure along with the same reference wave. If the resulting hologram is developed and illuminated by a reconstruction wave, then there would emerge from the hologram two object waves. One corresponding to the unstressed object and the other corresponding to the stressed object. These two object waves interfere to produce fringes. Thus on viewing through the hologram, one finds a reconstruction of the object superimposed with the fringe pattern. The shape and the number of fringes give the distribution of strain in the object. In *real time method*, the object is left in its original position throughout. The virtual image of the undisturbed object is made to overlap the object when it is stressed or disturbed. The distortions that appear as a fringe pattern are analysed (Fig. 8.19).

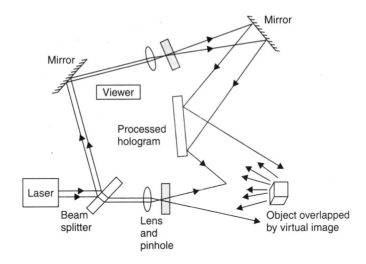

Fig. 8.19 Real time holographic interferometry

In *time average approach*, the photographic film is exposed for a relatively long duration during which time the vibrating object has executed a number of oscillations. The resulting hologram can be thought of as a superposition of a number of a multiple images. Bright areas correspond to undeflected or stationary nodal regions while contour lines trace out areas of constant vibrational amplitude.

Figure 8.20 shows the vibrational modes of a guitar.

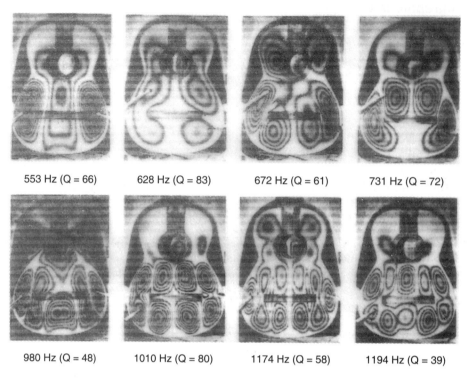

Fig. 8.20 Vibrational modes of Guitar using holographic interferometry

8.6.2 Holographic Microscopy

The magnification factor of the holographic image is given by the ratio of the wavelength of the reference wave to that of the wavelength of the reconstruction wave. Thus in holographic microscopy, the hologram is formed with a wave of longer wavelength while for reconstruction a wave of shorter wavelength is used.

8.6.3 Acoustic Holography

Holograms can be made with microwaves and acoustic waves and visible light can be used for reconstruction. They are helpful in producing pictures which are not visible. Acoustic holography uses sound waves instead of light waves to create a hologram. Acoustic holography requires coherent sound sources. The principal method is *liquid surface acoustic holography*. In the Fig. 8.21 there is an object in a body of water. An object reflects a coherent sound wave from the transmitter onto to the surface. A reference coherent sound beam transmitters directs the reference beam towards the surface. On the surface an interference pattern is produced leading to surface corrugations. The amplitude of the corrugations depends on the intensity of the coherent sound beams used. For low intensities the surface is practically smooth. A laser located above the surface is directed onto the surface. The reflected light is photographed. The plate becomes a recording of the hologram and can be illuminated with laser light to reconstruct the image of the object. However since the reconstruction is done with a light of lower wavelength, the image is enlarged in size and some aberrations appear which have to be corrected for.

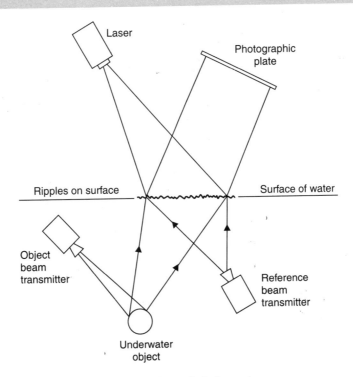

Fig. 8.21 Acoustic holography

8.6.4 Holographic Techniques for Pattern Recognition

Holography can be used for pattern recognition. In holography the complicated wavefront of an object is generated from a hologram by the simple wavefront of the reference beam. This process is reversible so that reference wave can be generated by the object wave. This principle

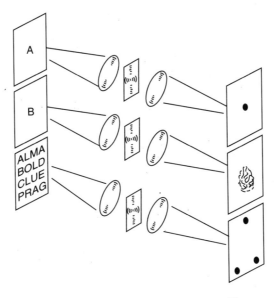

Fig. 8.22 Holographic pattern recognition

forms the basis for pattern recognition. This could be used to identify finger prints, postal addresses etc. For this purpose, one uses the Fourier transform hologram of the particular pattern to be recognized. Let us say, it is necessary to recognise the pattern in the form of the letter A (Fig. 8.22). The Fourier transform of the letter A is made. When this Fourier hologram of the letter A is illuminated from another letter A as shown in figures, a plane wave results which can be focused into a bright spot. The bright spot is easily recognised by the eye or can be detected photoelectrically. Alternately, if the Fourier hologram of the letter A is illuminated by any character other than A, a plane wave does not result and produce a focused spot of light. Thus, one can search for a given character from a matrix of characters. In this technique, essentially a hologram behaves like a filter and is often referred to as the *Vander Lugt filter*, after the person who discovered this technique.

8.6.5 Holographic Data Storage

In a thick, reflection hologram the spatial distribution of fringes is recorded throughout its entire thickness and is known as *volume hologram*. Volume hologram can be recorded in several types of media such as photoemulsion (intensity related changes in density of silver grains) or photochromic glass (intensity related absorption) or ferroelectric lithium niobate (intensity related changes in refractive index).

A volume hologram is essentially a three dimensional grating. The reconstruction process in such a hologram is very similar to diffraction of X-rays in crystals. One important feature of volume hologram is the interdependence of the wavelength and the scattering angle according to the Bragg's law. Thus different coloured images will be reconstructed at different scattering angles. Another significant property is that by successively altering the incident angle (or the wavelength) a single medium can store many holograms. This latter property makes such systems extremely appealing as densely packed memory devices. For example an 8 mm thick hologram has been used to store 550 pages of information each individually retrievable. These can also be used as data storage devices and hence are of much use in computer technology. Holography is employed in the production of photographic masks used to produce microelectronic circuits.

8.6.6 Holographic Diffraction Gratings

Holographic techniques have been employed to produce gratings. Diffraction gratings formed by recording an interference pattern in a suitable light sensitive medium are commonly called the holographic diffraction gratings. These are replacing the conventional ruled gratings, which are produced by drawing of lines on a glass plate by ruler having a diamond tip. Holographic gratings have several advantages over the ruled gratings. Besides being cheaper and simple to produce, they are free from periodic and random errors associated with ruled gratings. They exhibit much less scattered light, which is a great advantage in spectroscopy. Further, it is possible to produce much larger gratings of finer pitch. Holographic gratings can also be produced on substrates of varying shapes and gratings with curved grooves have unique focusing properties. This has opened up the possibility of efficient designing of spectrometers.

8.6.7 Holographic Optical Elements

A hologram can be used to transform an optical **wavefront** in much the same way as a lens. Zone plate, Gabor plate and holographic diffraction **gratings** are examples of holographic optical elements. The zone plate acts like a lens. A hologram **can** be designed for shaping an optical wavefront for a particular application or a specific purpose. *Thus holographic optical elements are holograms with predesigned interference patterns to shape the incident wavefront.* They are being increasingly used in the design of optical systems and have become important in optical engineering.

REFERENCES

1. B.A. Lengyel, *Introduction to Laser Physics,* John Wiley & Sons, NY, 1967.
2. M.J. Beesley, *Lasers and Their Applications,* Tylor and Francis Ltd., London, 1976.
3. O. Svelto, *Principles of Lasers,* Plenum Press, NY, 1976.
4. A.K. Ghatak and K. Thyagarajan, *Optical Electronics,* Cambridge Press, NY, 1989.
5. H.M. Muncheryan, *Laser and Optoelectronic Engineering,* Hemesphere Publishing Corporation, NY 1990.
6. W.T. Silfast, *Laser Fundamentals,* Cambridge Press, NY, 1996.
7. C.C. Davis, *Lasers and Electro-optics—Fundamentals and Engineering,* NY, 1996.

QUESTIONS

1. Distinguish between photography, stereography and holography.
2. Explain the principle of holography.
3. Write a short note on (*a*) Fourier hologram (*b*) Gabor hologram (*c*) Transmission hologram (*d*) reflection hologram (*e*) volume hologram and (*f*) copy hologram.
4. Explain the technique of holographic interferometry.
5. Write a short note on acoustic holography.
6. Discuss the use of holographic optical elements in optical engineering.
7. Explain the use of holography in pattern recognition.

9
Fiber Optics

9.1 INTRODUCTION

The rectilinear propagation of light forms the basis of geometrical optics. However, light can be made to travel along a curved path by means of optical fibers. This phenomenon is based on *total internal reflection*. The transmission of light in transparent cylinders by multiple total internal reflection is a fairly old and well-known phenomenon. The earliest recorded scientific demonstration of this phenomenon was given by John Tyndall at the Royal Society in England in 1870. He was able to demonstrate that light can be guided by total internal reflection inside a water jet flowing down from a tank (Fig. 9.1). The interest in this phenomenon described by Tyndall was dormant until 1927, when the possibility of using fibers for transmitting television pictures was tried out. However, these ideas were not pursued. A new burst of activity began in the early 1950's when the transmission of pictures along an aligned bundle of flexible glass fibers was carried out with remarkable success. This led to the development of *flexible fiberscope*. Thus the subject of fiber optics was established.

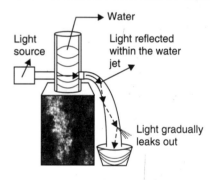

Fig. 9.1 Light is guided down a stream of water by total internal reflection

The use of optical fibers for communication received an added impetus with the advent of miniaturised semiconductor lasers and the availability of highly transparent glass fibers. Currently, optical fibers are being used to transmit voice, picture and digital data signals across the world. This has significant advantages as compared to transmission of signals by conventional coaxial cables. Fiber optics communication systems have got a tremendous capacity to carry information greater than that of electronic communication systems. Fiber optics is now a well-proven technology. The rapid strides in information technology involving world-wide computer networks have made fiber optics an indispensable technology. The global demand for new information services like the internet and e-commerce have fueled the demand for optical fibers capable of transmitting information at high speeds. Fiber optic communication systems supplement microwave and satellite communication systems. Apart from communication systems, optical fibers have found applications in various branches engineering as *sensors*. In medicine, it has become a useful tool for the examination of internal organs.

FIBER OPTICS

9.2 BASIC PRINCIPLES

9.2.1 Total Internal Reflection

The basic mechanism of light propagation along fibers can be understood using the principles of geometrical optics. When a light ray travelling in a medium of refractive index n_1 strikes a second medium of refractive index n_2, the angle of refraction θ is given by the Snell's law.

$$n_1 \sin i = n_2 \sin \theta$$, where i is the angle of incidence.

When a ray of light is travelling from an optically dense medium (high index of refraction) to a less dense medium (low index of refraction), it will bend away from the normal. It will not be refracted, if it strikes the surface at an angle equal to or greater than a particular angle called the *critical angle*. Instead it will be totally reflected at the surface, between the two media (Fig.9.2). *Total internal reflection is not possible when the ray of light travels from rarer medium to denser medium.*

Consider the case when $\theta = 90°$ for a particular value of $i = \theta_c$, then we have

$$n_1 \sin \theta_c = n_2 \sin 90°$$

$$\sin \theta_c = \frac{n_2}{n_1}$$

Thus for any ray whose angle of incidence is greater than this critical angle, total internal reflection occurs.

9.2.2 Evanescent Wave and Tunneling of Photons

On analysing the phenomenon of total internal reflection in a prism by treating light as an electromagnetic wave, one can prove that there is penetration of the wave, for a distance of the order of few wavelengths, beyond the interface. Therefore if the face of a second glass prism is kept parallel to the interface, (the gap between them being a few wavelengths) the light will *tunnel* through this barrier and generate a transmitted wave (Fig. 9.3). This phenomenon is used in the design of fiber optic couplers and fiber optic sensors.

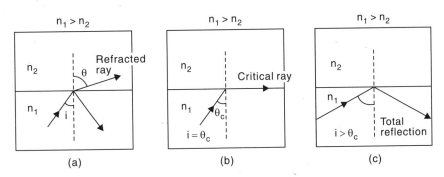

Fig. 9.2 Light ray travelling from a medium of higher refractive index to a medium of lower refractive index. (a) when $i < \theta_c$, it is refracted (b) when $i = \theta_c$, it traverses along the surface (c) when $i > \theta_c$, it is totally reflected

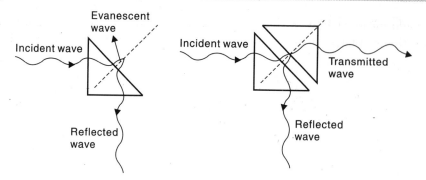

Fig. 9.3 Evanescent wave and tunneling of photons

9.3 OPTICAL FIBERS AND CABLES

9.3.1 Fiber Construction

An optical fiber is a very thin, flexible thread of transparent plastic or glass in which light is transmitted through multiple total internal reflection. It consists of a central cylinder or core surrounded by a layer of material called the cladding, which in turn is covered by a jacket. Light is transmitted within the core. The cladding keeps the light waves within the core because the refractive index of the cladding material is less than that of the core. The cladding also provides some strength to the core. The additional jacket protects the fiber from moisture and abrasion (Fig. 9.4).

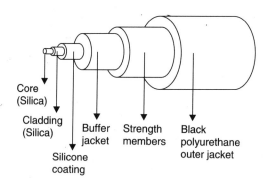

Fig. 9.4 Typical construction of an optical fiber

The core as well as the cladding is made of either glass or plastic. With these materials three major types of fibers are made: plastic core with plastic cladding, glass core with plastic cladding and glass core with glass cladding. In the case of plastics, the core can be polystyrene or polymethyl meta acrylate (PMMA), the cladding is generally silicone or teflon. The glass is made of silica. A small amount of components such as boron, germanium or phosphorus is added to change the refractive index of the fiber. In comparison with glass, plastic fibers are flexible and inexpensive.

9.3.2 Fiber Dimensions

Optical fibers are typically made in lengths of 1 km. Optical fibers can be joined using suitable connectors. Its outer diameter ranges from 0.1 to 0.15 mm. Core diameters range from 5 to 600 µm, whereas cladding diameters vary from 125 to 750 µm. To keep the light within the core, the cladding must have a minimum thickness of one or two wavelength of light. The protective jacket may add as much as 100 μm in diameter to the fiber's total diameter.

Ordinary glass is brittle. But the optical fibers which are made of glass are quite tough. They have a high tensile strength *i.e.*, ability to withstand hard pulling or stretching. The toughest optical fibers are as tough as steel wires of the same diameter. Yet these fibers can be easily bent unlike steel wires.

9.3.3 Fiber Fabrication

The fabrication of optical fibers is based on sophisticated technology. It involves the following three steps:

(a) Making of a glass rod or *preform* of highly purified glass with core and cladding structure.

(b) Drawing of fibers from this *preform*.

(c) Measuring the fiber characteristics for its performance in a communication system.

Preforms can be made in several different ways. Some of the widely used methods are described here.

Outside Vapour Deposition (OVD) Method

In this method ultrapure vapours of silicon tetrachloride and germanium tetrachloride are burnt in a flame producing fine soot of SiO_2 and GeO_2. Germanium oxide, Boron oxide, phosphorous pentoxide, titanium oxide and aluminum oxide are added as impurities to modify the refractive index. Typical chemical reactions that occur are represented by the following equations:

$SiCl_4\uparrow + O_2\uparrow \rightarrow SiO_2 + 2Cl_2\uparrow$ $SiCl_4\uparrow + 2H_2O\uparrow \rightarrow SiO_2 + 2Cl_2\uparrow + 2H_2\uparrow$

$GeCl_4\uparrow + O_2\uparrow \rightarrow GeO_2 + 2Cl_2\uparrow$ $GeCl_4\uparrow + 2H_2O\uparrow \rightarrow GeO_2 + 2Cl_2\uparrow + 2H_2\uparrow$

$TiCl_4\uparrow + O_2\uparrow \rightarrow TiO_2 + 2Cl_2\uparrow$ $2POCl_3\uparrow + 3H_2O\uparrow \rightarrow P_2O_5 + 3Cl_2\uparrow + 3H_2\uparrow$

$4POCl_3\uparrow + 3O_2\uparrow \rightarrow 2P_2O_5 + 6Cl_2\uparrow$ $2BBr_3\uparrow + 3H_2O\uparrow \rightarrow B_2O_3 + 3Br_2\uparrow + 3H_2\uparrow$

$4BBr_3\uparrow + 3O_2\uparrow \rightarrow 2B_2O_3 + 6Br_2\uparrow$

Excess of water vapour is removed using $SOCl_2$ by the following reactions:

$SOCl_2\uparrow + H_2O\uparrow \rightarrow SO_2\uparrow + H_2\uparrow + Cl_2\uparrow$

$6SOCl_2\uparrow + 6OH^-\uparrow \rightarrow 6SO_2\uparrow + 3H_2\uparrow + 6Cl_2\uparrow$

$Cl_2\uparrow + 2OH^-\uparrow \rightarrow 2HCl + O_2\uparrow$

The soot is deposited on a rotating rod or mandrel (Fig. 9.5). First the core glass is deposited and then the composition is changed to deposit the cladding. Finally the rod is removed and the remaining glass is heated and collapsed into a dense glass. This method of *outside vapour deposition* produces an extremely pure preform because all the material is deposited in the vapour phase.

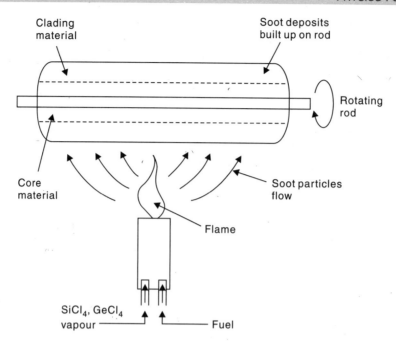

Fig. 9.5 Outside vapour deposition (OVD) technique

Modified Chemical Vapour Deposition (MCVD) Method

In a slight variation of the previous method, the glass vapour particles arising from the reaction of constituent metal halide gases and oxygen, flow through the inside of a revolving silica tube (Fig. 9.6). As the SiO_2 particles are deposited, they are sintered to a clear glass layer by an oxyhydrogen flame-torch, which travels back and forth along the tube. When the desired thickness of the glass is deposited, the vapour is shut off. The tube is heated strongly to cause it to collapse into a solid *preform*.

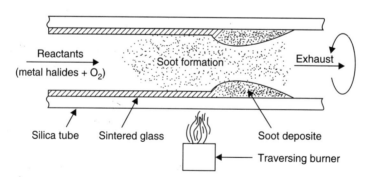

Fig. 9.6 Modified chemical vapour deposition (MCVD) technique

Plasma Activated Chemical Vapour Deposition (PCVD) Method

This process differs from MCVD in its method of heating the reaction zone. Instead of delivering heat from outside through a burner, PCVD uses microwaves to form ionized gas plasma inside the silica tube. A schematic of this process is shown in Fig. 9.7. Within the plasma, electrons acquire energy equivalent to that of 60,000K. The actual temperature inside the furnace is about 1500K. The highly energetic electrons recombine with ions, releasing considerable heat. This heat melts the soot particles obtained from the reaction between gaseous $SiCl_4$, $GeCl_4$, Fe_2Cl_6 and O_2. As a result, the deposition of the desired glass occurs directly on the target silica tube *without the formation of the soot*. A microwave resonator, within which the plasma zone exists, can move quickly along the target silica tube. Thus very thin layers are formed because there is no restriction caused by the size of the soot particles which are of the size of microns. This is the key advantage of this method. In this technique there is better control over the refractive index profile. The final steps of the PCVD process are similar to those of MCVD technique. The tube is collapsed, forming a glass *preform*.

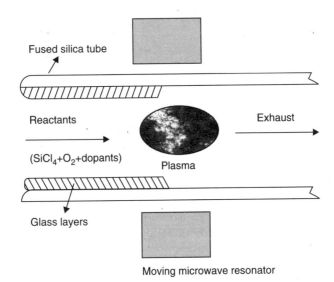

Fig. 9.7 Plasma activated chemical vapour deposition (PCVD) technique

Vapour-Axial Deposition (VAD) Method

In this technique, the deposition of the glass particles, which are obtained from the reaction among the gases in the heated zone occur at the bottom end of a seed rod. The seed rod rotates and moves upward as shown in Fig. 9.8. This deposition forms a porous preform, the upper end of which is heated in a ring furnace to produce a glass *preform*. The refractive index profile is controlled using many burners.

In all the above mentioned processes, the refractive index of the core and cladding can be modified using suitable dopants in the reaction zone. Typical variation of refractive index with various dopants and their concentration is shown in Fig. 9.9. The *preform* obtained by any of these methods is heated and drawn into a thin fiber as shown in Fig. 9.10. The fiber is coated with a protective plastic layer as it is drawn. Later on during packaging, these fibers are coiled for easier handling and protection.

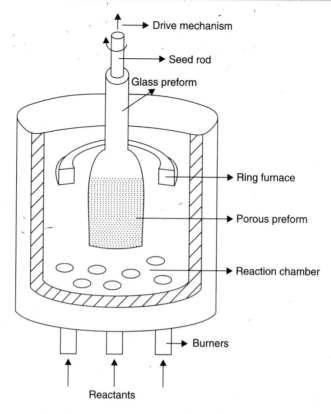

Fig. 9.8 Vapour axial deposition (VAD) technique

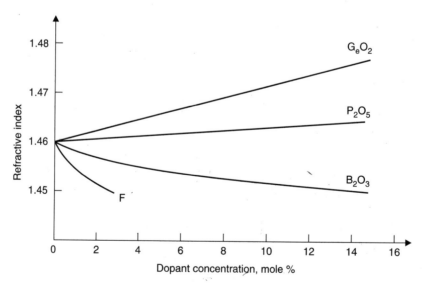

Fig. 9.9 Refractive index as a function of the dopant materials and their concentration

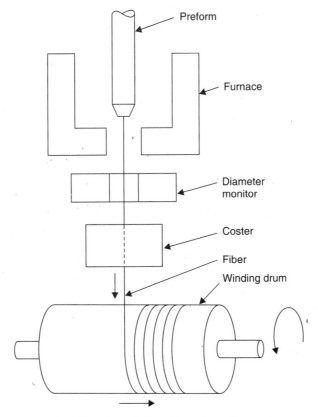

Fig. 9.10 Drawing of glass fibers from the preform

9.3.4 Fiber Cables

A fiber cable consists of many fibers just as an electric cable consists of many strands of copper wire. One reason for cabling fibers is to make them easier to handle. A single optical fiber with

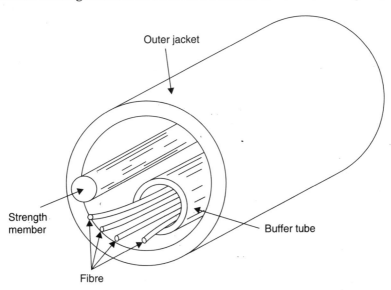

Fig. 9.11 Basic structure of a fiber optic cable

plastic coating has a diameter in the range of 250—950 µm and is difficult to handle. Further, most communication systems require more than one fiber. Thus cabling helps to handle multiple fibers. The design of the cable varies widely. Indoor cables are not subjected to wide temperature fluctuations and stresses. Hence, their design is simple. The cables which are buried under the ground as well as the sea require more elaborate and careful design. The basic structure of a fiber optic cable is shown in Fig. 9.11. A fiber cable consists of three elements:

(a) buffer tube (b) strengthening element and (c) protective element

A bare fiber or many fibers is put into a buffer tube. The buffer jacket falls into one of the three types. A tight buffer jacket consists of a hard plastic (nylon or teflon) and is in contact with the primary coated fiber. An alternative approach is to have loose buffering in an over sized cavity. In a variation of the loose buffering the oversized cavity is filled with moisture resistant soft compound.

The second element is the strength member. Its function is to release the fiber from the mechanical stresses acting on the fiber during installation and operation.

The third basic element is an outer jacket that assembles the entire construction and shields it from adverse environment. PVC (Polyvinylchloride) or PE (Polyethylene) sheaths are used for this purpose.

Some of the typical cable designs are shown in Fig. 9.12.

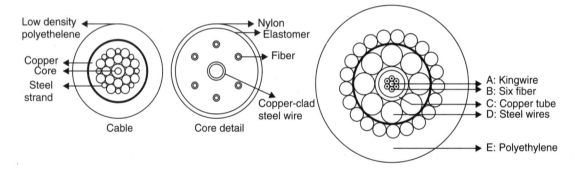

Fig. 9.12 Typical fiber optic cable structures

9.4 LIGHT PROPAGATION IN FIBERS

9.4.1 Numerical Aperture of an Optical Fiber

The numerical aperture is a measure of the light gathering capacity of the fiber. It is defined as the sine of the acceptance angle. *Acceptance angle is the maximum allowed angle that the incident ray can make with the fiber axis for transmission through the fiber*. Only rays of light which make angle less than the acceptance angle with the fiber axis are transmitted by the fiber.

Consider a ray which is incident on the entrance aperture of the fiber making an angle i with the axis (Fig. 9.13). Let the refracted ray make an angle θ with the axis. Assuming the outside medium to have a refractive index n_0 (which for most practical cases is unity), we get

$$\frac{\sin i}{\sin \theta} = \frac{n_1}{n_0} \quad \Rightarrow \quad \sin i = \frac{n_1}{n_0} \sin \theta \tag{9.1}$$

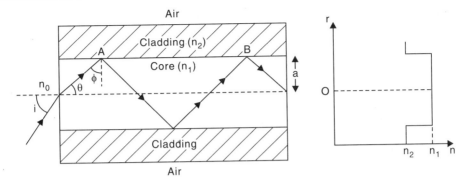

Fig. 9.13 Numerical aperture of an optical fiber. The refractive index profile is also shown.

If this ray has to suffer total internal reflection at the core cladding interface,

$$\sin \phi (= \cos \theta) \geq \frac{n_2}{n_1} \Rightarrow \sqrt{(1 - \sin^2 \theta)} \geq \frac{n_2}{n_1} \text{ or } (1 - \sin^2 \theta) \geq \left(\frac{n_2}{n_1}\right)^2$$

$$\Rightarrow \sin \theta \leq \sqrt{1 - \left\{\frac{n_2}{n_1}\right\}^2} \tag{9.2}$$

From eqns. (9.1) and (9.2)

$$\sin i \leq \frac{n_1}{n_0}\left[1 - \left(\frac{n_2}{n_1}\right)^2\right]^{1/2} = \left[\frac{n_1^2 - n_2^2}{n_0^2}\right]^{\frac{1}{2}} \tag{9.3}$$

If $(n_1^2 - n_2^2) \geq n_0^2$, then for all values of i, total internal reflection will occur since sine of an angle can never be greater than 1.

Assuming $n_0 = 1$, the maximum value of $\sin i$ for a ray to be guided is given by

$$\sin i_m = (n_1^2 - n_2^2)^{1/2}, \text{ when } (n_1^2 - n_2^2)^{1/2} < 1 \tag{9.4}$$

$$= 1 \text{ when } (n_1^2 - n_2^2)^{1/2} > 1$$

Thus, if a cone of light is incident on one end of the fiber, it will be guided through the fiber provided the semi-vertical angle of the cone is less than i_m. This angle, known as the acceptance angle is a measure of the light gathering power of the fiber. The numerical aperture (NA) of the fiber by the following equation

$$\text{NA} = \sqrt{n_1^2 - n_2^2} = \sqrt{(n_1 + n_2)(n_1 - n_2)} \approx n_1\sqrt{2\Delta}, \text{ where } \Delta = \frac{n_1 - n_2}{n_1} \tag{9.5}$$

Δ is known as the fractional change in the refractive index. In the last step we have assumed $n_1 \approx n_2$ which is indeed the case for all practical optical fibers. For a typical fiber, $n_2 = 1.458$, $\Delta = 0.01$ and the corresponding NA is = 0.2. Thus the fiber would accept light incident over a cone with a semi-angle $\sin^{-1}(0.2) = 11.5°$ about the axis. The light gathering capacity increases with numerical aperture since the acceptance angle is proportional to the numerical aperture.

9.4.2 Types of Optical Fibers

In optical fibers the refractive index of the core (n_1) is greater than the index of the cladding (n_2) i.e., $n_1 > n_2$. Depending the refractive index profile within the core (i.e., the variation of refractive index across the core) optical fibers are classified as *step-index fibers* and *graded index fibers*.

Step-index Fiber

In a step-index fiber the refractive index is constant throughout the core. In such a fiber the refractive index profile abruptly changes at the junction of the core and the cladding. Because of this abrupt change they are called step-index fibers. In a step-index fiber, the light paths within the core are zigzag, on account of multiple reflections within the fiber. Step-index fiber is essentially a reflective type fiber (Fig. 9.14).

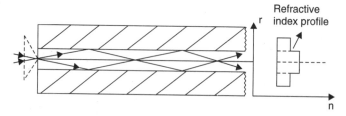

Fig. 9.14 Refractive index profile and light ray propagation in a step index optical fiber

Step-index fibers could be *either single mode* or *multimode*. Mode is an important mathematical and physical concept describing the propagation of electromagnetic waves. In the language of geometrical optics, mode refers to the various paths that light can take in a fiber. Multimode means that several paths are available. In wave optics mode has a deeper physical significance. Whenever a wave is spatially confined standing waves or modes result.

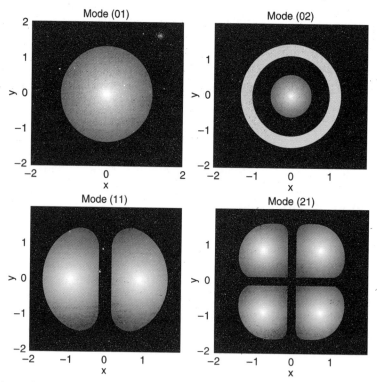

Fig. 9.15 Intensity distribution across the core of the fiber of lower order modes of step-index fiber

FIBER OPTICS

An optical fiber is an example of a wave being confined in a plane perpendicular to the direction of propagation. Such confinement results in various modes just as in the case of a stretched string whose ends are fixed. Intensity distribution across the core in some of the lower order modes in a step-index fiber are shown in Fig. 9.15. It depicts the variation of the intensity of light across the core.

The number of modes in a multimode fiber depends on numerical aperture, core radius, operating wavelength and refractive index profile. The number of modes in a step-index fiber is given by

$$N \approx \frac{V^2}{2} \qquad \text{where } V = \frac{2\pi a \text{NA}}{\lambda} \text{ is called the fiber parameter.}$$

Here a is the core radius, NA is the numerical aperture and λ is the operating wavelength. For a *single mode* step-index fiber, $V = \sqrt{2}$. The core radius is comparable to wavelength of light and is so small as to have only a single path for the propagation of light.

Graded Index Fiber

In a graded index fiber the refractive index varies continuously across the core. It is highest at the centre of the core and tapers off radially towards the outer edge as shown in Fig. 9.16a. The variation of refractive index as a function of the distance from the axis of the fiber is given by:

$$n^2(r) = n_1^2 \left[1 - 2\Delta \left(\frac{r}{a}\right)^q \right] ; 0 < r < a$$

$$n_2^2 = n_1^2 (1 - 2\Delta) ; r > a$$

where r corresponds to a cylindrical radial coordinates. n_1 represents the value of the refractive index on the axis (i.e., at $r = 0$) and n_2 represents the refractive index of the cladding; $q = 1$, $q = 2$ and $q = \infty$ correspond to the linear, parabolic and step-index profiles respectively. In graded index fiber light is continuously refracted along a curved path (Fig. 9.16 b). Hence, it is often referred to as refractive fiber. In a graded index fiber the signal distortion is minimum as will be discussed later on. The number of modes in a graded index fiber is given by:

$$M \approx \frac{V^2}{4} \qquad \text{(for a fiber with parabolic refractive index profile)}$$

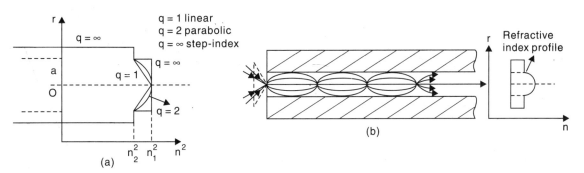

Fig. 9.16 (a) Refractive index profile and (b) Light ray propagation in a graded index optical fiber

Skew Rays and Meridional rays The geometrical paths in optical fibers can be classified as meridional rays and skew rays. The meridional rays intersect the fiber axis and reflect in the same plane without changing their angle of incidence. Thus the rays of light travel as though in a planar wave guide (Fig. 9.17a). These rays are guided if their angle θ with the fiber axis is smaller than the complement of the critical angle between the core-cladding interface. Skew rays lies in a plane offset from the fiber axis by a distance R. The plane of incidence intersects the core-cladding cylindrical boundary at an angle φ with the normal to the boundary and lies at a distance R from the fiber axis. The ray is identified by its angle θ with the fiber axis and by the angle φ of its plane (Fig. 9.17b). Thus when $R \neq 0$ and $\phi \neq 0$, then the ray is skewed. For meridional rays $\phi = R = 0$. The skewed ray reflects repeatedly into planes that make the same angle φ with the core-cladding boundary. In a graded index fiber it follows a helical trajectory confined within a cylindrical shell of radius R and core radius a.

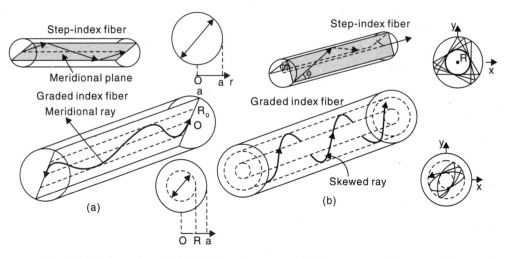

Fig. 9.17 Propagation of (a) Meridional rays and (b) skew rays. Projection of the paths across the fiber cross-section is also shown

9.5 SIGNAL DISTORTION OR DISPERSION IN OPTICAL FIBERS

Light is sent in the form of pulses in digital communication systems using optical fibers. The shape of these pulses gets distorted on account of various factors arising out of *dispersion of light*. There are four types of dispersion. They are *modal dispersion, material dispersion, waveguide dispersion* and *polarization dispersion*. The effect of dispersion is to distort the signal. This affects the rate at which the data is transmitted through the fiber.

9.5.1 Modal Dispersion

When light travels in a step index fiber, each ray of light may be reflected hundreds or thousands of times. The rays reflected at high angles (*i.e.*, the higher order modes) travel a greater distance than the low angle rays to reach the end of the fiber (Fig. 9.18). Because of this longer distance, the higher—angle rays arrive later than the lower angle rays. As a result, modulated light pulses broaden as they travel down the fiber causing signal distortion. The output pulses then no longer exactly match the input pulses. Since light is sent in the form of pulses, it is proper to speak of group velocity. This means that group velocities of various modes are different. Group velocities of various modes in a step-index fiber are shown in Fig. 9.19a. *The modal dispersion can be eliminated either by having a single mode fiber or by changing the refractive index profile of the core.*

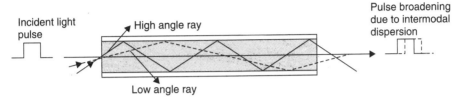

Fig. 9.18 Signal distortion due to modal dispersion in multimode fiber

In a graded index fiber, light rays travel at different speeds in different parts of the fiber because the refractive index varies across the core. Near the outer edge of the fiber the refractive index is lower. As a result, rays near the outer edge travel faster than rays in the centre of the core. Because of this all the rays arrive at the end of the fiber at approximately the same time. In effect light rays in these fibers are continuously refocused as they travel down the fiber (Fig. 9.16). This in effect reduces signal distortion. In a single mode step-index fiber the signal distortion is reduced, but the problem is one of aligning different fibers.

An estimate of the pulse broadening that occurs in step-index and graded index multimode fibers can be done as follows. For a step-index fiber, the group velocities of various modes varies from c/n_1 to $(c/n_1)(1-\Delta)$, while for a graded index fiber, the group velocities of various modes vary from c/n_1 to $(c/n_1)(1-\Delta^2/2)$ (Fig. 9.19b). If T_{max} and T_{min} is the travel time for the longest and shortest rays then the pulse width broadening σ_{modal}

$$\sigma_{modal} = T_{max} - T_{min} = \frac{L}{\left(\frac{c}{n_1}\right)(1-\Delta)} - \frac{L}{\frac{c}{n_1}} = \frac{n_1 L}{c}\left\{\frac{1}{(1-\Delta)} - 1\right\}$$

$$= \frac{n_1 L}{c}\left\{\frac{\Delta}{1-\Delta}\right\} \approx \frac{n_1 L \Delta}{c} \text{ (for step-index fiber)}$$

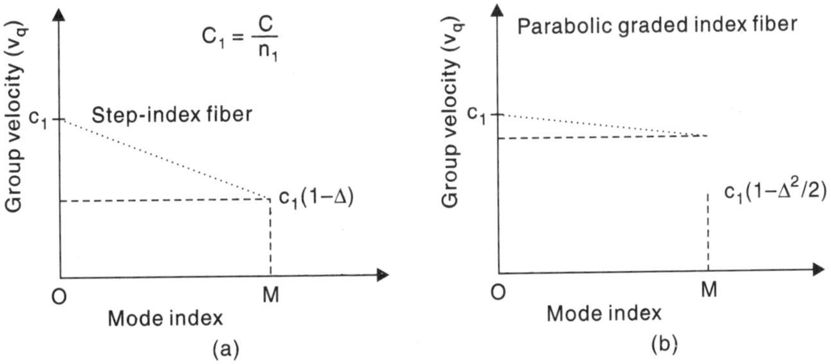

Fig. 9.19 Group velocities of various modes in (a) step-index fiber and (b) graded index fiber

$$\sigma_{modal} = T_{max} - T_{min} = \cfrac{L}{\left(\cfrac{c}{n_1}\right)\left(1-\cfrac{\Delta^2}{2}\right)} - \cfrac{L}{\cfrac{c}{n_1}} = \cfrac{n_1 L}{c}\left\{\cfrac{1}{\left(1-\cfrac{\Delta^2}{2}\right)}-1\right\}$$

$$= \frac{n_1 L}{c}\left\{\left(1-\frac{\Delta^2}{2}\right)^{-1}-1\right\} = \frac{n_1 L \Delta^2}{2c} \quad \text{(for gradex index fiber)}$$

The above equations clearly demonstrate how the pulse distortion is minimised in graded index fibers.

9.5.2 Material or Chromatic Dispersion

Material or chromatic dispersion refers to the variation of refractive index with wavelength. Whenever light undergoes refraction at a plane or a curved surface, dispersion effects occur. In a prism, it leads to the splitting of white light into various colours. In a lens, it gives rise to chromatic aberration or distortion of the image of a white object. These effects are shown in Fig. 9.20. Optical fiber is made of glass, which is a dispersive medium *i.e.*, its refractive index is a function of wavelength. In digital communication, the signal is sent in the form of pulses. A pulse is a wave packet, composed of a spectrum of components of different wavelengths each travelling at a different phase velocity. Therefore all the components of an input light pulse will not arrive simultaneously at the output. This leads to an increase in the pulse width as shown in Fig. 9.21. The group velocity is given by

$$\frac{\partial \omega}{\partial k} = v_p + k \cdot \frac{\partial v_p}{\partial k}$$

$$v_p = \omega/k, \text{ where } k = 2\pi/\lambda \text{ and } \omega = v_p k$$

$$v_g = v_p + \frac{2\pi}{\lambda}\frac{\partial(c/n)}{\partial(2\pi/\lambda)} = v_p + \frac{c}{\lambda}\frac{\partial(1/n)}{\partial(1/\lambda)} = v_p + \frac{\frac{\partial(1/n)}{\partial \lambda}}{\frac{\partial(1/\lambda)}{\partial \lambda}} = \frac{c}{n} + \left(\frac{c}{\lambda}\right)\frac{-\frac{1}{n^2}\cdot\frac{\partial n}{\partial \lambda}}{-\frac{1}{\lambda^2}}$$

$$\therefore \quad \frac{1}{n_g} = \frac{1}{n} + \frac{1}{\lambda}\frac{\lambda^2}{n^2}\frac{\partial n}{\partial \lambda} = \frac{1}{n}\left\{1 + \frac{\lambda}{n}\frac{\partial n}{\partial \lambda}\right\}$$

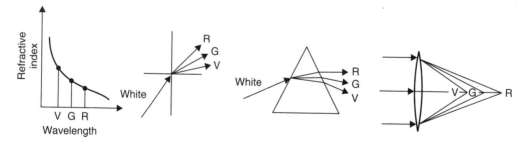

Fig. 9.20 Optical components made of dispersive media refract waves of different wavelengths (e.g., V = violet, G = green and R = red) by different angles. The refractive index for violet, green and red light is also shown.

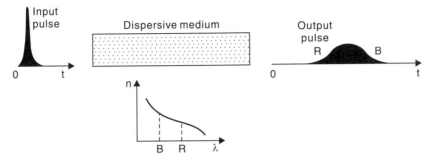

Fig. 9.21 A dispersive medium broadens a pulse of light because the different frequency components that constitute the pulse travel at different velocities. The low frequency component (long wavelength, denoted R) travels faster than the high-frequency component (short wavelength, denoted B) and arrives earlier.

i.e.,
$$n_g = n\left\{1 + \frac{\lambda}{n}\frac{\partial n}{\partial \lambda}\right\}^{-1} = n\left\{1 - \frac{\lambda}{n}\frac{\partial n}{\partial \lambda}\right\} = n - \lambda\frac{\partial n}{\partial \lambda}$$

This gives the relation between the *group refractive index* and the *phase refractive index*. Note that $(\partial n/\partial \lambda)$ has to be evaluated at the operating wavelength of the optical fiber. The typical variation of refractive index n and the group refractive index N with wavelength is shown in Fig. 9.22.

The temporal width of an optical pulse of spectral width σ_λ after travelling a distance L is

$$\sigma_{mat} = \frac{d(L/v_g)}{d\lambda}\sigma_\lambda = \frac{d(Ln_g/c)}{d\lambda}\sigma_\lambda = \frac{L\sigma_\lambda}{c}\frac{dn_g}{d\lambda}$$

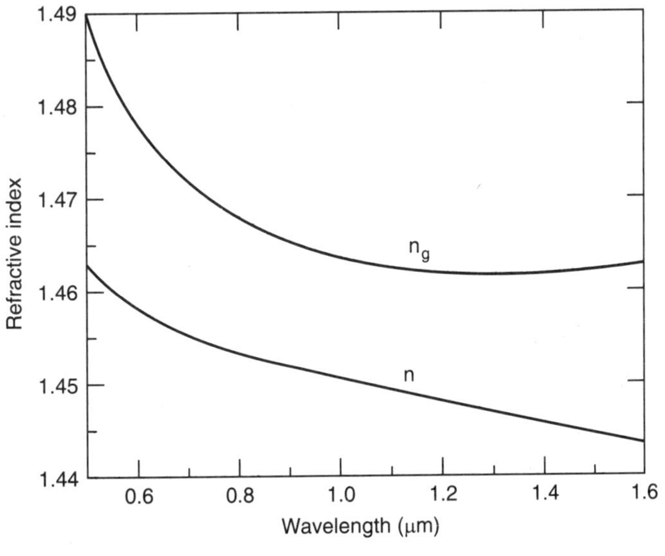

Fig. 9.22 Variation of refractive index n and the group index n_g with wavelength for fused silica

$$= -\frac{L\sigma_\lambda \lambda}{c}\frac{d^2n}{d\lambda^2} = |D_\lambda|\,\sigma_\lambda L \text{ where } D_\lambda = -\frac{\lambda}{c}\frac{d^2n}{d\lambda^2}$$

D_λ is called the *material dispersion coefficient.*

9.5.3 Waveguide Dispersion

In a single mode fiber, modal dispersion is not present. However the group velocity of the mode depends on the wavelength *even if the material dispersion is negligible.* This dependence, known as waveguide dispersion, results from the electromagnetic field distribution within the core and cladding of the fiber (Fig. 9.23). This depends on the ratio between the core radius and the wavelength (a/λ) or the V parameter. The V parameter is altered for the mode on account of the finite spectral width of the source. Hence the relative portions of optical power in the core and cladding are modified due to spectral width. Since the phase velocities of the light in the cladding and core are different, the group velocity of the mode is altered. Waveguide dispersion is particularly important in single mode fibers at wavelengths for which the material dispersion is small. A detailed analysis of the waveguide dispersion is beyond the scope of the book since it requires a knowledge of the propagation of electromagnetic waves in waveguides which is obtained by solving the Maxwell's equations.

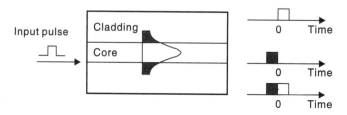

Fig. 9.23 Origin of waveguide dispersion

The waveguide dispersion is given by:

$$\sigma_{\text{waveguide}} = -\left(\frac{1}{2\pi\varepsilon_0}\right) V^2 \frac{d^2\beta}{dV^2}$$

where $\beta(=nk)$ is the propagation constant and V is the fiber parameter. The dependence of β on V for the lowest order mode is shown in Fig. 9.24. Since β varies nonlinearly with V, waveguide dispersion coefficient is itself a function of V and hence of λ. $\sigma_{\text{waveguide}}$ is found to be always negative.

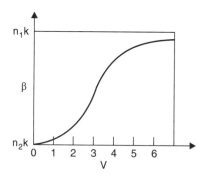

Fig. 9.24 Variation of propagation constant β with the fiber parameter V

In single mode fibers, the intermodal dispersion is absent. As a result, only the material and waveguide dispersion has to be considered. A plot of material and waveguide dispersion for a single mode fiber is shown in Fig. 9.25. The material and waveguide dispersion can be opposite in sign. They cancel out close to 1300 nm in silica fibers. These fibers also have low attenuation around this wavelength. However, this does not mean that the signal distortion is completely absent for an optical fiber operating at ~1300 nm. The total dispersion is zero only for a *single wavelength*. Since all the light sources emit a range of wavelengths, in practice it is impossible to eliminate signal distortion. Signal distortion can only be minimised.

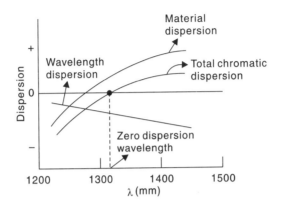

Fig. 9.25 Chromatic and waveguide dispersion in a single mode step-index fiber. Zero dispersion wavelength is shown.

9.5.4 Polarization Dispersion

Consider a single mode fiber of perfect circular symmetry. If the propagation of light is along the z-axis, the velocities of plane polarized light along x and y axes are the same. In real fibers the refractive index and hence the velocity of propagation for plane polarized light along x and y axes are not identical. This difference in refractive index is often referred to as *birefringence*. The birefringence could be geometrical in origin *i.e.*, the fiber is not perfectly circular throughout its length. This is referred to as *geometrical birefringence*. There could also be *stress induced birefringence*. Unequal stresses throughout the length of the fiber can cause changes in refractive index for light plane polarized along x and y axes. The pulse broadening due to birefringence is shown in Fig. 9.26. The time delay ΔT between light polarized along x and y axes is given by:

$$\Delta T = \left\{ \frac{L}{v_{gx}} - \frac{L}{v_{gy}} \right\}$$

where v_{gx} and v_{gy} correspond to the group velocities of plane polarized light along x and y axes.

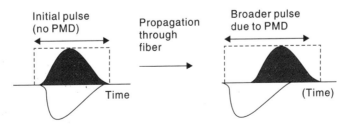

Fig. 9.26 Origin of polarization dispersion

In the above equation it is assumed that the fiber has constant birefringence. This is true in polarization-maintaining fibers. In conventional fibers, the birefringence varies randomly and the above equation is not valid. The time delay is found to be proportional to $\sqrt{L}$. In a conventional fiber, plane polarized light launched into the fiber quickly reaches a state of arbitrary polarization. The unknown polarization state is generally not a problem for optical communication systems whose receivers detect the total intensity directly. *However it is a serious problem in coherent optical communication systems and fiber optic sensors based on phase detection.*

Polarization Maintaining Fibers

Normal optical fibers are essentially insensitive to the state of polarization. For most of communication systems the state of polarization of light is immaterial. However, it is desirable to have optical fibers which maintain the state of polarization of light. This is necessary in the case of fiber optic sensors and coherent optical communication systems. It is possible to make fibers for which random fluctuations in the core shape and size are not the governing factors in determining the state of polarization. Such fibers are known as *polarization-maintaining fibers*. A large amount of birefringence is introduced intentionally in these fibers through design modifications so that small random birefringence fluctuations do not affect the state of polarization significantly. In a fiber having constant birefringence, there is a periodic power exchange between light polarized along x and y axes. The period referred to as the *beat length*, is given by (Fig. 9.27)

$L_B = \lambda/B$ where $B = (n_{gx} - n_{gy})$ denote the *birefringence*.

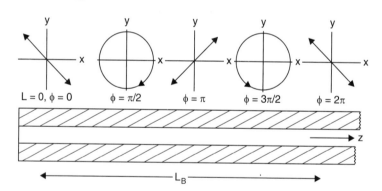

Fig. 9.27 Beat length in a polarization maintaining fiber

Polarization-maintaining fibers are constructed by designing asymmetries into the fiber. Examples include fibers with elliptical cores and fibers that contain nonsymmetrical stress-producing parts. The difference between the two types is subtle. True single-polarization (or polarizing) fiber has different attenuations for light polarized along x and y axes. Any one of the state of polarization is transmitted well, but the other is strongly attenuated, so that it is lost after transmission. Thus only one state of polarization remains in the end. Polarization states of light in fibers with elliptical cores is shown in Fig. 9.28a. If linearly polarized light enters a polarization-maintaining fiber with polarization parallel to one of its polarizing axes, the polarization will remain unchanged along the length of the fiber.

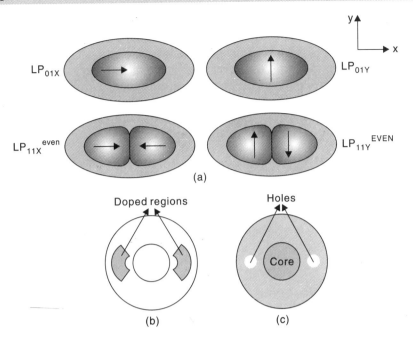

Fig. 9.28 (a) Polarization of light in lower modes of elliptic core fibers (b) Bow type fibers (c) PANDA fibers

Polarization-maintaining fiber maintains the polarization of light that originally entered the fiber by isolating the two orthogonal polarizations from each other even while they travel down the same single mode fiber. Light in the two states of polarization modes experiences about the same attenuation. However, the fiber manufacturing process deliberately introduces strain, which makes the refractive index differ for the two states of polarization. This means that the light with two states of polarization travel through the fiber at different speeds. In bow-type fibers there are stress induced regions on either side of the core by heavy doping Fig. 9.28. In PANDA fiber, holes are drilled on opposite sides of the core to produce high stress regions (Fig. 9.28c).

9.6 ATTENUATION

Light traveling through an optical fibre exhibits a power that decreases exponentially with distance as a result of absorption, scattering and other extraneous effects. Attenuation thus limits the magnitude of the optical power transmitted through the fiber. The attenuation coefficient α is usually defined in units of dB/km and is given by

$$\alpha = \frac{1}{L} 10 \log \frac{1}{\tau}$$

where $\tau = \dfrac{P(L)}{P(0)}$ is the transmission ratio and L is the fiber length.

The attenuation of light in optical fibres occurs on account of absorption of light and scattering of light.

9.6.1 Attenuation due to Absorption

Absorption of light could be *intrinsic* or *extrinsic*. The vibrations of the constituent Si-O atoms absorb light. Also silica has absorption in the ultraviolet region due to electronic transitions.

Both these are classified as *intrinsic* absorption. The extrinsic absorption of light is due to the presence of impurities in the fiber. Mainly absorption occurs due to the hydroxyl group O-H since very minute amounts of water vapour is present in the glass. The absorption also occurs due to the metallic ion impurities. Recent progress in the technology of fabricating glass fibers has made it possible to remove most metal impurities. However, hydroxyl group is very difficult to eliminate. The attenuation coefficient of fused silica (SiO_2) due to absorption is strongly dependent on wavelength as shown in Fig. 9.29. In the near-infrared region the attenuation is minimum and is the proper choice for optical communication.

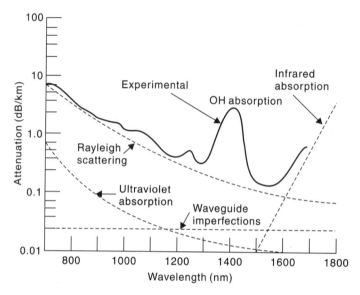

Fig. 9.29 Attenuation due to various loss mechanisms, as a function of wavelength in silica fiber

9.6.2 Attenuation due to Scattering

The attenuation of light also occurs due to scattering. Scattering involves a change in the direction of propagation of light and consequently leads to loss of power in the propagating direction. Scattering of light occurs mainly because of inhomogeneities in the fibre. *In a perfectly homogenous medium and in the absence of thermal motion, there is no scattering of light.* Two types of inhomogeneities are present in a fiber. One is structural and the other compositional. The structural inhomogeneity arises because silica glass is not crystalline and the basic molecular units (SiO_2) are connected in random fashion. The compositional inhomogeneity arises because the exact composition of glass may not be the same throughout the fiber. Because of these inhomogeneities, local variations in the refractive index occur and this leads to scattering and attenuation. *The scattered intensity is inversely proportional to fourth power of wavelength and increases for lower wavelengths* (Fig. 9.29). This type of elastic scattering of light by molecules is called the Rayleigh scattering. It is easy to demonstrate Rayleigh scattering in an optical fiber (Fig. 9.30). A tungsten halogen lamp emitting white light is coupled to two fibers one of length 1m and the other of length 1km. In the former case the emerging light looks white while in the latter case it is reddish. This is because in the fiber of long length the blue component of the white light would have undergone more scattering. Rayleigh scattering is responsible for the blueness of the sky. Sunlight gets scattered on account of the molecules of various gases present in the atmosphere. Since the blue light is scattered more than the red light sky appears blue. Attenuation due to Rayleigh scattering is intrinsic to the fiber itself. Attenuation also occurs due to other extrinsic factors as is evident in the ensuing discussion.

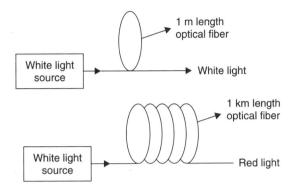

Fig. 9.30 Rayleigh scattering in a optical fiber

9.6.3 Attenuation due to Bending Losses

Small kinks or microbends can be formed when a fibre is pressed against a surface that is not perfectly smooth as shown in Fig. 9.31a. These also can lead to attenuation. To see the effect of microbending, consider the effect of a small angular change α in the direction of fibre on a ray traveling down the fiber (Fig. 9.31b). If the ray initially has an internal angle of incidence θ, then after discontinuity it will become $\theta' = \theta - \alpha$. If θ' turns out to be less than the critical angle, the ray will be refracted out of the fiber and contribute to loss of power. The effect of bending is to change the angle of incidence. Since different modes make different angles of incidence, one can also say that effectively microbending leads to conversion of energy from one mode to another.

It is reasonable to deduce that the losses will be greater for (a) bends with smaller radius of curvature and (b) for those modes which extend most into the cladding region. The loss can be generally represented by an absorption coefficient α_B given by the relation

$$\alpha_B = C \exp(-R/R_c)$$

where C is a constant, R is the radius of curvature of the bend and $R_c = a(\text{NA})^2$, where a is the core radius of the fiber. It is evident that bends with radii of curvature of the order of the fiber radius will give rise to heavy losses. However, the mechanical properties of the fiber usually

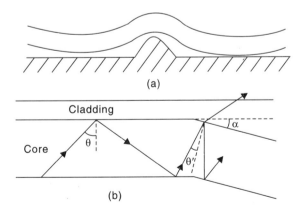

Fig. 9.31 (a) A microbend caused while the fiber rests on a small irregularity on an otherwise flat surface. The strain induced by such bending is also indicated. (b) In the presence of a small 'kink' of angle α in the fiber, the internal angles of the rays within the fiber will be altered. For the particular ray shown, θ changes to θ' where $\theta' = \theta - \alpha$.

cause the fiber to break long before such a small curvature bends can be achieved. So usually losses from such *macrobends* tend to be small.

9.6.4 Attenuation due to Coupling Losses

In many practical applications, the optical fibers have to be laid over very large distances. It can be even of the order of thousands of kilometers if it is a communication channel connecting the different continents. In such a situation, it becomes necessary to interconnect two fibers, which are usually of a kilometer length. When the fibers are interconnected, losses could occur due to mechanical misalignment. Three types of mechanical misalignments can occur and are shown in Fig. 9.32a. They are termed *lateral offset, longitudinal offset* and *angular offset*. Typical losses, which do occur as a result of these misalignments, are shown in Fig. 9.32b–d. Losses could also occur in the coupling of the optical fiber to the transmitter and the receiver.

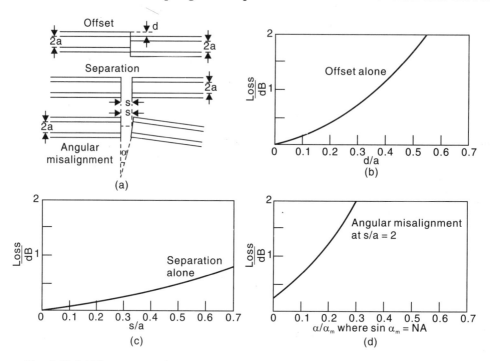

Fig. 9.32 (a) Three types of misalignment that can occur while coupling the two fibers (b)-(d) Typical losses associated with the three types of misalignment

Coupling and bending losses contribute significantly to attenuation. In addition, losses can also occur because the core-cladding surface may not be perfectly smooth. This could lead to non-specular reflection of light (*i.e.*, angle of incidence being not equal to angle of reflection). As a result, the criterion for total internal reflection may not be satisfied in subsequent reflections. Losses could also occur due to changes in the core size as well as slight changes in the refracting index profiles.

9.6.5 Low Attenuation and Low Dispersion Fibers

The need for low attenuation and low dispersion fibers has led to considerable research activity in the field of fiber technology. Attenuation is found to be minimum at ~1550nm (~0.16dB/km)

compared to 1300nm (~0.34dB/km). In the case of single mode fibers, the *zero dispersion* wavelength has been shifted from 1300 to 1550nm by slightly modifying the refractive index profile in the cladding. These are known as *Zero dispersion fibers*. Typical refractive index profiles and dispersion as a function of wavelength in such fibers are shown in Fig. 9.33. The wavelength of the light emitted by a LED or a laser diode can be varied by changing the band gap, which itself is a function of the composition of the semiconductor. Thus it is possible to arrive at an optimum wavelength of operation for fibers.

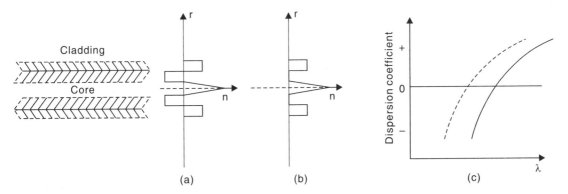

Fig. 9.33 Refractive index profiles for dispersion shifted fibers
(a) triangular profile multiple index design (b) segmented-core triangular profile design
(c) wavelength dependence of dispersion

Since the optical sources emit light over a range of wavelengths, it is better to have low dispersion over a range of wavelengths rather than at a single wavelength as in the case of *Zero dispersion fibers*. This has led to *Dispersion Flattened Fibers*. In such fibers zero dispersion occurs at two wavelengths and for the wavelengths in between these two wavelengths, the dispersion is very low. *Dispersion Flattened Fibers* can be obtained by modifying the refractive index profile of the cladding. These profiles along with wavelength dependence of dispersion is shown in Fig. 9.34. These can be thought of as double, triple and quadruple clad fibers.

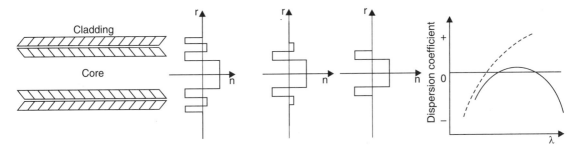

Fig. 9.34 Typical refractive index profiles for dispersion flat fibers
and wavelength dependence of dispersion

9.6.6 Holey and Photonic Crystal Fibers

In a conventional fiber light is guided by total internal reflection. In recent years novel approaches have been tried to confine and guide light in an optical fiber. These can be classified as *photonic band gap fibers* and *holey fibers*. In a photonic band gap fiber low index-core (usually air) of the fiber is surrounded by several layers of cladding, which has a periodic distribution of refractive index (Fig. 9.35). This sort of microstructure in the fiber forbids the propagation of photons of certain frequencies and allows the propagation of specific frequencies. In *holey fibers*, the core which is usually made of silica is surrounded by air holes running parallel to the core. The transmission properties depend on the hole size and the geometrical distribution of holes around the silica core (Fig. 9.36).

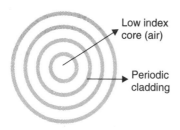

Fig. 9.35 Photon band gap fibers   **Fig. 9.36** Holey fibers

9.7 BASIC PRINCIPLES OF COMMUNICATION

Fiber optic communication systems are one of the many types of communication systems. Communication is sending and receiving information. Information could be in the form of voice, picture or data. In recent years computers have become an important component of communication systems. A communication system is defined as the collection of hardware and software that facilitates the exchange of information. For effective communication the following three fundamental characteristics are to be considered.

Delivery: The system must deliver data to the correct or the intended destination

Accuracy: The system must deliver data accurately (error free)

Timeliness: The system must deliver data in a timely manner without much delay.

9.7.1 Data Communication Components

There are five basic components in data communication system:

Message: This is the information to be communicated

Sender: The sender is the device that sends the message

Receiver: The receiver is the device that receives the message

Medium: The transmission medium provides the physical path that carries the message from the sender to the receiver.

Protocol: Protocol refers to a set of rules that coordinates the exchange of information. Both the sender and receiver should follow the same protocol to communicate data. Without the protocol, the sender and receiver cannot communicate with each other, just as two individuals who do not know a common language cannot communicate.

9.7.2 Data Transmission Mode

Data transmission mode refers to the direction of signal flow between two linked devices. There are three types of transmission modes *simplex, half-duplex and full-duplex*.

Simplex: Simplex transmission is unidirectional. The information flows in one direction across the path with no ability to support response in the other direction. Television broadcast can be considered as an example of this type.

Half-duplex: This is bi-directional but not at the same time. When one device is sending information the other can only receive it at that point of time. The most common example of this type is the wireless handsets (generally used by military and police personnel) where one user talks at a time and another listens. The listener hears a code word signaling the end of speaker's message and can respond only afterwards.

Full-duplex: Full-duplex transmission mode allows both the sender and the receiver to transmit and receive data simultaneously. Telephone network where two people converse by talking and listening at the same time belongs to this type.

9.7.3 Data Communication Measurement

The measurement of the quantity of data that can be passed down a communication link in a given time is done in terms of bandwidth. Bandwidth refers to the maximum volume of information that can be transferred over any communication link. The required bandwidth is proportional to the volume of information to be transmitted in a given time. On digital circuits, bandwidth is measured in *bits per* second (bps). The bandwidth falls into three categories:

Narrowband: In this there is a single transmission channel of 64kbps or less. The bandwidth required for transmitting a speech signal is 64 kbps. There can also be number of 64 kbps transmission ($N \times 64$ kbps) but not more than 1.544 Mbps.

Wideband: In wideband, the bandwidth capacity lies between 1.544 Mbps (also called $T1$ line) and 45 Mbps ($T3$ line).

Broadband: The bandwidth capacity in broadband is equal to 45 Mbps.

9.7.4 Transmission Media

Transmission media refers to the physical media through which communication signals (data and information) are transmitted. The information or a signal is transmitted from one device to another as electromagnetic fields. These signals can travel in vacuum, air or any other transmission medium. Voice signals are generally transmitted as electric current over metal cables. Radio waves are generally transmitted through air or space. Visible or infrared light is currently being used for communication through the fiber optic cable.

Guided Media

Guided transmission medium is also known as bound medium. Guided medium uses a cabling system that guides the data signals along the specific path. The data signals are confined within the cabling system. *Cabling* refers to transmission medium that consists of cables of various metals like copper, tin or silver. There are four types of guided media : *Open wire, twisted pair, coaxial cable* and *optical fiber*.

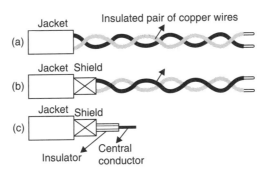

Fig. 9.37 Various guided media (a) unshielded twisted pair cable (b) shielded twisted pair cable (c) coaxial cable

Open wire: Open wire is traditionally used to describe the electrical wire system or power transmission wires strung along the power poles. No shielding or protection from noise interference is used. This can include multi-conductor cables or single wires. Thus medium is susceptible to a large degree of noise and interference. In addition, it also suffers from loss of energy and it can be easily tapped. Hence it is not suitable for long data transmission.

Twisted pair: In this kind of cabling, pairs of wires are twisted together which are surrounded by an insulating material and an outer layer called jacket. Each pair consists of two wires Fig. 9.37a. One wire is used for receiving data signal and the other is used for transmitting data signal. The wires are twisted in order to reduce noise and interference from external sources. Twisted pairs are used in a short distance communication (less than 100 m) and they are available in two forms.

Unshielded twisted pair (UTP) cable: This consists of two metal conductors (usually copper) that are insulated separately with their own coloured plastic insulation (Fig. 9.37a). UTP cables have a maximum transmission speed of up to 9600 bps.

Shielded twisted pair (STP) cable: STP cable has a metal foil or braided-mesh covering that covers each pair of insulated conductors (Fig. 9.37b). The metal foil is used to prevent infiltration of electromagnetic noise. This shield also helps to eliminate crosstalk.

Coaxial cable: Unlike twisted pairs that have two wires, coaxial cables have a single central conductor, which is made up of solid wire (usually copper) (Fig. 9.37c). This conductor is surrounded by an insulator over which a sleeve of metal mesh is woven. This metal mesh is again shielded by an outer covering of a thick material (PVC) known as jacket. Coaxial cable is very robust and is commonly used in Cable TV network. As compared to twisted pairs, it offers higher bandwidth. A coaxial cable is capable of transmitting data at a rate of 10Mbps.

Optical fiber: Optical fibers are made of glass or plastic. They are used to carry light signals. The typical optical fiber consists of a very narrow strand of glass called the core. Around the core is a concentric layer of glass called the cladding. A typical core diameter is 62.5 microns. Cladding has a diameter of 125 microns. The cladding is covered by the jacket. Optical fibers have large information carrying capacity.

9.7.5 Unguided Media

In unguided media data signals are not guided or bound to a fixed channel to follow. One of the common unguided media of transmission is air or vacuum in which *radio and microwaves* propagate.

Radio wave propagation: In radio wave propagation, the signal is carried over carrier waves which have frequencies in the range of radio frequency spectrum (1 kHz to 1 MHz). There are three types of RF (radio frequency) propagation, namely, *ground wave, ionospheric* and *line of sight*.

Ground wave propagation follows the curvature of the earth. They have carrier frequencies of up to 2MHz. AM radio is an example of ground wave propagation (Fig. 9.38a). In ionospheric propagation the signal is reflected by the ionosphere. It operates in the frequency region 30–85MHz (Fig. 9.38b).

Line of sight propagation transmits exactly in the line of sight. The receiving station must be in view of the transmitting station. It is sometimes called space waves or tropospheric propagation. It is limited by the curvature of the earth for ground based stations (~ 50 km). Examples of the line of sight propagation are FM radio, microwave and satellite (Fig. 9.38c).

Microwaves: Microwave transmission is line of sight transmission. The transmit station must be in visible contact with the receiver station. This sets a limit on the distance between stations depending on the local geography. Typically, the line of sight due to earth's curvature is only about 50km to the horizon. Therefore, repeater stations must be placed so that the data signal can travel further than the distance limit (Fig. 9.38d).

FIBER OPTICS

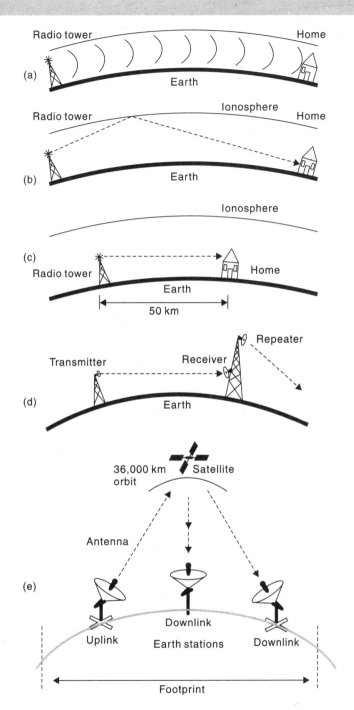

Fig. 9.38 Various unguided media (a) ground wave propagation (b) ionospheric propagation (c) line of sight propagation and (d) microwave transmission (e) satellite communication

Satellites: Satellite transmission is also a kind of line of sight transmission. Satellites are set in geostationary orbits directly over the equator at an altitude of 36,000 km above the earth's surface. Since the period of revolution of the satellite is equal to one day they appear to be stationary from any point on earth. The communication is carried through uplinks and downlinks. The uplink transmits the data to the satellite and the downlink receives the data from the satellite. Uplinks and downlinks are also called *earth stations* because they are located on the earth. The area shadowed by the satellite in which the information or data can be transmitted and received is called the *footprint* (9.38e).

9.7.6 Analog and Digital Data Transmission

Analog and digital transmission are the two modes of information transfer. In the analog case, the signal (for example electric current) varies continuously with time as shown in Fig. 9.39a. Familiar example is the audio output from a microphone. In contrast, the digital signal can take only a few discrete values. In the binary case only two values are possible. The simplest case of a binary digit signal is one in which the electric current is either on or off. These possibilities are called *bit 1 or bit 0*. (bit stands for the contracted form of *binary digit*). Each bit lasts for a certain period time T_b known as the bit period or the bit slot (Fig. 9.39b). The bit rate (B) is defined as the number of bits per second *i.e.*, $B = T_b^{-1}$.

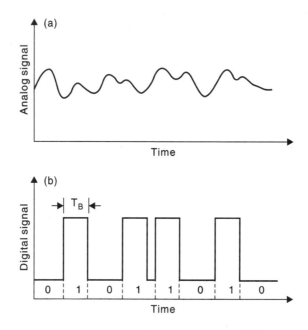

Fig. 9.39 Representation of (a) an analog signal and (b) a digital signal

A well-known example of the digital signal is provided by the computer data. Each letter of the alphabet together with other symbols (decimal numerals, punctuation marks etc.) is assigned a code number (ASCII code) in the range of 0 to 127, whose binary representation corresponds to a 7-bit digital signal. Both analog and digital signals are characterised by a bandwidth, which is a measure of the spectral contents (frequency range) of the signal. *The signal bandwidth represents the range of frequencies contained within the Fourier transform of the signal.*

The analog signal can be converted into a digital form by sampling it at regular intervals of time. Figure 9.43 shows how the sampling is done. The sampling rate is determined by the bandwidth of the analog signal. *According to the sampling theorem, the bandwidth limited signal can be fully represented by discrete samples, without any loss of information, if the sampling frequency f_s satisfies the Nyquist criterion $f_s > 2\Delta f$, where Δf is the bandwidth of the* *signal*. Thus for the speech signal whose bandwidth is 4 kHz, the sampling has to be done at 8 kHz *i.e.*, the value of the speech signal has to be specified at regular intervals of 1/8000 sec.

9.7.7 Modulation Techniques

Analog Signals

Some feature of the carrier wave has to be modulated by the signal for information transfer. Audible sound has frequency in the range 20 Hz to 4000 Hz. For transmitting the sound signals over a large distance as in the case of radio broadcasting, the audio signal is made to modulate a radio wave. Typically radio waves have frequencies in the range 10^4 to 10^6 Hz. Amplitude modulation and frequency modulation are the two techniques which are usually employed. In amplitude modulation the amplitude of the radio wave is varied in accordance with the signal (Fig. 9.40). The frequency of radio wave is modulated in *frequency modulation* (Fig. 9.41).

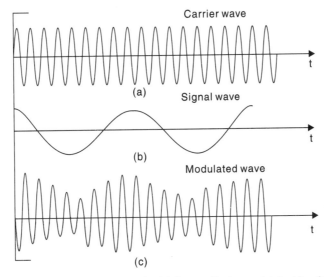

Fig. 9.40 When the carrier wave shown in (a) is amplitude modulated by the modulating wave in (b), one obtains an amplitude modulated wave as shown in (c)

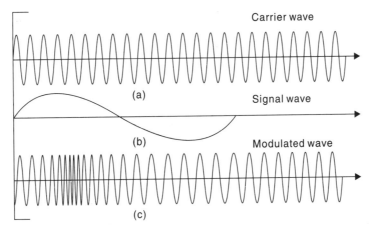

Fig. 9.41 When the carrier wave shown in (a) is frequency modulated by the modulating wave in (b), one obtains an frequency modulated wave as shown in (c)

Digital Signals

Several techniques of digital modulation are possible. Either the amplitude or the width or the position of the pulse could be modulated in accordance with the signal.

Pulse amplitude modulation (PAM): In the pulse modulation technique, the amplitude of the pulse is proportional to the value of the signal. The width of the pulse and the interval between the pulses is kept constant (Fig. 9.42a).

Pulse width modulation (PWM): In this technique, the width of the pulse is modulated in accordance with the signal value, while keeping the amplitude and the time interval between the pulses is kept constant (Fig. 9.42b).

Pulse position modulation (PPM): In this technique, the amplitude and the width of the pulse are constant. But the position of the pulse is altered within a time interval in accordance with the signal (Fig. 9.42c).

Pulse code modulation (PCM): This is a very important mode of transmission of the sampled values of the signal. In this scheme the signal value is first converted into a code formed by identical pulses. The signal value is first converted into the binary code and then the ones and zeros in the code are represented by the presence or absence of a pulse. In Fig. 9.43 the signal value is represented by three digits. But usually the coding system will have eight bits. Thus the signal value at a particular time is coded in the form of eight pulses. All the pulses have the same height, same width and same time interval between them. The eighth pulse is required for synchronisation of the receiver. The speech signal has to be sampled at 8kHz. Since each sample is replaced by 8 pulses, one would have 8×8 kHz = 64 kbps. Thus the information contained in the speech signal has to be transmitted at 64 kb/sec.

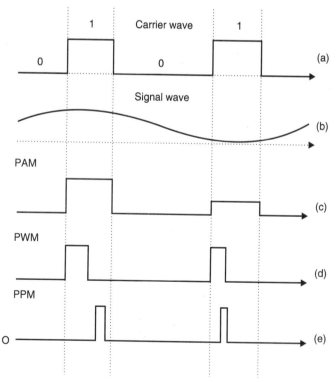

Fig. 9.42. (a) Carrier wave (b) Signal wave (c) Pulse amplitude modulation (PAM) (b) Pulse width modulation (PWM) (c) Pulse position modulation (PPM)

PCM technique requires more time interval for transmission of the same information than the PAM system, since corresponding to each pulse in PAM there are 8 pulses in PCM. This disadvantage of the PCM system is offset by the fact that a system employing PCM is more immune to noise and interference effects. This is because in the PCM system, the receiver has only to detect the presence or absence of the pulse, regardless of the amplitude value or the width. Thus all the external factors, which tend to distort the amplitude or the shape of the pulses, have no effect in the PCM technique.

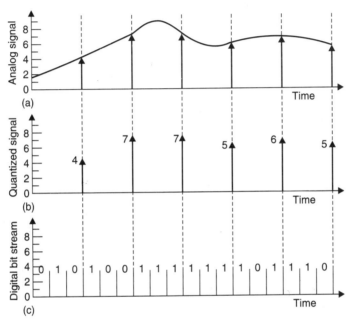

Fig. 9.43 Pulse code modulation (PCM). Pulse code waveform of sample signal depicted in
(a) The PCM using binary coding involving three bits is shown in
(b) The ones and zeros in the binary coding may be represented by the presence or absence of a pulse.

9.7.8 Error Detection

In digital communication, the presence and absence of a light pulse is assigned the digits 1 and 0 respectively. However, light signals are distorted by many mechanisms. This could lead to a change in pulse length, overlapping of pulses or even loss of pulses altogether. These possible distortions would make the receiving end misinterpret the signal. Two solutions to this problem exist. These are *error detection and digital encoding systems*. One way of overcoming the possible distortion problem is to devise a system by which the receiving end has at least a partial idea of what the signal should be. That way, any errors in signal will be apparent, and the receiver can act accordingly. Generally, the receiver will detect an error which is corrected or will ask for a complete retransmission of the signal.

Checking of errors is often accomplished using the *parity* system. *Parity refers to the number of 1s in a standard digital transmission*. Data is normally sent in groups (or words) of 1s and 0s. A single 1 or 0 is known as a *bit*, four *nibble* and eight a *byte*. Some communication systems may use 16, 32 or more bits in a word. Whatever the case, a word with an even number of 1s is said to have an *even parity*. An odd number of 1s is known as *odd parity*. In a parity error checking system, each word has an additional bit added to make it have a set parity. For example an odd parity system would add an extra bit (1 or 0) so that the total number of bits is odd. At the receiving end, the parity of each word can be checked and if a word has the wrong parity, an error in transmission is detected. Table 9.1 indicates how an even and odd parity would be achieved.

Table 9.1

Data	Even Parity Bit	Odd Parity Bit	Final Word	
			Even	Odd
1101	1	0	11011	11010
1111	0	1	11110	11111
1001	0	1	10010	10011
0001	1	0	00011	00010

Parity checking offers a simple way to check the errors in transmission. However, it has its limitations. It is possible that two bits get changed in a transmission. The parity of a word that contains an error may still be correct. To solve this problem, an even more advanced system is used. This system is known as *cyclic redundancy check* (CRC). The CRC uses a mathematical formula to detect errors. Each word (which is a number, in a sense) is passed through this formula and the result is transmitted along with the word. At the receiving end, the same process is repeated, and the result is compared with the transmitted result. Since the probability of the words being distorted so that they produce the same result are slim, CRC is a very effective method for checking for errors.

The second solution for detecting errors in transmission is the use of advanced digital encoding systems. These systems deviate from the pulse = 1 and no pulse = 0 method of encoding so that the possibility for error is reduced. There are many such encoding methods, but the most popular are non return to zero (NRZ), nonreturn to zero inverted (NRZI), Manchester, Miller and Biphase (Fig. 9.44).

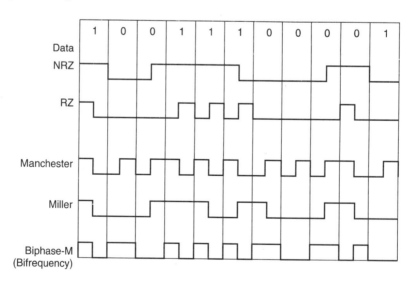

Fig. 9.44 NRZ, NRZI, Manchester and Biphase coding

NRZ (Non return to zero) coding: Signal level is low for a 0 bit and high for a 1 bit and does not return to zero between successive 1 bits.

RZ (Return to zero) coding: Signal level during the first half of a bit interval is low for a 0 bit and high for a 1 bit. Then it returns to zero for either a 0 or 1 in the second half of the bit interval.

Manchester coding: Signal level always changes in the middle of a bit interval. For a 0 bit, the signal starts out low and changes to high. For a 1 bit, the signal starts out high and changes to low. This means that the signal level changes at the end of a bit interval only when two successive bits are identical (*e.g.*, between two zeros).

Miller coding: For a 1 bit, the signal changes in the middle of a bit interval but not at the beginning or end. For a 0 bit, the signal level remains constant through a bit interval, changing at the end of it if followed by another 0 but not if it is followed by a 1.

Biphase-M or Bifrequency coding: For a 0 bit, the signal level changes at the start of an interval. For a 1 bit, the signal level changes at the start and at the middle of a bit interval.

NRZ coding is probably the most common in fiber systems, but each scheme has its advantages and disadvantages. Some including RZ, Manchester and bifrequency coding can make two transitions during a bit interval. This requires more system bandwidth, but improves performance.

9.7.9 Information Carrying Capacity

In amplitude modulation carrier wave frequency v_c is modulated with a sound wave of frequency of v_m. The modulated wave consists of waves of frequencies $(v_c - v_m)$ and $(v_c + v_m)$, in addition to the carrier wave frequency at v_c. The waves at frequency at $(v_c - v_m)$ and $(v_c + v_m)$ are said to lie in the lower side band and upper side band respectively. In general, the modulating wave may consist of frequencies lying from 0 to v_m. Thus on modulating, one obtains two bands of frequencies lying from $(v_c - v_m)$ to v_c (the lower side band) and v_c to $(v_c + v_m)$ (the upper side band). Since both the side bands contain the same information, one may transmit only of the side bands. Such a system is called the single side band transmission. One can extract the information out of the signal by again mixing the received wave with frequency v_c. Thus if the lower side band frequencies are mixed with the carrier frequency, waves having the frequencies $(2v_c - v_m)$ and v_m are obtained. The wave of frequency v_m is the signal of interest and thus recovered.

The frequency of speech signal may be anywhere between 0 to 4 kHz. If the speech signal is used to modulate a carrier wave of frequency 100 kHz, then the upper side band transmission will have waves whose frequencies lie between 100 to 104 kHz. And the lower band transmission will have waves whose frequencies lie between 96 kHz and 100 kHz. Since a band of 4 kHz must be reserved for one speech signal, for carrier frequencies between 100 kHz and 500 kHz, one can at most send (500–100 kHz)/4 kHz = 100 independent speech signals. Since the same band width of 4 kHz is required irrespective of the carrier wave frequency, in a higher carrier frequency channel between 10^{15} Hz and 5×10^{15} Hz, one can send 10^{12} independent speech signals. *Thus the information carrying capacity increases with the frequency of the carrier wave* (Table 9.2). Hence the importance of the optical communication over radio and microwave communication. The bandwidth of 4 kHz is required for the speech signal. But for music the bandwidth is about 20 kHz and for television it is about 6 MHz. Transmission of television signals to homes using copper cable is now the standard practice. In future, the copper wire will be replaced by the optical fiber. The optical fiber will allow reception of even more number of TV channels as well as telephone and computer data. Cable television uses *frequency division multiplexing* to transmit video data. Each channel broadcasts at a different frequency. TV sets use a tuning system to select the desired frequency.

Table 9.2 A comparison of number of telephone and TV channels that can be transmitted on different type of carriers

Carrier frequency	Frequency range	Usable bandwidth	Number of telephone channels	Number of TV channels
LW	30 kHz–300 kHz	10%	3	
MW	300 kHz–3 MHz	10%	25	
SW	3 MHz–30 MHz	10%	200	
VHF	30 MHz–300 MHz	10%	4000	
UHF	300 MHz–3 GHz	10%	10^4	10
Microwave	3 GHz–1 THz	10%	10^5	100
Optical	5×10^{13} Hz–10^{15} Hz	0.1%	10^8	10^5

9.8 MULTIPLEXING

In a network environment it is common that the transmission capacity of a medium linking two devices is greater than the transmission needs of the connected devices. Hence, the medium is shared so that it can be used fully. This is done by *multiplexing*. *Multiplexing* is a technique used for sending several signals simultaneously over a common medium (Fig. 9.45). The following types of multiplexing are employed.

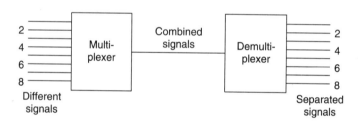

Fig. 9.45 Multiplexing signals

Frequency division multiplexing (FDM): In this technique, multiple signals are sent along the same channel by assigning different carrier frequency bands for each signal. Thus all the signals are overlapping in the time domain but non-overlapping in the frequency domain. The different signals are separated at the receiver by making use of filters, which transmit only the frequency band corresponding to that particular signal (Fig. 9.46).

Time division multiplexing (TDM): In this technique, the signals are sent along the same channel on a time sharing basis. In such a scheme all the signals occupy the same frequency band but are separated in the time domain (Fig. 9.47).

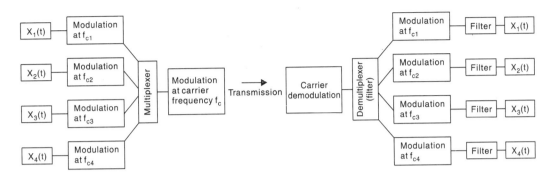

Fig. 9.46 Frequency division multiplexing (FDM)

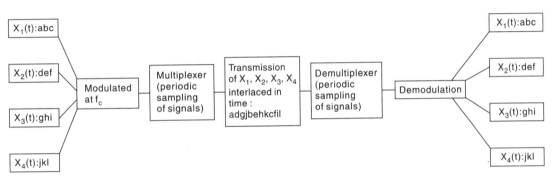

Fig. 9.47 Time division multiplexing (TDM)

Wavelength division multiplexing (WDM): In this technique two or more wavelengths are used to carry signals through the same channel. The signals are generated by separate light sources in the same transmitter, which are combined for transmission through a single fiber. Thus the same fiber could carry one signal at 1300 nm and another at 1550 nm (Fig. 9.48). Dense wavelength division multiplexing DWDM refers to WDM using large number of wavelengths.

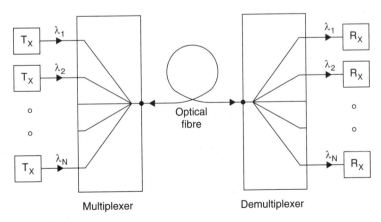

Fig. 9.48 Wavelength division multiplexing (WDM) A multichannel point to point fiber link. Separate transmitter-receiver pairs are used to send and receive the signal at wavelengths $\lambda_1, \lambda_2, \ldots \lambda_n$. Multiplexers combine the transmitter outputs and demultiplexers separate them at the receiving end

9.9 ASYNCHRONOUS AND SYNCHRONOUS TRANSMISSION

The digital data can be sent either serially (one bit at a time along a single channel) or parallely (group of bits along separate channels at the same time). There is only one way of sending data in parallel mode, but there are two subclasses of serial transmission, namely synchronous transmission and asynchronous transmission.

Asynchronous transmission. In asynchronous transmission the timing of the signal is not important. The information that is received or transmitted follows predefined pattern (Fig. 9.49a). As long as the patterns are followed, the receiving device can retrieve the information without any regard to the timing of the signal sent. However, a synchronizing pulse is necessary for the receiver to know when the data is coming and when it is ending. Hence each byte of information is preceded by a *start bit* (denoted by 0) and ended by a *stop bit* (denoted by 1). Therefore, the information is one byte *i.e.,* eight bits becomes ten bits, increasing the overheads. In addition, the transmission of each byte may be followed by a gap of varying duration, which can further help in synchronizing the information with the data stream or channel. As soon as receiver detects the *stop bit*, it ignores any received pulses until it detects the next bit. Asynchronous transmission is slower than the other forms of transmission but it is cheaper and an attractive choice for low-speed communication.

Synchronous transmission. In synchronous transmission the receiver's clock is synchronized with that of the transmitter (9.49b). Consequently there is no need for start and stop bits as in the case of asynchronous transmission. The advantage of synchronous transmission is *speed*. It is used for high speed communication.

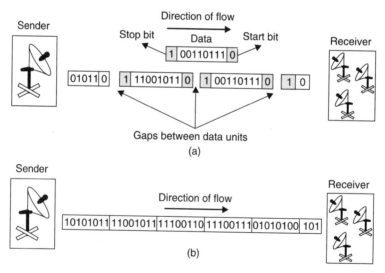

Fig. 9.49 Asynchronous and Synchronous transmission

9.10 TYPES OF COMMUNICATION SYSTEMS

Communication systems can be categorised according to the way the signals are transmitted. There are four types of transmission systems. They are (*a*) point to point transmission (*b*) point to multipoint transmission (broadcasting) (*c*) switched transmission and (*d*) network transmission. These days it is quite common to use personal computers to send and receive information.

Point to point transmission : This is the simplest type of fiber optic communication system. It provides two-way communication between any two terminals, each of which has a transmitter and a receiver (Fig. 9.50a).

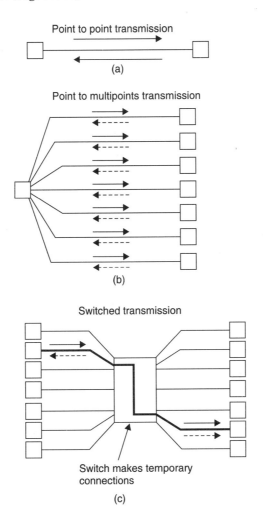

Fig. 9.50 Types of transmission systems (a) point to point (b) point to multipoint transmission (Broadcasting)

Point to multipoint transmission (Broadcasting): In this system, signal from one terminal (transmitter) is sent to many terminals (receivers). The other terminals could also be relay transmitters. This is the arrangement one has for broadcasting (Fig. 9.50b).

Switched transmission: Switched transmission allows any pair of terminals to send and receive signals directly. However, the connection between the sender and the receiver is temporary (Fig. 9.50c).

Network transmission: In networks, all terminals are interconnected for two-way transmission. A network can be as few as several computers on a small network or as large as the Internet, a worldwide network of computers. There are three categories of networks. *Local*

Area Networks (LAN), Metropolitan Area Network (MAN) and Wide Area Network (WAN). These categories are defined depending upon various factors like size of the network, the distance it covers and the type of link used in interconnections.

Local area network(LAN): A LAN or Local Area Network is a computer network that spans only a small geographical area (usually ~ km^2), such as office, home or building. In a local network, the connected computers have a network operating system installed onto them. One computer is designated as the file *server*, which stores all the software that controls the network along with the software that can be shared by the computers attached to the network. Other computers connected to the file server are called *work stations*. On most LANs, cables or optical fibers are used to connect computers. Usually, LAN offers a bandwidth of 10 to 100Mbps. They are generally limited to a maximum distance of only a few kilometers, although they are usually much shorter.

Metropolitan area network (MAN): A MAN, or Metropolitan Area Network, is a network of computers spread over a metropolitan area such as a city and its suburbs. It is usually reserved for metropolitan areas where the city bridges its local area networks with a series of backbones, making one large network for the entire city. It may be a single network such as cable television network or it may be a means of connecting a number of LANs. Note that, MAN may be operated by one organization (a corporate with several offices in one city) or its resources used by several organizations in the same city.

Wide area network (WAN): A WAN or Wide Area Network is a system of interconnecting many computers over a large geographical area such as cities, states, countries or even the whole world. These kinds of networks use telephone lines, satellite links and other long range communications technologies to connect. Such networks are designed to serve an area of hundreds or thousands of miles such as public and private packet switching networks and national telephone networks. For example, a company with offices in New Delhi, Chennai and Mumbai may interconnect the LANs at these places through a WAN. WAN is usually a network designed to connect small and intermediate sized networks together. The largest WAN in existence is the Internet.

Network Topologies

The term network topology refers to the way a network is laid out, either physically or logically.

Bus topology. Bus topology uses a common bus or backbone to connect all devices with terminators at both ends. The backbone acts as a shared communication medium and each mode (file server, workstations and peripherals) is attached to it with an interface connector. Whenever a message is to be transmitted on the network, it is passed back and forth along the cable, past the stations (computers) and between the two terminators, from one end of the network to the other. As the message passes each station, the station checks the message's destination address. If the address in the message matches the station's address, the station receives the message. If the addresses do not match, the bus carries the message to the next station and so on. Figure 9.51a shows how devices such as fileservers, workstations and printers are connected to the linear cable or the backbone.

Advantages

(a) Connecting a device to a linear bus is easy.

(b) Requires least amount of cabling and hence least expensive.

(c) Easy to extend the bus

Disadvantages

(*a*) Entire network shuts down if there is a failure in the backbone.

(*b*) Heavy traffic can slow down a bus because computers on such networks do not coordinate with each other to reserve time to transmit.

Ring topology: In ring topology, the devices are connected on a cable-ring without any terminated ends (Fig. 9.51*b*). Every node has exactly two neighbours for communication purposes. All messages travel through a ring in the same direction (clockwise or anticlockwise) until it reaches its destination. Each node in the ring incorporates a repeater. When a node receives signal intended for another device, its repeater regenerates the bits and passes them along the wire.

Advantages

(*a*) Ring topology is easy to install and reconfigure.

(*b*) Every computer is given equal access to the ring. Hence no single computer can monopolise the network.

Disadvantages

(*a*) Failure in any cable or node breaks the loop and can make the entire network non-operative.

(*b*) There is ceiling on maximum ring length and number of nodes.

Star topology: In star topology devices are not directly linked to each other, but they are connected via a centralized network component known as *hub* or *concentrator* (Fig. 9.51 *c*). The hub acts as a central controller and if a node wants to send data to another node, it boosts up the message and sends the message to the intended node. This topology uses twisted copper cable. However, coaxial cable or fiber optic cable can also be used.

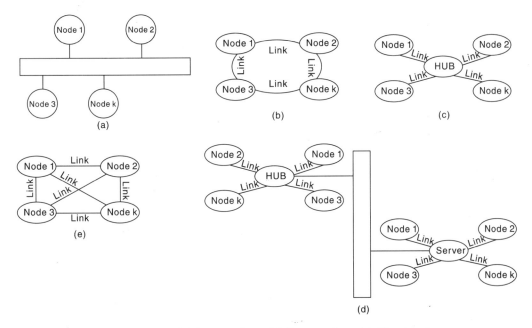

Fig. 9.51 (a) Bus topology (b) Star topology (c) Ring topology
(d) Tree topology (e) Mesh topology

Advantages

(a) Star topology is easy to install and wire.

(b) The network is not disrupted even if a node fails or is removed from the network.

(c) Fault detection and removal of faulty parts is easiest in star topology.

Disadvantages

(a) It requires a longer length of cable.

(b) If the hub fails nodes attached to it are disabled.

(c) The cost of the hub makes the network expensive as compared to bus and ring topology.

Tree topology: A tree topology combines characteristics of linear bus and star topologies (Fig. 9.51d). It consists of groups of star-configured workstations connected to a bus backbone cable. Not every node plugs directly to central hub. The majority of nodes connect to a secondary hub that in turn is connected to the central hub. Each secondary hub in this topology functions as the originating point of a branch to which other nodes connect.

Advantages

(a) The distance to which a signal can travel increases as the signal passes through a chain of hubs.

(b) Tree topology allows isolating and prioritising communications from different nodes.

(c) Tree topology allows for easy expansion of an existing network, which enable organizations to configure a network to meet their needs.

Disadvantages

(a) If the backbone line breaks the entire segment goes down.

(b) It is more difficult to configure and wire than other topologies.

Mesh topology: In a mesh topology, every node has a dedicated point to point link to every other node. Messages sent on a mesh network can take any of several possible paths from source to destination (Fig. 9.51e). A fully connected mesh network has $n(n-1)$ physical links to link n devices. In addition, to accommodate that many links, every device on the network must have $(n-1)$ communication (input/output) ports.

Advantages

(a) The use of large number of links eliminates network congestion.

(b) If one link becomes unusable, it does not disable the entire system.

Disadvantages

(a) The amount of required cabling is very large.

(b) As every node is connected to the other, installation and reconfiguration is very difficult.

(c) The amount of hardware required in this type of topology can make it expensive to implement.

9.11 SWITCHING

Switch: For establishing direct communication links among n devices, the number of direct communication channels required is $n(n-1)/2$. For large n, the number of channels required becomes enormous. In addition, channels remain idle for considerable lengths of time. A better solution is *switching*. A switch connects individual devices on a network so that they can communicate with each other. They inspect the data packets as they are received. They determine the source from which the data packet has arrived and forward it to the destination device. On a network, switching means routing traffic, by setting up temporary connections between two or more network points. This is done by devices located at different locations on a network, called *switches*. In a switched network, some switches are directly connected to the communicating devices while others are used for routing or forwarding information. Figure 9.52 depicts a switched network in which communicating devices are indicated as different nodes. and switches are labeled (I, II, III, IV, etc.). Note that multiple switches are used to complete the connection between any two communicating devices at a time.

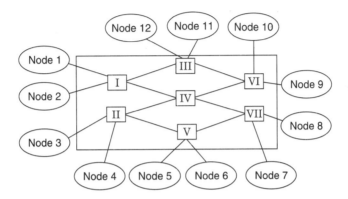

Fig. 9.52 Switched transmission

Circuit switching: When there is a need for a device to communicate with another device, circuit switching technique creates a fixed bandwidth channel (called a *circuit*) between the source and the destination. This circuit is reserved exclusively for a particular information flow and no other flow can use it. Other circuits are isolated from each other. Thus switches I, II and III are used to connect nodes 1 and 4 as shown in Fig. 9.53. The path connecting any two devices is fixed and not alterable. A common example of a circuit switched network is Public Switched Telephone Network (PSTN). In circuit switching, data is transmitted with no delay except the time needed for the signal to propagate. This method is simple and requires no special facilities. Therefore, it is well suited for low speed data transmission. However, it is less suited for data transmission, where the data comes in surges with idle gaps between them. When there is no flow of data, the capacity of incurring link is wasted. It is inflexible as well. Once the circuit is established, that is the only path taken by the flow of information whether or not it remains the most efficient.

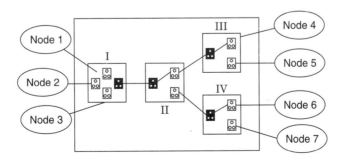

Fig. 9.53 Circuit switching

Packet switching: In Packet switching data is divided into *packets*. Packets are discrete units of potentially variable length blocks of data. Apart from data, these packets also contain a *header* with control information like the destination address, priority of the message, etc. These packets are passed by the source point to its local packet switching exchange (PSE). When the PSE receives a packet, it inspects the destination address contained in the packet. Each PSE contains a navigation directory specifying the outgoing links to be used for each network address. On receipt of each packet, the PSE examines the packet header information and then either removes the header or forwards the packet to another system. If the channel is not free, then the packet is placed in a queue until the channel becomes free. As each packet is received at each transitional PSE along the route, it is forwarded on to the appropriate link mixed with other packets. At the destination PSE, the packet is finally passed to its destination. All the packets traveling between the same two points (even those from a single message) need not necessarily follow the same route. Therefore, after reaching their destination, each packet is put into order by a packet assembler and disassembler (PAD). Figure 9.54 shows packet (1, 2, 3 and 4) from node 1 transmitted to node 4 via various routes. All the packets do not follow the same path. Hence at node 4 packets are not received in the same order in which it is sent at node 1. The machine at node 4 then assembles the arrived packets in order and retrieves the information. In packet switching the links are used for maximum advantage. It is ideal for data transfer.

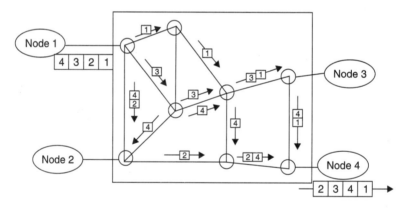

Fig. 9.54 Packet switching

Large volume of data transfer requires a series of packets to be sent. But this does not require the link to be dedicated in between the transmission of each packet. The packets belonging to data having high priority can be sent along with the packets belonging to the data having low priority. Hence, packet switching provides a much fairer and efficient sharing of the resources. Packet switching is widely used in data networks like the Internet.

Message switching: Message switching technique employs the *store and forward* mechanism (Fig. 9.55). In this mechanism, a special device (usually a computer system with large memory storage) in the network receives the message from a communicating device and stores it in its memory. Then it finds a free route and sends the stored information to the intended receiver. In this kind of switching, messages are always delivered to one device where they are stored and then rerouted to their destination. Message switching was common in 1960s and 1970s. However, presently this technique has virtually become obsolete because of the time delay in storing and forwarding the messages. This technique requires large capacity of data storage which is not cost effective.

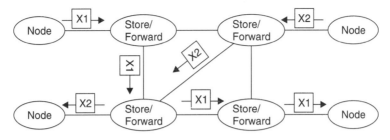

Fig. 9.55 Message switching

9.12 NETWORK DEVICES

Network Interface Card: Network interface card is the first contact between a machine and the network. It connects clients, servers and peripherals to the network via a port. Most network interfaces come as small circuit board that can be inserted onto one of the computer motherboard's slots. Alternatively, modern computers sometimes include the network interface as part of their main circuit boards (mother boards). Each network interface is associated with a unique address called its *media access control* (MAC) address. The MAC address helps in sending information to its intended destination.

Hub: A hub is a small box that connects individual devices on a network so that they can communicate with one another. The hub operates by gathering the signals from individual network devices, optionally amplifying the signals and then sending them onto all other connected devices. Amplification of the signal ensures that devices on the network receive reliable information. A hub can be thought of as the centre of a bicycle wheel, where the spokes (individual computers) meet.

Repeater: A repeater is an electronic device that operates on the physical layer of the OSI model. Signals that carry information within a network can travel a fixed distance before attenuation corrupts the data. A repeater installed on the link receives the signal, regenerates it and sends the refreshed copy back to the link.

Bridge: A bridge filters data traffic at a network boundary. It reduces the amount of traffic on a LAN by dividing it into two segments. It inspects the incoming traffic and decides whether to forward or discard it.

Router: A router is an essential network device for interconnecting two or more networks. Router's sole aim is to trace the best route for information travel. A router creates and/or maintains a table called a *routing table* that stores the best routes to certain destinations. Routers are critical components of the network.

Gateway: A gateway is an internet-working device, which connects two different network protocols together.

9.13 OSI COMMUNICATION MODEL

Communication Protocol: The communication between devices is visualised as a seven layer model (Fig. 9.56). It consists of seven separate but related layers, namely, *physical, data link, network, transport, session, presentation and application.* The *application, presentation and session layers* constitute the upper layers. Primarily these layers deal with application issues, and are implemented only in the software. The highest layer, *application*, is closest to the end user. The remaining four layers constitute the *lower layers*. A brief description of these layers is given here.

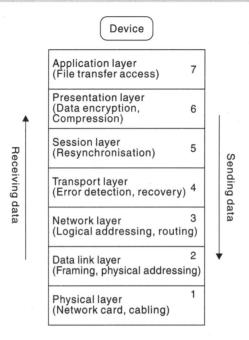

Fig. 9.56 OSI model

Physical Layer: The physical layer is the lowest layer of the OSI model. It defines the physical and electrical characteristics of the network. It defines the various types of copper or fiber optic cable as well as different wireless solutions. This layer communicates with the data link layer and regulates the transmission of a stream of bits over a physical medium. This layer also defines which transmission technique is used to send data over the cable.

Data Link Layer: The function of the data link layer is to transform the data into a *line* that is free of transmission errors and is responsible for node-to-node delivery. On the sender side, the data link layer divides the stream of bits from the network layer into manageable form known as *frames*. These data frames are then transmitted sequentially to the receiver. Data link layer also performs flow control of the frames in order to match the data transfer speed of the sender to the data processing speed of the receiver. On the receiver end, the data link layer detects and corrects any errors in the transmitted data, which it gets from the physical layer.

Network Layer: The network layer provides the physical routing of the data *i.e.*, it determines the path between the sender and the receiver. The outbound data is passed down from the transport layer, is encapsulated in the network layer's protocol and then sent to the data link layer for segmentation and transmission. This layer organizes frames from the data link layer into packets *i.e.*, the inbound data is de-fragmented in the correct order and then the assembled packet is passed to the transport layer. Network layer also manages traffic problems such as switching, routing and controlling the congestion of data packets. Network layer provides uniform addressing mechanisms so that more than one network can be interconnected.

Transport Layer: The basic function of the transport layer is to handle error recognition and recovery of the data packets. The transport layer establishes, maintains and terminates communication between the sender and the receiver. At the receiving end the transport layer ensures that all the packets have arrived and converts packets into the original message. The receiving transport layer sends receipt acknowledgments.

Session Layer: The session layer organises and synchronises the exchange of data between the sending and the receiving applications. This layer establishes the control between the two computers in a session, regulating which side transmits, when and for how much duration. The session layer lets each application at one end know the status of the other at the other end. An error in sending application is handled by the session layer in such a manner that the receiving application may know that the error has occurred. The session layer can resynchronise applications that are currently connected to each other. This may be necessary when communications are temporarily interrupted, or when an error has occurred that results in the loss of data.

Presentation Layer: The function of the presentation layer is to ensure that information sent from the application layer of one system would be readable by the application layer of another system. This is where application data is packed or unpacked, ready for use by the running application. This layer also manages security issues by providing services such as data encryption and compresses data so that fewer bits need to be transferred on the network.

Application Layer: The application layer is the entrance point that programs use to access the OSI model and utilise network resources. This layer represents the services that directly support applications. This OSI layer is closest to the end user.

9.14 FIBER OPTIC COMMUNICATION SYSTEM

9.14.1 Basic Layout

Figure 9.57 shows a generic block diagram of an optical communication system. It consists of a transmitter, a communication channel and a receiver. These three elements are common to all communication systems. Optical communication systems can be classified into two broad categories: *guided* and *unguided*. In *guided optical communication systems* the optical beam emitted by the transmitter remains spatially confined. This is achieved using optical fibers. Since all guided systems currently use optical fibers they are known as *fiber optical communication systems*. In the case of *unguided optical communication* systems, the optical beam emitted by the transmitter spreads in space, similar to radio or microwave waves. However, *unguided optical communication systems are less suitable for broadcasting applications, since the optical beams spread mainly in the forward direction (because of their small wavelength).* Their use generally requires accurate pointing between the transmitter and the receiver. That is why laser beams are used. LIDAR, which is the acronym for Light Detection And Ranging, is similar to RADAR. The basic elements of a fiber optic communication system are shown in Fig. 9.58. It consists of an

(a) *Optical transmitter,*

(b) *Optical fiber* and

(c) *Receiver.*

The other fiber optic devices which are used in communication systems are

(d) *Optical amplifiers*

(e) *Couplers*

Fig. 9.57 A generic block diagram of an optical communication system

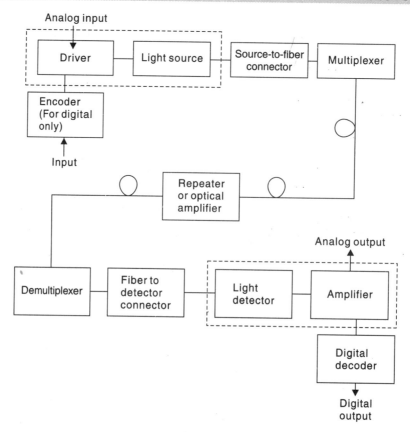

Fig. 9.58 Basic elements of a fiber optic communication system

(f) *Optical isolators*

(g) *Circulators* and

(h) *Fiber Bragg Gratings*

9.15 ADVANTAGES OF OPTICAL FIBERS

Following are the advantages of communication with optical fibers.

(a) *Extremely wide bandwidth:* This essentially means that greater volume of information or messages can be carried over in a fiber optic system. This is because the rate at which information can be transmitted is directly related to carrier frequency. Light has a frequency in the range of $10^{14} - 10^{15}$ Hz and microwave frequencies $10^8 - 10^{10}$ Hz. Therefore a transmission system that operates at the frequency of light can theoretically transmit information at a higher rate than systems that operate at radio frequencies or microwave frequencies. In recent years using optical fibers data transfer at the rate of Tbps ($\sim 10^{12}$ bps) has been demonstrated.

(b) *Smaller diameter and lighter weight cables:* Because of their lightweight and flexibility they can be more easily handled than copper cables.

(c) *Immunity to inductive interference:* Optical fibers are not metallic and hence they do not pick up electromagnetic waves. The result is noise free transmission. The fiber optic cables are immune to interference caused by lightning or other electromagnetic equipment.

(d) *Absence of cross talk:* In conventional communication circuits, signals often stray from one circuit to another, resulting in extraneous voice signals being heard in the background. This *cross-talk* is negligible in optical fiber cables even though they consist of many strands of optical fibers.

(e) *Low cost:* Optical fiber technology has the potential of delivering signals at low cost.

(f) *Safety:* Optical fibers are much safer than copper cables, since no electricity is conducted.

(g) *Longer life span:* The life span of optical fibers is expected to be 20–30 years in contrast to copper cables which have a life of 12–15 years.

(h) *Temperature resistant:* In contrast to copper cables, they have high tolerance to temperature extremes as well as corrosive liquids and gases.

(i) *Easy maintenance:* Optical fibers are more reliable and easy to maintain than copper cables.

9.15.1 Optical Transmitter

The role of an optical transmitter is to convert the electrical signal into optical form. The resulting optical signal is launched into the optical fiber.

The basic requirements for the light sources are:

(a) *Power:* The source power must be sufficiently high so that after transmission through the fiber the received signal is detectable with the required accuracy.

(b) *Small emitting area:* Light must be emitted in a small area comparable to the size of the core.

(c) *Linewidth:* The source must have a narrow spectral width so that the effect of chromatic dispersion in the fiber is minimized.

(d) *Speed:* It must be possible to modulate the source power at the desired rate.

(e) *Noise:* The source must be free of random fluctuations. This requirement is particularly strict for coherent communication system.

Other features include ruggedness, insensitivity to environmental changes such as temperature, reliability, low cost and long lifetime.

Light emitting diode, *laser diode* and *quantum well lasers* are used as light sources in fiber optic communication system.

Light emitting diode (LED): LED is a *p-n* junction operated in the forward bias (Fig. 9.59). Holes and electrons recombine across the depletion region to emit light. The semiconductor used for the optical source must have a direct band-gap. In an indirect band-gap material, part of the energy involved in electron-hole recombination is released as heat. In a direct band-gap semiconductor, all the energy is radiated. None of the single-element semiconductors has a direct band-gap. However compounds of elements III and V of the periodic table are direct band-gap materials. These so-called III-V semiconductors are made of a group III element (such as Al, Ga or In) and a group V element (such as P, As or Sb). Different binary, ternary and quaternary combinations of the elements such as GaAs, InAs, AlGaAs and InGaAsP are direct band materials. The wavelength of the light emitted is given by

$$E_g = h\nu = hc/\lambda \text{ or } \lambda = (hc/E_g)$$

By varying the composition, E_g can be varied and hence light of the required wavelength can be obtained.

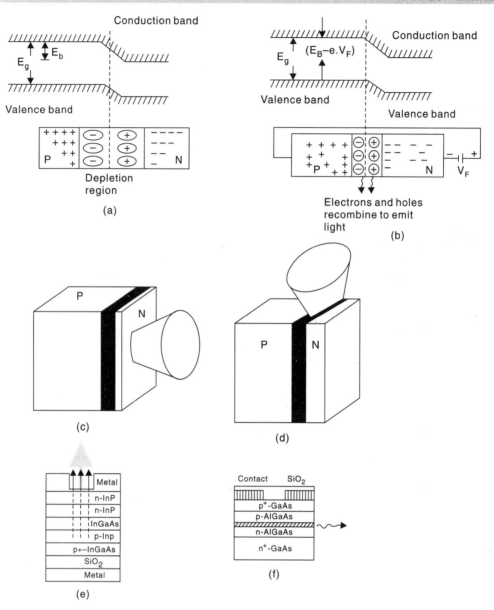

Fig. 9.59 (a) p-n Junction in equilibrium (b) p-n Junction forward biased (c) Surface emitting LED (d) Edge emitting LED (e) Structure of surface emitting LED (f) Structure of edge emitting LED

Light from a diode can be emitted along an edge or perpendicular to the surface. Typical structures of edge emitting diode and surface emitting diode are shown in Fig. 9.59e, f. These are heterojunction device in that they have three or more layers. The upper and lower layers with a lower index of refraction sandwiches the central layer. The change in the refractive index across the heterojunction serves to constrain some of the emitted light to the active region. Light emitted is incoherent and has large spectral width (Fig. 9.60a).

Laser diode: Laser diode is also a semiconductor *p-n* junction but it produces coherent light unlike LED. In LED light is produced is due to spontaneous emission whereas in a laser diode it is due to stimulated emission. Laser diode operates at high current densities. The lasing action is enhanced by having highly polished surfaces which act like mirrors as in the case of a laser. The advantage with laser diodes is the narrow line-width of the source (Fig. 9.60b). Further the active region forms a resonant cavity that resembles the Fabry-Perot cavity used for gas lasers (Fig. 9.61). If l is the length of the cavity then the allowed frequencies are given by

$$f_n = \frac{mc}{2ln}$$ where $m = 1, 2, 3, \ldots$ and n is the refractive index

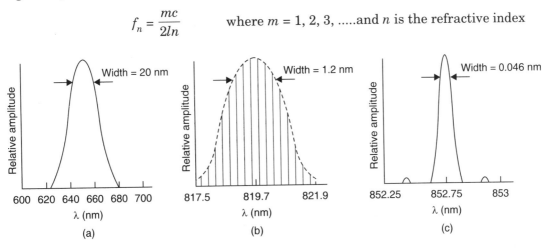

Fig. 9.60 Spectral width of (a) LED, (b) Laser diode and (c) DFB laser

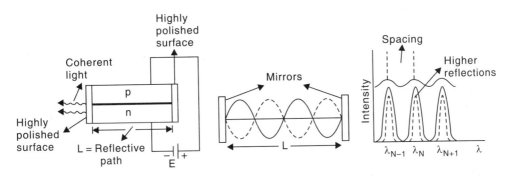

Fig. 9.61 Fabry Perot Cavity and the allowed modes.

The gain is however maximum at the central frequency

Distributed feedback laser and Distributed Bragg reflector: The spectral width of light emitted by a laser diode can be reduced further by making only one of the modes active and suppressing all the other modes. This is achieved by incorporating a periodic refractive index grating of period d. With such a grating only the mode satisfying the condition

$$\lambda = d \times n$$

will be present and all the other modes are suppressed on account of destructive interference between them.

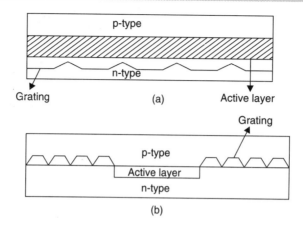

Fig. 9.62 Distributed Feedback laser and Bragg Reflector

Multi quantum well laser diodes: One of the major disadvantages of laser diodes is that the injection current densities needed are high. This disadvantage can be overcome by making the active region as thin as possible. At such small length (~nm) quantum effects begin to play a dramatic role. The band-gap energy of the active layer is smaller than the surrounding layers. The structure acts like a quantum well. MQW lasers employ a parallel stack of single quantum wells. A typical structure of a quantum well laser is shown in Fig. 9.63.

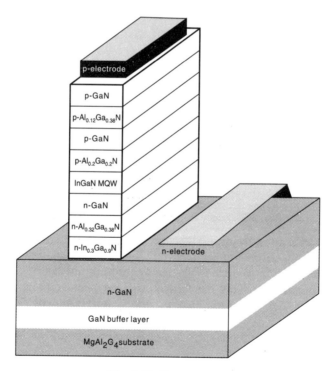

Fig. 9.63 Quantum well laser

The wavelength region used in fiber optic communication system is classified into different bands as given in Table 9.3.

Table 9.3 Nomenclature of different bands used in fiber optic communication system

Band	Descriptor	Wavelength range (nm)
O-band	Original	1260–1360
E-band	Extended	1360–1460
S-band	Short	1460–1530
C-band	Conventional	1530–1565
L-band	Long	1565–1625
U-band	Ultra-long	1625–1675

9.15.2 Optical Receiver

An optical receiver converts the optical signal received at the output end of the optical fiber back into electrical signal. Fiber to receiver coupling is similar to the source to fiber coupling as in the case of the surface emitting LED. Optical detectors are usually photodiodes. Photodiodes are *p-n* junctions made from indirect band-gap semiconductos. Semiconductors commonly used for making photodiodes are Si and Ge. Photodiodes must satisfy the following requirements.

Sensitivity: They must detect weak signals

Responsivity: They should respond uniformly to all wavelengths

Speed: They ought to be capable of operation at high speeds

Small size: They must have a small area comparable to that of fiber dimensions

Noise: They should not generate additional noise

Other features include ruggedness, cost and long lifetime.

Commonly employed optical detectors are *p-i-n* photodiode and the *avalanche diode* (APD). The function of the photodiode is the reverse of LED. Light falling on the depletion region generates electron hole pairs. These are swept away by applying a voltage.

p-i-n photodiode: It is made of a semiconductor *pn* junction operated in the reverse bias. It has lightly doped *n*-doped intrinsic layer sandwiched between the *p* and *n* type regions (Fig. 9.64). With the addition of the intrinsic layer, the active region of the diode is increased considerably. Light enters the diode through a tiny window about the same size as the fiber core. The electron-hole pairs formed are swept away by the applied potential giving rise to electric current.

Avalanche photodiode: This is more sensitive than the *p-i-n* photodiode. This is achieved by the addition of a high intensity electric field region. In this region the primary electron-hole pairs generated by the incident photons are accelerated. They collide with the atoms present in this region, thus generating more electron-hole pairs. This process of generating more electron-hole pairs is known as avalanche effect. Because of this the photocurrent generated by APD exceeds the current generated by a PIN device. Figure 9.65 denotes avalanche photodiode.

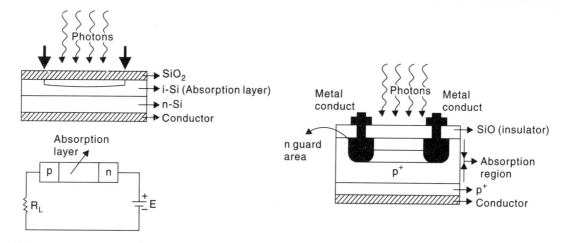

Fig. 9.64 A typical *p-i-n* photodiode which is used as a photodetector

Fig. 9.65 Avalanche photodiode

9.15.3 Optical Modulators

The input signal in the form of an analog or digital signal is applied to the driver of the optical source to obtain an optical signal. An *encoder* is used to convert an analog signal into a digital signal. LED or a laser diode is operated in the linear region of the forward characteristic to obtain the optical signal as shown in Figs. 9.66 and 9.67.

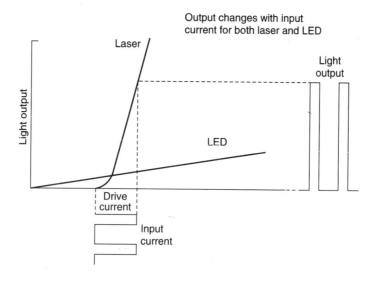

Fig. 9.66 Modulation of the intensity of the optical source. The light output of an LED or a laser diode, changes with the input binary current pulse.

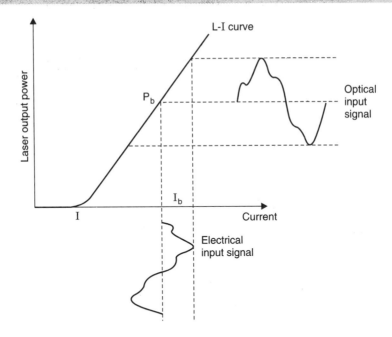

Fig. 9.67 Modulation of the intensity of the optical source for an analogue current signal

The modulation of light by an electric signal in LED or laser diode is known as *direct modulation*. There are other ways of modulating light. The most important of them is based on

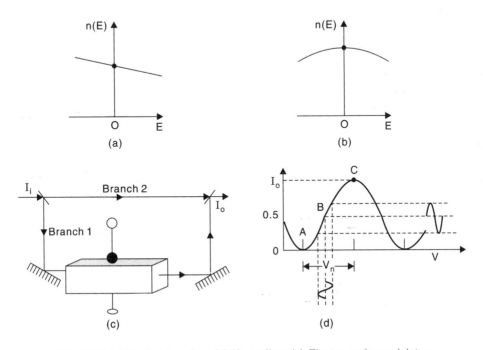

Fig. 9.68 (a) Pockels's effect (b) Kerr effect (c) Electro-optic modulator
(d) Intensity modulation by the applied voltage

electro-optic effect. This is based on the dependence of refractive index of a material on the applied electric field. Such materials are known as *electro-optic materials.* If the refractive index varies linearly with the electric field, it is referred to as *Pockel's* effect (Fig. 9.68a). In Kerr effect the refractive index depends on the square of the electric field (Fig. 9.68b). These effects are used to modulate the phase of the light beam. An electro-optic crystal is included in a Mach-Zhender interferometer as shown in Fig. 9.68c. The signal voltage modulates the refractive index of the material, which in turn modulates the phase or the path difference between the two beams. Thus the output intensity of the light beam is modulated, as shown in Fig. 9.68 d. An electro-optic modulator can be used as a switch. On the application of a voltage, a phase difference of π can be introduced between the two beams and the net transmittance is zero. In the absence of the voltage the two beams are in phase and there is light output.

9.15.4 Source to Fiber Coupling

The light signal from the LED or the laser diode has to be launched into the optical fiber, without any loss. In the case of the edge emitting LED, care must be taken to see that the light emitted by the LED is within the acceptance angle of the optical fiber as shown in Fig. 9.69. For the surface emitting LED or laser diode, the fiber is fixed onto the surface as shown in Fig. 9.70. Microlenses are also used for efficient coupling between the LED and the fiber. Typical coupling schemes using microlenses are shown in Fig. 9.71.

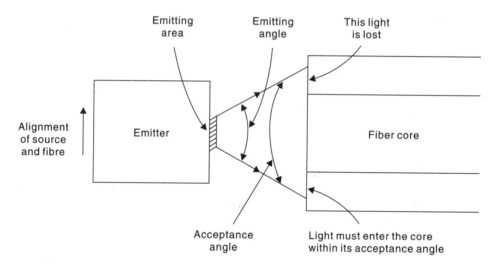

Fig. 9.69 Light transfer from an edge emitting LED to an optical fiber

9.15.5 Couplers and Switches

The guided light in an optical fiber penetrates into the cladding to a distance of ~ a few wavelengths (Fig. 9.71). This is known as the evanescent wave. This evanescent wave is responsible for the working of a fiber optic directional coupler. Such a device is formed if two fibers are laid side by side along some portion of their length in which cladding has been depleted to almost negligible thickness. If light is injected into port 1, it will split into ports 2 and 3 with negligible light appearing in port 4 (Fig. 9.72). The amount of light that will be distributed between ports 2 and 3 is a function of coupling length (L), core-to-core spacing (d) and refractive index n_2. The light exiting through port 3, for example can be varied by potting the fiber interaction region with a flexible elastomer of index n_2 close to that of the cladding. A change in either d or n_2 can be effected by the application of an external force field. Thus the power outputs across the ports can be varied.

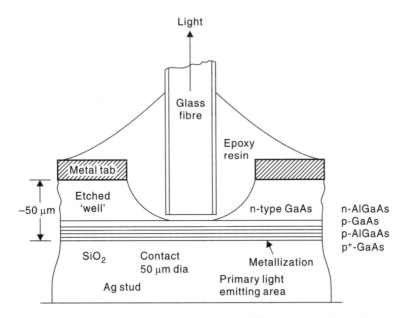

Fig. 9.70 Light coupling from a surface emitting LED or a laser diode with an optical fiber

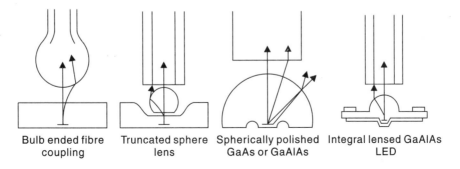

Fig. 9.71 Various lens coupling mechanisms of LED to optical fiber

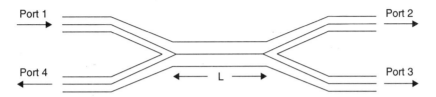

Fig. 9.72 Directional coupler

Couplers and switches direct the light beam that represents the various signals to their appropriate destinations. *Couplers* always operate on the incoming signal in the same manner. *Switches* are controllable couplers that can be modified by an external command. Examples of couplers are shown in Fig. 9.73. In the T-coupler, a signal at input port 1 reaches both output ports 2 and 3. The signal at the ports 2 and 3 reach the port 1. In the star coupler, the signal at any of the input ports reaches all the output ports. In the four port directional coupler, the signal at any of the input ports 1 or 2 reaches both the output ports 3 and 4. The signals from any of the output ports 3 or 4 in the opposite direction reaches both the ports 1 and 2. When operated as a switch, the four-port directional coupler is switched between the parallel state (1–3 and 2–4 ports) and the cross state (1–4 and 2–3 ports).

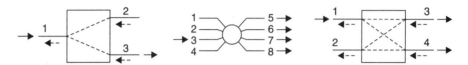

Fig. 9.73 T-coupler and Star coupler

Optical couplers can also be constructed by use of miniature beam splitters, lenses, graded-index rods, prisms, filters and gratings. The new technology is called macro-optics.

9.15.6 Optical Isolator

An optical isolator allows the light to pass through in one direction but blocks the transmission in the reverse direction. This is used in erbium doped fiber amplifier. Optoisolator is based on Faraday effect. Faraday effect is also utilised in the design of fiber optic based current and magnetic field sensor.

Faraday effect: Faraday discovered that when a block of glass is subjected to a strong magnetic field, it becomes optically active. When plane-polarized light is sent through glass in a direction parallel to the applied magnetic field, the plane of polarization is rotated (Fig. 9.74a). The angle of rotation is proportional to the distance traveled in the medium. The angle of rotation per unit length (ρ) is proportional to the component B of the magnetic field in the direction of propagation of light *i.e.*,

$$\rho = VB$$

where V is called the Verdet constant.

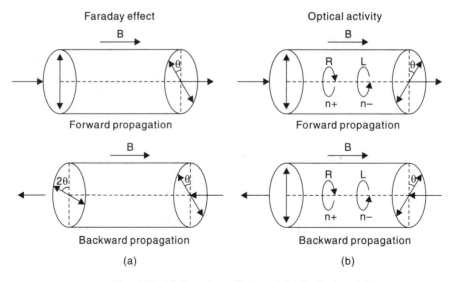

Fig. 9.74 (a) Faraday effect and (b) Optical activity

The sense of rotation is governed by the direction of the magnetic field. For V > 0, the rotation is in the direction of the right-handed screw pointing in the direction of the magnetic field. *In contrast to optical activity the sense of rotation does not reverse with the reversal of the direction of propagation* (Fig. 9.74b). This arises on account of the difference in velocities propagation for right circularly polarized and left circularly polarized light. When light travels through a Faraday rotator, reflects back onto itself and travels once more through the rotator in the opposite direction, it undergoes twice the rotation.

The principle of an optical isolator is shown in Fig. 9.75. The input unpolarised signal is passed through a polarizer which passes only light energy in the vertical state of polarization. It blocks the horizontally polarized light. The polarizer is followed by a Faraday rotator. *This rotates the plane of polarization by 45° regardless of the direction of propagation.* The Faraday rotator is followed by another polarizer which passes this polarized light. The reflected light from right to left passes through the polarizer and is rotated through 45° by the Faraday rotator. Thus the light is horizontally polarized and is blocked by polarizer 1.

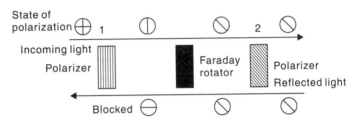

Fig. 9.75 Optoisolator

9.15.7 Circulators

A circulator is similar to an isolator, except that it has multiple ports, typically three or four as shown in Fig. 9.76. In a three-port circulator, an input signal on port 1 is sent out on port 2, an input signal on port 2 is sent on port 3 and an input signal on port 3 is sent out on port 1. Circulators are useful to construct optical add/drop elements. Circulators operate on the same principle as that of isolators.

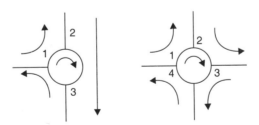

Fig. 9.76 Circulator

9.15.8 Fiber Amplifiers and Repeaters

An important component of the present day fiber optic communication systems is fiber amplifier or repeater and regenerator (Fig. 9.77). The fiber optic cable especially the ones used for communication between the continents could be hundreds or thousands of kilometers. An optical signal traveling along such a long optical fiber without attenuation and much distortion is not possible. This problem is overcome by using *repeaters and regenerators or fiber amplifiers* at short regular distances. With a semiconductor laser source of ~1 mW power, the maximum distance before regeneration is ~100 to 200 km.

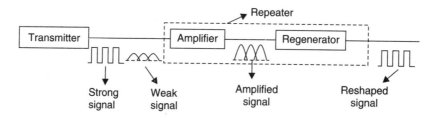

Fig. 9.77 Role of a repeater in a fiber optic system

A *repeater* detects the signal before it becomes too weak and then amplifies it. During this process of amplification, the noise in the signal also gets amplified. Hence the need for *regenerators*. *Regenerator* uses the input from the *repeater* to reconstruct the original signal. The output from the *regenerator* drives the light source coupled to the next fiber cable. In contrast to *repeaters and regenerators,* Fiber *amplifiers* directly amplify the optical signal. In a fiber amplifier the repeated conversion of optical to electrical and back to optical signal is avoided.

Erbium doped Fiber Amplifier: A fiber amplifier is essentially a laser where the active medium is the erbium doped silica glass fiber. Erbium provides a three level lasing scheme as shown in Fig. 9.78. Practical pump bands exist at wavelengths at 532 nm, 670 nm, 907 nm, 980 nm and 1480 nm. However, the latter three wavelengths comprise the most important pump bands. The laser radiation is at ~1.54 µm, which integrates well with the In GaAsP light source. Figure 9.79 shows the basic layout with the erbium-doped fiber spliced into the transmission fiber. Light from the laser is combined with the signal via the wavelength selective coupler. The laser light optically pumps the erbium atoms into the excited states and the signal induces stimulated emission at the signal wavelength. Optical isolators (antireflection coatings) prevent unwanted reflections and the filter blocks the laser light. The advantages of the erbium-doped fiber amplifier are that the gain is independent of the state of polarization and stable over 100°C temperature range. There are some disadvantages since the device operates only at 1.54 µm and the gain is not constant in the neighbourhood of this wavelength.

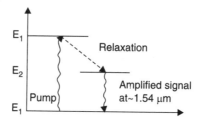

Fig. 9.78 Energy level diagram of erbium doped silica fiber laser amplifier. The energy levels are those of erbium ions

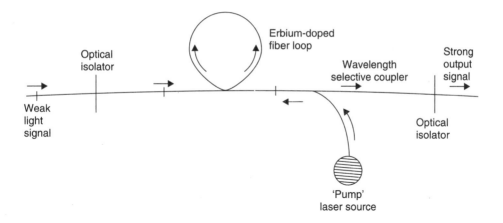

Fig. 9.79 Basic layout of an erbium doped fiber amplifier spliced (inserted) into the transmission fiber

Fiber Raman Amplifier

Spontaneous Raman scattering corresponds to inelastic scattering of light in the visible and ultraviolet region of the spectrum. It is analogous to Compton scattering of X-ray photons. When photons in the visible/ultraviolet region interact with the molecule, the scattered photon can have either higher or lower energy than the incident photons. This is known as the *Raman*

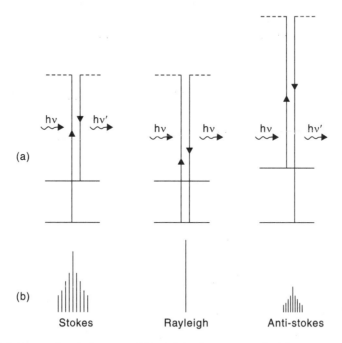

Fig. 9.80 (a) Energy level diagram of Rayleigh, Stokes and Anti-Stokes scattering and (b) corresponding frequency shifts observed in the spectra.

scattering. Rayleigh scattering refers to the elastic scattering *i.e.*, there is no change in the energy of the incident photon. There is only change in the direction of propagation of the photon. When the scattered photon has lesser energy than the incident photon it is referred to as Stokes Raman scattering. Anti-Stokes Raman scattering refers to the case when the scattered photon has greater energy than the incident energy. If the difference in the energy of the incident and the scattered photon corresponds to the vibrational energy of the molecule it is referred to as *Vibrational* Raman *scattering. Rotational Raman scattering* refers to the case when the difference in the incident and the scattered photon corresponds to the rotational energy of the molecule. The schematics of the Raman scattering is shown in Fig. 9.80a. Spontaneous Raman scattering is non-coherent and is weak in intensity compared to Rayleigh scattering. Stokes Raman scattering is more intense than AntiStokes Raman scattering by the virtue of the fact that molecules obey Maxwell-Boltzmann's distribution.

Stokes Raman shift = $h(\nu_0 - \nu_s) = \Delta E$ (vibrational or rotational energy)

Anti-Stokes Raman shift = $h(\nu_s - \nu_0) = -\Delta E$

Light corresponding to Rayleigh and Raman scattering are separated using a spectrometer. Figure 9.80b shows schematically the spectra of Rayleigh, Antistokes and Stokes Raman scattering.

Fiber Raman amplifier makes use of *stimulated Raman scattering. Stimulated Raman scattering* is a non-linear effect and occurs at high intensities. It is a threshold phenomenon. The observed Raman scattering is also coherent. Here the incident light of frequency ν_p (called the pump beam) induces a gain in the scattering medium at another frequency ν_s (called the signal beam), where $h(\nu_p - \nu_s)$ corresponds to the energy difference between two vibrational levels.

The basic configuration of FRA is shown in Fig. 9.81b. Both the pump beam at the frequency ν_p and the input signal frequency ν_s are injected into specific optical fiber serving as an optical amplifier, through an optical coupler. The pump wavelength λ_p is converted into a signal wavelength λ_s by SRS, thereby increasing the power at λ_s. When a high intense beam propagates through the fiber, weak Raman scattering is observed at the end of the fiber. Hence there is no need to use a separate fiber for the pump and the signal. In other words, if a suitable optical fiber is optically pumped by an appropriate source, the signal will get amplified as the two beams co-propagate along the fiber. In practice, both forward pumping (*i.e.*, the pump beam in the direction of the signal beam) and backward pumping (*i.e.*, the pump beam in the direction opposite to that of the signal beam) are possible. Since SRS is not a resonant phenomena it does not require population inversion. Since the spontaneous Raman scattering spectrum is broad, the corresponding gain spectrum of the Raman amplifier is also very broad. No matter what the wavelength of the pump light is, the fiber can act as an amplifier in the wavelength range corresponding to the spontaneous Raman scattering spectrum. Thus the signals in the 1310 nm can be amplified using a pump of wavelength 1240 nm and the signals at 1550 nm can be amplified using a pump of wavelength 1450 nm. The Raman scattering in a silica fiber is shown in Fig. 9.81a.

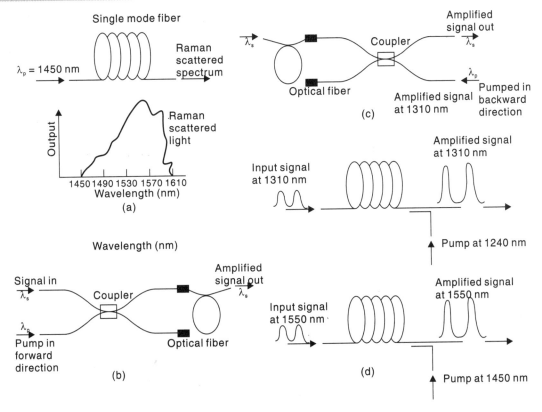

Fig. 9.81 (a) Raman scattering in silica fiber (b) Fiber Raman amplifier (forward pump) (c) Fiber Raman amplifier (backward pump) (d) Fiber Raman amplifier at (1310 nm and 1550 nm)

Fiber Brillouin Amplifier

Fiber Brillouin amplifier is essentially same as Fiber Raman amplifier. It is based on stimulated Brillouin scattering. Spontaneous Brillouin scattering is the inelastic scattering of light with acoustic phonons or acoustic waves (Fig. 9.82). It can also be thought of as the light beam undergoing Doppler shift due to scattering by acoustic wave which moves with the speed of sound. Thus by measuring the Brillouin shift one can estimate the speed of acoustic waves and hence the elastic constants. Stimulated Brillouin scattering is a nonlinear effect and has a threshold. The scattered beam is coherent unlike in the spontaneous Brillouin scattering. Here the incident light of frequency ν_p (called the pump beam) induces a gain in the scattering medium at another frequency ν_s (called the signal beam), where $h(\nu_p - \nu_s)$ corresponds to the energy of the acoustic phonon.

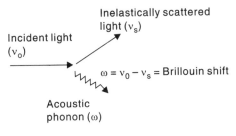

Fig. 9.82 Brillouin scattering

Figure 9.83 shows the configuration of Fiber Brillouin amplifier.

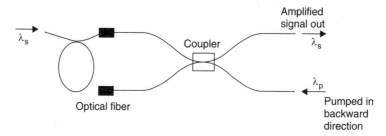

Fig. 9.83 Fiber Brillouin amplifer

Despite a formal similarity between SBS and SRS, SBS differs from SRS in three important aspects which affect the operation of Fiber Brillouin amplifiers.

(a) Amplification occurs only when the signal beam propagates in a direction opposite to that of the pump beam.

(b) Stokes shift for SBS is smaller (~10GHz) by three orders of magnitude compared with that of SRS and depends on the pump frequency.

(c) The Brillouin gain spectrum is extremely narrow with a bandwidth < 100 MHz.

9.15.9 Optical Fiber Gratings

Optical fiber gratings consists of periodic modulation of refractive index along the core of an optical fiber. There are two main types of fiber gratings, namely short period gratings and long period gratings. Short period gratings, also referred to as fiber Bragg gratings (FBG) have periods of the order of *half a* μm for operation around 1550 nm window, whereas long period gratings (LPG) have periods of *a few hundred* μm. Both types of gratings exhibit strong wavelength dependent characteristics. They have important applications in wavelength division multiplexed (WDM) optical fiber communication systems and networks. They are also used in fiber optic sensors.

Fiber optic gratings: Consider a periodic dielectric medium consisting of a number of alternate layers of higher $(n_0 + \Delta n)$ and $(n_0 - \Delta n)$ refractive indices with $\Delta n \ll n_0$ (Fig. 9.84). When light wave enters this medium, it undergoes minute reflections from every interface. If all the reflections are in phase, then the medium strongly reflects the incident wave. If the reflected waves are not in phase, then net reflection is weak. The phase difference between adjacent reflections is dependent on the wavelength. This implies that overall reflection from such a medium is very strongly wavelength dependent.

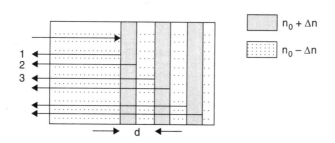

Fig. 9.84 Refractive index variation in a Fiber Bragg grating.

Assuming the thickness of each medium to be $d/2$, the phase difference between reflections 1 and 2 (Fig. 9.84) is (for $\Delta n \ll n_0$)

$$\Delta\phi = \pi - \frac{2\pi}{\lambda_0} n_0 d$$

The extra phase difference of π comes about due to the phase change on reflection suffered during the reflection from a denser medium. Similarly the phase difference between waves 2 and 3 is given by

$$\Delta\phi = \pi + \frac{2\pi}{\lambda_0} n_0 d$$

For, constructive interference between waves 1, 2, 3 etc. takes place:

$$\Delta\phi = \pm 2\pi$$

i.e., whenever the wavelength satisfies the following equation

$$\lambda_B = 2n_0 d$$

The above equation is referred to as the Bragg condition (reminiscent of X-ray diffraction from atomic planes in crystals) and the specific wavelength satisfying the above equation is called the Bragg wavelength. *Such a periodic index modulation within the fiber is referred to as an FBG.* Thus when broadband light or a set of wavelengths are incident on an FBG, then only the wavelength corresponding to λ_B gets reflected; the other wavelengths just get transmitted to the output (Fig. 9.85).

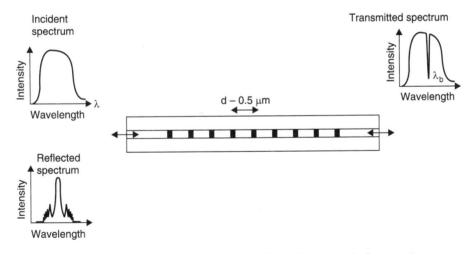

Fig. 9.85 Fiber Bragg grating (short period) and its transmission spectrum

In LPGs different modes propagating within the fiber get coupled. The specific cladding mode to which a core mode gets coupled depends on the fiber parameters as well as on the period of the grating. Because the cladding modes are lossy, this results in a loss peak in the transmission spectrum of the grating (Fig. 9.86). The magnitude of the loss peak, the spectral width and the position of the loss peak can all be controlled by an appropriate choice of the grating period, the length and Δn. Unlike FBGs, there is no reflected wave in the case of LPGs. Thus LPGs can be used to induce specific loss at a specific wavelength. Hence they can be used as filters in erbium-doped fiber amplifiers (EDFAs), in sensors and other devices.

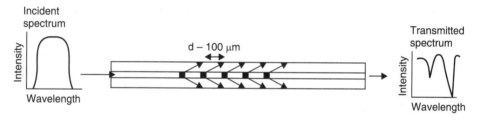

Fig. 9.86 Transmission spectrum of FBG (Long period) and its transmission spectrum

9.15.10 Add/Drop Multiplexer

In a wavelength division multiplexer it is required to add or drop a wavelength channel. This is possible only if the desired wavelength is transmitted or blocked. Figure 9.87 shows a typical configuration of an add/drop multiplexer based on FBG. An FBG, which reflects light at the desired wavelength that needs to be dropped is placed between two optical circulators. From among the incoming wavelength channels of the DWDM transmission system, the wavelength matching the FBG wavelength will get reflected and be routed to the dropped port. All other wavelength channels proceed in the forward direction along the link. At the same time, signal at the same wavelength can be added using the second circulator with the same grating reflecting the added channel into the link. Since an FBG reflects only the chosen wavelength and transmits all other channels, it is possible to add and drop multiple channels by using multiple FBG at the desired wavelengths instead of single FBG. The gratings used in add/drop multiplexers need to have high reflectivity so that there is no residual signal leading to crosstalk among the dropped and added channels. In WDM the wavelengths are separated by a grating and the graded index rod (Fig. 9.87b).

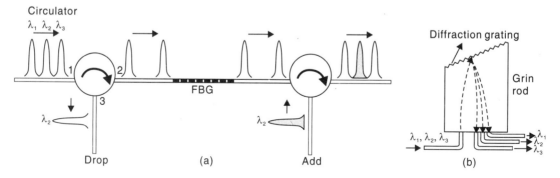

Fig. 9.87 (a) Add/drop wavelength multiplexer (b) GRIN rod with the grating for separating wavelengths

9.16 COHERENT OR HETERODYNE OPTICAL COMMUNICATION

The major attraction of coherent detection is that it avoids noise encountered in direct detection, making the receiver sensitive to fainter signals. The receiver sensitivity can be improved by up to 20dB compared with direct techniques. This enables to work with greater attenuation levels and consequently repeaters can be placed at much greater distances. The use of coherent detection allows an efficient use of the fiber bandwidth. Many channels can be transmitted simultaneously over the same fiber by using FDM with a channel spacing as small as 1–10 GHz. This also allows the use of frequency shift and phase shift modulation which does not work with direct detection. There are two types of coherent detection schemes. They are (a) homodyne and (b) heterodyne scheme. In the homodyne scheme the local-oscillator frequency is selected to coincide with the signal carrier frequency, so that the difference frequency generated is zero. In the case of heterodyne detection, the local oscillator frequency v_o is chosen to differ from the signal carrier frequency v_t, such that the intermediate frequency is in the microwave region. $v_{IF} = 1$ GHz.

Figure 9.88 shows the basic idea of coherent optical communication. The laser transmitter emits a frequency ω_1, which is modulated by the signal. The modulation can be in amplitude, phase or frequency. At the receiver, the light is mixed with light from a second laser at a nearby frequency ω_2. This gives rise to an intermediate frequency signal at the difference frequency $\pm (\omega_1 - \omega_2)$. The signal of this frequency can then be processed by the receiver electronics to give an output signal.

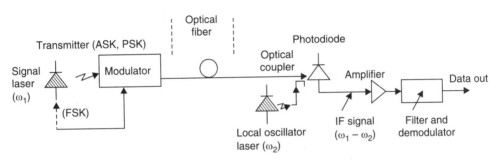

Fig. 9.88 Coherent optical detection

The several hundred GHz wide multimode spectrum of semiconductor laser must be reduced to a narrow linewidth of < 5 MHz. Polarization stability of the transmission media is an important consideration of fibers.

Let us consider the electric field of the transmitted signal to be a plane wave having the form

$$E_s = A_s \cos[\omega_s t + \phi_s(t)]$$

where A_s is the amplitude of the optical signal field, ω_s is the optical signal carrier frequency and $\phi_s(t)$ is the phase of the optical signal. To send information one can modulate either the amplitude, the frequency or the phase of the optical carrier. Thus one of the following three modulation techniques can be implemented.

1. Amplitude shift keying (ASK) or on-off keying (OOK). In this case ϕ is constant and the signal amplitude A_s takes one of two values during each bit period, depending on whether a 0 or a 1 being transmitted.

2. Frequency shift keying (FSK). In this case, the amplitude A_s is constant and $\phi_s(t)$ is either $\omega_1 t$ or $\omega_2 t$, where the frequencies ω_1 and ω_2 represent the binary signal values.

3. Phase shift keying (PSK). In this case, the information is conveyed by varying the phase with a sine wave $\phi_s(t) = \beta \sin \omega_m t$ where β is the modulation index and ω_m is the modulation frequency.

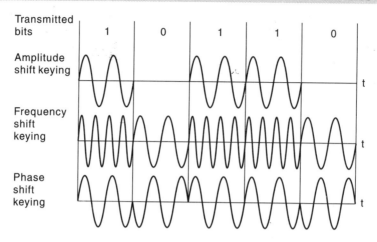

Fig. 9.89 ASK, FSK and PSK

The three modulation schemes are shown in Fig. 9.89.

9.17 FIBER SOLITON COMMUNICATION SYSTEMS

The word *soliton* was coined in 1965 to describe the particle like properties of pulse envelopes in *dispersive nonlinear* media under certain conditions. *The pulse envelope not only propagates undistorted but survives collisions just as particles do.* They were experimentally observed in optical fibers in 1980. Soliton was first observed in water wave. The existence of fiber soliton is the result of balance between *group velocity dispersion* and the *self phase modulation*.

Self phase modulation : In a single mode fibers on account of the narrow core diameter, the effective intensity of light within the fiber could be very large. At high intensities the refractive index is a function of the intensity of light and can be represented as

$n = n_0 + n_2 I$ where I is the intensity of light.

The increase in refractive index with intensity causes intensity dependent frequency shift within the optical pulse. There is shift to lower frequencies (a red shift) on the leading edge and a shift to higher frequencies (a blue shift) on the trailing edge. This is shown in Fig. 9.90b. The pulse gets compressed on the trailing edge and gets elongated on the leading edge. This is known as chirping. This phenomenon is called *self phase modulation* since the observed compression and elongation within the pulse arises on account the intensity dependent phase.

In the case of a single mode fiber due to chromatic dispersion there is broadening of the pulse. In addition there is also the chirping of the pulse. But this is of opposite sign to that observed in *self phase modulation* i.e., the pulse gets compressed on the leading edge and gets elongated on the trailing edge as shown in Fig. 9.90a. This occurs in the wavelength region above the zero dispersion wavelength.

Since chirping is of opposite sign for *self phase modulation* and *chromatic dispersion*, for some pulse shapes satisfying the required intensity, it is possible to generate an *optical soliton* (Fig. 9.90c). Such pulses will retain their initial shape even after traveling long distances in the optical fiber.

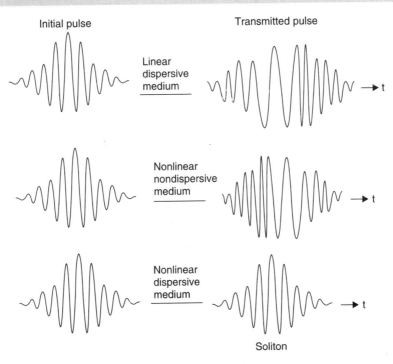

Fig. 9.90 (a) Chirping due to chromatic dispersion and (b) Self phase conjugation (c) Soliton

The evolution of fiber optic communication systems is given in Table 9.4.

Table 9.4 Evolution of fiber optic communication systems

Generation	Bit rate	Operating wavelength (μm)	Type of fiber	Loss (dB/km)	Repeater spacing (km)	Source	Detector
First (1978–82)	34 to 45 Mb/s	0.85	Multimode graded index	~3	8–15	GaAlAs	Si
Second (1980–84)	45 to 90 Mb/s	1.3	Multimode graded index	~1	~30	InGaAsP	Ge
Third (1982–87)	144 to 560 Mb/s	1.3	Single mode	~0.4	≥40	InGaAsP	PINFET
Fourth (1984–90)	≥ 1Gb/s	1.55	Single mode dispersion shifted	<0.3	≥100	DFB InGaAsP	(Ge) APD
Fifth (1990–1995)	≥ 5.5 Gb/s	1.55	Single mode dispersion shifted	<0.25	>100	DFB InGaAsP	(Ge)APD

9.18 FIBER SENSORS

9.18.1 Introduction

An important off-shoot of optical fiber technology is the emergence of the field of fiber optic sensors and devices. The advantages of fiber optic sensors are:

(a) sensed signal is immune to electromagnetic interference and radio frequency interference.

(b) intrinsically safe in explosive environments.

(c) highly reliable and secure with no risk of fire/ sparks.

(d) low volume and weight.

(e) as a point sensor they can be used to sense normally inaccessible regions without perturbation of the transmitted signals.

(f) potentially resistant to nuclear or ionizing radiation.

(g) can be easily interfaced with low-loss optical fiber telemetry and hence affords *remote sensing*.

(h) large bandwidth and hence offers possibility of multiplexing a large number of individually addressed point sensors in a fiber network or distributed sensing *i.e.*, continuous sensing along the fiber length.

(i) chemically inert and they can be readily employed to monitor chemical and biochemical process.

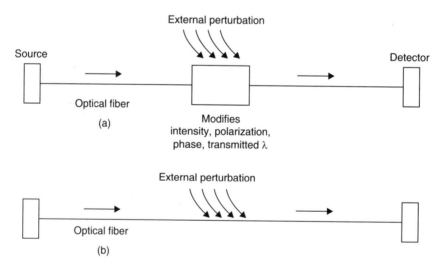

Fig. 9.91 (a) Extrinsic and (b) Intrinsic fiber optic sensor

A fiber optic sensor may be classified as either intrinsic or extrinsic (Fig. 9.91). In the intrinsic sensor, the physical parameter to be measured or the physical effect to be sensed modulates the transmission parameters of the sensing fiber. One or more of the physical properties of the guided light *e.g.*, intensity, phase, polarization and wavelength is modulated by the measurand. In an extrinsic fiber, the modulation of light takes place outside the fiber. The fiber merely acts as a conduit to transport the light signal to and from the sensor head. Four of the most common fiber optic sensing techniques are based on changes in intensity, phase, state of polarization and spectrum of the light. However, out of these four, the intensity and phase modulated ones offer the widest spectrum of optical fiber sensors. The advantage of intensity modulated sensors lie in their simplicity of construction and their compatibility with the multimode fibers. The phase modulated fiber optic sensors necessarily require an interferometric measurement set-up with associated complexity in construction.

9.18.2 Intensity Modulated Sensors

Sensor to detect the presence/absence of an object in an assembly line

Fiber sensors are used to detect changes in light intensity, which occur either inside or outside the fiber. For example a fiber might be used to collect light from a given point to see whether an object is present or not. Figure 9.92 shows an optical fiber used in an assembly line to test the presence of an object.

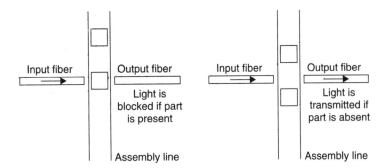

Fig. 9.92 Fiber optic probe checks for parts on an assembly line

Liquid Level Sensor

An optical fiber can be used to sense the liquid level in a closed tank. They are used in gasoline tanks or cryogenic fluid storage containers. A typical sensor is shown in Fig. 9.93. One fiber

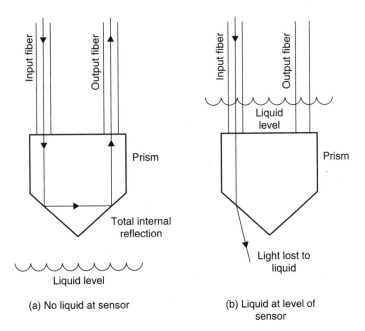

Fig. 9.93 Fiber optic liquid level sensor

delivers light to a prism mounted at a proper level. If there is no liquid in the tank, light from the fiber experiences total internal reflection at the base of the prism. This light is directed back into the collecting fiber. If the bottom of the prism is in contact with the liquid, total internal reflection cannot occur and no more light is reflected back into the fiber. When the light signal stops, the control system gets activated and can be designed to stop or start the liquid flow into the container as the case may be.

Displacement Sensor

Displacement sensor using optical fiber bundles have been designed. The basic design is shown in Fig. 9.94a. Light from the source travels along the fiber bundle and is incident on the surface, whose displacement has to be monitored. The reflected light from the surface is gathered from the other set of fiber bundles. The amount of light gathered by the fiber bundles is a function of the gap between the surface and the sensor head. The light traverses back the fiber bundles where it is converted into an electrical output using a photodetector. The variation of the electrical output with the gap distance is also shown in the Fig. 9.94b. The response in linear for a particular range of the gap values and the sensor can be used for measuring the displacement in this region.

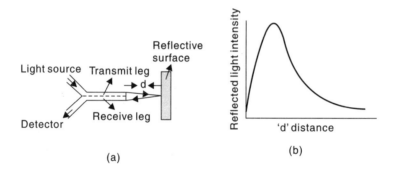

Fig. 9.94 (a) Displacement sensor and (b) Intensity versus gap distance

Pressure Sensor

Figure 9.95 shows how pressure can be measured by modifying the displacement sensor. Here the changes in displacement are caused by the movement of a diaphragm which is triggered by changes in pressure. The probe in the form of a reflecting diaphragm is positioned in front of the fiber. Light reflected from the diaphragm is detected by the photodetector. Changes in the external pressure cause the diaphragm to bend leading to changes in the numerical aperture. This leads to changes in light intensity transmitted by the fiber. Pressure changes upto 6Mpa can be measured with good accuracy. Such sensors are useful for monitoring pressure changes in arteries, bladder, urethra etc. These are also useful for pressure monitoring of gaseous reactants and products in chemical industries.

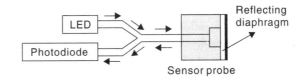

Fig. 9.95 Simple pressure sensor as a modified displacement sensor

Another design of a pressure sensor is shown in Fig. 9.96. The optical fiber is mounted between a pair of plates containing parallel grooves. When the pressure pushes the plate together, it causes microbends in the fiber. This, in turn, results in the leakage of light and reduces the light intensity transmitted by the fiber, which can be detected using a photodetector. Microbend sensors are not supposed to strain the fibers to breaking point. However, severe strains can break the fibers and that can be useful for sensing. The structural integrity of a dam can be monitored by embedding optical fibers in the concrete matrix. If the internal stresses develop the transmitted light intensity will decrease and finally drop to zero when the optical fiber snaps. Thus it can be used for early warning of a catastrophe. Pressure sensors are embedded in composites or other structures for similar reason.

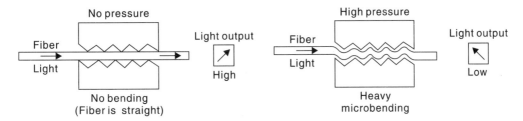

Fig. 9.96 Fiber optic pressure sensor. Increasing pressure on the plate causes microbending losses useful for sensing

An extrinsic fiber optic pressure sensor based on the photoelastic behaviour has also been designed (Fig. 9.97a). Photoelastic effect refers to the induced birefringence under the action of stress. Some materials like calcite and quartz exhibit double refraction or natural birefringence. In these materials, the incident light is split into ordinary and extraordinary ray. The ordinary and extraordinary rays are plane polarised. The two planes of polarisation are mutually perpendicular to one another. The velocity of the ordinary ray is isotropic while that of the extraordinary ray is anisotropic. The difference between the refractive indices of ordinary and extraordinary ray is known as *birefringence*. Materials like piezo-optic glass, polyurethane, epoxy resin exhibit birefringence only under the action of stress. This phenomenon is known as Photoelastic effect. When such a material is kept between crossed polaroids, no light can pass through. However, when the material is subjected to stress, on account of induced birefringence, light can pass through the material. The relative transmitted intensity is a function of stress as shown in the Fig. 9.97b. Hence, by measuring the transmitted intensity one can estimate the stress. Here, the multimode fibers which are used guide the light from the source to the photoelastic material and photoelastic material to the detector.

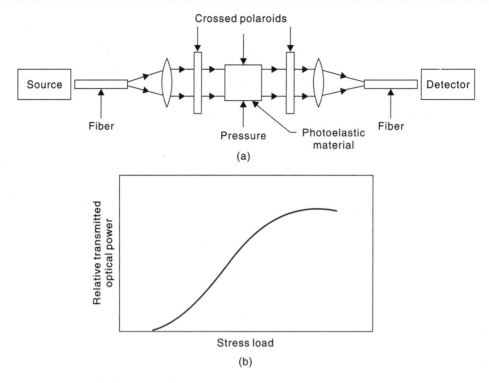

Fig. 9.97 Photoelastic pressure sensor using optical fiber

Temperature Sensor

Changes in temperature cause small variations in the refractive index. This can be used to design a temperature sensor using an optical fiber. Let us imagine a fiber in which the refractive index of the core and cladding vary with temperature in different ways (Fig. 9.98). At the temperature T_0, when the refractive index of core equals that of cladding, no light is transmitted by the fiber. Whereas for temperature $< T_0$, light transmission would occur.

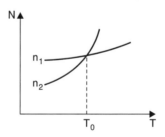

Fig. 9.98 Temperature variation of refractive index of core and the cladding

A black body tipped fiber optic radiation thermometer has also been devised for measuring the temperature. The basic design of the thermometer is shown in Fig. 9.99. It is based on the principle of estimating the temperature of a body by measuring the intensity of radiation emitted by it. It consists of a sensing probe made from a single crystal aluminium oxide (saphire) optical fiber 0.25 to 2 mm in diameter and 5–100 cm long. A cup shaped black body cavity is formed at the sensing tip coating platinum or iridium. A protective saphire film is also formed on the metal sensing film. For measuring temperature the sensing probe is kept in contact with the hot object. Saphire has very low thermal conductivity. As a result only radiation propagates along the fiber towards the detector where its intensity is measured. Assuming Stephan's law or Planck's model for the black body radiation, one can estimate the temperature of the hot body. The thermometer has to be calibrated to allow for the uncertainties in the emissivity of the hot body.

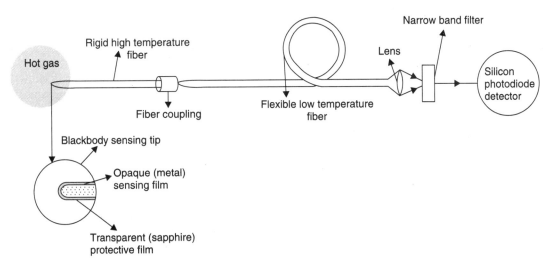

Fig. 9.99 Black body tippped fiber optic radiation thermometer

9.18.3 Phase Modulated Sensors

In these types of sensors the measurand (*e.g.*, temperature, pressure, strain, etc.) causes the phase modulation of the light wave guided within the fiber. Single mode fibers and highly monochromatic laser diodes are used as sources. Consider a monochromatic light of free space wavelength λ propagating through a single mode fiber of length L. The propagation constant β (*i.e.*, the phase change per unit length) of such a wave in the fiber is given by:

$$\beta = \frac{2\pi}{\lambda} n_1$$

where n_1 is the refractive index of the core. The total phase change after propagating through a length L of the fiber will be given by

$$\phi = \beta L = \frac{2\pi}{\lambda} n_1 L$$

∴
$$\Delta\phi = \frac{2\pi}{\lambda} (n_1 \Delta L + L \Delta n_1)$$

⇒
$$\frac{\Delta\phi}{\phi} = \frac{\Delta L}{L} + \frac{\Delta n_1}{n_1}$$

Here it is assumed that in principle the measurand can bring about change both in the length of the fiber as well as the refractive index of the core. The changes in phase can be detected by Mach Zhender interferometer.

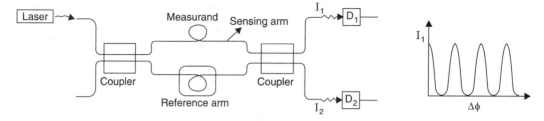

Fig. 9.100 Phase sensitive fiber optic sensor

The basic configuration of an all-fiber-Mach Zhender interferometric sensor is shown in Fig. 9.100. Light from a laser source is split equally by a coupler and sent to the sensing fiber and reference-fiber arms. Both these fibers are single mode fibers. The outputs of the two fibers are recombined at the second coupler. The sensing arm is in direct contact with the measurand, while the reference arm is shielded from external perturbation. The measurand acts on the sensing arm and changes the phase of the light wave either by changing its length, its refractive index or both. As the reference arm is not affected by the measurand, there appears a phase difference $\Delta\phi$ between the two waves arriving at the second coupler. The intensity I_1 and I_2 of light arriving at the two-output ports of the second coupler will depend on this phase difference between the two waves emerging from the sensing and the reference arms. It can be shown that

$$I_1 = (I_o/2)(1 + \cos \Delta\phi) = I_o \cos^2 (\Delta\phi/2)$$
$$I_2 = (I_o/2)(1 - \cos \Delta\phi) = I_o \sin^2 (\Delta\phi/2)$$

where $I_o = I_1 + I_2$ is the input intensity.

If the sensing and reference arms introduce identical phase shifts, $\Delta\phi$ will be zero and $I_1 = I_o$ and $I_2 = 0$. *i.e.,* the entire launched power appears at the output port 1. On the other hand, if $\Delta\phi = \pi$, $I_1 = 0$ and $I_2 = I_o$, the entire launched power appears at the port 2. For other values of $\Delta\phi$, the power will be divided into the two ports according to the above equations. In practice, the sensor is operated at the quadrature point corresponding to $\Delta\phi = \pi/2$. Thus a fixed bias of $\pi/2$ is induced and the phase change θ caused by the measurand is related to the output intensity. For this case

$$\Delta\phi = (\pi/2) + \theta$$

The above equations then become

$$I_2 = (I_o/2)[1 - \cos(\pi/2 + \theta)] = (I_o/2)[1 + \sin \theta]$$
$$= (I_o/2)[1 + \theta] \text{ for small } \theta$$

Similarly $\qquad I_1 = (I_o/2)[1 - \theta]$

Thus for small value of θ, I_1 and I_2 vary almost linearly with the phase change θ caused by the measurand.

Fiber Optic Gyroscope

Fiber optic gyroscope is used for sensing rotation, which is vital for an aircraft or missile. The basic idea is to send light traveling in opposite directions through a loop of single mode fiber. If the loop is perfectly still, the light traveling in opposite directions will travel exactly the same distance in one complete cycle. However if the ring has rotated, the two beams traveling in opposite directions travel different path lengths, which results in a difference in phase. This phase difference can be measured using interferometry and rotational speeds can be estimated. A fibre optic gyroscope is shown in Fig. 9.101. Assume that light takes a time t to travel around the loop of radius r. Let the loop turn through an angle θ. This rotation moves the starting point a distance $\Delta = r\theta$. Light going in the direction of rotation has to travel a distance $2\pi r + \Delta = 2\pi r + r\theta$, to get back to the starting point. The light traveling in the opposite direction has to travel only $2\pi r - r\theta$ or $2\pi r - r\theta$. Hence the path difference equals $2r\theta$ and the phase difference is $\phi = (2\pi/\lambda)r\theta$, where λ is the wavelength of light. Thus by measuring ϕ using interferometry θ can be estimated (θ/t) gives the rotational speed of the loop.

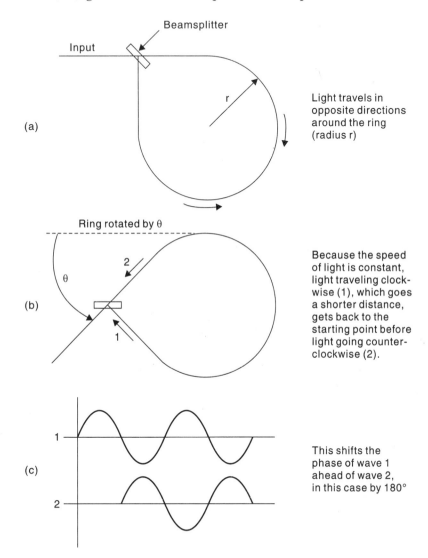

Fig. 9.101 Fiber-optic gyroscope

9.18.4 Polarization Sensitive Sensors

Current Sensor

The fiber optic current sensor is based on Faraday effect. Figure 9.102 shows a typical experimental set up for the measurement of current in the conductor using Faraday effect. A single mode polarization maintaining fiber of typically 500 m length is wound helically around the current carrying wire. The magnetic field due to the current results in the rotation of the plane of polarization of the light which is detected using a Wolfston prism. A Wolfston prism is made up of quartz or calcite. It is used to separate the horizontal and vertical components of polarization. Using this principle, currents greater than kiloamperes can be measured.

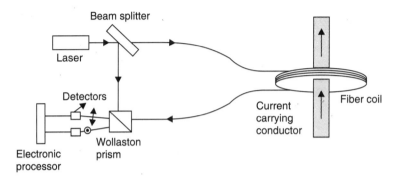

Fig. 9.102 Fiber optic current sensor

Birefringent based Fiber-optic Sensor

A birefringent optical fiber is a polarization maintaining fiber which is made up of a stress induced birefringent material. It is an intrinsic fiber optic sensor. Plane polarized light launched into such a fiber is transmitted without any loss in the state of polarization. However, on the application of a stress there is a change in the plane of polarization of light which can be measured (Fig. 9.103). In a balanced birefringent fiber optic sensor, two identical polarization maintaining fibers are used. One is treated as a reference and the other fiber is subjected to a stress. The change in the angle of plane of polarization with respect to the reference fiber is measured using a polarizer. The measured angle can be related to the applied stress.

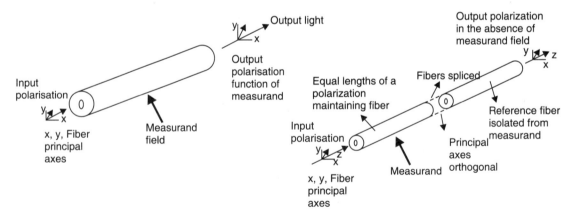

Fig. 9.103 Birefringence based fiber optic sensor for stress measurements

9.18.5 Evanescent Wave Based Refractive Index Sensors

This is based on the existence of the evanescent wave in an optical fiber. Figure 9.104a shows the typical construction of the sensor. In a single mode fiber, the cladding is removed on one side of the fiber and a side-polished transparent block is kept over it. If the refractive index of the material of the block is greater than the core refractive index, then the light in the core of the fiber escapes or leaks which can be detected. If the refractive index is less than that of the core of the fiber, the light is confined within the fiber and no light escapes. Thus the light transmission intensity shows a variation with the difference in the refractive indicies of the core and that of the block. This arrangement is ideally suited for the detection of any property which is based on the changes in refractive index. It is ideally suited for measuring the contamination of pure liquids because the presence of impurities gives rise to changes in refractive index. For measuring the refractive index of liquids a portion of the optical fiber whose cladding has been stripped off can be immersed in the liquid as shown in Fig. 9.105. If the refractive index of the liquid is greater than that of the core, the light will leak into the liquid. This will result in a decrease in the output light intensity.

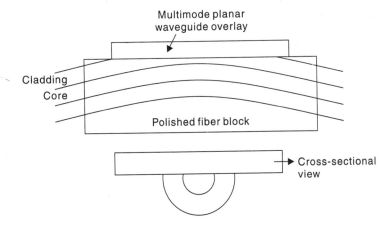

Fig. 9.104 Evanescent Wave sensor

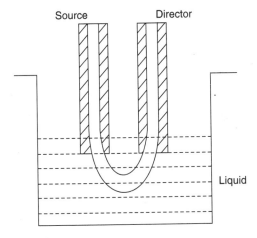

Fig. 9.105 U-tube shaped fiber optic sensor for refractive index measurements

9.18.6 Spectral Wavelength Sensitive Sensors (Fiber Bragg Grating Sensors)

The transmission characteristics of the fiber Bragg grating depends on the refractive index as well as the periodicity of the refractive index. In principle both these parameters can be affected by strain. Since both pressure and temperature induce strain in the optical fiber FBGs can be used as temperature and pressure sensors.

The Bragg wavelength of FBG is given by

$$\lambda_B = 2n_1 d$$

where d is the periodicity of the refractive index. For a longitudinal strain $\Delta\varepsilon$ applied to an FBG, the shift in $\Delta\lambda_B$ is given by:

$(\Delta\lambda_B)_{strain} = \lambda_B(1 - \rho)\,\Delta\varepsilon$ where ρ is the photoelastic constant of the material of the fiber. Photoelastic constant is measure of the change in refractive index with stress.

For a temperature change ΔT, the shift in λ_B is given by

$$(\Delta\lambda_B)_{temp} = \lambda_B(1 + \xi)\,\Delta T$$

where ξ is the thermo-optic coefficient of the material of the fiber.

Thermo-optic coefficient is a measure of the change in refractive index with temperature.

For a pressure change of ΔP, the corresponding shift in λ_B is given by

$$(\Delta\lambda_B)_{pressure} = \lambda_B \left[\frac{1}{d}\frac{\partial d}{\partial P} + \frac{1}{n}\frac{\partial n}{\partial P} \right] \Delta P$$

Typical configuration for use of fiber Bragg sensor is shown in Fig. 9.106.

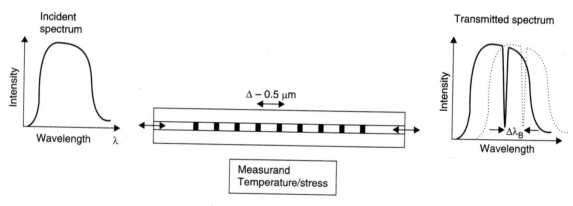

Fig. 9.106 Fiber Bragg sensor

9.18.7 Point and Distributed Sensors

Point sensors refer to the measurement carried out at a point in space. Distributed sensors refer to sensors which are used for measuring a physical parameter over a region of space. For example pressure measurements at various points in a dam or structure needs distributed sensors. For many applications temperature has to be measured at different points. This also needs distributed sensors. Optical fibers are ideally suited for distributed sensing. Two approaches have been tried out.

(a) the point sensors may be arranged in a desired network (Fig. 9.107a).

(b) a continuous length of a suitable configured optical fiber covering the region of interest is laid. The value of the measurand is obtained and displayed continuously as a function of the position of the fiber (Fig. 9.107b).

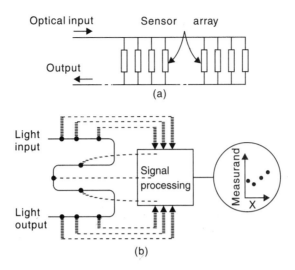

Fig. 9.107 Point and distributed fiber optic sensor

9.19 FIBER OPTICS IN MEDICINE AND INDUSTRY

The technology of fiber optics is intimately linked with its applications in medicine and industry. The initial efforts in fiber optics were concerned with the fabrication of fiberscope in endoscopy for the visualisation of internal portions of the human body. The lasers in combination with fiber optic probes not only permit the visualisation of internal portions of human body but also the selective cauterisation of tissues, if necessary.

A fiberscope is a device for forming an accurate image of the object using optical fibers. If the optical fibers are aligned properly *i.e.,* relative positions of the fibers in the input and output ends are the same, the bundle is said to be *coherent* (Fig. 9.108). If a particular fiber is illuminated at one of its ends, then there will be a bright spot at the other end of the same fiber at the same position. *Thus a coherent fiber bundle can transmit an image from one end to another.* In a fiberscope, two bundles of optical fibers are employed. One for guiding light from the source onto the object under examination. The other set of bundle gather light from the object and guide it to the detector (Fig. 9.109). The fiberscope is also useful in industry. It could be used to examine welds, nozzles and combustion chambers inside jet aircraft engines, which would be inaccessible for observations otherwise. Fiberscope not only permits the visualisation of internal parts of the human body but also the selective cauterisation of tissues using a laser beam.

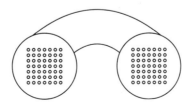

Fig. 9.108 Coherent fiber optic bundles

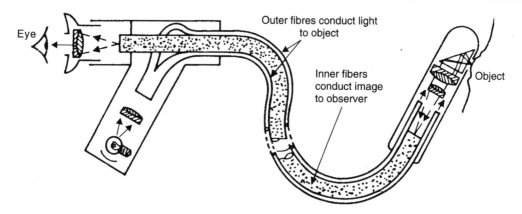

Fig. 9.109 Flexible fiberscope

REFERENCES

1. N.S. Kapany, *Fiber Optics, Principles and Applications,* Academic Press, New York. 1967.
2. A. Ghatak and K. Thyagarajan, *Introduction to Fiber Optics,* Cambridge University Press, Cambridge, 2004.
3. D.K. Mynabaev and L.L. Scheiner, *Fiber Optics Communications Technology.*
4. H. Kolimbiris, *Fiber Optics Communications,* Pearson Education Asia, New Delhi, 2004.
5. J.C. Palais, *Fiber Optics Communications,* Pearson Education Asia, New Delhi, 1998.

SOLVED EXAMPLES

1. An optical fiber had a diameter of 60 μm, a core index of 1.48 and a cladding index of 1.41. The wavelength of the light source is 0.8 μm. Determine the number of modes propagating in the fiber.
   Solution:

   We have, $\quad \text{NA} = \sqrt{n_1^2 - n_2^2}$

   Here $n_1 = 1.48$, $n_2 = 1.41$, $d = 60$ μm and $\lambda_0 = 0.8$ μm

   ∴ $\quad \text{NA} = \sqrt{1.48^2 - 1.41^2} = 0.450$

   $$V = \frac{\pi d \text{NA}}{\lambda_0} = \frac{\pi \times 60 \times 0.450}{0.8} = 106.08$$

   Number of modes $\approx \dfrac{V^2}{2} \approx \dfrac{106.8^2}{2} \approx 4.55 \times 10^3$.

2. The attenuation of light in optical fiber is 2.2 dB/km. What fraction of the initial intensity will remain after 2 km and 6 km.
   Solution: We have

   $$\alpha = \frac{10}{L} \log\left(\frac{I_t}{I_0}\right); \quad \text{Here } \alpha = 2.2 \text{ dB/km}$$

FIBER OPTICS

Case (a) $L = 2$ km

$$\therefore \quad -2.2 = \frac{10}{2} \log_{10}\left(\frac{I_t}{I_0}\right)$$

$$\Rightarrow \quad \log_{10}\left(\frac{I_t}{I_0}\right) = -\left(\frac{4.4}{10}\right) = -1.44 \quad \text{or} \quad \frac{I_t}{I_0} = 10^{-0.44} = 0.364$$

Case (b) $L = 6$ km

$$\therefore \quad -2.2 = \frac{10}{6} \log\left(\frac{I_t}{I_0}\right)$$

$$\Rightarrow \quad -2.2 = \frac{10}{6} \log\left(\frac{I_t}{I_0}\right) \quad \text{or} \quad \log\left(\frac{I_t}{I_0}\right) = -\frac{13.2}{10} = -1.32$$

or

$$\frac{I_t}{I_0} = 10^{-1.32} = 0.0479.$$

3. Calculate the numerical aperture and the angle of acceptance of a fiber having fractional refractive index of 0.05 and core refractive index of 1.48.

Solution: We have NA $= n_1 (2\Delta)^{1/2}$

Here $\quad n_1 = 1.48 \quad$ and $\quad \Delta = 0.05$

$\quad$ NA $= 1.48 (2 \times 0.05)^{1/2} = 0.468$

$\quad i_a = \sin^{-1} $ NA $= \sin^{-1} 0.468 = 27.9°$

4. Calculate the numerical aperture and angle of acceptance of an optical fiber whose core and cladding have refractive indices 1.45 and 1.40.

Solution: We have

$$\text{NA} = \sqrt{n_1^2 - n_2^2} \quad ; \text{Here } n_1 = 1.45 \text{ and } n_2 = 1.40$$

$$\therefore \quad \text{NA} = \sqrt{1.45^2 - 1.40^2} = 0.3774$$

$$i_a = \sin^{-1} \text{NA} = \sin^{-1} 0.3774 = 22°.$$

5. Find the loss specification of a fiber of length 500 m if the input power is 8.6 μW and the output power is 7.5 μW.

Solution: We have

$$\alpha = -\frac{10}{L} \log\left(\frac{I_t}{I_0}\right), \text{ Here } l = 500 \text{ m} = 0.5 \text{ km};$$

$$I_t = 7.5 \text{ μW and } I_0 = 8.6 \text{ μW}$$

$$\therefore \quad \alpha = -\frac{10}{0.5} \log\left(\frac{7.5 \times 10^{-6}}{8.6 \times 10^{-6}}\right) = -1.1 \text{ dB/km}.$$

6. The refractive index of the core is 1.5 and the fractional refractive index between the core and the cladding is 1.8%. Estimate (a) numerical aperture (b) acceptance angle (c) critical angle at the core-cladding interface (d) the velocity of light in the core and (e) the velocity of light in the cladding.

Solution: We have

$$NA = n_1(2\Delta)^{1/2}; \text{ Here } n_1 = 1.5 \text{ and } D = 1.8 \times 10^{-2}$$

$$\therefore NA = 1.5(2 \times 1.8 \times 10^{-2})^{1/2} = 0.285$$

$$i_a = \sin^{-1} NA = \sin^{-1} 0.285 = 16.6°$$

Also, $\Delta = \dfrac{n_1 - n_2}{n_1} = 0.018$, $\therefore n_1 - n_2 = 0.018\, n_1$ or $\dfrac{n_2}{n_1} = 0.982$

$$i_c = \sin^{-1}\left(\dfrac{n_2}{n_1}\right) = \sin^{-1} 0.982 = 79.1°$$

Velocity of light in the core

$$= \dfrac{c}{n_1} = \dfrac{3 \times 10^8}{1.5} = 2 \times 10^8 \text{ m/s}$$

Velocity of light in the cladding

$$= \dfrac{c}{n_2} = \left(\dfrac{c}{0.982 \times n_1}\right) = \dfrac{3 \times 10^8}{0.982 \times 1.5} = 2.04 \times 10^8 \text{ m/s}.$$

7. The optical power launched into a optical fiber is 1.5 mW. The fiber has attenuation of 0.5 dB/km. If the power output is 2 μW, calculate the fiber length.

Solution: We have

$$\alpha = -\dfrac{10}{L} \log\left(\dfrac{I_t}{I_0}\right)$$

Here $\alpha = -0.5$ dB/km; $I_t = 2 \times 10^{-6}$ W; $I_0 = 1.5 \times 10^{-3}$ W; $L = ?$

$$\therefore -0.5 = \dfrac{10}{L} \log\left(\dfrac{2 \times 10^{-6}}{1.5 \times 10^{-3}}\right) \Rightarrow L = -\dfrac{10}{0.5} \log\left(\dfrac{4 \times 10^{-3}}{3}\right) = 57.5 \text{ km}.$$

QUESTIONS

1. What is the principle behind the functioning of an optical fiber?
2. Write a short note on various methods of manufacturing optical fibers.
3. Write a short note on fiber optic cables.
4. Derive an expression for the numerical aperture of an optical fiber.
5. Describe the various types of optical fibers.
6. Discuss signal distortion in optical fibers.
7. Explain (a) modal dispersion (b) material dispersion (c) waveguide dispersion and (d) polarization dispersion.
8. Discuss various factors contributing to attenuation in optical fibers.
9. How are low attenuation and low dispersion fibers realised in practice?
10. Write a short note on fiber optic communication system.
11. What are the advantages using optical fibers in a communication system?
12. Discuss the functioning of an optical transmitter and optical receiver.
13. Write a short note on (a) LED (b) Laser diode (c) photodiode (d) p-i-n diode and (e) avalanche photodiode.

FIBER OPTICS

14. What is an optical modulator?
15. Discuss the functioning of a fiber amplifier.
16. Explain the working a fiber Bragg grating.
17. Explain how optical fibers can be used as sensors.
18. Write a short note on (a) intensity modulated fiber sensors (b) phase modulated fiber sensors.
19. Explain how optical fibers can be used to measure (a) changes in refractive index (b) changes in pressure (c) changes in temperature.
20. Explain the functioning of fiber Bragg grating sensors.

PROBLEMS

1. If an optical signal has lost 25% of the power after traveling through 250 m of the fiber, determine the loss specification of the fiber.
2. An optical fiber has a diameter of 60 μm and numerical aperture of 0.45. Calculate the number of modes for an operating wavelength of 10 μm.
3. The fractional refractive index of an optical fiber of diameter 80 μm is 0.03. The refractive index of the core is 1.48 and the operating wavelength is 0.8 μm. Calculate (a) numerical aperture (b) acceptance angle (c) critical angle for the core-cladding interface (d) velocity of light in the core and cladding and (e) number of modes in the fiber.
4. The refractive indices of the core and cladding are 1.45 and 1.48 respectively. Calculate the acceptance angle of the fiber. If the diameter of the fiber is 60 μm, find the number of modes the fiber can support at an operating wavelength of 1.5 μm. Also determine the velocity of propagation of light rays in the core and the cladding.

10
Matter and Radiation – Dual Nature

10.1 INTRODUCTION

The most important developments in physics in the twentieth century are those of quantum mechanics and theory of relativity. The theory of relativity revolutionised our concepts regarding space and time. The principles of quantum mechanics have led to advances in atomic physics, nuclear physics, particle physics, solid state physics and materials science. By the end of the nineteenth century, many physical phenomena had been understood on the basis of some fundamental laws. These laws of classical physics were those of mechanics, gravitation, electromagnetism and thermodynamics. The laws of mechanics and gravitation were formulated by Galileo and Newton. They could explain the motion of planets around the sun. The laws of electromagnetism developed by Coulomb, Ampere, Faraday and Maxwell culminated in the discovery of electromagnetic waves. The interference and diffraction of light could be understood by treating light as electromagnetic waves. The laws of thermodynamics and statistical mechanics enunciated by Clausius, Kelvin, Maxwell and Boltzmann could explain the nature of heat. It also led to the theoretical analysis of the performance of heat engines.

However, at the turn of the twentieth century, physics was in a state of turmoil. There were a plethora of experimental observations which could not be explained on the basis of classical physics. Experimental observations and theoretical studies which led naturally to the development of quantum mechanics are given in Table 10.1.

The period up to 1930 could be regarded as the first phase in the development of quantum mechanics. Subsequently, the principles of quantum physics have been applied to other branches of physics such as nuclear physics, particle physics and solid state physics. Quantum chemistry is the subject of molecular quantum mechanics. It has provided a better insight into the molecular structure and bonding. This has naturally lead to the better understanding of the material properties and the design of novel materials.

The subject of quantum field theory and quantum electrodynamics was developed during 1950-60. QED describes the way electrically charged particles interact with one another and with magnetic fields, through the exchange of virtual photons. The subject of quantum gravity deals with blending the theory of general relativity with principles of quantum mechanics. Needless to say that these subjects involve sophisticated mathematical techniques and are not always simple to comprehend. They have also raised fundamental questions about the nature of physical reality and the role of observer which appeal to people with a philosophical bent of mind.

The inadequacy of the laws of classical physics was first felt while trying to explain the observations related to black body radiation. The problem of black body radiation is of utmost importance. It was while trying to explain this, that Planck introduced the revolutionary idea of quantization of energy. It means that any physical system can have only discrete values of energy. This idea was made use of by Niels Bohr while proposing the model for the atom in which the electrons go round the nucleus in circular orbits. Sommerfeld showed that electrons could also go round the nucleus in elliptical orbits and that its energy had to be specified by additional quantum numbers. The electron orbits in space are also quantized. Apart from the quantization of energy, another crucial concept, which has played a major role in the development of quantum mechanics is the dual nature of matter and radiation. The behaviour of material particles such as electrons and neutrons can be described in terms of matter waves. Light can be thought of as a particle called photon or as an electromagnetic wave. In this chapter we shall study the various phenomena which led to these basic ideas of quantum mechanics.

Table 10.1 Important developments in modern Physics during 1900-1930

Year	Scientists	Topic
1896	Roentgen	X-rays
1896	Henri Becquerel, Marie Curie	Radioactivity
1900	Max Planck	Black body Radiation
1905	Albert Einstein	Photoelectric effect
1911	Rutherford	Model of the atom (Discovery of the atomic nucleus)
1913	Niels Bohr	Quantum model of the atom
1915	Sommerfeld	Elliptical orbits of electrons around the nucleus
1921	Stern and Gerlach	Spin of the electron and Space quantization
1923	Compton effect	Particle nature of X-rays
1924	de Broglie	Matter waves
1925	Pauli	Exclusion principle
1926	Schrodinger	Equation for matter waves
1926	Max Born	Probabilistic interpretation of wave function
1927	Heisenberg	Uncertainty Principle
1927	Davisson and Germer	Wave properties of electrons
1928	Paul Dirac	Relativistic wave equation for the electron

During the early years, quantum mechanics addressed itself to two fundamental questions that physics is concerned with viz., the nature of matter and radiation as well as the interaction between them. With advances in material synthesis and device fabrication, the importance of quantum mechanics in applied disciplines such as materials science, electronics, electrical engineering and applied physics has increased in the last couple of decades. An engineer can no longer afford to be ignorant of quantum mechanics. Devices such as semiconductor lasers, transistors, Josephson junctions, display systems and experimental techniques such as scanning microscopy, tunneling microscopy, atomic force spectroscopy used in the study of materials cannot be understood in terms of simple classical concepts. The subject of *nanoscience* and *nanoengineering* which involves the study and manipulation of materials and devices whose dimensions are ~10–100 nm makes the study of quantum mechanics an absolute necessity. At this length scale, atoms and molecules obey the rules of quantum mechanics. This chapter brings out the connection between the basic concepts of quantum mechanics and its applications. The advent of quantum mechanics does not make classical physics obsolete either. Classical physics still forms the bedrock on which the various branches of engineering are based. Quantum physics is primarily helpful in understanding the microscopic phenomena, where the usual common sense notions and intuition in dealing with macroscopic objects fail.

10.2 BLACK BODY RADIATION

10.2.1 Black body

A black body absorbs all the radiation, which is incident on it. The absorptivity, which is defined as the ratio of the energy absorbed to the incident energy, is thus equal to unity for a black body. According to Kirchhoff's laws, a black body is also a perfect emitter. The concept of a black body is an idealised one. It can be realised in practice as shown in Fig. 10.1(a). The inner wall of the enclosure is matt-black so that most of the radiation, which enters through the hole, is absorbed on reaching the wall. The sharp pointed-end just opposite to the aperture is to prevent the escape of radiation by normal reflection. The radiation, which is reflected from the walls, has very little chance of escaping through the hole as that too is absorbed in subsequent reflections within the cavity. The black body radiation is also known as cavity radiation, as the radiation is completely confined within the cavity. The black body radiation is characteristic of the temperature of the black body. It does not depend on the nature of the material of the black body. A black body radiator can be made by surrounding the enclosure with a heating coil (Fig. 10.1b). The radiation, which is emitted by any section of the wall, undergoes many reflections before it eventually emerges from the hole. Fig.10.2 shows the block diagram of the experimental arrangement used for analyzing the radiation. Since the radiation is emitted over a wide range of wavelengths it is necessary to employ different dispersing elements (prisms or gratings) and detectors to detect and measure the intensity. Thus in the visible region glass/quartz prisms and gratings are used as dispersing elements. The detector could be a photomultiplier tube. In the infrared region KBr/NaCl prisms are used as dispersing elements. The detector could be a semiconductor diode.

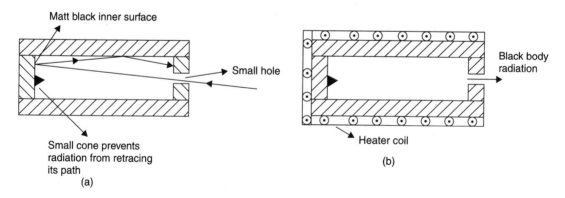

Fig. 10.1 (a) Approximate realisation of a black body (b) Black body emitter

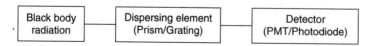

Fig. 10.2 Block diagram for the analysis of black body radiation

10.2.2 Energy Distribution in the Spectrum of the Black Body

Figure 10.3 illustrates the distribution of energy radiated by a *black body* amongst various wavelengths. E_λ is such that $E_\lambda.d\lambda$ represents the energy radiated per unit time per unit surface area of the black body in the wavelength interval λ to $\lambda + d\lambda$. It follows that the area under any particular curve is the *total* energy radiated per unit time per unit surface area at the corresponding temperature. The curves embody Wien's displacement law and Stefan's law.

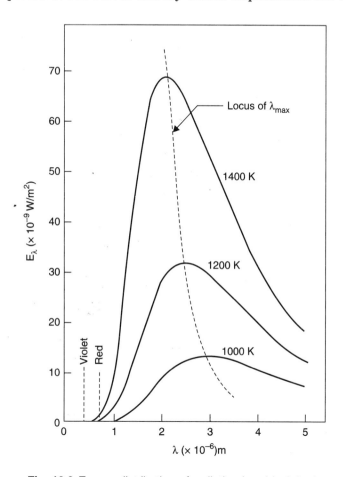

Fig. 10.3 Energy distribution of radiation in a black body

10.2.3 Wien's Displacement Law

The wavelength λ_{max} at which the maximum amount of energy is radiated decreases with the increase in temperature and is such that

$$\lambda_{max}T = \text{a constant} \qquad (10.1)$$

where T is the temperature of the *black body* in *Kelvin*.

Eqn. (10.1) is known as the Wien's displacement law. The value of the constant is found by experiment to be 2.98×10^{-3} metre-Kelvin (m-K).

The curves in Fig. 10.3 illustrate the well known observation of heating a metal like iron. As the temperature is increased one initially senses the heat. At this stage the iron is mainly emitting infrared radiation. At about 1200 K, it looks red hot because wavelengths emitted are in the red region of the spectrum. At higher temperatures the colour changes from red through yellow to white. When the iron is white hot, all the wavelengths in the visible region are emitted and eye perceives it as white. The intensity distribution of wavelengths emitted by the sun can be approximated to that of a black body at about 6000 K i.e., sun's surface temperature is about 6000 K. Some stars are much hotter and appear blue.

10.2.4 Stefan's Law

Stefan's law states that the total energy (E) radiated per unit time per unit surface area of a black body is proportional to the fourth power of the temperature of the body expressed in Kelvin. i.e.,

$$E = \sigma T^4 \tag{10.2}$$

where σ is a constant of proportionality known as Stefan's constant. The value is 5.67×10^{-8} Wm^{-2}K^{-4}. The value of E at any temperature T is equal to the area under the corresponding curve, i.e.,

$$E = \int_0^\infty E_\lambda \, d\lambda \tag{10.3}$$

If there are two black bodies at temperatures T_1 and T_2, the net rate of heat loss due to radiative transfer is

$$E_{net} = \sigma (T_1^4 - T_2^4) \text{ where } T_1 > T_2$$

In case of a non-black body the eqn. (10.2) is replaced by

$$E = \varepsilon \sigma T^4 \tag{10.4}$$

where ε is called the total emissivity of the body. It is defined as the ratio of the energy radiated by the body to the energy supplied. The value depends on the nature of the surface of the body and lies between 0 and 1. In general ε could be dependent on the wavelength. Normally ε_{av} is calculated over the range of wavelengths the energy is emitted. According to Kirchhoff's laws the emissivity and absorption coefficient are proportional. For a perfect black body the emissivity is equal to unity.

10.3 MODELS FOR THE BLACK BODY RADIATION

10.3.1 Energy Density and Emissive Power

Any model for black body radiation ought to explain the curve shown in Fig.10.3. This problem attracted the attention of brilliant minds since the model depends on the assumptions made about the nature of radiation. In the theoretical treatment of the problem of radiation, it is convenient to use *energy density* denoted by U_λ. The energy density is the energy contained in a unit volume of the *black body*. The energy density and the emissive power are related geometrically in a simple way. Let us determine the rate at which the radiant energy within the black body strikes the walls of the inner surface. Consider Fig. 10.4(a), in which dS is an element of wall surface and dV an element of volume in the black body cavity at a distance r from dS. Let U_λ denotes the radiation density which is considered to be uniform. The amount of radiation present in dV is $U_\lambda dV$. The surface area of a sphere of radius r surrounding dV is $4\pi r^2$. The projected area of dS on any portion of this sphere is $dS \cos \theta$. Hence the ratio $(dS \cos \theta)/4\pi r^2$ represents the fraction of the energy in dV that will ultimately strike dS by direct transmission. The energy from dV that impinges directly upon dS is, therefore,

MATTER AND RADIATION – DUAL NATURE

$$\Delta E = \frac{U_\lambda \, dV \, dS \cos\theta}{4\pi r^2} \tag{10.5}$$

Now assume that the volume element dV is a part of a hemispherical shell of radius r and thickness dr, centred upon dS. The radiation energy ΔE contributed to dS by this entire shell can then be obtained by considering a circular cone of radius $r \sin\theta$ for which θ is constant (Fig. 10.4b) and then integrating over the entire surface of the shell. The volume dV of this zone is $(2\pi r . \sin\theta). (dr). (r.d\theta)$ (i.e., $dV = 2\pi r^2 \sin\theta . d\theta . dr$). The integration is from $\theta = 0$ to $\theta = \pi/2$. On substituting the value of dV in eqn. (10.5), the expression for ΔE can be written as:

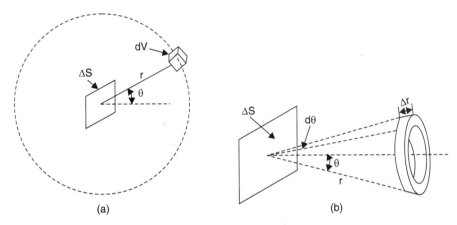

Fig. 10.4 Volume dV and surface ΔS elements used in deriving the relation between the intensity and the energy density of the black body radiation

$$\Delta E = \frac{U_\lambda \, dS. \, dr}{2} \int_0^{\pi/2} \sin\theta . \cos\theta . d\theta = \frac{U_\lambda . dS. \, dr}{4} \int_0^{\pi/2} \sin 2\theta . d\theta = \frac{U_\lambda . dS. \, dr}{4} \tag{10.6}$$

Since this energy arrives during the time interval $\Delta t = dr/c$, the rate at which the radiant energy falls on dS from all the directions is:

$$\frac{\Delta E}{\Delta t} = \frac{U_\lambda c. \, dS}{4} \tag{10.7}$$

Since the emissive power is defined as the radiant energy incident per unit area

$$E_\lambda = \frac{U_\lambda c}{4} \tag{10.8}$$

Hence energy density and emissive power can be used interchangeably so long as absolute values are not required.

10.3.2 Thermodynamics of Black Body Radiation—Wien's Law

According to Maxwell's electromagnetic theory, radiation in an enclosure exerts a pressure on its walls. The pressure is proportional to the energy density of the radiation. This is similar to a gas whose pressure is proportional to molar density i.e., $P = (n/V)RT$. Suppose radiation of a single wavelength λ_1 at a temperature T_1 is subjected to an adiabatic expansion, then Wien showed that

$$\frac{T_1}{T_2} = \frac{\lambda_2}{\lambda_1}$$

where T_2 is the final temperature and λ_2 is the corresponding wavelength of the radiation. The above equation can be written in the general form:

$$\lambda T = \text{constant} \qquad (10.9)$$

Wien also obtained another important result, namely that the monochromatic energy density and temperature before and after adiabatic expansion are related by the equation

$$\frac{U_{\lambda_1}}{U_{\lambda_2}} = \frac{T_1^5}{T_2^5}$$

The above equation may also be written in a general form:

$$\frac{U_\lambda}{T^5} = \text{constant} \qquad (10.10)$$

From eqns. (10.9) and (10.10) we have:

$$U_\lambda \lambda^5 = \text{constant} \qquad (10.11)$$

In order that eqns. (10.9) to (10.11) hold good simultaneously, $U_\lambda \lambda^5$ or U_λ / T^5 must be a function of the product λT and not functions dependent on λ and T separately. i.e.,

$$U_\lambda \lambda^5 = C\, f(\lambda T) \quad \text{or} \quad U_\lambda = \frac{C}{\lambda^5} f(\lambda T)$$

where C is a constant and the function $f(\lambda T)$ is, as yet, undetermined.

The first attempt to determine the function $f(\lambda T)$ was by Wien (1896). He assumed that the walls of the black body cavity consisted of oscillators of molecular size which emitted radiation. *Since the structure of the atom was unknown at that time, the nature of these oscillators was not well defined.* The frequency of radiation was assumed to be proportional to the kinetic energy of the oscillator. The intensity in any particular wavelength range is proportional to the number of oscillators with the requisite energy. *Wien's derivation was based on the conjecture that there is an analogy between the emissivity curves and the Maxwell's speed distribution curves for the molecules of an ideal gas** (Fig. 10.5). With these assumptions, Wien derived the following expression for U_λ:

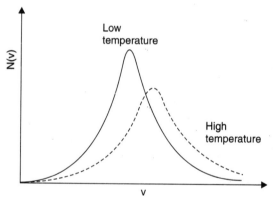

Fig. 10.5 Resemblance between the Maxwellian distribution of velocities in a gas and energy distribution inside a black body which prompted Wien to adopt the thermodynamic approach to solve the problem of black body radiation

$$U_\lambda = \frac{C_1}{\lambda^5} e^{(-C_2/\lambda k_B T)} \qquad (10.12)$$

where C_1 and C_2 are constants, k_B is the Boltzmann constant. This distribution has the required general forms from 0.5 μm to about 2 μm, but at longer wavelengths it predicts the values of the energy density which are too small (Fig. 10.6). The distribution formula also fails at high temperatures. According to eqn. (10.12), at any particular wavelength, U_λ should increase exponentially with temperature. This is not borne out by experiments. The experimental function is very satisfactory at short wavelengths.

---

*$N(v)\, dv$ gives the number of molecules having speeds between v and $v + dv$. $N(v)$ is given by:

$$N(v) = 4\pi \left(\frac{M}{2\pi RT}\right)^{3/2} v^2 e^{-Mv^2/2RT}$$

T is the gas temperature, M is the molar mass and R is the gas constant.

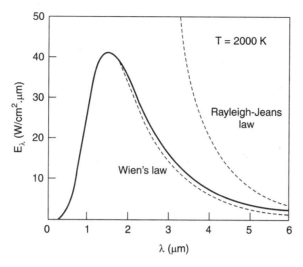

Fig. 10.6 The solid curve shows the experimental spectral radiance for radiation from a black body cavity at 2000 K. The predictions of the classical Rayleigh-Jeans law and Wein's law are shown as dashed lines

10.3.3 Classical Theory of Black Body Radiation: Rayleigh-Jeans Model

Another attempt to obtain a distribution law resulted in the Rayleigh-Jeans formula. This was derived on the basis of classical electrodynamics and Maxwell-Boltzmann statistics. Consider a hollow cavity with perfectly reflecting walls and for convenience assume that it is a cube. The material of the walls of the cavity can be thought to be made up of atomic oscillators which radiate electromagnetic waves. These waves constitute the thermal radiation. In such a cavity there can only be *standing electromagnetic waves.* If the average energy carried by each wave is known and if the number of standing waves having wavelengths between λ and $(\lambda + d\lambda)$ per unit volume is known, then the energy density of the radiation can be calculated. The number of standing waves per unit volume, having the wavelengths between λ and $(\lambda + d\lambda)$ can be shown to be $(8\pi/\lambda^4)\, d\lambda$, as described below.

Number of Modes in the Black Body Cacity

When a wave is confined within a cavity, standing waves are formed. This is on account of the reflection of waves at the inner walls of the cavity. For standing waves, only certain values of the wavelength are allowed. In the case of standing waves on a stretched string of length L fixed at both ends, the allowed values of wavelength is given by

$$\lambda = \frac{2L}{n} \quad \text{or} \quad k = \frac{2\pi}{\lambda} = \frac{2\pi n}{2L} = \frac{n\pi}{L} \tag{10.13}$$

where $n = 1, 2, 3 \ldots\ldots$ The standing wave with one loop corresponds to $n = 1$, the wave with two loops corresponds to $n = 2$ and so on. (Fig. 10.7).

The black body is a three-dimensional cavity and hence the number of modes in all the three directions has to be determined. Consider a black body in the shape of a cube of length L. Let $\mathbf{K}$ be the wave vector of the electromagnetic wave, then

$$\mathbf{K} = k_x \mathbf{i} + k_y \mathbf{j} + k_z \mathbf{k}$$

Here $\mathbf{i}, \mathbf{j}$ and $\mathbf{k}$ are unit vectors along the edges of the cube. k_x, k_y and k_z are the components along x, y and z directions. They have to satisfy the boundary condition analogous to eqn. (10.13) i.e.,

$$k_x = \frac{\pi n_x}{L}; \quad k_y = \frac{\pi n_y}{L}; \quad k_z = \frac{\pi n_z}{L} \quad (10.14)$$

$\mathbf{K}$ is a vector and hence

$$K^2 = k_x^2 + k_y^2 + k_z^2 \quad (10.15)$$

Let $\quad K = \dfrac{n\pi}{L} \quad$ or $\quad n = \dfrac{2L}{\lambda}$ as in eqn. (10.13) $\quad (10.16)$

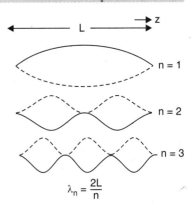

Fig. 10.7 Allowed wave patterns in a standing wave in one dimension

From equations (10.15) and (10.16) we have

$$\frac{\pi^2 n^2}{L^2} = \frac{\pi^2 n_x^2}{L} + \frac{\pi^2 n_y^2}{L} + \frac{\pi^2 n_z^2}{L} \quad \text{or} \quad n^2 = n_x^2 + n_y^2 + n_z^2 \quad (10.17)$$

Eqn. (10.17) can be used to calculate the number of modes in the black body cavity. Consider the n-space with coordinate axes representing n_x, n_y and n_z (Fig. 10.8). However, unlike the physical space, n_x, n_y and n_z can take only integer values. Eqn. (10.17) represents a sphere of radius n. Any point in this representation represents a mode. Also since one lattice point is shared by eight cubes, each point can be associated with a unit volume. Since n_x, n_y and n_z are all positive integers, the total number of modes N is given by:

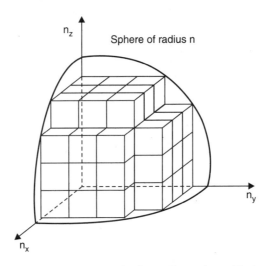

Fig. 10.8 The allowed standing waves in three dimensions. Each standing wave is represented by three integral numbers n_x, n_y and n_z

MATTER AND RADIATION – DUAL NATURE

$$N = \frac{\text{Volume of sphere of radius "}n\text{"}}{8 \times \text{unit cell volume}} = \frac{1}{8}\left(\frac{\frac{4\pi n^3}{3}}{1}\right)$$

$$= \frac{\pi n^3}{6} = \frac{1}{6}\pi\left(\frac{2L}{\lambda}\right)^3 = \frac{4\pi L^3}{3\lambda^3} = \frac{4\pi V}{3\lambda^3} \qquad (10.18)$$

Thus the number of modes N is a function of λ and volume $V (= L^3)$ of the cavity. For finding the number of modes lying between λ and $\lambda + d\lambda$, one has to differentiate eqn. (10.18) i.e.,

$$\frac{dN}{d\lambda} = \frac{4\pi V}{3}\left(\frac{-3}{\lambda^4}\right) = -\frac{4\pi V}{\lambda^4} \qquad (10.19)$$

Since the electromagnetic waves are transverse in nature, both horizontal and vertical polarization of the wave is possible. Hence the eqn. (10.19) has to be multiplied by a factor of 2. The *number of modes (dN)* per *unit volume* and having wavelengths lying between λ and $\lambda + d\lambda$ is given by:

$$dN = \frac{8\pi}{\lambda^4} d\lambda \qquad (10.20)$$

Average Energy of the Oscillator

A linear oscillator has two degrees of freedom, one corresponding to the kinetic energy and the other corresponding to the potential energy. From equipartition theorem it follows that the average energy of the oscillator is $(2 \times 1/2 k_B T)$ which is equal to $k_B T$.

The same result can be obtained through the following argument. Each oscillator has an energy ε which may take on any value between 0 and ∞. The average energy ε_{av} of the energy of the oscillator can be obtained from the classical mechanics. According to classical mechanics, when the oscillators are in thermal equilibrium with the radiation, the value of ε for any energy of the oscillator occurs with the relative probability of $\exp(-\varepsilon/k_B T)$, where k_B is the Boltzmann constant. The average energy ε_{av} is obtained by averaging overall values of ε, with this weight factor. i.e.,

$$\text{Average Energy} = \frac{\text{Energy of each oscillator} \times \text{Probability of occurrence}}{\text{Total number of oscillators}}$$

For convenience put $\beta = 1/k_B T$, then

$$\varepsilon_{av} = \frac{\int_0^\infty \varepsilon e^{-\beta\varepsilon} d\varepsilon}{\int_0^\infty e^{-\beta\varepsilon} d\varepsilon} = -\frac{d}{d\beta} \ln \int_0^\infty e^{-\beta\varepsilon} \cdot d\varepsilon = -\frac{d}{d\beta} \ln \left\{\frac{e^{-\beta\varepsilon}}{-\beta}\right\}_0^\infty$$

$$= -\frac{d}{d\beta} \ln \left(\frac{1}{\beta}\right) = \frac{d}{d\beta} \ln \beta = \frac{1}{\beta} = k_B T \qquad (10.21)$$

Calculation of the Energy Density

The energy density U_λ = (number of modes lying between λ and $\lambda + d\lambda$) × (Average energy of the oscillator)

$$U_\lambda d\lambda = \frac{8\pi k_B T}{\lambda^4} d\lambda \qquad (10.22)$$

The Rayleigh-Jeans formula agrees well with the experimental curves only for wavelengths much greater than 50 μm far beyond the scale shown in the Fig. 10.6.

According to eqn. (10.22) the energy radiated by a black body in a given range of wavelength increases rapidly as λ decreases and approaches infinity as the wavelength becomes very small. This is known as *ultraviolet catastrophe*. The experimentally observed curve is in complete disagreement with this conclusion. For very small wavelengths the energy density actually becomes vanishingly small. Further the total energy density would be

$$\int_0^\infty U_\lambda d\lambda = \int_0^\infty \frac{8\pi k T}{\lambda^4} d\lambda$$

and the integral is infinite for any value of T other than $T = 0$. *This means that the total energy radiated per unit area per unit time is infinite at all finite temperatures.* This conclusion is false, because the total energy radiated at any temperature is actually finite. Thus Rayleigh-Jeans formula also fails to account for the observed dependence of energy density of radiation on temperature.

10.3.4 Quantum Theory of Black Body Radiation: Planck's Model

The problem of the spectral distribution of thermal radiation was successfully solved by Planck in 1901 by means of a revolutionary hypothesis. Planck postulated that an atomic oscillator does not have an energy that can take on any value from zero to infinity, but restricted to have energy values equal to 0 or ε or 2ε or 3ε.......or $n\varepsilon$ where ε is a finite amount or quantum of energy and n is an integer. *This was a radical idea since according to classical physics the energy of an oscillator can vary only continuously and can take any value from zero to infinity.* The average energy of an oscillator is obtained in an analogous way to that used in deriving the Rayleigh-Jeans law except that *sums* are used instead of *integrals*. The average energy of the oscillator is now given by:

$$\varepsilon_{av} = \frac{\sum_{n=0}^{n=\infty} n\varepsilon e^{-\beta n\varepsilon}}{\sum_{n=0}^{n=\infty} e^{-\beta n\varepsilon}} = -\frac{d}{d\beta} \ln \sum_{n=0}^{n=\infty} e^{-\beta n\varepsilon} \qquad (10.23)$$

The expression within the summation sign is a geometric series with a common ratio of $e^{-\beta\varepsilon}$. Its first term is 1.

$$\therefore \quad \sum_{n=0}^{\infty} e^{-\beta n\varepsilon} = \frac{1 \times \left[1 - \left(e^{-\beta\varepsilon}\right)^n\right]}{\left(1 - e^{-\beta\varepsilon}\right)} = \frac{1 - e^{-\beta n\varepsilon}}{1 - e^{-\beta\varepsilon}} = \left(1 - e^{-\beta\varepsilon}\right)^{-1} \quad \text{(as } n \to \infty\text{)} \qquad (10.24)$$

From eqns. (10.23) and (10.24)

$$\therefore \quad \varepsilon_{av} = -\frac{d}{d\beta} \ln\left(1 - e^{-\beta\varepsilon}\right)^{-1} = \frac{d}{d\beta} \ln\left(1 - e^{-\beta\varepsilon}\right) = \frac{\varepsilon e^{-\beta\varepsilon}}{\left(1 - e^{-\beta\varepsilon}\right)} = \frac{\varepsilon}{\left(e^{\beta\varepsilon} - 1\right)} = \frac{\varepsilon}{e^{\varepsilon/k_B T} - 1}$$

$$(10.25)$$

The energy density is the given by:

$$U_\lambda d\lambda = \frac{8\pi}{\lambda^4} \frac{\varepsilon}{e^{\varepsilon/k_B T} - 1} d\lambda \qquad (10.26)$$

MATTER AND RADIATION – DUAL NATURE

Defining, $\varepsilon = h\nu = hc/\lambda$, the above expression can be cast in the form:

$$U_\lambda d\lambda = \frac{8\pi hc}{\lambda^5} \frac{1}{\left(e^{hc/\lambda k_B T} - 1\right)} d\lambda \qquad (10.27)$$

This is Planck's law for the energy density of the black body radiation and it fitted the experimental results very well (Fig. 10.9). It is interesting to compare Planck's formula with those of Wien and Rayleigh-Jeans.

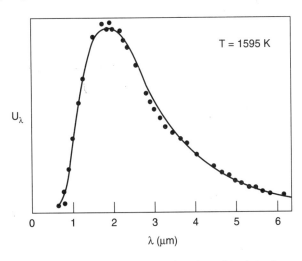

Fig. 10.9 Planck's radiation law fitted to experimental data for a black body cavity radiation at 1595 K

At long wavelengths

$$\frac{hc}{\lambda k_B T} \ll 1 \Rightarrow e^{hc/\lambda k_B T} \approx 1 + \frac{hc}{\lambda k_B T}$$

Substituting the above expression in eqn. (10.27), we get

$$\therefore \quad U_\lambda d\lambda = \frac{8\pi hc}{\lambda^5} \cdot \frac{1}{\left(\frac{hc}{hkT}\right)} \cdot d\lambda = \frac{8\pi}{\lambda^4} \cdot kT \, d\lambda$$

which is the same as Rayleigh-Jeans formula.

At short wavelengths

$$\frac{hc}{\lambda k_B T} \gg 1 \Rightarrow \left(e^{hc/\lambda k_B T} - 1\right) \approx e^{hc/\lambda k_B T}$$

Substituting the above expression in eqn. (10.27), we get

$$\therefore \quad U_\lambda d\lambda = \frac{8\pi hc}{\lambda^5} \cdot \frac{1}{e^{hc/\lambda k_B T}} \cdot d\lambda = \frac{8\pi hc}{\lambda^5} \cdot e^{-hc/\lambda k_B T} \cdot d\lambda = \frac{C_1}{\lambda^5} e^{-C_2/\lambda T} \cdot d\lambda$$

where $\quad C_1 = 8\pi hc, \, C_2 = \dfrac{hc}{k_B}$

which is the same as Wien's law.

Thus Wien's and Rayleigh-Jeans fomulae can be thought of as limiting cases of Planck's formula.

Stefan's Law

It can also be shown that Planck's radiation law leads to Stefan's law. An expression for Stefan's constant can be obtained as follows:

From eqn. (10.8) we have $E_\lambda = \dfrac{c}{4} \cdot U_\lambda$

The total energy radiated by the black body at a temperature T is given by

$$E = \int_0^\infty E_\lambda \cdot d\lambda = \int_0^\infty \dfrac{c}{4} \cdot U_\lambda \cdot d\lambda = \int_0^\infty \dfrac{c}{4} \cdot \dfrac{8\pi hc}{\lambda^5} \cdot \dfrac{1}{e^{hc/\lambda k_B T} - 1} \cdot d\lambda$$

Put $\dfrac{hc}{\lambda k_B T} = x$, so that $\dfrac{hc}{x k_B T} = \lambda$ and $d\lambda = -\dfrac{hc}{k_B T} \cdot \dfrac{dx}{x^2}$

$$\therefore E = \int_\infty^0 \dfrac{c}{4} \cdot \dfrac{8\pi hc}{\left(\dfrac{hc}{x k_B T}\right)^5} \cdot \dfrac{1}{e^x - 1} \cdot \left(-\dfrac{hc}{k_B T} \cdot \dfrac{dx}{x^2}\right)$$

$$= \int_0^\infty \dfrac{c}{4} \cdot \dfrac{x^5 (k_B T)^5 \cdot 8\pi hc}{(hc)^5} \left(\dfrac{hc}{k_B T}\right) \cdot \dfrac{1}{e^x - 1} \cdot \dfrac{dx}{x^2} = \dfrac{2\pi k_B^4 T^4}{h^3 c^2} \int_0^\infty \dfrac{x^3 \, dx}{e^x - 1}$$

Using the value of the definite integral

$$\int_0^\infty \dfrac{x^3 \, dx}{e^x - 1} = \dfrac{\pi^4}{15}$$

We get $\qquad E = \dfrac{2\pi^5 k_B^4}{15 c^2 h^3} T^4$

This is the Stefan's familiar T^4 law where Stefan's constant σ is given by:

$$\sigma = \dfrac{2\pi^5 k_B^4}{15 c^2 h^3} \qquad\qquad (10.28)$$

Wien's Displacement Law

Planck's formula leads to Wien's displacement law as well.

We know that $\qquad E_\lambda = U_\lambda \cdot \dfrac{c}{4} = \dfrac{8\pi hc}{\lambda^5} \dfrac{1}{\left(e^{hc/\lambda k_B T} - 1\right)} \dfrac{c}{4} = \dfrac{2\pi hc^2}{\lambda^5} \dfrac{1}{\left(e^{hc/\lambda k_B T} - 1\right)}$

Put $\dfrac{hc}{\lambda k_B T} = x$ so that $E_\lambda = \dfrac{2\pi k_B^5 T^5}{h^4 c^3} \dfrac{x^5}{(e^x - 1)} = \dfrac{2\pi k_B^5 T^5}{h^4 c^3} g(x) \qquad (10.29)$

Clearly the function $g(x)$ describes the universal shape of the wavelength spectrum for a black body at any temperature.

According to Wien's law $\lambda_{max} T = $ constant. Hence to arrive at this value we have to calculate $dE/d\lambda$ then equate it to zero.

Further, $\dfrac{dE}{d\lambda} = 0$ when $\dfrac{dg}{dx} = 0$

Now
$$\frac{dg}{dx} = \frac{x^4}{e^x - 1}\left(5 - \frac{xe^x}{e^x - 1}\right) = \frac{x^4}{e^x - 1}\left[5 - \frac{x}{1 - e^{-x}}\right] = 0 \Rightarrow 1 - e^{-x} = \frac{x}{5}$$

This is a transcendental equation, and can be solved by plotting the graph of $y = x/5$ as well $y = 1 - e^{-x}$ and locating the point of intersection. It turns out that $dg/dx = 0$, when $x_m = 4.965$. The parameters x_{max} and λ_{max} are related by the equation,

$$x_{max} = \frac{hc}{\lambda_{max} k_B T} = 4.965 \quad \therefore \quad \lambda_m T = \frac{hc}{k_B \times 4.965} = 2.897 \times 10^{-3} \text{ K-m}$$

$$\Rightarrow \quad \frac{h}{k_B} = \frac{2.897 \times 10^{-3} \times 4.965}{c} = \frac{14.384 \times 10^{-3}}{c} \tag{10.30}$$

This is an excellent agreement with the experimental observations.

Determination of Planck's Constant and Boltzmann Constant:

From eqns. (10.28) and (10.30), the Stefan's constant is given by:

$$\sigma = \frac{2}{15} \frac{\pi^5}{c^2}\left(\frac{k_B}{h}\right)^3 k_B = \frac{2}{15} \frac{\pi^5}{c^2}\left(\frac{c}{14.384 \times 10^{-3}}\right)^3 k_B$$

σ was known from the experiment to be 7.061×10^{-15} ergs/cm^3 K^4. With known values of σ and c, Planck obtained $k_B = 1.346 \times 10^{-23}$ J/K and $h = 6.55 \times 10^{-34}$ J.sec. This was the first calculation of Planck's constant. It was also the best calculation of Boltzmann constant. More recent and more precise experimental value of h and k_B by other methods give $h = 6.626 \times 10^{-34}$ J.sec and $k_B = 1.381 \times 10^{-23}$ J/K.

It is necessary to realise the revolutionary nature of Planck's theory. The importance of Planck's idea in the development of quantum mechanics is immense. According to Planck, the energy of an oscillator can vary only by discrete amounts. If this is so, then the emission and absorption of radiation must be discontinuous processes. Emission can take place when an oscillator makes a transition from a high energy state (E_2) to a lower energy state (E_1) by an amount, which is an integral multiple of 'hv'. i.e., $nh\nu = E_2 - E_1$. We are thus lead to the idea that an oscillator can have only discrete energy values or levels. Energies intermediate between these allowed values never occur. This discreteness of emission and absorption of radiation is in direct contrast to classical theory of radiation. According to classical theory, radiation is absorbed or emitted in a continuous manner when an electric charge distribution is accelerated or decelerated. The idea of discrete energy levels was utilized by Bohr while proposing his model for an atom.

10.4 PYROMETRY

10.4.1 Basic Principles

From an engineering point of view, the study of black body radiation is important because it forms the basis for *pyrometry*. *Pyrometry* refers to the *non-contact* method of measurement of high temperature, usually above 500°C and is routinely used in industry. Generally, in thermometry, the body whose temperature has to be measured is in physical contact with the thermometer. In *pyrometry*, the intensity of radiation emitted by the object is the basis of determining its temperature. Hence the body whose temperature has to be measured need not be in physical contact with the thermometer. A body at a temperature T emits radiation over a wide range of wavelengths. The intensity of the radiation is given by the Planck's formula. *Pyrometry* is based on sampling the energies in certain bandwidths (or ranges of wavelengths) of this spectrum. The Planck's formula can be used to estimate the temperature. Fig. 10.10a shows the energy emitted by the black body at various temperatures and Fig. 10.10b shows the wavelength range over which 90% of the total power is found for various temperatures.

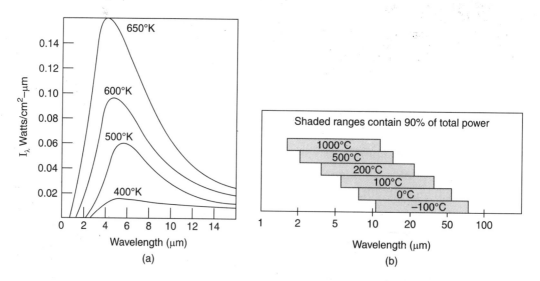

Fig. 10.10 (a) Emissive power of black body at various temperatures and (b) the wavelength region at different temperatures where 90% of the radiation power is emitted

Pyrometers make use of photodetectors for measuring the intensity of radiation. Photodetectors are of two general types, viz., *photon detectors* and *thermal detectors*. In *photon detectors the temperature of the sensing element does not change much on exposure to the incident radiation*. Typical photon detectors are the photodiodes. They are basically p-n junctions operated in the reverse bias. Light that falls on the depletion region of the junction generate electron-hole pairs. This is referred to as photogeneration. The photocurrent is proportional to the incident light intensity. Typical characteristic of the photodiode is shown Fig. 10.11. Photodiodes can be used in the *photoconductive* or *photovoltaic* mode. In the photovoltaic mode, used for solar cells, external bias is not applied. The photocurrent passes through an external load to generate power. In photoconductor mode, diode is subject to reverse bias. To improve the efficiency of photodiode a *p-i-n* diode or an avalanche diode is used (Fig. 10.12). A PIN diode has an insulating layer sandwiched between the *p* and the *n* regions. This results in efficient absorption of photons and results in more number of electron-hole pairs. In avalanche diode a *p*-layer and an insulating layer is sandwiched between highly *p*-doped and the *n*-junction (Fig.10.13). This results in steep barrier field inside the *p-n* junction. Consequently electron hole pairs are accelerated and produce more number of electron-hole pairs due to impact ionization.

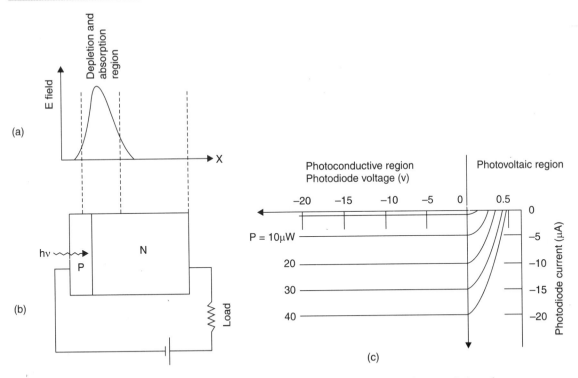

Fig. 10.11 (*a*) *p-n* photodiode showing the depletion (*b*) *I-V* characteristics of the photodiode indicating the photoconductive and photovoltaic region

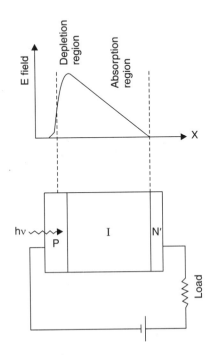

Fig. 10.12 *p-i-n* photodiode showing the combined absorption and depletion region

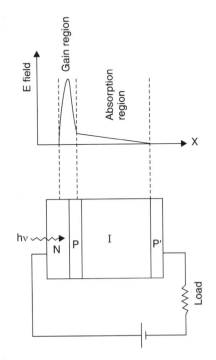

Fig. 10.13 Avalanche photodiode showing the high electric field (gain) region and carrier pair multiplication in the gain region

Thermal detectors are based on the temperature rise produced in a sensing element. When the energy radiated from a body (whose temperature has to be measured) is focused onto the sensing element, the sensing element gets heated. The sensing element could be a thermocouple, a thermopile, a thermistor or a pyroelectric material. A thermocouple is a thermoelectric device that converts thermal energy into electrical energy. Essentially it consists of two junctions of metals A and B, which are at different temperature (Fig. 10.14(a)). The *emf* developed depends on the difference in temperatures and also depends on the metals used. Usually one of the junctions is kept at 0°C and acts as the reference. The other junction is kept in physical contact with the body whose temperature has to be measured. In order to obtain a higher output, two or more thermocouples may be connected in series and this is referred to as *thermopile* (Fig. 10.14b). A typical thermopile is shown in Fig. 10.14(c). It consists of many pairs of junctions surrounded by an annular ring of mica which serves as both electrical and thermal insulation. The radiation is focused at the center and the resulting *emf* is measured. The rise in temperature can be estimated by measuring the *emf*. It is possible to estimate the intensity of radiation falling at the center by knowing the thermal capacity of the system. *Thermistors* are materials whose resistance varies with temperature. They are usually semiconductors or metals. The temperature variation of resistance of a typical metal and a semiconductor is shown in Fig.10.15a. The thermistor is kept in physical contact with the body whose temperature has to be determined. The thermistor resistance is measured using Wheatstone's bridge (Fig.10.15b). For measuring the radiation intensity, the radiation is focused onto the thermistor and resultant change in resistance is measured. The intensity of radiation falling on the thermocouple can be determined by knowing the thermal capacity of the system and rise in temperature.

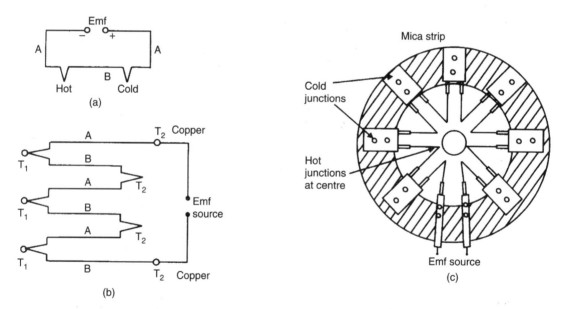

Fig. 10.14 (*a*) Simple thermocouple (*b*) Thermocouple in series
(*c*) A typical configuration employed in practice

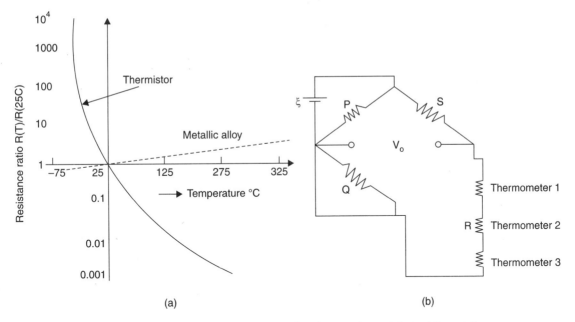

Fig. 10.15 (a) Temperature variation of resistance in metals and thermistor
(b) Wheatstone bridge arrangement for measuring resistance

In *infrared pyrometry* pyroelectric materials are used as sensing elements. Pyroelectric materials have spontaneous polarization that decreases with an increase of temperature (Fig. 10.16(a)). Pyroelectric elements thus act like thermal detectors that respond to changes in temperature by developing surface charges. These are similar to piezoelectric materials, which develop charges due to mechanical strain. In Fig 10.16(b) is shown a pyroelectric crystal with a thin metal film electrode. One of the electrodes faces the incident radiation. Pyroelectric materials are dielectrics and the electrode/crystal sandwich is a capacitor. The capacitance of

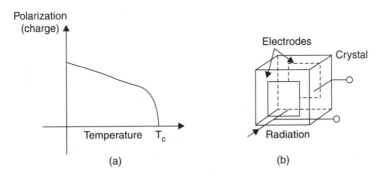

Fig. 10.16 (a) Variation of spontaneous polarization versus temperature in a pyroelectric material
(b) Electroded surfaces of the pyroelectric material on which the radiation is incident

this capacitor varies with the change in temperature. Thus this capacitor can be viewed as a radiation sensitive charge generator. Like piezoelectric charges, the pyroelectric charges also get neutralised by the surroundings. Hence a mechanical chopper is used for periodically chopping the radiation, that is incident on the pyroelectric material. This produces a steady pyroelectric current, which can be amplified and measured. A typical pyroelectric detector is shown in Fig. 10.17(a). The electrical equivalent circuit of the pyroelectric detector is shown in Fig. 10.17(b). Pyroelectric materials normally used are triglycine sulphate, lithium tantalate and organic polymers like polyvinyl fluoride and polyvinylidene fluoride films. Pyroelectric vidicon is routinely used for thermal imaging (Fig. 10.28).

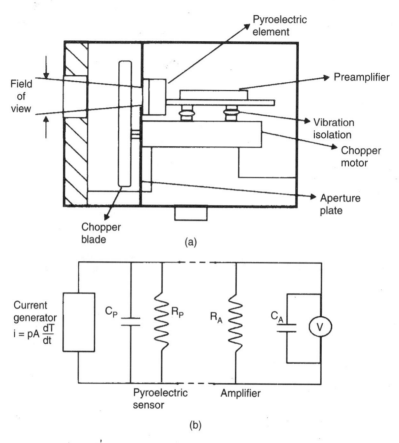

Fig. 10.17 (a) Typical pyroelectric detector and (b) Its electrical equivalent circuit.

A number of advantages accrue when the radiation coming from the target to the detector is interrupted periodically or chopped at a fixed frequency. When high sensitivity is needed, amplification is a must and high gain ac amplifiers are easier to construct than the dc counterparts. This is the main reason for using choppers. Additional benefits related to ambient-temperature compensation and reference-source comparison also may be obtained. Figure 10.18a shows radiation from a target focused onto the detector using a mirror. The radiation is chopped before it reaches the detector. The output voltage of the detector circuit is shown in Fig.10.18b. The difference between the target and chopper radiation levels is amplified (Fig.10.18c). If the chopper radiation level is considered as a known reference value, the target radiation and thus its temperature may be inferred. To provide a dc output signal related to target temperature and suitable for recording or control purposes, the ac amplifier is followed by a phase sensitive demodulator and filter circuit. The necessary synchronizing signal for the demodulator may be generated by placing magnetic proximity pickup near the chopper blade.

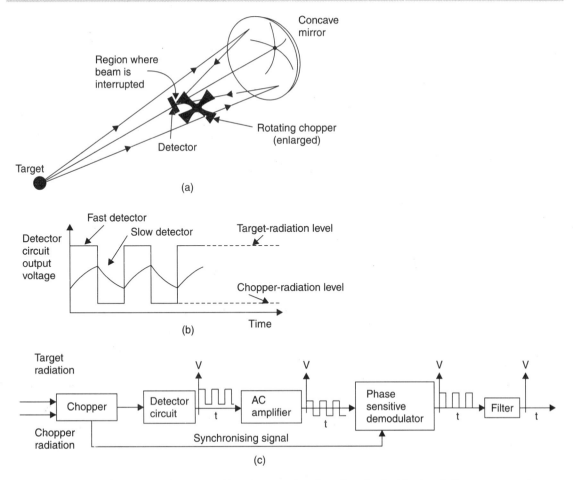

Fig. 10.18 Phase sensitive detector by chopping the incident radiation

10.4.2 Classification of Pyrometers

Pyrometers may be classified by the set of wavelengths used for making measurement. *A total radiation pyrometer* absorbs energy at *all wavelengths (such as infrared or visible wavelengths)*. It is also referred to as the *broad band pyrometer*. *An optical pyrometer* measures energy at *one specific wavelength*. A variant of this approach, the *two-colour pyrometer* compares the energy at *two specific wavelengths*. The most commonly used type is the *infrared pyrometer*, which determines the temperature from measurements over a range of infrared wavelengths. The *infrared pyrometer* has altered the traditional perception of *pyrometry* as a strictly high temperature technique because bodies having temperature close to the room temperature radiate most of their energy in the infrared wavelengths. However, the distinguishing feature of *pyrometry* is that it is a *non-contact* method. The methods of fiber optics also make it possible for the radiation to be guided along the fibers. Thus the intervening medium between the object whose temperature has to be measured and the sensing element need not be vacuum.

10.4.3 Total Radiation Pyrometry

Fig. 10.19 shows in a simplified form, the method of operation of a *total radiation pyrometer*. The radiation from the surface, whose temperature has to be measured, is focused onto the sensing element. A mirror system can also be used for focusing. The sensing element could be a thermal detector or a photon detector. On being exposed to the radiation, a balance is quickly established between the energy absorbed by the sensing element and that dissipated by conduction through the leads as well as emission to surroundings. The equilibrium temperature attained by the sensor is a measure of the temperature of the hot body. Depending on the nature of the sensing element, its change in temperature can be measured, either by measuring the *thermo-emf* if it is a thermocouple or photocurrent if it is a photodiode. Particular attention must be given to the optical system of the radiation pyrometer. Appropriate optical glasses must be selected to pass the necessary range of wavelengths. Pyrex glass may be used for the range of 0.3 to 2.7µm, fused silica for 0.3 to 3.8µm and calcium fluoride for 0.3 to 10µm. Radiation pyrometers are used ideally in applications where the source approaches black body conditions. If the emissivity of the source is not known precisely, the temperature measured is not reliable. To overcome this ambiguity in the measurement of temperature, the pyrometer is calibrated for the particular application it is used. Calibration consists of comparing the pyrometer reading with that of some standard device such as a thermocouple.

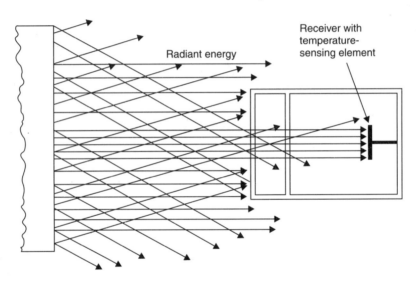

Fig. 10.19 Total radiation pyrometer

10.4.4 Optical Pyrometry

In optical pyrometry, the radiant intensity at one or two specific wavelengths is measured by the use of appropriate filters. The intensity is found either by visual comparison or by quantitative measurement using a sensing element. *The disappearing filament pyrometer* is an example of the visual comparison type (Fig. 10.20). The intensity of an electrically heated element is varied to match the source intensity at a particular wavelength. While in use, the pyrometer is directed at the distant test object whose temperature has to be determined. The objective lens focuses the source in the plane of the lamp filament. The eyepiece is adjusted so that the filament and the source appear superimposed and are in focus to the observer. In general, the filament will be either hotter than or colder than the unknown source as shown in Fig. 10.21(a) and Fig. 10.21(b). When the battery current is adjusted, the filament may be made to disappear as shown in Fig.10.21(c). The current indicated by a milliammeter to obtain this condition may then be used as the temperature read out. A red filter is generally used to obtain approximately monochromatic conditions. An absorption filter is used so that the incident intensity is attenuated. Thus the filament may be operated at reduced power and its life prolonged.

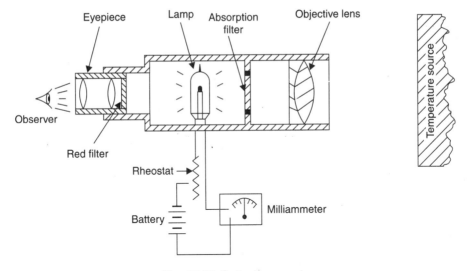

Fig. 10.20 Optical pyrometer

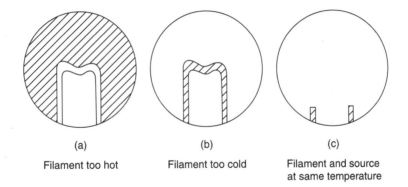

(a) Filament too hot (b) Filament too cold (c) Filament and source at same temperature

Fig. 10.21 Appearance of the filament against the background of the radiation from the body whose temperature is measured (a) when the filament temperature is less than that of the body (b) when the filament temperature is more than that of the body (c) when the temperature of the filament is equal to that of the body

Two-Colour Optical Pyrometry

Errors due to inaccurate values of source emissivity are important in optical pyrometer. *Two-colour pyrometry* is an optical technique that minimises the influence of emissivity. *The two-colour pyrometer* measures source intensity at two wavelengths λ_1 and λ_2. If the emissivity is independent of wavelength or if the wavelengths are nearly equal, then the ratio of measured intensities depends only on the temperature. From the Planck's formula we have

$$\frac{E(\lambda_1)}{E(\lambda_2)} = \left(\frac{\lambda_2}{\lambda_1}\right)^5 \frac{\left(e^{hc/\lambda_2 k_B T} - 1\right)}{\left(e^{hc/\lambda_1 k_B T} - 1\right)}$$

Thus by measuring $E(\lambda_1)$, $E(\lambda_2)$ and knowing λ_1 and λ_2, the temperature T can be evaluated from the above formula.

10.4.5 Fiber Optic Radiation Thermometer

Optical fibers can be used for transmitting radiation emitted by the object. This property of fibers have been utilized to develop pyrometry, when the intervening medium between the object and the detector poses problems for the transmission of radiation. This technique has been used to measure the temperature of hot gases in gas turbine combustors (Fig. 10.22). The optical fiber is made of single crystal sapphire (Al_2O_3), typically a fraction of a millimeter in diameter and a meter long. A cup shaped black body cavity is formed at the sensing tip by evaporating a metallic film of platinum or iridium. A protective sapphire film is also formed over the metal sensing film. When the probe is immersed in a hot gas, it attains the temperature of the hot gas. The thermal conductivity of the sapphire is small and hence the fiber does not get heated along its length. The radiation is transmitted by the fiber which is coupled to a photodiode. The photodiode produces an output which is proportional to the intensity of the radiation.

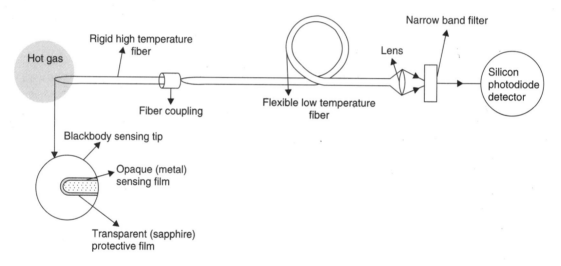

Fig. 10.22 Fiber optic radiation thermometer

10.4.6 Infrared Imaging Systems–IR Thermography

Infrared pyrometry is very useful for thermal imaging or infrared thermography. All objects around us emit electromagnetic radiation. Around ambient temperatures this is predominantly infrared radiation. IR is invisible to eye. But with the aid of a suitable detector, IR can be converted into a visible image. Variation in the surface temperature of the object can be visualized from the thermal image of an object. This means that deviation from the normal temperature can be detected from a distance. Thermography makes use of the infrared spectral band of the electromagnetic radiation. The infrared band is further subdivided into four smaller bands, the boundaries of which are arbitrarily chosen. They include the near infrared (0.75-3μm), the middle infrared (3-6μm), the far infrared (6-15μm) and the extreme infrared (15-100μm). The most commonly used band for commercial infrared imaging is between 0.75-15μm.

The intensity of infrared radiation is dependent on the absolute temperature of the body and is given by the Planck's law. The basic factors affecting the thermal measurement include (*a*) emissivity, (*b*) surroundings and (*c*) atmosphere. For a black body the emissivity is 1. Other materials and surfaces have emissivities ranging from 0.1 to 0.95. Emissivity is a critical parameter for quantitative measurement of the temperature. It is important to have the object surroundings free from thermal radiation sources. Otherwise the radiation from these sources would also be reflected by the object under examination leading to erroneous values. The effects of atmosphere are of importance when the object is far away. *The atmosphere not only attenuates the radiation from the target but also alters the spectral characteristics.* However, these effects are negligible in cases where the object under investigation is located quite close and the atmosphere is uncontaminated with vapours, smoke, fog, hot gases etc. Occasionally, one may need to make some critical measurements on an object in the presence of hot air/gases as in the case of furnaces. In such cases, suitable high temperature gas filters are used along with the appropriate correction factors to take into account the ambient temperature and attenuation by these filters. In case where the objects are situated at a large distance as in the case of airborne thermography, atmospheric absorption plays a very important role. The atmospheric absorption is quite a complex process and in these cases, mathematical modeling is resorted to for estimating the temperature.

Infrared imaging systems have the function of providing a television like visual display. The display consists of various shades of colours which represent different temperature levels. The block diagram of a non-contact thermography system is shown in Fig. 10.23. It basically consists of an infrared scanner, monitor, control unit and a computation unit for field applications. The output can also be stored in a modified video thermal recorder which can be analysed later using a personal computer with image processing facilities. The infrared scanner essentially consists of an optical system, scanning mechanism, infrared detector and associated electronics. The optical system collimates the incoming infrared radiation into the detector. The commonly used materials for mirrors and prisms in the optical system are germanium, silicon, sapphire, barium fluoride and arsenic trisulphide. The scanning mechanism scans the surface within the field of view. Various models of scanning are possible. In Fig. 10.24 is shown a typical scanning system. Here the scanning of the target surface is accomplished by focusing radiation on a plane mirror oscillating about a horizontal axis. This scans the line of sight vertically over the target surface. The image from the plane mirror is focused on a rotating octagonal prism which provides the horizontal scanning. A InSb detector produces an electrical signal proportional to the incident flux. Scanning can also be done using two octagonal prisms rotating perpendicular to each other at a high speed. The advent of personal computers has revolutionised the field of thermography. Thermograms can be subjected to image processing and refinement to obtain the minute details not otherwise visible. The non-contact nature of the sensing method, together with the display of the entire surface temperature distribution over an object, gives infrared imaging systems unique application possibilities.

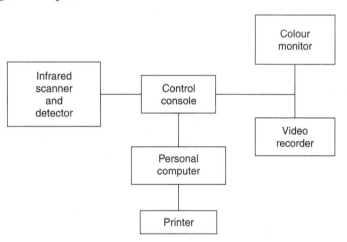

Fig. 10.23 Block diagram of the apparatus used for thermal imaging

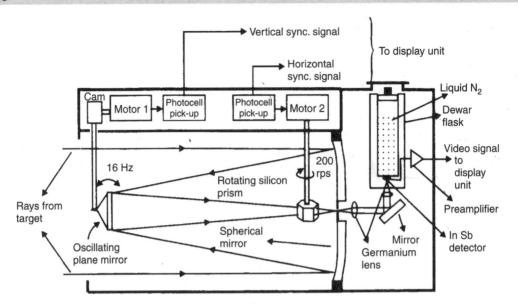

Fig. 10.24 Infrared camera

It is also common to use CCDs (Charge Coupled Devices) as thermal detectors. An individual photodetector registers the intensity of the radiation falling on it. In contrast an array of large number of photodetectors can register radiation intensity from many spatial points. Such an array is ideal for recording the image of an object. The image of an object may be focused onto the array or alternately the image may be scanned. The most important technology used for imaging applications is based on CCDs. The basic device upon which CCD technology is based is the metal-insulation-semiconductor (MIS) or metal-oxide-semiconductor (MOS) capacitor. In Si technology, the MOS structure is used. Typical MOS structure is shown in Fig. 10.25a. It is made by depositing a metal electrode on a thermally oxidised p-type silicon substrate. A positive bias is applied to the gate to create a depletion region below the metal electrode. The depletion region acts like a potential well in which the minority electron carriers are trapped (Fig. 10.25b). *This well can be thought of as a partially filled bucket with electron fluid.* On being exposed to radiation, electron-hole pairs are generated. The holes are swept away from the oxide-semiconductor interface due to the electric field while the electrons get accumulated in the well (Fig. 10.25c). The number of electrons and hence the amount of charge in the well is proportional to the intensity of radiation. In CCD an array of MOS capacitors are connected. By applying appropriate voltages on the adjacent metal electrodes, it is possible to generate depletion layers of various dimensions in the semiconductor underneath. Such a result is equivalent to produce a series of potential wells where electrons may be stored and transferred from one well to the other by appropriately altering the voltages on the metallic electrodes (Fig. 10.26(a-d)). In Fig. 10.27 is shown a CCD array where different amounts of charge are stored in each pitzel. On applying potential suitably, one pitzel row is shifted to parallel register. It is read out pitzel by pitzel at the output node. The stored charge in each pitzel is thus fed into an amplifier and digitized. In a CCD array, an array of MOS capacitors are isolated from a readout register by a transfer gate. When this gate is opened all the charges accumulated in the individual capacitors are transferred in parallel into the readout register, which has one transfer cell opposite each MOS capacitor. The transfer register is shielded from the incident light as depicted in Fig. 10.26b. Once all the cells are empty the array can be re-exposed. Rather than serial readout it is also possible to have full frame CCDs in which the whole frame is transferred to an adjacent storage array leaving the main array free to collect a new signal flux. The cells can be as small as 6 μm. High resolution CCD array detectors are available in 256 × 256 to 1024 × 1024 pitzels within 4.8 mm × 6.4 mm area. Thus the pitzel size is ~ μm. (Fig. 10.27). The main advantages of CCDs are that they have a very low noise and can accept low input signal levels. The major disadvantages are the speed and the cost.

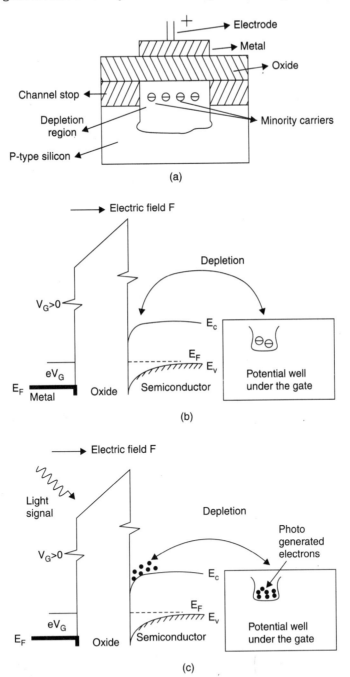

Fig. 10.25 (a) MOS capacitor (b) Band profile of a MOS capacitor (c) Effect of photon is to create a pocket of electrons under the gate

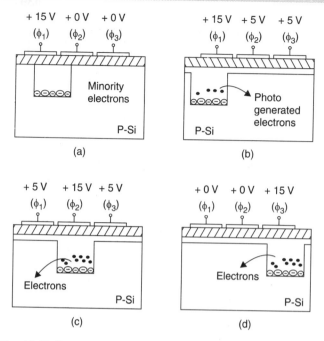

Fig. 10.26 Transfering electrons by applying potential in a CCD array

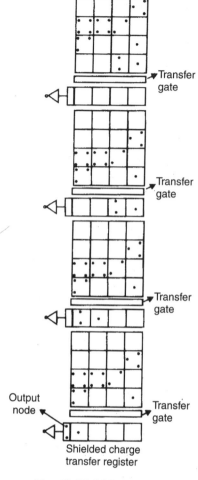

Fig. 10.27 CCD array

Figure 10.28 shows a pyroelectric vidicon used for thermal imaging. The thermal image of the scene is projected via an infrared lens on to a thin disc shaped pyroelectric crystal, usually TGS ~ 25μm thick. The front surface is covered with an infrared absorbing electrically conducting film, which becomes then an equipotential surface. Depending on the radiation intensity, distribution of charges takes place on the surface. This in turn gives rise to changes in the surface potential at various points on the surface. During the tube operation the electron beam scans the surface. This is achieved by the application of orthogonal electric fields to the beam of electrons as in the case of a TV tube. The pyroelectric

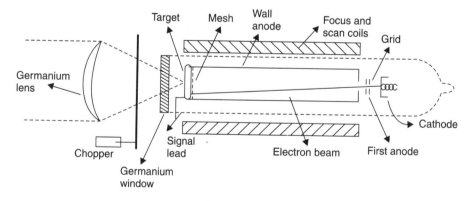

Fig. 10.28 Pyroelectric vidicon

crystal can be thought of as a two dimensional array of capacitors. The electron beam discharges these capacitors at each point while scanning. The charging current is amplified and is used to modulate the intensity of the TV display in synchronization with the electron beam.

Infrared Imaging in Industry

Thermal imaging is used in testing buildings for heat leakage and in measuring temperature distribution in electronic equipment. In petroleum industry petroleum has to be distilled out at different temperatures to obtain light gasoline, naphtha, kerosene and gas oil. Hence thermal imaging is used in the monitoring of stack temperature, maintenance of plant equipment such as reaction towers, refining furnaces, ducts and piping, detection of corrosion in oil tank shell and measurement of oil levels etc. Thermal imaging has been extensively applied for condition monitoring of furnace tubes, gas and fluid transfer lines and evaluation of heat resistant linings in refractory furnaces. This technique is ideally suited for the wear determination of refractories in blast furnaces, hot blast stoves and steel stoves and the inspection of rotary kiln lining and estimation of temperature within the kiln. Thermograhic inspection of the entire electrical power systems can be performed. Stator lamination insulation, core insulation, slip ring temperature measurement etc., of turbogenerators are also made using thermography. Application of thermography for the inspection of transmission lines, substations and distribution systems has become a regular feature in many countries abroad. Thermal imaging can also be applied for location of loose contacts on busbar joints of switchyards and switchgears, location of improper jointing of lugs in cable joints, finding irregularities in distribution boards, detection of hot spots in isolators due to presence of dirt or moisture which could lead to corona and checking the adequacy of insulation. In many materials changes in stress lead to changes in temperature. Consequently changes in temperature as recorded by thermograms can be utilized for stress analysis. In meteorology, thermal imaging using satellites and aircrafts is routinely done to monitor the surface temperature of the earth on land and sea. It is also used for detecting forest fires.

Infrared Imaging in Medicine

Measurements of body surface temperature indicate that the surface temperature varies from point to point depending on the external factors as well as internal metabolic processes and blood circulation near the skin. Since the variations in these internal processes may be symptomatic of abnormal conditions, many researchers have attempted to accurately measure the surface temperature of the body and relate it to pathologic condition. For example it has been found that most breast cancers could be characterised by an elevated skin temperature in the regions of the cancer. The surface temperature above a tumor is typically about 1°C higher than that above nearby normal tissue, indicating that a very sensitive temperature-measuring device has to be used. Thermography has also been used to study the circulation of blood in the head. Differences in temperature distribution between the left and right sides can indicate circulatory problems. Thermography has had considerable success in reducing leg amputation in diabetics. The blood supply in a diabetic's leg is usually adequate. But if the tissue breaks down and an ulcer is formed, the need for blood in the leg may double. Consequently a hot spot on the foot is usually formed which can be detected using thermography.

Infrared Imaging in Military

In missile guidance, the missile is designed to *home* on the infrared radiation emitted by the target, often the hot jet exhaust of the target aircraft's engine. An infrared scanning system in the missile locates the target and produces error signals that steer the missile onto the target. For satellite altitude sensing, the infrared sensors are able to distinguish between the radiation intensity from the earth, the moon or a planet from the background in space. These generate accurate orientation signals for satellite altitude control.

10.5 PARTICLE LIKE BEHAVIOUR OF LIGHT

10.5.1 Photoelectric Effect

Electrons are emitted from the surface of a metal on being exposed to visible and ultraviolet light. This phenomenon is called the *photoelectric effect*. The photoelectric effect can be demonstrated by shining light on the surface of a metal plate in an evacuated chamber (Fig. 10.29). The chamber also has another plate, which is connected to the battery. This plate is kept at a voltage negative with respect to the metal plate, which emits the electrons. The flow of current through the plates is measured using an ammeter. The current stops flowing when the light supply is cut off.

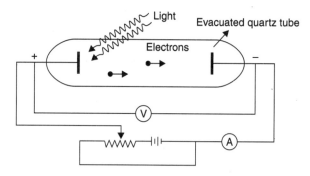

Fig. 10.29 Experimental observation of the photoelectric effect

Following are the experimental observations:

(a) The magnitude of the photoelectric current as a function of potential difference between the plates is shown in Fig. 10.30a. The current continues to flow even when the potential difference is negative. This suggests that the electrons have sufficient kinetic energy to overcome the electrostatic repulsion. The photocurrent decreases with increasing V and becomes zero at $V = V_0$. The magnitude of potential difference V_0 at which the current is just zero is called the *stopping potential*. All the electrons that are emitted from the emitter surface do not have the same energy. (This is because all the electrons in a metal do not have the same energy. The other reason is that it is easier to knock off the electrons, which are located near the surface than those deep inside the metal). This accounts for the decreasing photocurrent with V.

If v_m is the maximum velocity of the electrons emitted at the emitter, then their kinetic energy at the emitter is $mv_m^2/2$, where m is the mass of the electron. When the potential difference between the plates is negative, the kinetic energy of electron is equal to the work done in overcoming this electrostatic repulsion i.e., $mv_m^2/2 = eV_0$ where e is the magnitude of electronic charge. Thus the stopping potential is a measure of the maximum kinetic energy that the photoelectrons can have.

(b) The photoelectric current starts flowing as soon as the light supply is turned on. There is virtually no time lag between the shining of light and the emission of electrons. More precise measurments indicate that the time lag is less than 3×10^{-9} secs.

MATTER AND RADIATION – DUAL NATURE

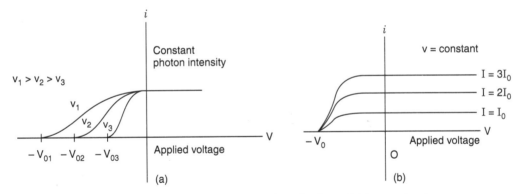

Fig. 10.30 Photoelectric current as a function of voltage (a) increasing frequency at constant intensity (b) increasing intensity at constant frequency

(c) For a light of given frequency, the stopping potential is independent of the intensity light (Fig. 10.30b). When the frequency of the light used is held fixed, the current is directly proportional to the intensity of light (Fig. 10.31).

(d) The photoelectric current vs voltage for different frequencies is shown in Fig. 10.30a. Higher the frequency, greater is the stopping potential.

(e) The stopping potential V_0 is dependent on the frequency (and hence wavelength) of the light used (Fig. 10.32). The relation between V_0 and ν is given by

$$V_0 = \alpha \nu + \beta \qquad (10.31)$$

where α and β are constant. β depends on the material while α does not. The linear relation between the stopping potential and frequency is shown in Fig. 10.32 for cesium, lithium and silver.

(f) There is a threshold frequency and only if the frequency of light is greater than the threshold frequency, the photoelectric effect is observed.

(g) In photoelectric effect when plane polarized light is used, there is a preferential direction for electron emission. Thus the polarization properties of light have to be closely

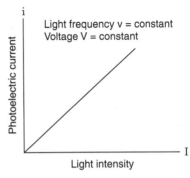

Fig. 10.31 Photocurrent as a plot of light intensity at constant frequency and constant retarding potential

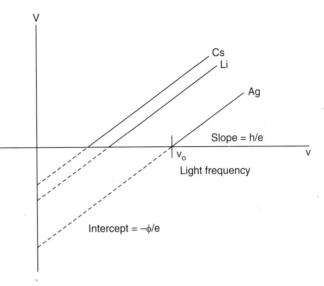

Fig. 10.32 Stopping potential as a function of frequency

connected with some particle properties. This is taken care of by ascribing *spin* to photons. Photons are particles with spin angular momentum of $h/2\pi$.

10.5.2 Failure of the Classical Theory

The classical theory in which light is treated as an electromagnetic wave is unable to explain the experimental observations for the following reasons:

(a) A metal being electrically neutral has both positive and negative charges. Metals are good conductors and hence the electrons are free to move inside the metal. A metal is best visualized as a sea of valence electrons in which the positively charged nuclei are embedded. To remove an electron from the metal, it needs to be pulled away from the positive charge and this requires work to be done. When a metal is exposed to an electromagnetic wave, the electrons experience a force, gain kinetic energy and hence can be ejected from the surface of the metal. The intensity of light is proportional to the square of the amplitude of the electric field intensity. Hence electrons are expected to gain more energy with an increase in the intensity of light. *In other words, classical theory predicts the stopping potential to be dependent on intensity of light which is not observed.*

(b) Further, the classical theory cannot account for the threshold frequency. Electromagnetic wave of any frequency with sufficient intensity ought to be able to generate photoelectrons, which is also not experimentally observed.

(c) According to classical physics, the transfer of energy from the electromagnetic field to the electron should take a long time, of the order of seconds (see solved example 1). This means that the photoelectric effect is not instantaneous and thus contradicts the experimental observations.

10.5.3 Einstein's Quantum Theory

The photoelectric effect was explained by Einstein in 1905. He suggested that a light beam should not be considered as a wave supplying energy to electrons in a continuous manner. Instead, it should be thought of as consisting of particles of light called photons. Each photon has a definite energy equal to $h\nu$, where h is the Planck's constant and ν is the frequency of the light beam. Photons travel with the velocity of light c and have *zero rest mass*. The energy of a light beam having n photons is $nh\nu$ and n can assume only integral values. According to this theory, when a metal is exposed to light, electrons undergo collisions with photons. During this process they absorb energy from the photon and are ejected from the metal. The collision is instantaneous so that there is no time lag between the emission of electrons and its collision with photon. The number of photons increase with the intensity of light. Hence the probability of collisions of photons with electrons also increases. Thus photoelectric current is a function of intensity of light. However, in each collision an electron can acquire only an energy $h\nu$. *It can be shown that the probability of an event in which an electron absorbs an energy equal to $2h\nu$ by simultaneously colliding with two photons is negligibly small.* Not all the energy $h\nu$ supplied to electron appears as kinetic energy. This is because the electrons use some energy for getting themselves released from the metal. The minimum energy required to remove an electron from the metal is called the *work function*. If ϕ is the *work function*, then

$$\text{Kinetic energy} = E_{max} = \frac{1}{2} m v_{max}^2 = h\nu - \phi$$

But
$$E_{max} = eV_0, \quad \therefore \quad eV_0 = h\nu - \phi$$

$$\Rightarrow \quad V_0 = \frac{h\nu}{e} - \frac{\phi}{e} \quad (10.32)$$

Comparing eqns. (10.31) and (10.32) we have

$$\alpha = h/e \quad \text{and} \quad \beta = -\phi/e$$

Thus α is independent of the material, while β depends on the nature of the metal.

This theory can explain the origin of threshold frequency ν_{th}. If $h\nu_{th}$ is the energy of photon then

$$\phi = h\nu_{th} \tag{10.33}$$

If energy of the photon is less than ϕ, photoelectrons cannot be liberated since energy of the photon is insufficient to impart the necessary energy to the electron to overcome the electrostatic attraction due to the positive charges. Increasing the intensity of light results in an increase in the number of photons and not the intrinsic energy of the photons. This explains why stopping potential is not a function of the intensity of light. For the same reason, increasing the intensity of photons having frequency less than the threshold frequency, will not result in photoelectric emission.

10.5.4 Contact Potential Between Two Metals

Work function of a metal is defined as the *minimum energy* needed to eject an electron from the surface of the metal. Let ϕ_A and ϕ_B be the work functions of metals A and B. When metals A and B are brought into contact, electrons from metal A flows into that of B, much the same way as a fluid flows from a higher level to a lower level (Fig. 10.33). As a result metal A is left electropositive with respect to metal B. A potential difference arises across the metallic surfaces in contact. The contact potential V_c is given by

$$eV_c = \phi_B - \phi_A$$

The validity of this relation has been confirmed by experiment.

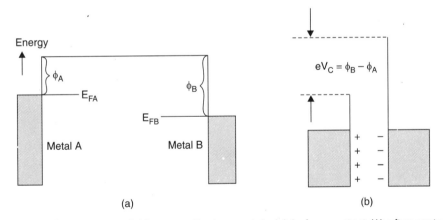

Fig. 10.33 Contact potential between the two metals (a) before contact (b) after contact

10.5.5 Light Detection Using Photoelectric Effect

Photoelectric effect is an important phenomenon which played a crucial role in clarifying the quantum nature of radiation. The effect is also utilized in many devices involving detection of light and measurement of its intensity. A photoelectric detector is also referred to as a photomultiplier tube (PMT). A PMT consists of a photoemissive cathode and a large number of anodes (Fig. 10.34). The cathode is so chosen that it exhibits photoelectric effect in the visible region. On being illuminated with light, the photoemissive cathode emits electrons. The collision of photons with electrons in the metal is a statistical process. Every photon interacting with an electron need not result in the emission of photoelectron. This is taken care by introducing a parameter called the quantum efficiency. Quantum efficiency is the ratio of number of photoelectrons emitted from the surface to the number of photons incident on the surface. The photocurrent obtained with a single anode is very small. Hence in a PMT a number of anodes is used. In the multiple anode arrangement, the electrons released from one dynode are attracted to the neighbouring dynode because of its higher positive potential. In that process, the electron gets accelerated and on striking the second anode a large number of secondary electrons are released. Two or three low energy secondary electrons are emitted for each incident electron. This produces an electron multiplying effect. The electrons from the last dynode stage are collected by the positive anode A. The final current is a measure of the illuminance at the cathode. The merits of the PMT are its satisfactory stability, low threshold and sensitivity. Sensitivity can be increased by application of large voltage or additional amplifiers. The response of the PMT is ~ 10^{-8} s. Hence it can be used to measure the intensity of short light pulses. Typical characteristic of the PMT is shown in Fig. 10.35. In some PMTs, light is gathered on an end window. A semitransparent photocathode film at the back of the glass window releases electrons. In such PMTs, the spectral response of the emissive material as well as the transparency of the glass window has to be taken into account.

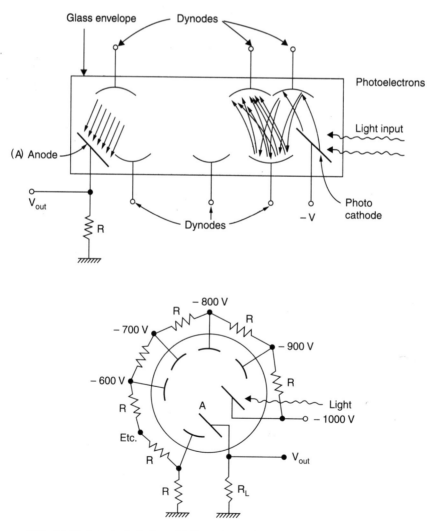

Fig. 10.34 A typical photomultiplier tube which is used for detecting light

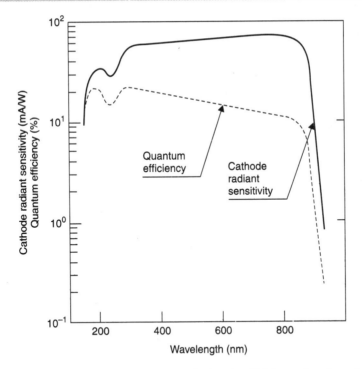

Fig. 10.35 Quantum efficiency and radiant sensitivity of a PMT as a function of wavelength

10.5.6 Photoelectron Spectroscopy in Materials Science

Metals exhibit photoelectric effect with visible and ultraviolet light. Many materials exhibit electron emission when X-rays are incident on them. By studying the energy of the photoelectrons, it is possible to obtain, information about the electronic structure of the material. Photoelectron spectroscopy is one of the routine tools employed by materials scientists in the analysis of the material. Photoelectron spectroscopy enables the identification of all the elements in the periodic table with the exception of hydrogen and helium. It also enables the determination of the *oxidation state* of an element and the *type of species* to which it is bonded. Photoelectron spectroscopy with X-rays is often referred to as X-ray Photoelectron Spectroscopy (XPS) or Electron Spectroscopy for Chemical Analysis (ESCA). Electron spectroscopy with ultraviolet light is called Ultraviolet Photoelectron Spectroscopy (UPS).

Photoelectron spectrometer has the following components: (*a*) X-rays or ultraviolet source, (*b*) a sample holder and (*c*) analyzer. The photoelectrons are deflected by an electrostatic field in such a way that the electrons travel in a curved path. By varying the field, electrons of various kinetic energies can be focused on the detector. The deflection plates in an electron spectrometer may be cylindrical or spherical or hemispherical. Fig. 10.36 shows a spectrometer with cylindrical electrodes. The relation between the plate voltage V_1 and V_2 and the kinetic energy E_k of the electrons is given by

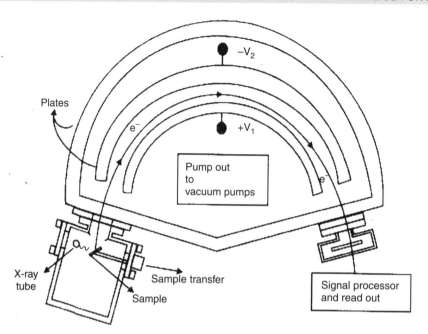

Fig. 10.36 Photoelectron spectrometer

$$V_2 - V_1 = 2E_k R \log\left(\frac{R_1}{R_2}\right)$$

where R_1 and R_2 are the radii of the two plates and R is their average value. Typically the pressure in the electron spectrometer is $\sim 10^{-5}$ torr. Detectors are solid state devices, which consist of glass doped with lead or vanadium. When a potential of several kilovolts is applied across these materials, a cascade or pulse of $10^6 - 10^8$ electrons are produced for each incident photoelectron. The pulses are then counted electronically.

10.6 COMPTON EFFECT

In photoelectric effect, a photon interacts with an electron and in the process gives its *entire* energy to the electron. An electron can also scatter an incident photon. In this situation the photon loses only *a part* of its energy to the electron. Therefore the scattered photon emerges with a wavelength larger than that of the incident photon. This difference in wavelength depends on the angle of scattering. This phenomenon which results in the change of wavelength of photon on scattering is called *Compton effect*. This was first observed with X-rays. Figure 10.37 shows a typical set-up for observing the Compton effect. X-rays from a tube having a molybdenum target is incident on the graphite. In this case the scattering angle is 90°. The scattered X-rays are detected using a suitable detector. The observed distribution of scattered wavelengths shows Compton scattering by electrons at $\lambda' = \lambda_o + \lambda_c$ and Thomson scattering (elastic scattering) by atoms at $\lambda' = \lambda_o$.

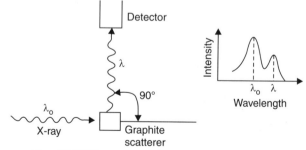

Fig. 10.37 X-ray scattering at 90°. The observed distribution of scattered wavelengths shows Compton scattering by electrons at $\lambda' = \lambda_o + \lambda_c$ and Thomson scattering by atoms at $\lambda' = \lambda_o$

MATTER AND RADIATION – DUAL NATURE

10.6.1 Compton Shift

The relation between the change in wavelength and angle of scattering can be derived as follows. Figure 10.38 indicates the energy and momentum of the X-ray photon and electron involved in the scattering process, before and after collision.

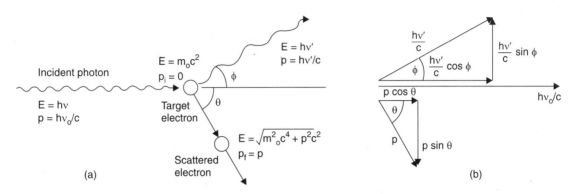

Fig. 10.38 (*a*) The scattering of a photon by an electron is called the Compton effect energy and momentum are conserved in such a event. As a result the scattered photon has less energy (longer wavelength) than the incident photon (*b*) Vector diagram of the momenta and their components of the incident, and scattered photons and the scattered electron

An incident X-ray beam of frequency v_0 travels along the *x*-axis. This corresponds to photons of energy hv_0 and momentum hv_0/c. The photons are scattered by an electron, which is initially at rest. Before the collision the total energy is $hv_0 + m_0c^2$ where m_0 is the rest mass of the electron. After collision, photon of frequency v travels along a direction making an angle ϕ with the *x*-axis. The electron has a velocity v along a direction θ with *x*-axis. The requirement that energy be conserved gives

$$hv_0 + m_0c^2 = hv + \frac{m_0c^2}{\sqrt{1-\left(\frac{v^2}{c^2}\right)}}$$

i.e.,
$$hv_0 = hv + m_0c^2\left[\frac{1}{\sqrt{1-\left(\frac{v^2}{c^2}\right)}} - 1\right] \qquad (10.34)$$

In this derivation relativistic expressions for energy transfer are used since electron velocities can be comparable to that of light.

The requirement that momentum be conserved gives two equations, one for the *x*-component of the momentum, and the other for *y*-component.

$$x\text{-component}: \frac{hv_0}{c} = \frac{hv}{c}\cos\phi + \frac{m_0v}{\sqrt{1-\left(\frac{v^2}{c^2}\right)}}\cos\theta \qquad (10.35)$$

$$y\text{-component}: 0 = \frac{h\nu}{c} \sin\phi - \frac{m_0 v}{\sqrt{1-\left(\frac{v^2}{c^2}\right)}} \sin\theta \qquad (10.36)$$

To solve these equations it is convenient to set $\beta = v/c$ and rewrite the last three eqns, (10.34) to (10.36). This gives

$$h\nu_0 = h\nu + m_0 c^2 \left[\frac{1}{\sqrt{1-\beta^2}} - 1\right] \qquad (10.37)$$

$$\frac{h\nu_0}{c} = \frac{h\nu}{c} \cos\phi + \frac{m_0 \beta c}{\sqrt{1-\beta^2}} \cos\theta \qquad (10.38)$$

$$0 = \frac{h\nu}{c} \sin\phi - \frac{m_0 \beta c}{\sqrt{1-\beta^2}} \sin\theta \qquad (10.39)$$

For a particular angle of scattering ϕ, the last three equations contain three unknowns, ν, β and θ, and expressions for these quantities can be found by solving these three equations. For comparison with experiment, however, it is more convenient to calculate the shift in wavelength. Let $\lambda_0 = c/\nu_0$ and $\lambda = c/\nu$ be the wavelengths of the incident and scattered radiation respectively. Then eqns. (10.37-10.39) become:

$$\frac{h}{\lambda_0} - \frac{h}{\lambda} + m_0 c = \frac{m_0 c}{\sqrt{1-\beta^2}} \qquad (10.40)$$

$$\frac{h}{\lambda_0} - \frac{h}{\lambda} \cos\phi = \frac{m_0 \beta c}{\sqrt{1-\beta^2}} \cos\theta \qquad (10.41)$$

$$\frac{h}{\lambda} \sin\phi = \frac{m_0 \beta c}{\sqrt{1-\beta^2}} \sin\theta \qquad (10.42)$$

Squaring and adding eqns. (10.41) and (10.42):

$$\frac{h^2}{\lambda_0^2} + \frac{h^2}{\lambda^2} - \frac{2h^2 \cos\phi}{\lambda_0 \lambda} = \frac{m_0^2 \beta^2 c^2}{1-\beta^2} = \frac{m_0^2 c^2}{1-\beta^2} - m_0^2 c^2 \qquad (10.43)$$

Similarly squaring eqn. (10.40):

$$\frac{h^2}{\lambda_0^2} + \frac{h^2}{\lambda^2} - \frac{2h^2}{\lambda_0 \lambda} + 2m_0 ch\left(\frac{1}{\lambda_0} - \frac{1}{\lambda}\right) + m_0^2 c^2 = \frac{m_0^2 c^2}{1-\beta^2} \qquad (10.44)$$

Subtracting eqn (10.43) from eqn. (10.44):

$$\frac{2h^2}{\lambda_0 \lambda}(\cos\phi - 1) + 2m_0 ch\left(\frac{1}{\lambda_0} - \frac{1}{\lambda}\right) = 0$$

$$\Rightarrow \qquad \Delta\lambda = \lambda - \lambda_0 = \frac{h}{m_0 c}(1 - \cos\phi) = \lambda_C(1 - \cos\phi) \qquad (10.45a)$$

where λ_C is known as the Compton shift. On substituting the values for h, m_0 and c:

$$\lambda_C = 0.0242 \text{ Å} = 2.42 \text{ pm}.$$
$$\Delta\lambda = 0.0242(1 - \cos\phi) \text{ Å} = 2.42(1 - \cos\phi) \text{ pm}. \qquad (10.45b)$$

Eqn. (10.45) states that when an incident X-ray of wavelength λ_o is scattered through an angle ϕ by a free electron, the wavelength λ of the scattered X-ray should be greater than that of the incident X-ray by the amount $0.0242\,(1-\cos\phi)$ Å (*i.e.*, in units of Angstroms) or $2.42(1-\cos\phi)$ pm. For a given value of the scattering angle ϕ, the shift in wavelength is independent of the wavelength of the incident radiation. For $\phi = 90°$, $\Delta\lambda = 2.42$ pm, which agrees very well with the observed value of 2.36 pm. The predicted dependence of $\Delta\lambda$ on the angle ϕ was also verified by experiment.

10.6.2 Kinetic Energy of Recoil Electrons

The kinetic energy T of the recoil electron is equal to the loss in the energy of the scattered X-ray photon and is given by :

$$T = h\nu_o - h\nu = \frac{hc}{\lambda_0} - \frac{hc}{\lambda} = \frac{hc}{\lambda\lambda_0}(\lambda - \lambda_0) = \frac{hc}{\lambda\lambda_0}(\Delta\lambda) \qquad (10.46)$$

From eqns. (10.45(a)) and (10.46)

$$T = \frac{h^2}{\lambda\lambda_0 m_0}(1-\cos\phi) = \frac{h^2(1-\cos\phi)}{\lambda_0(\lambda_0+\Delta\lambda)m_0} = \frac{h^2(1-\cos\phi)}{\lambda_0 m_0\left[\lambda_0 + (1-\cos\phi)\dfrac{h}{m_0 c}\right]}$$

Multiplying both numerator and denominator by $\left(\dfrac{\nu_0}{c}\right)$, we get

$$T = h\nu_0\left[\frac{(1-\cos\phi)\alpha}{1+(1-\cos\phi)\alpha}\right] \quad \text{where} \quad \alpha = \frac{h\nu_0}{m_0 c^2} \qquad (10.47)$$

The recoil electrons have been detected and their measured energies agree with the values predicted by the theory. The theory just developed explains the observed shift in wavelength, but does not account for the presence of the unshifted line (Thomson scattering or elastic scattering). In the above derivation it has been assumed that the electrons are free. Actually, electrons are bound to atoms more or less tightly, and a certain amount of energy is needed to detach an electron from the atom. If this energy is small, the electron acts like a free electron and eqn. (10.35) is valid. If the collision is such that the electron is not detached from the atom, but remains bound, the rest mass m_o of the electron in eqn. (10.35) must be replaced by the mass of the atom, which is several thousand times greater. The calculated value of $\Delta\lambda$ then becomes too much small to be detected. A photon, which collides with a bound electron, therefore, does not have its wavelength changed, and this accounts for the presence of the unshifted spectral line.

In the expression for Compton shift, the only trace of the relativity theory that appears is the zero subscript denoting the rest mass of the electron. It would seem as though the same effect would have been obtained if the problem had been set up without taking into account the relativistic effects. This is actually the case so far as the shift in wavelength is concerned. The more refined details of the Compton effect, however, demand a rigorous relativistic treatment. Hence, even the more elementary problems are usually set up relativistically.

It ought to be noted that $\Delta\lambda/\lambda \ll 1$ for $\lambda \gg \lambda_c$. Thus for light of a wavelength much larger than 10^{-12} m, one cannot expect and appreciable shift with wavelength. When $\lambda \sim \lambda_c$ then photon has an energy $\sim$MeV, the energy transfer of the electron is $\sim$MeV. In such a situation, as already stated, one has to use relativistic expressions for energy. The Compton effect provides another evidence for the particle nature of electromagnetic radiation.

Inelastic scattering of photons in the visible region of the electromagnetic spectrum has also been observed. In the visible region, the analogous of Compton scattering is Raman scattering and the one corresponding to Thomson scattering is Rayleigh scattering. In Raman scattering, the photon loses or gains a part of the energy equivalent to a vibrational, rotational or electronic energy states of the scattering system. If the photon loses in energy it is called Stokes Raman scattering and if it gains energy it is termed anti-Stokes Raman scattering (Fig. 10.39). In Rayleigh scattering the energy of the incident photon is unchanged and thus corresponds to elastic scattering. Raman scattering has been observed in isolated molecules as well as in liquid and solid phase of matter. It is a powerful technique to probe the vibrational, rotational and electronic states of the materials. Rayleigh scattering is important in optical fibers since it contributes to attenuation of the light beam through the fiber.

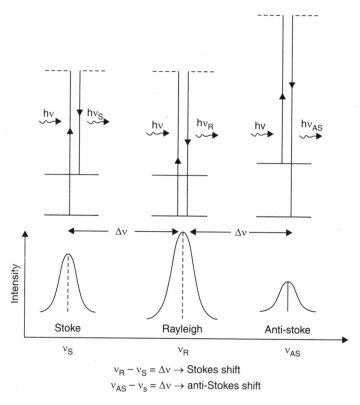

Fig. 10.39 Raman scattering as an analogue of Compton scattering

10.7 WAVELIKE BEHAVIOUR OF PARTICLES

10.7.1 de Broglie's Hypothesis-Matter Waves

The quantum theory of light is radically different from the wave theory of light. The wave theory was developed to understand the phenomenon of interference and diffraction. It is impossible to explain these phenomena by assuming light to be photons. It appears that both these physical models are required to explain all the phenomena associated with light. It is paradoxical that light behaves like particles in some phenomena and like waves in some others. *This is referred to as the dual nature of light.* According to the special theory of relativity, matter and radiation are interconvertible. Hence de Broglie conjectured that similar to light, all material particles have particle as well as wave like property. He postulated that particle-wave nature is a general property of all objects. He suggested that every object can be associated with a wave, often called the *matter wave* whose wavelength is given by

$$\lambda = \frac{h}{p} = \frac{h}{mv} \qquad (10.48)$$

where h is the Planck's constant and p is the momentum of the particle. The equation is called de Broglie relation and forms the basis for the mathematical development of quantum mechanics.

For a particle moving with speeds comparable to that of light, relativistic expression for momentum has to be used for calculating the de Broglie wavelength. According to the theory of special relativity, for a particle moving with velocity comparable to that of light, the total energy E, momentum p and the de Broglie wavelength are given by:

$$E^2 = p^2c^2 + m_0^2c^4 \Rightarrow p = \frac{\sqrt{E^2 - m_0^2c^4}}{c}$$

$$\Rightarrow \qquad \lambda = \frac{hc}{\sqrt{E^2 - m_0^2c^4}} \qquad (10.49)$$

Here m_0 is the rest mass of the particle. Note that for a photon the rest mass is zero and $\lambda = hc/E = hc/h\nu = c/\nu$.

10.7.2 Statistical Interpretation of Matter Waves

The physical significance of the amplitude of the matter waves is one of the key ideas in quantum mechanics. In general, a wave is a disturbance varying in time and space. The magnitude of the disturbance is given by the amplitude of the wave. In sound waves the amplitude corresponds to pressure variations while in light waves it corresponds to variations in electric field and magnetic field. *However in matter waves the amplitude does not correspond to any physical parameter like pressure, electric field or magnetic field. Matter wave is an abstract concept. The square of the amplitude of a matter wave gives the probability of locating the particle in space.* Thus if $\psi(x)$ denotes the wave function of a particle moving along x-axis, then $\psi^2(x)\,dx$ denotes the probability of finding the particle between x and $x + dx$. Similarly if $\psi(x, y, z)$ denotes the wave function of a particle in three dimensional space, $\psi^2(x, y, z)\,dx, dy. dz.$ denotes the probability of finding the particle in a volume element $dv = dx.dy.dz$ at (x, y, z).

The square of the amplitude in the case of sound and light wave is related to their intensity. The intensity of light is the energy per unit area per second. Light energy is equal to the number of photons multiplied by the energy of each photon. Thus the probability of finding a photon in an area is proportional to the light intensity.

10.7.3 X-ray Diffraction

The crucial experiments to verify de Broglie's hypothesis was conducted with electrons by C. J. Davisson and C.H. Germer in 1931. By then it was well known that X-ray photons do behave like waves. The wavelength of X-rays is of the order of Angstroms (~ 10^{-10} m). The distance between the lattice planes in a crystal is also of the order of Angstroms. The X-rays when scattered by the atoms in a crystal are expected to exhibit diffraction effects. X-ray diffraction in single crystals as well as the powdered cyrstalline specimens is shown in Figs. 10.40 and 10.41. The phenomenon is similar to that of light exhibiting diffraction when it passes through a diffraction grating. X-rays when diffracted by single crystals give rise to discrete spots known as *Laue spots* on the film. With powders one observes continuous concentric circles. Such patterns arise because X-rays when scattered by the atoms of the crystal exhibit constructive interference only in some directions and satisfy what is popularly known as the Bragg condition.

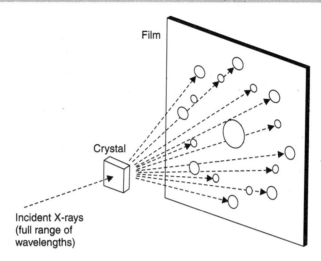

Fig. 10.40 X-ray scattering by a single crystal. An interference maximum (dot) appears on the film whenever a set of crystal planes happen to satisfy the Bragg condition for a particular wavelength. These dots are often called Laue spots or Laue pattern or Laue photograph

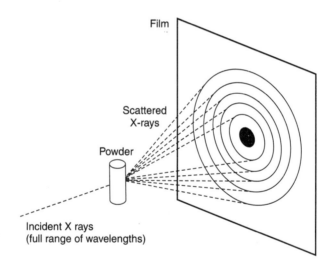

Fig. 10.41 X-ray scattering by a powdered sample. Crystallites in the powder have all possible orientations. The diffraction pattern is in the form of concentric circles

Bragg's law of X-ray diffraction

Fig. 10.42 shows a family of planes, with *interplanar spacing d*. Rays 1 and 2 get reflected from the first and second planes respectively. Reflections from the other planes are not shown for the sake of clarity. The angle made by the X-ray beam with the plane is called the *Bragg angle* and is denoted by θ. Note that it is the compliment of the angle of the incidence which is defined as the angle between the normal and the directions of the incident X-ray beam. At each plane X-rays obey the laws of reflection *i.e.*, the angle of incidence is equal to the angle of reflection. There is a path difference associated with rays 1 and 2. Simple geometry shows that the path difference is $2d \sin \theta$. Thus the condition for intensity maxima is given by:

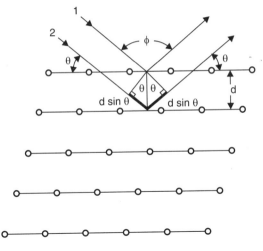

Fig. 10.42 Bragg's law

$$2d \sin \theta = n\lambda \tag{10.50}$$

This is known as *Bragg's law*. The magnitude of the intensity of scattered X-rays is determined by the atomic number of the particular element. X-rays are scattered by the electron cloud surrounding the nucleus of atoms. Hence elements with a large atomic number scatter X-rays better than the elements with lower atomic number. X-ray diffraction is a powerful technique for locating the positions of atoms in the cyrstal. Crystal structure analysis is routinely done using X-rays.

10.7.4 Electron Diffraction

Electrons also should exhibit diffraction effects similar to X-rays, if the wavelength of the waves associated with electrons is of the order of Angstroms. If an electron is accelerated through a voltage V_o, its kinetic energy is eV_o. Its momentum (p) and hence its de Broglie wavelength is given by:

$$p = mv = \sqrt{2m\left(\frac{1}{2}\right)mv^2} = \sqrt{2meV_o}$$

$$\lambda = \frac{h}{p} = \frac{h}{\sqrt{2meV_0}} = \frac{12.3}{\sqrt{V_0}} \times 10^{-10} \text{ m} \tag{10.51}$$

The de Broglie wavelength of ~ 150 eV electrons is of the order of 1A. Hence observation of diffraction effects with electrons having energies of the order of 100 – 200 eV ought to be possible. This argument motivated Davisson and Germer to observe electron diffraction in crystals. The experimental set-up for observing diffraction effects with electrons is shown in Fig. 10.43(a). Electrons from a hot filament are accelerated through various potential differences to produce beams of different energy. The electron beam is directed onto a surface cut in a single crystal of nickel. The scattered electrons are observed at different angles by varying the position of the detector with respect to the direction of the incident beam. The intensity of the scattered electrons is plotted as a function of scattering angle ϕ for each beam energy as shown in Fig. 10.43(b). The angle ϕ is the angle between the incident and the Bragg reflected beam. Fig. 10.43(b) is a radial plot, the length of the radius vector at any angle ϕ is denoted by its length. Thus the intesity of reflected electrons is maximum for certain angles of ϕ. These observations indicate that scattering of electrons from the crystal target is just like the diffraction of X-rays by the same cyrstal. The observations can be explained by applying matter-wave hypothesis to electrons and recognising the effect as due to Bragg reflection of electrons.

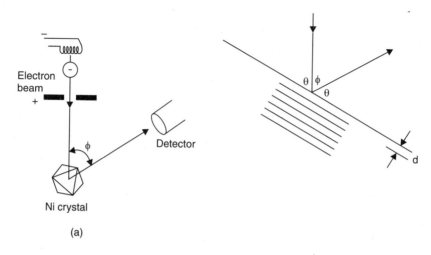

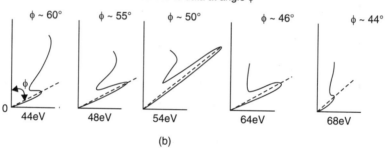

Fig. 10.43 (a) Davisson Germer experiment–electron diffraction from a Nickel crystal
(b) Variation of the intensity of the scattered electrons as a function of the scattering angle

Nickel belongs to the cubic structure with a cubic edge length D equal to 0.215 nm. There are many crystallographic planes from which electron diffraction can be observed. However we shall consider the one which is inclined at angle α with the horizontal as shown in Fig. 10.44. The interplanar distance $d = D \sin \alpha$. The relationship between the incident electron beam and the nickel cyrstal is shown in Fig. 10.43. In the Bragg's law, 2θ is the angle between the incident and exit beams.

$$\therefore \quad \phi = \pi - 2\theta = 2\alpha \;\Rightarrow\; 2\theta = \pi - 2\alpha \;\text{ or }\; \theta = \frac{\pi}{2} - \alpha$$

$$\therefore \quad \sin \theta = \sin\left(\frac{\pi}{2} - \alpha\right) = \cos \alpha$$

Hence the Bragg condition reduces to:

$$n\lambda = 2d \sin \theta = 2d \cos \alpha = 2D \sin \alpha \cos \alpha = D \sin 2\alpha = D \sin \phi$$

$$\therefore \quad \lambda = \frac{D \sin \phi}{n} = \frac{1.23}{\sqrt{V_0}}\text{nm} \;\text{ or }\; \sin \phi = \frac{1.23}{D\sqrt{V_0}}\text{nm for } n = 1$$

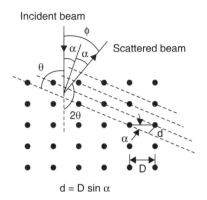

Fig. 10.44 Lattice of Ni crystal showing the Bragg planes

For nickel the interatomic distance $D = 0.215$ nm. Hence one would expect the peak at $\phi \approx 50°$ for electrons accelerated through 56 V.

Electron diffraction was observed in transmission geometry by Thomson. Laue spots (Fig. 10.40) and circular diffraction rings (Fig. 10.41) were observed with electron beam in thin samples of silver, gold and platinum.

10.7.5 Electron Microscopy

Apart from being a key idea of primary importance in quantum mechanics, de Broglie's hypothesis provided impetus for the design of *electron microscope,* which is routinely used for the structural analysis of materials. An electron microscope is similar to optical microscope except that a beam of electrons is employed instead of light. It is a device for obtaining pictures of high resolution employing a beam of electrons and is more efficient than an optical microscope.

In addition to the magnifying power, the microscope is characterized by *resolving power*. *Resolving power* is the ability of the microscope to produce separate images of two neighbouring points separated by a distance d on the object. The resolving power of a microscope is given by:

$$d = \frac{1.22\lambda}{2\sin\alpha}$$

Where λ is the wavelength of light used and α is the angle subtended by the object at the microscope. Hence resolving power can be increased by using light of shorter wavelengths. Since electrons behave like waves, in principle it is possible to obtain higher resolving power with electrons provided suitable lenses can be designed for focusing electrons. Assume that the electron is accelerated through a voltage V_o, According to eqn. (10.49) in the relativistic limit the de Broglie wavelength is given by:

$$\lambda = \frac{hc}{\sqrt{E^2 - m_o^2 c^4}} = \frac{hc}{\left[\left(eV_o + m_o c^2\right)^2 - m_o^2 c^4\right]^{1/2}} = \frac{h}{\left[2m_o eV_o\left(1 + \frac{eV_o}{2m_o c^2}\right)\right]^{\frac{1}{2}}}$$

Table 10.2 summarises the de Broglie wavelength of electrons for various accelerating voltages. Note that $m_o c^2$ is the rest mass of the electron and is approximately equal to 0.512 MeV. For low voltages *i.e.*, when $eV_o \ll m_o c^2$, the above expression reduces to non-relativistic expression of eqn. (10.50)

Table 10.2 de Broglie wavelengths of electrons accelerated through various voltages

Accelerating Voltage (kV)	Non-relativistic (λ) nm	Relativistic (λ) nm	Mass ($\times m_o$)	Velocity ($\times 10^8$ m/s)
100	0.00386	0.00370	1.196	1.644
120	0.00352	0.00335	1.235	1.759
200	0.00273	0.00251	1.391	2.086
300	0.00223	0.00197	1.587	2.330
400	0.00193	0.00164	1.783	2.484
1000	0.00122	0.0087	2.957	2.823

Thus the de Broglie wavelength of accelerated electrons is more than thousand times smaller than that for visible light. Hence resolving power of about a thousand times smaller than the optical microscope is possible. A perusal of the table reveals that with an electron microscope one can obtain resolving power of the order of a fraction of an angstrom.

The construction of electron microscope became possible owing Busch who demonstrated in 1926-27 that a suitably shaped magnetic field could be used as a lens in electron microscope. The production, propagation and focusing of electrons properly belongs to the subject of electron optics. This is discussed in the chapter on diffraction of light.

10.7.6 Neutron Diffraction

The de Broglie hypothesis was later verified for neutrons as well. It was found that neutrons do exhibit the phenomenon of Bragg diffraction (a wave property) on being scattered by atoms in a crystal. This is possible when the de Broglie wavelength of the neutrons is also of the order of Angstroms. The de Broglie relation can be expressed as:

$$\lambda = \frac{h}{p} = \frac{h}{mv} = \frac{h}{\sqrt{2mE}}$$

where E is the kinetic energy of the neutron.

Assuming neutron to be similar to monoatomic gas,

$$E = \frac{3}{2} k_B T$$

$$\therefore \quad \lambda = \frac{h}{\sqrt{2mE}} = \frac{h}{\sqrt{3mk_B T}} = 0.145 \times 10^{-9} \text{ m at } T = 300 \text{ K}$$

Neutrons with these de Broglie wavelengths are obtained from the nuclear reactors. The neutrons which are released during the fission, thermalise by repeated scattering in a graphite moderator that is cooled to room temperature. These neutrons can be utilised to observe diffraction effects in crystals.

X-rays are electromagnetic waves and hence the X-ray diffraction involves the electromagnetic interaction between the X-rays and the electron cloud that surround the nucleus. Neutrons being neutral particles are not scattered by electrons due to electric fields. Neutron has a magnetic moment and hence the interaction involved between the neutrons and the atoms is magnetic in nature. Atoms of paramagnetic and ferromagnetic materials possess a magnetic moment. The contribution to this magnetic moment arises mainly because

MATTER AND RADIATION – DUAL NATURE

of motion of spinning electrons around the nucleus. The contribution of the nucleus to the magnetic moment is negligible. Neutron diffraction is therefore employed to determine the crystal structure especially of magnetic materials. It is also utilised to locate the accurate position of light atoms like hydrogen in the unit cell, since the X-ray scattering from hydrogen atoms is weak.

A simple example to illustrate the advantages of using neutrons vis-à-vis X-rays for probing the atomic structure is provided by metallic chromium. The lattice is cubic (bcc). The structure is such that atoms at the body-centre location have magnetic moments opposite to those at the corners (Fig. 10.45). The structure can be regarded as consisting of two interpenetrating simple cubic lattices of Cr atoms having antiparallel magnetic moments. Therefore, although the atoms are identical as far as X-rays are concerned, they are different from the viewpoint of neutrons, on account of the difference in the orientation of the magnetic moments. Hence by neutron diffraction these two different sublattices can be identified. In X-ray diffraction it is impossible to distinguish these two sublattices. Thus neutron diffraction complements X-rays diffraction in the analysis of structure of crystals.

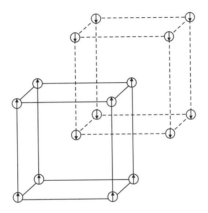

Fig. 10.45 In the case of an antiferromagnetic crystal, where the spins of one set of atoms are aligned antiparallel to those of the other set, neutron diffraction detects two interpenetrating simple cubic lattices on account of the magnetic interaction of the neutron with the atoms. X-ray diffraction would see only a single bcc lattice

10.7.7 Young's Double Experiment with Neutrons

Young's double slit experiment demonstrated the wave properties of light. The conventional Young's double slit experimental set up is shown in Fig. 10.46. A light source illuminates a tiny pin hole S. Light diverging from this pin hole falls on a barrier containing two narrow rectangular apertures S_1 and S_2, which are very close to each other and separated by a distance d. Spherical wavefronts travelling from S_1 and S_2 are coherent since S_1 and S_2 lie on the same wavefront emitted by light passing through S. An interference pattern is obtained on the screen which is at a distance D from the two slits. The wavelength (λ) of the light wave is related to the fringe width (β) by the relation:

$$\lambda = \frac{\beta d}{D}$$

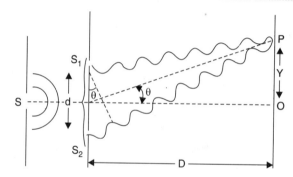

Fig. 10.46 Young's experiment with photons

The one important criterion for the observation of the interference pattern is that the two slits ought to be sufficiently close enough and be illuminated by a single source. If the two slits are separately illuminated by two independent sources, then no interference pattern is observed. This is because the slits S_1 and S_2 ought to lie on the same wave front. According to Huygen's principle, every point on the wavefront acts as a source of light. Hence the light waves emitted at S_1 and S_2 are in phase and thus S_1 and S_2 act as coherent sources. At any point on the screen, the amplitude of the light wave is given by the algebraic sum of the amplitudes of the coherent waves emitted by S_1 and S_2. The intensity is the square of the resultant amplitude. Points on the screen where the path difference between the two waves is an integral multiple of the wavelength corresponds to maximum brightness. Points on the screen where the path difference between the two waves is an integral multiple of half the wavelength correspond to minimum brightness.

Young's double slit experiment can also be performed with the source of neutrons. Figure 10.47a shows the experimental set up. Thermal neutrons from the nuclear reactor are the source. However, to get a monoenergetic neutrons (corresponding to monochromatic source of light) these neutrons are incident on a crystal and by choosing the scattering angle for Bragg reflection, it is possible to obtain a specific wavelength. In the experiment neutrons of kinetic energy 0.00024 eV which corresponds to a de Broglie wavelength of 1.85 nm were passed through a gap of diameter 148 μm in a material that absorbs virtually all of the neutrons incident on it. In the center of the gap is a boron wire of diameter 104 μm. Boron has a high absorption for neutrons. The neutrons can pass on either side of the wire through slits of width 22 μm. The intensity of neutrons that pass through this double slit was observed by sliding another slit across the beam and measuring the intensity of neutrons passing through this *scanning slit*. The intensity pattern observed is shown in Fig. 10.47b. The wavelength of neutrons can be estimated from the slit separation using the equation used in Young's double slit experiment. Estimating the spacing and hence the fringe width from Fig. 10.47b, to be about 75 μm, the wavelength is found to be:

$$\lambda = \frac{d\beta}{D} = \frac{(126 \, \mu m)(75 \, \mu m)}{5 \, m} = 1.89 \text{ nm}$$

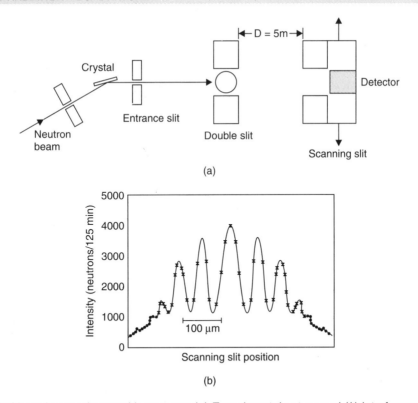

Fig. 10.47 Young's experiment with neutrons (a) Experimental set up and (b) interference pattern

The result agrees very well with the de Broglie wavelenth of 1.85 nm selected for the neutron beam. Young's double slit experiment has been performed with atoms such as helium as well. Thus the wave property has now been established to be a universal attribute of matter.

10.7.8 Mystery of Single Particle Interference

Young's double slit experiment is a classic experiment which is regarded as a conclusive evidence for the wave nature of light (Fig. 10.46). The intensity pattern does not depend on the intensity of the light source at all. Suppose one perform the double slit experiment with a very weak light source, emitting very few photons per second such that there is only one photon at any time in the apparatus. Then it is observed that after long intervals of time still the same intensity pattern is observed. It is impossible to account for the appearance of the interference pattern, by assuming the single photon passes either through S_1 or S_2. Also a single photon cannot split into two. This observation cannot be accounted on the basis of classical notions. *It can only be explained on the basis of quantum mechanics which assumes that a single photon somehow passes through both the slits simultaneously!* Needless to say that it is contrary to our common sense. Somehow a single photon interferes with itself to give rise to the interference pattern. This is the mystery of single particle interference.

Another experiment in support of wave nature of light is polarization. When polarized light passes through a sheet of Polaroid, the intensity of light emerging out of the Polaroid is given by

$$I = I_o \cos^2 \alpha$$

Where α is the angle between the direction of the electric field of incident polarized light and the axis of the Polaroid. (Fig. 10.48). This result also is independent of the light intensity used. If a very weak source is used and at any time only one photon is present in the apparatus, still the $\cos^2 \alpha$ form is obtained when the experiment is carried out over long intervals of time. This is incomprehensible since the single photon passes through the Polaroid or does not pass through. It is not possible for the single photon to split into two so that a part of it goes through in each event.

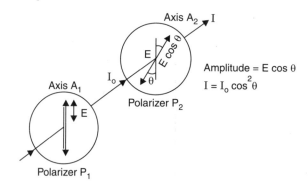

Fig. 10.48 Malthu's law-single photon interference using polariser

10.7.9 Principle of Complementarity

The particle aspect of matter requires the matter to be *localised*. A particle is an object occupying a certain definite volume in space at any given time. Its position at this time can also be specified. On the other hand, a matter wave always extends over a certain region. The two views are thus basically different. It is this difference which causes difficulties when one attempts to interpret a given experimental observation in terms of both waves and particles. *In a given experiment, if matter or radiation exhibits the particle character, then it is impossible to prove the wave character in the experiment. Similarly if the wave nature of either matter or radiation is exhibited in an experiment then it is impossible to prove their particle nature. This idea put forth by Niels Bohr is referred to as the principle of complementarity.*

10.8 PHASE VELOCITY AND GROUP VELOCITY OF WAVES

10.8.1 Introduction

The two seemingly different concepts, namely of particles and waves are not necessarily contradictory. The particle behaviour of matter can be understood in terms of *localised waves* or *wave packets*, which exist only in a small region of space. Consider a free particle of momentum p and energy E moving along the positive x-axis. Let us associate with the particle a *wave packet*. Wave packet is a wave confined to a small region, moving with the same velocity as that of the particle. A snapshot of the wave packet at a given time is shown in the Fig. 10.49(a). *The wave packet has to be necessarily confined to a small region since the square of the wave function denotes the probability of finding the particle.* The wave is quite different from a pure sinusoidal wave, which extends over the entire region of space.

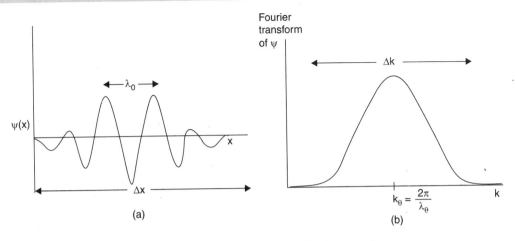

Fig. 10.49 A wave packet and its Fourier transform

10.8.2 Wave Packet and Fourier Analysis

A wave packet can be constructed by superposing several sinusoidal waves. This basic idea is due to the French mathematician Joseph Fourier (1768 – 1830). According to the theorem which bears his name, any periodic wave shape, however complicated can be built from sinusoidal waves. The frequencies of these sinusoidal waves are harmonics of a fundamental frequency. By choosing the amplitudes and phases of these harmonic waves and summing them up, the wave shape can be generated. This process is known as *Fourier synthesis*. A specification of the strengths (or amplitudes) of various harmonics is often referred to as spectrum. Hence this process is known as *harmonic analysis or spectrum analysis*. The wavelength of the constituent sinusoidal waves and their amplitude are shown in Fig. 10.49*b*. The Fourier spectra is illustrated for sawtooth, square and triangular waveshapes (Fig. 10.50). The Fourier synthesis of a sawtooth wave considering first six harmonics is shown in Fig. 10.51.

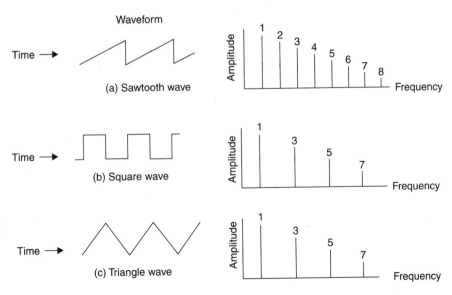

Fig. 10.50 Fourier analysis of different waveshapes (*a*) sawtooth wave (*b*) square and (*c*) triangle wave

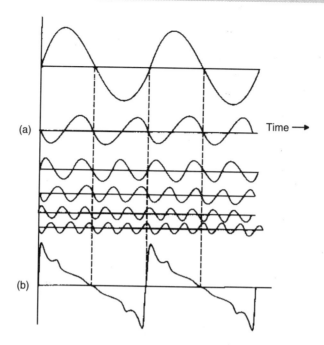

Fig. 10.51 Fourier synthesis of a sawtooth wave by considering the first six harmonics

In a wave packet, the superposition of constituent waves gives rise to constructive interference in a small region. The maximum peak produced by such a constructive interference moves with a velocity called the *group velocity* rather than the *phase velocity i.e.,* the velocity of an individual sinusoidal wave. We will first derive an expression for the group velocity of a wave packet.

Consider a single sinusoidal wave. The phase velocity associated with the wave is given by

$$v_p = \frac{\omega}{k} \tag{10.52}$$

Now consider a waveform consisting of two sinusoidal waves 1 and 2 of wavelengths λ and $\lambda + d\lambda$ (Fig. 10.52). The two waves travel with phase velocities v and $v + dv$ respectively. The two sinusoidal waves can be represented as

$$y_1(x, t) = A \sin(kx - \omega t) = A \sin[k(x - v_p t)] \tag{10.53}$$
$$y_2(x, t) = A \sin[(k + dk)x - (\omega + d\omega)t] \tag{10.54}$$

The sum of the two waves is

$$Y = y_1 + y_2 = A \sin(kx - \omega t) + A \sin[(k + dk)x - (\omega + d\omega)t]$$

Making use of the formula,

$$\sin A + \sin B = 2 \sin \frac{A+B}{2} \cos \frac{A-B}{2}$$

and making the approximation $2k + dk \approx 2k$ and $2\omega + d\omega \approx 2\omega$, we get

$$Y = 2A \cos\left[\frac{x.dk - t.d\omega}{2}\right] \sin(kx - \omega t)$$

$$= 2A \cos\left[\frac{dk}{2}\left(x - t\frac{d\omega}{dk}\right)\right] \sin(kx - \omega t) \tag{10.55}$$

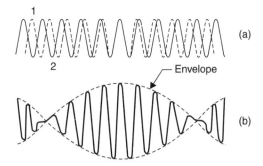

Fig. 10.52 A wave packet formed by the superposition of two waves

As can be seen, the second term on the right hand side here is the same as the term in eqn. (10.53). The total amplitude of the superposed waves is shown in the Fig. 10.51. As in eqn (10.53), the first term $\cos[(xdk - td\omega)/2]$ represents the envelope (shown by dashed curves in this figure) which modulates the amplitude of $\sin(kx - \omega t)$. The envelope is the wave packet formed by the two sinusoidal waves. The group velocity v_g of the envelope is obtained by examining the first term of equation (10.55) and is given by

$$v_g = \frac{d\omega}{dk} \qquad (10.56)$$

When the number of waves forming a wave packet is increased, the width of the envelope reduces and the height increases (Fig. 10.53). In other words, the wave packet becomes narrower, leading to a better localization. The group velocity of this wave packet would still be the same as that given by eqn. (10.56). Its energy, which is proportional to the square of its amplitude, is clearly localised in a small region.

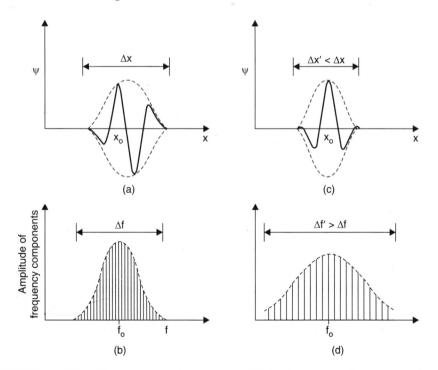

Fig. 10.53 The width of the wavepacket decreases with the increase in the number of waves

Thus effectively a wave packet is the sum of several sinusoidal waves of different wavelength. If all these sinusoidal waves travel with a speed 'v' then the wave packet will also travel with the speed 'v'. A medium in which sinusoidal waves of different wavelengths travel with the same speed is called a *non-dispersive medium*. In such a medium, the wave packet travels without changing its shape. A non-disperse medium is more of an exception than a rule. Most media are dispersive. In such a medium, the velocity of the wave depends on the wavelength of the wave. A wave packet traveling in a dispersive medium changes its shape as it moves (Fig. 10.54).

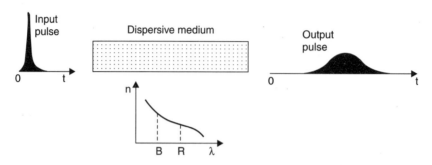

Fig. 10.54 Change in the shape of the wave packet as it passes through a dispersive medium

The above discussion can be best appreciated by considering the relation between the group velocity (v_g) and phase velocity (v_p). We have

$$v_g = \frac{d\omega}{dk}; \quad v_p = \frac{\omega}{k}; \quad k = \frac{2\pi}{\lambda} \quad \therefore \quad \frac{dk}{d\lambda} = -\frac{2\pi}{\lambda^2}$$

We have $\omega = v_p k$

$$\therefore \quad v_g = \frac{d\omega}{dk} = \frac{d}{dk}(v_p k) = v_p + k \cdot \frac{dv_p}{dk} = v_p + k \cdot \frac{dv_p}{d\lambda} \frac{d\lambda}{dk}$$

i.e.,
$$v_g = v_p + k \cdot \frac{dv_p}{d\lambda}\left(-\frac{\lambda^2}{2\pi}\right) = v_p - \lambda \frac{dv_p}{d\lambda} \quad (10.57)$$

Clearly in a non-dispersive medium $dv_p/d\lambda = 0$ and hence $v_g = v_p$.

10.8.3 Phase Velocity and Group Velocity of de Broglie Waves

We will show that for both non-relativistic and relativistic case the group velocity of the wave packet represents the velocity of the particle.

Non-relativistic case

An in the case of the photon, one can assign a frequency ν such that the energy of the particle $E = h\nu$. Its de Broglie wavelength is $\lambda = h/p$.

$$v_g = \frac{d\omega}{dk} = \frac{d(2\pi\nu)}{d(2\pi/\lambda)} = \frac{d(\nu)}{d(1/\lambda)} = \frac{d(h\nu)}{d(h/\lambda)} = \frac{dE}{dp} = \frac{d(p^2/2m)}{dp} = \frac{p}{m} = v$$

which is the velocity of the particle.

Relativistic case

The same result can be proved for relativistic case also as shown below. According to the theory of relativity the total energy E associated with a body of rest mass m_o moving with a velocity v is given by

MATTER AND RADIATION – DUAL NATURE

$$E = mc^2 = \frac{m_0 c^2}{\sqrt{1-\dfrac{v^2}{c^2}}} = h\nu$$

where a frequency ν has been assigned to the particle as in the case of a photon.

$$\therefore \quad \omega = 2\pi\nu = \frac{2\pi mc^2}{h} = \frac{2\pi m_0 c^2}{h\sqrt{1-\dfrac{v^2}{c^2}}}$$

where ω is the angular frequency of the de Broglie wave. The wave vector

$$k = \frac{2\pi}{\lambda} = \frac{2\pi}{(h/p)} = \frac{2\pi m v}{h} = \frac{2\pi m_0 v}{h\sqrt{1-\dfrac{v^2}{c^2}}} \qquad (\because \lambda = h/p)$$

The group velocity of the de Broglie wave is given by

$$v_g = \frac{d\omega}{dk} = \frac{d\omega/dv}{dk/dv} = c^2 \frac{\dfrac{d}{dv}\dfrac{1}{\sqrt{1-\dfrac{v^2}{c^2}}}}{\dfrac{d}{dv}\dfrac{v}{\sqrt{1-\dfrac{v^2}{c^2}}}} = \frac{c^2\left(-\dfrac{1}{2}\right)\left(1-\dfrac{v^2}{c^2}\right)^{-3/2}\left(-\dfrac{2v}{c^2}\right)}{\left(1-\dfrac{v^2}{c^2}\right)^{-1/2}+v\left(-\dfrac{1}{2}\right)\left(1-\dfrac{v^2}{c^2}\right)^{-3/2}\left(-\dfrac{2v}{c^2}\right)}$$

$$= \frac{v\left(1-\dfrac{v^2}{c^2}\right)^{-3/2}}{\left(1-\dfrac{v^2}{c^2}\right)^{-1/2}+\dfrac{v^2}{c^2}\left(1-\dfrac{v^2}{c^2}\right)^{-3/2}} = v$$

Thus the group velocity of the de Broglie wave is the velocity of the particle itself.
The phase velocity v_p of the de Broglie waves is given by

$$v_p = \frac{\omega}{k} = \frac{2\pi mc^2/h}{2\pi mv/h} = \frac{c^2}{v} \tag{10.58}$$

Thus the phase velocity of the de Broglie waves exceeds both the velocity of the body as well as that of light.

10.8.4 Phase and Group Velocity of Light Waves

The relation between the phase velocity and the group velocity is valid for *all types of waves*. It is useful to apply these concepts for the case of light waves, since in fiber optic communication light is sent in the form of *pulses or wave packets*. It is customary to express phase and group velocities in terms of the refractive index of the medium. The phase refractive index and the group refractive indices are defined by the relations:

$$n = \frac{c}{v_p}, \quad N = \frac{c}{v_g} \tag{10.59}$$

where n and N are the phase and group refractive indices respectively. From eqns. (10.57) and (10.59)

$$\frac{c}{N} = \frac{c}{n} - \lambda \frac{d\left(\frac{c}{n}\right)}{d\lambda}$$

$$\Rightarrow \quad \frac{1}{N} = \frac{1}{n} - \lambda \frac{d\left(\frac{1}{n}\right)}{d\lambda} = \frac{1}{n} - \lambda \frac{d(n^{-1})}{dn} \frac{dn}{d\lambda}$$

$$\Rightarrow \quad = \frac{1}{n} - \lambda \left(-\frac{1}{n^2}\right) \frac{dn}{d\lambda} = \frac{1}{n}\left[1 + \frac{\lambda}{n}\frac{dn}{d\lambda}\right]$$

$$\Rightarrow \quad N = n\left[1 + \frac{\lambda}{n}\frac{dn}{d\lambda}\right]^{-1} \approx n\left[1 - \frac{\lambda}{n}\frac{dn}{d\lambda}\right] \approx n - \lambda \frac{dn}{d\lambda} \quad (10.60)$$

The velocity with which a light pulse or a light wave packet travels in an optical fiber is given by $v_g = c/N$. In a non-dispersive medium, the refractive index is not a function of wavelength i.e., $dn/d\lambda = 0$. Hence the group velocity is the same as the phase velocity. i.e., all the components of the wave packet travel with the same velocity. However, in a dispersive medium the components of the wave packet travel with different velocities. As a result pulse shape also gets altered during the transit through the fiber. This has lot of consequences in fiber optic communication. The distortion of pulses leads to overlapping of pulses in digital communication and consequent loss of information content (Fig. 10.55).

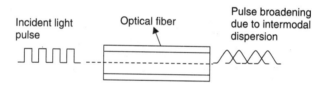

Fig. 10.55 Distortion of light pulses in optical communication

10.9 HEISENBERG'S UNCERTAINTY PRINCIPLE

10.9.1 Uncertainty Principle as Consequence of Dual Nature of Matter and Radiation

The fact that an object exhibits both particle-like and wave-like aspects has far reaching consequences. One such very important result is the uncertainty principle formulated by W. Heisenberg. According to this principle it is impossible to know both the exact position and exact momentum of an object at the same time.

A particle is represented by a wave packet or a group of waves. If the wave packet is narrow, then the particle's position can be specified to a great accuracy. However, a narrow wave packet can arise only from the superposition of a large number of sinusoidal waves of different wavelengths. Since the wavelength $\lambda = h/mv$, it means that the particle's momentum is not very precise. A single sinusoidal wave has a precise wavelength but then that can represent only a particle which is completely delocalised i.e., the position of the particle is not defined. The wave packet is represented by $\psi(x)$, the wave function.

In the language of mathematics, the summation is replaced by integration:

$$\psi(x) = \int_0^\infty g(k) \sin(kx) \, dk \quad (10.61)$$

where $g(k)$ denotes the amplitude of the sine function in the momentum space. ($p = h/\lambda = hk/2\pi$). Integration denotes summation over all the possible wavelengths. *$g(k)$ is known as the Fourier*

MATTER AND RADIATION – DUAL NATURE

transform of the function $\psi(x)$ Fig. 10.49b is a pictorial representation of this idea. The relation between Δx and the wave number spread Δk depends upon the shape of the wave packet. The minimum value of the product $\Delta x \cdot \Delta k$ occurs when the envelope of the group has the familiar bell shape of a Gaussian function (Fig. 10.56).

$$\psi(x) = \exp\left(-\frac{x^2}{(\Delta x)^2}\right)$$

Thus allowing that the wave packet is represented by a Gaussian distribution, then

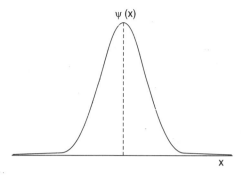

Fig. 10.56 A Gaussian wave packet

$$\Delta x \cdot \Delta k \geq \frac{1}{2}.$$

$$k = \frac{2\pi}{\lambda} = \frac{2\pi p}{h} \quad \text{(de Broglie's hypothesis)}$$

$$\therefore \quad \Delta k = \frac{2\pi}{h}\Delta p \quad \Rightarrow \quad \Delta x \cdot \Delta p_x \geq \frac{h}{4\pi}$$

Similarly, $\quad \Delta y \cdot \Delta p_y \geq \dfrac{h}{4\pi} \quad$ and $\quad \Delta z \cdot \Delta p_z \geq \dfrac{h}{4\pi}$ \hfill (10.62)

Wave Packets in Time Domain

So far we have thought of wave packets in space. One can also examine the dependence of wave packet in time, at a particular point in space. Similar to eqn. (10.61) we have

$$\psi(t) = \int_0^\infty g(\omega)\sin\omega t \, d\omega \tag{10.63}$$

Thus wave packet in the time domain can be thought of as a superposition of a large number of sinusoidal waves of different frequencies (Fig. 10.57). Once again if the wave packet is Gaussian we have

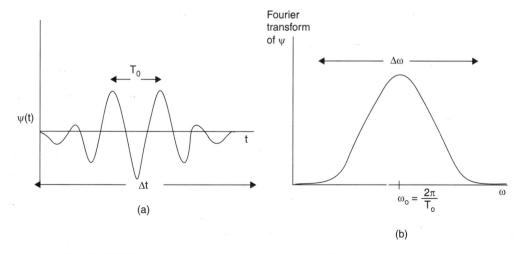

Fig. 10.57 Wave packet as a function of time and its frequency components

$$\Delta t \Delta \omega \geq \frac{1}{2}. \quad \text{But} \quad E = \frac{h}{2\pi}\omega \quad \therefore \quad \Delta E = \frac{h}{2\pi}\Delta\omega$$

$$\Rightarrow \quad \Delta E.\Delta t \geq \frac{h}{4\pi} \tag{10.64}$$

This shows that there is an uncertainty in the measurement of ΔE as well. Δt refers to the uncertainty in the time the particle spends in that energy state. Δt is a measure of the life time of the energy state. *Energy is frequency multiplied by the Planck's constant i.e., $E = h\nu$. Hence the wavepacket in time domain can be thought of as a superposition of different energy states.*

Diffraction of Electrons at a Single Slit

Heisenberg's uncertainty principle can also be understood by considering the diffraction of electrons at a single slit. Consider a beam of electrons moving upward (along y-axis) with a speed v_o (Fig. 10.58). They pass through a narrow slit of width Δx and undergo diffraction because of their wave like nature. The angle corresponding to the first minimum in the diffraction intensity is given by

$$\sin \theta = \frac{\lambda}{\Delta x} \approx \frac{\Delta v_x}{v_0}$$

$$\Rightarrow \quad \Delta x.\Delta v_x \approx \lambda v_0$$
$$\therefore \quad \Delta x.\Delta m v_x = \Delta x.\Delta p_x \approx m\lambda v_0 \approx \lambda p \approx h$$

By narrowing down the slit width, one can locate the position of electron with great accuracy. However, narrower the slit, greater is the value of θ. Hence the uncertainty in momentum increases.

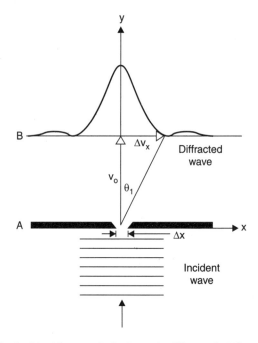

Fig. 10.58 An incident beam of electrons is diffracted at the slit in screen. As the screen is made narrower, the diffraction pattern becomes wider

Diffraction of Light at a Grating

Heisenberg's uncertainty principle can also be illustrated by considering the diffraction of light at a grating (Fig. 10.59). A grating can be regarded as a series of multiple slits. Assume a beam of light of diameter l is collimated by a slit and impinges on the grating. The grating constant of the grating is d. Thus the number of grating lines which are important for the diffraction are $N = l/d$. The resolving power of a grating is its ability to separate two waves of different wavelengths and is given by

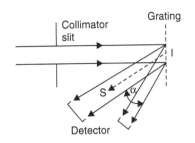

Fig. 10.59 Diffraction of light at a grating

$$\frac{\Delta \lambda}{\lambda} = \frac{1}{N} = \frac{d}{l} \tag{10.65}$$

The uncertainty in the position (Δs) of the photon is equal to the diameter of the beam

$$\therefore \quad \Delta s = l \tag{10.66}$$

The momentum associated with the photons is given by

$$p = \frac{h}{\lambda}$$

$$\Rightarrow \quad dp = -\frac{h}{\lambda^2} d\lambda = -p\frac{d\lambda}{\lambda}$$

$$\Rightarrow \quad \Delta p = -p\frac{\Delta \lambda}{\lambda} \tag{10.67}$$

From eqns. (10.65), (10.66) and (10.67)

$$\Delta p \Delta s = -p\frac{\Delta \lambda}{\lambda} \Delta s = -\frac{h}{\lambda} \frac{1}{N} \frac{l}{\alpha} = \frac{h}{\lambda} \cdot \frac{d}{l} \cdot l = h \cdot \frac{d}{\lambda} \tag{10.68}$$

In order to observe diffraction d and λ must at least be of the same order of magnitude. Hence we have

$$\Delta p \Delta s \approx h.$$

10.9.2 Uncertainty Principle as a Consequence of Observation

Gamma Ray Microscope

The uncertainty principle can also be thought as arising due to the act of observation. The very act of observation disturbs the object of measurement. Hence the uncertainty in the physical parameters of measurement becomes inevitable. Two examples are provided here to illustrate this point of view.

Figure 10.60 shows the experimental arrangement in which an electron is viewed through a microscope. The electron can be observed only if the photon scattered by it reaches the observer's eye. However, when the photon bounces off the electron into the microscope, it transfers momentum to the electron. The magnitude of the momentum of the photon cannot be known precisely because the photon may have been scattered anywhere within the angular aperture 2θ subtended by the objective lens at the electron. Hence the uncertainty in momentum of electron (Δp_x) is

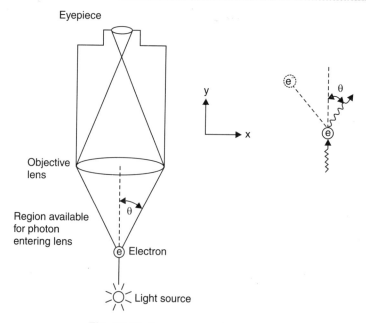

Fig. 10.60 Gamma ray microscope

i.e.,
$$\Delta p_x = 2p \sin \theta = \frac{2h}{\lambda} \sin \theta \quad (10.69)$$

The resolving power of the microscope is given by

$$\Delta x \approx \frac{\lambda}{\sin \theta} \quad (10.70)$$

This is because the microscope's image of the point object is not a point object, but a diffraction pattern. The resolving power determines the ultimate accuracy with which the electron can be located. From eqns. (10.69) and (10.70)

$$\Delta p_x \cdot \Delta x \approx 2h$$

Counting the Number of Waves Crossing a Point

Figure. 10.61 a train of ocean waves passing a buoy that is anchored at a fixed point. An observer on the buoy could measure the frequency (number of waves passing the buoy per unit time) at time t_0 by counting the number of crests and troughs passing the buoy between the times $t_0 - \Delta t$ and $t_0 + \Delta t$, dividing by 2 to obtain the number of waves in time $2\Delta t$, and obtain the frequency (ν), which is defined as the number of waves in unit time:

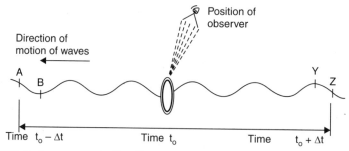

Fig. 10.61 A diagram illustrating the uncertainty in determining the frequency by counting the number of waves passing a point during a period of time

$$\nu = \text{Number of crests plus number of troughs} / (2 \times 2\Delta t) \quad (10.71)$$

This measurement is an average for a period of time $2\Delta t$ in the neighbourhood of t_0, we may describe it as being for the time t_0 with uncertainty Δt. There is also an uncertainty in the frequency. The crest A (Fig 10.61) might or might not be counted, and similarly the trough Z. Hence there is an uncertainty of about 2 in the number of crests plus the number of troughs, and hence of about $1/2\Delta t$ in the frequency:

$$\Delta \nu = \frac{2}{2 \times 2\Delta t} = \frac{1}{2\Delta t} \quad (10.72)$$

We may rewrite this equation as $\Delta \nu \times \Delta t = \frac{1}{2}$. A more detailed discussion based on an error-function definition of $\Delta \nu$ and Δt leads to the uncertainty equation for frequency and time in its customary form:

$$\Delta \nu \times \Delta t = \frac{1}{2\pi} \quad (10.73)$$

By use of quantum theory this equation can be at once converted into the uncertainty equation for energy and time for photons. The energy of a photon with frequency ν is $h\nu$. The uncertainty in frequency $\Delta \nu$ when multiplied by h is the uncertainty in energy ΔE.

$$\Delta E = h \Delta \nu.$$

By substituting this relation in equation we obtain the energy-time uncertainty equation.

$$\Delta E \times \Delta t = \frac{h}{2\pi} \quad (10.74)$$

It has been found by analysis of many experiments on the basis of quantum theory that this relation holds for any system. Only by making the measurement of the energy of any system over a long period of time, the error in the measured energy of the system can be made small.

10.9.3 Heisenberg's Uncertainty Principle for Other Physical Variables

There are similar uncertainty relations involving several other pairs of measurable physical quantities. One such relation is the energy – time uncertainty principle. If ΔE is the uncertainty in the measurement of energy and Δt is the time requirement for measurement, then it can be shown that

$$\Delta E \, \Delta t \geq h/4\pi$$

We can verify this relation for a free particle. Consider a free particle with a momentum p in the x-direction. The energy of the particle E is $p^2/2m$.

$$\therefore \quad \frac{dE}{dp} = \frac{p}{m} \quad \text{or} \quad \Delta E = \frac{p \Delta p}{m}$$

But $\quad \Delta p \Delta x \geq \dfrac{h}{4\pi} \quad \text{or} \quad \Delta p \geq \dfrac{h}{4\pi \Delta x} \quad \text{or} \quad \Delta E \geq \dfrac{p}{m} \dfrac{h}{4\pi \Delta x}$

$$\Delta E \geq v \cdot \frac{h}{4\pi \Delta x} \quad \text{But} \quad \frac{\Delta x}{\Delta t} = v \quad \therefore \quad \Delta E \geq \frac{h}{4\pi \Delta t} \quad \text{or} \quad \Delta E \Delta t \geq \frac{h}{4\pi}$$

Since we are considering a free particle, we may wonder why energy is uncertain. This is because the measurement disturbs the particle, leading to an uncertainty in its energy.

Similarly we have uncertainty relations involving angular displacements and angular momentum.

$$\Delta\theta_x \cdot \Delta L_x \geq h/4\pi \ ; \ \Delta\theta_y \cdot \Delta L_y \geq h/4\pi; \ \Delta\theta_z \cdot \Delta L_z \geq h/4\pi$$

where $\Delta\theta_x$, $\Delta\theta_y$ and $\Delta\theta_z$ is the uncertainty in the angular displacements about x, y and z axes respectively and ΔL_x, ΔL_y and ΔL_z is the uncertainty in the angular momentum about the x, y and z axes respectively.

10.10 APPLICATIONS OF HEISENBERG'S UNCERTAINTY PRINCIPLE

10.10.1 Radar Communication

Consider the problem of designing an antenna for launching an information-carrying microwave beam from the earth to a communication satellite as shown in Fig. 10.62. The beam-spread angle θ can be ascribed to the fact that the photons (considered here as particles) are confined initially to a distance

$$\Delta x \sim 2R$$

corresponding to the dish diameter $2R$. Since $\Delta x \Delta p > h/4\pi$, their transverse momentum spread is,

$$\Delta p_x > h/8\pi R$$

The spreading angle θ can be taken as:

$$\frac{\theta}{2} \approx \frac{\Delta p}{p}$$

$$\Rightarrow \quad \theta \approx \frac{2\Delta p}{p} \approx \frac{2}{\left(\frac{h}{\lambda}\right)} \frac{h}{8\pi R} \approx \frac{\lambda}{4\pi R}$$

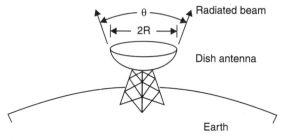

Fig. 10.62 Heisenberg's uncertainty principle and the antenna

This is of course a well-known result. It states that the spread angle is directly proportional to the wavelength. Hence the need to use waves of shorter wavelength for communication in the line of sight.

10.10.2 Information Theory (Shannon's Theorem)

Information theory was developed by Shannon and others. It deals with methods to quantify or measure information. It also concerns with the rate at which information can be sent from one point to the other. The most basic piece of information about an event is a yes or no, *i.e.*, whether the event has occurred or not. In the tossing of coin there is the possibility of appearance of head or tail. Information may correspond to truth or falsity of a statement. In the case of a binary digit it corresponds to the digit 0 and 1. The basic unit of information is called the binary digit or bit. It is the basis of the binary number system.

Consider a system of N equi-probable events. Assume that only one of those events have to be specified. Then according to information theory, the number of bits required to determine one of the N events is

$$I = \log_2 N \qquad (10.75)$$

Using Heisenberg's uncertainty principle it is possible to arrive at the maximum rate at which this information can be transmitted. Consider the example where information is transmitted via a sequence of sinusoidal electromagnetic pulses as shown in Fig. 10.63. The average power is taken as p, so that the average energy per pulse is $p\Delta t$. The transmitted information is coded into the amplitude of the pulses so that each predetermined value of

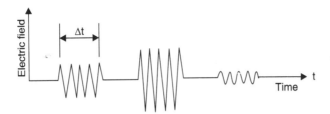

Fig. 10.63 Shannon's theorem

pulse energy is associated with a message. Here N corresponds to the number of distinguishable energy values that the pulse can take. The information content is given by eqn (10.75):

$$I_{pulse} = \log_2\left(\frac{p.\Delta t}{\Delta E}\right) \qquad (10.76)$$

where the energy resolution ΔE is the smallest pulse energy increment that can be measured. Since the time available for measuring the energy of a given pulse is Δt, the energy resolution ΔE is given according to the uncertainty principle by

$$\Delta E \approx \frac{h}{2\pi \Delta t} \qquad (10.77)$$

From eqns. (10.76) and (10.77)

$$I \approx \log_2\left(\frac{p(\Delta t)^2\, 2\pi}{h}\right) \qquad (10.78)$$

The average information transmission rate (C) is equal to the average information content per pulse multiplied by the number of pulses per second (i.e., $1/\Delta t$.)

$$C \approx \left(\frac{1}{\Delta t}\right)\log_2\left(\frac{p(\Delta t)^2\, 2\pi}{h}\right) \qquad (10.79)$$

It is clear that C increases as Δt is made smaller. The limit for this procedure is reached when Δt becomes comparable to the oscillation period f^{-1}. Further, each pulse is a wave packet in the time domain and in the frequency domain it will have a band width B. Assuming that

$$B \approx f \approx (\Delta t)^{-1}$$

C can be cast in the form:

$$C \approx B\left(\log_2\left(\frac{p}{hfB}\right)\right) \qquad (10.80)$$

Shannon's classical result is

$$C = B(\log_2(1 + SNR)) \qquad (10.81)$$

where SNR is the signal to noise ratio.

Eqn. (10.80) describes a fundamental limit to the rate at which information can be transmitted. The quantity hfB has the dimension of power and by analogy with eqn. (10.81) can be thought of as a noise *contaminating* the signal power p. Since it results from the application of the uncertainty principle, one can think of hfB as the noise power. It arises because of the wave-particle duality of the electromagnetic pulse.

10.10.3 Size of an Atom

Heisenberg's uncertainty principle can be utilised to arrive at the size of an atom. Let us assume that an electron is confined to a region $\sim a$. According to uncertainty principle,

$$p \sim \Delta p \sim \frac{h}{a}$$

The kinetic energy is $\quad \text{K.E.} = \dfrac{p^2}{2m} \approx \dfrac{h^2}{2ma^2}$

The potential energy $\quad \text{P.E.} = -\dfrac{e^2}{a}$

$\therefore\quad$ The total energy $\quad \text{T.E.} \approx \dfrac{h^2}{2ma^2} - \dfrac{e^2}{a}$

The ground state of the atom will correspond to a minimum value of E for which dE/da must be zero. Thus

$$\therefore\quad \frac{dE}{da} = -\frac{2h^2}{2ma^3} + \frac{e^2}{a^2} = 0$$

$$\Rightarrow \quad a = a_o = \frac{h^2}{me^2} = 0.5 \times 10^{-10}\,\text{m}$$

where a_o represents the value of a corresponding to the minimum in the total energy. Thus the finite size of the atom is due to the uncertainty principle. The above equation indicates that the atomic dimensions are of the order of Angstroms which is indeed correct. The minimum energy is given by

$$E = \frac{h^2}{2ma_o^2} - \frac{e^2}{a_o} = -\frac{me^4}{2h^2} = -13.6\,\text{eV}$$

which is indeed the ground state energy of the hydrogen atom.

10.10.4 Zero Point Energy of the Harmonic Oscillator

An important feature of quantum mechanics is its prediction of the zero energy of the harmonic oscillator. The energy of the one dimensional harmonic oscillator is given by

$$E = \frac{p_x^2}{2m} + \frac{1}{2}m\omega^2 x^2$$

Where $\omega = 2\pi\nu$ is the angular frequency of the oscillator. Assume the particle to be confined to a region $\sim a$; i.e., $\Delta x \sim a$. According to the uncertainty principle,

$$\Delta p_x \sim \frac{h}{4\pi a}$$

Thus the energy of the oscillator is given by

$$E = \frac{h^2}{32\pi^2 ma^2} + \frac{1}{2}m\omega^2 a^2$$

The lowest energy state of the oscillator (which is called the ground state) will correspond to the minimum value of E for which

$$\frac{dE}{da} = 0$$

$$\frac{dE}{da} = -\frac{h^2}{16\pi^2 ma^3} + m\omega^2 a = 0 \Rightarrow a \approx a_o \approx \left\{\frac{h}{4\pi m\omega}\right\}^{\frac{1}{2}}$$

The minimum energy is given by

$$E = \frac{h^2}{32\pi a^2 m a_o^2} + \frac{1}{2} m w^2 a_o^2 = \frac{h^2}{8\pi^2 a} \frac{1}{2m} + \frac{1}{2} m\omega^2 \frac{h}{4\pi m\omega} = \frac{h\omega}{4\pi} = \frac{1}{2} h\nu$$

This minimum energy is called the ground state energy of the linear harmonic oscillator. The quantity (1/2) $h\nu$ is called the *zero point energy and is a consequence of the uncertainty principle.*

10.10.5 Possibility of Electron's Existence Inside the Nucleus

Many nuclei exhibit β-radioactivity *i.e.*, they emit electrons. Naturally the question arises whether electrons can exist within the nucleus. This was an important question to be resolved during the early days of nuclear physics. By applying uncertainty principle, it is possible to show that electrons cannot exist within the nucleus. If an electron were to exist within the nucleus, the uncertainty in its position Δx cannot be greater than the nuclear diameter. The nuclear diameter d is of ~10^{-15} m. Hence the uncertainty in the momentum of the electron is

$$\Delta p \geq \frac{h}{2\pi d}$$

Hence the momentum of the electron (p)

$$p \approx \frac{h}{2\pi d} \approx \frac{6.63 \times 10^{-34}}{2\pi \times 10^{-15}} \approx 1.05 \times 10^{-19} \text{ kg-ms}^{-1}.$$

According to the theory of relativity the kinetic energy of the electron is given by:

$$\text{K.E} + m_o c^2 = \sqrt{p^2 c^2 + (m_o c^2)^2} \Rightarrow \text{K.E} = \sqrt{p^2 c^2 + (m_o c^2)^2} - m_o c^2$$

where m_o is the rest mass of the electron, p is the momentum of electron and c is the velocity of light. On substituting the numerical values for m_o, c, h and p

$$\text{K.E} \approx 200 \text{ MeV}.$$

However, the experimental measurements give the kinetic energy of electrons as ~MeV. This shows that electrons cannot be the constituents of the nucleus. The observed electron emission is attributed to the decay of the neutron.

10.10.6 Spectral Linewidth

The energy levels of a system are quantised. They are usually classified as vibrational, rotational and electronic levels. Vibrational levels correspond to different vibrational energy of molecules; rotational levels correspond to different rotational motion and electronic levels involve the electrons. Transitions between the various energy levels lead to either absorption or emission of radiation. Let E_i and E_f be the initial and final states of the system. Then the frequency of radiation absorbed or emitted is given by:

$$h\nu = |E_i - E_f|$$

Let E_o be the ground state of the system (*i.e.*, $E_o = 0$). On absorbing a photon of energy $h\nu$, the system undergoes a transition to a level corresponding to energy E. However it does not stay there indefinitely. It stays for a time Δt, which is called the lifetime of that level.

$$E = h\nu$$

$$\lambda = \frac{c}{\nu} = \frac{hc}{h\nu} = \frac{hc}{E}$$

$$\Rightarrow \qquad \frac{d\lambda}{dE} = -\frac{hc}{E^2} = -\frac{E\lambda}{E^2} = -\frac{\lambda}{E}$$

$$\Rightarrow \qquad \frac{\Delta\lambda}{\lambda} = -\frac{\Delta E}{E} \approx \frac{h}{2\pi \Delta t E}$$

This suggests that greater the linewidth, shorter the time the system spends in the excited state and vice versa. For a perfectly monochromatic source, $\Delta\lambda = 0$, and consequently $\Delta t = \infty$.

10.11 SIGNIFICANCE OF UNCERTAINTY PRINCIPLE

According to classical mechanics the position and velocity of any macroscopic body can be determined to the desired accuracy. Measurement errors are associated only with the apparatus. The position and velocity can be absolutely determined without any errors with a perfectly designed apparatus. The measurements are independent of the observer. Hence in classical mechanics it is conventional to describe the motion by a trajectory. In quantum mechanics on account of the uncertainty principle, the motion of a particle cannot be described by a trajectory. In Bohr's model the electron describes circular orbit around the nucleus. However, in later developments of quantum mechanics this notion is given up. Instead the motion of electron is described in terms of orbitals. These do not denote trajectories. Instead they represent the probability of finding the electron around the nucleus. Figure 10.64 shows the s, p and d orbitals.

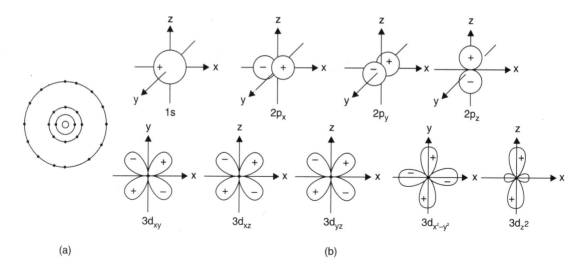

(a)   (b)

Fig. 10.64 Orbits and orbitals

Also according to classical mechanics, if at a particular instant of time the position and velocity are known, it is possible to accurately predict the position and velocity at a later time by making use of Newton's laws of motion. This is referred to as classical determinism. However in quantum mechanics, on account of uncertainties in measurements, it is only possible to predict the time evolution of motion of the particle in terms of probabilities. Some physicists attribute this impossibility to *certain hidden variables* in the system. According to them, the probabilistic estimates of the measurements are some sort of averages over these hidden variables.

MATTER AND RADIATION – DUAL NATURE

Heisenberg's uncertainty principle has attracted the attention of best minds on account of its radical views regarding scientific measurements. It defines inherent limits of measurements. Even with very careful prolonged measurements done with perfect instruments, one cannot measure better than the limits set by this principle. This also questions the objective method scientific enquiry. If the observer himself disturbs the system, the notion of an objective reality is debatable. The existence of an objective reality not affected by the observer is taken for granted in classical mechanics.

Uncertainty principle also allows for small violations of conservation laws of energy and momentum. These violations are not paradoxical since they cannot be measured. Suppose one wishes to observe deviation from exact energy conservation by an amount ΔE. The uncertainty principle requires that the time during which this violation takes place is of the order of $\Delta t \approx h/2\pi\Delta E$. To observe in this time interval, one needs a clock ticking at intervals less than Δt. i.e. trains of pulses of frequency $\nu \geq 1/\Delta t$. But quanta in this wave train have energies $E = h\nu = h/\Delta t = 4\pi\Delta E$. Hence the quanta have sufficient energy to disturb the system. Thus any attempt to measure the energy ΔE must disturb the system by at least as much as the uncertainty in Δt.

REFERENCES

1. K. Krane, *Modern Physics*, John Wiley, New York, 1998.
2. A. Beiser, *Concepts of Modern Physics*, McGraw-Hill, New York, 1995.
3. J. Bernstein, P.M. Fishbane and S. Gasiorowicz, *Modern Physics*, Pearson Education, New Delhi, 2001.
4. S.T. Thornton and A. Rex, *Modern Physics for Scientists and Engineers*, Saunders College Publishing, New York, 2000.

SOLVED EXAMPLES

Black Body Radiation:

1. Luminosity of Rigel star in Orion Constellation is 17000 times that of sun. If surface temperature of sun is 6000 K, calculate the temperature of star.

   Solution:

   The energy radiated by a black body per unit area per sec is

   $$E = \sigma T^4$$

   If E_1 and E_2 are luminosities of sun and the star respectively and T_1, T_2 their respective temperatures, then

   $$E_1 = \sigma T_1^4 \quad \text{and} \quad E_2 = \sigma T_2^4$$

   $\therefore$
   $$\frac{E_1}{E_2} = \frac{T_1^4}{T_2^4} \quad \Rightarrow \quad T_2 = T_1 \left[\frac{E_2}{E_1}\right]^{\frac{1}{4}}$$

   Here $\dfrac{E_2}{E_1} = 17000 \quad \therefore \quad T_2 = 6000(17000)^{1/4} = 68512$ K

2. Calculate the energy radiated per minute from the filament of an incandescent lamp at 2000 K if the surface area is 5×10^{-5} m² and its emissivity is 0.85.

Solution:

The energy radiated per unit area per second by a body of relative emissivity e at temperature T.
$$E = e\sigma T^4$$
∴ Total energy radiated by surface area A in time t is
$$Q = EAt = \varepsilon\sigma T^4 At$$
$$= 0.85 \times 5.7 \times 10^{-8} \times (2000)^4 \times (5 \times 10^{-5}) \times 60$$
$$Q = 2326 \text{ Joules.}$$

3. The peak emission from a black body at a certain temperature occurs at a wavelength of 900 nm. On increasing its temperature the total radiation emitted increases by 81 times. At the initial temperature when the peak radiation from the black body is incident on a metal surface it does not cause any photoemission from the surface. After the increase of temperature the peak radiation from the black body caused photoemission. To bring these photoelectrons to rest, a potential equivalent to the excitation energy between the $n = 2$ to $n = 3$ Bohr levels of hydrogen atom is required. Find the work function of the metal.

Solution:

The radiant energy emitted by a black body $E = \sigma T^4$.

When energy radiated increases by 81 times, let the temperature increases to T_2 from initial temperature T_1. Then
$$T_2 = T_1 \left(\frac{E_2}{E_1}\right)^{\frac{1}{4}} = T_1(81)^{\frac{1}{4}} = 3T_1$$

From Wien's displacement law, $\lambda_m T = $ constant
∴
$$900 \text{ nm} \times T_1 = \lambda_{m2} \times T_2$$
$$\lambda_{m2} = \left(\frac{T_1}{T_2}\right) \times 900 = (9000/3) = 300 \text{ nm}$$

Energy of incident photon corresponding to this wavelength is given by:
$$\frac{hc}{\lambda} = \frac{6.63 \times 10^{-34} \times 3 \times 10^8}{300 \times 10^{-9} \times 1.6 \times 10^{-19}} = 4.14 \text{ eV}$$

Excitation energy from Bohr's formula is given by:
$$Rhc\left[\frac{1}{n_1^2} - \frac{1}{n_2^2}\right] = \frac{1.097 \times 10^7 \times 6.63 \times 10^{-34} \times 3 \times 10^8}{1.6 \times 10^{-19}}\left[\frac{1}{2^2} - \frac{1}{3^2}\right] = 1.89 \text{ eV}$$

This excitation energy is equal to the kinetic energy of the electrons.

Work function = Excitation energy − Kinetic energy
$$= 4.14 \text{ eV} - 1.89 \text{ eV} = 2.25 \text{ eV.}$$

4. Radiant energy from the sun strikes the earth at a rate of 1.4×10^3 watt/m². Calculate the temperature at the surface of sun. (Radius of sun = 7×10^8 m, Radius of earth's orbit = 1.5×10^{11} m).

MATTER AND RADIATION – DUAL NATURE

Solution:

Let R be the radius of the sun and T its surface temperature. Let E be the power radiated per unit area of the sun. Treating the sun as a black body; the power radiated by the sun is given by

$$EA = \sigma T^4 (4\pi R^2).$$

If r is radius of earth's orbit, then this radiated energy will be spread over a sphere of surface area $4\pi r^2$. Hence this energy received per unit time per unit surface area of earth is given by:

$$\frac{\sigma T^4 (4\pi R^2)}{4\pi r^2} = \frac{\sigma T^4 R^2}{r^2} = 1.4 \times 10^3 \text{ W/m}^2 = S \text{(Solar constant)}$$

$$\therefore \quad T = \left[\frac{Sr^2}{\sigma R^2}\right]^{\frac{1}{4}} = \left[\frac{1.4 \times 10^3 \times (1.5 \times 10^{11})^2}{5.7 \times 10^{-8} \times (7 \times 10^8)^2}\right] = 5800 \text{ K}.$$

5. How much faster does a cup of tea cool by 1° when at 373 K than when it is at 303 K? Consider the tea as a black body, the room temperature being 293 K.

Solution:

According to Stefan's law of radiation

$$E = \sigma (T^4 - T_0^4)$$

Where T is the temperature of a black body and T_0 the temperature of the surroundings.

Let us consider the two cases: $T_1 = 373\ K$ and $T_2 = 303\ K$

E is the energy radiated per unit area per second from the hot body and hence proportional to the rate of cooling. In the first case

$$E_1 = \sigma (T_1^4 - T_0^4) = \sigma (373^4 - 293^4)$$

In the second case

$$E_2 = \sigma (T_2^4 - T_0^4) = \sigma (303^4 - 293^4)$$

Now E_1/E_2 will indicate how much faster the body cools in the first case than in the second case.

$$\frac{E_1}{E_2} = \frac{373^4 - 293^4}{303^4 - 293^4} = 11.3.$$

6. A body which has a surface area 5.00 cm² and a temperature of 727 °C radiates 300 joules of energy each minute. What is its emissivity?

Solution:

The total energy radiated by the body of emissivity ε, surface area A at temperature T is given by

$$Q = E A t = \varepsilon \sigma (T^4 - T_0^4) A t \approx \varepsilon \sigma T^4 A t \text{ for } T \gg T_o$$

where T_o is the temperature of surroundings and t is the time over which the radiation is emitted. Both T and T_o have to be expressed in Kelvin. Solving for ε we get:

Solving we get, $\varepsilon = \dfrac{Q}{\sigma T^4 At} = \dfrac{300}{5.67 \times 10^{-8} \times (1000)^4 \times (5 \times 10^{-4}) \times 60} = 0.18$.

7. The outer surface of a sphere of copper of radius 5 cm is coated black. How much time is required for the block to cool down from 1000 K to 300 K? Density of copper = 9×10^3 kg/m^3 and specific heat = 4 kJ/kg.

Solution:

When the copper block is coated black, it behaves like a perfect black body. The rate of heat energy radiated out at any instant is given by

$$\dfrac{dQ}{dt} = \sigma A T^4 \tag{1}$$

where A is the surface area and T is the temperature of black body at that instant. T changes continuously with time as black body cools. The heat given out by a block of mass m and specific heat c is given by

$$dQ = mc\,dT \tag{2}$$

From eqns. (1) and (2),

$$\therefore \quad \dfrac{dT}{dt} = -\left(\dfrac{\sigma A}{mc}\right) T^4 \tag{3}$$

Negative sign is used to show that temperature falls as time increases.

From eqn. (3) we have

$$dt = -\left(\dfrac{mc}{\sigma A}\right)\left(\dfrac{dT}{T^4}\right)$$

Hence, time required for the block to cool down from T_1 to T_2 is given by

$$\int_0^t dt = -\dfrac{mc}{\sigma A} \int_{T_1}^{T_2} \dfrac{dT}{T^4} \Rightarrow t = \dfrac{mc}{3\sigma A}\left[\dfrac{1}{T_2^3} - \dfrac{1}{T_1^3}\right]$$

If ρ be the density of copper and r be the radius of sphere, then

or $$t = \dfrac{\left(\dfrac{4\pi r^3 \rho}{3}\right)}{3\sigma(4\pi r^2)}\left[\dfrac{1}{T_2^3} - \dfrac{1}{T_1^3}\right] = \dfrac{\rho rc}{9\sigma}\left[\dfrac{1}{T_2^3} - \dfrac{1}{T_1^3}\right]$$

where ρ is the density of copper and r is the radius of the sphere.

Here $r = 5$ cm $= 5 \times 10^{-2}$ m, $\rho = 9 \times 10^3$ kg/m^3, $c = 4 \times 10^3$ J/kg, $T_1 = 1000 = 10^3$ K, $T_2 = 300$ K and $\sigma = 5.67 \times 10^{-8}$ J/m^2 sK^4.

Substituting these values and solving we get $t = 127 \times 10^3$ sec.

8. A wire of length 1.0 m and radius 10^{-3} m is carrying a heavy current and is assumed to radiate as a black body. At equilibrium its temperature is 900 K while that of the surroundings is 300K. The resistivity of the material of the wire at 300K is $\pi^2 \times 10^{-8}$ Ωm and its temperature coefficient of resistance is 7.8×10^{-3} per °C. Find the current in the wire.

MATTER AND RADIATION – DUAL NATURE 481

Solution:

According to Stefan's law
$$E = \sigma(T^4 - T_0^4) A = \sigma(T^4 - T_0^4) \times 2\pi rl$$
$$= (5.68 \times 10^{-8})[(900)^4 - (300)^4] [2\pi \times 10^{-3} \times 1.0]$$
$$= 5.68 \times 6480 \times 2\pi \times 10^{-3} \text{ watt} \quad (1)$$

The resistivity of wire at 900 K is given by
$$\rho_{900} = \rho_{300}(1 + \alpha \Delta T)$$
$$= \pi^2 \times 10^{-8}[1 + 7.8 \times 10^{-3} \times 600]$$
$$= \pi^2 \times 5.68 \times 10^{-8} \text{ ohm-meter} \quad (2)$$

The resistance of the wire at 900 K is given by
$$R_{900} = \rho_{900} \times (l/\pi r^2)$$
$$= (\pi^2 \times 5.68 \times 10^{-8}) \times [1.0/\pi(10^{-3})^2]$$
$$= \pi \times 5.68 \times 10^{-2} \text{ ohm.} \quad (3)$$

Now $\quad P = I^2 R_{900} = I^2 (\pi \times 5.68 \times 10^{-2}) \text{ watt} \quad (4)$

In steady state $\quad P = E$

$\therefore \quad I^2 (\pi \times 5.68 \times 10^{-2}) = 5.68 \times 6480 \times 2\pi \times 10^{-3}$

$\Rightarrow \quad I^2 = 648 \times 2 = 1296 \quad \text{or} \quad I = 36 \text{ amp.}$

9. Calculate the energy density of black body radiation at 1000K using both Rayleigh-Jeans law and Planck's law for wavelengths in the range (a) 1 – 1.1mm and (b) 100 – 101nm. Comment on the results obtained in both the cases.

Solution:

Case (a) The energy densities according to Rayleigh-Jeans and Planck's law corresponding to 1 – 1.1mm wavelength are given by:

$$U_\lambda d\lambda = \frac{8\pi k_B T}{\lambda^4} d\lambda = \frac{8\pi(1.38 \times 10^{-23}) \times 1000}{(10^{-3})^4} \times (0.1 \times 10^{-3})$$

$$= 3.47 \times 10^{-11} \text{ J/m}^3$$

$$U_\lambda d\lambda = \frac{8\pi hc}{\lambda^5} \times \frac{1}{\left(e^{hc/\lambda k_B T} - 1\right)} \times d\lambda$$

Let us first calculate the exponential term

$$\frac{hc}{\lambda k_B T} = \frac{6.63 \times 10^{-34} \times 3 \times 10^8}{10^{-3} \times 1.38 \times 10^{-23} \times 10^3} = 14.39 \times 10^{-3} = 0.01439$$

$\therefore \quad e^{hc/\lambda k_B T} = e^{0.01439} = 1.01449$

$\therefore \quad U_\lambda d\lambda = \dfrac{8\pi \times 6.63 \times 10^{-34} \times 3 \times 10^8}{(10^{-3})^5} \times \dfrac{1}{(1.01449 - 1)} \times (0.1 \times 10^{-3})$

$$= 3.44 \times 10^{-11} \text{ J/m}^3$$

Case (b) The energy densities according to Rayleigh-Jeans law and Planck's law corresponding to 100 – 101 nm are given by:

$$U_\lambda d\lambda = \frac{8\pi k_B T}{\lambda^4} d\lambda = \frac{8\pi \times (1.38 \times 10^{-23}) \times 1000}{(100 \times 10^{-9})^4} \times (1 \times 10^{-9}) = 3.47 \text{ J/m}^3$$

$$U_\lambda d\lambda = \frac{8\pi hc}{\lambda^5} \times \frac{1}{\left(e^{hc/\lambda k_B T} - 1\right)} \times d\lambda$$

Let us first calculate the exponential term

$$\frac{hc}{\lambda k_B T} = \frac{6.63 \times 10^{-34} \times 3 \times 10^8}{(100 \times 10^{-9}) \times 1.38 \times 10^{-23} \times 10^3} = 14.4$$

$$\therefore \quad e^{hc/\lambda k_B T} = e^{14.4} = 1.79 \times 10^6$$

$$\therefore \quad U_\lambda d\lambda = \frac{8\pi \times 6.63 \times 10^{-34} \times 3 \times 10^8}{(100 \times 10^{-9})^5} \times \frac{1}{(1.79 \times 10^6 - 1)} \times (1 \times 10^{-9})$$

$$= 2.79 \times 10^{-6} \text{ J/m}^3$$

Thus for shorter wavelengths the energy densities predicted by Rayleigh-Jeans law and Planck's law are considerably different while for longer wavelengths the energy densities predicted by them are same.

Photoelectric Effect:

10. The energy required to remove an electron from sodium is 2.3 eV. Does sodium shows a photo-electric effect for light with $\lambda = 680$ nm?

    Solution:

    Given $\quad W = 2.3 \text{ eV} = 2.3 \times 1.6 \times 10^{-19}$ joules.

    If λ_0 is threshold wavelength for sodium, then $W = (hc/\lambda_0)$

    or $\quad \lambda_0 = \frac{hc}{W} = \frac{6.63 \times 10^{-34} \times 3 \times 10^8}{2.3 \times 1.6 \times 10^{-19}} = 538$ nm

    From definition, threshold wavelength is the longest wavelength capable of showing photoelectric effect. Hence the light with $\lambda = 680$ nm cannot show photo-electric effect when incident on the surface of sodium.

11. Calculate the value of the retarding potential needed to stop the photo-electrons ejected from a metal surface of work function 1.2 eV with light of frequency 5.5×10^{14} Hz.

    Solution:

    According to Einstein's photo-electric equation, if V_s is stopping or retarding potential, then

    $$eV_s = h\nu - W$$

    Here $\quad e = 1.6 \times 10^{-19}$ coulomb, $\nu = 5.5 \times 10^{14}$ Hz

$$W = 1.2 \text{ eV} = 1.2 \times 1.6 \times 10^{-19} \text{ joules}$$

$$\therefore \quad 1.6 \times 10^{-19} V_s = 6.6 \times 10^{-34} \times 5.5 \times 10^{14} - 1.2 \times 1.6 \times 10^{-19}$$

or $\quad 1.6 \times 10^{-19} V_s = 1.71 \times 10^{-19}$

i.e., $\quad V_s = \dfrac{1.71 \times 10^{-19}}{1.6 \times 10^{-19}} = 1.06 \text{ volt}$

12. Find the frequency of light which ejects electrons from a metal surface. Photo-electrons are fully stopped by a retarding potential of 3V. The photo-electic threshold occurs at 6×10^{14} Hz.

    Solution:

    According to Einstein's photo-electric equation

    $$E_k = h\nu - W$$

    If V_s is retarding or stopping potential and ν_o the threshold frequency, then the above equation becomes

    $$eV_s = h\nu - h\nu_o \text{ or } h\nu = eV_s + h\nu_o$$

    or $\quad \nu = \dfrac{eV_s}{h} + \nu_o = \dfrac{(1.6 \times 10^{-19} \times 3)}{6.63 \times 10^{-34}} + 6 \times 10^{14} = 1.324 \times 10^{15}$ Hz.

13. Light of wavelength 200nm falls on an aluminum surface. In aluminum 4.2 eV is required to remove an electron. (*i*) What is the kinetic energy in electron volts of (*a*) the fastest, (*b*) the slowest emitted photo-electrons, (*ii*) What is the stopping potential? (*iii*) What is the cut off wavelength for aluminum?

    Solution:

    Energy corresponding to incident photon

    $$h\nu = \dfrac{hc}{\lambda} = \dfrac{6.6 \times 10^{-34} \times 3 \times 10^8}{200 \times 10^{-9}} = 9.9 \times 10^{-19} \text{ J} = \dfrac{9.9 \times 10^{-19}}{1.6 \times 10^{-19}} = 6.2 \text{eV}$$

    (*i*) (*a*) The kinetic energy of fastest electrons

    $$E_k = h\nu - W$$
    $$E_k = 6.2 \text{ eV} - 4.2 \text{ eV} = 2 \text{ eV}.$$

    (*b*) The kinetic energy of slowest electrons is zero. As the emitted electrons have all possible energies from 0 to certain maximum value E_k.

    (*ii*) If V_s is the stopping potential, then $E_k = eV_s$. Since the electrons have a maximum kinetic energy of 2 eV, the stopping potential is 2V.

    (*iii*) If λ_0 is the cut off wavelength for aluminum, then

    $$W = (hc/\lambda_0) \text{ or } \lambda_0 = (hc/W)$$

    i.e., $\quad \lambda_0 = \dfrac{hc}{W} = \dfrac{6.63 \times 10^{-34} \times 3 \times 10^8}{4.2 \times 1.6 \times 10^{-19}} = 300$ nm.

14. One milliwatt of light of wavelength 4560 λ is incident on a cesium surface. Calculate the photo-electric current liberated, assuming a quantum efficiency of 0.5%. The work function for cesium = 1.93 eV.

Solution:

Energy of each photon of incident light is

$$\varepsilon = h\nu = \frac{hc}{\lambda} = \frac{6.62 \times 10^{-34} \times 3 \times 10^8}{4560 \times 10^{-10}} = 4.32 \times 10^{-19} \text{ joules}$$

Power incident on the surface, $P = 10^{-3}$ watt. Number of photons incident on the surface of the metal per second is given by:

$$N = \frac{P}{E} = \frac{10^{-3}}{4.32 \times 10^{-19}} = 2.32 \times 10^{15}$$

Since the quantum efficiency is 0.5%, only 0.5% of incident photons release photoelectrons. Hence the number of electrons released per second is given by:

$$N_e = \frac{0.5}{100} \times 2.32 \times 10^{15} = 1.16 \times 10^{3}$$

Photoelectric current $= N_e e = 1.16 \times 10^{13} \times 1.6 \times 10^{-19} = 1.856 \times 10^{-6}$ amps

$= 1.856$ μA.

15. A stopping potential of 0.82 volt is required to stop the emission of photoelectrons from the surface of a metal by light of wavelength 400 nm. For light of wavelength 300 nm, the stopping potential is 1.85 volt. (*i*) Find the value of Planck's constant. (*ii*) At stopping potential, if the wavelength of the incident light is kept fixed at 400 nm, but the intensity of light increased by two times, will any photoelectric current be obtained? Give reasons for your answer.

Solution:

(*i*) $$\frac{hc}{\lambda_1} = W + eV_1$$

$$\frac{hc}{\lambda_2} = W + eV_2$$

Subtracting $\frac{hc}{\lambda_1} - \frac{hc}{\lambda_2} = e(V_1 - V_2) \Rightarrow h = \frac{e(V_1 - V_2)\lambda_1 \lambda_2}{c(\lambda_2 - \lambda_1)}$

Here $\lambda_1 = 400$ nm $\lambda_2 = 300$ nm $V_1 = 0.82$V $V_2 = 1.85$ V

∴ $$h = \frac{1.6 \times 10^{-19} \times (0.82 - 1.85) \times 400 \times 10^{-9} \times 300 \times 10^{-9}}{3 \times 10^8 \times (300 - 400) \times 10^{-9}}$$

$= 6.592 \times 10^{-34}$ J–s

(*ii*) No, because stopping potential does not depend on intensity of incident light.

MATTER AND RADIATION – DUAL NATURE

16. Radiation of wavelength 180 nm ejects photo-electrons from a plate whose work function is 2.0 eV. If a uniform magnetic field of flux density 5.0×10^{-5} tesla (or weber/m^2) is applied parallel to the plate, what should be the radius of the path followed by electrons ejected normally from the plate with maximum energy?

Solution:

According to Einstein's photoelectric equation,

$$E_k = h\nu - W = \frac{hc}{\lambda} - W = \frac{6.63 \times 10^{-34} \times 3 \times 10^8}{180 \times 10^{-9}} - 2 \times 1.6 \times 10^{-19}$$

$$= 7.8 \times 10^{-19} \text{ J}$$

∴ Velocity v of electron corresponding to maximum kinetic energy E_k is given by

$$v = \sqrt{\frac{2E_k}{m}} = \sqrt{\left(\frac{2 \times 7.8 \times 10^{-19}}{9.1 \times 10^{-31}}\right)} = 1.30 \times 10^6 \text{ m/s}.$$

Radius of circular path r in a magnetic field of induction B is given by

$$Bev = \frac{mv^2}{r} \quad \text{or} \quad r = \frac{mv}{eB} = \frac{9.1 \times 10^{-31} \times 1.30 \times 10^6}{1.6 \times 10^{-19} \times 5.0 \times 10^5} = 0.149 \text{ m}$$

Compton Effect:

17. A 6.2 keV X-ray photon falling on a carbon block is scattered by Compton collision and its frequency is shifted by 0.5%. Through what angle is the photon scattered? How much energy is imparted to the electron?

Solution:

The wavelength of the incident X-rays is given by:

$$\lambda_0 = \frac{hc}{E_o} = \frac{6.63 \times 10^{-34} \times 3 \times 10^8}{6.20 \times 10^3 \times 1.6 \times 10^{-19}} = 2.01 \times 10^{-10} \text{ m}$$

Loss in energy of the incident photon is equal to

$$\Delta E = \frac{0.5}{100} \times 6.2 \times 10^3 \text{ eV} = 31 \text{ eV} = 0.031 \text{ keV}$$

The energy of the scattered photon is given by

$$E = E_o - \Delta E = 6.2 \text{ keV} - 0.031 \text{ keV} = 6.169 \times 10^3 \text{ eV}$$

The wavelength of the scattered photon is given by :

$$\lambda = \frac{hc}{E} = \frac{6.63 \times 10^{-34} \times 3 \times 10^8}{6.169 \times 10^3 \times 1.6 \times 10^{-19}} = 2.02 \times 10^{-10} \text{ m}$$

The Compton shift $(\lambda - \lambda_0) = \Delta\lambda = (2.02 - 2.01) \times 10^{-10} = 10^{-14}$ m

The angle through which the X-ray is scattered can be obtained using the formula

$$\Delta\lambda = \frac{h}{mc}(1 - \cos\phi)$$

$$\cos\phi = 1 - \frac{mc\Delta\lambda}{h} = 1 - \frac{9.1 \times 10^{-31} \times 3 \times 10^8 \times 10^{-14}}{6.63 \times 10^{-34}} = 0.587$$

$$\Rightarrow \quad \phi = 54.1°.$$

18. X-rays with a wavelength of 100 pm are Compton scattered from a carbon target. The scattered radiation is viewed at 90° to the incident direction. (a) What is the Compton shift and (b) What is the kinetic energy imparted to the electron?

Solution:

The Compton shift is given by:

$$\Delta\lambda = \frac{h}{mc}(1 - \cos\phi), \quad \text{Here } \phi = 90°$$

$$\therefore \quad \Delta\lambda = \frac{h}{mc} = \frac{6.63 \times 10^{-34}}{9.1 \times 10^{-31} \times 3 \times 10^8} = 2.43 \text{ pm}$$

The wavelength of the scattered photon = (100 – 2.43)pm = 97.57 pm

The energy lost by the X-ray photon is equal to the energy imparted to the electron and is given by:

$$\Delta E = \frac{hc}{\lambda_o} - \frac{hc}{\lambda} = \frac{hc(\lambda - \lambda_o)}{\lambda\lambda_o} = \frac{6.63 \times 10^{-34} \times 2.43 \times 10^{-12}}{100 \times 10^{-12} \times 97.57 \times 10^{-12}}$$

$$= 4.72 \times 10^{-17} \text{ J} = 295 \text{ eV}.$$

19. If E is the energy of the Compton scattered photon and E_o that of the incident photon, show that

$$\frac{\Delta E}{E} = \frac{E_o}{mc^2}(1 - \cos\phi).$$

Solution:

We have the expression for Compton shift as

$$\lambda - \lambda_o = \frac{h}{mc}(1 - \cos\phi)$$

Expressing in terms of E and E_o, we get

$$\frac{hc}{E} - \frac{hc}{E_o} = \frac{h}{mc}(1 - \cos\phi) \quad \Rightarrow \quad \frac{E_o - E}{EE_o} = \frac{(1 - \cos\phi)}{mc^2}$$

or

$$\frac{\Delta E}{E} = \frac{E_o}{mc^2}(1 - \cos\phi).$$

20. At what scattering angle will the 100 keV X-rays incident on a target will get Compton scattered with an energy of 90 keV?

Solution:

We have the relation

$$\frac{\Delta E}{E} = \frac{E_o}{mc^2}(1 - \cos\phi) \quad \text{or} \quad \cos\phi = 1 - \frac{\Delta E}{E} \times \frac{mc^2}{E_o}$$

Here $E_o = 100$ keV, $E = 90$ keV, $\Delta E = 10$ keV

and

$$mc^2 = \frac{9.1 \times 10^{-31} \times (3 \times 10^8)^2}{1.6 \times 10^{-19} \times 10^3} = 512.44 \text{ keV}$$

MATTER AND RADIATION – DUAL NATURE

$$\therefore \qquad \cos\phi = 1 - \frac{10}{90} \times \frac{512.44}{100} = 0.431 \Rightarrow \phi = 64.5°$$

Matter waves:

21. Calculate the de Broglie wavelength of the following:

    (a) A 10 electron volt electron, (b) A hydrogen molecule moving with a velocity of 2200 m/sec and (c) A golf ball of 45 g moving with a velocity of 22 m/sec.

    Solution:

    (a) The de Broglie wavelength is given by:

    $$\lambda = \frac{h}{p} = \frac{h}{\sqrt{2m(KE)}} = \frac{h}{\sqrt{2 \times 9.1 \times 10^{-31} \times 10 \times 1.6 \times 10^{-19}}} = 0.388 \text{ nm}$$

    (b) The gram molecular weight of hydrogen molecule is 2g. Hence the mass of the hydrogen molecule is equal to

    $$m = \frac{2 \times 10^{-3}}{6.023 \times 10^{23}} \text{ kg} = 3.319 \times 10^{-27} \text{ kg}$$

    The de Broglie wavelength is given by

    $$\lambda = \frac{h}{p} = \frac{h}{mv} = \frac{6.63 \times 10^{-34}}{3.319 \times 10^{-27} \times 2200} = 0.091 \text{ nm}$$

    (c) The de Broglie wavelength of the golf ball is given by

    $$\lambda = \frac{h}{p} = \frac{h}{mv} = \frac{6.63 \times 10^{-34}}{45 \times 10^{-3} \times 22} = 6.7 \times 10^{-34} \text{ m.}$$

22. Assuming neutrons to be a monatomic gas, calculate their de Broglie wavelength at 300 K. The mass of neutron is 1.0087 amu.

    Solution:

    The de Broglie wavelength of neutrons is given by

    $$\lambda = \frac{h}{p} = \frac{h}{\sqrt{2m(K.E)}} = \frac{h}{\sqrt{2m\left(\frac{3}{2}k_B T\right)}} = \frac{h}{\sqrt{3mk_B T}}$$

    $$\therefore \qquad \lambda = \frac{6.63 \times 10^{-34}}{\sqrt{3 \times 1.0087 \times 1.66 \times 10^{-27} \times 1.38 \times 10^{-23} \times 300}} = 0.145 \text{ nm.}$$

23. Show that the de Broglie wavelength of an electron accelerated through V volts is ~ $1.227\, V^{-1/2}$ nm. Hence find the voltage through which an electron has to be accelerated to have a de Broglie wavelength of 0.1 nm.

    Solution:

    The de Broglie wavelength of the electron is given by

    $$\lambda = \frac{h}{p} = \frac{h}{\sqrt{2m(KE)}} = \frac{h}{\sqrt{2meV}} = \frac{6.63 \times 10^{-34}}{\sqrt{2 \times 9.1 \times 10^{-31} \times 1.6 \times 10^{-19} \times V}} = \frac{1.227}{\sqrt{V}} \text{ nm.}$$

If the de Broglie wavelength is 0.1 nm we have

$$0.1 \text{ nm} = \frac{1.227 nm}{\sqrt{V}} \Rightarrow V = \left(\frac{1.227}{0.1}\right)^2 = 150.55 \text{ volts}$$

Hence the electron has to be accelerated through 150.55 volts.

24. Calculate the de Broglie wavelength, phase velocity and group velocity for a neutron of energy 0.04 eV given the mass of neutron to be 1.675×10^{-27} kg.

Solution:

The de Broglie wavelength of the neutron is given by

$$\lambda = \frac{h}{p} = \frac{h}{\sqrt{2m(KE)}} = \frac{6.63 \times 10^{-34}}{\sqrt{2 \times 1.675 \times 10^{-27} \times 0.04 \times 1.6 \times 10^{-19}}} = 0.143 \text{ nm}$$

The velocity of the neutron is given by:

$$v = \frac{p}{m} = \frac{h}{\lambda m} = \frac{6.63 \times 10^{-34}}{0.143 \times 10^{-9} \times 1.675 \times 10^{-27}} = 2.76 \times 10^3 \text{ m/s}$$

$$v_g = v = 2.76 \times 10^3 \text{ m/s}$$

$$v_p = \frac{c^2}{v_g} = \frac{(3 \times 10^8)^2}{2.76 \times 10^3} = 3.26 \times 10^{13} \text{ m/s}$$

The energy of the photon having the same de Broglie wavelength is given by:

$$E = h\nu = \frac{hc}{\lambda} = \frac{6.63 \times 10^{-34} \times 3 \times 10^8}{0.143 \times 10^{-9}} = 1.39 \times 10^{-15} \text{ J} = 8.69 \text{ keV}$$

25. The frequency of the tidal wave is $\sqrt{gk}$, where g is the acceleration due to gravity and k is the wave vector. Calculate the group velocity for wave packet with $\lambda = 10$ m.

Solution:

$$\nu = (gk)^{\frac{1}{2}} \quad \text{or} \quad \omega = 2\pi\nu = 2\pi(gk)^{\frac{1}{2}}$$

$$\therefore \quad v_g = \frac{d\omega}{dk} = \pi\left(\frac{g}{k}\right)^{\frac{1}{2}} = \pi\left(\frac{\lambda g}{2\pi}\right)^{\frac{1}{2}} = \left(\frac{\lambda g \pi}{2}\right)^{\frac{1}{2}} = 12.41 \text{ m/s for } \lambda = 10 \text{ m}.$$

26. Electrons are accelerated by 344 volt and are Bragg reflected from a set of parallel planes in a crystal. The first reflection maximum occurs when the glancing angle is 60°. Determine the interplanar distance.

Solution:

We have from Bragg's law:

$$2d \sin \theta = n\lambda = \frac{nh}{p} = \frac{nh}{\sqrt{2m(KE)}} = \frac{nh}{\sqrt{2meV}}$$

MATTER AND RADIATION – DUAL NATURE

$$d = \frac{nh}{2 \sin \theta \times \sqrt{2meV}} = \frac{1 \times 6.63 \times 10^{-34}}{2 \times \sin 60° \times \sqrt{2 \times 9.1 \times 10^{-31} \times 1.6 \times 10^{-19} \times 344}}$$

$$= 0.038 \text{ nm}.$$

27. A neutron beam of kinetic energy 0.04 electron volt is diffracted by a set of parallel planes in a crystal with interplanar spacing of 0.314 nm. Calculate the glancing angle at which first order Bragg spectrum will be observed. (neutron mass = 1.675×10^{-27} kg).

Solution:
We have from Bragg's law

$$2d \sin \theta = n\lambda = \frac{nh}{p} = \frac{nh}{\sqrt{2m(KE)}}$$

$$\therefore \quad \sin \theta = \frac{nh}{2d\sqrt{2m(KE)}} = \frac{1 \times 6.63 \times 10^{-34}}{2 \times 0.314 \times 10^{-9} \times \sqrt{2 \times 1.675 \times 10^{-27} \times 0.04 \times 1.6 \times 10^{-19}}}$$

$$= 0.2334 \quad \Rightarrow \quad \theta = 13.5°.$$

28. Davisson Germer experiment was repeated with a beam of 0.25 eV neutrons. Neutrons reflected back from a crystal surface is found to produce a first order maximum at an angle of 70° with the initial beam direction. Determine the interplanar distance for the crystal.

Solution:
The angle between the incident and the reflected beam is
$\phi = 180° - 2\theta = 70°$. Hence the Bragg angle is 55°. We have from Bragg's law

$$2d \sin \theta = n\lambda = \frac{nh}{p} = \frac{nh}{\sqrt{2m(KE)}}$$

$$\therefore \quad d = \frac{nh}{2 \sin \theta \sqrt{2m(KE)}}$$

$$= \frac{1 \times 6.63 \times 10^{-34}}{2 \sin 55° \sqrt{2 \times 1.675 \times 10^{-27} \times 1.6 \times 10^{-19}}} = 3.5 \times 10^{-11} \text{ m}.$$

Uncertainty Principle:

29. An electron has a speed of 4×10^5 metre/sec, accurate to 0.01%. With what accuracy can we locate the position of the electron?

Solution:
Momentum of the electron, $p = mv = 9.1 \times 10^{-31} \times 4 \times 10^5 = 36.4 \times 10^{-26}$ kg-m/sec
Since the uncertainty in velocity is 0.01% (= 10^{-4}), the uncertainty in momentum will also be 0.01%.
Thus, $\quad \Delta p = (10^{-4}) \times 36.44 \times 10^{-26} = 3.64 \times 10^{-25}$ kg-m/s
From uncertainty principle the uncertainty in position Δx is given by:

$$\Delta x \geq \frac{h}{2\pi \Delta p} \geq \frac{6.63 \times 10^{-34}}{2\pi \times 3.6 \times 10^{-25}} \geq 0.292 \text{ nm}.$$

30. If the uncertainty in the location of a particle is equal to its de Broglie wavelength, what is the % uncertainty in its velocity?

Solution:

$$\Delta x \Delta p \geq \frac{h}{4\pi}. \quad \text{Since } \Delta x = \lambda = \frac{h}{p}, \text{ we have}$$

$$\frac{h}{p} \times \Delta p \geq \frac{h}{4\pi} \quad \Rightarrow \quad \frac{\Delta p}{p} \geq \frac{1}{4\pi} \geq 7.95\%.$$

Hence the % uncertainty in its velocity is also equal to 7.95.

31. A nucleon is confined to a nucleus of diameter 5×10^{-14} m. Calculate the minimum uncertainty in the momentum of the nucleon. Also calculate the minimum kinetic energy of the nucleon.

Solution:

$$\Delta p \geq \frac{h}{4\pi \Delta x} \geq \frac{6.63 \times 10^{-34}}{4\pi \times 5 \times 10^{-14}} \geq 1.05 \times 10^{-21} \text{ kg.m/s}$$

Hence the minimum momentum p_{min} should be ~ 1.05×10^{-21} kg.m/s

$$\therefore \quad E_{min} \approx \frac{(p_{min})^2}{2m} \approx \frac{(1.05 \times 10^{-21})^2}{2 \times 1.675 \times 10^{-27}} \approx 0.33 \times 10^{-15} \text{ J} \approx 2 \text{ keV}$$

32. In the ruby laser the metastable state from which the laser transition arises has a life time of 1 ms. Calculate the natural line width of the laser transition whose wavelength is 694.5 nm.

Solution:

We have from the uncertainty principle

$$\Delta E \Delta t \geq \frac{h}{4\pi} \Rightarrow h.\Delta \nu.\Delta t \geq \frac{h}{4\pi} \Rightarrow \Delta \nu.\Delta t \geq \frac{1}{4\pi}$$

Here Δt refers to the life time of the energy state. Also

$$c = \nu \lambda \Rightarrow 0 = \nu.d\lambda + \lambda.d\nu \Rightarrow d\lambda = -\frac{\lambda.d\nu}{\nu} = -\frac{\lambda^2}{c} \times \frac{1}{4\pi \Delta t}$$

$$\therefore \quad d\lambda = \left| -\frac{(694.5 \times 10^{-9})^2}{3 \times 10^8} \times \frac{1}{4\pi(10^{-3})} \right| = 1.28 \times 10^{-10} \text{ nm}.$$

QUESTIONS

Black body radiation
1. Define emissive power and emissivity.
2. Derive the relation between energy density and intensity of radiation.
3. How is black body radiation generated in the laboratory?
4. Explain Stefan's law and Wien's displacement law.

MATTER AND RADIATION – DUAL NATURE

5. Derive an expression for the number density of modes as a function of wavelength for electromagnetic waves confined in a cavity.
6. Derive Rayleigh-Jeans law and explain it limitations.
7. What is ultraviolet catastrophe?
8. Explain the assumptions of the Planck's model of black body radiation.
9. Explain the revolutionary nature of Planck's hypothesis and its significance for modern developments in physics.
10. Derive Planck's formula for black body radiation.
11. Show how Planck's formula reduces to Rayleigh-Jeans formula at long wavelengths.
12. Show how Planck's formula reduces to Wien's law at short wavelengths.
13. Derive Stefan's law from Planck's formula.
14. Derive Wein's displacement law from Planck's formula.
15. Describe the technique of pyrometry.
16. Explain the principle underlying on infrared imaging system.
17. Discuss the applications of infrared imaging in engineering and medicine.
18. Light from stars may appear reddish or bluish. What information would one infer from this observation?
19. Why is it difficult to produce an incandescent bulb with a visible spectrum similar to sunlight?
20. If Planck's constant were smaller than it is, would quantum phenomena be more or less conspicuous than they are now?

Photoelectric effect

21. Explain photoelectric effect.
22. Discuss the salient experimental observations of photoelectric effect.
23. Define work function of a metal.
24. Explain why classical electromagnetic theory fails to explain photoelectric effect.
25. Could a sufficiently powerful AM radio signal produce a photoelectric effect?
26. (a) When a surface is illuminated with monochromatic light, why is there a maximum kinetic energy for photoelectrons? (b) For a given frequency greater than the threshold frequency, why is there a range of kinetic energies for the emitted electrons?
27. If the intensity of light is fixed, does the number of photoelectrons depend on frequency?
28. The existence of a photoelectric work function is not contrary to classical physics. Since the work function is equal to hf_o, why isn't the existence of a cutoff frequency also acceptable classically?
29. Explain Einstein's theory of photoelectric effect.
30. Explain the functioning of a photomultiplier tube.
31. Explain how photoelectric effect can be used for light detection.
32. Discuss the use of photoelectron spectroscopy in the study of materials.

Compton effect

33. Explain Compton effect.
34. Derive an expression for the Compton shift.
35. Explain why classical electromagnetic theory cannot explain Compton effect.
36. What are the similarities between Compton effect and Raman effect?
37. In what way(s) are the photoelectric effect and the Compton effect (a) similar, (b) different?
38. Why does the Compton effect not occur with visible light?
39. In a Compton effect, why is $\Delta\lambda$ independent of the material? Why is it independent of λ?
40. In the Compton effect, why is it preferable to use short wavelengths for the incident radiation?

Matter waves

41. What are matter waves? Explain how are they different from sound and light waves?
42. State de Broglie hypothesis.
43. Derive Bragg's law and explain how it demonstrates the wave nature of X-rays.
44. Describe Davisson Germer experiment and explain how it validates de Broglie hypothesis?
45. Explain why an electron microscope has a better resolving power than that of an optical microscope.
46. Discuss neutron diffraction and its advantages vis a vis X-ray diffraction in materials science.
47. Explain single particle interference.
48. Explain the principle of complimentarity.
49. Define a wave packet.
50. Derive the relation between phase velocity and group velocity.
51. Show that the group velocity of a matter wave is equal to the velocity of the material particle.
52. Show that the phase velocity of a matter wave is always greater than the velocity of light.
53. Show that the product of group and phase velocities of a wave packet is equal to the square of the velocity of light. Why is the wave nature of matter not more apparent in our daily observations?
54. How many experiments can you recall that support the wave theory of light?, the particle theory of light?, the wave theory of matter?, the particle theory of matter?
55. Is an electron a particle? Is it a wave? Explain your answer, citing relevant experimental evidence.
56. Is equation for the de Broglie wavelength $\lambda = h/p$, valid for a relativistic particle? Justify your answer.

Uncertainty principle

57. Explain Heisenberg's uncertainty principle.
58. Explain how Heisenberg's uncertainty principle is a natural consequence of dual nature of matter.
59. Explain the physical significance of wave function.
60. Deduce uncertainty principle by considering the diffraction of electrons at a single slit.
61. Deduce uncertainty principle by considering the diffraction of electrons in a grating.
62. Deduce uncertainty principle by considering a gamma ray microscope.
63. Discuss the significance of uncertainty principle for information theory.
64. Show how Shannon's theorem in information theory is a natural consequence of Uncertainty principle.
65. Based on uncertainty principle show that electrons cannot exist within the atomic nucleus.
66. Based on uncertainty principle deduce the zero point energy of a harmonic oscillator.
67. Based on uncertainty principle deduce the size of hydrogen atom.
68. Explain how uncertainty principle leads to natural spectral width of a radiative transition.
69. Deduce an expression for spectral line width based on uncertainty principle.
70. Discuss uncertainty principle in relation to experimental errors in classical mechanics.
71. What do de Broglie waves and electromagnetic waves have in common that distinguishes them from other types of waves?
72. Discuss the analogy between (a) wave optics and geometrical optics and (b) wave mechanics and classical mechanics.
73. Does a photon have a de Broglie wavelength? Explain.
74. Discuss similarities and differences between a matter wave and an electromagnetic wave.
75. Can a de Broglie wavelength associated with a particle be smaller than the size of the particle? Larger? Is there any relation necessarily between such quantities?

76. If, in the de Broglie formula $\lambda = h/mv$, we let $m \to \infty$, do we get the classical result for particles of matter?
77. Considering electrons and photons as particles how are they different from each other?
78. Would you expect de Broglie waves to exhibit a Doppler effect?
79. In what ways is Bohr's model of the hydrogen atom not compatible with quantum mechanics?
80. The wave function gives us information regarding only probabilities, yet the predictions of wave mechanics are quite precise. Reconcile these two statements.
81. Is the Heisenberg uncertainty principle not more readily apparent in our daily observations?
82. (a) Give examples of how the process of measurement disturbs the system being measured. (b) Can the disturbances be taken into account ahead of time by suitable calculations?
83. You measure the pressure in a tyre, using a pressure gauge. The gauge, however, bleeds a little air from the tyre in the process, so that the act of measuring changes the property that you are trying to measure. Is this an example of the Heisenberg uncertainty principle? Explain.
84. "The energy of the ground state of an atomic system can be precisely known, but the energies of its excited states are always subject to some uncertainty." Can you explain this statement on the basis of the uncertainty principle?
85. "If an electron is localized in space, its momentum becomes uncertain. If it is localized in time, its energy becomes uncertain." Explain this statement.

PROBLEMS

Black body radiation

1. Show that the unit of Planck's constant is the same as that of angular momentum.
2. (a) Show that the energy, E, of a photon (in eV) can be written in the form $E = 1250\lambda^{-1}$, where the wavelength λ is in nanometers, (b) What is the range in energy of photons in the visible region from 400 nm to 700 nm?
3. Find the energy (in eV) of photons of the following wavelengths or frequencies: (a) visible light at 550 nm; (b) an FM radio wave at 100 MHz; (c) an AM radio wave at 940 kHz; (d) an X ray at 0.071 nm.
4. The dissociation energy of CO is 11 eV. What is the minimum frequency of radiation that could break this bond? (b) The maximum wavelength of radiation capable of dissociating the O_2 molecule in 175 nm. What is the binding energy in eV?
5. The C-C bond has dissociation energy 2.8 eV. What is the longest wavelength of radiation that could break this bond? To what part of the spectrum does it belong?
6. The intensity of solar radiation incident on the earth's atmosphere is 1.34 kW/m². Assuming it is monochromatic at 550 nm (yellow), how many photons/m².s does this involve.
7. What is the wavelength of the peak in black body radiation at the following temperatures: (a) The 3 K cosmic background radiation that is a remnant of the "big bang" that created the universe, (b) a tungsten filament at 3000 K, and (c) a fusion reaction at $10^7 \, K$?
8. The peak in the radiation from the sun occurs at 470 nm. (a) What is the surface temperature of the sun? (b) What would be the surface temperature of a star whose thermal radiation peaked at 350 nm?
9. For what range of temperatures does the wavelength of the peak in black body radiation vary through the visible range (400 nm – 700 nm)?
10. (a) What is the frequency of a photon whose energy is twice the rest energy of an electron? (b) What would be the linear momentum of the photon?
11. With a pupil diameter of 5 nm, the eye can detect 8 photons/s at 500 nm. What is the required power of a point source at the distance of (a) the moon; (b) Alpha-centauri, 4.2 light-years away?
12. Estimate the net radiated intensity for the following:
    (a) A hot coal at 2000 °C in a room of 20 °C, (b) A person with a skin temperature of 34 °C in air at 10 °C, (c) The earth's surface at 22 °C radiating into space at – 270 °C.

13. Given that the sun's surface temperature is 5760 K. Find the total power radiated into space (taken to be at 0 K). The sun's radius is 6.96×10^8 m.
14. A heater filament has a radius of 2 mm and a length of 20 cm. If its temperature is 2000 K, what is the net radiated power?
15. What is the wavelength of the peak in the black body radiation of a body at 300 K?
16. Assuming the Planck radiation formula in terms of wavelength, express the formula in terms of frequency.
17. Find the wavelength and frequency of a 100 MeV photon.
18. A 1.00 kW radio transmitter operates at a frequency of 880 kHz. How many photons per second does it emit?
19. Under favourable circumstances the human eye can detect 1.0×10^{-18} J of electromagnetic energy. How many 600-nm photons does this represent?
20. A detached retina is being "welded" back in place using 20-ms pulses from a 50-W laser operating at a wavelength of 632 nm. How many photons are there in each pulse?
21. A 100-W bulb converts 5% of the electrical energy input to visible light. Assume the light has a wavelength of 600 nm and the bulb is a point source. (a) What is the number of photons emitted per second? (b) If the eye can detect 20 photons/s, at what distance would the bulb be visible? Take the pupil diameter to be 3 mm.

Photoelectric effect

22. The threshold wavelength for photoelectric emission in tungsten is 230 nm. What wavelength of light must be used in order for electrons with a maximum energy of 1.5 eV to be ejected?
23. The threshold frequency for photoelectric emission in copper is 1.1×10^{15} Hz. Find the maximum energy of the photoelectrons (in eV) when light of frequency 1.5×10^{15} Hz is directed on a copper surface.
24. A silver ball is suspended by a string in a vacuum chamber and ultraviolet light of wavelength 200 nm is directed to it. What electrical potential will the ball acquire as a result?
25. The work function for lithium is 2.3 eV. (a) What is the maximum kinetic energy of photoelectrons when the surface is illuminated with light of wavelength 400 nm? (b) If the stopping potential is 0.6 V, what is the wavelength?
26. Radiation of wavelength 200 nm is incident on mercury, which has a work function of 4.5 eV. What is (a) the maximum kinetic energy of the ejected electrons, and (b) the stopping potential?
27. When radiation of wavelength 350 nm is incident on a surface, the maximum kinetic energy of the photoelectrons is 1.2 eV. What is the stopping potential for a wavelength of 230 nm?
28. The work function for potassium is 2.25 eV. A beam with a wavelength of 400 nm has an intensity of 10^{-9} W/m^2. Find (a) the maximum kinetic energy of the photoelectrons, (b) the number of electrons emitted per meter squared per second from the surface assuming 3 % of the incident photons are effective in ejecting electrons.
29. 1.5 mW of 400 nm light is directed at a photoelectric cell. If 0.10 per cent of the incident photons produce photoelectrons, find the current in the cell.
30. Light of wavelength 400 nm is shone on a metal surface. The work function of the metal is 2.50 eV. (a) Find the retarding voltage at which the photoelectron current disappears; (b) Find the speed of the fastest photoelectrons.
31. A metal surface illuminated by 8.5×10^{14} Hz light emits electrons whose maximum energy is 0.52 eV. The same surface illuminated by 12.0×10^{14} Hz light emits electrons whose maximum energy is 1.97 eV. From these data find Planck's constant and the work function of the surface.
32. The work function of a tungsten surface is 5.4 eV. When the surface is illuminated by light of wavelength 175 nm, the maximum photoelectron energy is 1.7 eV. Find Planck's constant from this data.
33. Show that it is impossible for a photon to give up all its energy and momentum to a free electron in free space. Hence show that the photoelectric effect can take place only when photons strike electrons in a metal.

34. When violet light of wavelength 420 nm illuminates a surface, the stopping potential of the photoelectrons is 2.4 V. What is the threshold frequency for this surface?
35. When a metal is illuminated with light of frequency f, the maximum kinetic energy of the photoelectrons is 1.3 eV. When the frequency is increased by 50%, the maximum kinetic energy increases to 3.6 eV. What is the threshold frequency for this metal?
36. The following data on wavelengths and stopping potentials were obtained from an experiment on the photoelectric effect.

λ (nm)	500	450	400	350	300
V_o (V)	0.37	0.65	1.0	1.37	2.0

Plot a graph and from it determine (a) h/e; (b) the threshold frequency.

Compton effect

37. A beam of X rays with an energy of 30 keV undergoes Compton scattering. A scattered photon emerges at 50° relative to the incoming beam; (a) Find the modified wavelength, (b) What is the kinetic energy of the scattered electron?
38. An X-ray beam has an energy of 40 keV. Find the maximum possible kinetic energy of Compton scattered electrons.
39. A 0.071 nm wavelength X ray is scattered by a carbon target. It suffers a 0.02% shift in wavelength. At what angle to its original direction does it emerge?
40. The wavelength of a photon is equal to the Compton wavelength. What is its energy?
41. A 30 keV beam of X rays is Compton scattered through 37°. (a) What is the shift in wavelength? (b) What is the energy of the scattered photon?
42. The fractional shift experienced by a beam of Compton-scattered radiation is $\Delta\lambda/\lambda = 0.03$ %. What is the energy of the incident photon if it is scattered through 53°?
43. An X-ray of wavelength 0.08 nm is scattered by 70° by a block of carbon. What is the Compton shift in wavelength? (b) What is the kinetic energy of the scattered electron?
44. X-rays with an energy of 50 keV are scattered by 45°. Find the frequency of the scattered photons.
45. A beam of X-rays of wavelength of 0.08 nm undergoes Compton scattering by a target. Calculate the shift in wavelength if the scattered photon is deflected by (a) 30°, (b) 90°, (c) 150°.
46. In a Compton-scattering experiment, the scattered photon has an energy of 130 keV and the scattered electron's kinetic energy is 45 keV. Find (a) the wavelength of the incident photons, (b) the angle Y through which the photon is scattered, and (c) the angle ϕ at which the electron moves off.
47. X-rays scattered by a crystal are assumed to undergo no change in wavelength. Show that this assumption is reasonable by calculating the Compton wavelength of a Na atom and comparing it with the typical X-ray wavelength of 0.1 nm.
48. A monochromatic X-ray beam whose wavelength is 55.8 pm is scattered through 46°. Find the wavelength of the scattered beam.
49. A beam of X-rays is scattered by a target. At 45° from the beam direction the scattered X-rays have a wavelength of 2.2 pm. What is the wavelength of the X-rays in the direct beam?
50. An X-ray photon whose initial frequency was 1.5×10^{19} Hz emerges from a collision with an electron with a frequency of 1.2×10^{19} Hz. How much kinetic energy was imparted to the electron?
51. An X-ray photon of initial frequency 3×10^{19} Hz collides with an electron and is scattered through 90°. Find its new frequency.
52. Find the energy of an X-ray photon which can impart a maximum energy of 50 keV to an electron (a) Find the change in wavelength of 80-pm X-rays that are scattered 120° by a target. (b) Find the angle between the directions of the recoil electron and the incident photon. (c) Find the energy of the recoil electron.

53. A photon of frequency γ is scattered by an electron initially at rest. Verify that the maximum kinetic energy of the recoil electron is

$$KE_{max} = (2h^2\gamma^2/m_oC^2)(1 + h\gamma/m_oC^2).$$

54. In a Compton-effect experiment in which the incident X-rays have a wavelength of 10.0 pm, they scattered at a certain angle having a wavelength of 10.5 pm. Find the momentum (magnitude and direction) of the corresponding recoil electrons.

55. A photon whose energy equals the rest energy of the electron undergoes a Compton collision with an electron. If the electron moves off at an angle of 40° with the original photon direction, what is the energy of the scattered photon?

Matter waves

56. A non-relativistic particle is moving three times as fast as an electron. The ratio of their de Broglie wavelengths is 1.813×10^{-4}. By calculating its mass, identify the particle.

57. A photon in free space has energy of 1.5 eV and an electron, also in free space, has a kinetic energy of that same amount. (a) What are their wavelengths? (b) Repeat for an energy of 1.5eV.

58. Electrons are accelerated through a potential difference of 25.0 kV. Find the de Broglie wavelength of such electrons (a) using the classical expression for momentum and (b) taking relativity into account.

59. What accelerating voltage would be required for electrons in an electron microscope to obtain the same ultimate resolving power as that which could be obtained from a gamma ray microscope using 1360 keV gamma rays?

60. The highest achievable resolving power of microscope is limited only by the wavelength used; that is, the smallest detail that can be separated is about equal to the wavelength. Suppose one wishes to "see" inside an atom. Assuming the atom to have a diameter of 100 pm this means that we wish to resolve the detail of separation about 100pm. (a) If an electron microscope is used, what minimum energy of electrons is needed? (b) If a light microscope is used, what minimum energy of photons is needed? (c) Which microscope seems more practical for this purpose? Why?

61. The 32-GeV electron accelerator at Stanford provides an electron beam of small wavelength, suitable for probing the fine details of nuclear structure by scattering experiments. What is this wavelength and how does it compare with the size of an average nucleus?

62. Consider a balloon filled with (monatomic) helium gas at 18°C and 1.0 atm pressure. Calculate (a) the average de Broglie wavelength of the helium atoms and (b) the average distance between the atoms. Can the atoms be treated as particles under these conditions?

63. Compare the de Broglie wavelengths of an electron and a proton if they have (a) the same speed, and (b) the same energy.

64. An electron is accelerated from rest by a potential difference of 120 V. What is its de Broglie wavelength?

65. Calculate the de Broglie wavelengths of (a) an electron and (b) a photon if the kinetic energy of the electron is equal to the energy of the photon, which is 2eV.

66. A bullet of mass 41 g travels at 960 m/s. (a) What wavelength can we associate with it? (b) Why does the wave nature of the bullet not reveal itself through diffraction effects?

67. Calculate the wavelength of a 1.00 keV (a) electron, (b) photon, and (c) neutron.

68. The wavelength of the yellow spectral emission line of sodium is 589 nm. At what kinetic energy would an electron have the same de Broglie wavelength?

69. If the de Broglie wavelength of a photon is 0.113 pm, (a) what is the speed of the proton and (b) through what electric potential would the proton have to be accelerated from rest to acquire this speed?

70. Singly charged sodium ions are accelerated through a potential difference of 325 V. (a) What is the momentum acquired by the ions? (b) Calculate their de Broglie wavelength.

71. Find the de Broglie wavelength of a proton moving at (a) 10^3 m/s and (b) 10^6 m/s.

MATTER AND RADIATION – DUAL NATURE

72. A thermal neutron, which has a kinetic energy of 0.04 eV at 300 K, plays an important role in the fission of uranium in a nuclear reactor. What is its de Broglie wavelength?

73. A 1-g pellet moves at 10 m/s. For what slit width would the first diffraction be minimum at 0.5°? Is this a practical experiment?

74. A photon and an electron each has a de Broglie wavelength of 5 nm. Compare their energies in eV.

75. For what energy (in eV) is the wavelength of a photon (a) 10^{-10} m, (b) 10^{-15} m?

76. At what speed would the de Broglie wavelength of an electron equals that of yellow light, which is 600 nm?

77. Through what potential difference must a proton be accelerated for it to have a de Broglie wavelength of 0.1 pm?

78. The electrons in the Davisson-Germer experiment were accelerated by a potential difference of 75 V. What is the corresponding angle ϕ for the first order peak?

79. The resolving power of a microscope, which tells us the smallest detail that can be distinguished, is approximately equal to one wavelength. At what speed is the de Broglie wavelength of an electron equal to 0.1 nm, the approximate size of an atom?

80. At what speed would the de Broglie wavelength of an electron equals the Bohr radius, which is 0.0533 nm? (b) Compare the speed found in part (a) with the speed of the electron in the ground state, as predicted by the Bohr model.

81. An electron is attracted to a proton that is held at rest. Assuming the electron starts from rest at infinity, find its de Broglie wavelength when it is 0.1 nm from the proton.

82. Thermal neutrons, whose kinetic energy is 0.04 eV, pass through two slits separated by 0.1 mm. What is the expected separation between like fringes on a screen 2 m from the slits?

84. What is the de Broglie wavelength of an electron with energy of 200 MeV? Use relativistic expressions and ignore the rest energy (0.5 MeV) of the electron.

85. Calculate the linear momentum of the electron in the ground state of the hydrogen atom according to Bohr's model. (b) If the uncertainty in the momentum is $\Delta p = 2p$, find the uncertainty in the position and compare it with the Bohr radius.

86. What is the frequency of an X-ray photon whose momentum is 1.1×10^{-23} kg.m/s?

87. How much energy must a photon have if it is to have the momentum of a 10 MeV proton?

88. If the particles listed below all have the same energy, which has the shortest wavelength: electron, neutron and proton?

89. What common expression can be used for the momentum of either a photon or a particle?

90. Do electron diffraction experiments give different information about crystals than can be obtained from X-ray diffraction experiments? From neutron diffraction experiments? Give examples.

91. A neutron crystal spectrometer utilizes crystal planes of spacing d = 73.2 pm in a beryllium crystal. What must be the Bragg angle θ so that only neutrons of energy K = 4.2eV are reflected? Consider only first-order reflections.

92. A beam of thermal neutrons from a nuclear reactor falls on a crystal of calcium fluoride, the beam direction making an angle θ with the surface of the crystal. The atomic planes parallel to the crystal surface have an inter-planar spacing of 54.64 pm. The de Broglie wavelength of neutrons in the incident beam is 11.00pm. For what values of θ will the first three orders of Bragg-reflected neutron beams occur?

93. In the experiment of Davisson and Germer (a) at what angles would the second and third diffracted beams corresponding to a strong maximum in Fig and occur, provided they are present? (b) At what angle would the first order diffracted beam occur if the accelerating potential were changed from 54 to 60 V?

94. A potassium chloride crystal is cut so that the layers of atomic planes parallel to its surface having a spacing of 314 pm between adjacent lines of atoms. A beam of 380-eV electrons is

incident normally on the crystal surface. Calculate the angles ϕ at which the detector must be positioned to record strongly diffracted beams of all orders present.

95. A beam of low-energy neutrons emerges from a reactor and is diffracted from a crystal. The kinetic energies of the neutrons are contained in a band of width ΔK centered on kinetic energy K. Show that the angles for a given order of diffraction are spread over a range $\Delta\theta$ given in degrees by $\Delta\theta = (90/\pi) (\tan \theta) (\Delta K/K)$, where θ is the diffraction angle for a neutron with kinetic energy K.

Uncertainty principle

96. Using a rotating shutter arrangement, you listen to a 540 Hz standard tuning fork for 0.23s. What approximate spread of frequencies is contained in this acoustic pulse?

97. The signal from a television station contains pulses of full width $\Delta t \approx 10$ ns. Is it feasible to transmit television in the AM broadcasting band, which runs from about 500 to 1600 kHz?

98. A nucleus in an excited state will return to its ground state, emitting a gamma ray in the process. If its mean lifetime is 8.7 ps in a particular excited state of energy 1.32 MeV, find the uncertainty in the energy of the corresponding emitted gamma-ray photon.

99. An atom in an excited state has a lifetime of 12 ns; in a second excited state the lifetime is 23 ns. What is the uncertainty in energy for a photon emitted when an electron makes a transition between these two states?

100. A microscope using photons is employed to locate an electron in an atom to within a distance of 12 pm. What is the minimum uncertainty in the momentum of the electron located in this way?

101. Imagine playing baseball in a universe where Planck's constant was 0.60J-s. What would be the uncertainty in the position of a 0.50-kg baseball moving at 20m/s with an uncertainty in velocity of 1.2 m/s? Why would it be hard to catch such a ball?

102. Find the uncertainty in the location of a particle, in terms of its de Broglie wavelength λ, so that the uncertainty in its velocity is equal to its velocity.

11
X-rays

11.1 INTRODUCTION

The discovery of X-rays has been crucial to the development of modern physics. In 1895 Wilhelm Roentgen found that a highly penetrating radiation is produced when fast electrons impinge on metals. X-rays were found to

(a) travel in straight lines

(b) be unaffected by electric and magnetic fields

(c) pass readily through opaque materials

(d) cause phosphorescent substances to glow and

(e) affect photographic plates

The particle nature of X-rays was demonstrated by Compton effect. The wave nature of X-rays was evident on observing diffraction effects in crystals. A single crystal acts like a three-dimensional grating. The interplanar spacing in crystals is of the order of the wavelength of X-rays. Hence X-rays are diffracted by crystals. The X-ray wavelength lies in the range 0.01 to 10 nm. X-rays thus paved the way for the acceptance of the idea of dual nature of radiation. X-rays also confirmed the basic features of the model of the atom proposed by Niels Bohr. X-ray studies enabled H.G.Moseley to develop the concept of atomic number. This in turn led to the proper understanding of the arrangement of elements in the periodic table. Earlier Mendeleyev had arranged the elements in the periodic table on the basis of their atomic weight. Moseley's work clearly showed that atomic number was a more rational basis for arranging the elements in the periodic table.

X-rays on account of their penetrating power have found applications in industry, medicine and materials science.

11.2 PRODUCTION OF X-RAYS

X-rays are produced when fast energetic electrons strike a metal. Figure 11.1 shows the diagram of an evacuated tube, used for producing X-rays. It consists of an anode and a cathode which are connected to a high voltage dc power supply of the order of tens of kilo volts. A filament, which emits electrons heats the cathode by thermionic emission. The inside of the X-ray tube is evacuated so that the electrons do not lose energy through collisions. The high potential difference V between the cathode and the anode accelerates the electrons towards the anode. The anode, also known as *target* is a metallic element of high atomic number such as iron, cobalt, tungsten or molybdenum. The face of the target is at an angle relative to the electron beam. When fast electrons strike the metallic target, X-rays are produced. They pass through the side of the tube.

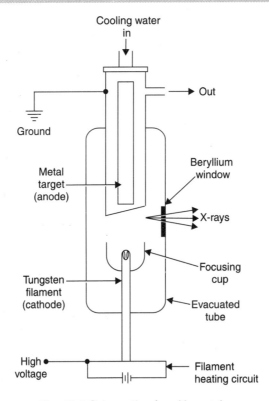

Fig. 11.1 Schematic of an X-ray tube

Because of the bombardment of electrons the anode gets heated and has to be cooled using water flow. Heat dissipation can also be achieved by having a rotating anode in which the electrons do not strike the same part of the anode (Fig. 11.2). In such tubes electrons are focused to strike a small spot ($\sim$ mm^2) on the rim of the anode disc. The heat generated is radiated away and the disc has sufficient thermal capacity to prevent it from getting unduly heated. Anodes of this type rotate at $\sim$ 10000 rpm and the rotating disc shaft spins as a result of magnetic coupling through the sealed glass envelope.

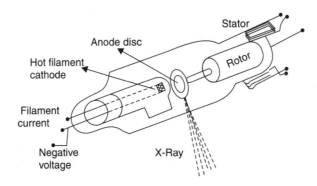

Fig. 11.2 X-ray tube with a rotating anode

An analysis of the intensity of X-rays as a function of wavelength reveals the following features.

(a) X-ray spectrum is continuous on which are superposed sharp peaks. The sharp peaks are known as the characteristic X-ray spectrum since the position of the peaks depends on the nature of the anode element (Fig. 11.3).

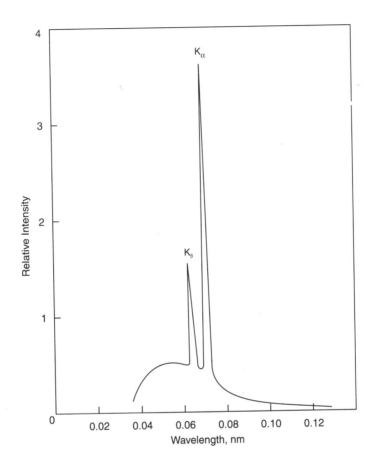

Fig. 11.3 X-ray spectrum

(b) For the same anode material, the broad intensity peak in the continuous spectrum shifts towards shorter wavelengths, with increasing accelerating potential (Fig.11.4).

(c) At any accelerating potential, there is a cut off wavelength λ_{min}. X-rays have only wavelengths $> \lambda_{min}$. Increasing 'V' decreases λ_{min}. λ_{min} or ν_{max} corresponds to the case when the entire kinetic energy of the electron is converted into a single photon.

The continuous spectrum is explained as arising due to the deceleration of electrons in the metallic target. The characteristic spectrum is explained on the basis of Bohr's atomic model.

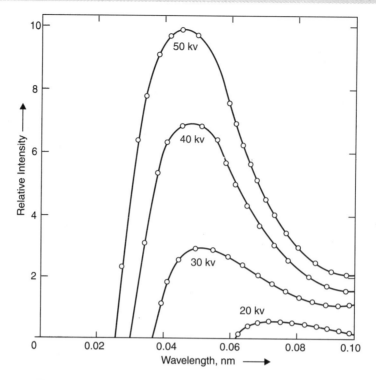

Fig. 11.4 X-ray spectrum for various accelerating voltages

11.3 ORIGIN OF X-RAY SPECTRUM

The white radiation or the continuous X-ray spectrum (also known as Bremsstrahlung) is due to slowing down of high-speed electrons as they pass close to the nuclei of the atoms within the target of the X-ray tube. According to classical electromagnetic theory, radiation is emitted whenever electrons are accelerated or decelerated (Fig.11.5). If the electron loses all its energy in the target, then the maximum frequency or minimum wavelength of photon is given by:

$$eV = h\nu_{max} = \frac{hc}{\lambda_{min}}; \text{ or } \lambda_{min} = \frac{hc}{eV} = \frac{1.24 \times 10^{-6}}{V} \qquad (11.1)$$

Sometimes a fast electron strikes an innermost electrons in the target atom (such as K-electron, L-electron or M-electron) and knocks it out of its orbit and frees these electrons. The vacancy thus created in any shell is filled almost immediately, when an electron from an outer shell of the target atom falls into it. In such a process a characteristic X-ray photon is emitted (Fig.11.5). The wavelength of the X-ray photon can be obtained using Bohr's formula with a minor modification. Bohr's model is applicable for a single electron going round the nucleus. However, in X-ray emission due to electronic transition, the electron undergoing the transition is under the influence of a nucleus which is screened by the other electrons in the orbit. Hence the effective charge on the nucleus is not the same as Ze. In the modified Bohr's formula this is taken care of by using an effective charge Z^*e for the nucleus. $Z^* = (Z - b)$ where b is the screening constant. Thus the wavelength of the X-ray photon is given by:

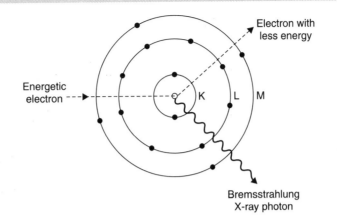

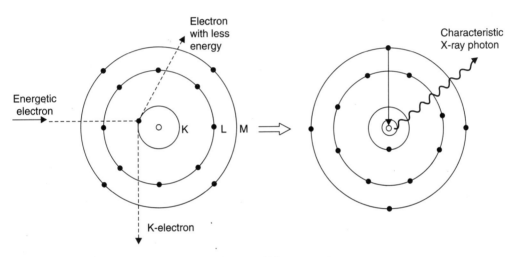

Fig. 11.5 Origin of X-ray spectrum

$$\frac{1}{\lambda} = R(Z^*)^2 \left[\frac{1}{n_1^2} - \frac{1}{n_2^2} \right] \text{ where } R \text{ is the Rydberg constant.} \qquad (11.2)$$

The nomenclature of different X-ray lines arising from various transitions is shown in Fig. 11.6.

For, K_α line, $n_1 = 1, n_2 = 2$; K_β line, $n_1 = 1, n_2 = 3$; K_γ line, $n_1 = 1, n_2 = 4$
L_α line, $n_1 = 2, n_2 = 3$; L_β line, $n_1 = 2, n_2 = 4$; L_γ line $n_1 = 2, n_2 = 5$
M_α line, $n_1 = 3, n_2 = 4$; M_β line, $n_1 = 3, n_2 = 5$; M_γ line, $n_1 = 3, n_2 = 5$

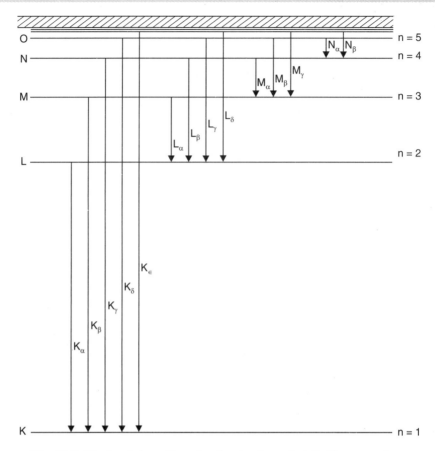

Fig. 11.6 Electronic transitions leading to characteristic X-ray spectrum

11.4 MOSELEY'S LAW

Moseley determined the wavelengths of the characteristic X-ray lines in a number of elements. He found that for a given spectra line such as K_α, the plot of $\sqrt{\text{(frequency)}}$, versus atomic number yielded a straight line (Fig. 11.7). Prior to Moseley's work, an element's place in the periodic table was assigned on the basis of its mass. Due to Moseley's work, *atomic number* became important and the characteristic X-ray lines became a universally accepted signature of an element.

The Moseley's plot finds its natural explanation in the Bohr's atomic model. In the Bohr atomic model consider the transition of electron from an initial state (n_2) to the final state (n_1). The expression for the $(n_2 > n_1)$ wavelength of light is

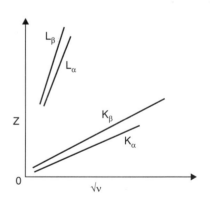

Fig. 11.7 Moseley plot

$$\frac{1}{\lambda} = R(Z^*)^2 \left\{ \frac{1}{n_1^2} - \frac{1}{n_1^2} \right\}$$

where R is the Rydberg constant and is given by $R = 1.097 \times 10^{-7} \text{m}^{-1}$.

The frequency of the X-ray characteristic emission is given as

$$\nu = cR(Z^*)^2 \left\{ \frac{1}{n_1^2} - \frac{1}{n_1^2} \right\} = cR(Z-b)^2 \left\{ \frac{1}{n_1^2} - \frac{1}{n_1^2} \right\} \qquad (11.3)$$

The above equation can be cast in the form:

$$Z - b = \frac{\sqrt{\nu}}{\sqrt{cR \left\{ \frac{1}{n_1^2} - \frac{1}{n_1^2} \right\}}} \quad \text{or} \quad Z = \frac{\sqrt{\nu}}{\sqrt{cR \left\{ \frac{1}{n_1^2} - \frac{1}{n_1^2} \right\}}} + b \qquad (11.4)$$

The above equation predicts a linear relation between Z and $\sqrt{\nu}$, with different slopes for each series. The intercept gives the screening constant. Thus in K-series K_α and K_β spectral lines with $n_2 = 2$ and 3 have corresponding slopes $\sqrt{(4/3cR)}$ and $\sqrt{(9/8cR)}$. The L-series include L_α, L_β and L_γ for $n_2 = 3, 4$ and 5 with slopes $\sqrt{(36/5cR)}$, $\sqrt{(16/3cR)}$ and $\sqrt{(100/21cR)}$. Figure 11.7 shows Z versus $\sqrt{\nu}$ graphs whose linear behaviour agrees with the predictions based on Bohr's atomic model. It is found that $b_K \approx 1$ and $b_L \approx 7.4$.

11.5 SYNCHROTRON RADIATION AS X-RAY SOURCE

In recent years, synchrotron radiation or synchrotron light source is used as a source of X-rays. According to classical electromagnetic theory, a charged particle such as an electron when subject to acceleration emits radiation. When the velocity of the electron is comparable to that of light (i.e., $\beta = (v/c) \approx 1$), the emitted radiation is highly intense and collimated (Fig. 11.8). In a synchrotron, electrons are accelerated to very great energy ($\sim GeV$) by the application of electric and magnetic fields. Electrons experience a centripetal acceleration when they move in circular orbits in what are called storage rings (Fig. 11.8). As a consequence they emit radiation whose power is given by:

$$P = \frac{88.47 E^4 I}{R} = 2.65 B E^4 I$$

where E is the energy of electrons in GeV, I is the current in amps, R is the radius of curvature of the electron orbit and B is the magnetic field in kilo Gauss.

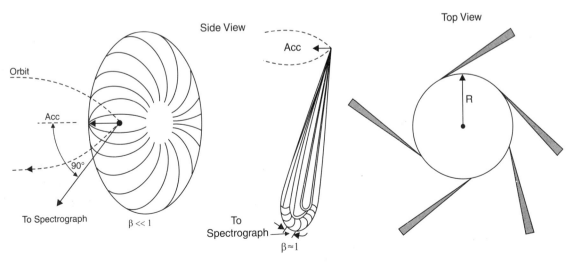

Fig. 11.8 Synchrotron light source. The pattern of energy radiated also shown.

The radiation is distributed over a wide range of wavelengths as shown in Fig. 11.9. The advantage of synchrotron light source is its high intensity and collimation. The disadvantage is that it is very expensive and can only be a national facility.

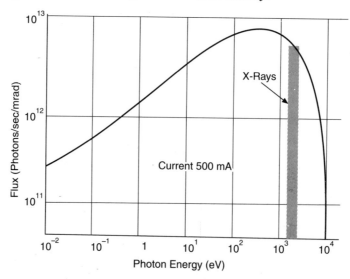

Fig. 11.9 Spectrum of synchrotron light source

11.6 INTERACTION OF X-RAYS WITH MATTER

When X-rays pass through matter, they interact with it in a complex manner. The effect of this interaction is the attenuation of the X-ray radiation. The X-ray intensity, when it passes through matter can be written as

$$I_x = I_o e^{-\mu x}$$

where I_o is the intensity of the incident beam, I_x is the intensity after traversing a distance x and μ is the linear attenuation or absorption coefficient. The linear absorption coefficient has units of m^{-1}. The half value layer (HVL) for an X-ray beam is the thickness of a given material that will reduce the beam intensity by one-half. The HVL is related to the linear absorption coefficient by the relation

$$HVL(\text{in units of cm}) = \frac{ln_e 2}{\mu} = \frac{0.693}{\mu} \qquad (11.5)$$

The attenuation of X-rays depends on the density of the material as well. Hence the linear absorption coefficient is normalized with respect to the density of the material. The above equation can be recast in the following form:

$$I = I_0 e^{-\left(\frac{\mu}{\rho} \times \rho x\right)} = I_0 e^{-\mu_m (\rho x)}$$

The quantity ρx has the units of grams per cm^2 and is called the area density. μ_m is called the mass attenuation coefficient and has the units of cm^2/gram. The mass attenuation coefficient emphasizes that the mass is primarily responsible for attenuating X-rays. In terms of area density, the HVL is given by

$$HVL(\text{in units of } gram/cm^2) = \frac{ln_e 2}{\mu_m} = \frac{0.693}{\mu_m} = \frac{0.693}{\mu} \times \rho$$

Thus HVL (in units of gram/cm^2) = HVL (in units of cm) $\times$ density $\qquad (11.6)$

X-RAYS

Major factors for the attenuation of X-rays are

(a) Photoelectric emission
(b) Compton scattering and
(c) Pair production

All these processes are shown in Fig. 11.10.

(a) In photoelectric effect, the incident energy is transmitted to an orbital electron. As a result the electron is no longer bound to the atom and get released with a particular velocity. Photoelectric emission is predominant around 20 keV to 100 keV.

(b) In Compton scattering, the incident X-ray photon is inelastically scattered by free electron. Due to this collion process, the incident X-ray photon loses energy and consequently there is an increase in its wavelength. Compton scattering predominates in the region 100 keV to 200 MeV.

(c) In pair production the incident X-ray photon directly interacts with the nucleus of the atom giving rise to an electron-positron pair. Pair production dominates above 6 MeV.

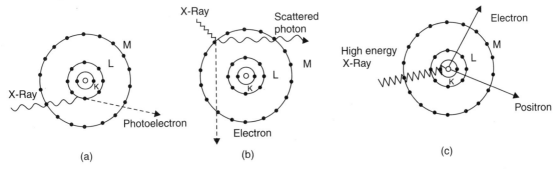

Fig. 11.10 X-rays lose energy in three ways (a) in photoelectric effect all of the photon's energy is given to the photoelectron, (b) in Compton effect, some energy is given to an electron and some goes into a scattered photon, (c) in pair production, a high energy photon is converted into an electron and positron

The mass absorption coefficient of water for these processes as a function of energy is shown in Fig. 11.11.

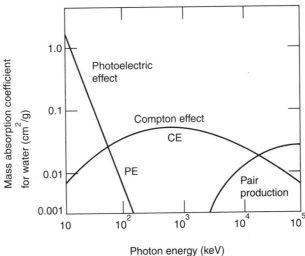

Fig. 11.11 Mass attenuation coefficient for water at different energies in the range 10 to 10^5 keV

11.7 DETECTION OF X-RAYS

Early X-ray equipment employed photographic emulsions for detection and measurement of radiation. For reasons of convenience, speed and accuracy, modern instruments are equipped with detectors that convert the X-ray radiant energy into an electrical signal. Three types of transducers are used. These are gas filled detectors, semiconductor based detectors and scintillation detectors. Gas filled detectors and semiconductor based detectors are based on the property of X-rays to ionize the medium in which they travel. Scintillation detectors are based on the property of X-rays to produce luminescence in some materials.

11.7.1 Gas Filled Detectors

Inert gases such as argon, xenon or krypton, which are ordinarily insulators, become conducting when X-rays pass through them. This is because X-rays produce a large number of positive gaseous ions and electrons, which enhance the conductivity. This phenomenon is used to detect and measure the energy of X-rays. Three types of X-ray detectors viz., ionization chambers, proportional counters and Geiger tubes are based upon this principle. A typical gas-filled detector is shown schematically in Fig. 11.12. Radiation enters the chamber through a transparent window of mica, beryllium, aluminum or mylar. Each X-ray photon knocks off the outermost electrons *of the gas atom. Each photoelectron has a large kinetic energy, which is equal to the difference between the X-ray photon energy and the binding energy of the energy in the argon atom. The photoelectron then loses this excess kinetic energy by ionizing several hundred additional atoms of the gas.* Under the influence of an applied potential, the mobile electrons migrate towards the central wire anode while the slower-moving cations are attracted towards the cylindrical metal cathode. In the absence of the applied potential, the ions tend to recombine with electrons and no current results.

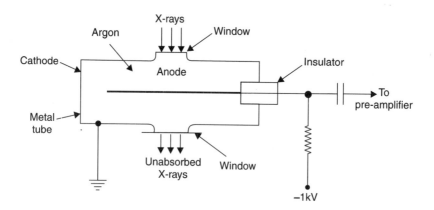

Fig. 11.12 Gas filled detector

Figure 11.13 shows the effect of applied potential upon the number of electrons that reach the anode of a gas-filled detector for each X-ray photon entering the detector. Three characteristic voltage regions are indicated. At potentials less than V_1, the accelerating force on the ion pairs is low and the rate at which the positive and negative species separate is insufficient to prevent partial recombination. As a consequence, the number of electrons reaching the anode is smaller than the number produced initially by incoming radiation.

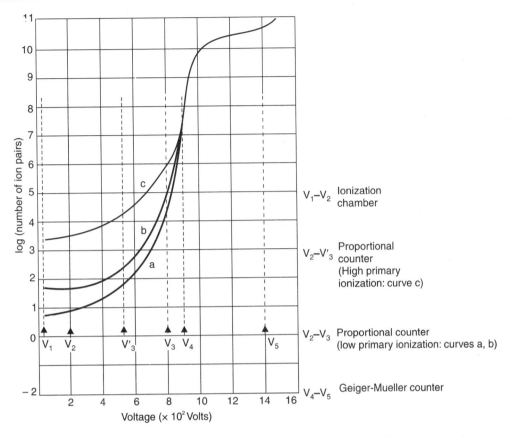

Fig. 11.13 Effect of applied potential on the number of electrons that reach anode

Ionization chambers are operated in the voltage range from V_1 to V_2 (Fig. 11.13). In this region the number of electrons reaching the anode is reasonably constant and represents the total number formed by a single photon. Here the currents are small (typically 10^{-13} to 10^{-16} A) and relatively independent of applied voltage. Ionization chambers are not employed in X-ray spectrometry because of their lack of sensitivity.

Proportional counters operate in the region between V_2 and V_3. In this region the number of electrons increases rapidly with applied potential. This increase is the result of secondary ion-pair production caused by collisions between the accelerated electrons and gas molecules. Amplification of the ion current results. Typically the pulse produced by a photon is amplified by a factor of 500 to 10,000. But the number of positive ions produced is so small that the dead time is only about 1μs. In general, the pulses from a proportional counter tube must be amplified before being counted. The number of electrons per pulse (the pulse height) produced in the proportional region depends directly on the energy (and thus the frequency) of the incoming radiation. Hence it can be used for measuring the energy of the incident photon. A proportional counter can be made sensitive to a restricted range of X-ray frequencies with a *pulse height analyzer,* which counts a pulse only if its amplitude falls within certain limits. Proportional counters have been widely used as detectors in X-ray spectrometers. For highly energetic X-rays, which produce a large primary ionization, proportional counters can be operated in the region $V_3' - V_2$.

Geiger Muller Counters operate in the region V_4 to V_5. In this region, amplification of the electrical pulse is enormous. Here the gas amplification is greater than 10^9. Each photon produces an avalanche of electrons and cations. The resulting currents are thus large and relatively easy to detect and measure. But the space charge which consists of heavy positively charged gas ions migrate towards the cathode very slowly. Because of this effect, the number of electrons reaching the anode is independent of the type and energy of incoming radiation and is governed instead by the geometry and the pressure of the tube. The net effect is a momentary pulse of current followed by an interval during which the tube does not conduct. Before conduction can again occur, this space charge must be dissipated by the migration of the cations to the walls of the chamber. During the *dead time*, when the tube is non-conducting, response to radiation is impossible. The dead time thus represents the upper limit in the response capability of the tube. Typically, the dead time of a Geiger tube is in the range from 50 to 200 μs. Geiger tubes are usually filled with argon. A low concentration of an organic substance, often alcohol or methane (as quenching gas) is also present to minimise the production of secondary electrons when the cations strike the chamber wall. This helps to reduce the dead time. With a Geiger tube, radiation intensity is determined by counting the pulses of current. The device is applicable to all types of nuclear and X-ray radiation. However, it lacks the large counting range of other detectors because of its relatively long dead time. Its use in X-ray spectrometers is limited by this factor.

11.7.2 Semiconductor Detectors

Semiconductor detectors have assumed a major importance as detectors of X-ray radiation. These are basically photodiodes with a very wide depletion region. In a normal photodiode, the photon on traversing through the depletion region, is absorbed and gives rise to electron-hole pairs. The photodiode can then be operated either in the *photovoltaic* or *photoconductive* mode. However, in the photodiode used for X-ray detection, provision ought to be made for the high penetration of X-rays. Hence *p-i-n* diode is employed. *P-I-N* diode is essentially a *pn* junction with a wide intrinsic region between the *p* and *n* semiconductors. Depending on the X-ray energy to be measured, the width of the intrinsic region is varied. These diodes are made by diffusing Li in Si or Ge in the presence of electric fields at high temperature. Hence they are sometimes called *lithium drifted silicon detectors* Si (Li) or lithium drifted germanium detectors Ge (Li). Figure 11.14 illustrates one form of a lithium drifted detector, which is fashioned from a wafer of crystalline silicon. Three layers exist in the crystal, a *p*-type semiconducting layer that faces the X-ray source, a central intrinsic zone and an *n*-type layer. The outer surface of the *p*-type layer is coated with a thin layer of gold for electrical contact. Often, it is also covered with a thin beryllium window which is transparent to X-rays. The signal output is taken from an aluminum layer which coats the *n*-type silicon. This output is fed into a preamplifier with an amplification factor of about 10. The preamplifier is frequently a field-effect transistor which is made an integral part of the detector.

The intrinsic layer of a silicon detector functions in a way that is analogous to argon in the gas filled detector. Initially, the absorption of a photon results in the formation of a highly energetic photoelectron, which then loses its kinetic energy by elevating several thousand electrons in the silicon to a conduction band. A marked increase in conductivity results. When a potential is applied across the crystal, a current pulse accompanies the absorption of each photon. In common with a proportional detector, the size of the pulse is directly proportional to the energy of the absorbed photons. In contrast to the proportional detector, however, secondary amplification of the pulse does not occur. Generally these detectors are operated at $77K$ to reduce noise. Germanium is used in place of silicon to give lithium-drifted detectors that are particularly useful for detection of radiation that is shorter in wavelength than $0.3A$. These detectors must be cooled at all times. Germanium detectors that do not require lithium drifting have been produced from very pure germanium. These detectors, which are called *intrinsic germanium detectors,* need to be cooled only during use.

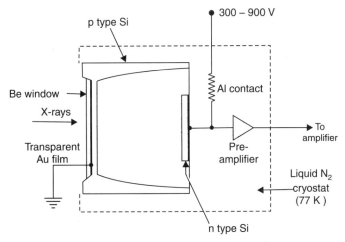

Fig. 11.14 Semiconductor detector

11.7.3 Scintillation Detectors

X-rays produce luminescence when incident on some materials known as *phosphors*. This property is made use of in scintillation detectors. In the earliest applications, the technique involved the manual counting of flashes that resulted when individual photons or radiochemical particles struck a zinc sulphide screen. The tedium of counting individual flashes by eye led Geiger to the development of gas-filled detectors. These were not only more convenient and reliable but also more responsive to radiation. The advent of the photomultiplier tube and better phosphors has reversed the trend. Scintillation counting has again become one of the important methods for radiation detection.

The most widely used modern scintillation detector consists of a transparent crystal of sodium iodide that has been activated by the introduction of perhaps 0.2% thallium iodide. Often, the crystal is shaped as a cylinder that is 3 to 4 inches in each dimension. One of the plane surfaces then faces the cathode of a photomultiplier tube (Fig. 11.15). As the incoming radiation traverses the crystal, its energy is first lost to the scintillator. This energy is subsequently released in the form of photons of fluorescent radiation. Several thousand photons with a wavelength of about 400 nm are produced by each primary particle or photon over a period of about 0.25 µs (the *dead time*). The *dead time* of a scintillation counter is thus significantly smaller than the *dead time* of a gas-filled detector.

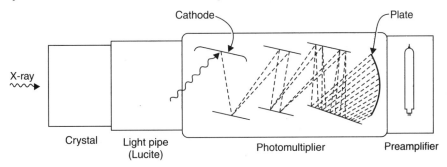

Fig. 11.15 Scintillation detector

The flashes of light produced in the scintillation crystal are transmitted in the photocathode of a photomultiplier tube. They are in turn converted to electrical pulses that can be amplified and counted. An important characteristic of scintillators is that the number of photons produced in each flash is approximately proportional to the energy of the incoming radiation. Thus, the incorporation of a pulse-height analyzer to monitor the output of a scintillation counter forms the basis of energy-dispersive photometers.

In addition to sodium iodide crystals, a number of organic scintillators such as stilbene, anthracene and terphenyl have been used. In crystalline form, these compounds have dead times of 0.01 and 0.1 μs. Organic liquid scintillators such as a solution of p-tetraphenyl in toluene have also been used in scintillation detectors.

11.8 X-RAY RADIOGRAPHY

Ever since the discovery of X-rays by Roentgen in 1895, they have been put into numerous applications in the fields of research, medicine and industry. Roentgen's experiments revealed the extraordinary penetrating power of X-rays. Radiography progressed little beyond laboratory curiosity for another 20 years. The invention of Coolidge X-ray tube in 1913, providing greater output of X-rays at higher energies made industrial radiography a practical proposition. Radiography is now found to be most satisfactory for finding internal defects such as porosity and voids. It is also suitable for detecting changes in material composition, for thickness measurement and for locating unwanted or defective components hidden from view in assembled parts.

The following properties of X-rays are relevant for industrial radiography:

(a) Rectilinear propagation.

(b) Differential absorption in matter.

(c) Ability to affect photographic plates.

(d) Fluorescence.

(e) Liberation of photoelectrons.

11.8.1 Basic Principles

The basic principle of radiographic inspection is shown in Fig 11.16. The X-ray source emits radiation that travels in straight lines and penetrates the object under investigation. Once the radiation has passed through the material, its image (as received on a recording plane opposite to the source) is used to evaluate the condition of the part being inspected. Films or fluoroscopic screens are made use of for recording the image. If film is used it must be chemically developed in order to obtain the image.

The image of the object will appear on the film provided there is a sufficient difference in the radiation intensities received by the film under the defect as compared to that received through the remainder of the material i.e., the absorption coefficient of X-rays passing through the defect should be considerably different from those passing through the remainder of the material. X-ray radiographic technique is most helpful in locating internal defects such as inclusions, blowholes, porosity and big voids.

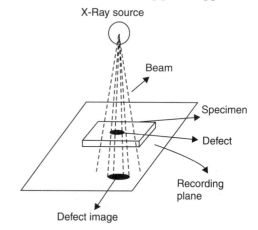

Fig. 11.16 Layout for radiographic inspection

Visualization of X-rays

An important component of the X-ray radiography is the visualization of the X-ray image. X-rays cannot be detected by the human eye and hence indirect methods of visualization must be used to depict the intensity distribution of X-rays that have passed through the object of interest. Three different techniques are in common use.

Fluoroscopy

The term fluoroscopy is synonymous with real-time radiography and electronic radiography. Roentgen actually discovered X-rays when he noticed that certain metal salts glowed in the dark when exposed to this radiation. This phenomenon is known as *fluorescence*. The brightness of this *fluorescence* is a function of the radiation intensity. Cardboard pieces coated with such metal salts were first used exclusively to visualize X-ray image. The fluoroscopic image obtained in this way is rather faint. And the X-ray intensity necessary to obtain a reasonably bright image is of such a magnitude that it can be harmful to the observer. If the radiation intensity is reduced to a safer level, the fluoroscopic image becomes so faint that it must be observed in a completely darkened room and after the eyes of the observer have adapted to the dark for 10 to 20 minutes. Because of these inconveniences, direct fluoroscopy now has only limited use.

Image intensifiers

Basic equipment for conventional fluoroscopy consists of a source of radiation, a fluoroscopic conversion screen, mirrors and a viewing port. The faint image of the fluoroscopic screen can be made brighter with the help of an electronic image intensifier as shown in Fig.11.17. The intensifier tube contains a fluorescent screen, the surface of which is coated with a suitable material to act as a photocathode. The electron image thus obtained is projected onto a phosphor screen at the other end of the tube by means of an electrostatic lens system. The resulting brightness gain is due to the acceleration of the electrons in the lens system and the fact that the output image is smaller than the primary fluorescent image. The gain can reach an overall value of several hundred. It not only allows the X-ray intensity to be decreased but makes it possible to observe the image in a normally illuminated room. The intensifying tube, however, is rather heavy and requires a special suspension. To get the basic real time image an object is placed between the source of radiation and fluoroscope screen that converts the transmitted radiation to visible light. A specifically coated mirror then reflects the visible image to a viewing port that lets the interpreter view the object.

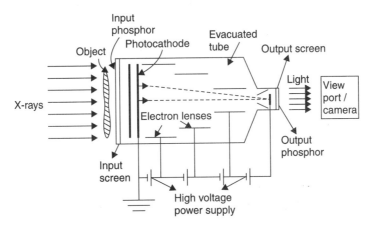

Fig. 11.17 X-ray fluoroscopy with an image intensifier tube

X-ray films

X-rays affect the photographic plate in a fashion similar to that of visible light. After processing in a developing solution, a film exposed to X-rays, will show the image of the X-ray intensity. The sensitivity of this effect can be increased by the use of *intensifying screens*, which are similar to the fluoroscopic screens described above (Fig.11.18). The screen is brought into close contact with the film surface so that the film is exposed to the X-rays as well as to the light from the fluorescence of the screen. X-ray films, with or without intensifying screens, are packaged in light-tight cassettes with which one side is made of thin plastic that can easily be penetrated by the X-rays.

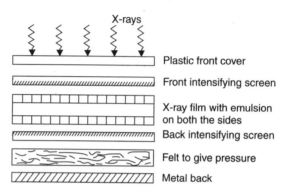

Fig. 11.18 X-ray film with intensifying screens

11.8.2 Real Time Radiography

Real time radiography is similar to conventional radiography. However, a major difference is in viewing the image. During film radiography, the image is viewed in a static mode. During real time radiography, the image is interpreted generally at the same time as the radiation passes through the object. This is the dynamic mode. Another difference of real time image is that a positive image is normally presented, whereas the X-ray film gives a negative image. Real time radiography has the advantages of high speed and low cost of inspection.

11.8.3 Applications of Industrial Radiography

Radiography can be used to inspect most types of solid material both ferrous and non-ferrous alloys as well as non-metallic materials and composites. It can be used to inspect the condition and proper placement of components, for liquid level measurements in sealed components, etc. The method is used extensively for castings, weldments and forgings when there is a critical need to ensure that the object is free from internal flaws. Radiography is well suited to the inspection of semiconductor devices for detection of cracks, broken wires, unsoldered connections, foreign material and misplaced components whereas other methods are limited in ability to inspect semiconductor devices.

11.8.4 Limitations of Industrial Radiography

Like other NDT methods, radiographic inspection method has certain limitations. Certain types of flaws are difficult to detect. For example, cracks cannot be detected unless they are parallel to the radiation beam. Tight cracks in thick sections usually cannot be detected at all, even when properly oriented. Minute discontinuities such as inclusions in wrought material, flakes, microporosity and microfissures cannot be detected unless they are sufficiently large in size. Laminations are nearly impossible to detect with radiography because of their unfavourable orientations.

The defect or discontinuity must be parallel to the radiation beam, or sufficiently large, to register on the radiograph. A defect usually must be at least 2% of the thickness of the material before it can register on a radiograph with sufficient contrast, thus to be detected.

Certain areas in many items cannot be radiographed because of the geometric considerations involved. Often it is difficult, if not impossible, to position the film and source of radiation so as to obtain a radiograph of the area desired.

Compared to other NDT methods of inspection, radiography is expensive. When portable X-ray or gamma source is used, capital costs can be relatively low. Inspection of thick section is a time consuming process. Radioactive sources also limit the thickness that can be inspected, primarily because high activity sources require heavy shielding for protection of personnel. Protection of personnel for not only those engaged in radiographic work but also those in the vicinity of radiographic inspection size are of major importance. Safety requirements impose both economic and operational constraints on the use of radiography for inspection.

11.8.5 X-ray Radiography in Medicine

X-rays are used in medicine both for diagnosis and therapy.

X-ray Diagnostics with Still Picture

The use of X-rays as a diagnostic tool is based on the fact that various components of the body have different absorption coefficients for these rays. When X-rays from a point source penetrate a body section, the internal structure of the body absorbs varying amounts of the radiation. The radiation that leaves the body, therefore, has a spatial intensity variation that is an image of the internal structure of the body. When this intensity distribution is visualized using a photograph, a shadow image is generated that can show organs which differ in their absorption coefficients. Bones and foreign bodies, especially metallic ones show up well on these images because they have a much higher absorption coefficient than the surrounding tissue (Fig. 11.19). Air filled cavities also show up well because these cavities have much lower absorption coefficient than the surrounding medium. Body organs which differ very little in the absorption coefficients do not show up well on the X-ray image. In some cases this problem can be overcome by using some of the special techniques.

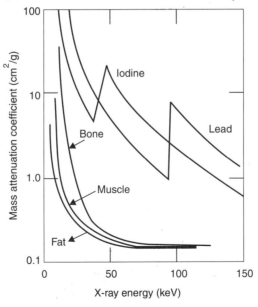

Fig. 11.19 Mass attenuation coefficient of various tissues

Special X-ray Techniques in Medicine

The previous section described the general principle of obtaining X-ray images, but often special techniques must be used to obtain usable images from certain body structures.

Use of Grids

While obtaining X-ray image it is assumed that X-rays travel in straight lines. In actual practice, X-rays do get scattered and no longer travel in a straight line. If the body section examined is very thick and if the X-rayed area is large, the scattered X-rays can cause blurring of the X-ray image. This effect can be reduced by the use of a grid. This device consists of a grid-like structure made of thin lead strips that is placed directly in front of the X-ray film (Fig. 11.20). Like a venetian blind that lets sun rays through only when they strike parallel to the slots of the blind, the grid absorbs the scattered X-rays while those traveling in straight lines pass. In order to keep the grid from throwing its own shadow on the film, it has to be moved by a motorised drive during the exposure of the film.

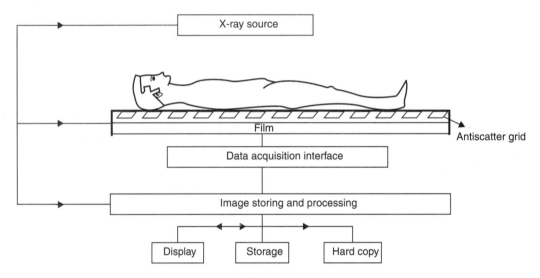

Fig. 11.20 X-ray film and the grid structure

Contrast Media

While foreign bodies and bones absorb X-rays much more readily than the soft tissue, the organs and soft tissue structures of the body show very little difference in X-ray absorption. In order to make their outlines visible on the X-ray image, it may be necessary to fill them with a *contrast medium* prior to taking the X-ray photo. Thus for example the structures of the gastrointestinal tract can be visible with the help of *barium sulphate* given orally or as an enema, which has higher X-ray absorption than the surrounding tissue. In the *pneumoencephalogram*, the ventricles of the brain are made visible by filling them with air, whose absorption coefficient is less than the surrounding brain structure. Other body structures and organs can also be visualized by filling them with suitable contrast media.

Angiography

In angiographic procedures, the outlines of blood vessels are made visible on the X-ray image by injecting a bolus of contrast medium directly into the bloodstream in the region to be investigated. Because the contrast medium is rapidly diluted in the blood circulation, an X-ray photo or a series of such photos must be taken immediately after the injection. This procedure is often performed automatically with the help of a power-operated syringe and an electrical cassette changer. Thus we have *cardioangeography* (to detect malfunctioning of the heart), *renal angiography* (to detect malfunctioning of the kidneys) and *cerebroangiography* (to detect tumors in the brain).

Three Dimensional Visualization–X-ray Tomography

The basic limitation of X-ray images is the fact that they are two-dimensional presentations of three-dimensional structures. Therefore an organ located in front or behind the target organ frequently obscures details in the image of the target organ. In *stereoradiagraphy* two X-ray photos are taken from different angles, which, when viewed in a stereo viewer, gives a sense of depth in an X-ray image. In *tomography* (from the Greek work *tomos* meaning slice or section) the X-ray photo shows the structure of only a thin slice of a section of the body. Several photos representing slices taken at different levels permit three-dimensional visualization. Tomographic X-ray photos can be obtained by moving the X-ray tube and the film cassette in opposite directions during the exposure of the film (Fig. 11.21). This procedure causes the image of the structures above and below a certain plane to be blurred by the motion, whereas structures in this plane are imaged without distortion. Special tomography machines that scan body sections with a thin X-ray beam have been developed. The image of the section is reconstructed from a large number of such scans with the help of a digital computer.

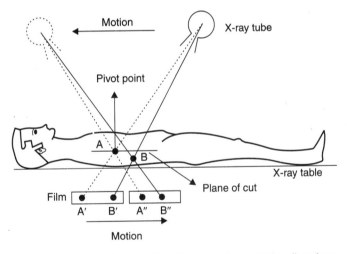

Fig. 11.21 X-ray tube and the film move in opposite directions

The simple linear motion of the tomographic unit is useful for many situations. But it can introduce confusing artifacts. An *axial tomograph* is an image of a slice across the body and is taken by rotating the X-ray tube and the film around the patient (Fig. 11.22). Axial tomography has been improved by *Computerised Axial Tomography* (CAT). In a CAT scan, an X-ray tube produces two narrow beams and scan linearly across the patient's head and the intensities of the transmitted X-ray beams are recorded by two detectors moving with the beams, on the other side of the patient (Fig.11.23). The data from the linear scan are stored in the memory of the computer. The tube and the detector are then rotated 1° and the process is repeated. After 180 scans, which require about 4 min, the computer analyses the data to determine the image of the target organ. CAT scanners that can scan any portion of the body have been developed.

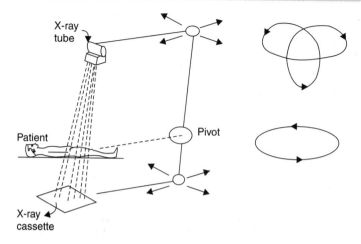

Fig. 11.22 Axial tomography

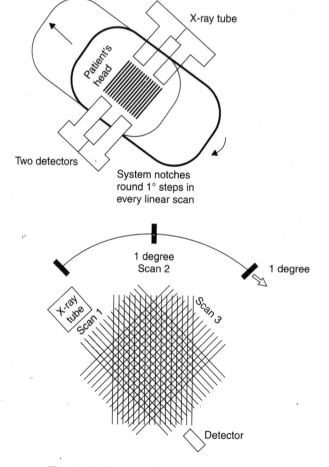

Fig. 11.23 Computerised axial tomography

X-ray Therapy
This involves the destruction of cancerous tissues and tumors using X-rays. The energy of the X-rays used and the irradiation time must be carefully chosen. It is also necessary to identify the boundary of the cancerous tissue prior to irradiation using a scanning device. Special lead shields of suitable forms are used to protect the surrounding healthy tissues. However, some of the healthy tissues around the unwanted ones may be destroyed. Normally the healthy tissues will regenerate.

11.9 CRYSTAL STRUCTURE DETERMINATION USING X-RAYS

11.9.1 Introduction
The determination of crystal structure involves the estimation of unit cell parameters and the position of atoms within the unit cell. The dimensions of the unit cell are of the order of Angstroms (*i.e.*, 10^{-10}m). Hence it is necessary to use probes whose wavelength is of this order. X-rays, thermal neutrons and electrons accelerated through a suitable potential are therefore employed as probes for crystal structure determination. On account of the translational periodicity in three dimensions, a crystal acts like *a three dimensional grating* for these probes. Diffraction data, which consists of the angles of diffraction and the intensity of the diffracted beams from different planes of atoms, is collected using an instrument called the *diffractometer*. The data is analysed using the mathematical technique of *Fourier transforms* to determine the crystal structure. Over the years, several novel techniques have been developed to determine the crystal structure. Nowadays, even the structure of complicated biomolecules such as proteins and nucleic acids is determined routinely using this method.

X-ray diffraction should not be confused with X-ray radiography. In X-ray radiography one obtains an image of an object, which is inhomogeneous as far as X-ray transmission is concerned *i.e.,* different portions of the object have different absorption and transmission coefficients for the X-ray beam. This property is made use of in industry for non-destructive testing of materials as well as in medicine for diagnostic purposes.

11.9.2 Miller Indices
It is necessary to name the planes and directions in a crystal lattice to describe the diffraction in crystals. This is done using *Miller indices*. Miller indices are a set of lowest integers which denote the orientation of a plane as well as direction within the unit cell. They are also useful in describing the anisotropic nature of physical properties of crystals. In general, physical properties of crystals are anisotropic. (*i.e.,* they depend on the direction of measurement with reference to the crystallographic axes). For example, the elastic modulus of *bcc* ion is greater along the body diagonal than along the cube edge. On the contrary, the magnetic permeability of iron is greatest in a direction parallel to the edge of the unit cell rather than along the body diagonal. Certain planes of atoms in a crystal are also significant. Metals deform along planes in which atoms are most tightly packed.

Indices of Directions
All parallel directions use the same label or indices. Therefore to label a direction, a line parallel to the given direction which passes through the origin $O(0, 0, 0)$ is chosen. The direction is specified by the coordinates of any point $P(x, y, z)$ on that line. However, since there is an infinite number of points on any line, a point with the lowest set of integers is chosen.

The procedure for arriving at the Miller indices for a direction OP is as follows:
(*a*) x, y and z axes are chosen to be right handed.
(*b*) Coordinates of P are expressed as fractions of the unit cell parameters viz.,
   $[x/a, y/b, z/c]$

(c) The coordinates are multiplied by the LCM to obtain the set of least integers. Thus if the point P has fractional coordinates (1/5, 1/3, 1/2), then its Miller indices are [6 10 15].

Thus [111] direction passes from the origin (0, 0, 0) through point (1, 1, 1). Note however, that this direction passes through the point (1/2, 1/2, 1/2) etc. Likewise, the direction [112] passes through the point (1/2, 1/2, 1) or the point (2, 2, 4). But for the sake of simplicity, integer notation is used. The Miller indices for direction viz., u, v and w are enclosed in the square brackets [uvw]. Further, the indices [uvw] may be negative which is designated with an over bar. Thus [11$\bar{1}$] will correspond to the direction joining the origin to the point (1, 1, –1). Also while specifying the direction as [uvw] *no comma* is written between u and v or v and w. Figure 11.24 shows the indices for various directions in a cubic unit cell.

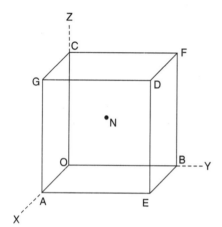

Positions	Coordinates	Direction →	Miller Indices
N (Centre of Cube)	(1/2, 1/2, 1/2)	ON	[1 1 1]

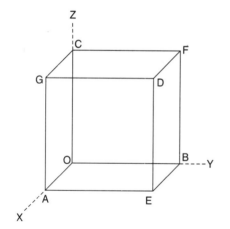

Positions	Coordinates	Direction →	Miller Indices
O	(0, 0, 0)		
A	(1, 0, 0)	OA	[1 0 0]
B	(0, 1, 0)	OB	[0 1 0]
C	(0, 0, 1)	OC	[0 0 1]
D	(1, 1, 1)	OD	[1 1 1]
E	(1, 1, 0)	OE	[1 1 0]
F	(0, 1, 1)	OF	[0 1 1]
G	(1, 0, 1)	OG	[1 0 1]

(Contd.)

X-RAYS

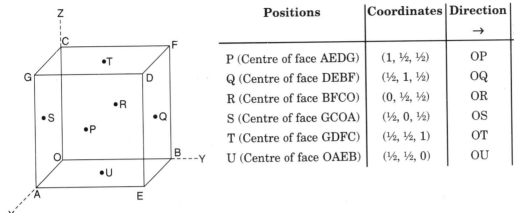

Positions	Coordinates	Direction $\rightarrow$	Miller Indices
P (Centre of face AEDG)	(1, ½, ½)	OP	[2 1 1]
Q (Centre of face DEBF)	(½, 1, ½)	OQ	[1 2 1]
R (Centre of face BFCO)	(0, ½, ½)	OR	[0 1 1]
S (Centre of face GCOA)	(½, 0, ½)	OS	[1 0 1]
T (Centre of face GDFC)	(½, ½, 1)	OT	[1 1 2]
U (Centre of face OAEB)	(½, ½, 0)	OU	[1 1 0]

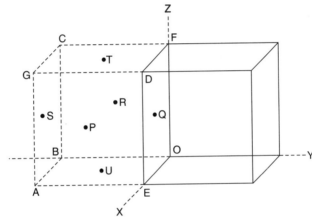

Positions	Coordinates	Direction $\rightarrow$	Miller Indices
P (Centre of face AEDG)	(1, − ½, ½)	OP	[2 $\bar{1}$ 1]
Q (Centre of face DEOF)	(½, 0, ½)	OQ	[1 0 1]
R (Centre of face OFCB)	(0, − ½, ½)	OR	[0 $\bar{1}$ 1]
S (Centre of face GCBA)	(½, − 1, ½)	OS	[1 $\bar{2}$ 1]
T (Centre of face GDFC)	(½, − ½, 1)	OT	[1 $\bar{1}$ 2]
U (Centre of face OBAE)	(½, − ½, 0)	OU	[1 $\bar{1}$ 0]

Fig. 11.24 Miller indices for directions in a cubic system

Angle between directions: In certain calculations it is necessary to calculate the angle between two different directions. If $[u_1 v_1 w_1]$ and $[u_2 v_2 w_2]$ are the two directions, then the vectors representing these directions are given by

and
$$A = u_1 i + v_1 j + w_1 k$$
$$B = u_2 i + v_2 j + w_2 k$$

where **i**, **j** and **k** are unit vectors along x, y and z-axes. Then,

$$A \cdot B = |A||B| \cos \theta$$

$$\Rightarrow \quad \theta = \cos^{-1} \frac{A \cdot B}{|A||B|} = \frac{u_1 u_2 + v_1 v_2 + w_1 w_2}{\sqrt{u_1^2 + v_1^2 + w_1^2} \sqrt{u_2^2 + v_2^2 + w_2^2}} \quad (11.7)$$

The above formula is applicable only for orthonormal axes.

Families of directions: In the cubic crystal, the following directions are identical since they depend on our choice of x, y and z labels on the axes.

Thus [111], [1$\bar{1}$1], [11$\bar{1}$], [$\bar{1}$11], [$\bar{1}\bar{1}$1], [$\bar{1}$1$\bar{1}$], [1$\bar{1}\bar{1}$] and [$\bar{1}\bar{1}\bar{1}$] indicate the body diagonals of the cube.

Any directional property will be identical for these four opposing pairs. Therefore, it is convenient to identify this family of directions as <111>, rather than writing the eight separate indices. Note that the closure symbols are *angle brackets* and again *no commas* are used.

Miller Indices of the Planes

The procedure to arrive at the Miller indices for a plane is as follows.

(a) x, y, z axes are chosen to be right handed system of axes.

(b) *intercepts of the plane* along the axes (*i.e*, X, Y and Z) are expressed in *units of lattice constants i.e.,* ($X/a, Y/b, Z/c$). These will be fractions since the planes are shown within the unit cell.

(c) Reciprocal of these fractional intercepts are reduced to smallest integers (hkl)

(d) (hkl) denotes a family of parallel planes rather than a single plane.

(e) Note that the Miller indices are inversely proportional to the intercept. A plane which is parallel to a coordinate axis has Miller index zero corresponding to that axis.

(f) The notation {hkl} denotes a family of equivalent planes. Thus the planes (100), (010) and (001) are equivalent. So are the planes (110), (101), (011).

In a cubic system the Miller indices of a plane are the same as the direction perpendicular to the plane. Figure 11.25 shows some examples of the use of the Miller indices to define planes.

X-RAYS 523

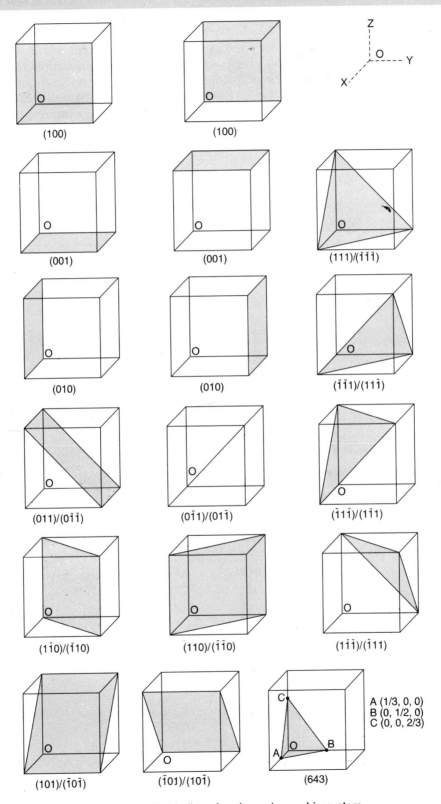

Fig. 11.25 Miller indices for planes in a cubic system

11.9.3 Interplanar Distance

The distance between two consecutive parallel planes in a given unit cell is of considerable interest since it determines the angles of diffraction. The discussion is limited to the unit cells which are expressed in terms of the orthogonal coordinate axes, so that simple cartesian geometry is applicable. Let ox, oy and oz be the three mutually perpendicular coordinate axes. A plane (hkl) parallel to the plane passing through the origin makes intercepts on the three axes at A, B and C respectively as shown in Fig. 11.26. OP ($= d$, the interplanar spacing) is the normal to the plane drawn from the origin. It makes angles α, β and γ with the three axes.

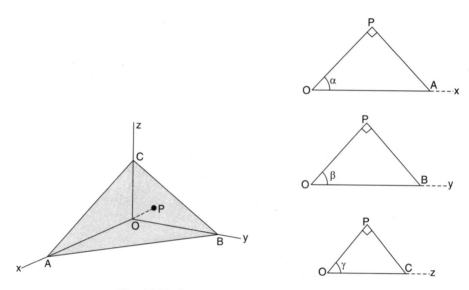

Fig. 11.26 Calculation of interplanar distance

By definition of Miller indices

$$OA = \frac{a}{h}, \quad OB = \frac{b}{k} \quad \text{and} \quad OC = \frac{c}{l}$$

From $\triangle OPA$, $\triangle OPB$ and $\triangle OPC$

$$\cos \alpha = \frac{OP}{OA} = \frac{d}{(a/h)}, \quad \cos \beta = \frac{OP}{OB} = \frac{d}{(b/k)}$$

and

$$\cos \gamma = \frac{OP}{OC} = \frac{d}{(c/l)}$$

But

$$\cos^2 \alpha + \cos^2 \beta + \cos^2 \gamma = 1$$

Hence

$$\frac{d^2}{(a/h)^2} + \frac{d^2}{(b/k)^2} + \frac{d^2}{(c/l)^2} = 1$$

$$\Rightarrow \quad d^2 \left[\frac{h^2}{a^2} + \frac{k^2}{b^2} + \frac{l^2}{c^2} \right] = 1$$

So that

$$d = \left[\frac{h^2}{a^2} + \frac{k^2}{b^2} + \frac{l^2}{c^2} \right]^{-1/2} \quad \text{or} \quad d = \frac{1}{\sqrt{\left(\frac{h}{a}\right)^2 + \left(\frac{k}{b}\right)^2 + \left(\frac{l}{c}\right)^2}} \quad (11.8)$$

This is a general formula and is applicable to the *primitive lattices* of orthorhombic, tetragonal and cubic systems.

(*i*) In tetragonal system; $a = b \neq c$. Hence eqn (11.8) reduces to

$$d = \left[\frac{h^2 + k^2}{a^2} + \frac{l^2}{c^2}\right]^{-1/2}$$

(*ii*) In cubic system $a = b = c$, Hence eqn (11.8) reduces to

$$d = \frac{a}{\left(h^2 + k^2 + l^2\right)^{1/2}}$$

11.9.4 Interplanar Distances in SC, BCC and FCC Lattices

Simple Cubic

In a simple cubic system, the lattice parameters $a = b = c$. The lattice points are only at the corners of the unit cell. Thus the interplanar distance can be determined by simply using eqn (11.8). They are shown in Fig. 11.27*a*. The interplanar distances corresponding to three low index planes (100), (110) and (111) are given by (Fig. 11.27*a*):

$$d_{100} = a, \quad d_{110} = \frac{a}{\sqrt{2}} \quad \text{and} \quad d_{111} = \frac{a}{\sqrt{3}}$$

Hence their ratio is

$$d_{100} : d_{110} : d_{111} := 1 : \frac{1}{\sqrt{2}} : \frac{1}{\sqrt{3}} \tag{11.9}$$

Body Centered Cubic

In this case the lattice points are at the eight corners and the centre of the unit cell. Because of the presence of an additional point at the body centre, the ratio of interplanar distances for three low index planes (100), (110) and (111) are different. There appears an additional plane half way between (100) planes. This is true of (111) planes also (Fig. 11.27*b*). No new planes appear in between (110) planes when compared with the simple cubic lattice. Therefore, the interplanar distances for the low index planes in the body centered cubic system are:

$$d_{100} = \frac{1}{2}(d_{100})_{sc} = \frac{a}{2}$$

$$d_{110} = (d_{110})_{sc} = \frac{a}{\sqrt{2}}$$

$$d_{111} = \frac{1}{2}(d_{111})_{sc} = \frac{a}{2\sqrt{3}}$$

Hence their ratio is

$$d_{100} : d_{110} : d_{111} = \frac{1}{2} : \frac{1}{\sqrt{2}} : \frac{1}{2\sqrt{3}} = 1 : \sqrt{2} : \frac{1}{\sqrt{3}} \tag{11.10}$$

Face Centered Cubic

In this case, the lattice points are at eight corners and at the six face centres of the unit cell. Fig. 11.27c shows the appearance of additional planes halfway between (100) planes. This is true of (110) planes as well. No new planes appear in between (111) planes. Therefore, the interplanar distances for three low index planes are given as:

$$d_{100} = \frac{1}{2}(d_{100})_{sc} = \frac{a}{2}$$

$$d_{110} = \frac{1}{2}(d_{110})_{sc} = \frac{a}{2\sqrt{2}}$$

$$d_{111} = (d_{111})_{sc} = \frac{a}{\sqrt{3}}$$

Hence their ratio is

$$d_{100} : d_{110} : d_{111} = \frac{1}{2} : \frac{1}{2\sqrt{2}} : \frac{1}{\sqrt{3}} = 1 : \frac{1}{\sqrt{2}} : \frac{2}{\sqrt{3}} \quad (11.11)$$

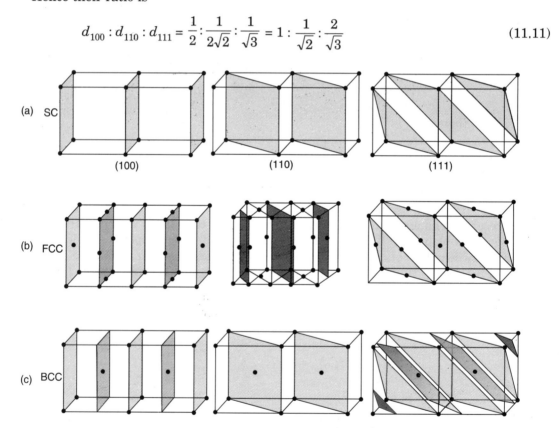

Fig. 11.27 Interplanar distances for SC, FCC and BCC systems

Note that (100) planes in simple cubic, (110) planes in body centered cubic and (111) planes in face centered cubic systems have large interplanar distances. These are also planes of the closest packing.

For convenience, the interplanar distances for all simple lattices are provided in Table 11.1

Table 11.1 Interplanar distances in various crystal systems

System	d_{hkl}
Cubic	$a(h^2 + k^2 + l^2)^{-1/2}$
Tetragonal	$\left[\dfrac{h^2 + k^2}{a^2} + \dfrac{l^2}{c^2}\right]^{-1/2}$
Orthorhombic	$\left[\dfrac{h^2}{a^2} + \dfrac{k^2}{b^2} + \dfrac{l^2}{c^2}\right]^{-1/2}$
Hexagonal	$\left[\dfrac{4/3(h^2 + hk + k^2)}{a^2} + \dfrac{l^2}{c^2}\right]^{-1/2}$
Rhombohedral	$\dfrac{a(1 + 2\cos^3\alpha - 3\cos^2\alpha)^{1/2}}{[(h^2 + k^2 + l^2)\sin^2\alpha + 2(hk + kl + lh)(\cos^2\alpha - \cos\alpha)]^{1/2}}$
Monoclinic	$\left[\dfrac{h^2/a^2 + l^2/c^2 + (2hl\cos\beta)/ac}{\sin^2\beta} + \dfrac{k^2}{b^2}\right]^{-1/2}$
Triclinic	$\left[\dfrac{h}{a}\begin{vmatrix} h/a & \cos\gamma & \cos\beta \\ k/b & 1 & \cos\alpha \\ l/c & \cos\alpha & 1 \end{vmatrix} + \dfrac{k}{b}\begin{vmatrix} 1 & h/a & \cos\beta \\ \cos\gamma & k/b & \cos\alpha \\ \cos\beta & l/c & 1 \end{vmatrix} + \dfrac{l}{c}\begin{vmatrix} 1 & \cos\gamma & h/a \\ \cos\gamma & 1 & k/b \\ \cos\beta & \cos\alpha & l/c \end{vmatrix} \middle/ \begin{vmatrix} 1 & \cos\gamma & \cos\beta \\ \cos\gamma & 1 & \cos\alpha \\ \cos\beta & \cos\alpha & 1 \end{vmatrix}\right]^{-\frac{1}{2}}$

11.9.5 Bragg's Law

Figure 11.28 shows a family of planes (hkl), with *interplanar spacing* d_{hkl}. Rays 1 and 2 get reflected from the first and second planes respectively. Reflections from the other planes are not shown for the sake of clarity. The angle made by the X-ray beam with the plane is called the *Bragg angle* and is denoted by θ_{hkl}. Note that it is the compliment of the angle of the incidence which is defined as the angle between the normal and the directions of the incident X-ray beam. At each plane X-rays obey the laws of reflection i.e., the angle of incidence is equal to the angle of reflection. There is a path difference associated with rays 1 and 2. Simple geometry shows that the path difference is $2d_{hkl}\sin\theta_{hkl}$. Thus the condition for intensity maxima is given by:

$$2d_{hkl}\sin\theta_{hkl} = n\lambda \qquad (11.12)$$

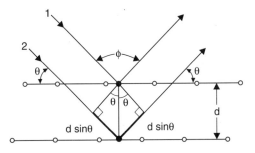

Fig. 11.28 Bragg's law

This is known as *Bragg's law*. Note that the angle of diffraction which is defined as the angle between the incident ray and the diffracted ray is 2θ.

11.9.6 Laue Diffraction

Laue's approach is more general and can be shown to be equivalent to Bragg's law. When a beam of X-rays of wavelength λ strikes a row of equally spaced atoms, each atom scatters waves in all directions. These reinforce in certain directions to produce various orders of diffraction spots. The condition of reinforcement is that the path difference between the rays diffracted by two adjacent atoms in the row must be an integral multiple of the wavelength.

For simplicity, consider a one dimensional row of atoms with interatomic distance a (Fig 11.29). Suppose that AB is the incident plane wavefront making an angle α_0 with a row of atoms, and CD is the diffracted plane wavefront leaving an angle α with the same row of atoms. Then the path difference between the two consecutive rays is

$$\Delta = (AC - BD) = a\,(\cos \alpha - \cos \alpha_0)$$

The diffracted beam is observed only if this path difference is an integral multiple of λ

i.e., $\qquad a\,(\cos \alpha - \cos \alpha_0) = e\lambda \qquad (11.13)$

where $e = 0, 1, 2\ldots$ is any integer giving the order of diffraction.

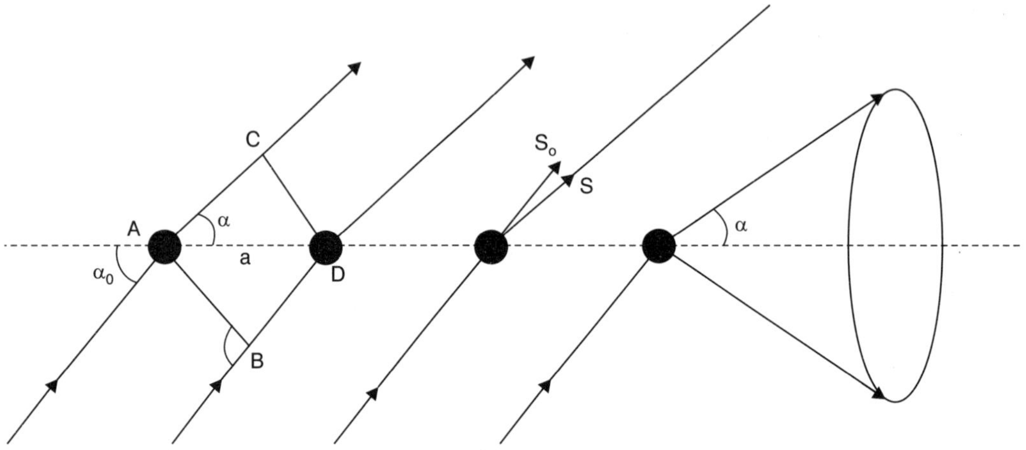

Fig. 11.29 Laue diffraction

This equation will be satisfied by all the diffracted beams lying on the concentric cone with respect to the line of atoms and has the semi apex angle α (Fig 11.29). Thus, for any given angle of incidence there will be a series of concentric cones surrounding the row of atoms, where each cone represents various orders of diffraction. (Fig. 11.30). Assuming S_o and S as the unit vectors, respectively in the directions of the incident and diffracted beam and a as the translation vector along the x-axis, in the vector notation eqn.(11.13) can be written as

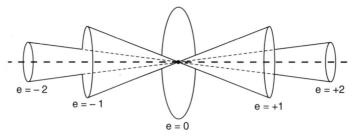

Fig. 11.30 X-rays are diffracted in a cone

$$a \cdot (S - S_o) = e\lambda \tag{11.14}$$

A crystal is a three dimensional periodic arrangement of atoms. Thus for the diffraction to occur from a simple space lattice with a unit cell defined by the primitive translations a, b and c, the following three equation must be satisfied simultaneously:

$$\begin{aligned} a(\cos\alpha - \cos\alpha_o) &= a \cdot (S - S_o) = e\lambda \\ b(\cos\beta - \cos\beta_o) &= b \cdot (S - S_o) = f\lambda \\ c(\cos\gamma - \cos\gamma_o) &= c \cdot (S - S_o) = g\lambda \end{aligned} \tag{11.15}$$

where β_o, β is the angle made by the incident and the scattered X-ray beam with the row of atom along b-axis and γ_o and γ is the angle made by the incident and the scattered beam with the row of atoms along the c-axis. f and g are integers.

These three equations together are called Laue equations. Thus for a crystal, there are three sets of cones, one each around a, b and c-rows of atoms and the most intense diffracted beam will be directed along the intersection of three sets of cones. These points are known as the Laue spots. (Fig. 11.31).

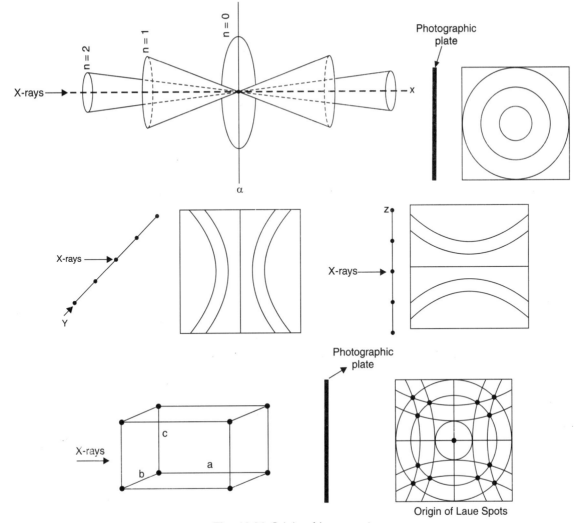

Fig. 11.31 Origin of Laue spots

It is to be noted that for a fixed λ (*i.e.*, a monochromatic beam) and an arbitrary direction of incidence (*i.e.*, fixed S_0) it is not possible to find a direction S which satisfies the Laue equations simultaneously and hence in general no diffraction should be observed. This actually means that the diffraction will occur only for a particular angle of incidence.

11.9.7 Comparison of Bragg and Laue Equations

Consider the case of a simple cubic lattice with $a = b = c$ and assume the angle between incident and diffracted beam to be equal to 2θ. Then squaring and adding the three Laue equations we can write

$$a^2 (\cos^2 \alpha + \cos^2 \beta + \cos^2 \gamma + \cos^2 \alpha_0 + \cos^2 \beta_0 + \cos^2 \gamma_0)$$
$$- 2(\cos \alpha \cos \alpha_0 + \cos \beta \cos \beta_0 + \cos \gamma \cos \gamma_0) = \lambda^2 (e^2 + f^2 + g^2) \qquad (11.16)$$

From the theorems in solid geometry,

$$\cos^2 \alpha + \cos^2 \beta + \cos^2 \lambda = \cos^2 \alpha_0 + \cos^2 \beta_0 + \cos^2 \lambda_0 = 1 \text{ and}$$
$$\cos \alpha \cos \alpha_0 + \cos \beta \cos \beta_0 + \cos \gamma \cos \gamma_0 = \cos 2\theta$$

Making use of these results, eqn (11.16) reduces to

$$2(1 - \cos 2\theta) = \frac{\lambda^2}{a^2} (e^2 + f^2 + g^2) \qquad (11.17)$$

or
$$\sin^2 \theta = \frac{\lambda^2}{4a^2} (e^2 + f^2 + g^2)$$

Similarly, squaring eqn(11.12) and making use of eqn (11.8):

$$\sin^2 \theta = \frac{n^2 \lambda^2}{4a^2} (h^2 + k^2 + l^2) \qquad (11.18))$$

Now, comparing eqns.(11.17) and (11.18), we have

$$e = nh, f = nk \text{ and } g = nl$$

This means that a diffracted beam defined in the Laue's treatment by the integers e, f, g may be interpreted as the nth order diffraction from a set of (hkl) planes in the Bragg's treatment. The order of diffraction n is simply equal to the largest common factor of the numbers e, f and g. This proves the equivalence of Bragg and Laue equations and suggests that the Bragg equation is a consequence of the more general Laue equations.

11.10 BASICS OF X-RAY SCATTERING

11.10.1 X-ray Scattering by a Single Electron

X-rays are electromagnetic waves. Hence they are scattered by the electron cloud surrounding the nucleus. In the classical picture, under the action of X-rays the electron cloud is set into oscillations which re-radiate energy. This re-radiated energy corresponds to the scattered X-rays.

Consider an electron at O, which is undergoing a periodic oscillation with an acceleration amplitude a. According to the classical theory, the amplitude of the radiated electromagnetic wave traveling in the direction OP is given by:

$$E = \frac{ea \sin \phi}{4\pi \varepsilon_0 r c^2} \qquad (11.19)$$

which is perpendicular to OP and in the plane defined by OP and a (Fig. 11.32a). In Fig.11.32b a parallel beam of X-rays traveling along OX encounters an electron at O.

X-RAYS

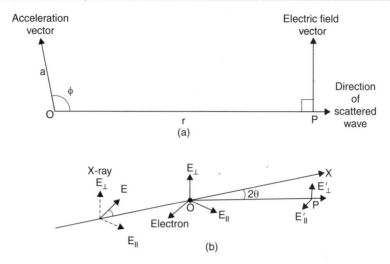

Fig. 11.32 (a) Electric field due to an accelerating electron
(b) Electric field vector of scattered and incident X-ray

It is now required to determine the amplitude of the scattered wave at P when the electron at O is set into oscillations. The amplitude of the electric vector E of the incident radiation is perpendicular to OX. It can be resolved into components $E_\perp$ and $E_\parallel$ perpendicular to and in the plane OXP. The electron will have corresponding components of acceleration of amplitude.

$$a_\perp = \frac{eE_\perp}{m} \quad \text{and} \quad a_\parallel = \frac{eE_\parallel}{m} \qquad (11.20)$$

From eqns (11.19) and (11.20), the electric field vector components of the scattered wave at P is given by:

$$E'_\perp = \frac{e^2 \sin(90°)}{4\pi\varepsilon_o rc^2 m} E_\perp = \frac{e^2}{4\pi\varepsilon_o rc^2 m} E_\perp$$

$$E'_\parallel = \frac{e^2 \sin(90° - 2\theta)}{4\pi\varepsilon_o rc^2 m} E_\parallel = \frac{e^2 \cos 2\theta}{4\pi\varepsilon_o rc^2 m} E_\parallel \qquad (11.21)$$

If the intensity of incident radiation is I_o and if this radiation is unpolarised then

$$I_{2\theta} \propto (E'_\perp)^2 + (E'_\parallel)^2 \propto \left(\frac{e^2}{4\pi\varepsilon_o rc^2 m}\right)^2 (E_\perp^2 + \cos^2 2\theta E_\parallel^2) \qquad (11.22)$$

But $\qquad E_\perp^2 = E_\parallel^2 \propto \dfrac{I_o}{2}$

$\therefore \qquad I_{2\theta} \propto \dfrac{I_o}{r^2} (1 + \cos^2 2\theta) \left(\dfrac{e^2}{4\pi\varepsilon_o mc^2}\right)^2$

The intensity of the scattered radiation, defined as the power per unit solid angle scattered through an angle 2θ is given by:

$$I_{2\theta} \cdot r^2 \propto I_o (1 + \cos^2 2\theta) \left(\frac{e^2}{4\pi\varepsilon_o mc^2}\right)^2 \qquad (11.23)$$

The factor $(1/m^2)$ in eqn (11.23) shows that the intensity is inversely proportional to the square of the mass of the scatterer. *Hence electrons scatter X-rays better than the nucleus.*

11.10.2 X-ray Scattering by an Atom

Eqn (11.23) shows that the scattering intensity is proportional to the square of the electronic charge. The scattering is proportional to the number of electrons around the nucleus and hence to the atomic number of the element. Since the wavelengths of X-rays are comparable to the size of the atom, the interference of scattered radiation from different parts of the atom should be included in the theory. The *scattering factor f* for an atom is defined to be

$$f = \frac{\text{Amplitude of the wave scattered by all the electrons in an atom}}{\text{Amplitude of the wave scattered by a single electron}}$$

The superposition of scattered waves from various electrons within an atom can be shown to be:

$$f = \int_0^\infty \rho(r) \frac{\sin kr}{kr} r^2 \, dr \quad \text{where } \rho(r) \text{ is the electron density and } k = \frac{2\pi}{\lambda} \sin \theta$$

Figure 11.33 shows the scattering factors for a few atoms. The atomic scattering factor f is a measure of the total contribution of all the electrons surrounding a nucleus in scattering the incident radiation. It is a function of $\sin \theta$. If θ is small, f approaches the atomic number of the element because all the electrons scatter X-rays which are nearly in phase. As θ increases, f decreases because the waves from individual electrons are more out of phase. In practice, values of f are tabulated as functions of $\lambda^{-1} \sin \theta$. Elements with high atomic number scatter X-rays better. This is the reason why it is difficult to locate the position of hydrogen atoms with X-rays. The problem is overcome either by deuteration or by using neutrons.

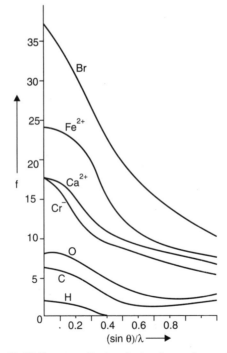

Fig. 11.33 X-ray scattering factor for various atoms

11.10.3 X-ray Scattering from a Crystal

While calculating the intensity of scattered waves from a crystal, scattering from a unit cell is considered. Since unit cell is the building block of a crystal, the scattered intensity will be equal to the sum of the intensities of the X-rays scattered by all the unit cells. The intensity of diffracted X-ray beam depends primarily on two properties of the crystal.

(a) The atomic scattering factor of each atom and

(b) The position of each atom in the unit cell.

Since the atoms are located at different positions, the phase of the scattered waves from various atoms will be different. In some directions, on account of destructive interference, scattered intensity turns out to be zero. In some other directions, scattered intensity is maximum on account of constructive interference of the scattered waves from various atoms in the unit cell.

The detailed calculation of position of atoms from the observed X-ray intensities at different Bragg angles is beyond the scope of the book. Only Laue method and powder method will be discussed here.

11.10.4 Laue Method

In this method a *single crystal* is exposed to a polychromatic beam of X-rays. X-rays are diffracted at angles and wavelengths for which the Laue equations are satisfied. A flat film receives the diffracted beams and the diffraction pattern is composed of a series of spots that shows the symmetry of the crystal. The origin of Laue spots is shown in Fig. 11.31. The Laue method is used extensively for the rapid determination of crystal orientation and crystal symmetry. In the former case, the crystal is oriented in a goniometer and rotated until a desired direction is found, as indicated by the X-ray pattern. The Laue method is almost never used for crystal structure determination because different orders may be reflected from a single plane on account of the polychromatic nature of the X-ray beam. This hinders the accurate measurement of X-ray intensity from a particular set of planes.

11.10.5 Powder Method

Powder method is a very simple technique and dispenses with the use of single crystals. This is a great advantage since it takes considerable time and effort to grow single crystals. Powder is a large collection of randomly oriented microscopic crystallites. Some of the crystallites in the sample will be so oriented to give Bragg reflections at certain angles. In practice, the sample is rotated in the X-ray beam to make orientations completely random. As shown in Fig. 11.34 the diffraction occurs 2θ from the direction of the incident beam. The diffraction pattern from a set of planes will trace a cone around the incident beam. A portion of this beam is recorded on the photographic plate. *The powder diffraction pattern of a compound is characteristic of the compound. Hence, like the finger print, it is routinely used for ready identification of materials.*

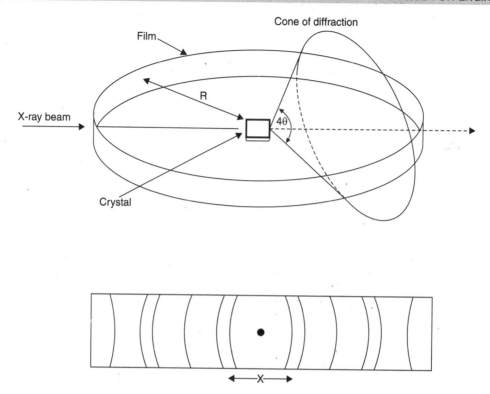

Fig. 11.34 Powder diffraction pattern

11.10.6 Modern X-ray Diffractometer

X-ray diffractometer is routinely used these days to collect the X-ray diffraction data. It consists of a crystal holder and a detector. The crystal can be rotated about three mutually perpendicular axes (ω, χ and ϕ). In perfect alignment, the three points, the source (S), the crystal position(C) and the ideal detector position (D) define a plane. The axis passing through the crystal and perpendicular to this plane is the main axis of the instrument. The detector rotates around the axis on a circle of constant radius. The angle subtended at the crystal by the source and the detector is $180° - 2\theta$, where 2θ is twice the Bragg angle (Fig. 11.35). When the crystal is rotated through an angle α, the detector turns through an angle 2α in accordance with the principle of geometrical optics. Detectors are usually based on semiconductors. Many of them are equipped with CCD cameras (charge coupled devices), which enable a visual display of the diffraction pattern. Also, computer programs are now commonly used to automate the determination of even quite complicated crystal structures from the measured intensity and diffraction angles.

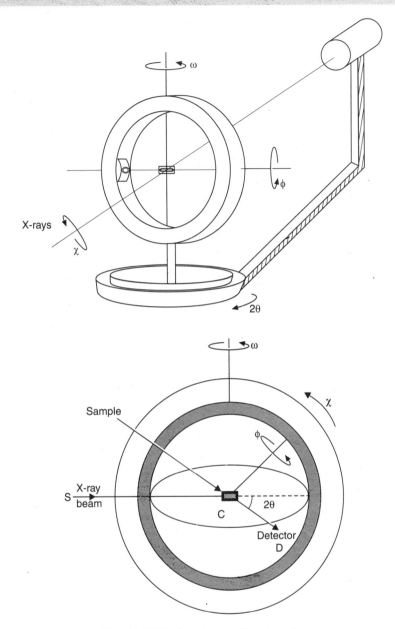

Fig. 11.35 Modern X-ray diffractometer

11.10.7 X-ray Diffraction from Cubic Crystals

The analysis of X-ray diffraction patterns is complicated for crystals with low symmetry. However, it is simple for the case of cubic crystals as described below.

The analysis of the powder diffraction pattern for cubic crystals is given here to illustrate the method. For deducing the interplanar distances, Miller indices have to be assigned to the diffraction pattern. This is called *indexing*. In the case of cubic crystals the Bragg equation can be written as:

$$\sin\theta = \frac{n\lambda}{2d} = \frac{n\lambda}{2\left[\dfrac{a}{\sqrt{h^2+k^2+l^2}}\right]}$$

$\Rightarrow \qquad \sin^2\theta = \dfrac{\lambda^2 J}{4a^2}$ where $J = h^2 + k^2 + l^2$ for $n = 1$ \hfill (11.24)

The quantity J takes only integral values as seen from Table 11.2.

Table 11.2

(hkl)	J	(hkl)	J	(hkl)	J	(hkl)	J
100	1	210	5	310	10	321	14
110	2	211	6	311	11	400	16
111	3	220	8	222	12	322	17
200	4	221 or 300	9	320	13		

Note that a few J values, 7 and 15 are missing, since they cannot be equal to the sum of the squares of three integers. Thus indexing the reflections of cubic crystals is in principle straight forward. The measured $\sin^2\theta$ values should form an arithmetic sequence except for a few missing numbers. The difference between two consecutive terms is $\lambda^2/4a^2$. The value of J and the corresponding Miller indices may be determined simply by reading the sequence. Hence the unit cell dimension a is easily determined. Figure 11.36 shows a plot of $\sin^2\theta$ vs. J.

Following factors are important in the analysis of the powder diffraction pattern.

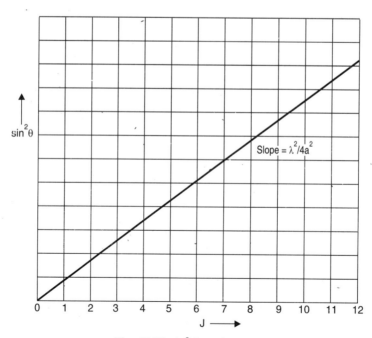

Fig. 11.36 $\sin^2\theta$ vs J curve

(a) Systematic Absences

Reflections from some planes are absent. Table 11.3 summarises the conditions of systematic absences for FCC and BCC crystal which contains identical atoms. No such restriction exists for a simple cubic crystal. The derivation of these conditions is beyond the scope of the book. Figure 11.37 shows the expected diffraction pattern for the three cubic lattices.

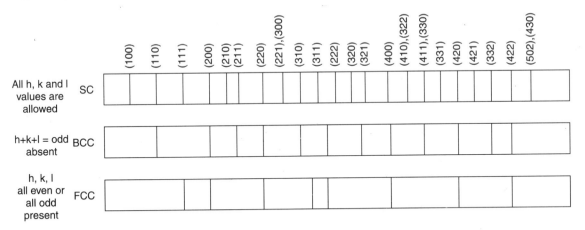

Fig. 11.37 Expected powder diffraction pattern for SC, BCC and FCC

Table 11.3

Simple Cubic	Face Centered Cubic	Body Centered Cubic
No condition	Reflections present only if h, k, l are *all odd or all even*	Reflections absent if $h + k + l = odd$

(b) Scattering Powers of Atoms

An important factor in understanding the powder pattern is the difference in the scattering power of different atoms or ions. Both NaCl and KCl crystallise in the *fcc* lattice. But the observed powder diffraction pattern is different. Figure. 11.38 shows the observed diffraction pattern of NaCl and KCl. The K^+ and Cl^- ions have the same number of electrons (18 each) and their scattering strength is identical. Hence the scattering from the A planes of Cl^- ions destructively interfere with scattering from the B planes of K^+ ions. The Na^+ ions scatter significantly less than Cl^- ions. Hence the intensity cancellation between the A and B planes is not complete and weak reflections are present for which there are no counterparts in the KCl pattern.

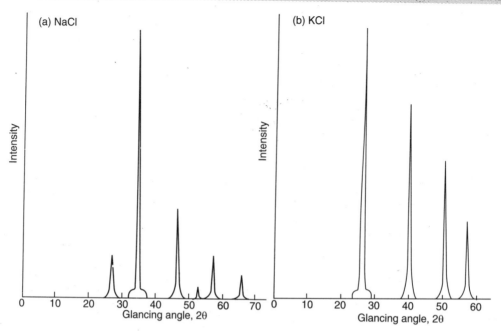

Fig. 11.38 Diffraction pattern for NaCl and KCl

(c) Other Considerations

There are, however, some important practical considerations. Reflections for which θ is too close to 0° or 180°, which are the first members of the series, are difficult to measure. Indexing the powder patterns for crystals of lower symmetry poses special difficulties. For these samples, powder patterns may still be used for their identification. But structure determination calls for single crystal studies. These factors have to be taken into account for a correct interpretation.

11.11 X-RAY TECHNIQUES IN MATERIALS SCIENCE

X-ray techniques in materials science can be classified as those based on X-ray diffraction and X-ray fluorescence. In materials science and engineering, the ultimate aim is to correlate the structure and properties of the materials. X-ray diffraction is used for a precise determination of structure (*i.e.*, to locate atomic positions within the unit cell and to measure the unit cell parameters). Several X-ray techniques have been developed for structural studies both for single crystals and polycrystalline samples. X-ray diffraction has become an indispensable tool in the study of materials. Materials also exhibit fluorescence on being exposed to X-rays. An analysis of the luminescence spectra can be used to characterize the electronic structure of the material.

11.11.1 Techniques Based on X-ray Diffraction

X-ray diffraction techniques are used to

(a) prove the crystallinity of the sample
(b) to orient the crystals
(c) to determine the perfection of crystals
(d) to determine the chemical composition through lattice parameter determination
(e) to determine the particle size
(f) to determine thermal amplitudes and
(g) to determine the crystal structure

X-ray diffraction from a crystal is different from that of a liquid or glass. X-ray diffraction spots are very sharp in the case of crystals while they are broad in the glassy state. This is utilised to characterize a sample as glassy or crystalline (Fig. 11.39).

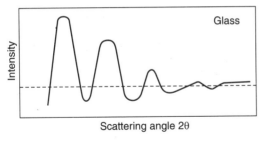 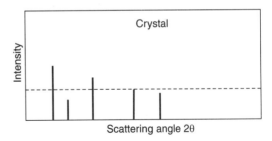

Fig. 11.39 X-ray diffraction from the crystalline and glassy state

It is well known that the lattice constants are a function of the particle size (Fig. 11.40). Consequently, the interplanar spacing and hence the Bragg reflection angle depends on the particle size. Hence an accurate measurement of the Bragg angle can lead to an estimation of the particle size.

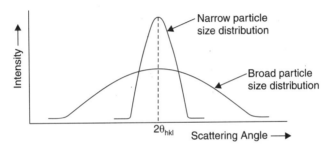

Fig. 11.40 Lattice constant versus particle size

In the approach of Bragg or Von Laue, it was assumed that the atoms were at rest. This is strictly not true on account of the thermal motion of atoms. The effect of thermal motion is to increase the width of the Bragg reflections. This is known as *thermal diffuse scattering*. A measurement of the width of the X-ray diffraction lines can be used to estimate the thermal amplitudes (Fig. 11.41)

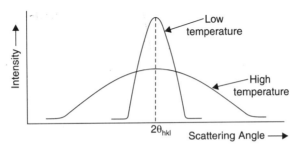

Fig. 11.41 Thermal diffuse scattering of X-rays

A crystal refers to the state of solid in which there is a perfect translational and orientational order of atoms or molecules constituting the solid. However, in practice this perfect nature expected of a crystalline state is destroyed by imperfections of various kinds. These defects could be interstitial and substitutional. They could be twins or domains. X-ray topography is a technique used often in materials science for the visualization of disclocations, twins, domain walls, inclusions, impurities present in the crystal volume. Essentially defects create non-uniform of microscopic strain. X-ray topography records long range distortion fields. These strain fields affect the diffracted intensity. This gives rise to a contrast in the image. Thus essentially X-ray topography is a study of the fine structure of a Bragg spot which contains information about the deviations from the perfect crystalline state *i.e.* the defect structure. Fig. 11.42 shows the typical image of a Bragg spot of a crystal with defects. The strain in the crystal results in a non-uniform intensity within the spot.

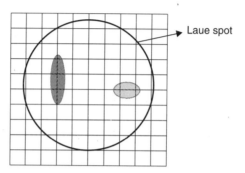

Fig. 11.42 X-ray topograph

In semiconducting binary compounds the lattice constant is a function of composition (Fig. 11.43). Measurement of lattice constant can lead to an estimation of chemical composition of the crystal. Since the powder diffraction pattern is characteristic of the compound, it is invariably used to characterize the sample.

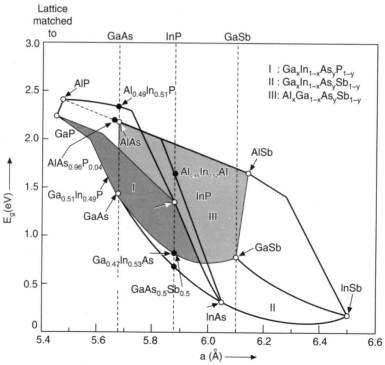

Fig. 11.43 Lattice parameter as a function of composition in binary semiconductors

11.11.2 Techniques Based on X-ray Fluorescence

X-ray fluorescence (XRF) is one of the widely used of all analytical methods for the qualitative and quantitative identification of elements having atomic numbers greater than oxygen (>8). Fluorescence and phosphorescence refer to the emission of light by materials after being irradiated by X-rays. These two phenomena are often referred to by the more general term *photoluminescence*.

The energy levels of atoms and molecules can be classified as electronic, vibrational and rotational. The electronic energy levels denote the possible energy states for the electron while vibrational and rotational levels correspond to the energy states of vibration and rotation of molecules. A typical energy level scheme for an atom/molecule is shown in Fig. 11.44. Note that $(\Delta E)_{electronic} > (\Delta E)_{vibrational} > (\Delta E)_{rotational}$. On being irradiated with X-rays the atom/molecule is excited to various energy levels. Every energy level is characterized by a life time which denotes the time the atom or molecule spends in that energy state. Generally the lifetime of an atom/molecule excited by absorption of radiation is brief because several relaxation processes exist permitting its return to the ground state. While undergoing transition to the ground state atom/molecule can emit radiation or the difference energy can be given to the medium as heat. These transitions are termed as *radiative transition and non-radiative transition* respectively. Figure 11.45 denotes these transitions. Non-radiative relaxation involves the loss of energy in a series of small steps. The loss of energy occurs on account of collisions with other molecules. A minute increase in temperature of the system results. Vibrational relaxation is such an efficient process that the average lifetime of an excited vibrational state is only about 10^{-15}s. Non-radiative relaxation between the lowest vibrational level of an excited electronic state and the upper vibrational level of another electronic state can also occur. This type of relaxation, which is sometimes called *internal conversion* is also depicted in Fig. 11.45. Internal conversion is much less efficient than vibrational relaxation since the average life time of an electronic state is between 10^{-5} and 10^{-9}s. The mechanisms by which this type of relaxation occurs are not fully understood but the net effect is again a tiny rise in the temperature of the medium.

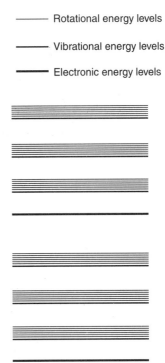

Fig. 11.44 Energy level scheme showing the electronic, vibrational and rotational levels

Radiant emission then occurs as the excited atom or molecule returns to the ground state. The wavelength of the emitted radiation is larger than that used for exciting the atom/molecule. Typical luminescence spectra of the atom/molecule is shown in Fig. 11.45. Fluorescence occurs more rapidly then phosphorescence and is generally completed after 10^{-5}s (or less) from the time of excitation. Phosphorescence emission takes place over periods longer than 10^{-5}s and may indeed continue for minutes or hours after the irradiation has ceased. Fluorescence and phosphorescence are most easily observed at 90° angle to the excitation beam.

Resonance fluorescence describes a process in which the emitted radiation is identical in frequency to the radiation in excitation. The lines labeled 1 and 2 in Fig. 11.45 (a) and (c) illustrate the type of fluorescence. Here the species is excited to the energy states E_1 or E_2 by radiation having an energy of $(E_1 - E_2)$ or $(E_2 - E_1)$. After a brief period, emission of radiation of identical energy occurs, as depicted in Fig. 11.45(c). Resonance fluorescence is most commonly produced by atoms in gaseous state, which do not have vibrational energy states superimposed on electronic energy levels.

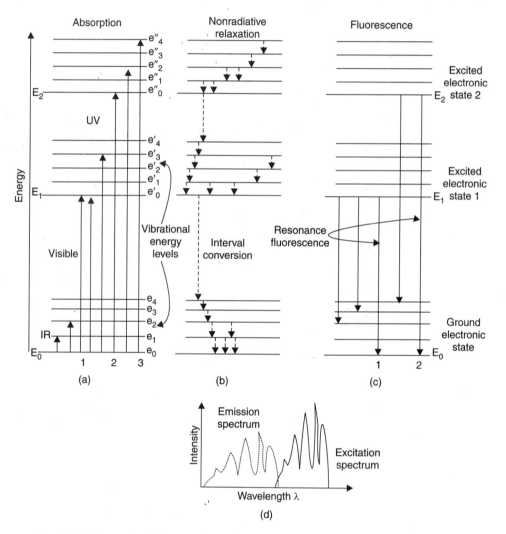

Fig. 11.45 Energy level diagram for a fluorescent molecule depicting the excitation as well as fluorescence spectra. (*a*) Absorption, (*b*) Non-radiative relaxation, (*c*) Fluorescence, (*d*) Luminescence spectra

Non-resonance fluorescence is brought about by irradiation of molecules in solution or in the gaseous state. As shown in Fig. 11.45, absorption of radiation promotes the molecules into any of the several vibrational levels associated with the two excited electronic levels. The lifetimes of these excited vibrational states are, however, only of the order of 10^{-15}s, which is much smaller than the lifetimes of the excited electronic states (10^{-8}s). Therefore, on the average, vibrational relaxation occurs before electronic relaxation. As a consequence, the energy of the emitted radiation is smaller than that of the radiation absorbed by an amount equal to the vibrational excitation energy. For example, for the absorption labeled 3 in Fig. 11.45a, the absorbed energy is equal to $(E_4 - E_o)$ whereas the energy of the fluorescent radiation is given by $(E_2 - E_o)$. Thus, the emitted radiation has a lower frequency or longer wavelength than the radiation that excited fluorescence. This shift in wavelength to lower frequencies is sometimes called the *Stokes shift*. Clearly, both resonance and non-resonance radiation can accompany fluorescence of molecules, although the latter tends to predominate because of the much larger number of vibrationally excited states.

XRF can be utilized as a diagnostic tool because every element has its own characterisitic fluorescence, which can be detected. By measuring the Stokes shift as well as the intensity of fluorescence, it is possible to carry out both the qualitative and quantitative elemental analysis. A typical experimental setup for measuring fluorescence is shown in Fig. 11.46. Nearly all fluorescence instruments employ double beam optics as shown in the figure in order to compensate for fluctuations in the power of the source. The sample beam first passes through an excitation filter or monochromator, which transmits radiation that will excite fluorescence but excludes or limits the emitted radiation. Fluorescence emanates from the sample in all directions but is most conveniently observed at right angles to the excitation beam. The emitted radiation reaches a photomultiplier tube after passing through a second filter or monochromator that isolates a fluorescence peak for measurement. The reference beam passes through an attenuator that reduces its power to approximately that of the fluorescence radiation. The signals from the reference and sample photomultiplier tubes are then fed into a difference amplifier whose output is visually displayed and recorded.

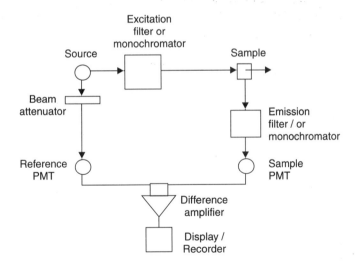

Fig. 11.46 X-ray fluorescence spectrometer

XRF spectrometry is perhaps the most powerful tool available to the chemist for the rapid quantitative determination of all but lightest elements in complex samples. It is used for

(a) quality control in the manufacture of metals and alloys.

(b) quantitative determination of lead and bromine in aviation gasoline samples.

(c) quantitative determination of calcium, barium and zinc in lubricating oils

(d) direct determination of the pigments in paint samples and

(e) analysis of atmosphere pollutants.

REFERENCES

1. F.C. Phillips, *An Introduction to Crystallography*, ELBS, London, 1963.
2. L.V. Azaroff, *Introduction to solids*, TMH, New Delhi, 1960.
3. C. Kittel, *Introduction to Solid State Physics*, 7th edn., John Wiley, New York, 2004.
4. W.D. Callister, *Materials Science and Engineering—An Introduction*, John Wiley, New York, 1997.
5. W.F. Smith, *Principles of Materials Science and Engineering*, McGraw-Hill, 1990.
6. N.W. Ashcroft and N.D. Mermin, *Solid State Physics*, Saunders College Publishing, Harcourt Brace College Publishers, New York, 1976.
7. S. Elliott, *The Physics and Chemistry of Solids*, John Wiley, New York, 1998.
8. A.R. West, *Solid State Chemistry and its Applications*, John Wiley, New York, 1987.

SOLVED EXAMPLES

1. A voltage of 25kV is applied across an X-ray tube and a current of 10 mA flows through it. Find the number of electrons striking the target per second and the minimum wavelength in the X-ray spectrum.

   Solution:

   The number of electrons striking the target is given by

   $$n = \frac{\text{Current}}{\text{Charge on the electron}} = \frac{10 \times 10^{-3}}{1.6 \times 10^{-9}} = 6.25 \times 10^{16}$$

   The minimum wavelength of the emitted X-rays is given by:

   $$\lambda_{min} = \frac{hc}{eV} = \frac{(6.63 \times 10^{-34}) \times (3 \times 10^{8})}{(1.6 \times 10^{-19}) \times (25 \times 10^{3})} = 4.97 \times 10^{-11} \text{ m}$$

2. The cutoff wavelength from an X-ray tube operating at 50 kV is diffracted from NaCl crystal. The first order maxima is observed at a Bragg angle of 26°. Calculate the interplanar spacing and the angle through which the receiver of the diffractometer must be rotated to observe for the second order maxima.

   Solution:

   The cutoff wavelength is given by:

   $$\lambda_{min} = \frac{hc}{eV} = \frac{(6.63 \times 10^{-34}) \times (3 \times 10^{8})}{(1.6 \times 10^{-19}) \times (50 \times 10^{3})} = 2.48 \times 10^{-11} \text{ m}$$

   The interplanar spacing (d) is calculated using the Bragg's law:

   $$d = \frac{n\lambda}{2 \sin \theta_1} = \frac{1 \times (2.48 \times 10^{-11})}{2 \times \sin 26°} = 2.83 \times 10^{-11} \text{ m}$$

   The Bragg angle for the second order reflection is given by:

   $$\sin \theta_2 = \frac{n\lambda}{2d} = \frac{2 \times (2.48 \times 10^{-11})}{2 \times (2.83 \times 10^{-11})} = 0.876$$

   $$\Rightarrow \qquad \theta_2 = \sin^{-1} 0.876 = 61.2°$$

The change in the angle of incidence is (61.2° − 26°) = 35.2°. Hence the receiver arm of the diffractometer must be turned through an angle of 2 × 35.2°. = 70.4°.

3. An X-ray beam of wavelengths from 95pm to 140pm is incident on a family of reflecting planes with a spacing of 275pm at a glancing angle of 45°. At what wavelengths will these planes produce intensity maxima in their reflections.

Solution:

We have from Bragg's law

$$\lambda = \frac{2d \sin \theta}{n} = \frac{2 \times (275 \text{ pm}) \times \sin 45°}{n} = \frac{389 \text{ pm}}{n}$$

Only for $n = 3$ and 4 the value of λ lies within the range of the incident beam. The corresponding wavelengths for $n = 3$ and $n = 4$ are 130pm and 97.3pm respectively. Hence Bragg reflections are observed only for these two wavelengths.

4. Which element has K_α line of wavelength 0.180×10^{-9}m?

Solution:

The wavelength of the K_α line is given by:

$$\frac{1}{\lambda} = R(Z^*)^2 \left[\left(\frac{1}{1}\right)^2 - \left(\frac{1}{2}\right)^2 \right] = R(Z^*)^2 \left(\frac{3}{4}\right) \text{ where } R = 1.097 \times 10^7 \text{ m}^{-1}$$

Here $\lambda = 0.180 \times 10^{-9}$m. Hence Z^* is given by

$$Z^* = \left[\frac{4}{3\lambda R}\right]^{\frac{1}{2}} = \left[\frac{4}{3 \times 0.180 \times 10^{-10} \times 1.097 \times 10^7}\right]^{\frac{1}{2}} = 26$$

$$Z^* = Z - 1 \quad \text{or} \quad Z = Z^* + 1 = 26 + 1 = 27$$

The target element is cobalt.

5. The calcium target contaminated with an impurity element emits faint K_α X-ray line of wavelength 0.194 nm. The K_α line of calcium has energy of 3.69 keV. Identify the impurity element.

Solution:

From Moseley's law we have the relation:

$$\frac{\sqrt{\nu_{Ca}}}{\sqrt{\nu_{imp}}} = \frac{Z_{Ca} - 1}{Z_{imp} - 1} = \frac{\sqrt{E_{Ca}}}{\sqrt{E_{imp}}}$$

The energy E_{imp} corresponding to the X-ray line of the impurity element is given by:

$$E_{imp} = \frac{hc}{\lambda} = \frac{(6.63 \times 10^{-34}) \times (3 \times 10^8)}{0.194 \times 10^{-9} \times (1.6 \times 10^{-19})} \text{ eV} = 6.41 \text{ keV}$$

Making use of Moseley's law we get

$$Z_{imp} = \left(\frac{\sqrt{E_{imp}}}{\sqrt{E_{Ca}}}\right)(Z_{Ca} - 1) + 1 = \left(\frac{\sqrt{6.41}}{\sqrt{3.69}}\right)(20 - 1) + 1 = 26$$

The impurity element is therefore iron.

6. In an orthorhombic crystal, a plane cuts intercepts 2a, −3b and 6c along the three crystallographic axes. Determine the Miller indices of the plane and sketch the plane.

Solution:

Following the procedure for determining the Miller indices we have:

Intercepts	2a	−3b	6c
÷ unit cell edges	2	−3	6
Taking reciprocals	1/2	−1/3	1/6
× L.C.M	3	−2	1
Coordinates of A, B, C	(1/3,0,0)	(0, −1/2,0)	(0,0,1)

The required Miller indices of the plane are (3 $\bar{2}$ 1). The plane is shown in Fig.11. 47.

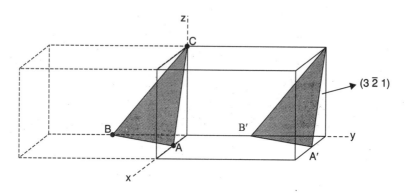

Fig. 11.47

7. Determine the Miller indices of a plane which is parallel to the x-axis and whose intercepts are in the ratio 3 : 2 along the y- and z-axes in a cubic crystal. Also sketch the plane.

Solution:

Since the plane is parallel to the x-axis, its intercept along x-axis is ∞. Miller index corresponding to x-axis is 0. Let OB and OC be the intercepts along the y- and z-axes. OY: OZ = 3:2 or OZ= (2/3) OY. Following the procedure for determining the Miller indices:

Intercepts	∞	OY	OZ
÷ unit cell edges	∞	1	2/3
Taking reciprocals	0	1	3/2
× L.C.M	0	2	3
Coordinates of A, B, C	(∞,0,0)	(0,1,0)	(0,0,2/3)

The required Miller indices of the plane are (0 2 3). The plane is shown in Fig.11.48.

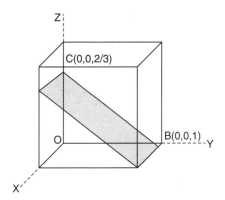

Fig. 11.48

8. An orthorhombic crystal has cell parameters $a = 0.121$ nm, $b = 0.184$ nm and $c = 0.197$ nm. If a plane with Miller indices (2 3 1) cuts an intercept of 0.121 nm along the x-axis, find the intercepts along the y-and z-axes.

   Solution:

   Miller indices are in the inverse ratio of intercepts. Hence the *fractional intercepts OA, OB* and *OC* are in the ratio given by:

   $$OA : OB = 3 : 2 \text{ and } OA : OC = 1 : 2$$

   ∴ $OB = (2/3) OA = (2/3) \times 0.184 \text{ nm} = 0.123 \text{ nm}$

   $OC = 2 \times OA = 2 \times 0.197 = 0.394 \text{ nm}$

9. Calculate the interplanar distance between the (123) planes and (246) planes of an orthorhombic cell with $a = 0.82$ nm, $b = 0.94$ nm and $c = 0.75$ nm

   Solution:

   The interplanar distance is given by:

   $$d_{hkl} = \left[\left(\frac{h}{a}\right)^2 + \left(\frac{k}{b}\right)^2 + \left(\frac{l}{c}\right)^2\right]^{-1/2}$$

   Here $a = 0.82$ nm, $b = 0.94$ nm and $c = 0.75$ nm

   For (123) planes, $h = 1$, $k = 2$ and $l = 3$. On substituting the values of the relevant parameters we get $d_{123} = 0.11$ nm. $d_{246} = d_{123}/2 = 0.055$ nm.

10. X-rays of wavelength 0.12 nm are found to undergo second order reflection at a Bragg angle of 28° from LiF crystal. What is the interplanar spacing of the reflecting planes of the crystal?

    Solution:

    We have the Bragg equation $2d \sin \theta = n\lambda$

    Here $\theta = 28°$, $\lambda = 0.12$ nm and $n = 2$.

    ∴ $$d = \frac{n\lambda}{2 \times \sin \theta} = \frac{2 \times 0.12 \text{ nm}}{2 \times \sin 28°} = 0.26 \text{ nm}$$

11. An X-ray beam of wavelength (λ_1) undergoes a first order Bragg reflection at a Bragg angle of 23°. X-rays of wavelength 97pm undergo third order reflection at a Bragg angle of 60°. Assuming that the two beams are reflected from the same set of planes, calculate (a) interplanar spacing and (b) λ_1.

Solution:

We have the Bragg equation: $2d \sin\theta = n\lambda$

For case (i): $n = 1$, $\theta = 23°$ and $\lambda = \lambda_1$

$\therefore \quad 2 \times d \times \sin 23° = \lambda_1$ \hfill (1)

case (ii): $n = 3$, $\theta = 60°$ and $\lambda = 97$ pm

$\therefore \quad 2 \times d \times \sin 60° = 3 \times 97$ pm \hfill (2)

Dividing eqn (1) by eqn (2)

$$\lambda_1 = \frac{3 \times 97 \times \sin 23°}{\sin 60°} = 130 \text{ pm}$$

From eqn(2) $\quad d = \dfrac{3 \times 97}{2 \times \sin 60°} = 168$ pm

12. X-rays are incident from (111) planes of a single cubic crystal with a lattice parameter of 0.2 nm. The first order maxima is observed in the direction of 87° to the incident ray. Calculate the Bragg angle and the wavelength of the X-rays.

Solution:

The angle between the incident and the diffracted beam is 2θ, where θ is the Bragg angle.

Here $2\theta = 87°$ $\Rightarrow$ $\theta = 43.5°$. Further, for the simple cubic the interplanar distance is given by:

$$d = \frac{a}{\sqrt{h^2 + k^2 + l^2}} = \frac{0.2 \text{ nm}}{\sqrt{1^2 + 1^2 + 1^2}} = \frac{0.2 \text{ nm}}{\sqrt{3}}$$

Making use of the Bragg equation we get:

$$\lambda = 2d \sin\theta = \frac{2 \times 0.2 \times \sin 43.5°}{\sqrt{3}} = 0.159 \text{ nm}$$

13. Electrons are accelerated by 844 volts and are Bragg reflected from a crystal. The first order reflection maximum occurs at a Bragg angle of 58°. Determine the interplanar spacing.

Solution:

The de Broglie for an electron accelerated through a potential of V_o (= 844 volts) volts is given by:

$$\lambda = \frac{h}{p} = \frac{h}{\sqrt{2m(K.E)}} = \frac{h}{\sqrt{2meV_o}} = \frac{6.63 \times 10^{-34}}{\sqrt{2 \times 9.1 \times 10^{-31} \times 1.6 \times 10^{-19} \times 844}}$$

$= 0.420 \times 10^{-10}$ m

We have the Bragg equation $2d \sin\theta = n\lambda$. Here $\lambda = 0.420 \times 10^{-10}$ m, $\theta = 58°$ and $n = 1$

$$d = \frac{n\lambda}{2 \sin\theta} = \frac{1 \times 0.420 \times 10^{-10}}{2 \times \sin 58°} = 0.248 \times 10^{-10} \text{ m}$$

14. Thermal neutrons at 300K are incident on (111) plane of a simple cubic crystal. The second order Bragg reflection occurs at an angle of 30°. Calculate the lattice constant.

Solution:
The de Broglie wavelength of the neutron is given by:

$$\lambda = \frac{h}{p} = \frac{h}{\sqrt{2m(\text{K.E})}} = \frac{h}{\sqrt{2m(3k_B T/2)}} = \frac{h}{\sqrt{3mk_B T}}$$

$$= \frac{6.63 \times 10^{-34}}{\sqrt{3 \times 1.804 \times 10^{-27} \times 1.38 \times 10^{-23} \times 300}}$$

$$= 1.45 \times 10^{-10} \text{ m}.$$

We have $2d \sin\theta = n\lambda$

Here $d = \dfrac{a}{\sqrt{3}}$, $\theta = 30°$, $n = 2$ and $\lambda = 1.45 \times 10^{-10}$ m

$$\therefore a = \frac{\sqrt{3}\lambda}{2} = \frac{\sqrt{3} \times 1.45 \times 10^{-10}}{2} = 1.26 \times 10^{-10} \text{ m}$$

15. A powder diffraction photograph of the element polonium gave lines at the following values of 2θ in degrees when 71pm Mo X-rays were used: 12.1, 17.1, 21.0, 24.3, 27.2, 29.9, 34.7, 36.9, 40.9 and 42.8. Identify the unit cell and determine its dimensions.

Solution:
We have the relation

$$\sin^2\theta = \left(\frac{\lambda}{2a}\right)^2 (h^2 + k^2 + l^2) = J\left(\frac{\lambda}{2a}\right)^2$$

which can be used to estimate the cell parameters after indexing. The experimental data can be cast in the following form:

2θ	12.1	17.1	21.0	24.3	27.2	29.9	34.7	36.9	38.9	40.9	42.8
θ	6.05	8.55	10.5	12.2	13.6	15.0	17.4	19.5	19.5	20.5	21.4
$\sin^2\theta$	.0111	.0221	.0332	.0447	.0553	.0670	.0894	.101	.111	.123	.133
÷.0111]	1	2	3	4	5	6	8	9	10	11	12
(hkl)	(100)	(110)	(111)	(200)	(210)	(211)	(220)	(300)	(310)	(311)	(222)

From the table we have

$$.0111 = \left(\frac{\lambda}{2a}\right)^2 \Rightarrow a = \left(\frac{\lambda}{2}\right)\left(\frac{1}{\sqrt{.0111}}\right) = \frac{71}{2 \times \sqrt{.0111}} = 337 \text{ pm}$$

16. An X-ray diffractometer chart for an element which has either the *bcc* or *fcc* crystal structure shows diffraction peaks at the following 2θ angles: 40, 58, 73, 86.8, 100.4, and

114.7. The wavelength of the incoming X-ray used was 0.154 nm. Determine the cubic structure of the element and its lattice constant. Also identify the element.

Solution:

The diffraction data can be cast in the following format for analysis:

2θ	θ	$\sin\theta$	$\sin^2\theta$
40	20	0.3420	0.1120
58	29	0.4848	0.2350
73	36.5	0.5948	0.3538
86.8	43.4	0.6871	0.4721
100.4	50.2	0.7683	0.5903
114.7	57.35	0.8420	0.7090

The ratio of the $\sin^2\theta$ values for the first and second angles is calculated:

$$\frac{\sin^2 20°}{\sin^2 29°} = \frac{0.117}{0.235} = 0.498 \approx 0.5$$

The crystal structure is *bcc* since this ratio is 0.5. If the ratio had been 0.75, the structure would have been *fcc*. For determining the lattice constant we have the formula:

$$a^2 = \frac{\lambda^2 \times \sqrt{h^2+k^2+l^2}}{4\sin^2\theta} \Rightarrow a = \frac{\lambda}{2}\sqrt{\frac{h^2+k^2+l^2}{\sin^2\theta}}$$

Miller indices for the first set of principal diffracting planes for the BCC structure is (111). The corresponding $\sin^2\theta = 0.117$. Here $\lambda = 0.154$ nm. Therefore

$$a = \frac{0.154\,\text{nm}}{2}\sqrt{\frac{1^2+1^2+0^2}{0.117}} = 0.318\,\text{nm}$$

The element is *tungsten* since this element has lattice constant of 0.316 nm and crystallizes in the *bcc* structure.

17. The results of powder X-ray diffraction show that the diffraction peaks occur at the following 2θ angles. Determine the crystal structure, the indices of the plane corresponding to each peak and the lattice parameter of the material ($\lambda = 0.07107$ nm).

Peak no.	1	2	3	4	5	6	7	8
2θ	20.7	28.72	35.36	41.07	46.19	50.90	55.28	59.42

Solution:

We will analyse the diffraction data by drawing up the following table.

Peak no.	1	2	3	4	5	6	7	8
2θ	20.7	28.72	35.36	41.07	46.19	50.90	55.28	59.42
θ	10.35	14.36	17.68	20.535	23.095	25.45	27.64	29.71
$\sin^2\theta$	0.0308	0.0615	0.0922	0.1230	0.1539	0.1847	0.2152	0.2456
$\div 0.0308$	1	2	3	4	5	6	7	8
$h^2 + k^2 + l^2$	2	4	6	8	10	12	14	16
(hkl)	110	200	211	220	310	232	321	400

On analysing the powder diffraction data as tabulated in the table, we find that $\sin^2\theta \div 0.0308$, gives values 1, 2, 3, 4, 5, 6, 7 and 8. If the material were to be simple cubic, 7 should not be present. Hence the pattern must really be 2, 4, 6, 8, 10, 12, 14 and 16. The corresponding (hkl) values are given in the table. The material must correspond to *bcc* structure. For *bcc* structure $(h + k + l)$ must be either even or odd.

Consider peak number 8. We have

$$d_{400} = \frac{\lambda}{2\sin\theta} = \frac{0.07107}{2\times\sin(29.71)} = 0.071699 \text{ nm}$$

$$a = d_{400}\sqrt{h^2 + k^2 + l^2} = 0.71699 \times 4 = 0.2868 \text{ nm}.$$

18. In a neutron diffractometer, the Bragg angle for neutrons reflected from (111) planes of Ni crystal is found to be 28°30′. Calculate the effective temperature of the neutrons. Ni has *fcc* structure with a lattice constant of 0.352 nm.

Solution:

The interplanar distance d is given by:

$$d = \frac{a}{\sqrt{h^2 + k^2 + l^2}} = \frac{0.352\,\text{nm}}{\sqrt{1^2 + 1^2 + 1^2}} = 0.203 \text{ nm}$$

Making use of the Bragg equation we have

$$\lambda = 2d\sin\theta = 2 \times 0.203 \times \sin 28.5 = 0.194 \text{nm}$$

For neutrons, K.E $= \frac{1}{2}mv^2 = \frac{p^2}{2m} = \frac{3}{2}k_B T \Rightarrow p = \sqrt{3mk_B T} = \frac{h}{\lambda}$ (de Broglie's principle)

$$\therefore \quad T = \frac{h^2}{3mk_B\lambda^2} = \frac{(6.626\times 10^{-34})^2}{3\times 1.67\times 10^{-27} \times 1.38\times 10^{-23} \times (0.194\times 10^{-9})^2}$$

$$= 169 \text{ K}$$

19. Calculate the Bragg angle at which electrons accelerated from rest through a potential difference of 80 volts will be diffracted from the (111) planes of a *fcc* crystal of lattice parameter 0.35 nm.

Solution:

The de Broglie wavelength of the electron is given by:

$$\lambda = \frac{h}{\sqrt{2meV_o}} = \frac{6.626 \times 10^{-34}}{\sqrt{2 \times 9.1 \times 10^{-31} \times 1.6 \times 10^{-19} \times 80}} = 0.137 \text{ nm}$$

The interplanar distance is given by:

$$d_{111} = \frac{a}{\sqrt{h^2 + k^2 + l^2}} = \frac{0.35 \text{ nm}}{\sqrt{3}} = 0.202 \text{ nm}$$

The Bragg angle is given by:

$$\theta = \sin^{-1}\left(\frac{\lambda}{2d}\right) = \sin^{-1}\left(\frac{0.137 \times 10^{-9}}{2 \times 0.202 \times 10^{-9}}\right) = 19°\ 40'.$$

20. Copper atoms are 0.2552 nm in diameter and form *fcc* structure. X-rays of wavelength 0.152 nm is used for the analysis of two samples of copper. For sample A the first order Bragg reflection from (111) planes occurred at an angle of 21°00′, while for sample B the first order Bragg reflection from (111) planes was at 21°23′. Give an explanation for the difference between the samples.

Solution:

For an *fcc* crystal the lattice parameter is given by $a = D\sqrt{2}$ where D is the diameter of the atom.

$$\therefore \quad a = 0.2552 \times \sqrt{2} = 0.3609 \text{ nm}$$

For sample A, $\quad d_{111} = \dfrac{\lambda}{2\sin\theta} = \dfrac{0.152}{2\sin 21°0'} = 0.212 \text{ nm}$

$$a = d_{111}(h^2 + k^2 + l^2)^{1/2} = 0.212\sqrt{3} = 0.3673 \text{ nm}$$

For sample B, $d_{111} = \dfrac{\lambda}{2\sin\theta} = \dfrac{0.152}{2\sin 21°23'} = 0.208$

$$a = d_{111}(h^2 + k^2 + l^2)^{1/2} = 0.208\sqrt{3} = 0.361 \text{ nm}$$

The lattice parameter of sample B is the same as that of pure copper. Therefore sample B is a high purity copper. The lattice parameter of sample A is 1.75% greater than that of pure copper. Sample B is not pure and the presence of impurity atoms has caused strain in the crystal lattice.

21. X-rays of wavelength 0.171 nm is incident on a cubic crystalline metal. The first two Bragg reflections occur at angle of 30°00′ and 35°17′ respectively. Determine whether the crystal type is *bcc* and *fcc*. Also find the lattice parameter and the atomic diameter.

Solution:

If the metal is *bcc* the first two Bragg reflections will be from (110) and (200) planes respectively. Whereas if the metal is *fcc* the first two reflecting planes will be (111) and (200). Assuming that the metal is *bcc*.

$$d_{110} = \frac{\lambda}{2\sin\theta} = \frac{0.171}{2\sin 30°\,0'} = 0.171 \text{ nm}$$

and
$$d_{200} = \frac{\lambda}{2\sin\theta} = \frac{0.171}{2\sin 35°\,17'} = 0.148 \text{ nm}$$

The lattice parameter, $a = d\sqrt{h^2 + k^2 + l^2} = 0.171\sqrt{2} = 0.242$ nm or

$$a = 0.148\sqrt{4} = 0.296 \text{ nm}.$$

The two values of lattice parameter are not same. Hence the metal is not bcc.

Assuming the metal is fcc, $a = 0.171\sqrt{3} = 0.296$ nm and $a = 0.148\sqrt{4} = 0.296$ nm which is consistent.

Atomic diameter $= \dfrac{a}{\sqrt{2}} = \dfrac{0.296}{\sqrt{2}} = 0.2093$ nm.

22. Bragg reflections may occur from the following planes in *bcc* crystals: (110), (200), (211), (220), (310), (222), (321), (400), (411), (420), (322), and (422). Which of these planes will give reflections when X-rays of wavelength 0.154 nm is incident on a crystal of chromium? Assume the diameter of the chromium atoms to be 0.2494 nm.

Solution:

We have the relation:

$$\sin\theta = \frac{\lambda}{2d}, \text{ but } \sin\theta \leq 1.$$

$\therefore \quad \dfrac{\lambda}{2d} \leq 1$ or $d \geq \dfrac{\lambda}{2} = \dfrac{0.154 \text{ nm}}{2} = 0.077$ nm

But $\quad d = \dfrac{a}{\sqrt{h^2 + k^2 + l^2}} \quad \therefore \quad \dfrac{a}{\sqrt{h^2 + k^2 + l^2}} \geq 0.077$ nm

$\Rightarrow \quad \sqrt{h^2 + k^2 + l^2} \leq \dfrac{a}{0.077}$

$$\leq \frac{2d}{0.077\sqrt{3}} \leq \frac{2 \times 0.228}{0.077\sqrt{3}} = 3.74$$

$\therefore \quad h^2 + k^2 + l^2 \leq 13.98.$

The highest possible values of (hkl) are (222). Bragg reflections will occur from the first planes, including (222).

QUESTIONS

1. Describe how X-rays are produced.
2. Explain the origin of continuous and characteristic X-ray spectrum.
3. Write a short note on synchrotron radiation or synchrotron light source. Discuss its advantages.
4. Explain Moseley's law.
5. Explain the various mechanisms by which X-rays interact with matter.
6. Write a short note on (a) X-ray gas detectors, (b) X-ray semiconductor detectors and (c) X-ray scintillation detector.
7. Discuss the principle of X-ray radiography.

8. Describe the use of X-ray radiography in industry.
9. Discuss the limitations of X-ray radiography.
10. Describe the use of X-ray radiography in medicine.
11. Define Miller indices for designating (a) direction (b) plane (c) family of directions and (d) family of planes.
12. Briefly describe the procedure for arriving at the Miller indices of (a) direction and (b) plane.
13. Derive an expression for the interplanar distance between parallel planes(hkl).
14. Derive the ratio $d_{100}: d_{110}: d_{111}$ for (a) simple cubic, (b) face centered cubic and (c) body centered cubic crystals.
15. Describe the classical picture of X-ray scattering from an atom.
16. Describe the classical picture of X-ray scattering from a crystal.
17. Define scattering factor for an atom.
18. Derive Bragg's law
19. Derive Laue's equation for X-ray diffraction.
20. Compare Bragg's equation with those of Laue and show that they are equivalent.
21. Explain the origin of Laue spots.
22. What are the factors which contribute to the intensity of diffracted X-ray beam?
23. What are systematic absences? Explain.
24. Describe the four circle X-ray diffractometer.
25. Discuss the technique of X-ray diffraction using powders.
26. What are the various factors to be considered in interpreting the powder X-ray diffraction data?
27. How do you distinguish between the simple cubic, body centered cubic and face centered cubic using powder X-ray diffraction data?
28. Discuss the utility of X-ray diffraction techniques in characterizing materials.
29. Describe the X-ray fluorescence techniques in characterizing materials.

PROBLEMS

1. An X-ray tube operates at 50 kV. Calculate the shortest wavelength of the X-ray produced.
2. What is the minimum X-ray wavelength produced in Bremsstrahlung by electrons that have been accelerated through 250 kV.
3. Electrons are accelerated through a potential difference of 30 kV. Calculate the maximum frequency of radiation emitted. Calculate the Bragg angle for first and second order maxima when this radiation is diffracted from a set of planes separated by 0.1nm.
4. A polychromatic X-ray beam having wavelengths in the range 10 pm to 500 pm is diffracted from (111) planes of a simple cubic crystal with a lattice parameter of 0.1 nm at a Bragg angle of 45°. Calculate the X-ray wavelengths for which the Bragg reflection is observed.
5. Calculate the L_α lines for elements with atomic numbers 45, 60 and 75.
6. The resolution of a spectrograph is 1 pm. Would it be able to separate the K_α lines for platinum and gold.
7. Calculate K_α and K_β X-ray wavelengths for He and Li.
8. Cobalt target is bombarded with electrons and the wavelength of the characteristic spectra are measured. A second fainter characteristic line is also found because of an impurity in the target. The wavelengths of the K_α lines are 178.9 pm and 143.5 pm due to cobalt and the impurity element respectively. What is the impurity element?
9. An X-ray beam of wavelength 'A' undergoes a first order Bragg reflection from a crystal when its angle of incidence to a crystal face is 23° and an X-ray beam of wavelength 97 pm undergoes third order reflection when its angle of incidence to that same face is 60°. Assuming that the two beams are reflected from the same family of planes find the interplanar distance and A.

X-RAYS

10. Sketch the following planes and directions within a cubic unit cell:
    (a) [101]   (b) [010]   (c) [12$\bar{2}$]
    (d) [301]   (e) [$\bar{2}$01]   (f) [2$\bar{1}$3]
    (g) [0$\bar{1}\bar{1}$]   (h) (102)   (i) (002)
    (j) [1$\bar{3}$0]   (k) ($\bar{2}$12)   (l) (31$\bar{2}$).

11. Sketch the following planes and directions within an orthorhombic unit cell:
    (a) [110]   (b) [$\bar{2}\bar{2}$1]   (c) [410]
    (d) [0$\bar{1}$2]   (e) [$\bar{3}\bar{2}$1]   (f) (1$\bar{1}$1)
    (g) (11$\bar{1}$)   (h) (01$\bar{1}$)   (i) (030)
    (j) ($\bar{1}$21)   (k) (11$\bar{3}$)   (l) (04$\bar{1}$).

12. Determine the Miller indices of the plane that passes through three points having the coordinates:
    (a) (0, 0, 1) ; (1, 0, 0) ; (1/2, 1/2, 0)
    (b) (1/2, 0, 1) ; (1/2, 0, 0) ; (0, 1, 0)
    (c) (1, 0, 0) ; (0, 0, –1/4); (1/2, –1, 0)
    (d) (1, 0, 0) ; (0, –1, 1/2) ; (1, 1/2, 1/4)

13. A cubic plane has the following axial intercepts $a = 1/3$, $b = -1/2$ and $c = 1/4$. What are the Miller indices of the plane?

14. A plane which is parallel to the z-axis has intercepts along the x and y-axes in the ratio 3:2. Determine its Miller indices and sketch the plane.

15. A plane which is parallel to the x-axis has negative intercepts along both y and z-axes in the ratio 3:4. Determine its Miller indices and sketch the plane.

16. Calculate the energy of X-rays that produce first order Bragg reflection at 20° when incident on crystal planes separated by a distance of 0.2 nm.

17. A beam of X-rays is incident on a sodium chloride crystal. They get Bragg reflected at an angle of 8° 35′ from a set of parallel planes with a separation of 0.28 nm. What is the wavelength of the X-rays? At what angles would the second and third order Bragg reflections occur?

18. A beam of X-rays of wavelength 0.1 nm is incident on planes of crystal having bcc structure. Calculate the lattice constant if the first order Bragg reflection is observed at 10°.

19. Calculate the wavelength of neutrons that have reached thermal equilibrium with their surroundings at 100°C.

20. Draw (110) plane in the cubic unit cell. Indicate the <111> direction that lie on this plane.

21. In an orthorhombic cell, the unit cell parameters are $a = 0.127$ nm, $b = 0.214$ nm and $c = 0.151$ nm. Determine the intercepts of the (213) on x and y-axes if it has an intercept of 0.151nm along the z-axis.

22. Determine the Miller indices of a plane that makes an intercept of 0.3nm, 0.4nm and 0.5nm on the orthorhombic crystal whose $a : b : c = 1 : 2 : 5$.

23. An X-ray diffractometer recorder chart of an element which has either the bcc or fcc crystal structure showed diffraction peaks at the following 2θ values: 44.39°, 64.578°, 81.717° and 98.141°. Given the wavelength of the X-ray beam to be 0.1541 nm, determine the crystal structure and the lattice constant of the element.

24. An X-ray diffractometer recorder chart for an element which has either bcc or fcc crystal structure showed diffraction peaks at the following 2θ values: 42.171°, 61.160°, 77.079° and 92.146°. Given the wavelength of the X-rays to be 0.1541 nm, determine the crystal structure and the lattice constant of the element.

25. For which set of crystallographic planes will a first order diffraction peak occur at a diffraction angle of 46.21° for *bcc* iron when monochromatic radiation having a wavelength of 0.071 nm is used?

26. The metal rubidium has a *bcc* crystal structure. If the angle of diffraction for (321) set of planes occur at 27.00° (first order reflection) when monochromatic X-radiation having a wavelength of 0.0711 nm is used. Compute (*a*) the interplanar spacing for this set of planes, and (*b*) the atomic radius for the rubidium atom.

27. The metal iridium has an *fcc* crystal structure. If the angle of diffraction for the (221) set of planes occurs at 69.22° (first order reflection) when monochromatic X-ray radiation having a wavelength of 0.1542 nm is used, compute (*a*) the interplanar spacing for this set of planes and (*b*) the atomic radius for the iridium atom.

28. Using X-radiation of wavelength 154.2 pm metallic copper lattice (*fcc*) produces reflections from the (111) and (200) planes. If the density of copper is 8995 kg/m^3, at what angles will the first order reflections occur?

29. The smallest observed Bragg diffraction angle of the third order reflection from the (111) planes of potassium crystal is 6.613°, when X-radiation of wavelength 70.926pm is used. Given that the potassium crystallizes in the *bcc*, determine the unit cell length and the density of the crystal.

30. Lead is known to crystallise in one of the cubic structures. A powder sample of lead gives Bragg reflections at the following angles:15.66°, 18.17°, 26.13°, 31.11°, 32.71° and 38.59° with *X*-rays of wavelength 154.433 pm. Determine the type of the unit cell and its length.

31. The density of tantalum at 20°C is 16.69 g/cc and its unit cell is a cube. Given the first five observed Bragg reflection angles are 19.31°, 27.88°, 34.95°, 41.41° and 47.69°, find the type of the unit cell and its length. Assume the wavelength of *X*-rays to be 154.433 pm.

32. The density of silver at 20°C is 10.50 g/cc. and its unit cell is a cube. Given that the first five observed Bragg diffraction angles are 19.10°, 22.17°, 32.33°, 38.82° and 40.88°, find the type of the unit cell and its length. Assume the wavelength of the X-rays to be 154.433 pm.

33. Chromium crystallises in the *bcc* structure with a density of 7.20 g/cc. Calculate the length of the unit cell and the distance between the successive (110), (200) and (111) planes. Assume the wavelength of the X-rays to be 154.433 pm.

34. The unit cell of topaz is orthorhombic with *a* = 839 pm, *b* = 879 pm and c = 465 pm. Calculate the value of the Bragg diffraction angles from the (110), (101), (111) and (222) planes. Assume the wavelength of the X-rays to be 154.433 pm.

35. The distance between the (100) planes of a nickel substrate, whose surface in a (100) plane is 351pm. Calculate the minimum accelerating potential so that the electrons can diffract from the crystal. Calculate the kinetic energy and the de Broglie wavelength of these electrons.

12
Basic Quantum Mechanics

12.1 INTRODUCTION

In contrast to classical mechanics, quantum mechanics deals with the physics of atomic and nuclear systems. The atomic phenomena cannot be explained in terms of classical notions. The basic non-classical ideas on which the quantum mechanics is founded are the following:

(a) the particle aspect of radiation

(b) the wave aspect of particles

(c) the discrete nature of energy and momentum of atomic systems and

(d) the Heisenberg's uncertainty principle.

An important idea associated with the quantum mechanics is that the behaviour of a material particle can be described by a de Broglie wave which is represented by wave function Ψ. The wave function *does not represent* any physical wave or a physical quantity. It ought to be understood as a *mere mathematical description* of a particle, which enables us to calculate its actual behaviour in a convenient way. Ψ is a function of x, y, z and t. Though Ψ does not correspond to any physical quantity, the square of the wave function $\Psi\Psi^*$ (because Ψ in general could be a complex function) denotes the probability of finding a particle at a certain location.

Quantum mechanics is an abstract subject involving fair amount of mathematics. Since it is counter-intuitive and violates *common sense* notions the early steps towards its understanding are not easy. None the less, an appreciation of the subject is a must for an engineer since the physics of semiconductor devices is based completely on quantum mechanics. Many of the physical properties of materials are better explained on the basis of quantum mechanics. Quantum computation and quantum cryptography are subjects in their nascent state which are likely to have a profound impact on technology. The subject of nanoscience and nanotechnology heavily depends on the ideas of quantum mechanics. This chapter is intended to be a brief and elementary introduction to the subject.

Newton's laws form the foundation of classical mechanics. Similarly, Schrödinger equation forms the basis of quantum mechanics. *We cannot really derive Schrödinger's equation, just as we cannot derive Newton's equations. It is just a convenient mathematical description of the behaviour of material particles. Its validity and usefulness lie in the experimental verification of the results obtained from them.*

12.2 WAVE EQUATION IN CLASSICAL MECHANICS

In classical mechanics, while describing the wave phenomena, the notion of wave function $y(x, t)$ is introduced. Thus, in the case of sound wave, $y(x, t)$ represents pressure and for light wave $y(x, t)$ represents the electric or magnetic field vector. Any travelling sinusoidal wave of wavelength λ (= $2\pi/k$) and angular frequency ω is described by a function $sin\ (kx \pm \omega t)$ or $cos\ (kx \pm \omega t)$ or in general by $e^{i(kx \pm \omega t)}$. Note that –sign indicates a forward moving wave while the + sign corresponds to a backward moving wave.

Consider $y(x, t) = Ae^{i(kx - \omega t)}$ where $\omega = 2\pi\nu$ and $k = 2\pi/\lambda$ $\hspace{2cm}$ (12.1)

$$\frac{\partial y}{\partial x} = (+ik)Ae^{i(kx - \omega t)}$$

$$\frac{\partial^2 y}{\partial x^2} = (+ik)^2 Ae^{i(kx - \omega t)}$$

$$= -k^2 Ae^{i(kx - \omega t)} = -k^2 y \hspace{2cm} (12.2)$$

Similarly,

$$\frac{\partial y}{\partial t} = (-i\omega)e^{i(kx - \omega t)}$$

$$\frac{\partial^2 y}{\partial t^2} = (-i\omega)^2 e^{i(kx - \omega t)}$$

$$= (-\omega^2 y) \hspace{2cm} (12.3)$$

From eqns. (12.2) and (12.3)

$$\frac{\partial^2 y}{\partial x^2} = \frac{k^2}{\omega^2} \cdot \frac{\partial^2 y}{\partial t^2} = \frac{1}{v^2} \cdot \frac{\partial^2 y}{\partial t^2} \hspace{2cm} (12.4)$$

where v is the velocity of propagation of the wave and is given by $v = \nu\lambda$.

12.3 SCHRÖDINGER WAVE EQUATION

We assume that Schrödinger's equation is analogous to classical wave equation *i.e.*, the wave function obeys the above form of eqn. (12.4). Hence, we have

$$\frac{\partial^2 \Psi}{\partial x^2} = \frac{1}{v^2} \cdot \frac{\partial^2 \Psi}{\partial t^2} \hspace{2cm} (12.5)$$

Here, v represents the velocity of particle or it represents the group velocity.

$$\omega = 2\pi\nu = \frac{2\pi E}{h} = \frac{E}{\hbar} \text{ where } \hbar = \frac{h}{2\pi}$$

E is the total energy of particle. Let us consider a particle whose motion is confined only along the x-axis. We assume that $\Psi = \psi(x)e^{-i\omega t}$ *i.e.*, the wave function is thought of as a product of two functions, one depending only on x and the other only on t

$$\frac{\partial \Psi}{\partial x} = \frac{d\psi}{dx} \cdot e^{-i\omega t}$$

$$\frac{\partial^2 \Psi}{\partial x^2} = \frac{d^2\psi}{dx^2} \cdot e^{-i\omega t} \hspace{2cm} (12.6)$$

$$\frac{\partial \Psi}{\partial t} = (-i\omega)\psi \cdot e^{-i\omega t}$$

$$\frac{\partial^2 \Psi}{\partial t^2} = -\omega^2 \psi \cdot e^{-i\omega t} \hspace{2cm} (12.7)$$

From eqns. (12.5), (12.6) and (12.7)

$$\frac{\partial^2 \psi}{\partial x^2} e^{-i\omega t} = -\frac{\omega^2}{v^2} \psi \cdot e^{-i\omega t}$$

$$\frac{d^2 \psi}{dx^2} + \frac{4\pi^2}{\lambda^2} \psi = 0 \qquad (\because \ v = \nu\lambda) \tag{12.8}$$

But Kinetic energy $= K.E = \dfrac{p^2}{2m}$ or $p^2 = \left(\dfrac{h}{\lambda}\right)^2 = 2m(K.E)$ $\quad \left(\because \lambda = \dfrac{h}{p}\right)$

If E is the total energy and V is the potential energy, then $K.E = E-V$

$$\therefore \qquad E - V = \frac{h^2}{2m\lambda^2} \quad or \quad \frac{1}{\lambda^2} = \frac{2m(E-V)}{h^2} \tag{12.9}$$

From eqns. (12.8) and (12.9)

$$\frac{d^2 \psi}{dx^2} + \frac{8\pi^2 m(E-V)\psi}{h^2} = 0$$

i.e.,
$$\frac{d^2 \psi}{dx^2} + \frac{2m}{\hbar^2}(E-V)\psi = 0 \tag{12.10}$$

This is known as the time independent Schrödinger wave equation.

Alternate Method : There is an alternate way of arriving at Schrödinger's wave equation. Assume Ψ for a free particle moving along $+ x$ direction to be of the form

$$\Psi = A e^{i(kx - \omega t)} \tag{12.11}$$

But $E = \hbar\omega$ and $p = \hbar k$.

$$\therefore \qquad \Psi = A e^{\frac{i}{\hbar}(px - Et)}$$

$$\frac{\partial^2 \Psi}{\partial x^2} = \left(\frac{ip}{\hbar}\right)^2 \Psi = -\frac{p^2}{\hbar^2} \Psi$$

$$\Rightarrow \qquad p^2 = -\hbar^2 \frac{\partial^2 \Psi}{\partial x^2} \tag{12.12}$$

Similarly, $\qquad \dfrac{\partial \Psi}{\partial t} = \left(-i\dfrac{E}{\hbar}\right)\Psi$

or $\qquad E\Psi = -\dfrac{\hbar}{i} \cdot \dfrac{\partial \Psi}{\partial t} = i\hbar \dfrac{\partial \Psi}{\partial t} \tag{12.13}$

At speeds small compared to that of light, the total energy E of the particle is the sum of its kinetic energy and potential energy V

or
$$E = \frac{p^2}{2m} + V(x, t)$$

or
$$E\Psi = \frac{p^2\Psi}{2m} + V\Psi \tag{12.14}$$

From eqns. (12.12), (12.13) and (12.14),

or
$$i\hbar \frac{\partial \Psi}{\partial t} = \frac{\hbar^2}{2m} \frac{\partial^2 \Psi}{\partial x^2} + V\Psi \tag{12.15}$$

This is what is called the *time dependent* Schrödinger's equation.

$$\Psi = Ae^{i(kx - \omega t)} = \psi(x)e^{-i\omega t} = \psi(x) \cdot e^{-\left(\frac{iEt}{\hbar}\right)} \tag{12.16}$$

$$\therefore \quad \frac{\partial \Psi}{\partial t} = -\frac{iE}{\hbar} \psi(x) \cdot e^{-\left(\frac{iEt}{\hbar}\right)} \tag{12.17}$$

and
$$\frac{\partial^2 \Psi}{\partial x^2} = \frac{d^2 \psi}{dx^2} \cdot e^{-\left(\frac{iEt}{\hbar}\right)} \tag{12.18}$$

From eqns. (12.15), (12.16), (12.17) and (12.18)

$$E\psi e^{-\left(\frac{iEt}{\hbar}\right)} = -\frac{\hbar^2}{2m} \frac{d^2 \psi}{dx^2} \cdot e^{-\left(\frac{iEt}{\hbar}\right)} + V\psi \cdot e^{-\left(\frac{iEt}{\hbar}\right)}$$

i.e.,
$$E\psi = -\frac{\hbar^2}{2m} \cdot \frac{d^2 \psi}{dx^2} + V\psi$$

or
$$\frac{d^2 \psi}{dx^2} + \frac{2m}{\hbar^2} (E - V)\psi = 0 \tag{12.19}$$

This is known as the *time independent* Schrödinger equation.

12.4 PROPERTIES OF WAVE FUNCTIONS

The product $\psi\psi^* dV$ denotes the probability of finding the particle in a volume element dV at (x, y, z). Hence,

$$\int \psi\psi^* dV = 1$$

where the integration is carried out over the entire volume. This is referred to as the normalizing condition.

Following are the important implications of the probabilistic interpretation of the wave function:

(a) The wave function must be *single valued*. The probabilistic interpretation of the wave function would be physically absurd if ψ is a multi-valued function (Fig.12.1).

(b) The wave function should not be infinite over a finite region. Otherwise, the normalization condition wont be satisfied (Fig.12.2a). However, the wave function can be a delta function. This corresponds to the case of

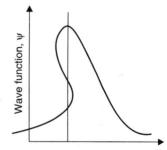

Fig. 12.1 A wave function that is not single valued

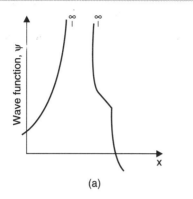 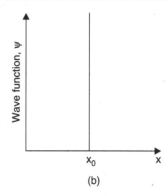

Fig. 12.2(*a*) A wave function must not be infinite over a finite range because it is not square integrable; (*b*) However, it can be a delta function *i.e.*, $f(x_0)= 1$ for $x = x_0$ and $f(x) = 0$ for all other x. This indicates the localization of particle at $x = x_0$

localization of the particle at a particular point in space (Fig.12.2*b*).

(*c*) Since, the wave function has to obey a second order differential equation, it must be continuous everywhere (Fig.12.3). Note that $f(x)$ is continuous at $x = a$, if
$$\operatorname*{Lt}_{x \to a^-} f(x) = \operatorname*{Lt}_{x \to a^+} f(x) = f(a).$$

(*d*) Its first derivative also must be continuous, except at the ill-behaved regions of the potential (Fig.12.4). This is a weaker condition because there are systems with certain ill-behaved potential energies. A particle in a potential box of infinite height is an example of this kind because the potential changes from zero to infinity.

(*e*) An important idea in quantum mechanics is that the expectation values of physical quantities can be calculated using the wave function. This is discussed in the operator formalism of Schrödinger equation.

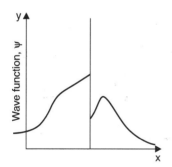 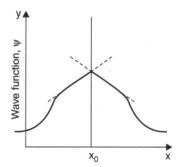

Fig. 12.3 Discontinuous wave function

Fig. 12.4 Wave function that is discontinuous in slope

12.5 OPERATOR FORMALISM OF SCHRÖDINGER EQUATION

12.5.1 Definitions

An operator $\hat{A}$ is any mathematical entity which operates on any function. The simplest example of an operator is to take $\hat{A}$ itself to be a function of x. Thus $\hat{A}(x)$ denotes the operation of mutiplication by 'x'. Thus we might have

$$\hat{A}(x) f(x) = x \times f(x)$$

A less trivial example is that of differentiation. Thus

$$\hat{A}\left(\frac{d}{dx}\right) f(x) = \frac{df}{dx}$$

To each operation $\hat{A}\left(x, \frac{d}{dx}, \frac{d^2}{dx^2}\right)$ belongs a set of numbers a_n and a set of functions $u_n(x)$ defined by the equation:

$$\hat{A}\left(x, \frac{d}{dx}, \frac{d^2}{dx^2}\right) u_n(x) = a_n u_n(x)$$

where a_n is an eigen value and $u_n(x)$ is the corresponding eigen function.

This is known as the eigen value equation.

The commutator of two operators $\hat{A}$ and $\hat{B}$ is defined to be $= \left[\hat{A}, \hat{B}\right] = \hat{A}\hat{B} - \hat{B}\hat{A}$.

12.5.2 Operators in Schrödinger Representation

In Schrödinger representation, the position, momentum and energy of a particle are represented by the following operators:

$$\hat{x} \to x$$

$$\hat{p} \to -i\hbar \frac{d}{dx} \quad \text{where} \quad \hbar = \frac{h}{2\pi}$$

$$\hat{p}^2 = \hat{p} \cdot \hat{p} = -\hbar^2 \frac{d^2}{dx^2}$$

$$E \to i\hbar \frac{d}{dt}$$

The Hamiltonian is defined as corresponding to the total energy i.e.,

$$H = K.E + P.E = \frac{p^2}{2m} + V$$

$$\therefore \quad \hat{H} = -\frac{\hbar^2}{2m} \frac{d^2}{dx^2} + V$$

Thus, the Schrödinger equation is an eigen value equation:

$$\hat{H}\psi = E\psi$$

BASIC QUANTUM MECHANICS

$$\left[-\frac{\hbar^2}{2m}\frac{d^2}{dx^2}\right]\psi = E\psi$$

$$\Rightarrow \quad \frac{d^2\psi}{dx^2} + \frac{2m}{\hbar^2}(E-V)\psi = 0$$

This is the time independent Schrödinger wave equation.

In the operator formalism $\hat{H}\psi = E\psi$ where $\hat{H}$ is the Hamiltonian operator.

$$\therefore \quad \left[-\frac{\hbar^2}{2m}\frac{d^2}{dx^2} + V\right]\Psi = i\hbar\frac{\partial\Psi}{\partial t}$$

This is the time dependent Schrödinger equation.

12.5.3 Expectation Values of Observations

The observations are represented by operators $\hat{A}$.

The possible results of an observation $\hat{A}$ are the corresponding eigen values a_n.

An observation $\hat{A}$ on a system in an eigen state $u_n(x)$ certainly leads to the result a_n.

The average value of repeated observation $\hat{A}$ on a set of systems, each one in an arbitrary state is

$$\bar{a} = \frac{\int_{-\infty}^{+\infty}\Psi^*(x)\hat{A}\left(x,\frac{\partial}{\partial x}\right)\Psi(x)dx}{\int_{-\infty}^{+\infty}\Psi^*(x)\Psi(x)dx} \quad \text{where } \psi^* \text{ is the complex conjugate of } \psi.$$

12.5.4 Commutation Relations

The commutation relations incorporate the Heisenberg's uncertainty principle and are given by

$$[\hat{x},\hat{p}_x] = i\hbar; \quad [\hat{x},\hat{p}_y] = 0; \quad [\hat{x},\hat{p}_z] = 0$$
$$[\hat{y},\hat{p}_x] = 0; \quad [\hat{y},\hat{p}_y] = i\hbar; \quad [\hat{y},\hat{p}_z] = 0$$
$$[\hat{z},\hat{p}_x] = 0; \quad [\hat{z},\hat{p}_y] = 0; \quad [\hat{z},\hat{p}_z] = i\hbar$$

12.6 FREE PARTICLE

We will solve the Schrödinger equation for a free particle. The particle is not subjected to any potential and moves in +ve x-direction. The potential energy $V = 0$ and Schrödinger equation assumes the form

$$\frac{d^2\psi}{dx^2} + \frac{2m}{\hbar^2}E\psi = 0 \qquad (12.20)$$

This is a differential equation which is similar to that of an undamped simple harmonic vibration.

The solution is of the form $\psi(x) = Ae^{i\alpha x}$. (12.21)

The total wave function will however be $\psi(x)e^{-i\omega t}$

From eqns. (12.20) and (12.21)

$$(i\alpha)^2 \psi + \frac{2m}{\hbar^2} E\psi = 0 \quad i.e., \quad -\alpha^2 \psi + \frac{2m}{\hbar^2} E\psi = 0$$

or
$$\alpha = +\sqrt{\frac{2mE}{\hbar^2}} \tag{12.22}$$

Since the particle propagates along +ve x-axis, α is taken to be +ve.

From eqn. (12.22)

$$E = \frac{\hbar^2 \alpha^2}{2m} = \frac{p^2}{2m} \tag{12.23a}$$

Hence, the momentum of the particle is given by

$$p = \hbar \alpha \tag{12.23b}$$

Equation (12.23) is the classical expression for the energy of a free particle. We also note that E is continuous and all the values of energy are allowed.

The group velocity and the phase velocity of the particle can be calculated as follows:

$$v_g = \frac{dE}{dp} = \frac{2p}{2m} = v$$

$$v_p = \frac{\omega}{k} = \frac{\hbar \omega}{\hbar k} = \frac{E}{p} = \frac{p^2}{2m \times p} = \frac{p}{2m} = \frac{v}{2} = \frac{v_g}{2} \tag{12.24}$$

Thus, for a free particle the phase velocity is equal to half of the group velocity.

Since $v_g v_p = c^2$, it follows that $v_p = \dfrac{c}{\sqrt{2}}$ (12.25)

12.7 PARTICLE CONFINEMENT IN POTENTIAL WELLS

12.7.1 Introduction

In semiconductor devices and lasers it is common to have electrons being confined in a potential well. The potential well could be of one, two or three dimensions. In such cases, electron energies are quantized. The distribution of electrons among these allowed energy levels plays an important role in the design and function of these devices.

12.7.2 One Dimensional Potential Well of Infinite Height (Quantum Wire)

Consider a particle which is confined within a potential well of infinite height (Fig.12.5). Let the width of the potential well be 'L'. The particle cannot escape from this potential well which means that $\psi = 0$ for $x \leq 0$ and $x \geq L$. The potential energy inside the well is constant and assumed to be zero (*i.e.*, $V = 0$ for $0 < x < L$ and $V = \infty$ all other x). Hence, the Schrödinger's equation for the particle inside the well is

$$\frac{d^2 \psi}{dx^2} + \frac{2m}{\hbar^2} E\psi = 0 \quad \text{or} \quad \frac{d^2 \psi}{dx^2} = -\frac{2m}{\hbar^2} E\psi \tag{12.26}$$

BASIC QUANTUM MECHANICS

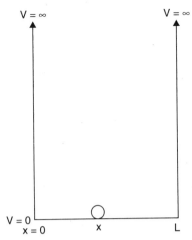

Fig. 12.5 Particle in a one-dimensional box of infinite height

This equation is similar to the differential equation governing the simple harmonic motion. Hence, the solution of eqn. (12.26) is

$$\psi = A \cos kx + B \sin kx \qquad (12.27)$$

Taking the first and second derivates, we get

$$\frac{d\psi}{dx} = -Ak \sin kx + Bk \cos kx \qquad \text{and}$$

$$\frac{d^2\psi}{dx^2} = -Ak^2 \cos kx - Bk^2 \sin kx = -k^2 \psi \qquad (12.28)$$

From eqns. (12.26) and (12.28)

$$k^2 = \frac{2m}{\hbar^2} E \quad \text{or} \quad E = \frac{\hbar^2 k^2}{2m} \qquad (12.29)$$

The constants A and B in eqn. (12.27) have to be determined by means of boundary conditions

$$\psi = 0 \text{ at } x = 0, \quad \therefore \quad 0 = A + (B \times 0) \Rightarrow A = 0 \qquad (12.30)$$
$$\psi = 0 \text{ at } x = L, \quad \therefore \quad 0 = B \sin k\alpha \Rightarrow kL = n\pi, \text{ where } n = 1, 2, 3 \qquad (12.31)$$

From eqns. (12.27), (12.30) and (12.31), we conclude that the the wave function is of the form
$$\psi_n = B \sin \frac{kx}{L} = B \sin \frac{n\pi x}{L} \qquad (12.32a)$$

The constant B is evaluated by applying normalizing condition. *i.e.*,

$$\int_0^L \psi \psi^* \, dx = 1$$

i.e., $\qquad \int_0^L B^2 \sin^2 kx \cdot dx = 1 \Rightarrow \frac{B^2}{2} \int_0^L 2 \sin^2 kx \cdot dx = 1$

i.e., $\qquad \frac{B^2}{2} \int_0^L (1 - \cos 2kx) \, dx = 1 \Rightarrow \frac{B^2}{2} \left\{ \int_0^L dx - \int_0^L \cos 2kx \cdot dx \right\} = 1$

i.e., $\qquad \frac{B^2}{2} \left[x - \frac{\sin 2kx}{2k} \right]_0^L = 1 \Rightarrow \frac{B^2 L}{2} = 1 \qquad (\because kL = n\pi)$

$$\Rightarrow \qquad B = \pm \sqrt{\frac{2}{L}} \qquad (12.32b)$$

From eqns. (12.32a) and (12.32b), we have

$$\psi_n = \pm \sqrt{\frac{2}{L}} \sin \frac{n\pi x}{L} \qquad (12.33)$$

A plot of ψ and $\psi\psi^*$ for different values of n are shown in Figs. 12.6 and 12.7. We see that standing waves are created between the walls of the potential well. The results are analogous to the modes of a stretched string fixed at its ends. This is on account of the wave nature of the particle.

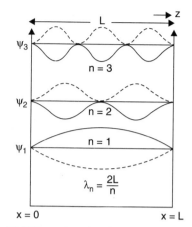

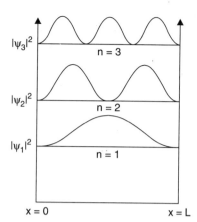

Fig. 12.6 Wave function for the first three energy states for a particle in a one dimensional potential well of infinite height

Fig. 12.7 Probability distribution ($\psi\psi^*$) for particle in a one dimensional well of infinite height corresponding to first three energy states

Of special interest is the behaviour of the function $\psi\psi^*$ i.e., the probability of finding the electron at a certain place within the well. In the classical case, the particle travels back and forth between the walls. Its probability function is therefore equally distributed along the whole length of the well. In quantum mechanics, the deviation from the classical case is most pronounced for $n = 1$. In this case, $\psi\psi^*$ is largest in the middle of the well and vanishes at the boundary. For higher n values i.e., for higher energies, the values of $\psi\psi^*$ approach those of classical values (Fig.12.8). This is in accordance with the Bohr's correspondence principle. This principle states that for large quantum numbers, the quantum behaviour ought to correspond to that predicted by classical mechanics.

From eqns. (12.29) and (12.31)

$$E_n = \frac{\hbar^2 k^2}{2m} = \frac{\hbar^2 n^2 \pi^2}{2mL^2} = \frac{n^2 h^2}{8mL^2} \quad \text{where} \quad n = 1, 2, 3, \ldots \quad (12.34)$$

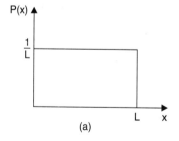
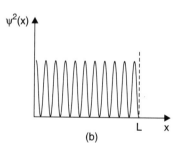

Fig. 12.8 (a) Classical Probability density for one dimensional potential well which is uniform throughout the well (b) Probability distribution according to quantum mechanics for a particle in an energy state with high quantum number

BASIC QUANTUM MECHANICS

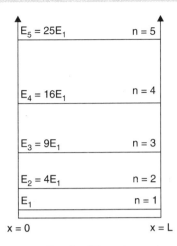

Fig. 12.9 Energy corresponding to different quantum states for a particle in a potential well of infinite height.

Thus, unlike in the case of a free particle the particle in the well can have only discrete energy values (Fig. 12.9). *Further, the ground state of the particle is not zero.* This is because the uncertainty in the position of the particle is L and consequently there is minimum momentum associated with the particle.

Density of States

The density of states is an important parameter which determines the electrical and optical properties of quantum dots and wires. It denotes the number of available energy states per unit length per unit energy interval around energy E. If we denote the density of states as $\rho(E)$, then $\rho(E)dE$ denotes the number of energy states (dN) available for the electron between energies E and $E + dE$.

$$\rho = \frac{dN}{dE} = \frac{dN}{dn} \times \frac{dn}{dE} \tag{12.35}$$

The first factor can be calculated with reference to Fig. 12.10. Since, there is only one energy state per every quantum number, the first factor is equal to unity. Hence,

$$\rho = \frac{dN}{dE} = 1 \times \frac{dn}{dE}$$

From eqn. (12.34), we get

$$\frac{dE}{dn} = \frac{nh^2}{4mL^2} = \frac{h\sqrt{8mL^2E}}{4mL^2} = \frac{h\sqrt{E}}{L\sqrt{2m}}$$

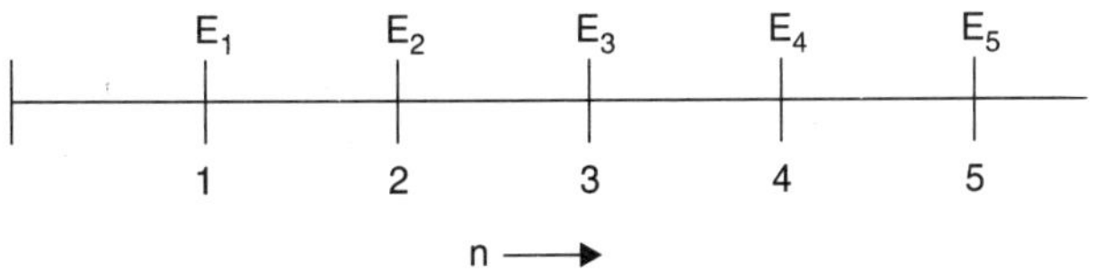

```
      E₁      E₂      E₃      E₄      E₅
   ───┼───────┼───────┼───────┼───────┼───
      1       2       3       4       5
                      n ⟶
```

Fig. 12.10 Calculation of density of states for quantum wire

$$\therefore \quad \rho = \frac{dN}{dE} = \frac{dn}{dE} = \frac{L\sqrt{2m}}{h\sqrt{E}}$$

The density of states is defined for the well with unit length. Hence, assuming L = 1, we get

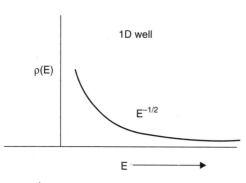

Fig. 12.11 Plot of density of states as a function of energy for a quantum wire

$$\rho = \frac{dN}{dE} = \frac{\sqrt{2m}}{h\sqrt{E}} \qquad (12.36)$$

Thus, the density of energy states is directly proportional to the square root of the mass of the particle and inversely proportional to square root of energy. This is shown in Fig. 12.11. The limiting case when the width of the well $L \to 0$, corresponds to the case of quantum dot. Since L can never be zero, it is often referred to as the quasi one dimensional quantum dot. Note that for a perfect quantum dot both the potential and the wave functions are delta functions.

12.7.3 Particle in a Two-Dimensional Box of Infinite Height (Quantum Well)

Consider a particle which is confined within a square potential well of infinite height (Fig. 12.12). Let the width of the potential well be 'L'. The potential $V(x, y)$ is defined by the equation:

$$V(x, y) = \begin{cases} 0 & \text{for} \quad 0 < x < L \quad \text{and} \quad 0 < y < L \\ \infty & \text{otherwise} \end{cases} \qquad (12.37)$$

The Schrödinger equation for the particle inside the well is given by

$$\frac{\partial^2 \psi}{\partial x^2} + \frac{\partial^2 \psi}{\partial y^2} = -\frac{2m}{\hbar^2} E\psi \qquad (12.38)$$

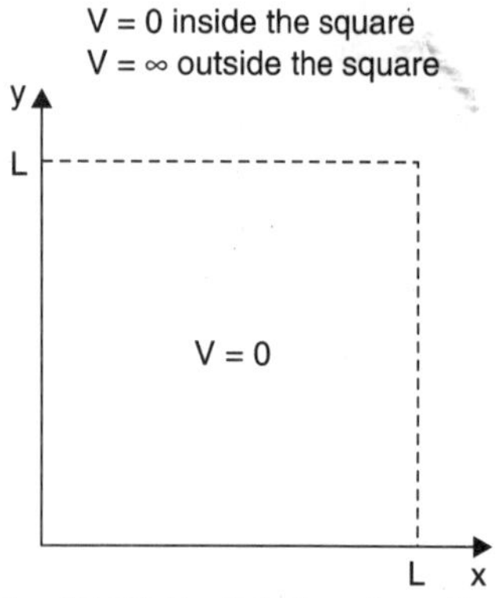

Here, ψ is a function of x and y. To solve this equation ψ is written as a product of two functions, each of which depends only on one of the variables x or y.

i.e., $$\psi(x, y) = X(x)\, Y(y) \qquad (12.39)$$

$$\frac{\partial^2 \psi}{\partial x^2} = Y \frac{\partial^2 X}{\partial x^2}; \quad \frac{\partial^2 \psi}{\partial y^2} = X \frac{\partial^2 Y}{\partial y^2} \qquad (12.40)$$

From eqns. (12.38), (12.39) and (12.40)

$$Y \frac{\partial^2 X}{\partial x^2} + X \frac{\partial^2 Y}{\partial y^2} = -\frac{2mE}{\hbar^2} XY$$

Fig. 12.12 Particle in a two-dimensional potential well of infinite height (quantum well)

Dividing throughout by XY

$$\frac{1}{X} \cdot \frac{\partial^2 X}{\partial x^2} + \frac{1}{Y} \cdot \frac{\partial^2 Y}{\partial y^2} = \frac{-2mE}{\hbar^2} \qquad (12.41)$$

BASIC QUANTUM MECHANICS

The first term is a function of only x and the second is a function of only y, while the sum of the two is a constant. This can be true only if each term is a constant. Let

$$\frac{1}{X} \cdot \frac{\partial^2 X}{\partial x^2} = -k_x^2 \quad \text{and} \quad \frac{1}{Y} \cdot \frac{\partial^2 Y}{\partial y^2} = -k_y^2 \tag{12.42}$$

$$\therefore \quad k_x^2 + k_y^2 = \frac{2mE}{\hbar^2} \tag{12.43}$$

Eqn. (12.42) may be written as

$$\frac{\partial^2 X}{\partial x^2} + k_x^2 X = 0 \quad \text{and} \quad \frac{\partial^2 Y}{\partial y^2} + k_y^2 Y = 0 \tag{12.44}$$

Following the same procedure as in the case of one dimensional case we get

$$X = \pm \sqrt{\frac{2}{L}} \sin \frac{n_x \pi x}{L} \quad \text{and} \quad Y = \pm \sqrt{\frac{2}{L}} \sin \frac{n_y \pi y}{L} \tag{12.45}$$

where, $k_x = \dfrac{n_x \pi}{L}; k_y = \dfrac{n_y \pi}{L}$. Note that n_x and n_y are integers. \hfill (12.46)

From eqns. (12.39) & (12.45) the total wave function

$$\psi = XY$$

$$\psi = XY = \pm \frac{2}{L} \sin \frac{n_x \pi x}{L} \cdot \sin \frac{n_y \pi y}{L}. \tag{12.47}$$

A plot of these functions is shown in Fig.12.13.

From eqns. (12.43), and (12.46) we have

$$E = \frac{\hbar^2}{2m}(k_x^2 + k_y^2) = \frac{\hbar^2}{2m}\left(\frac{n_x^2 \pi^2}{L^2} + \frac{n_y^2 \pi^2}{L^2}\right) = \frac{h^2}{8mL^2}(n_x^2 + n_y^2) = \frac{h^2 R^2}{8mL^2} \tag{12.48}$$

where $\quad R^2 = n_x^2 + n_y^2$

Thus, in a quantum well the energy state is characterized by two quantum numbers n_x and n_y. One feature in two dimensions is the appearance of degenerate states. *Degenerate states correspond to energy states having the same energy but different quantum numbers.* Thus, quantum states (1, 2) and (2, 1) have the same energy.

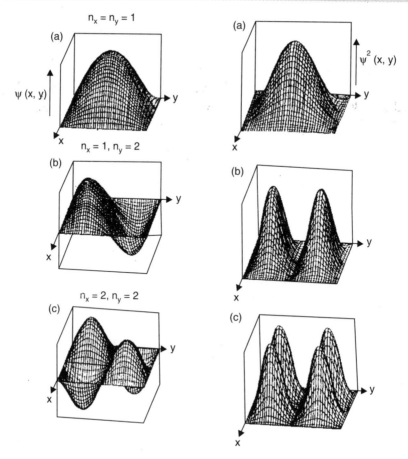

Fig. 12.13 Plot of wave function and probability distribution for the first few states for a particle in two-dimensional potential well of infinite height

Density of States

The density of states for the two-dimensional case is obtained as follows:

$$\rho = \frac{dN}{dE} = \frac{dN}{dR} \times \frac{dR}{dE} \qquad (12.49)$$

The first factor can be calculated by referring to Fig. 12.14.

In eqn. (12.48) $R^2 = n_x^2 + n_y^2$ represents a cricle of radius R.

Each point at the vertex of the square represents an energy state. Effectively, there is one point associated with unit area of the square. Hence, the number (N) of energy states is equal to the area of the quadrant. Since, both n_x and n_y are positive, only positive quadrant of the circle is considered.

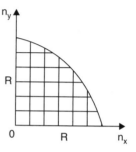

Fig. 12.14 Calculation of density of states for a two-dimensional quantum well. Every vertex of the square indicates an energy state

BASIC QUANTUM MECHANICS

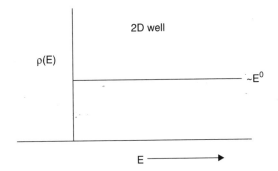

Fig. 12.15 Plot of density of states as a function of energy for a quantum well

$$N = \frac{\pi R^2}{4}, \quad \therefore \quad \frac{dN}{dR} = \frac{\pi R}{2} \tag{12.50}$$

From eqn. (12.48) we have

$$\frac{dE}{dR} = \frac{h^2 R}{4mL^2} \tag{12.51}$$

From eqns. (12.49), (12.50) and (12.51)

$$\rho = \frac{\pi R}{2} \times \frac{4mL^2}{h^2 R} = \frac{2m\pi L^2}{h^2} \tag{12.52}$$

Since ρ is defined for unit area, $L^2 = 1 \quad \therefore \quad \rho = \dfrac{2m\pi}{h^2}$

Thus, the density of states is a constant for two-dimensional case. This is shown in Fig. 12.15.

12.7.4 Particle in a Three-dimensional Cubical Box (Bulk Solid)

Consider a particle of mass 'm' constrained to move in a cubical box of length 'L' (Fig.12.16). The potential energy V is equal to zero within the box and in infinite at the boundary i.e.,

$$V(x, y, z) = 0 \begin{cases} 0 < x < L \\ \text{for } 0 < y < L \\ 0 < z < L \end{cases} \tag{12.53}$$

$$= \infty \text{ otherwise}$$

The three-dimensional Schrödinger equation is

$$\nabla^2 \psi + \frac{2mE}{\hbar^2} \psi = 0$$

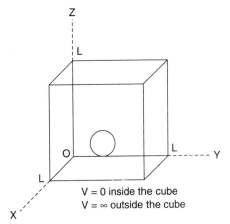

Fig. 12.16 Particle in a three-dimensional cubic potential well of length L.

i.e.,
$$\frac{\partial^2 \psi}{\partial x^2} + \frac{\partial^2 \psi}{\partial y^2} + \frac{\partial^2 \psi}{\partial z^2} + \frac{2mE}{\hbar^2}\psi = 0 \tag{12.54}$$

Here ψ is a function of x, y & z. To solve this equation ψ is written as a product of three functions, each of which depends only one of the variables x, y or z

i.e.,
$$\psi(x, y, z) = X(x)\, Y(y)\, Z(z) = XYZ \tag{12.55}$$

$$\frac{\partial^2 \psi}{\partial x^2} = YZ\frac{\partial^2 X}{\partial x^2};\quad \frac{\partial^2 \psi}{\partial y^2} = XZ\frac{\partial^2 Y}{\partial y^2}\ \text{and}\ \frac{\partial^2 \psi}{\partial z^2} = XY\frac{\partial^2 Z}{\partial z^2} \tag{12.56}$$

From eqns. (12.54), (12.55) and (12.56)

$$YZ\frac{\partial^2 X}{\partial x^2} + XZ\frac{\partial^2 Y}{\partial y^2} + XY\frac{\partial^2 Z}{\partial z^2} = -\frac{2mE}{\hbar^2}XYZ$$

Dividing throughout by XYZ, we get

$$\frac{1}{X}\cdot\frac{\partial^2 X}{\partial x^2} + \frac{1}{Y}\cdot\frac{\partial^2 Y}{\partial y^2} + \frac{1}{Z}\cdot\frac{\partial^2 Z}{\partial z^2} = -\frac{2mE}{\hbar^2} \tag{12.57}$$

The first term is a function of x only, the second, a function of y only and the third, a function of z only, while the sum of the three is a constant. This can be true only if each term is a constant.

$$\frac{1}{X}\cdot\frac{\partial^2 X}{\partial x^2} = -k_x^2;\ \frac{1}{Y}\cdot\frac{\partial^2 Y}{\partial y^2} = -k_y^2\ \text{and}\ \frac{1}{Z}\cdot\frac{\partial^2 Z}{\partial z^2} = -k_z^2 \tag{12.58}$$

$$\therefore\quad k_x^2 + k_y^2 + k_z^2 = \frac{2mE}{\hbar^2} \tag{12.59}$$

Eqn. (12.58) may be written as

$$\frac{\partial^2 X}{\partial x^2} + k_x^2 X = 0;\ \frac{\partial^2 Y}{\partial y^2} + k_y^2 Y = 0;\ \text{and}\ \frac{\partial^2 Z}{\partial z^2} + k_z^2 Z = 0 \tag{12.60}$$

Following the same procedure as in the case of one and two-dimensional case, we get

$$X = \pm\sqrt{\frac{2}{L}}\sin\frac{n_x \pi x}{L};\quad Y = \pm\sqrt{\frac{2}{L}}\sin\frac{n_y \pi y}{L}\ \text{and}$$

$$Z = \pm\sqrt{\frac{2}{L}}\sin\frac{n_x \pi z}{L} \tag{12.61}$$

where,
$$k_x = \frac{n_x \pi}{L},\ k_y = \frac{n_y \pi}{L}\ \text{and}\ k_z = \frac{n_z \pi}{L} \tag{12.62}$$

Note that n_x, n_y and n_z are integers.

From eqns. (12.55) & (12.61) the total wave function

$$\psi = XYZ$$

$$\psi = \pm\sqrt{\frac{8}{v}}\sin\frac{n_x \pi x}{L}\cdot\sin\frac{n_y \pi y}{L}\cdot\sin\frac{n_z \pi z}{L}\ \text{where}\ v = L^3 \tag{12.63}$$

From eqns. (12.59), and (12.62) we have

$$E = \frac{\hbar^2}{2m}(k_x^2 + k_y^2 + k_z^2)$$

$$= \frac{\hbar^2}{2m}\left(\frac{n_x^2\pi^2}{L^2} + \frac{n_y^2\pi^2}{L^2} + \frac{n_z^2\pi^2}{L^2}\right)$$

$$= \frac{h^2}{8mL^2}\left(n_x^2 + n_y^2 + n_z^2\right) = \frac{h^2 R^2}{8mL^2} \text{ where } R^2 = n_x^2 + n_y^2 + n_z^2 \quad (12.64)$$

The smallest allowed energy in a three-dimensional potential well is occupied by a particle for $n_x = n_y = n_z = 1$. For the next higher energy there are three different possibilities for combining the n-values, viz., $(n_x, n_y, n_z) = (1,1,2), (1,2,1),$ or $(2,1,1)$. *These states which have the same energy but different quantum numbers are the degenerate states.* The example just given describes a threefold degenerate energy state.

Density of States

The density of states for the three-dimensional case is obtained as follows. As in the case of two-dimensional case, the density of states is given by eqn. (12.49).

The first factor can be calculated by referring to Fig. 12.17.

$R^2 = n_x^2 + n_y^2 + n_z^2$ represents a sphere. n_x, n_y and n_z are positive integers. Hence only that part of the sphere in which all the numbers are positive is shown. Every corner of the cubical mesh in side the sphere represents an energy state. Effectively, there is one point associated with unit volume of the cube since each point is shared by eight cubes. Thus, the number of energy states N is equal to one-eighth the volume of the sphere.

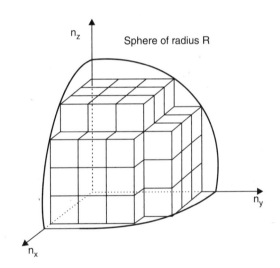

Fig. 12.17 Calculation of density of states for a particle in three-dimensional infinite potential well. Every corner of the cubical mesh denotes an energy state

i.e., $N = \frac{1}{8} \times \left(\frac{4\pi R^3}{3}\right) = \frac{\pi R^3}{6}$

$$\therefore \quad \frac{dN}{dR} = \frac{\pi R^2}{2} \quad (12.65)$$

From eqn. (12.64) we have

$$\frac{dE}{dR} = \frac{Rh^2}{4L^2 m} \quad (12.66)$$

Using eqns. (12.49), (12.64), (12.65) and (12.66)

$$\therefore \quad \rho = \frac{dN}{dE} = \frac{dN}{dR} \times \frac{dR}{dE} = \frac{\pi R^2}{2} \times \frac{4L^2 m}{Rh^2} = \frac{2\pi L^2 mR}{h^2}$$

$$= \frac{2\pi L^2 m}{h^2} \times \frac{L\sqrt{8mE}}{h} = \frac{4\sqrt{2}\pi v m^{3/2}}{h^3} E^{1/2}$$

where $v = L^3$ = volume of the cube. The density of states (ρ) is defined per unit volume. Hence, $v = 1$

$$\therefore \quad \rho = \frac{4\sqrt{2}\pi m^{3/2}}{h^3} E^{1/2} \text{ for unit volume of the material} \tag{12.67}$$

Thus, the density of states is parabolic for three-dimensional case. This is shown in Fig.12.18.

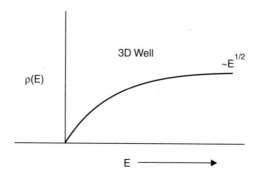

Fig.12.18 Plot of density of states for a particle in a three-dimensional infinite potential well

Calculation of Fermi Energy for a Metal at Absolute Zero

The result obtained in eqn. (12.67) can be used to calculate the Fermi energy of a metal. In the quantum theory of metals, electron inside the metal is treated as a particle in 3-D box. The electrons obey the Fermi-Dirac distribution which is given by:

$$f(E) = \frac{1}{e^{\frac{E-E_F}{k_B T}} + 1}$$

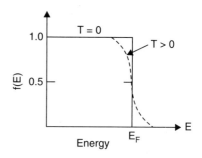

$f(E)$ denotes the probability of an electron having energy E at a temperature T. A plot of this function is shown in Fig. 12.19. Note that at absolute zero, all the energy levels below the Fermi energy are filled. Since electrons obey Fermi-Dirac distribution, each level can accommodate only two electrons possessing opposite spin. Thus, Fermi energy is the maximum energy that an electron can have at absolute zero. Hence, if n is the number of electrons per unit volume of the metal. Then

Fig. 12.19 Fermi Dirac distribution function

$$n = \int_0^{E_F} 2 \times \rho(E) dE = \int_0^{E_F} \frac{8\sqrt{2}\pi m^{3/2}}{h^3} E^{1/2} dE = \frac{8\sqrt{2}\pi m^{3/2}}{h^3} \times \frac{2E_F^{3/2}}{3}$$

BASIC QUANTUM MECHANICS

Rewriting the above equation we get

$$E_F = \frac{h^2}{8m}\left(\frac{3n}{\pi}\right)^{2/3}$$

12.7.5 Potential Well of Finite Height

Let us consider a potential well of finite height V_0 and having a width $2a$ (Fig.12.20a). In semiconductor technology a good approximation to such quantum wells is produced when a narrow-bandgap material of width $W = 2a$ is sandwiched between a large-bandgap material (Fig. 12.20b). Let the particle has energy E less than V_0 (i.e., $E < V_0$).

For the sake of convenience the physical space is divided into three regions.

$$x < -a \quad \text{Region 1}$$
$$-a < x < +a \quad \text{Region 2}$$
$$x > +a \quad \text{Region 3}$$

According to classical mechanics, when the particle strikes the sides of the well, it bounces off without entering regions 1 and 3 (Fig. 12.21). In quantum mechanics, the particle also bounces back and forth, but now it has a certain probability of penetrating into regions I and III even though $E < V_0$. *Thus the wave function does not become zero at the boundary.* This is in contrast to the case of potential well of infinite height where the wave function becomes zero at the boundaries.

In region 1 (i.e., $x < -a$) and region 3 (i.e., $x > +a$) Schrödinger equation is

$$\frac{d^2\psi}{dx^2} + \frac{2m(V_0 - E)}{\hbar^2}\psi = 0 \qquad (12.68)$$

In region 2 (i.e., $-a < x < +a$) the Schrödinger equation is

$$\frac{d^2\psi}{dx^2} + \frac{2mE}{\hbar^2}\psi = 0 \qquad (12.69)$$

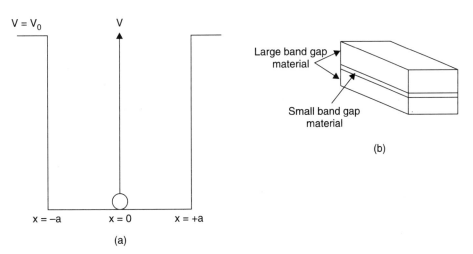

Fig. 12.20 (a) Narrow band gap material sandwiched between two large band gap materials
(b) The corresponding finite potential experienced by the particle

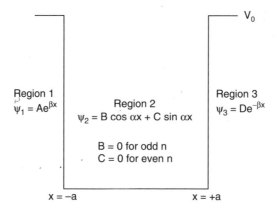

Fig. 12.21 Forms of wave functions in the three regions of interest

Note that in region 2, the potential is zero. Further the particle is bound and we are interested in solving the Schrödinger equation for $E < V_0$.

Eqns. (12.69) and (12.68) can be rewritten as

$$\frac{d^2\psi}{dx^2} + \alpha^2 \psi = 0 \quad \text{for} \quad -a < x < +a \quad \text{where} \quad \alpha = \sqrt{\frac{2mE}{\hbar^2}} \tag{12.70}$$

$$\frac{d^2\psi}{dx^2} + \beta^2 \psi = 0 \quad \text{for} \quad x < -a \quad \text{and} \quad x > +a \quad \text{where} \quad \beta = \sqrt{\frac{2m(V_0 - E)}{\hbar^2}} \tag{12.71}$$

Let ψ_1, ψ_2 and ψ_3 denote the wave functions in region 1, 2 and 3. The solutions to eqns. (12.70) and (12.71) are given by:

$$\psi_1 = A e^{\beta x} \tag{12.72}$$
$$\psi_2 = B \cos \alpha x + C \sin \alpha x \tag{12.73}$$
$$\psi_3 = D e^{-\beta x} \tag{12.74}$$

A, B, C and D are constants to be determined by boundary conditions.

The form of the solutions in region 1 and 3 are chosen to satisfy the following conditions:

(a) Both ψ_1 and ψ_3 must be finite everywhere.

(b) Further ψ_1 should tend to 0 as $x \to -\infty$ and ψ_3 should tend to 0 as $x \to \infty$. Note that β is > 0.

(c) In region 2, the form of ψ_2 chosen is similar to that of a particle in a potential well of infinite height.

(d) Both ψ and $d\psi/dx$ must be continuous at $x = +a$ and $x = -a$. i.e.,

At $x = -a$, $\quad \psi_1 = \psi_2$

$$\left(\frac{d\psi_1}{dx}\right)_{x=-a} = \left(\frac{d\psi_2}{dx}\right)_{x=-a}$$

At $x = +a$, $\quad \psi_2 = \psi_3$

$$\left(\frac{d\psi_2}{dx}\right)_{x=+a} = \left(\frac{d\psi_3}{dx}\right)_{x=+a}$$

BASIC QUANTUM MECHANICS

Applying these boundary conditions we get,

$$Ae^{-\beta a} = B\cos(-a\times\alpha) + C\sin(-a\times\alpha) = B\cos\alpha a - C\sin\alpha a \qquad (12.75)$$

$$\beta Ae^{-\beta a} = -B\alpha\sin(-a\times\alpha) + C\alpha\cos(-a\times\alpha) = B\alpha\sin\alpha a + C\alpha\cos\alpha a \qquad (12.76)$$

$$De^{-\beta a} = B\cos(+a\times\alpha) + C\sin(+a\times\alpha) = B\cos\alpha a + C\sin\alpha a \qquad (12.77)$$

$$-\beta De^{-\beta a} = -B\alpha\sin(+a\times\alpha) + C\alpha\cos(+a\times\alpha) = -B\alpha\sin\alpha a + C\alpha\cos\alpha a \qquad (12.78)$$

Adding eqns. (12.75) and (12.77) we get

$$(A+D)e^{-\beta a} = 2B\cos\alpha a \qquad (12.79)$$

Subtracting eqn. (12.78) from (12.76)

$$\beta(A+D)e^{-\beta a} = 2\alpha B\sin\alpha a \qquad (12.80)$$

Subtracting eqn. (12.75) from (12.77)

$$(D-A)e^{-\beta a} = 2C\sin\alpha a \qquad (12.81)$$

Adding eqns. (12.78) and (12.76)

$$-\beta(D-A)e^{-\beta a} = 2\alpha C\cos\alpha a \qquad (12.82)$$

Dividing eqn. (12.80) by eqn. (12.79) and multiplying by 'a'

$$\beta a = \alpha a \tan\alpha a \qquad (12.83a)$$

Dividing eqn. (12.82) by eqn. (12.81) and multiplying by 'a'

$$-\beta a = \alpha a \cot\alpha a \qquad (12.83b)$$

Symmetric and Anti-symmetric Wave Functions

Equations (12.83a) and (12.83b) lead to symmetric and anti-symmetric wave functions respectively. The wave functions with *even parity* are symmetric with respective to the origin i.e., $\psi(x) = \psi(-x)$. The wave functions with *odd parity* are anti-symmetric with respect to the origin. i.e., $\psi(x) = -\psi(-x)$.

From eqns. (12.81) and (12.82) we have

$$-\beta.2C.\sin\alpha a = 2\alpha C\cos\alpha a$$

Substituting for $\sin\alpha a$ from eqn. (12.83a) in the above equation we get

$$-\beta.2C\left(\frac{\beta\cos\alpha a}{\alpha}\right) = 2\alpha C\cos\alpha a$$

$$\Rightarrow \qquad 2C(\alpha^2 + \beta^2)\cos\alpha a = 0$$

$$\Rightarrow \qquad C = 0 \qquad (12.84a)$$

From eqns. (12.81) and (12.82) we get, $A = D \qquad (12.84b)$

The form of wave functions can be obtained from eqns. (12.72), (12.73), (12.74), (12.84a) and (12.84b). They are:

$$\psi_1 = Ae^{\beta x};\ \psi_2 = B\cos\alpha x \text{ and } \psi_3 = Ae^{-\beta x}$$

$$\therefore \qquad \psi_1(-x) = Ae^{\beta(-x)} = Ae^{-\beta x} = \psi_3(x)$$

$$\psi_2(-x) = B\cos\alpha(-x) = B\cos\alpha x = \psi_2(x)$$

$$\psi_3(-x) = Ae^{-\beta(-x)} = Ae^{\beta x} = \psi_1(x)$$

Thus, the wave functions are symmetric about the origin. Hence they are of even parity.

From eqns. (12.79) and (12.80) we have

$$\beta.2B.\cos\alpha a = 2\alpha B\sin\alpha a$$

Substituting for cos αa from eqn. (12.83b)

$$\beta 2B \left(\frac{-\beta}{\alpha}\right) \sin \alpha a = 2\alpha B \sin \alpha a$$

$\Rightarrow \qquad 2B(\alpha^2 + \beta^2) \sin \alpha a = 0$

$\Rightarrow \qquad\qquad\qquad\qquad B = 0$ \hfill (12.84c)

From eqns. (12.79) and (12.80) we get $A = -D$ \hfill (12.84d)

Again the form of wave functions can be obtained from eqns. (12.72), (12.73), (12.74), (12.84c) and (12.84d). They are:

$$\psi_1 = Ae^{\beta x};\ \psi_2 = C\sin \alpha x\ \text{and}\ \psi_3 = -Ae^{-\beta x}$$

$\therefore \qquad \psi_1(-x) = Ae^{\beta(-x)} = Ae^{-\beta x} = -\psi_3(x)$

$\qquad \psi_2(-x) = C \sin \alpha(-x) = -C \sin \alpha x = -\psi_2(x)$

$\qquad \psi_3(-x) = -Ae^{-\beta(-x)} = -Ae^{\beta x} = -\psi_1(x)$

Thus, the wave functions are anti-symmetric about the origin. Hence they are of odd parity.

Energy Eigen Values

Plots of eqns. (12.83a) and (12.83b) are shown in Fig.12.22a and Fig.12.22b. These equations can be solved by numerical techniques. Thus, values of α and β can be obtained. However, from eqns. (12.70) and (12.71), α and β also have to satisfy the equation:

$$(\alpha a)^2 + (\beta a)^2 = \frac{2mV_0 a^2}{\hbar^2} = r^2 \text{ where } r \text{ is the radius of the circle.} \qquad (12.85)$$

Therefore, the desired energy values can be obtained by constructing the intersection of the curves shown in Fig.12.22 with the circle defined by eqn. (12.85).

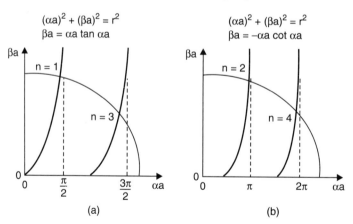

Fig. 12.22 (a) Plot of $\beta a = \alpha a \tan(\alpha a)$ for odd n integer
(b) Plot of $\beta a = -\alpha a \cot(\alpha a)$ for even n integer. The intersection of the curve $(\alpha a)^2 + (\beta a)^2 = r^2$ is also shown for both the cases

However, for the case of wave functions with even parity (Fig. 12.23), the tan function of eqn. (12.83a) passes through the origin. Hence, at least one solution exists for any value of V_0.

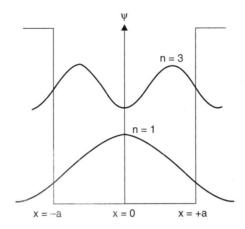

Fig. 12.23 Plot of $\psi(x)$ for the $n = 1$ and $n = 3$ excited states of a particle in a one dimensional well of finite height

In the case of wave functions with odd parity (Fig. 12.24), the radius of the circle has to be larger than the minimum value so that the two curves intersect. *i.e.*, the potential well must have a certain depth for the given width a and mass m to permit a solution.

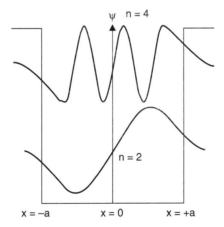

Fig. 12.24 Plot of $\psi(x)$ for the $n = 2$ and $n = 4$ excited states of a particle in a one dimensional well of finite height

In the limit $mV_0 a^2 \to \infty$, the intersection points are given by

$$\tan \alpha a = \infty \text{ corresponding to } \alpha a = \frac{(2n-1)\pi}{2} \text{ or } \alpha(2a) = 2(n-1)\pi$$

where $n = 1, 2, 3, \ldots$

and $-\cot \alpha a = \infty$ corresponding $\alpha a = n\pi$ or $\alpha(2a) = 2n\pi$

where $n = 1, 2, 3,$

or we can combine both the conditions to give $\alpha(2a) = n\pi$ where $n = 1, 2, 3, \ldots$

Note that the last condition corresponds to eqn. (12.31) of the potential well of infinite height. As in the case of particle in an infinite well, the energy spectrum of the particle is given by:

$$E_n = \frac{n^2 h^2}{8m(2a)^2} - V_0 = \frac{n^2 h^2}{32ma^2} - V_0$$

Hence the energy levels E_n are lower for each n than they are for a particle in an infinite well. Even n integers correspond to even parity states or symmetric wave functions and odd n integers correspond odd parity states or anti-symmetric wave functions. Fig. 12.23 shows the plot of symmetric wave functions for $n = 2$ and $n = 4$ states and Fig. 12.24 shows the plot of anti-symmetric wave functions for $n = 1$ and $n = 3$ states. Note that the wave functions penetrate the barrier region and thus there is a finite probability of finding the particle outside the well. This is known as *barrier penetration* or *tunneling*. This is discussed in greater detail in the next section.

12.7.6 Bohr's Atomic Model and Wave Function of Electron

It is relevant here to discuss the Bohr's postulate regarding the quantization of angular momentum of an electron going round the nucleus. The waves associated with an orbiting electron have to be standing waves. If this were not to be the case, the wave would be out of phase with itself after each revolution. After a large number of revolutions in the orbit, all possible phases would be obtained and the wave amplitude would be zero on account of destructive interference. This can only be avoided if the perimeter of the orbit is an integral multiple of λ (Fig. 12.25) i.e.,

$$2\pi r = n\lambda = \frac{nh}{p} = \frac{nh}{mv} \quad \text{(From de Broglie hypothesis)}$$

$$\Rightarrow \qquad mvr = \frac{nh}{2\pi}$$

which is the same as Bohr's postulate regarding quantization of angular momentum of electron.

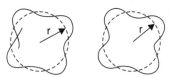

Fig. 12.25 (a) The wavelength associated with the electrons does not fit into the orbit
(b) The wave fits correctly satisfying the quantum condition $2\pi r = n\lambda$

12.8 BARRIER PENETRATION AND TUNNELING

In many semiconductor devices and other physical phenomena particles encountering a potential barrier are common. The behaviour of particles in such cases does not conform to the usual ideas of classical physics. It has to be understood on the basis of quantum mechanics. According to quantum mechanics, even though the particles may have energies much less than those of barriers, they may *penetrate* and *tunnel* through barriers. Such a behavior of particles is not only verified by experiments but also forms the basis for the functioning of a number of semiconductor and superconductor devices.

12.8.1 Step Barrier or Step Potential

Consider a potential step such that $V = 0$ for $x < 0$ and $V = V_0$ for $x > 0$. The particle moves in the positive x-direction. To solve the Schrödinger equation for the particle it is customary to divide the physical space into two regions 1 and 2. They correspond to $x < 0$ and $x > 0$ (Fig.12.26). In region 1, $V = 0$ and region 2, $V = V_0$. It is necessary to consider two cases viz., when the energy E of the particle is greater than V_0 ($E > V_0$) and when the energy E of the particle is lesser than V_0 ($E < V_0$).

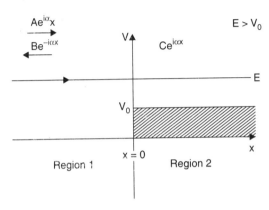

Fig. 12.26 Particle encountering a step potential ($E > V_0$)

Case (1) When the energy of the particle is greater than that of the barrier height. ($E > V_0$)

The Schrödinger equation for regions 1 and 2 are given by:

Region 1

$$\frac{d^2\psi_1}{dx^2} = -\frac{2mE}{\hbar^2}\psi_1 = -\alpha^2\psi_1 \quad \text{where } \alpha = \sqrt{\frac{2mE}{\hbar^2}} \quad (12.86)$$

Note that $\alpha = (2\pi/\lambda_1)$ where λ_1 is the de Broglie wavelength of the particle in region 1.

Region 2

$$\frac{d^2\psi_2}{dx^2} = -\frac{2m(E-V_0)}{\hbar^2}\psi_2 = -\beta^2\psi_2 \quad \text{where } \beta = \sqrt{\frac{2m(E-V_0)}{\hbar^2}} \quad (12.87)$$

Note that $\beta = (2\pi/\lambda_2)$ where λ_2 is the de Broglie wavelength of the particle in region 2. Further β is > 0 because E is > 0.

In region 1, the incident wave and reflected waves are possible. Hence, solution to eqn. (12.86) is of the form

$$\psi_1 = Ae^{i\alpha x} + Be^{-i\alpha x} \quad \text{(Region 1)} \quad (12.88)$$

Note that the first term denotes the wave propagating in the +ve x-direction while the second term denotes the wave moving in the –ve x-direction.

In region 2, the general solution to eqn. (12.87) is of the form

$$\psi_2 = Ce^{i\beta x} + De^{-i\beta x} \quad \text{(Region 2)} \quad (12.89)$$

However, in region 2, there is no possibility of a wave propagating in the –ve x-direction. Hence,

$$\psi_2 = Ce^{i\beta x} \quad \text{(Region 2)} \quad (12.90)$$

ψ_1 and ψ_2 must obey the following boundary conditions.

At $x = 0$, $\quad\quad\quad \psi_1 = \psi_2 \Rightarrow A + B = C$ \hfill (12.91)

$$\left(\frac{d\psi_1}{dx}\right)_{x=0} = \left(\frac{d\psi_2}{dx}\right)_{x=0} \Rightarrow A(i\alpha)e^{i\alpha x} + B(-i\alpha)e^{i\alpha x} = C(i\beta)e^{i\beta x} \quad \text{at } x = 0$$

$\Rightarrow \quad\quad\quad i\alpha(A - B) = i\beta C \quad \text{or} \quad \alpha(A - B) = \beta C$ \hfill (12.92)

Solving for A, B and C from eqns. (12.91) and (12.92):

$$B = \left(\frac{\alpha - \beta}{\alpha + \beta}\right) A \quad \text{and} \quad C = \left(\frac{2\alpha}{\alpha + \beta}\right) A \quad\quad (12.93)$$

Hence, the wave functions in region 1 and 2 are given by:

$$\psi_1 = Ae^{i\alpha x} + A\left(\frac{\alpha - \beta}{\alpha + \beta}\right) e^{-i\alpha x} \quad\quad (12.94)$$

$$\psi_2 = A\left(\frac{2\alpha}{\alpha + \beta}\right) e^{i\beta x} \quad\quad (12.95)$$

The constant A can be evaluated by using the normalizing condition:

$$\int_{-\infty}^{0} \psi_1 \psi_1^* \, dx + \int_{0}^{\infty} \psi_2 \psi_2^* dx = 1$$

A plot of ψ_1 and ψ_2 is shown in Fig.12.27.

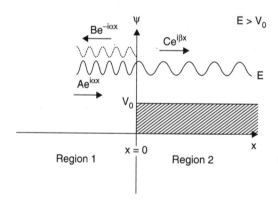

Fig. 12.27 Plot of ψ_1 and ψ_2 when the particle energy is greater than the potential height

Transmission coefficient:

It is required to calculate the transmission coefficient and reflection coefficient of the particle. This is done by calculating the probability flux per second. The probability is given by the square of the amplitude of the wave function. *The probability of finding the particle at a point also depends on the velocity of the particle.* If the velocity of particle is low the probability of finding it at that point is low and *vice versa*. Hence, for calculating the transmission and reflection coefficients the square of the amplitude of the wave function is multiplied by the corresponding velocity.

The transmission and reflection coefficients (T and R) can be calculated using eqn (12.93) and are given by:

$$T = \frac{v_2(C^*C)}{v_1(A^*A)} = \frac{p_2(C^*C)}{p_1(A^*A)} = \frac{\hbar\beta(C^*C)}{\hbar\alpha(A^*A)} \quad (\because p_1 = \hbar\alpha; p_2 = \hbar\beta)$$

$$= \frac{\beta}{\alpha}\left(\frac{2\alpha}{\alpha+\beta}\right)^2 = \frac{4\alpha\beta}{(\alpha+\beta)^2} = \frac{4\left(\dfrac{\beta}{\alpha}\right)}{\left[1+\dfrac{\beta}{\alpha}\right]^2} \quad (12.96)$$

$$R = \frac{v_1(B^*B)}{v_1(A^*A)} = \left(\frac{\alpha-\beta}{\alpha+\beta}\right)^2 = \left[\dfrac{1-\dfrac{\beta}{\alpha}}{1+\dfrac{\beta}{\alpha}}\right]^2 \quad (12.97)$$

From eqns. (12.96) and (12.97), $R + T = 1$ as expected. From eqns. (12.86) and (12.87)

$$\left(\frac{\beta}{\alpha}\right)^2 = 1 - \left(\frac{V_0}{E}\right)$$

A plot of R and T as a function of (β/α) is shown in Fig. 12.28.

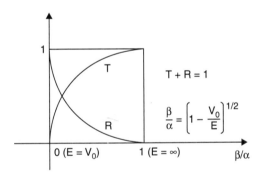

Fig. 12.28 Plot of reflection and transmission coefficients as a function of particle energy (β/α)

According to classical mechanics, a particle possessing energy greater than that of the barrier does not get reflected at all. However, this is contradicted by quantum mechanics. From Fig.12.28 there is a finite probability of a particle getting reflected even when its energy is greater than that of the barrier. The reflection coefficient becomes very small and tends to zero only when the energy of the particle is very much greater than the barrier energy.

Case 2. When the energy of the particle is less than that of the barrier. $(E < V_0)$

The Schrödinger equation for regions 1 and 2 (Fig.12.29) are given by:

Region 1

$$\frac{d^2\psi_1}{dx^2} = -\frac{2mE}{\hbar^2}\psi_1 = -\alpha^2\psi_1 \quad \text{where } \alpha = \sqrt{\frac{2mE}{\hbar^2}} \quad (12.98)$$

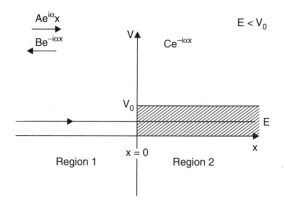

Fig. 12.29 Particle encountering a step potential ($E < V_0$)

Region 2

$$\frac{d^2\psi_2}{dx^2} = -\frac{2m(E-V_0)}{\hbar^2}\psi_2 = -\beta^2\psi_2 \quad \text{where } \beta = \sqrt{\frac{2m(E-V_0)}{\hbar^2}} \quad (12.99)$$

Further β^2 is < 0 because E is $< V_0$. Let $\beta = i\gamma$ where γ is > 0

In region 1, the incident wave and reflected waves are possible. Hence, solution to eqn. (12.99) is of the form

$$\psi_1 = Ae^{i\alpha x} + Be^{-i\alpha x} \quad \text{(Region 1)} \quad (12.100)$$

In region 2 the general solution to eqn. (12.99) is of the form

$$\psi_2 = Ce^{i\beta x} = Ce^{i(i\gamma x)} = Ce^{-\gamma x} \quad \text{(Region 2)} \quad (12.101)$$

A, B and C can be obtained by replacing $\beta = i\gamma$ in eqn. (12.93).

$$B = \left(\frac{\alpha - i\gamma}{\alpha + i\gamma}\right)A \quad \text{and} \quad C = \left(\frac{2\alpha}{\alpha + i\gamma}\right)A \quad (12.102)$$

Hence, the wave functions in region 1 and 2 are given by:

$$\psi_1 = Ae^{i\alpha x} + A\left(\frac{\alpha - i\gamma}{\alpha + i\gamma}\right)e^{-i\alpha x} \quad (12.103)$$

$$\psi_1 = A\left(\frac{2\alpha}{\alpha + i\gamma}\right)e^{-\gamma x} \quad (12.104)$$

The constant A can be evaluated by using the normalizing condition:

$$\int_{-\infty}^{0} \psi_1 \psi_1^* \, dx + \int_{0}^{\infty} \psi_2 \psi_2^* \, dx = 1$$

BASIC QUANTUM MECHANICS

A plot of ψ_1 and ψ_2 is shown in Fig.12.30. Note that ψ_2 is an exponentially decaying function of x. The distance over which the amplitude reduces to $(1/e)$ of the maximum amplitude is given by $(1/\gamma)$

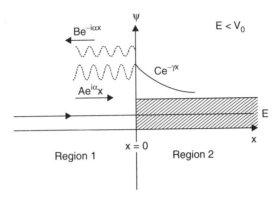

Fig. 12.30 Plot of ψ_1 and ψ_2 when the particle energy is less than the potential height

Reflection coefficient:
The reflection coefficient can be calculated from eqn. (12.97) and is given by:

$$R = \frac{v_2(B*B)}{v_1(A*A)}\left(\frac{\alpha - i\gamma}{\alpha + i\gamma}\right)\left(\frac{\alpha + i\gamma}{\alpha - i\gamma}\right) = 1 \tag{12.105}$$

From eqns. (12.98) and (12.99)

$$\left(\frac{\beta}{\alpha}\right)^2 = \left(\frac{i\gamma}{\alpha}\right)^2 = 1 - \left(\frac{V_0}{E}\right) \Rightarrow \left(\frac{V_0}{E}\right) - 1 = \left(\frac{\gamma}{\alpha}\right)^2 \tag{12.106}$$

A plot of R as a function of (E/V_0) is shown in Fig. 12.31.

Thus as expected from classical mechanics, the particle gets completely reflected at the barrier when its energy is lower than that of the barrier. However, the novel thing predicted by quantum mechanics is the penetration of the barrier. According to classical mechanics, the particle can never be found in region 2. Hence, region 2 is called the *forbidden region*. According to quantum mechanics region 2 is accessible

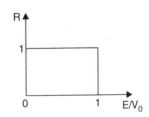

Fig. 12.31 Plot of R as a function of (E/V_0)

to the particle. The particle penetrates into the barrier. The penetration distance is proportional to $(1/\gamma)$. The penetration distance decreases with the mass of the particle and the height of the barrier above the energy of the particle. Macroscopic bodies have such large masses that their penetration depth is almost zero whatever be the height of the barrier. Hence, for all practical purposes they are not found in the classically forbidden region. An electron or a proton, on the other hand, may penetrate into a forbidden region to an appreciable extent. For example an electron which is accelerated through a potential difference of 1.0V, and which has acquired a kinetic energy of 1eV, incident on a potential barrier equivalent to 2eV, will have a wave function that decays to (1/e) of its maximum amplitude after 0.20 nm. This is comparable to diameter of one atom. Hence, penetration can have important effects on processes at surfaces, such as electrodes and for all events on an atomic scale.

It ought to be pointed out here that the penetration of barriers by waves is not uncommon in classical physics. When light gets totally internally reflected in a prism there is penetration of light wave upto a distance of the order of few wavelengths, beyond the interface. Therefore, if the face of a second glass prism is kept parallel to the interface, (the gap between them being not more than a few wavelengths) the light will *tunnel* through this barrier and generate a transmitted wave (Fig 12.32). The wave is called the *evanescent wave*. This phenomenon is used in fiber optic couplers (Fig. 12.33) and fiber optic sensors (Fig. 12.34).

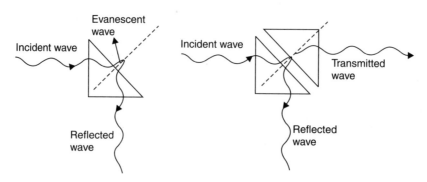

Fig. 12.32 Tunneling of photons when two prisms are kept close to each other

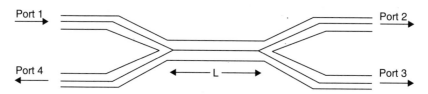

Fig. 12.33 Fiber optic coupler

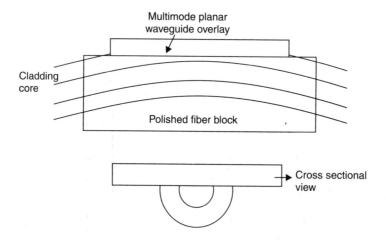

Fig. 12.34 Evanescent wave based fiber optic sensor

BASIC QUANTUM MECHANICS

Heisenberg's uncertainty principle shows that the wave like properties exhibited by an entity in penetrating classically excluded regions are really not in conflict with its particle like properties. Consider an experiment capable of proving that the particle is located somewhere in the region $x > 0$. Since, the probability density for $x > 0$ is appreciable only in the range of penetration depth, Δx is ~ penetration depth (γ^{-1})

$$\therefore \quad \Delta p \geq \frac{\hbar}{2\Delta x} \geq \frac{\hbar}{2} \times \sqrt{\frac{2m(V_0 - E)}{\hbar^2}} \geq \frac{\sqrt{2m(V_0 - E)}}{2}$$

Hence, the energy of the particle is uncertain by an amount

$$\Delta E \geq \frac{(\Delta p)^2}{2m} \geq \frac{V_0 - E}{4} \approx V_0 - E$$

Thus, it is no longer possible to say that the total energy of the particle is definitely less than the barrier height V_0.

12.8.2 Square Barrier or Square Potential

Consider a particle encountering a square barrier of width a (Fig.12.35). The potential step is such that $V = 0$ for $x < 0$ as well as $x > +a$ and $V = V_0$ for $0 < x < +a$. The particle moves in the positive x-direction. To solve the Schrödinger equation for the particle it is customary to divide the physical space into regions 1, 2 and 3. They correspond to $x < 0$; $0 < x < +a$ and $x > +a$. In regions 1 and 3, $V = 0$ and region 2, $V = V_0$. It is necessary to consider two cases viz., when the energy E of the particle is greater than V_0 (i.e., $E > V_0$) and when the energy E of the particle is lesser than V_0 ($E < V_0$).

Case 1. When the energy of the particle is greater than V_0 ($E > V_0$)

The Schrödinger equation for regions 1, 2 and 3 are given by:

Region 1

$$\frac{d^2\psi_1}{dx^2} = -\frac{2mE}{\hbar^2}\psi_1 = -\alpha^2\psi_1 \quad \text{where } \alpha = \sqrt{\frac{2mE}{\hbar^2}} \tag{12.107}$$

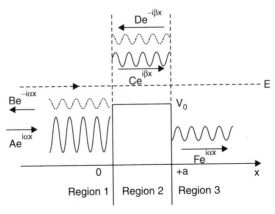

Fig. 12.35 Wave function for a particle encountering a square barrier of width a ($E > V_0$)

Region 2

$$\frac{d^2\psi_2}{dx^2} = -\frac{2m(E-V_0)}{\hbar^2}\psi_2 = -\beta^2\psi_2 \quad \text{where } \beta = \sqrt{\frac{2m(E-V_0)}{\hbar^2}} \quad (12.108)$$

Region 3

$$\frac{d^2\psi_3}{dx^2} = -\frac{2mE}{\hbar^2}\psi_3 = -\alpha^2\psi_3 \quad \text{where } \alpha = \sqrt{\frac{2mE}{\hbar^2}} \quad (12.109)$$

The solutions of the Schrödinger equation in the three regions are:

$$\psi_1 = Ae^{i\alpha x} + Be^{-i\alpha x} \quad \text{(Region 1)}$$
$$\psi_2 = Ce^{i\beta x} + De^{-i\beta x} \quad \text{(Region 2)} \quad (12.110)$$
$$\psi_3 = Fe^{i\alpha x} \quad \text{(Region 3)}$$

Note that in regions 1 and 2, both forward and backward moving waves are possible, while in region 3 only the forward moving wave is possible. It is convenient to define the parameter g which is characteristic of the barrier. It is defined as:

$$(a\alpha)^2 - (a\beta)^2 = \frac{2mV_0 a^2}{\hbar^2} = g^2 \quad (12.111)$$

Eqn. (12.111) represents a hyperbola (Fig.12.36). The permitted values of α (and hence E) comprise a positive unbounded continuum. For each such value of α, there are wave functions ψ_1, ψ_2 and ψ_3. The coefficients A, B, C, D and F have to be determined from the boundary conditions: It is customary to calculate (B/A), (C/A), (D/A) and (F/A). The value of A can be determined the normalizing condition. ψ_1, ψ_2 and ψ_3 must obey the following boundary conditions.

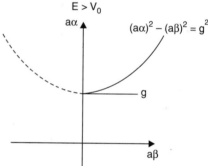

Fig. 12.36 Plot of $(\alpha a)^2 - (\beta a)^2 = g^2$ (hyperbola)

At $x = 0$, $\quad \psi_1 = \psi_2 \Rightarrow A + B = C + D$

$$(12.112)$$

$$\left(\frac{d\psi_1}{dx}\right)_{x=0} = \left(\frac{d\psi_2}{dx}\right)_{x=0} \Rightarrow \alpha(A - B) = \beta(C - D) \quad (12.113)$$

At $x = a$, $\quad \psi_2 = \psi_3 \Rightarrow Ce^{i\beta a} + De^{-i\beta a} = Fe^{i\alpha a} \quad (12.114)$

$$\left(\frac{d\psi_2}{dx}\right)_{x=a} = \left(\frac{d\psi_3}{dx}\right)_{x=a} \Rightarrow \beta(Ce^{i\beta a} - De^{-i\beta a}) = \alpha Fe^{i\alpha a} \quad (12.115)$$

Solving for A and B from eqns. (12.112) and (12.113) we get

$$A = \left(\frac{\alpha+\beta}{2\alpha}\right)C + \left(\frac{\alpha-\beta}{2\alpha}\right)D \qquad (12.116)$$

$$B = \left(\frac{\alpha-\beta}{2\alpha}\right)C + \left(\frac{\alpha+\beta}{2\alpha}\right)D \qquad (12.117)$$

Solving for C and D from eqns. (12.114) and (12.115)

$$C = \left(\frac{\alpha+\beta}{2\beta}\right) F e^{i\alpha a} e^{-i\beta a} \qquad (12.118)$$

$$D = \left(\frac{\beta-\alpha}{2\beta}\right) F e^{i\alpha a} e^{i\beta a} \qquad (12.119)$$

Substituting for C and D in eqns. (12.116) and (12.117) we get

$$\frac{B}{A} = \frac{(\alpha^2-\beta^2)(1-e^{2i\beta a})}{(\alpha+\beta)^2 - (\alpha-\beta)^2 e^{2i\beta a}} \qquad (12.120)$$

$$\frac{F}{A} = \frac{4\alpha\beta e^{i(\beta-\alpha)a}}{(\alpha+\beta)^2 - (\alpha-\beta)^2 e^{2i\beta a}} \qquad (12.121)$$

Making use of eqn. (12.121) and substituting for α and β from eqns. (12.107) and (12.108). The transmission coefficient (T) is given by:

$$T = \left(\frac{FF^*}{AA^*}\right) = \frac{4E(E-V_0)}{V_0^2 \sin^2 \beta a + 4E(E-V_0)} \qquad (12.122)$$

A plot of transmission coefficient as a function of E is shown in Fig.12.37. Making use of eqn. (12.120) and substituting for α and β from eqns. (12.107) and (12.108) the reflection coefficient is given by:

$$R = \left(\frac{BB^*}{AA^*}\right) = \frac{V_0^2 \sin^2 \beta a}{V_0^2 \sin^2 \beta a + 4E(E-V_0)} \qquad (12.123)$$

According to classical physics, the particle should not get reflected at all when its energy is greater than that of the barrier height. However, quantum mechanics predicts finite probability of reflection. Note that $T + R = 1$.

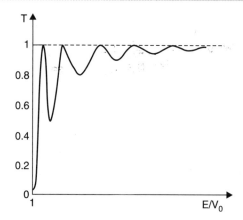

Fig. 12.37 Plot of transmission coefficient as function of particle energy $(E > V_0)$

Also note that $T = 1$, when $\sin(\beta a) = 0$ or $\beta a = n\pi$, $n = 1, 2, 3, 4......$

or
$$\frac{2\pi a}{\lambda_2} = n\pi \text{ or } \lambda_2 = \frac{2a}{n} \text{ or } a = n\left(\frac{\lambda_2}{2}\right) \tag{12.124}$$

i.e., when the width of the region is an integral multiple of half the de Broglie wavelength inside region 2

This behaviour presents resonant phenomena which does not occur in classical physics for particles. This situation does occur in wave optics. Total transmission of light through thin refracting layers is observed.

The condition for perfect transmission could also be expressed as follows:
$$\beta^2 a^2 = n^2 \pi^2$$

$$\frac{2m(E - V_0)a^2}{\hbar^2} = n^2 \pi^2$$

$$\Rightarrow \qquad (E - V_0) = n^2 \left(\frac{h^2}{8ma^2}\right) = n^2 E_1 \tag{12.125}$$

where E_1 is the ground state energy of a one dimensional box of width a and infinite height. Thus perfect transmission occurs when the difference between the energy of the particle and that of the barrier is equal to the integral square multiple of the ground state energy of the particle confined in an infinite well of width equal to the width of the barrier.

Case 2. When the energy of the particle is less than V_0 ($E < V_0$)

When the energy of the particle is less than that of the barrier height, the problem is analyzed in the same manner. However, note that in the region 2, the wave vector β is imaginary and hence the wave function in region 2 is an exponential function (Fig.12.38).

$$\beta = \sqrt{\frac{2m(E - V_0)}{\hbar^2}} = i\gamma \qquad \text{where } \gamma \text{ is } > 0 \tag{12.126}$$

In such a case,

$$\sin \beta a = \frac{e^{i\beta a} - e^{-i\beta a}}{2i} = \frac{e^{i(i\gamma)a} - e^{-i(i\gamma)a}}{2i} = \frac{-(e^{\gamma a} - e^{-\gamma a})}{2i} = i \sinh \gamma a \tag{12.127}$$

BASIC QUANTUM MECHANICS

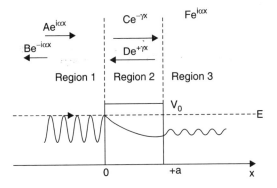

Fig. 12.38 Wave function for a particle encountering a square barrier of width a ($E < V_0$)

Here also it is convenient to define the parameter g which is characteristic of the barrier. It is defined as:

$$(\alpha a)^2 - (\beta a)^2 = (a\alpha)^2 - (i\gamma a)^2 = (\alpha a)^2 + (\gamma a)^2 = \frac{2mV_0 a^2}{\hbar^2} = g^2 \qquad (12.128)$$

This is the locus of a circle of radius g. The permitted values of a comprise a positive bounded continuum (Fig.12.39)

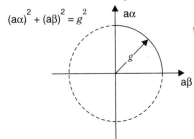

Fig. 12.39 Plot of $(\alpha a)^2 + (\gamma a)^2 = g^2$ (circle)

The transmission and reflection coefficients can be got using eqns. (12.124), (12.125) and (12.127)

$$R = \frac{V_0^2 \sinh^2 \gamma a}{V_0^2 \sinh^2 \gamma a + 4E(V_0 - E)} \qquad (12.129)$$

$$T = \frac{4E(V_0 - E)}{V_0^2 \sinh^2 \gamma a + 4E(V_0 - E)}$$

$$= \frac{4 \times (\alpha\gamma)^2 \times (\hbar^2/2m)^2}{V_0^2 \sinh^2 \gamma a + 4 \times (\alpha\gamma)^2 \times (\hbar^2/2m)^2} \qquad (12.130)$$

A plot of T versus γ is shown in Fig.12.40. The tunneling probability can be reduced to a simple expression under the following approximation $a >> d$. i.e., the barrier width is very much greater than the penetration depth. The tunneling is dominated by the exponential term ($e^{\alpha\gamma}$) and the polynomial term ($\alpha^2\gamma^2$) can be neglected.

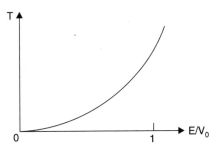

Fig. 12.40 Plot of tunneling probability as function of particle energy ($E < V_0$)

Let $\gamma = \dfrac{1}{2d} = \sqrt{\dfrac{2m(V_0 - E)}{\hbar^2}}$

$$= \frac{\sqrt{2m(V_0 - E)}}{\hbar}$$

$$\sinh(\gamma a) = \sinh\left(\frac{a}{2d}\right) \approx \frac{e^{\gamma a}}{2}$$

Substituting for α and γ from eqns. (12.107) and (12.126) in eqn. (12.130) we get

$$\therefore \quad T \approx \frac{4 \times (\alpha\gamma)^2 \times (\hbar^2/2m)^2}{V_0^2 \sinh^2 \gamma a} \approx \frac{16E(V_0 - E)}{V_0^2} e^{-2\gamma a}$$

Considering only the dominant exponential term we have

$$T \approx e^{-2\gamma a} \approx \exp\left(-\frac{a\sqrt{8m(V_0 - E)}}{\hbar}\right) \tag{12.131}$$

12.8.3 Barrier of Arbitrary Shape

Consider a particle encountering a potential of arbitrary shape as shown in Fig.12.41. The potential can be divided into small regions as shown in the figure. The total tunneling probability through the barrier can be taken as the product of the tunneling probability T_i through each region. *This is valid only if the potential is smoothly varying compared to the de Broglie wavelength of the particle.*

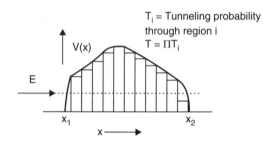

Fig. 12.41 Particle encountering a potential of arbitrary shape

$$T = \prod_i T_i \approx \exp\left[-\frac{\sqrt{8m}}{\hbar} \int_{x_1}^{x_2} [V(x) - E]^{1/2}\, dx\right] \tag{12.132}$$

12.9 APPLICATIONS OF TUNNELING

12.9.1 Field Emission Devices

Field emission devices are based on electron emission from a metal under the action of an electric field. This is also referred to as *cold emission.* Important technologies such as laptop computer displays, flat television displays and cockpit displays depend upon flat panel display technology.

Electrons need to have a minimum energy to escape from the surface of a metal. This is referred to as the work function (W) of the metal. In the quantum theory of metals, electron is treated as a particle in a three-dimensional box. Its energy is quantized. Fermi energy refers to the maximum energy that an electron can have at absolute zero. To a good approximation its variation with temperature can be neglected. The electron energy levels in a metal are shown in Fig. 12.42a. The Fermi energy of the electron is taken to be the reference. Electrons can be removed from the metal through irradiation of ultraviolet and X-rays. This is the familiar photoelectric effect. Electrons can also be emitted by heating the metal. This is known as *thermoionic emission.* Thermoionic emission is used in X-ray tubes and some of the electronic devices.

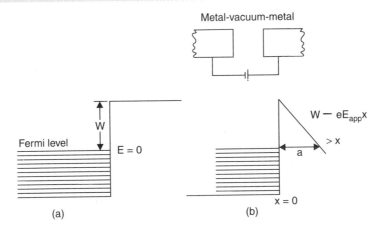

Fig. 12.42 Potential seen by an electron on the application of an electric field (E_{app})

Electrons can be removed at room temperature by the application of an electric field. This is because the external field changes the potential seen by the electron. This is shown in Fig.12.42b. The electrons in the metal on the left side see a triangular potential barrier. If the applied field is strong enough, they can tunnel through the barrier into the vacuum region, causing a current flow.

The tunneling probability is given by:

$$T = Ce^{\int_0^a \frac{\sqrt{8m(W - eE_{app}x)}}{\hbar} \cdot dx}$$

where a is the width of the barrier and is given by

$$a = \frac{W}{eE_{app}}$$

Let
$$I = \int_0^a \frac{\sqrt{8m(W - eE_{app}x)}}{\hbar} \cdot dx$$

Let $(a - x) = y^2$ so that $dx = -2\, y\, dy$

$$\therefore \quad I = \frac{\sqrt{32meE_{app}}}{\hbar} \int_0^{\sqrt{a}} y^2\, dy = \frac{4\sqrt{2}}{3} \sqrt{\frac{Wa^2}{\hbar^2}}$$

$$\therefore \quad T \approx C \exp(-I) \cong C \exp\left[-\frac{4\sqrt{2}a}{3\hbar} \sqrt{W}\right] \quad (12.133)$$

This is called the Fowler-Nordheim formula and explains the emission only qualitatively. When electrons are emitted, the metal acquires a positive charge which retards the emission. This retarding effect has not been taken into account in this derivation. This derivation also neglects the surface imperfections which change the electric field locally.

Figure 12.43 shows the display systems based on field emission devices. The metal or semiconductor emitters are formed into sharp tips to enhance the electric field produced on applying the potential. The voltage across each tip is controlled by the driver circuitry. The phosphor screen is separated from the tips by a very small gap (~ μm), so that the entire system is paper thin. Each pitzel is illuminated by its own set of tips, allowing a thin, low power consuming system.

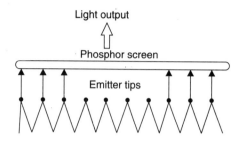

Fig. 12.43 Display system for field emission devices

12.9.2 Scanning Tunneling Microscope

The barrier tunneling of matter waves is best demonstrated in the scanning tunneling microscope. STM is a device for examining the surface of materials and is a useful tool in materials science. In a STM, a fine needle tip is scanned mechanically over the surface of the sample being investigated as shown in the Fig 12.44. Electrons from the sample tunnel through the gap between the sample and the needle and are recorded as a *tunnel current*. The tunnel current is a function of the gap between the sample and the needle. Since the surface is uneven *tunnel current* varies during the scan. Thus, from the variation of the tunneling current across the surface, the topography of the surface can be determined. STM is also operated in constant current mode. In this mode, a mechanism is provided that automatically moves the needle up or down during the scan, so as to keep the tunnel current and thus the gap constant. The needle's vertical position can then be displayed on a screen as a function of its location, producing a three-dimensional plot of the surface.

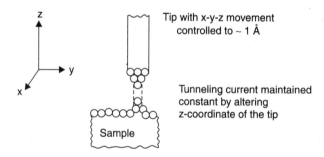

Fig. 12.44 Scanning tunneling microscope

12.9.3 Ohmic Contacts

In many semiconductor devices it is necessary to have electrode contacts which are ohmic. *i.e.*, the voltage varies linearly with current. An ohmic contact is a low resistance junction providing conduction in both directions between a metal and a semiconductor. Unlike a Schottky diode which is a rectifying metal-semiconductor contact, ohmic contact is non-rectifying. When a metal is deposited on a semiconductor a potential barrier is produced for the electron flow as shown in Fig.12.45a. The barrier is referred to as Schottky barrier. The height of the Schottky barrier is determined by the nature of the semiconductor surface and the metal. Typically, the barrier height $\phi_i = eV_{bi}$ is equal to the half of the band gap of the semiconductor. The width of the barrier is also called the depletion depth since in this region mobile charge carriers are

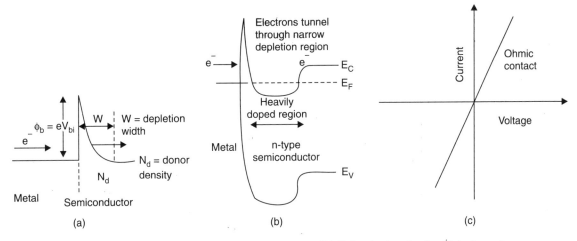

Fig. 12.45 Ohmic contact (a) Schottky barrier (b) Schottky barrier for high dopant concentration (c) I-V characteristic

swept away by the electric field. If N_d is the doping density in the semiconductor, the depletion width on the semiconductor side is

$$W = \sqrt{\frac{2\varepsilon V_{bi}}{eN_d}} \quad \therefore \quad \frac{W}{\sqrt{V_{bi}}} \alpha \frac{1}{\sqrt{N_d}} \qquad (12.134a)$$

where ε is the dielectric constant of the semiconductor. The depletion width is inversely proportional to the square root of the dopant density.

If the semiconductor is heavily doped near the interface region the depletion width could be made extremely narrow. The depletion width could be made so small that the electrons tunnel through the barrier with ease (Fig.12.45b). If A is the area of contact and R is its resistance, the specific resistance r_c of the ohmic contact is given by

$$r_c = R.A$$

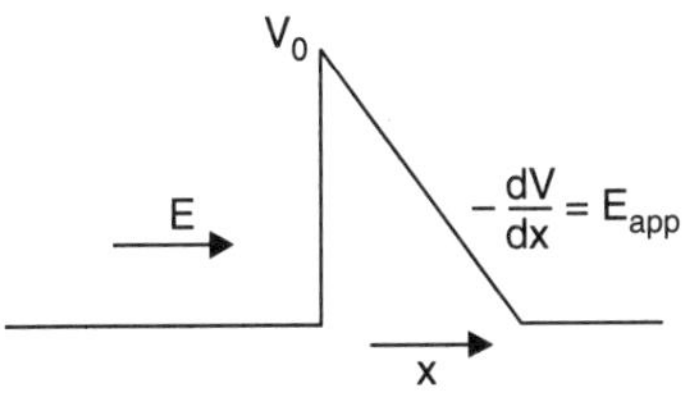

Note that the resistance of the contact is inversely proportional to the area. The figure of merit for an ohmic contact is given by $(\partial J/\partial V)_{V=0}$, where J is the current density and V is the applied voltage. It ought to be as small as possible. For an ohmic contact, at constant voltage, the current is inversely proportional to the contact resistance. The tunneling probability is directly proportional to the tunneling current. For a triangular potential barrier, the tunneling probability T for an applied field E_{app} is given by (Fig.12.46):

Fig. 12.46 Tunneling probability for a triangular barrier

$$T = \exp\left(\frac{-(V_0 - E)^{3/2}}{E_{app}}\right) \approx \exp\left(\frac{-V_{bi}^{3/2}}{E_{app}}\right) \qquad (12.134b)$$

For an ohmic contact

$$r_c \propto T^{-1} \Rightarrow \ln(r_c) \propto -\ln T \propto \frac{V_{bi}^{3/2}}{E_{app}} \tag{12.134c}$$

But $\quad E_{app} = \dfrac{V_{bi}}{W} \quad \therefore \quad \ln(r_c) \propto W\sqrt{V_{bi}} \tag{12.134d}$

From Eqns. (12.134a) and (12.134d)

$$\ln(r_c) \propto \frac{V_{bi}}{\sqrt{N_d}} \tag{12.134e}$$

Thus, the contact resistance can be reduced by using a low Schottky barrier height and doping the semiconductor as heavily as possible. The current-voltage relation of a good ohmic contact is highly linear (Fig.12.45c).

12.9.4 Zener Diode

Zener diode is a heavily doped *p-n* junction which is operated in the reverse bias. It is used as a voltage stabilizer. *On account of heavy doping the depletion region is very thin.* The energy band diagram of the *pn* junction in equilibrium and when it is in reverse bias is shown in Fig.12.47a. The I-V characteristic of the zener diode is shown in Fig.12.47b. At a particular voltage called the zener breakdown voltage, the current increases very rapidly. This feature of the characteristic enables it to be used as a voltage stabilizer. At the zener breakdown voltage the band alignment across the junction is such that electrons can tunnel across the barrier and hence the tunneling current is maximum.

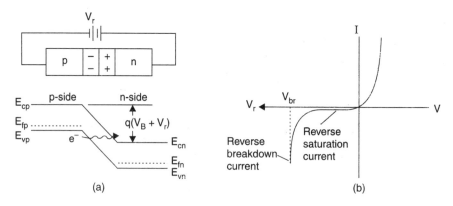

Fig. 12.47 Band diagram of *pn* junction in reverse bias and *I-V* characteristics

12.9.5 Tunnel Diode or Esaki Diode

The tunnel diode is also a heavily doped *pn* junction and operates in the forward bias. In the tunnel diode the energy bands are so aligned such that it exhibits *negative resistance, i.e.,* for a particular region of bias the current decreases with voltage. The operation of the device can be understood with respect to Fig.12.48. At a small forward bias, some of the electrons on the *n*-side can tunnel into the holes on the *p*-side. As the bias is increased, the band alignment is such that more and more of the electrons find holes to tunnel into and the current increases as shown. Beyond the point marked B, the empty states on the *p*-side move higher in energy with respect to the electron energies on the *n*-side. Hence, the tunneling current decreases.

Eventually, at point C, there are no empty states (holes) available for the electrons to tunnel into. This causes the current to decrease to a minimum value. At higher voltages, the current increases because the electrons and holes are able to overcome the reduced barrier due to thermal energy.

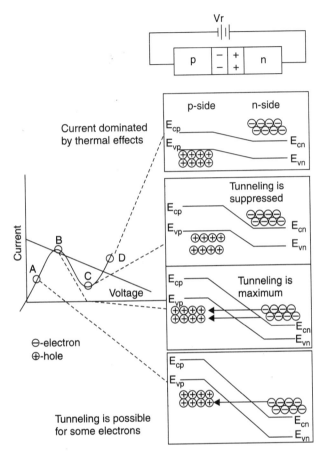

Fig. 12.48 Band diagram for a tunnel diode and the I-V characteristics

12.9.6 Resonant Tunneling

Resonant tunneling is important in semiconductor heterostructures. A typical resonant tunneling structure is shown in Fig.12.49a. It consists of regions A and B where the particle is *free* enclosing a series of potential barriers and wells. An important resonant tunneling structure is the double-barrier resonant tunneling structure which has form as shown in Fig.12.49b. The quantum well-enclosed by the two barriers has *quasi bound* states. If the barriers were infinitely thick, these levels would simply be the bound states in a quantum well. However, because of the finite barrier thickness, the quasi bound states have wave functions that *leak* out of the well region. When the energy of the particle approaches one of those quasi bound energies, the transmission coefficient approaches unity. This is shown in Fig.12.50.

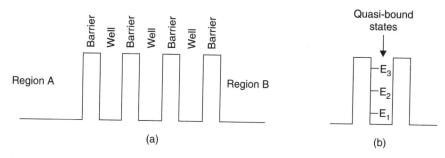

Fig. 12.49 (a) Particle encountering a series of potential barriers and potential wells (b) Double potential barrier enclosing a potential well

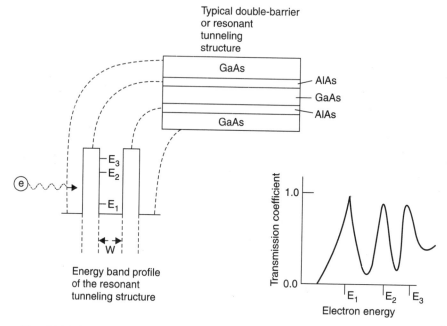

Fig. 12.50 Typical double heterostructure and its transmission characteristics

Semiconductor diodes incorporating the double barrier resonant structure are known as *resonant tunneling diodes*. The barrier height and width allows only the energy state E_1 to lie within the barrier height. The energy band diagram of a resonant tunneling diode is shown in Fig.12.51a. The typical I-V characteristic of the resonant tunnel diode is shown in Fig.12.51b. For small applied voltages, the tunneling probability of electrons with energies near the Fermi level energy is very small and as a result very little current flows through the structure. At point B, when the Fermi energy lines up with the quasi-bound state, a maximum amount of current flows through the structure, since the tunneling probability reaches unity. Further increasing the bias results in the energy band profile (point C), where the current through the structure decreases with increasing bias. Applying a larger bias results in a strong thermionic emission current and thus the current increases substantially.

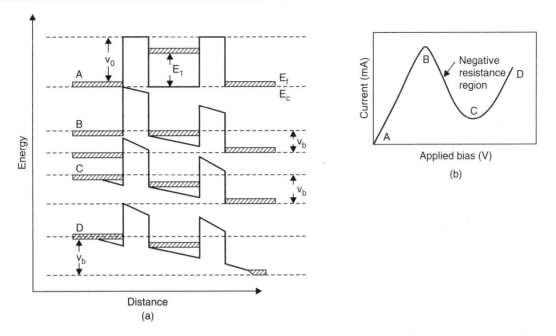

Fig. 12.51 Energy band alignment in resonant tunnel diode and its *I-V* characteristics

12.9.7 Josephson Junction and SQUID

Josephson junction refers to a thin insulating layer separating two superconductors (Fig.12.52a). In superconductors the current is carried by *cooper pairs*. A cooper pair refers to two electrons of opposite momentum and spin. A remarkable feature of such a junction is that a super current can flow even in the presence of the insulating layer. This is due to the tunneling of cooper pairs across the insulating layer. However, once the super current exceeds a certain value, a potential difference appears across the junction. *i.e.,* the insulating junction no longer behaves like a superconductor. The super current is given by

$$J = J_0 \sin \delta_0 \qquad (12.135)$$

Where J_0 is dependent upon tunneling probability and δ_0 refers to the phase difference between the wave functions of cooper pairs on either side of the junction. This is referred to as the *dc* Josephson effect.

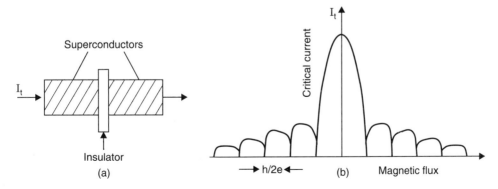

Fig. 12.52 (*a*) Josephson junction (*b*) Effect of magnetic field on the super current

When a voltage V is applied across a Josephson junction, the phase difference δ_0 increases with time at the rate

$$\nu = \frac{d\delta_0}{dt} = \frac{2eV}{h} \quad \text{or} \quad \omega = \frac{2eV}{\hbar} \tag{12.136}$$

As a result, the super current varies sinusoidally with time. This is called the *ac* Josephson effect. The value of $2e/h$ is 483.5979THz/volt. Since, ν is proportional to V and can be measured accurately (for instance by measuring the frequency of the electromagnetic radiation emitted by the junction) the *ac* Josephson effect enables very precise voltage determination. This effect is also used as a voltage standard. One volt is defined as that potential difference which when applied across the Josephson junction produces electromagnetic wave of frequency 483.5879THz.

The phase of the cooper pair can be changed by the application of magnetic field. Hence, interesting effects are observed with Josephson junctions. If a magnetic field is applied perpendicular to the junction, the value of the critical super current drops to zero whenever the flux through the junction is a multiple of flux quantum *i.e.*, $nh/2e$ (Fig.12.52b).

SQUID is an acronym for superconducting quantum interference device. It consists of two Josephson junctions in parallel (Fig.12.53a). Let the junction enclose a magnetic flux Φ. In such a device the total super current can be shown to be

$$J_{total} = 2J_0 \sin \delta_0 \cos\left(\frac{e\Phi}{\hbar}\right) \tag{12.137}$$

The current can be seen to very with Φ and has a maxima when

$$\frac{e\Phi}{\hbar} = n\pi \quad \text{or} \quad \Phi = n\left(\frac{h}{2e}\right) \tag{12.138}$$

Fig.12.53b shows the variation of current with the magnetic field. The control of current through a Josephson loop by a magnetic field is the basis of many important superconducting devices. It is important to note that the magnetic flux needed to alter the current through a the loop is very small. Hence SQUID is an extremely sensitive device for measuring changes in magnetic flux.

SQUID can be used in either a *dc* or *ac* configuration. In the *dc* configuration the Josephson loop encloses the flux Φ to be detected. The operation depends upon the fact that the maximum *dc* super current as well as the I-V relations depend upon the flux Φ. Note that the current is modulated by the applied magnetic field. This results in voltage getting modulated by the applied flux. This is shown in Fig.12.54. The *dc* device uses a constant current source in which case the voltage across the device oscillates with changes in the flux through the loop.

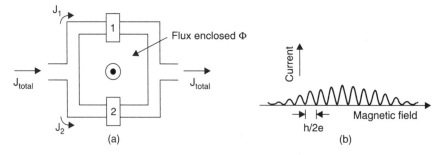

Fig. 12.53 (a) SQUID (b) Variation of tunneling super current as a function of magnetic field

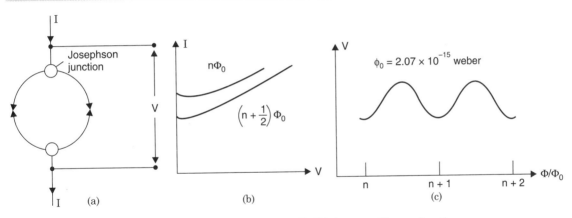

Fig. 12.54 (a) dc operation of SQUID (b) *I* versus *V* as a function of ϕ (c) *V* as a function of ϕ

The *rf* SQUID consists of a single Josephson junction incorporated into a superconducting loop and the circuit operates with an *rf* bias. The SQUID is coupled to the inductance of an LC circuit excited at its resonant frequency. Here the current induced in the inductor depends on the magnetic field enclosed by the Josephson loop. The *rf* voltage across the circuit versus the *rf* current is shown in Fig. 12.55 and oscillates with the applied flux.

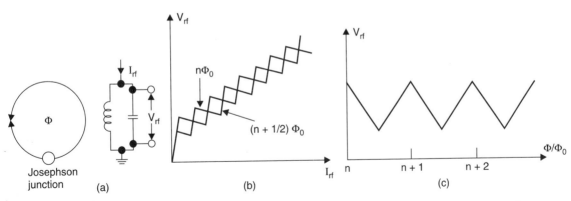

Fig. 12.55 (a) ac operation of SQUID (b) *V* versus *I* as a function of ϕ (c) *V* as a function of ϕ

Due to the extreme sensitivity of SQUID, the device (in various configurations) finds use in biomagnetism, geophysical exploration, gravitational experiments, Hall effect and many other fields.

12.9.8 Ammonia Maser

Maser is an acronym for microwave amplification by stimulated emission of radiation. The principle of maser is the same as that of laser. The radiation wavelength corresponds to that of microwaves and hence the name maser. Maser action in ammonia was observed in ammonia prior to the laser action in ruby laser.

Ammonia molecule is a pyramidal molecule with nitrogen forming the apex and the three hydrogens forming the base (Fig.12.56a). Symmetry considerations demand that the apex of the pyramid can be either above or below the plane formed by the three hydrogens. However, the two positions of nitrogen are separated by a potential barrier (Fig.12.56b).

According to classical mechanics, based on the strength of the N-H bonds, the nitrogen atom does not have sufficient energy to overcome this barrier. However, according to quantum mechanics the nitrogen atom can tunnel through the barrier and the nitrogen atom oscillates with a frequency of about 2.4×10^{10} Hz. This corresponds to a photon of energy 1 µeV. The transition is between two energy states E_1 and E_2. They correspond to the configurations in which the ammonia molecule has anti-parallel dipole moments. Radiation of this frequency is observed for ammonia and thus confirms the phenomena of tunneling. In the ammonia maser the population inversion and hence the stimulated emission is observed for this transition.

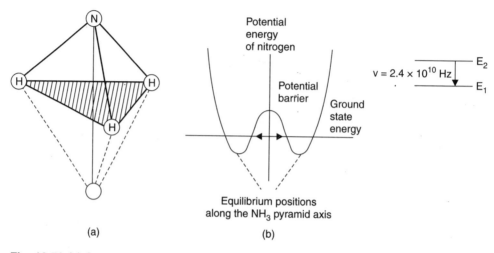

Fig. 12.56 (a) Structure of ammonia molecule (b) Potential barrier and tunneling of N atom

Since the energy separation of the levels is so small in thermal equilibrium the number of molecules in these two states are practically the same. However, when the beam of ammonia molecules is passed through an inhomogeneous electric field a separation of the molecules in the two levels can be achieved (Fig. 12.57). The force experienced by a molecule carrying a dipole moment µ is $\mu \times \partial E/\partial z$ where the gradient in the electric field is along the z-direction. This is because the dipole moments in these two states are anti-parallel. On passing through the inhomogeneous field, only molecules predominantly populated in E_2 comes out. This beam is passed through a cavity tuned to the required frequency of 2.4×10^{10} Hz. This results in stimulated emission of radiation at the same frequency. Power outputs of $\sim 10^{-10}$ W can be obtained with a line width as small as 10^{-2} Hz

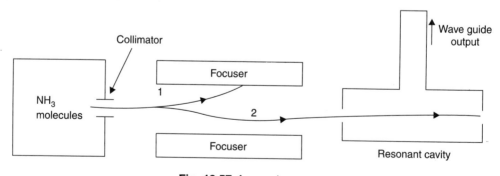

Fig. 12.57 Ammonia maser

12.9.9 Radioactive Decay

Many heavy nuclei exhibit α-radioactivity *i.e.,* they emit an alpha particle (helium nucleus). The radioactivity is described by the equation:

$N(t) = N(0)e^{-\lambda t}$, where the decay constant λ is related to the half life $t_{1/2}$ by the relation:

$$t_{1/2} = \frac{0.693}{\lambda}$$

The kinetic energies of alpha particles vary from 4 to 8MeV, while half life times vary over a factor of 10^{13}. The half life time is related to the kinetic energy (E) by the relation:

$$\ln t_{1/2} = A - \frac{B}{\sqrt{E}} \qquad (12.139)$$

where A and B are constants (Fig.12.58). On account of the strong nuclear force the alpha particle is trapped inside the nucleus. Classically, it does not have sufficient energy to surmount the barrier. However, the alpha particle can tunnel through the barrier.

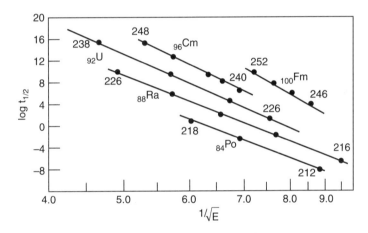

Fig. 12.58 Plot of half life versus the energy of the alpha particles

The experimental data of alpha radioactivity is explained on the basis of tunneling. Inside the nucleus an alpha particle feels the strong, short range attractive nuclear force as well as the repulsive Coulomb force. The shape of the potential is shown in Fig.12.59. The nuclear force dominates inside the nuclear radius r_0 and the potential can be approximated by a square well. At large distances from the nucleus, the potential energy is the electrostatic energy between a helium ion and the nucleus

Thus the general potential is of the form:

$$V(r) = -V_0, \text{ for } r < r_0$$

$$= \frac{2z'e^2}{4\pi\varepsilon_0 r}, \text{ for } r > r_0$$

Note that in the expression for Coulomb potential z' is the atomic number of the nucleus after the decay. The expression denotes the potential energy between the alpha particle and the nucleus after decay.

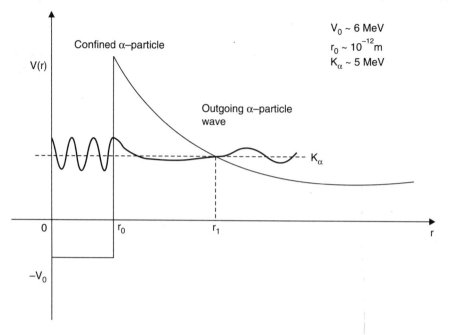

Fig. 12.59 Potential seen by an alpha particle. The square well corresponds to the nuclear force and the other to the columbic repulsion.

The tunneling probability is therefore given by eqn.(12.131):

$$\ln T = -\frac{-\sqrt{8m}}{\hbar}\int_{r_0}^{r_1}\left[\frac{2Z'e^2}{4\pi\varepsilon_0 r}-E\right]^{1/2}dr$$

The lower limit is $r = r_0$ and the upper limit is at $r = r_1$ where $V(r_1) = E$. Hence,

$$E = \frac{2Z'e^2}{4\pi\varepsilon_0 r_1} \quad \text{or} \quad r_1 = \frac{2Z'e^2}{4\pi\varepsilon_0 E} \qquad (12.140)$$

Using these limits we can evaluate the tunneling probability T is given by:

$$\ln T = \frac{-\sqrt{8mE}}{\hbar}\int_{r_0}^{r_1}\left(\frac{r_1}{r}-1\right)^{1/2}dr$$

Let $r = r_1 \cos^2\theta$ so that $\cos\theta = \sqrt{\frac{r}{r_1}}$ and $dr = -2r_1\cos\theta\sin\theta\, d\theta$

$$\therefore \quad \ln T = r_1 \int_0^{\theta_0} 2\sin^2\theta\, d\theta \quad \text{where } \theta_0 = \cos^{-1}\sqrt{\frac{r_0}{r_1}}$$

$$= r_1[\theta_0 - \sin\theta_0 \cos\theta_0]_0^{\theta_0}$$

$$= r_1\left[\cos^{-1}\sqrt{\frac{r_0}{r_1}} - \sqrt{\frac{r_0}{r_1}}\sqrt{1-\frac{r_0}{r_1}}\right]$$

$$\approx r_1 \left[\frac{\pi}{2} - \sqrt{\frac{r_0}{r_1}} - \sqrt{\frac{r_0}{r_1}} \right] = r_1 \left[\frac{\pi}{2} - 2\sqrt{\frac{r_0}{r_1}} \right] \text{ since } r_0 << r$$

Substituting for r_1 from eqn. (12.140) and $m = 4 \times 1.67 \times 10^{-27}$ kg, $e = 1.6 \times 10^{-19}$ C and $\varepsilon_0 = 8.87 \times 10^{-12}$ F/m we get

$$\ln T = 2.97 \, (Z')^{1/2} \, r_0^{1/2} - 3.95 Z' \left(\frac{1}{\sqrt{E}} \right) \quad (12.141)$$

where E is the units of MeV. Thus, tunneling is able to account for the experimental data on α-decay.

12.10 ELECTRONS IN A PERIODIC POTENTIAL OF A CRYSTAL—BAND THEORY OF CRYSTALLINE SOLIDS

12.10.1 Introduction

The behaviour of an electron in a crystalline solid can be analyzed using Schrödinger equation. The analysis leads to the concept of energy bands in solids. In a crystalline solid the atoms/molecules/molecular groups are arranged periodically in all the three dimensions. One dimensional periodic potential due to a row of atoms is shown in Fig. 12.60. To solve the Schrödinger equation for the potential shown is rather complicated. Hence a simplified model of the same potential, known as the Kronig-Penney model will be considered. In this analysis, the energy of electrons is shown to be confined to only particular energy bands. The bands are separated by forbidden energy regions. An electron can go over to the next band only if it possesses energy greater than the energy gap, separating the bands. Within a particular allowed band electrons behave much the same as free electrons. The band theory of solids has been most successful in explaining some of the anomalies predicted by the free electron model and also can account for the differing electrical properties of conductors, semiconductors and insulators. What determines the conduction properties of a particular material is whether the electronic states within an allowed energy band are empty or full. The analysis also brings out the quantum mechanical justification for introducing the concept of *hole*. It will be shown that the properties of a material with an almost filled band are identical to those of a material containing a few positive charge carriers called *holes* in an otherwise empty band. The behaviour of electrons and holes can be described by attributing an *effective mass* which takes care of the internal periodic field experienced by the electron.

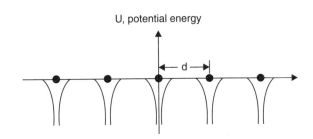

Fig. 12.60 Electrons in a periodic potential

12.10.2 Kronig-Penney Model

In the Kronig-Penney model, the periodic potential seen by the electrons is assumed to be a simple potential of the following form:

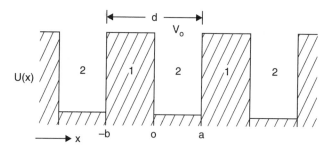

Fig. 12.61 Kronig-Penny model

$$U(x) = 0 \quad \text{for} \quad 0 \le x \le a$$
$$= V_0 \quad \text{for} \quad -b \le x \le 0 \tag{12.142}$$

It consists of a barrier of height V_0 and width b separated by a potential well of width a (Fig.12.61). The potential is repeated periodically with a periodicity of $d(= a + b)$. Region 2 corresponds to the potential well while region 1 corresponds to that of the barrier. The Schrödinger equation for the two regions can be written as:

Region 2 $(0 \le x \le a)$

$$\frac{d^2\psi_1}{dx^2} = -\frac{2mE}{\hbar^2}\psi_1 = -\alpha^2\psi_1 \quad \text{where} \quad \alpha = \sqrt{\frac{2mE}{\hbar^2}} \tag{12.143}$$

Region 1 $(-b \le x \le 0)$

$$\frac{d^2\psi_2}{dx^2} = -\frac{2m(E-V_0)}{\hbar^2}\psi_2 = -\beta^2\psi_2 \quad \text{where} \quad \beta = \sqrt{\frac{2m(E-V_0)}{\hbar^2}} \tag{12.144}$$

Further β is > 0 because E is > 0.

Equations (12.143) and (12.144) need to be solved simultaneously, a task which can be achieved only with considerable mathematical effort. Bloch showed that the solution of this type of equation has the following form:

$$\psi(x) = U(x)e^{ikx} \tag{12.145}$$

where $U(x)$ is a periodic function which possesses the periodicity of the lattice in the x direction. *Therefore, ψ has no longer a constant amplitude A as in equation (12.21) but has an amplitude which is modulated by the periodicity of the lattice. $\psi(x)$ in this form is known as the Bloch function.* Of course, ψ is different for various directions in the crystal lattice.

Since, the potential is periodic, the electron wave function satisfies the Bloch's theorem and is of the form:

$$\psi(x + d) = e^{i\phi}\psi(x) \quad \text{where} \quad \phi = k_x d \tag{12.146}$$

In the region, $-b < x < a$, the electron wave function has the form:

$$\psi_1 = Ae^{i\beta x} + Be^{-i\beta x}, \quad \text{for} \quad -b \le x \le 0 \tag{12.147}$$
$$\psi_2 = Ce^{i\alpha x} + De^{-i\alpha x}, \quad \text{for} \quad 0 \le x \le a \tag{12.148}$$

In the following period, $a < x < a + d$ according to eqn. (12.146), the wave function has the form:

$$\psi_1 = e^{i\phi}(Ae^{i\beta(x-d)} + Be^{-i\beta(x-d)}) \quad \text{for} \quad a \le x \le d \tag{12.149}$$

$$\psi_2 = e^{i\phi}(Ce^{i\alpha(x-d)} + De^{-i\alpha(x-d)}) \quad \text{for} \quad a \leq x \leq a+d \tag{12.150}$$

Applying the boundary conditions that the wave functions ought to be continuous at $x = 0$ and $x = a$, we have

At $x = 0$,
$$\psi_1 = \psi_2 \tag{12.151}$$

$$\left(\frac{d\psi_1}{dx}\right)_{x=0} = \left(\frac{d\psi_2}{dx}\right)_{x=0} \tag{12.152}$$

At $x = a$, \hfill (12.153)

ψ_1 (i.e., region 1 of next period) = ψ_2 (i.e., region 2 of previous period)

i.e., ψ_1 of eqn. (12.147) = ψ_2 of eqn. (12.150)

$$\left(\frac{d\psi_1}{dx}\right)_{x=a} = \left(\frac{d\psi_2}{dx}\right)_{x=a} \tag{12.154}$$

Corresponding to the boundary conditions specified in eqns. (12.151) to (12.154) we have the following equations:

$$A + B = C + D \tag{12.155}$$

$$\beta(A - B) = \alpha(C - D) \tag{12.156}$$

$$e^{i\phi}(Ae^{-i\beta b} + Be^{i\beta b}) = Ce^{i\alpha a} + De^{-i\alpha a} \quad (\because \; a - d = -b) \tag{12.157}$$

$$\beta e^{i\phi}(Ae^{-i\beta b} - Be^{i\beta b}) = \alpha(Ce^{i\alpha a} - De^{-i\alpha a}) \tag{12.158}$$

Equations (12.155) to (12.158) can be put in the following form:

$$A + B - C - D = 0 \tag{12.159}$$

$$\beta A - \beta B - \alpha C + \alpha D = 0 \tag{12.160}$$

$$e^{i\phi}e^{-i\beta b}A + e^{i\phi}e^{i\beta b}B - Ce^{i\alpha a} - De^{-i\alpha a} = 0 \tag{12.161}$$

$$\beta e^{i\phi}e^{-i\beta b}A - \beta e^{i\phi}e^{i\beta b}B - \alpha e^{i\alpha a}C + \alpha e^{-i\alpha a}D = 0 \tag{12.162}$$

Non-trivial solutions for A, B, C and D are obtained only if the coefficients formed by their determinant is zero. i.e.,

$$\begin{vmatrix} 1 & 1 & -1 & 1 \\ \beta & -\beta & -\alpha & \alpha \\ e^{i\phi}e^{-i\beta b} & e^{i\phi}e^{i\beta b} & -e^{i\alpha a} & -e^{-i\alpha a} \\ \beta e^{i\phi}e^{-i\beta b} & -\beta e^{i\phi}e^{i\beta b} & -\alpha e^{i\alpha a} & \alpha e^{-i\alpha a} \end{vmatrix} = 0$$

In the steps that follow the properties of determinants are made use for simplification.

Adding columns 1 and 2 as well as 3 and 4, we get

$$\begin{vmatrix} 1 & 2 & 0 & -1 \\ \beta & 0 & 0 & \alpha \\ e^{i\phi}e^{-i\beta b} & e^{i\phi}(e^{i\beta b} + e^{-i\beta b}) & -(e^{i\alpha a} + e^{-i\alpha a}) & -e^{-i\alpha a} \\ \beta e^{i\phi}e^{-i\beta b} & -\beta e^{i\phi}(e^{i\beta b} - e^{-i\beta b}) & -\alpha(e^{i\alpha a} - e^{-i\alpha a}) & \alpha e^{-i\alpha a} \end{vmatrix} = 0$$

Adding columns 1 and 4 as well as 2 and 3, we get

$$\begin{vmatrix} 1 & 2 & 0 & 0 \\ \beta & 0 & 0 & \alpha+\beta \\ e^{i\phi}e^{-i\beta b} & e^{i\phi}(e^{i\beta b}+e^{-i\beta b}) & e^{i\phi}(e^{i\beta b}+e^{-i\beta b})-(e^{i\alpha a}+e^{-i\alpha a}) & e^{i\phi}e^{-i\beta b}-e^{-i\alpha a} \\ \beta e^{i\phi}e^{-i\beta b} & -\beta e^{i\phi}(e^{i\beta b}-e^{-i\beta b}) & -\beta e^{i\phi}(e^{i\beta b}-e^{-i\beta b})-\alpha(e^{i\alpha a}-e^{-i\alpha a}) & \beta e^{i\phi}e^{-i\beta b}+\alpha e^{-i\alpha a} \end{vmatrix} = 0$$

On expanding the determinant we get,

$$\begin{vmatrix} 0 & 0 & \alpha+\beta \\ e^{i\phi}(e^{i\beta b}+e^{-i\beta b}) & e^{i\phi}(e^{i\beta b}+e^{-i\beta b})-(e^{i\alpha a}+e^{-i\alpha a}) & e^{i\phi}e^{-i\beta b}-e^{-i\alpha a} \\ -\beta e^{i\phi}(e^{i\beta b}-e^{-i\beta b}) & -\beta e^{i\phi}(e^{i\beta b}-e^{-i\beta b})-\alpha(e^{i\alpha a}-e^{-i\alpha a}) & \beta e^{i\phi}e^{-i\beta b}+\alpha e^{-i\alpha a} \end{vmatrix} +$$

$$(-2)\begin{vmatrix} \beta & 0 & \alpha+\beta \\ e^{i\phi}e^{-i\beta b} & e^{i\phi}(e^{i\beta b}+e^{-i\beta b})-(e^{i\alpha a}+e^{-i\alpha a}) & -e^{-i\alpha a}+e^{i\phi}e^{-i\beta b} \\ \beta e^{i\phi}e^{-i\beta b} & -\beta e^{i\phi}(e^{i\beta b}-e^{-i\beta b})-\alpha(e^{i\alpha a}-e^{-i\alpha a}) & \beta e^{i\phi}e^{-i\beta b}+\alpha e^{-i\alpha a} \end{vmatrix} = 0$$

On further expanding the determinants we get

$$(\alpha+\beta)e^{i\phi}\begin{vmatrix} (e^{i\beta b}+e^{-i\beta b}) & e^{i\phi}(e^{i\beta b}+e^{-i\beta b})-(e^{i\alpha a}+e^{-i\alpha a}) \\ -\beta(e^{i\beta b}-e^{-i\beta b}) & -\beta e^{i\phi}(e^{i\beta b}-e^{-i\beta b})-\alpha(e^{i\alpha a}-e^{-i\alpha a}) \end{vmatrix} +$$

$$(-2)\begin{vmatrix} \beta & 0 & \beta \\ e^{i\phi}e^{-i\beta b} & e^{i\phi}(e^{i\beta b}+e^{-i\beta b})-(e^{i\alpha a}+e^{-i\alpha a}) & e^{i\phi}e^{-i\beta b} \\ \beta e^{i\phi}e^{-i\beta b} & -\beta e^{i\phi}(e^{i\beta b}-e^{-i\beta b})-\alpha(e^{i\alpha a}-e^{-i\alpha a}) & \beta e^{i\phi}e^{-i\beta b} \end{vmatrix} +$$

$$(-2)\begin{vmatrix} \beta & 0 & \alpha \\ e^{i\phi}e^{-i\beta b} & e^{i\phi}(e^{i\beta b}+e^{-i\beta b})-(e^{i\alpha a}+e^{-i\alpha a}) & -e^{-i\alpha a} \\ \beta e^{i\phi}e^{-i\beta b} & -\beta e^{i\phi}(e^{i\beta b}-e^{-i\beta b})-\alpha(e^{i\alpha a}-e^{-i\alpha a}) & \alpha e^{-i\alpha a} \end{vmatrix} = 0$$

The second determinant is zero since the first and second columns are identical. Hence, we have to expand the first and the last determinants only. On doing so and rearranging the terms we get

$$-(\alpha+\beta)\alpha e^{i\phi}(e^{i\beta b}+e^{-i\beta b})(e^{i\alpha a}-e^{-i\alpha a}) - (\alpha+\beta)\beta e^{i\phi}(e^{i\beta b}-e^{-i\beta b})(e^{i\alpha a}+e^{-i\alpha a}) +$$

$$-2\alpha\beta e^{-i\alpha a}e^{i\phi}(e^{i\beta b}+e^{-i\beta b}) + 2\alpha\beta e^{-i\alpha a}(e^{i\alpha a}+e^{-i\alpha a}) + 2\beta^2 e^{-i\alpha a}e^{i\phi}(e^{i\beta b}-e^{-i\beta b}) +$$

$$+ 2\alpha\beta e^{-i\alpha a}(e^{i\alpha a}-e^{-i\alpha a}) + 2\alpha\beta e^{2i\phi}e^{-i\beta b}(e^{i\beta b}-e^{-i\beta b}) + 2\alpha^2 e^{i\phi}e^{-i\beta b}(e^{i\alpha a}-e^{-i\alpha a}) +$$

$$+ 2\alpha\beta e^{2i\phi}e^{-i\beta b}(e^{i\beta b}+e^{-i\beta b}) - 2\alpha\beta e^{i\phi}e^{-i\beta b}(e^{i\alpha a}+e^{-i\alpha a}) = 0$$

Multiplying the above equation throughout by $e^{-i\phi}$ and making use of the relations
$(e^{i\theta}+e^{-i\theta}) = 2\cos\theta$ and $(e^{i\theta}-e^{-i\theta}) = 2i\sin\theta$, we get

$$\cos\phi = \cos\alpha a \cos\beta b - \frac{\alpha^2+\beta^2}{2\alpha\beta}\sin\alpha a \sin\beta b \quad \text{for} \quad E > V_0 \qquad (12.163)$$

For the case of $E < V_0$ the same analysis can be carried out by putting
$\beta = i\gamma$ where γ is given by:

$$\gamma = \sqrt{\frac{2m(V_0 - E)}{\hbar^2}} = -i\beta.$$ Note that $\sin \beta b = i \sin h\, \gamma b$ and $\cos \beta b = \cosh \gamma b$

The equation corresponding to (12.163) is given by:

$$\cos \phi = \cos \alpha a \cosh \gamma b - \frac{\alpha^2 - \gamma^2}{2\alpha\gamma} \sin \alpha a \sinh \gamma b \quad \text{for } E < V_0 \tag{12.164}$$

The energy E which appears in eqns. (12.163) and (12.164) through α, β and γ is physically allowed only if $-1 < \cos \phi < 1$.

Consider the case, $E < V_0$, denoting the right hand side of eqn. (12.164) as $f(E)$, we get

$$f(E) = \cos\left(a\sqrt{\frac{2mE}{\hbar^2}}\right) \cosh\left(b\sqrt{\frac{2m(V_0 - E)}{\hbar^2}}\right)$$

$$+ \frac{V_0 - 2E}{2\sqrt{E(V_0 - E)}} \sin\left(a\sqrt{\frac{2mE}{\hbar^2}}\right) \sinh\left(b\sqrt{\frac{2m(V_0 - E)}{\hbar^2}}\right) \tag{12.165}$$

This function must lie between -1 and $+1$ since it is equal to $\cos \phi$ i.e., $\cos k_x d$. We wish to find the relation between E and ϕ or E and k_x. In general this can be found using a numerical method employing computer algorithm indicated in Fig.12.62. This would give the value of ϕ for every allowed value of E. The typical result is shown in Fig.12.63a. We see from the figure that $f(E)$ remains between $+1$ and -1 only for certain regions of energies. These allowed energies from the allowed bands are separated by band gaps. The Fig.12.63a can be recast in the form of Fig.12.63b indicating the allowed energies and the corresponding k values. The energies between E_2 and E_1 form the first allowed band, the energies between E_4 and E_3 form the second band while the energies between E_2 and E_3 correspond to energy gap.

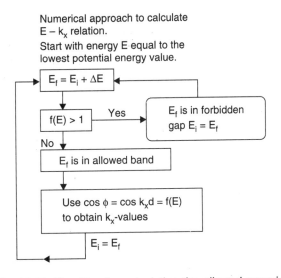

Fig. 12.62 Algorithm for calculating the allowed energies

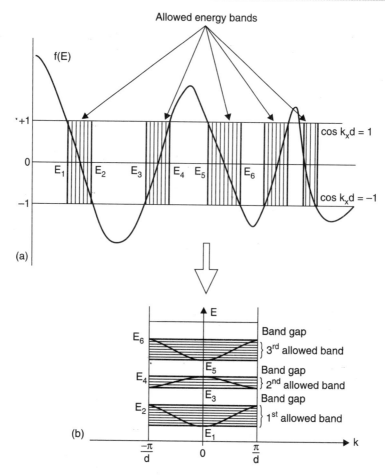

Fig. 12.63 (a) Graphical solution to obtain energy levels. The function f(E) is plotted as a function of E. Only energies for which f(E) lies between +1 and –1 are allowed
(b) The allowed and forbidden bands are plotted in the E versus k relation using the results of (a)

A simplification of the eqn. (12.164) is possible under the following assumption. In the limit $b \to 0$ and $V_0 \to \infty$ while bV_0 remains finite and very small, $\cosh(b\gamma) \approx 1$ and $\sinh(b\gamma) \approx b\gamma$. Further neglecting α^2 in comparison to γ^2 since $V_0 >> E$, we get

$$\cos\phi = \cos\alpha a \cdot 1 + \frac{\gamma^2 - \alpha^2}{2\alpha\gamma}(\gamma b)(\sin\alpha a) \Rightarrow \cos\alpha a + \frac{\gamma^2 ab}{2}\left(\frac{\sin\alpha a}{\alpha a}\right) = \cos k_x \cdot d$$

i.e.,
$$\left(\frac{mV_0 ba}{\hbar^2}\right)\frac{\sin\alpha a}{\alpha a} + \cos\alpha a = \cos k_x d$$

i.e.,
$$P \frac{\sin\alpha a}{\alpha a} + \cos\alpha a = \cos k_x d \text{ where } P = \frac{mV_0 ba}{\hbar^2} \quad (12.166)$$

This is the desired relation which provides the allowed solutions to the Schrödinger equations (12.143) and (12.144). We notice that the boundary conditions lead to an equation with trigonometric functions. Therefore, only certain values of α are possible. This in turn means that because of eqn. (12.143) only certain values for the energy E are defined.

Equation (12.166) can be used to discuss two extreme cases.

Case 1, When the barrier strength $V_0 b$ or P is very small

If the potential barrier strength becomes smaller and smaller and finally disappears completely, P goes toward zero, and one obtains from eqn. (12.165) equation

$$\cos \alpha a = \cos ka \qquad (12.167)$$

or $\alpha = k$. Hence, it follows from eqn. (12.143), that

$$E = \frac{\hbar^2 k^2}{2m}$$

This corresponds to free electrons, which is identical to eqn. (12.23)

Case 2. When the barrier strength or P is very large

If the potential barrier strength is very large, P approaches infinity. However, because the left-hand side of eqn. (12.165) has to stay within the limits + 1 and – 1 *i.e.*, it has to remain finite, it follows that

$$\frac{\sin \alpha a}{\alpha a} \to 0.$$

That is, $\sin \alpha a \to 0$. This is only possible if $\alpha a = n\pi$ \qquad (12.168)

From eqns. (12.168) and (12.144) it follows that

$$E = \frac{n^2 h^2}{8ma^2}$$

which is the result of eqn. (12.34) corresponding to the energy levels of a particle in a potential of infinite height.

In summary if the electrons are strongly bound, *i.e.*, if the potential barrier is very large, one obtains sharp energy levels. This corresponds to the electron in the potential field of a single ion. If the electron is not bound, it can have continuous energy. If the electron moves in a periodic potential field one obtains energy bands (solid). The widening of the energy levels into energy bands and the transition into a quasi-continuous energy region is the essence of band theory of solids. This widening occurs because the atoms increasingly interact as their mutual distance decreases. This is depicted in Fig.12.64.

It is impossible for an electron to possess energy corresponding to that in a forbidden band. However, an interesting consequence of the analysis is that, within the allowed energy bands, traveling wave solutions exist that are not attenuated. This implies that there is no electron scattering in the uniform lattice of a perfect crystal and within an allowed band an electron can move in a completely unrestricted manner. However, in real solids on account of thermal vibrations and impurities, the perfect periodicity of the lattice is destroyed. Hence electrons get scattered and attenuated. This gives rise to Joule heating.

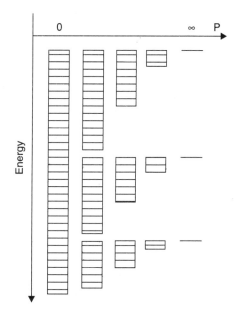

Fig. 12.64 The formation of bands as a function of the strength *P* of the barrier potential

12.11 PERIODIC NATURE OF ENERGIES AND THE BRILLOUIN ZONE

Note that $\phi = k_x d$ term on the left hand side equation (12.164) appears as the argument of a cosine function which is periodic in $2n\pi$. Hence, if $k_x d$ corresponds to a certain allowed electron energy, then $k_x d + 2n\pi$ is also allowed. *This reflects the periodic nature of the lattice.* It is customary to show the $E - k$ relation for the smallest k-values. Each region between the discontinuities in eqn. (12.164) is called a *Brillouin zone*. The discontinuities occur at the edges of the Brillouin zones. This is shown in Fig.12.65.

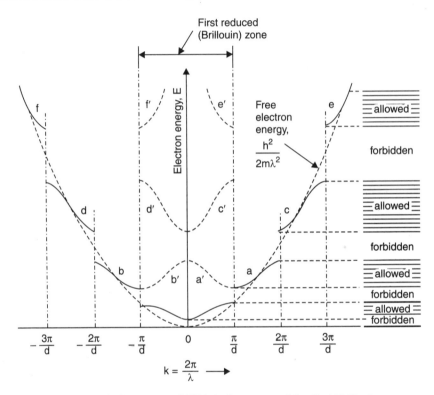

Fig. 12.65 Periodic nature of E Vs k diagram and the first Brillouin zone

12.12 STANDING WAVES AT THE BAND EDGES AND THE ORIGIN OF BAND GAPS

When $\phi = kd = n\pi$, $e^{ikd} = (-1)^n$ the wave functions ψ_1 and ψ_2 are not traveling waves. Instead they are standing waves. This is because the amplitude of the forward moving and the backward moving waves are the same as will be evident from the following argument. This actually satisfies the condition for Bragg reflection.

Equations (12.155) to (12.158) corresponding to the boundary conditions can be written as:

$$A + B = C + D \tag{12.169}$$

$$\beta(A - B) = \alpha(C - D) \tag{12.170}$$

$$(Ae^{-i\beta b} + Be^{i\beta b}) = Ce^{i\alpha a} + De^{-i\alpha a} \tag{12.171}$$

$$\beta(Ae^{-i\beta b} - Be^{i\beta b}) = \alpha(Ce^{i\alpha a} - De^{-\alpha a}) \tag{12.172}$$

Eqn. (12.169) can be written as
$$\alpha(A + B) = \alpha(C + D) \tag{12.173}$$
From eqns. (12.173) and (12.170)
$$A(\alpha + \beta) + B(\alpha - \beta) = 2\alpha C \quad \text{or} \quad e^{i\alpha a}[A(\alpha + \beta) + B(\alpha - \beta)] = 2\alpha e^{i\alpha a} C \tag{12.174}$$
Multiplying eqn. (12.171) by α and adding to eqn. (12.172) we get
$$A(\alpha + \beta)e^{-i\beta b} + B(\alpha - \beta)e^{i\beta b} = 2\alpha e^{i\alpha a} C \tag{12.175}$$
Equating eqns. (12.174) and (12.175) and rearranging the terms we get
$$\frac{A}{B} = \frac{(\alpha - \beta)(e^{i\beta b} - e^{i\alpha a})}{(\alpha + \beta)(e^{i\alpha b} - e^{i\beta b})}$$

$$\therefore \quad \frac{AA^*}{BB^*} = \frac{A^2}{B^2} = \frac{(\alpha - \beta)^2(e^{-i\beta b} - e^{i\alpha a})e^{-i\beta b} - e^{i\alpha a})}{(\alpha + \beta)^2(e^{i\alpha b} - e^{i\beta b})(e^{-i\alpha a} - e^{i\beta b})}$$

$$= \frac{(\alpha - \beta)^2\,[2 - (e^{i(\beta b - \alpha a)} + e^{-i(\beta b - \alpha a)})]}{(\alpha + \beta)^2\,[2 - (e^{i(\beta b + \xi a)} + e^{-i(\beta b + \alpha a)})]} = \frac{(\alpha - \beta)^2\,[1 - \cos(\beta b - \alpha a)]}{(\alpha + \beta)^2\,[1 - \cos(\beta b + \alpha a)]}$$

$$= \frac{(\alpha - \beta)^2(1 - \cos\alpha a \cos\beta b - \sin\alpha a \sin\beta b)}{(\alpha + \beta)^2(1 - \cos\beta b \cos\alpha a + \sin\beta b \sin\alpha a)} \tag{12.176}$$

At the boundary, $\cos\phi = 1$. Hence, from eqn. (12.164) we get
$$1 = \cos(\alpha a)\cos(\beta b) - \frac{\alpha^2 + \beta^2}{2\alpha\beta}\sin(\beta a)\sin(\beta b) \tag{12.177}$$

From eqns. (12.176) and (12.177)
$$A^2 = B^2 \quad \text{or} \quad A = +B \quad \text{or} \quad A = -B$$
Similarly, it can be shown that $C = +D$ or $C = -D$
Hence, the wave functions can be written as
$$\psi(+) = e^{\frac{i\pi x}{d}} + e^{\frac{i\pi x}{d}} = 2\cos\left(\frac{\pi x}{d}\right)$$

$$\psi(-) = e^{\frac{i\pi x}{d}} - e^{\frac{i\pi x}{d}} = 2i\sin\left(\frac{\pi x}{d}\right)$$

This form results from Bragg reflection. We can form two standing waves. They are labeled (+) or (−) according as whether or not they change sign when $-x$ is substituted for x.

The two standing waves $\psi(+)$ and $\psi(-)$ pile up electrons at different regions as shown in Fig.12.66. Hence, the two waves have different values of the potential energy. This is the origin of energy gap. The charge density is given by $\psi^*\psi$. For a pure traveling wave the charge density is constant throughout. The charge densities $\rho(+)$ and $\rho(-)$ for the two waves are given by:

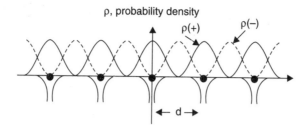

Fig. 12.66 Standing waves of electrons with at the edge of the Brillouin zone ($k = \pi/d$)

$$\rho(+) \propto \cos^2\left(\frac{\pi x}{d}\right) \quad \text{and} \quad \rho(-) \propto \sin^2\left(\frac{\pi x}{d}\right)$$

For $\rho(+)$ the electrons get piled up just above the ion cores while for $\rho(-)$ they get piled up in between the ion cores. $\rho(+)$ is less than that of the traveling wave while $\rho(-)$ is higher than that of the traveling wave. The difference in their potential energies corresponds to the energy gap. The magnitude of the energy gap can be shown to be equal to

$$E_g = \frac{\hbar^2}{2m}(\alpha^2 - \beta^2) \tag{12.178}$$

The case of Kronig-Penny model refers to a periodic potential which is linear and infinite. It can be made more realistic by making it three-dimensional. Since in bulk samples the surface atoms do not contribute much it is assumed that the potential function lies on a circle of radius which is very large compared to the distance between the ion sites d. In such a case, the wave function repeats itself after N unit cells. This is shown in Fig. 12.67. This is often referred to as the cyclic boundary condition. Thus, in the expression kd = $N\pi$, N refers to the number of unit cells and it is usually large. In each of the Brillouin zones there are N values for k.

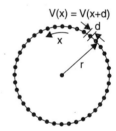

Fig. 12.67 Cyclic boundary condition

12.13 BAND THEORY OF ELECTRICAL CONDUCTION IN SOLIDS

Quantum mechanics has been successful in explaining the various properties of solids. We will discuss only the electrical conduction in solids. According to quantum theory, when an allowed band is completely filled with electrons, the electrons in the band cannot carry current. Electrons are fermions and they need empty energy states to occupy. On applying the electric field, the electron gains energy. But there are no available energy levels for the electron. Hence, electrons cannot conduct current. On the other hand, a material with a partial filled band can be a good conductor and this is the situation with a metal.

In the case of both semiconductors and insulators, the band is completely filled at 0K. The band which is completely filled is called the valence band and the next band is called the conduction band. In the case of a semiconductor, the gap between the filled band and the next completely unfilled band is very small. It is 0.72eV for Ge and 1.11eV for Si. At a temperature T, the electrons have a thermal energy of ~ $T \times 8.6 \times 10^{-5}$ eV. Thus, at room temperature, some of them get excited to the conduction band. This number is ~7×10^{10} per cm^3 for Si and ~ 2.5×10^{13}. These number densities account for the measured conductivities. Such materials are known as *intrinsic semiconductors*. The band gap for an insulator is large. Hence, very large electric fields are needed to excite the electrons into the conduction band. The energy bands for a metal, semiconductor and an insulator is shown in Fig.12.68.

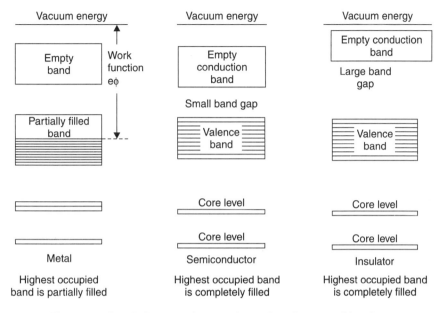

Fig. 12.68 Band diagram of a metal, semiconductor and insulator

A semiconductor acts like an insulator at sufficiently low temperatures. Its conductivity begins to increase with the increase of temperature. In semiconductors the conductivity can be drastically changed by doping. A doped semiconductor is known as an *extrinsic semiconductor*. Doping provides additional energy levels in the otherwise forbidden energy gap. Donor levels are close to the conduction band while the acceptor levels are near the valence band (Fig.12.69).

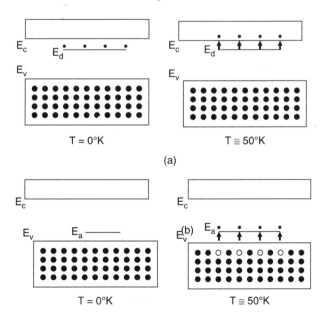

Fig. 12.69 Additional energy levels in the band gap of an extrinsic semiconductor
(a) n-type at $T = 0$ and 50 K (b) p-type at $T = 0$ and 50 K.

For insulators the energy gap is much larger than those of semiconductors. For diamond it is 6eV. Note also that the photons in the visible spectrum do not have sufficient energy to excite the valence electrons to the conduction band. Hence, diamond is transparent in visible light.

The metals have partially filled bands. They have very high conductivity because of the very large number of electrons can participate in the current transport. Empty energy levels in the partially filled band allow the electrons to gain energy from the electric field. Unlike semiconductors, it is difficult to alter their conductivity.

12.13.1 Velocity and Effective Mass of Electrons in a Solid

A consequence of the de Broglie hypothesis is the new definition of mass of a particle. In terms of its wave nature. The kinetic energy of the particle is given by

$$E = \frac{p^2}{2m} = \frac{(\hbar k)^2}{2m} = \frac{\hbar^2 k^2}{2m}$$

$$\Rightarrow \quad \frac{dE}{dk} = \frac{\hbar^2 k}{m} \quad \text{and} \quad \frac{d^2 E}{dk^2} = \frac{\hbar^2}{m}$$

$$\Rightarrow \quad m = \hbar^2 \left[\frac{d^2 E}{dk^2}\right]^{-1} \tag{12.179}$$

The above expression for the effective mass can also be obtained in the following way. On applying an electric field E_{app}, the electron gains an energy dE. This is given by

$$dE = eE_{app}\, dx = eE_{app}\, \frac{dx}{dt}\, dt = eE_{app}\, v_g\, dt \tag{12.180a}$$

But,

$$dE = \hbar d\omega = \hbar\, \frac{d\omega}{dk}\, dk = \hbar v_g\, dk \tag{12.180b}$$

From eqns. (12.180a) and (12.180b)

$$eE_{app} v_g\, dt = \hbar v_g\, dk \quad \Rightarrow \quad \hbar\, \frac{dk}{dt} = \frac{dp}{dt} = eE_{app} \tag{12.180c}$$

This is in the form of Newton's law of motion. Further, we have

$$v_g = \frac{d\omega}{dk} = \frac{1}{\hbar}\frac{dE}{dk}$$

$$\Rightarrow \quad \frac{dv_g}{dt} = \frac{1}{\hbar}\frac{d^2 E}{dt\, dk} = \frac{1}{\hbar}\frac{d^2 E}{dk\, dt} = \frac{1}{\hbar}\frac{d^2 E}{dk^2}\frac{dk}{dt} = \frac{1}{\hbar^2}\frac{d^2 E}{dk^2} eE_{app}$$

We have $\dfrac{dv}{dt} = \dfrac{eE_{app}}{m^*}$ (in the particle picture) $= \dfrac{dv_g}{dt}$ \hfill (12.181)

$$\therefore \quad \frac{1}{m^*} = \frac{1}{\hbar^2}\frac{d^2 E}{dk^2} \quad \Rightarrow \quad m^* = \hbar^2 \left(\frac{d^2 E}{dk^2}\right)^{-1}$$

The electron moves under the influence of internal forces, exerted on it by the ions of the lattice and an external force, exerted on it by the applied electric field E_{app}. We can use eqn. (12.180c) to discuss the motion *in terms of the external force alone* since that equation is in the form of Newton's law. The effects of the internal forces are contained in the equation. They appear only in the effective mass m^*, which has values different form the actual electron mass in free space.

Thus, mass of an electron wave packet can be defined in terms of the second derivative of E vs k curve. The group velocity of the wave packet is given by

$$v_g = \frac{d\omega}{dk} = \frac{1}{\hbar}\left(\frac{dE}{dk}\right) = \frac{dE}{dp}$$

The variation of the group velocity of the electron, (dE/dK) and its effective mass is shown in Fig.12.70.

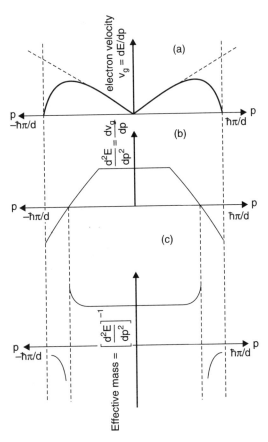

Fig. 12.70 Variation of group velocity, dE/dk and effective mass in the first Brillouin zone

Note that the velocity of the electron becomes zero at the band edge. This is in agreement with the finding the electronic wave functions become standing waves at the top and the bottom of the band. Note also that for some k values the mass of the electron is also negative. The negative mass has to be understood in the following sense.

Consider an electron having E and k values corresponding to the point a on the E–k curve (Fig.12.71). If an electric field is applied in the direction shown, the electron will get accelerated and move to the right on the diagram from point a to some point b, where both its energy and its velocity have increased. This conventional behaviour corresponds to *positive effective mass*. Now, consider an electron at the upper end of a band, at c. When the field is applied, electron moves to the point d, where its energy is increased but its velocity has decreased. The electron appears to have been decelerated by the previously accelerating force and this can be explained as saying that the electron has a negative mass. Negative effective masses occur whenever E–k curve is concave downwards and that the electron mass is positive whenever the curve is concave upwards.

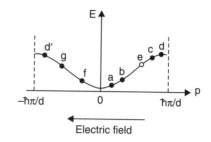

Fig. 12.71 E-p diagram of a hypothetical solid

The direction of acceleration of an electron is determined by the sign of both its effective mass and its charge. An alternative way of accounting for the properties of an electron with negative mass is to consider it as a particle with a positive mass but having a positive charge $+e$. i.e.,

$$\text{Acceleration} = \frac{(-e)E_{app}}{(-m_e^*)} = \frac{(+e)E_{app}}{(+m_h^*)}$$

Thus, when electrons at the top of a band are acted on by an applied field the resulting currents correspond to the movement of particles with a positive charge $+e$ and a positive effective mass m_h. These are called as *holes*.

Alternate Definition of Hole

Semiconductors are materials in which the valence band is full of electrons and the conduction band is empty at 0K. At finite temperatures some of the electrons leave the valence band and occupy the conduction band. The valence band is then left with some unoccupied states. Consider the situation shown in Fig.12.72 where an electron with the k_e vector is missing from the valence band. When all the valence band states are occupied, the sum over all wave vector states is zero; i.e.,

$$\sum k_i = 0 = k_e + \sum_{i \neq e} k_i$$

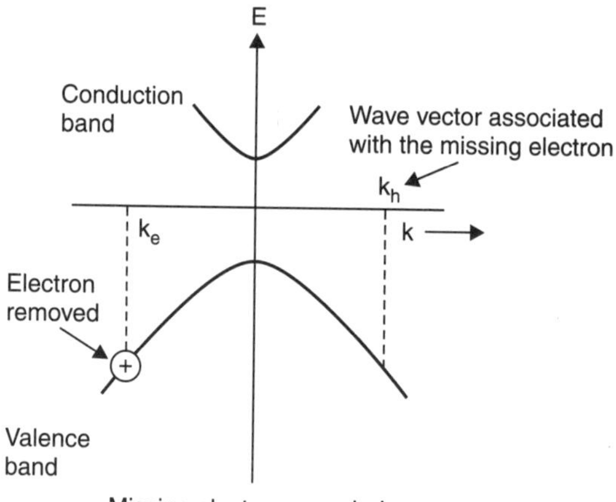

This result is just an indication that there are as many positive k states occupied as there are negative ones. When the electron having wave vector k_e is missing the total wave vector is

$$\sum_{i \neq e} k_i = -k_e$$

Fig. 12.72 Wave vector of the missing electron and the hole

BASIC QUANTUM MECHANICS

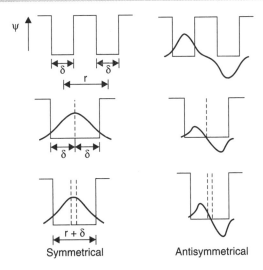

Fig. 12.73 Two atom systems and the possible wave functions

The missing state is called a hole and the wave vector of the system $-k_e$ is attributed to it. It is important to note that the electron is missing from the state k_e and the momentum associated with the hole is at $-k_e$. The position of the hole is depicted as that of the missing electron. But in reality the hole wave vector is $-k_e$ as shown in Fig.12.72.

i.e., $$k_h = -k_e$$

Note that the hole is a representation for the valence band with a missing electron. If the electron is not missing the valence band electrons cannot carry any current. However, if an electron is missing the current flow is allowed. If an electric field is applied, all the electrons move in the direction opposite to the electric field. This results in the unoccupied state moving in the direction of the electric field. *The hole responds as if it had a positive charge*. It therefore responds to external electric and magnetic fields E and B respectively, according to the equation of motion

$$\hbar \frac{dk_h}{dt} = e\,[E + v_h \times B]$$

where $\hbar k_h$ and v_h are the momentum and the velocity of the hole.

Thus the equation of motion of holes is that of particles with a positive charge e. The mass of a hole has a positive value, although the effective electron mass in its valence band is negative. When we discuss the conduction band properties of semiconductors or insulators we refer to electrons but when we discuss the valence band properties, we refer to holes. This is because in the valence band only missing electrons or holes lead to charge transport and current flow.

Energy Bands in Solids—An Alternate Approach

The basic results obtained from Kronig Penney model could also be obtained by examining the how energy levels of individual atoms get modified when they are brought together. Fig.12.73 shows the wave functions of two isolated atoms when they are brought together. The possible energy values for the ground state as a function of interatomic distance are shown in Fig.12.74. The ground state energy is split into two and the difference in energy decreases as the interatomic distance increases. Similarly in the case of three atoms, the wave functions and the energy state splittings for the ground state are shown in Figs.12.75 and Fig.12.76. The extension of the argument is straight forward and if N atoms are brought together the energy level is split into N levels. This gives rise to a band (Fig.12.77). Typically, in a solid, the number of interacting atoms is ~10^{22} and typically the width of the band is ~ eV. Since, in this case 10^{22} levels have to be accommodated in an energy range that is only ~ eV, the individual levels in a band are necessarily very closed spaced together. The allowed energy levels within a band are therefore said to be *quasi continuous*. The band formed from the overlap of *s* orbitals is called the *s band*. If the atoms have available *p* orbitals the same procedure gives rise to *p* band. If the atomic orbitals lie higher in energy than *s* orbitals, then the *p band* will be higher than the *s band*. Thus, there may arise a band gap.

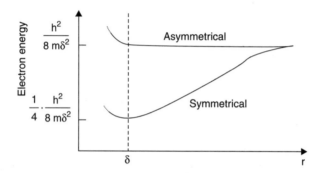

Fig. 12.74 Energy of the lowest states as a function of distance

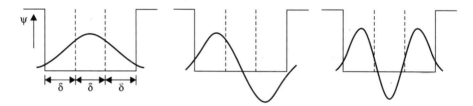

Fig. 12.75 Three atom systems and possible wave functions

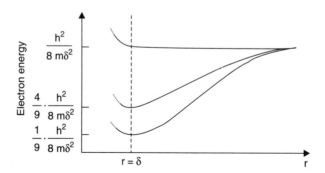

Fig. 12.76 Energy as a function of atomic separation

BASIC QUANTUM MECHANICS

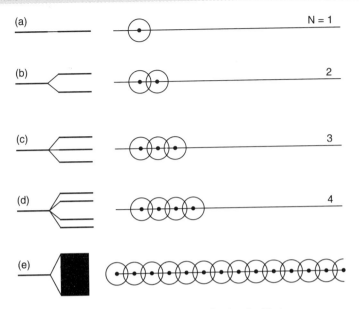

Fig. 12.77 Formation of bands due to *N*-atoms

This approach can be appreciated by considering the case of diamond. The electronic structure of a single carbon atom is $1s^2\, 2s^2\, 2p^2$. Thus, its inner principal shell is filled but there are only four electrons in its outer shell and there are four vacancies in the outer subshell. If we first of all consider a *gas* of such atoms with the interatomic spacing or lattice constant a so large that no interaction occurs between them, then the energy levels of each atom are shown in Fig.12.78 (i). As the carbon atoms are brought into closer proximity (*i.e.*, a is reduced) level splitting occurs as described which results in bands of allowed energies at (ii). For even closer spacings as at (iii), the bands can overlap. Eventually as a is reduced still further, the energies of the outer shell electrons can lie in one of the two bands, separated by a forbidden gap as in (iv)

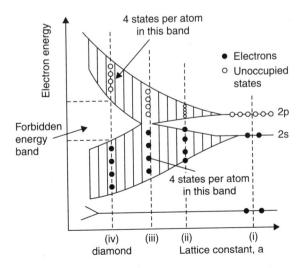

Fig. 12.78 Energy bands for carbon with varying atomic spacing

12.14 ANALOGUE OF KRONIG-PENNEY MODEL

An insight into the Kronig-Penney model can be obtained by considering similar problems in wave propagation. Consider the coaxial transmission line as shown in Fig.12.79. The voltage at any point x on the line is given by

$$\frac{d^2V}{dx^2} + \omega^2 \mu_0 \varepsilon_r \varepsilon_0 \, V = 0$$

If the line is air filled but periodically loaded at intervals of a with dielectric discs of relative permittivity ε_r and thickness w, then the voltage equation becomes similar in form to eqns. (12.143) and (12.144) where $\omega^2 \varepsilon_0 \mu_0 = \beta^2$ and $\omega^2 \varepsilon_0 \varepsilon_r \mu_0 = -\alpha^2$. Hence, in the circuit analogy, voltage V is equivalent to electron energy E and frequency bands are analogous to electron energy bands. The ω versus k diagram for the loaded line can be obtained by solving eqns. using the appropriate boundary conditions. This results in a series of stop and pass bands of frequency that occur when $ka = \pi, 2\pi, 3\pi, 4\pi, 5\pi$ ... The stop bands arise because when $ka = n\pi$ the electromagnetic waves reflected from the successive discs add in phase as shown in Fig.12.79. Then even if individual reflections are weak, their combined effect is to produce total reflection. Hence, for this condition no traveling wave solution exists for the voltage and only standing wave solutions are possible.

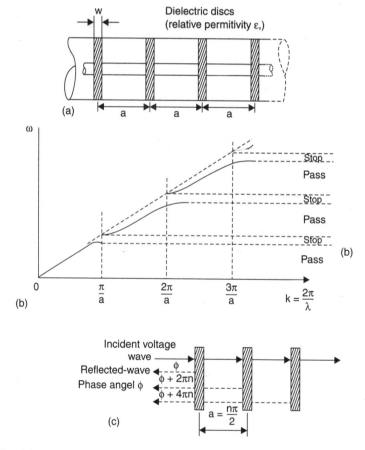

Fig. 12.79 (a) Periodically loaded transmission line (b) ω-k diagram and (c) reflections occurring when $ka = n\pi$

BASIC QUANTUM MECHANICS

REFERENCES

1. R.L. Liboff, *Introductory Quantum Mechanics,* Pearson Publications, New Delhi, 2003.
2. R. Eisberg and R. Resnick, *Quantum Physics of Atoms, Molecules, Solids, Nuclei and Particles,* John Wiley, New York, 2002.
3. S. Gasiorowicz, *Quantum Physics,* John Wiley, New York, 2005.
4. E. Merzbacher, *Quantum Mechanics,* John Wiley, New York, 1999.
5. J. Singh, *Modern Physics for Engineers,* John Wiley, New York, 2003.

SOLVED EXAMPLES

1. A particle of mass 1.67×10^{-27} kg is confined to the second excited state in a one dimensional potential well of infinite height and width $L = 0.1$nm. Calculate (a) its energy (b) its momentum and (c) the probability of finding the particle between 0 and $L/3$.

    Solution:

    (a) The particle is in the second excited state. Hence, the quantum number $n = 3$.

    The energy (E) of the particle is given by:

    $$E = \frac{n^2 h^2}{8mL^2} = \frac{3^2 \times (6.63 \times 10^{-34})^2}{8 \times (1.67 \times 10^{-27}) \times (0.1 \times 10^{-9})^2} = 29.34 \times 10^{-21} \text{ Joules}$$

    (b) The de Broglie wavelength of the particle in n=3 state is given by:

    $$\lambda = \frac{2L}{n} = \frac{2 \times (0.1 \times 10^{-9})}{3} = 6.6 \times 10^{-11} \text{ m}$$

    $$\therefore \quad p = \frac{h}{\lambda} = \frac{6.63 \times 10^{-34}}{6.6 \times 10^{-11}} = 10^{-23} \text{ kg.ms}^{-1}$$

    (c) The probability (P) of finding the particle between $x = 0$ and $x = L/3$ is given by:

    $$= P \int_0^{L/3} \psi_3^2 dx = \frac{2}{L} \int_0^{L/3} \sin^2\left(\frac{3\pi x}{L}\right) = \frac{1}{L} \int_0^{L/3} \left[1 - \cos\left(\frac{6\pi x}{L}\right)\right] dx$$

    $$= \frac{1}{L}\left[\int_0^{L/3} dx - \int_0^{L/3} \cos\left(\frac{6\pi x}{L}\right) dx\right] = \frac{1}{L}\left[\frac{L}{3} - 0\right] = \frac{1}{3}$$

2. An electron is confined to a one dimensional infinite potential of width 1nm. Calculate the wavelength of the radiation emitted when it undergoes a transition from $n = 6$ state to the $n = 2$ state.

    Solution:

    The wavelength of the radiation emitted is given by:

    $$\lambda = \frac{hc}{E} = \frac{hc}{(E_5 - E_2)} = \frac{hc \times 8mL^2}{h^2(6^2 - 3^2)} = \frac{8mcL^2}{27h}$$

    $$= \frac{8 \times (9.1 \times 10^{-31}) \times (3 \times 10^8) \times (10^{-9})^2}{27 \times (6.63 \times 10^{-34})} = 123\text{nm}$$

3. An electron is confined to a one dimensional infinite potential of width 1nm. Calculate the uncertainty in its momentum in the ground state.

Solution:

The uncertainty in the position of the electron is 1nm. Hence, the uncertainty in its momentum (Δp) is given by:

$$\Delta p \geq \frac{\hbar}{2 \times \Delta x} \geq \frac{6.63 \times 10^{-34}}{4\pi \times (10^{-9})} \geq 0.53 \times 10^{-15} \text{ kgms}^{-1}$$

4. An electron is confined in a potential well of width 0.5nm and height 15eV. How many states the well can accommodate?

   Solution:

   The energy corresponding to energy state with a quantum number n is given by:

   $$E_n = \frac{n^2 h^2}{8mL^2} - V_0 \quad \text{or} \quad n_{max} = \sqrt{\frac{8mL^2(E_{max} + V_0)}{h^2}}$$

   But the maximum energy the electron can have is equal to the height of the well.

   $$n_{max} = \frac{4L\sqrt{m \times V_0}}{h} = \frac{4 \times (0.5 \times 10^{-9}) \times \sqrt{9.1 \times 10^{-31} \times (15 \times 1.6 \times 10^{-19})}}{6.63 \times 10^{-34}} = 4.45$$

   But the maximum energy the electron can have is equal to the height of the well.

   $$n_{max} = \frac{4L\sqrt{m \times V_0}}{h} = \frac{4 \times (0.5 \times 10^{-9}) \times \sqrt{9.1 \times 10^{-31} \times (15 \times 1.6 \times 10^{-19})}}{6.63 \times 10^{-34}} = 4.45$$

   Hence, the maximum quantum number possible is 4.

5. An electron traveling with a speed of 10^5m/s encounters a step barrier whose height is twice that of its kinetic energy. Estimate the penetration distance of the electron.

   Solution:

   The kinetic energy of the electron is given by:

   $$E = \frac{mv^2}{2} = \frac{(9.1 \times 10^{-31}) \times (10^5)^2}{2} = 4.56 \times 10^{-21} \text{ J}$$

   The height of the barrier is therefore $2 \times 4.56 \times 10^{-21}$ J $= 9.12 \times 10^{-12}$ J
   The wave vector in the barrier region is given by:

   $$\gamma = \sqrt{\frac{2m(V_0 - E)}{\hbar^2}} = \sqrt{\frac{2m(2E - E)}{\hbar^2}} = \frac{\sqrt{2mE}}{\hbar}$$

   $$= \frac{2\pi\sqrt{2 \times 9.1 \times 10^{-31} \times 4.56 \times 10^{-21}}}{6.63 \times 10^{-34}} = 0.086 \times 10^{10} \text{ m}^{-1}$$

   The wave function in the barrier region is $\psi_2 = Ce^{-\gamma x}$

   Hence, the distance (d) over which the amplitude of the wave function reduces to e^{-1} of its value is given by:

   $$d = \frac{1}{\gamma} = \frac{1}{0.086 \times 10^{10}} = 1.16 \text{ nm}$$

6. An electron with energy of 2eV strikes a potential barrier of width 0.3nm and height 20eV. Calculate the tunneling probability assuming that the barrier width is much larger than the penetration distance.

Solution:

The tunneling probability is given by:

$$T = \frac{16E(V_0 - E)}{V_0^2} e^{-2\gamma a} \quad \text{where} \quad \gamma = \frac{\sqrt{2m(V_0 - E)}}{\hbar}$$

Here, $\gamma = \dfrac{2\pi\sqrt{2 \times (9.1 \times 10^{-31}) \times (20 - 2) \times (1.6 \times 10^{-19})}}{6.63 \times 10^{-34}} = 2.17 \times 10^{10} \text{ m}^{-1}$

$\therefore \quad T = \dfrac{16 \times 2 \times (20 - 2)}{20^2} \times exp[-2 \times (2.17 \times 10^{10}) \times (0.3 \times 10^{-9})] = 3.17 \times 10^{-6}$

QUESTIONS

1. Mention the key ideas which form the basis of quantum mechanics.
2. Derive the time independent Schrödinger wave equation.
3. Derive the time dependent Schrödinger wave equation.
4. Discuss the mathematical properties of wave functions which satisfy Schrödinger wave equation.
5. Discuss the physical significance of wave function.
6. What are position, momentum and Hamiltonian operators?
7. Write down Schrödinger wave equation using operator formalism.
8. Define the expectation value of an operator.
9. What are commutation relations? Explain their significance.
10. Solve the Schrödinger wave equation for a free particle. Obtain the wave function and an expression for its energy.
11. Solve the Schrödinger wave equation for a particle confined in a one dimensional potential of width L and infinite height. Obtain an expression for its energy and wave function.
12. Solve the Schrödinger wave equation for a particle confined in a two-dimensional square potential of width L and infinite height. Obtain an expression for its energy and wave function.
13. Solve the Schrödinger wave equation for a particle confined in a three-dimensional cubic potential of width L and infinite height. Obtain an expression for its energy and wave function.
14. What are degenerate energy states? Explain with examples.
15. Define density of states for a particle confined in a one-dimensional potential well of infinite height. Obtain an expression for the same.
16. Define density of states for a particle confined in a two-dimensional square potential well of infinite height. Obtain an expression for the same.
17. Define density of states for a particle confined in a three-dimensional cubic potential well of infinite height. Obtain an expression for the same.
18. Solve the Schrödinger wave equation for a particle encountering a one dimensional step potential of height V_0, when the energy of the particle is greater than that of the height of the barrier. Obtain an expression for the transmission and reflection coefficient.
19. Solve the Schrödinger wave equation for a particle encountering a one dimensional step potential of height V_0, when the energy of the particle is less than that of the height of the barrier. Obtain an expression for the transmission and reflection coefficient.
20. What is an evanescent wave? Discuss its role in fiber optic couplers and fiber optic sensors.
21. Solve the Schrödinger wave equation for a particle encountering a one dimensional of width a potential of height V_0, when the energy of the particle is greater than that of the height of the barrier. Obtain an expression for the transmission and reflection coefficient.

22. Solve the Schrödinger wave equation for a particle encountering a one dimensional of width a potential of height V_0, when the energy of the particle is less than that of the height of the barrier. Obtain an expression for the tunneling probability.
23. Discuss the role of tunneling in field emission devices.
24. Discuss the role of tunneling in ohmic contacts.
25. Discuss the role of tunneling in tunnel diode.
26. Discuss the role of tunneling in resonant tunnel diode.
27. Discuss the role of tunneling in zener diode
28. Discuss the role of tunneling in Josephson junction.
29. What is SQUID? Discuss the ac and dc operation of SQUID magnetometer.
30. Discuss the role of tunneling in alpha radioactivity.
31. Discuss the role of tunneling in ammonia maser.
32. Discuss the Kronig Penney model and show how it leads to the origin of energy bands in a solid.
33. Explain the origin of energy gap by considering the Brag reflection of electrons at the zone boundary.
34. Distinguish between a metal, a semiconductor and an insulator.
35. Discuss the salient features of electrical conduction in a metal and a semiconductor.
36. Write a short note on the effective mass of an electron.
37. Discuss the concept of *hole* on the basis of band theory of solids.

Problems

1. A proton is trapped in a one-dimensional infinite potential well of length 10^{-14}m. (a) what are the first two energy levels? (b) What is the frequency of the photon emitted when the proton makes a transition from the upper level to the ground state? In what part of the electromagnetic spectrum does it lie?
2. An electron moves within a one-dimensional infinite potential well of length of 0.1nm. (a) Calculate the energies of the ground state and the first excited state. (b) What is the wavelength of the photon emitted when the electron undergoes a transition from the first excited state to the ground state?
3. An electron in an infinite potential well has an energy of 5 eV in the $n = 4$ level. What is the width of the well?
4. What is the photon energy required to transfer an electron from ground state to the second excited state in an infinite potential well of width 0.2 nm. To which part of the electromagnetic spectrum does the photon belong?
5. What is the minimum speed of an electron in an infinite potential well of width 0.1 mm?
6. The ground state energy of an electron in an infinite potential well is 20 eV.
   (a) What is the energy of the first excited level? (b) What is the length of the well?
7. Suppose an electron was trapped with an infinite potential well whose length is 10^{-14} m, which is an approximate size of a nucleus. (a) Calculate the ground state energy of the electron; (b) Calculate the potential energy of the electron at this distance. What can you say about the possibility of electrons being with the nucleus?
8. An electron is in an infinite potential well of length 0.2 nm. What is the uncertainty in its linear momentum?
9. A proton is confined to a nucleus of radius 2×10^{-14} m. (a) Estimate the uncertainty in its linear momentum. (b) If the linear momentum were equal to the uncertainty found in (a) What would be the kinetic energy in MeV?
10. Consider the ground state wave function for a particle in an infinite potential well that extends from $x = 0$ to $x = L$. What is the probability of finding the particle from $x = L/4$ to $3L/4$?
11. An impenetrable box extends from $x = -L/2$ to $x = L/2$. What are the normalized wave functions for the three lowest energy states?
12. Show that for a particle in the nth state of a one-dimensional impenetrable box of Length L,
$$\langle x^2 \rangle = (1/3 - 1/2n^2\pi^2) L^2$$

BASIC QUANTUM MECHANICS

13. What must be the width of an infinite well such that a trapped electron in the $n = 3$ state has an energy of 4.70 eV?

14. (a) Calculate the smallest allowed energy of an electron confined to an infinitely deep well with a width equal to the diameter of an atomic nucleus (about 1.4×10^{-14}m). (b) Repeat for a neutron. (c) Compare these results with the binding energy (several MeV) of protons and neutrons inside the nucleus. On this basis, should we expect to find electrons inside nuclei?

15. The ground state energy of an electron is an infinite well is 2.6 eV. What will the ground-state energy be if the width of the well is doubled?

16. (a) Calculate the fractional difference between two adjacent energy levels of a particle confined in a one-dimensional well of infinite depth. (b) Discuss the result in terms of the correspondence principle.

17. (a) Calculate the separation in energy between the lowest two energy levels for a container 20cm on a side containing argon atoms. (b) Find the ratio with the thermal energy of the argon atoms at 300 K. (c) At what temperature does the thermal energy equal the spacing between these two energy levels? Assume, for simplicity, that the argon atoms are trapped in a one-dimensional well 20 cm wide. The molar mass of argon is 39.9 g/mol.

18. Assume the electron in a metal to be a particle confined in an three-dimensional box of infinite height. Calculate the energies of the lowest five distinct states for a conduction electron moving in a cubical crystal of edge length $L = 250$ nm.

19. Consider an electron trapped in an infinite well whose width is 98.5 pm. If it is in state with $n = 15$, what are (a) its energy? (b) The uncertainty in its momentum? (c) The Uncertainty in its position?

20. Where are the points of (a) maximum and (b) minimum probability for a particle trapped in an infinitely deep well of length L if the particle is in the state n?

21. A particle is confined between rigid walls separated by a distance L, (a) Show that the probability P that it will be found within a distance $L/3$ from one wall is given by
    $$P = (1/3)(1 - (\sin(2\pi n/3)/(2\pi n/3)))$$
    Evaluate the probability for (b) $n = 1$, (c) $n = 2$, (d) $n = 3$, and (e) under the assumption of classical physics.

22. A particle is confined between rigid walls located at $x = 0$ and $x = L$. For the $n = 4$ energy state, (a) sketch the probability density curve for the particle's location. Calculate the approximate probabilities of finding the particle within a region $\Delta x = 0.0003L$ when (b) $\Delta x = L/8$ and (c) at $x = 3L/16$.

23. Consider a finite square-well potential of width 3×10^{-15}m that a particle of mass GeV/c^2. How deep does this potential well need to be to contain three energy states?

24. Calculate the density of states for an electron confined to a one dimensional potential well of width 1nm and infinite height.

25. Calculate the density of states for a proton confined to a two-dimensional square potential well of width 1fm and infinite height.

26. Calculate the density of states for an electron confined to a three-dimensional cubic potential well of width 1cm and infinite height.

27. Calculate the transmission probability of an alpha particle of energy $E = 5$MeV through a square barrier of width 13 fm and height 15MeV.

28. A 1 eV electron has a 10^{-4} probability of tunneling through a 2.5eV potential barrier. What is the probability of a 1eV proton tunneling through the same barrier?

20. Consider the penetration of a step potential function of height 2.4eV by an electron whose energy is 2.1eV. Determine the relative probability of finding the electron at a distance of (a) 1.2nm beyond the barrier and (b) 5nm beyond the barrier, compared to the probability of finding the incident particle at the barrier edge.

30. Evaluate the transmission coefficient for an electron of energy 2.2eV impinging on a potential barrier of height 6eV and thickness 0.1nm. Repeat the calculation for a barrier thickness of 1nm.

13
Quantum Information and Quantum Computation

13.1 INTRODUCTION

Quantum information and quantum computation is an emerging interdisciplinary field involving quantum physics, theoretical computer and information science. In the past two decades the subject of computation and information processing using *quantum computers* has generated much interest among physicists, computer scientists and information theorists. The present day computers which are in use may be called as *classical computers*. Their design is based on semiconductor technology. The physics of semiconductors is based on quantum mechanics. However, this does not make these computers truly *quantum mechanical*. In quantum computers, information is stored and manipulated using *single isolated atoms or photons or electrons*. Since atoms, molecules and photons obey the laws of quantum mechanics, quantum computers perform computation and process information according to the laws of quantum mechanics. Their operation and design is qualitatively different from those of classical computers. In classical computers these tasks are done using *an ensemble of atoms*.

Quantum computers are found to be more efficient than their classical counterparts in performing computational tasks such as factorizing a number or carrying out searches in a data base. Quantum computer networks can efficiently transmit and receive information in *unusual* ways. The novel features of these computers have immense potential in the science of cryptography. Cryptography is the science of maintaining secrecy and security in communication. Its utility in military and business applications is self-evident. With the proliferation of information, carrying out efficient searches in a data base is an important technological problem. These factors have been the main driving force for the emergence of quantum computation and quantum information. This novel and challenging discipline also has turned out to be a laboratory toy kit for understanding the basic and unique aspects of quantum mechanics. It is an interdisciplinary subject involving several disciplines as shown in Fig.13.1.

Traditionally computation has been a branch of mathematics. That physics should be related to computation might appear strange. But this notion loses its force when we realize that storage and manipulation of information is done using a *physical device*. The present day computers have a long history. Initially computers were built using mechanical and electromechanical devices such as gears and relays. Later with the development of science of electronics they were built with vacuum tubes as well as transistors. The present day computers rely on very large scale integrated circuits (VLSI). The important component is the silicon chip of area 0.25mm thick and surface area of ~1cm². This contains millions of transistors which act as logical gates. The present trend in silicon technology is to cram more and more transistors onto to the silicon chip and increase the clock speed of the microprocessors. The increase in the number of transistors on a silicon chip obeys Moore's law (Fig.13.2). This states that approximately in every two years the density of transistors doubles. By 2010, one expects the

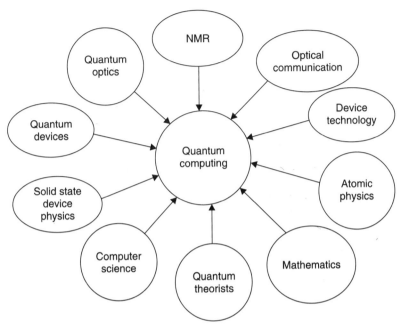

Fig. 13.1 Disciplines that contribute to quantum computation and quantum information processing

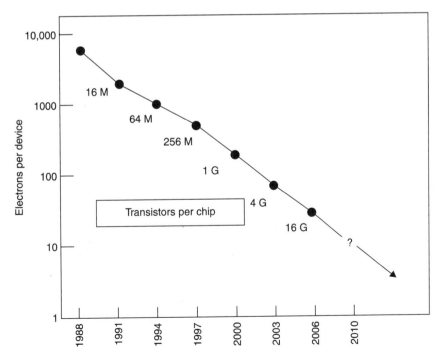

Fig. 13.2 Moore's law—Growing number of transistors per chip with time is shown. Also shown is the declining trend in the number of electrons in a transistor

number of electrons to be as small as ~10 per device. *A physical device of such microscopic dimensions may not behave the same way as a device of bigger dimensions.* The present day research in the field of nanotechnology does suggest that materials behave differently at such minute length scales. Quantum computers can be thought of as the logical extension of this trend in technology, where data storage and manipulation is done at the *atomic scale.*

An analogy with the relation of physics with geometry might help to reinforce this shift in paradigm regarding the relation between physics and computation. Euclidean geometry is not based on experimental observations, though it might have originated from the practical necessities of Greek architecture. Euclidean geometry can be thought of as a pure result of human thought. Its approach consists in assuming a number of basic postulates and proving various theorems using deductive logic. Later on there have been several geometries of space including the one due to Riemann. *However, Einstein in his theory of general relativity demonstrated how the Riemannian geometry of space could be related to the distribution of masses contained in it.* It is only natural that the study of information storage, processing and computation be linked to the study of physical devices which store information. From an engineering perspective, mastery of the principles of physics and material science is needed to develop the state-of-art computer hardware.

The subject of quantum computers has motivated physicists to think computationally about quantum physics and impelled computer scientists as well as information theorists to understand computation through the *magic glass of quantum mechanics.* It has provided a *new paradigm* for physicists, computer scientists and information theorists for exciting and challenging research. Simulation of a phenomena or working of a physical device is routinely done in the present day scientific and engineering research. Richard Feynman, the legendary figure of the twentieth century physics had anticipated the advent of quantum computers. He showed that simulation of quantum systems cannot be done efficiently on a classical computer. He found that to simulate a quantum system, the number of computational steps rises exponentially both with the size of the quantum system and with the amount of time over which the system's behavior is tracked. On the other hand, a quantum computer does not pose such a difficulty. It is hoped that simulation on quantum computers could provide useful tools for physics and technology of nanoscale devices.

Quantum mechanics is weird and mysterious. It defines our common sense notions about nature. It predicts a number of counter intuitive effects that have been verified experimentally. It is worth recording the opinions of two prominent physicists who played a major role in the development of quantum mechanics. Niels Bohr said that *any one who can contemplate quantum mechanics without getting dizzy has not properly understood it.* Einstein called it *black magic calculus.* Though Einstein made use of the quantum theory to explain photoelectric effect, he never reconciled himself with the philosophy of quantum physics.

13.2 BASIC IDEAS

Some of the basic ideas which are central to quantum computers are *quantum superposition, single particle interference, qubit, quantum entanglement and EPR states.*

13.2.1 Quantum Superposition and Single Particle Interference

The principle of quantum superposition is an extension of the usual principle of superposition but for a single particle. This principle can be best appreciated by considering the following experiments.

Young's Double Experiment

In Young's double slit experiment, two closely spaced pin holes are illuminated by a source of light (Fig. 13.3). On a far off screen an interference pattern is obtained. The phenomenon is explained as due to the interference of light from the two coherent sources at S_1 and S_2. According to Huygens's theory the points S_1 and S_2 lie on the same wavefront and consequently act as coherent sources. If the two slits S_1 and S_2 are illuminated independently, one does not observe the interference pattern. *It is interesting to note that the interference pattern is formed even when there is only one photon in the apparatus at any time.* Assume that the experiment is performed by the source emitting a single photon at a time, such that there is only one photon in the apparatus. The interference pattern can be perceived when the experiment is performed over long periods of time. A necessary condition for quantum interference is that the experiment is performed in such a way that there is no way of knowing which of the slits the photon has traveled. Since material particles also exhibit wave like behavior, the Young's double experiment can be repeated with electrons and neutrons as well. Similar results are obtained. This result cannot be explained in the usual way. It is explained by assuming that somehow the particle travels both the paths at the same time! *i.e.*, it is in a superposition of two experimental paths.

$$\begin{matrix} \text{Single Particle} \\ \text{Interference} \end{matrix} = \left| \begin{matrix} \text{Passage through} \\ \text{the upper slit} \end{matrix} \right\rangle + \left| \begin{matrix} \text{Passage through} \\ \text{the lower slit} \end{matrix} \right\rangle$$

Let us now consider an alternate way of demonstrating single particle interference. In Fig.13.4a a single photon is directed at a half silvered mirror (beam splitter). The beam splitter reflects half the intensity of light impinging on it and allows the rest to pass through. The photon detectors in the two possible exit paths detect the photon with equal probability. Hence it is natural to conclude that during any one run of the experiment photon has traveled one of the paths since it cannot be split into two. *However this assumption is not true.* Mysteriously photon takes both the paths at once! This is best demonstrated by another experimental set up as shown in Fig.13.4b. With this one can observe astonishingly pure phenomenon of *single particle interference*. Suppose that a particular photon travels horizontally after striking the first mirror. By comparison with the previous experiment set up, one finds that the photon is detected at the two detectors with equal probability. The same would be observed if the photon were on the vertical path. Hence if the photon really did take a single path through the apparatus, both the detectors would detect with equal probability. *However this does not happen. In this experimental arrangement the photon always strikes detector A and never detector B.*

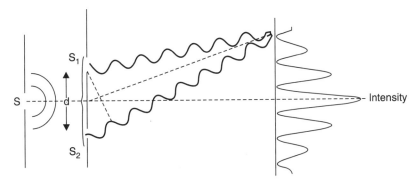

Fig. 13.3 Young's double slit experiment. Source emits only one photon at a time. At any time there is only one photon in the apparatus

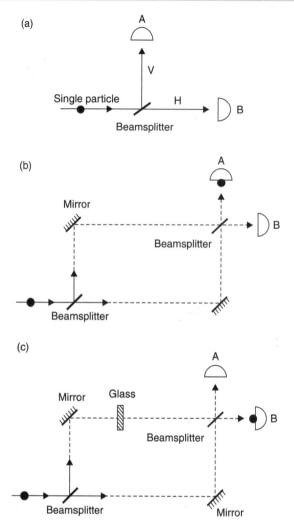

Fig. 13.4 Single particle interference using (a) single beam splitter (b) two beam splitters and (c) with a thin glass plate

A slight change in the path length could be introduced by having a glass plate as shown in Fig.13.4c. *Then the photon is detected by detector B and never by detector A.*

The inevitable conclusion is that photon must in some sense have traveled both the paths at once. If either of the two paths is blocked by an absorbing screen, then both detectors A and B would detect the photon with equal probability. In other words, blocking off either path illuminates the detector B. But when both the paths are open, somehow the photon is subjected to an influence which prevents from reaching B. This *mysterious occult influence* must travel along the other path at the speed of light, exactly as a photon would. When a thin glass is introduced, the exact arrival times between the photon and the influence is not exact and hence is detected by detector B.

This property of quantum interference—that the motion of particles that we detect appears to be affected by invisible influences applies not only to photons but also to other particles and physical systems. At the microscopic level, Nature behaves in an incomprehensible way and the sense of mystery deepens.

13.2.2 Qubit

In classical computer, the data is stored in a *bit*. At any instant of time, the bit can take the value 0 or 1. The corresponding parameter in quantum computer is *qubit*. However, there is a fundamental difference between the two. A qubit can be in a superposition state corresponding to logical states 0 and 1 just like a single photon. These states are represented as $|0\rangle$ and $|1\rangle$. Qualitatively speaking, if white and black colour correspond to the states $|0\rangle$ and $|1\rangle$, then a qubit can be imagined to have any shade of grey. Also qubit obeys the laws of quantum mechanics. The physical state of a qubit is denoted by ψ.

Geometrical Representation of a Qubit

It is convenient to represent the state of a qubit in what is known as *Bloch sphere*. Bloch sphere is a sphere of unit radius and the state of a qubit is represented by a vector in this sphere as shown in Fig.13.5

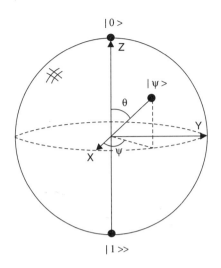

Fig. 13.5 Geometrical representation of a qubit in Bloch sphere

i.e.,
$$|\psi\rangle = \cos\frac{\theta}{2}|0\rangle + e^{i\phi}\sin\frac{\theta}{2}|1\rangle \qquad (13.1)$$

Note for $\phi = 0$ and $\theta = 0$, the state $|\psi\rangle$ corresponds to $|0\rangle$ and is along $+z$-axis

for $\phi = 0$ and $\theta = 180°$, the state $|\psi\rangle$ corresponds to $|1\rangle$ and is along $-z$-axis.

When $\theta = 90°$, the vector is the x–y plane. For

$\phi = 90°$, $|\psi\rangle = \dfrac{1}{\sqrt{2}}(|0\rangle + i|1\rangle)$, is a superposition state along $+y$-axis

$\phi = -90°$, $|\psi\rangle = \dfrac{1}{\sqrt{2}}(|0\rangle - i|1\rangle)$, is a superposition state along $-y$-axis

$\phi = 0°$, $|\psi\rangle = \dfrac{1}{\sqrt{2}}(|0\rangle + |1\rangle)$, is a superposition state along x-axis

$\phi = 180°$, $|\psi\rangle = \dfrac{1}{\sqrt{2}}(|0\rangle - |1\rangle)$, is a superposition state along $-x$-axis

For a classical computer, the two logical states 0 and 1 are represented by the poles of a sphere as shown in Fig.13.5. In contrast, the state of a qubit can be represented by any point on the sphere. Since there are infinite points on the sphere, a qubit in principle has more capacity to store information compared to a classical bit. *Bloch sphere represents the state of only one qubit. There is no generalization of Bloch sphere for multiple qubits.*

Physical Realisation of a Qubit

The physical realization of qubit is shown in Fig.13.6. It could be

(*a*) horizontal and vertical polarization of a single photon

(*b*) right and left circular polarization of a single photon

(c) up and down spin states of an electron
(d) two energy states of an atom
(e) two energy levels of a quantum dot
(f) beam splitter modes of two particles

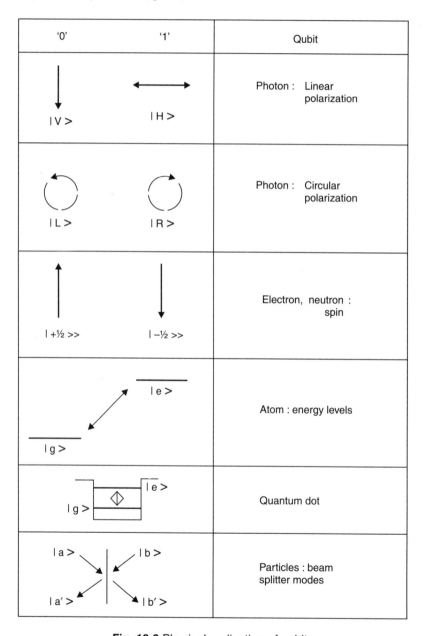

Fig. 13.6 Physical realization of qubits

Measurement of the State of a Qubit

In general a qubit will be in a superposition state of $|0\rangle$ and $|1\rangle$. However, on measurement a qubit will be found either in $|0\rangle$ or $|1\rangle$, whatever might have been its state prior to

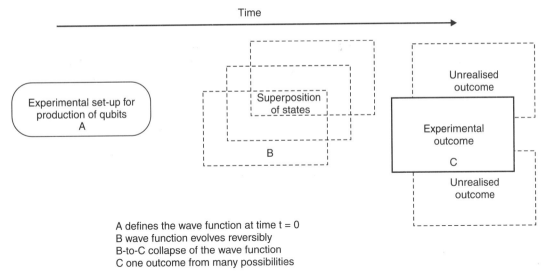

A defines the wave function at time t = 0
B wave function evolves reversibly
B-to-C collapse of the wave function
C one outcome from many possibilities

Fig. 13.7 Quantum behaviour of a qubit. Qubit is in a superposition state prior to measurement. One of the states is measured in an experiment

measurement. *The superposition state collapses to the state $|0\rangle$ or $|1\rangle$ during measurement* (Fig.13.7). According to quantum mechanics, the experimental outcome is probabilistic. It is impossible to predict the outcome in any single isolated event. It requires several measurements to arrive at the state prior to measurement. A superposition state contains hidden information that is not accessible to measurement. The subject of quantum tomography deals with the method of determining the state of qubit prior to measurement. Let the superposition state of the qubit be represented by the expression:

$$|\psi\rangle = a|0\rangle + b|1\rangle \qquad (13.2)$$

where a and b can in general be complex numbers. If a^* and b^* represent their complex conjugates, then, aa^* and bb^* indicate the probability of measuring the logical state $|0\rangle$ and $|1\rangle$. i.e., the square of the modulus of the coefficients of states $|0\rangle$ and $|1\rangle$ denote the probability of measurement of these states respectively.

During computation the state of a qubit changes. These changes are reversible. However, whenever a measurement is done, irreversibility comes into picture since on measurement the qubit irreversibly collapses to state $|0\rangle$ or $|1\rangle$. Irreversibility is avoided only in the special case when the qubit is actually in the pure state $|0\rangle$ or $|1\rangle$ before measurement. Thus any measurement will corrupt the data. Obviously, one cannot infer the state of a qubit prior to measurement from the outcome of a single measurement. This is not a question of experimental competence. It is the property of nature. This fragility of the quantum state is exploited in quantum cryptography. Sending information encoded in individual qubits guarantees that no eavesdropper can read it in transit without leaving evidence of tampering.

Another important property of the qubit is that its state cannot be copied. This is known as *no cloning theorem*. This is not at all true of a classical bit. The state of a classical bit can be inferred from measurement and also copied to another register.

The state of a qubit is represented by a vector. A vector can be represented using different sets of basis vectors. Hence, the state of a qubit can also be represented using different basis states. The basis vectors chosen to represent states $|0\rangle$ and $|1\rangle$ are arbitrary. Assume we have two photons whose polarization we are measuring in a chosen basis. We identify the horizontal polarization in this base as $|0\rangle$ and the vertical as $|1\rangle$. We can now choose a new conjugate basis which is rotated with respect to the first one at 45° in the plane of polarization. The states now are represented by:

$$|0'\rangle = \frac{(|0\rangle + |1\rangle)}{\sqrt{2}} \quad |1'\rangle = \frac{(|0\rangle - |1\rangle)}{\sqrt{2}} \tag{13.3}$$

These new bases are obtained by putting $\theta = 90°$ and $\phi = 0°$; $\theta = 90°$ and $\phi = 180°$ in eqn. (13.1). Conjugate bases cannot be used *at the same time* in an experiment. But the possibility to switch alternately between various bases is utilized in many experimental methods for qubit measurement.

13.2.3 Qubit and Stern Gerlach Experiment

The electron spin is a good example of a qubit. Spin is an abstract concept. This concept was introduced to explain the spectral features of elements. The spectral lines get split when atoms are excited in the presence of a magnetic or electric field. These phenomena are known as Zeeman effect and Stark effect respectively. In classical physics the spin angular momentum is associated with the rotational motion of a rigid body about its axis. The concept of electron as a tiny ball of charge spinning on its axis is useful. By analogy with the motion of planets, electrons were imagined to go round the nucleus and have a spinning motion as well. However the Stern Gerlach experiment described below does not support such a classical view. In fact the behavior of electron spin in a magnetic field is best described by a *qubit*.

Stern Gerlach experiment is one of the well known experiments in modern physics. The experiment demonstrates *space quantization*, a key concept in quantum mechanics. It is a prototype of measurement fundamental to a proper understanding of quantum mechanics and qubits. A schematic diagram of the apparatus is shown in Fig.13.8*a*. Metallic silver is vapourized in an electrically heated oven. These silver atoms escape from a small hole in the oven wall. They are collimated using a thin slit and enter the external vacuum of the apparatus. Silver atoms are electrically neutral but possess a magnetic moment. The collimated beam passes through the pole pieces of an electromagnet and is detected on the screen. The pole faces are shaped to make the field as non-uniform as possible. It is observed that the beam is split into two in the presence of this non-uniform magnetic field.

Theory

A dipole of magnetic moment µ experiences only a torque in a *uniform magnetic field*. However in a non-uniform magnetic field, it experiences a net force in addition to a torque, since the forces on the north and south poles of the dipole are not equal in magnitude (Fig.13.8*b*). The net force acting on the dipole can be calculated as follows:

The potential energy $U(\theta) = -\vec{\mu} \cdot \vec{B_z} = -\mu B_z \cos\theta = -\mu_z B_z$

where θ is the angle between the dipole and the magnetic field along the z-direction. μ_z is the component of the dipole moment along the z-direction. In a non-uniform magnetic field, the force experienced is given by:

$$F(\theta) = -\frac{\partial U}{\partial z} = \mu_z \frac{\partial B_z}{\partial z} \tag{13.4}$$

The deflecting force is determined by the gradient of the magnetic field and not the actual magnetic field itself. The force is in the positive or negative z-direction depending on the orientation of the dipole with respect to the field. Thus atoms on leaving the magnet will spread out according to their vertical components of the magnetic moment. In classical theory all angles are possible. Hence one would expect a smear of silver atoms along a vertical line. The height of the line would be proportional to the magnetic moment as well as the gradient of the magnetic field. The failure of classical theory was obvious when Stern and Gerlach observed two spots as shown in Fig. 13.8a. The original beam of silver atoms had split into two beams.

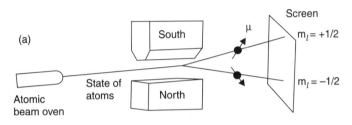

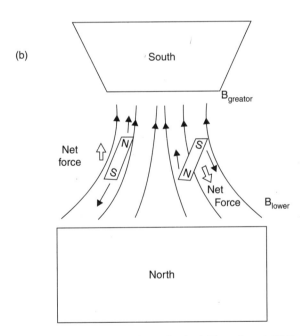

Fig. 13.8 (a) Schematic diagram of the Stern Gerlach Experiment (b) Force acting on a magnetic dipole in a non-uniform magnetic field

In any atom, both the spin and orbital angular momentum of each electron contribute to the total magnetic moment of the atom. Of the 47 electrons in the silver atom, 46 form closed shells, with zero angular momentum and magnetic moment. It is only the spin of the last electron, in the $l = 0$ (s orbital), that contributes to the angular momentum. In the s shell the magnetic moment due to orbital motion is zero. Hence the magnetic moment of silver atom is due to the spin of the 47th electron. Experimental observations with hydrogen atom are also similar, since the magnetic moment of the hydrogen atom is also equal to the magnetic moment due to the spin of the lone electron. Experimental observations with other atoms which have other values of spin magnetic moment prove that the magnetic dipole can have only discrete orientations with respect to the magnetic field. In general an atom with spin I is split into $(2I + 1)$ beams.

Apart from space *quantization*, the experiments with cascaded Stern Gerlach apparatus bring out the peculiar behavior of electrons. Figure 13.9 shows a cascaded Stern Gerlach apparatus.

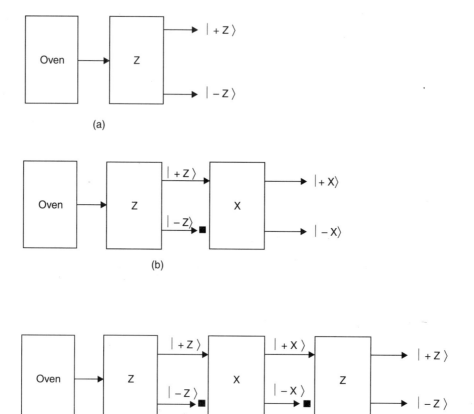

Fig. 13.9 Abstract schematic of the Stern Gerlach Experiment with electrons (*a*) single stage -magnetic field along z-direction (*b*) two stage -magnetic field along z and later along x and (*c*) three stage cascaded measurements magnetic field along z, x and z directions

The magnetic field in the first is along the z-axis while in the second it is along the x-axis. On passing through the first electron beam is split into two beams of equal intensity. The $|-Z\rangle$ output beam form the first SG apparatus is blocked, while the $|+Z\rangle$ output is sent through the second apparatus whose magnetic field is along the x-axis. A magnetic dipole which is oriented along the z-axis ought not to experience any force if the magnetic field is along the x-axis. If the electrons were to behave classically, the beam ought not to split and one has to observe a single peak on the screen. However, again two peaks of equal intensity are observed. If the electron beam with $|-X\rangle$ is blocked and the beam with $|+X\rangle$ is passed through a third SG apparatus, then again surprisingly it is split into $|+Z\rangle$ and $|-Z\rangle$ states with equal intensity. Thus somehow $|+Z\rangle$ state consists of $|+X\rangle$ and $|-X\rangle$ states and $|+X\rangle$ state consist of equal portions of $|+Z\rangle$ and $|-Z\rangle$ states. *This observation suggests that the electron is in a superposition state prior to measurement and collapses into pure states after measurement.* This behavior is similar to that of a qubit. The qubit model provides a simple explanation of the experimental observations.

Let $|0\rangle$ and $|1\rangle$ be the states of the qubit with reference to the z-axis. Then

$$|0\rangle \to |+Z\rangle \qquad (13.5a)$$
$$|1\rangle \to |-Z\rangle$$

Then comparing with eqn. (13.1)

$$\frac{(|0\rangle + |1\rangle)}{\sqrt{2}} \to |+X\rangle$$

$$\frac{(|0\rangle - |1\rangle)}{\sqrt{2}} \to |-X\rangle \qquad (13.5b)$$

Note that +x-axis corresponds to θ = 90° and φ = 0°, while –x-axis corresponds to θ = 90° and φ = 180°. The similarity between eqns. (13.2) and (13.5b) is obvious.

The z-axis SG apparatus measures the spin (*i.e.*, the state of the qubit) in the computational basis $|0\rangle$ and $|1\rangle$. And the x-axis SG apparatus measures with respect to the basis $|+x\rangle$ and $|-x\rangle$.

From eqns. (13.5a) and (13.5b) we have:

$$|+Z\rangle = |0\rangle = \frac{(|+X\rangle + |-X\rangle)}{\sqrt{2}}$$

$$|+X\rangle = |1\rangle = \frac{(|+Z\rangle + |-Z\rangle)}{\sqrt{2}} \qquad (13.6)$$

Hence in measurements carried out along the z-axis and the x-axis, one would expect the beam to be split into two. Feynman has analyzed the results of the SG experiments with atomic having spin 1. In such a case the beam is split into 3. In general an atom with an angular momentum l is split into $(2l + 1)$ beams.

The experiments suggest that one cannot specify the direction of the electron spin prior to measurement. What all one can say is that after measurement it is in a direction along or opposite to the magnetic field. Here the magnetic field defines the direction of measurement.

13.2.4 Qubit Registers

A collection of qubits is called the quantum register. Two qubit registers are the most useful. In quantum computers many of the interesting features occur with two qubits. Suppose we had two classical bits the possible configurations are 00, 01, 10 and 11. It can store only the numbers 0, 1, 2 and 3. However, with two qubits the state of qubit register is represented as:

$$|\psi\rangle = a_1|00\rangle + a_2|01\rangle + a_3|10\rangle + a_4|11\rangle \qquad (13.7)$$

These four states define the various possibilities. Thus a two qubit register can store all four numbers simultaneously.

An interesting feature with two qubit register is the occurrence of *entangled states*. They are also known as Bell states or EPR states. These states enormously enhance the power of computation and communication as will be explained in next section. Computation and information processing using *entangled states* do not have any parallel in classical computation and information processing. The Bell or EPR states are discussed in section 13.3.

The general state of a 3 qubit register is represented by:

$$|\psi\rangle = a_1|000\rangle + a_2|001\rangle + a_3|010\rangle + a_4|011\rangle + a_5|100\rangle + a_6|101\rangle + a_7|110\rangle + a_8|111\rangle \tag{13.8}$$

A three bit register in a classical computer can represent numbers 0, 1, 2, 3, 4, 5, 6 and 7. However, a three qubit register holds all these numbers in a superposition state at any instant of time. This might appear paradoxical but is no more mysterious than both 0 and 1 states being held in a single qubit or a photon traveling simultaneously the two paths in the experiment described in section 13.2.

In general the general state of a n-qubit register is a superposition involving 2^n terms and can be represented by:

$$|\psi\rangle = \sum_{x=0}^{2^n-1} a_x |x\rangle \tag{13.9}$$

Where x within the $|\rangle$ is written in the binary notation. *Thus in general a register of L qubits can store upto 2^L numbers at the same time in superposition*. However, this is true so long as the contents of the register are not read. If a measurement is made to look for the contents of the register it will show one of these 2^L numbers with a certain probability given by the appropriate coefficient. Thus the probability of reading a number 2^n is given by the coefficient $a_n a_n^*$.

13.3 QUANTUM ENTANGLEMENT

Entanglement is one of the distinct properties of quantum systems (together with quantum superposition and probabilistic measurement among others) that make quantum information processing so different from classical information technology. This phenomenon refers to the *joint state of two or more qubits*. It describes the correlations between them. *Mathematically an entangled state cannot be written as a simple product of two particle states*. There are four possible Bell states for a pair of two-state particles. The two states of a qubit are $|0\rangle$ and $|1\rangle$. Let the qubits be labeled by A and B. The four states are:

$$|\psi^+\rangle = \frac{(|0\rangle_A |1\rangle_B + |1\rangle_A |0\rangle_B)}{\sqrt{2}} \tag{13.10a}$$

$$|\psi^-\rangle = \frac{(|0\rangle_A |1\rangle_B - |1\rangle_A |0\rangle_B)}{\sqrt{2}} \tag{13.10b}$$

$$|\phi^+\rangle = \frac{(|0\rangle_A |0\rangle_B + |1\rangle_A |1\rangle_B)}{\sqrt{2}} \tag{13.10c}$$

$$|\phi^-\rangle = \frac{(|0_A\rangle |0\rangle_B - |1\rangle_A |1\rangle_B)}{\sqrt{2}} \tag{13.10d}$$

Each Bell state represents a coherent superposition of two possibilities. In $|\psi^+\rangle$ and $|\psi^-\rangle$, the states of two qubits are different and are said to be anticorrelated. In $|\phi^+\rangle$ and $|\phi^-\rangle$, the states of the two qubits are the same and are said to be correlated. Measurements which enables one to distinguish between $|\psi^+\rangle, |\psi^-\rangle, |\phi^+\rangle$ and $|\phi^-\rangle$ are known as Bell measurements.

In the state $|\psi^+\rangle$, for example, particle A can be in state $|0\rangle$ and particle B in state $|1\rangle$ or vice versa, but there is no way of knowing which particle is in which state. All that is defined is the fact that the two qubits are in different states. This means that all the information is distributed between two qubits and that none of the individual qubits carries any specific information. This is the essence of entanglement and is one of the novel and counterintuitive features of quantum mechanics. Assume that the two qubits A and B are entangled. Knowing the state of qubit A i.e., $|1\rangle$ or $|0\rangle$, allows one to know the state of qubit B. Even more amazing is the result due to quantum superposition. When not measured qubit is in a general superposition state such as described by eqn. (13.2). The state of A is decided at the time of measurement. This is somehow communicated to qubit B, which assumes the state opposite to that of qubit A. This is a real phenomenon which Einstein called *spooky action at a distance.* The mechanism of transmission of this information is yet to be explained by any theory. Quantum entanglement allows qubits that are separated by incredible distances to interact with each other instantaneously. It appears as if the particles are communicating faster than the speed of light. But the special theory of relativity is not violated since no information is exchanged. No matter how great is the distance between them, qubits A and B remain entangled as long as they are isolated. Imagine that a pair of entangled qubits is generated and two persons Alice and Bob are in possession of each one of them. Let Alice possess the qubit A and Bob possess the qubit B. Then whenever Alice makes a measurement of the state of qubit A, Bob will instantaneously come to know of it even without making any measurements although they may be separated by any distance!

This peculiar property of entangled states has a long history. It was pointed out by Einstein, Podosky and Rosen in 1935. The puzzle has two key elements viz., *locality* and *reality*. Locality means that no physical action can instantly go from Alice's apparatus to Bob's. Consequently Alice performing the measurement on qubit A does not influence the state of the qubit possessed by Bob. *Reality* refers to the assumption that physical properties exist independent of observation. According to quantum mechanics observations not only disturb what has been measured but they also produce it. For example the experimenter can make light behave like a particle or a wave by choosing the experiment of photoelectric effect or Young's double slit experiment. Prior to the experiment, the light is in a supersposition state of a particle and a wave. The act of experiment decides whether it is a particle or a wave. Note that according to classical mechanics, physical properties have definite values independent of observation. This nondeterministic character of quantum world is inaccessible *information*. There have been repeated attempts to replace the quantum theory by a statistical theory. According to these there exist hidden variables whose values prescribe the values of all observables for any particular object. The hidden variables are unknown to the experimenter. This gives rise to the probabilistic nature of the quantum theory. This probabilistic nature of quantum mechanics would then be analogous to the statistical mechanics where one can imagine that the motion of all particles in principle can be known. In this example, the assumption of *reality* requires that there must be *some element* in the physical world that allows Alice to know Bob's results. Following this line of reasoning and assuming the validity of both locality and reality, the late John Bell investigated possible correlations for a thought experiment in which Alice and Bob choose bases that are at oblique angles. For three arbitrary angles, α, β and γ the following inequality must be fulfilled:

$$N(1_\alpha, 1_\beta) \leq N(1_\alpha, 1_\gamma) + N(1_\beta, 0_\gamma) \tag{13.11}$$

Where $N(1_\alpha, 1_\beta)$ is the number of times Alice obtains '1' with her apparatus at orientation β and Bob obtains '1' with orientation 'α' and so on. (Fig.13.10)

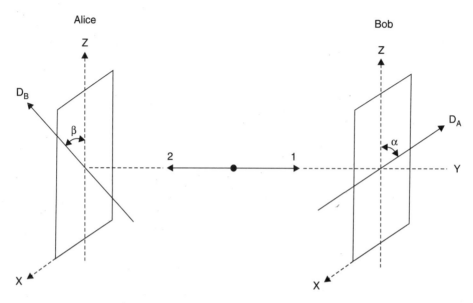

Fig. 13.10 Experimental geometry for verifying Bell's inequalities

Quantum mechanics predicts that

$$N(1_\alpha, 1_\beta) = (1/2)N_0 \cos^2(\alpha - \beta) \text{ and} \tag{13.12a}$$

$$N(1_\beta, 0_\gamma) = (1/2)N_0 \sin^2(\beta - \gamma) \tag{13.12b}$$

Where N_0 is the number of entangled pairs emitted by the source. This inequality is clearly violated if

$(\alpha - \beta) = (\beta - \gamma) = 30°$ when $(\alpha - \gamma) = 60°$. This is because

$$N(1_\alpha, 1_\beta) = \frac{1}{2} N_0 \cos^2 30° = \frac{1}{2} \times N_0 \times \frac{3}{4} = \frac{3N_0}{8}$$

$$N(1_\alpha, 1_\gamma) = \frac{1}{2} N_0 \cos^2 60° = \frac{1}{2} \times N_0 \times \frac{1}{4} = \frac{N_0}{8}$$

$$N(1_\beta, 0_\gamma) = \frac{1}{2} N_0 \sin^2 30° = \frac{1}{2} \times N_0 \times \frac{1}{4} = \frac{N_0}{8}$$

This has been confirmed experimentally many times and implies that one of the assumptions underlying Bell's inequality—e.g., locality or reality—must be in conflict with quantum mechanics. These experiments are usually viewed as evidence for non-locality, though this is by no means the only possible explanation. Bell's inequality is the basis of the two-particle scheme for quantum cryptography. To show that Bell's inequality is violated in a two-particle experiment, it is necessary to perform a statistical experiment. The violation cannot be demonstrated with a single measurement. However, the situation is very different with three-particle entangled states such as the GHZ states. There the contradiction between the Einstein-Podolsky-Rosen assumptions and quantum mechanics arises for individual events.

QUANTUM INFORMATION AND QUANTUM COMPUTATION

Entanglement lies at the heart of microscopic world as described by quantum mechanics. It need not be confined to just two particles. It is possible to have entanglement of three qubits. Three particle entangle states are known as Greenberger-Horne-Zeilinger (GHZ) states. In an entangled state of several particles, measurements on one particle can affect all the others even if they are too far apart for a casual influence to propagate between them. Entangled states offer the possibility to encode information in a complete new way. Entanglement makes possible quantum teleportation, quantum end correction and quantum dense coding.

Production of Entangled States

Entangled pairs can be generated in a number of ways.

(a) Photons with entangled polarization can be generated by passing through a calcite crystal (Fig.13.11).

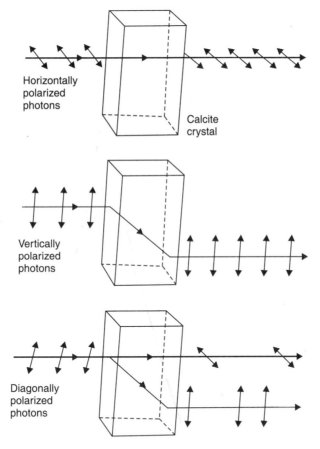

Fig. 13.11 Entangled photons in a calcite crystal will be in a superposition of horizontally and vertically polarized plates

(b) They can also be generated by passing a single photon through a nonlinear crystal. On passing ultraviolet light through a nonlinear crystal it is possible to obtain infrared light. This phenomenon known as parametric down conversion is the reverse of second harmonic generation. In this process a single ultraviolet photon is converted into two infrared photons. The two infrared photons are polarized mutually perpendicular to each other. They are emitted in a direction such that both the momentum and energy are conserved in the process (Fig.13.12).

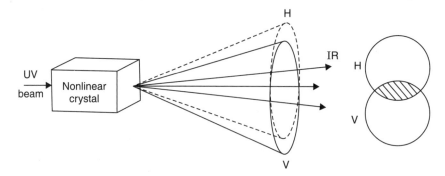

Fig. 13.12 Entangled photons in a nonlinear crystal

(c) Entangled photons can also be generated using a beam splitter. A beam splitter divides the incident light into two beams of equal intensity. When two photons are incident on a beam splitter, it is possible for both the photons to emerge in either the same beam (left) or in different beams (right). In general, bosons emerge in the same beam while fermions emerge in different beams. The situation is more complex for entangled states. Two photons in the Bell state (ψ^-) emerge in different beams. Therefore if detectors are placed in two outward directions register photons at the same time, the experimenter knows that they are in entangled state (Fig.13.13).

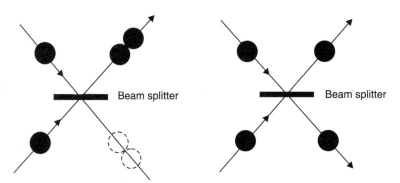

Fig. 13.13 Entangled photons using a beam splitter

(d) In *entanglement swapping* two pairs of entangled particles are used to establish an entanglement between two particles that have never interacted earlier. (Performing a Bell measurement of one half of pair A and one half of pair B transfers the entanglement to the other halves of the pair (Fig.13.14). In this scheme, one ends up with the ability to affect the quantum state of a particle by performing a measurement on another particle that has never had any interaction with the first. It is an astonishing and puzzling—but nonetheless experimentally proven-feature that demonstrates new potential for communication technologies in the quantum world.

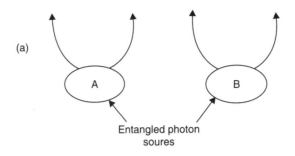

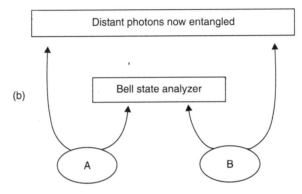

Fig. 13.14 Entanglement swapping

(e) Two energy states of an atom E_1 and E_2 can be used for storing information. Atom is state E_1 corresponds to bit 0 and when in state E_2 corresponds to bit 1. To write 0 bit one does nothing. To write 1, the atom is excited to state E_2. Let τ be the time taken to flip the bit from 0 to 1. If the radiation is applied for a time less than $\tau/2$, the atom will be in a superposition state.

13.4 COMPUTATION – BASIC IDEAS

13.4.1 Computation and Its Complexity

Theoretical computer science and computation can be regarded as a branch of abstract mathematics. Algorithm is central to computation. A computer algorithm is a procedure for carrying out a particular computation. *Computational complexity* deals with the study of difficulties associated in solving different computational problems and relative merits of algorithms. From a mathematical point of view, all existing computers are capable of performing precisely the same set of computational tasks. The difference between one computer and the other is in speed, memory capacity and input/output mode. Computer scientists find it convenient to analyze an algorithm using a universal computer, often known as the Turing machine named after Turing, one of the pioneers of computer science. A Turing machine is an abstract toy machine and is shown in Fig. 13.15. Not long after Von Neumann developed a theoretical model

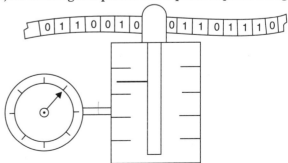

Fig. 13.15 Turing machine

put together in a practical fashion all the components necessary for a computer. Figure 13.16 shows the typical block diagram of the modern computer. The wide variety of computers that have evolved over the past couple of decades can be thought of as different realizations of the universal Turing machine.

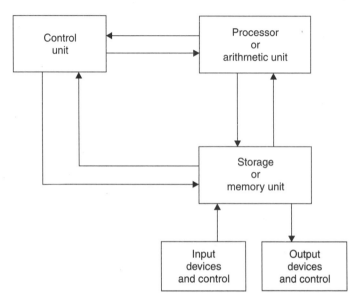

Fig. 13.16 Block diagram of a digital computer

The Turing machine has a lot of significance to computer scientist. Turing showed that it is possible to construct a universal computer which can simulate the action of any other computer in the following sense. The machine has a fixed set of internal states and fixed design. An indicator dial shows the internal state of the Turing machine at any instant of time. The Turing machine reads one binary symbol at a time supplied on a tape. The machine's action on reading a given symbol s depends only on that symbol and the internal state G. The action consists in overwriting a new symbol s' on the current tape location, moving the internal state to G' and moving the tape one place in the same direction d (left or right). The internal construction of the machine can therefore be specified by a finite fixed list of rules of the form:

$$s, G \to s', G', d$$

One special internal state is the *halt state*. Once the machine is in this state the machine does not run further. An input program on the tape is transformed into an output result provided on the tape. The various internal states G are defined for each algorithm or the computational problem. Turing had not made any assumptions about the physical device used for storing and manipulating information in the form of bits. Also he had implicitly assumed classical notions of measurement. *In a classical computer it is possible to measure the state of a bit and copy it, whereas such an operation is not feasible in a quantum computer.* Nor he had imagined that information can be stored in a *superposition state*. When physical devices function according to the laws of quantum mechanics, *Quantum Turing machine* has to be used. In 1985, Deutch described a universal quantum computer. The capabilities of such a machine are found to be quite different from that of classical computer. By abstracting away the physics of the machine, the universal quantum computer becomes the study of universal Turing machine.

Computational complexity is the study of time and memory requirements to solve computational problems. Time and memory requirements are often referred to as *resources* required to perform an algorithm. There are rigorous ways of defining what makes an algorithm efficient. The memory and time resources are usually the criterion for evaluating an algorithm. The input size is the amount of information (measured in bits) needed to specify the input. For example, a number N requires $\sim \log_2 N$ bits of storage in a digital computer. A problem is regarded as easy, tractable or feasible if an algorithm for solving the problem using polynomial resource exists. It is termed as hard, intractable or infeasible if the best possible algorithm requires exponential resources. Many computational problems are formulated as decision problems *i.e.*, problems with YES or NO answers. Is a number prime or not? etc.

Computer scientists define the efficiency of algorithms for obvious reasons. An algorithm is a computational procedure that takes the input data and produces the desired result after time 't'. While running the algorithm, the computer may use up to s bits of memory space to store temporary information. The running time of an algorithm is usually measured as the number of fundamental steps needed to obtain the result. The fundamental step can be taken to mean a floating point operation. The memory space is usually taken as the maximum amount of memory used at any time during the completion of algorithm. An algorithm that minimizes the running time or memory space is preferred.

A polynomial time algorithm is an algorithm whose worst case running time is $\sim(L^k)$, where L is the size of the input and k is a constant. An exponential time algorithm is any algorithm that is not polynomial time. Similar definitions can be made for polynomial space and exponential space algorithm. Broadly speaking polynomial or exponential behavior will determine whether or not an algorithm is efficient. For example the algorithms used for addition or multiplication are efficient, since the algorithms are polynomial. Where as the converse problem of factorization is more difficult and comes under exponential. To provide a numerical example it is much easier to compute $127 \times 229 = 29083$, then to compute $? \times ? = 29083$. A plot of computational time required for factorization and multiplication problem as a function of input is shown in Fig. 13.17. In general, polynomial time algorithms are said to be efficient whereas exponential ones are said to be intractable.

Computation problems are also classified as decision problems. A decision problem is one where the output is an answer YES or NO. The complexity class P is the set of all decision problems that are solvable in polynomial time. The class NP is the class of decision problems for which a YES answer can be found in polynomial time using some additional information. Factorization of a number does not belong to the category of P, while addition and multiplication of numbers falls within the class.

Any algorithm can be simulated efficiently using a Turing machine. A Turing machine has an infinite memory. In analog computation the representation of information is continuous instead of zeroes and ones. Certain types of analog computation can efficiently solve problems believed to have no efficient solution on a Turing machine. However, when realistic assumptions are made about the presence of inherent noise in electronic circuits, they cannot solve problems efficiently. Some algorithms make use of random number generator. Such algorithms may not have an efficient solution on a Turing machine. However it can give a probable answer. For example an algorithm can determine whether a number is probably prime or else composite with certainty.

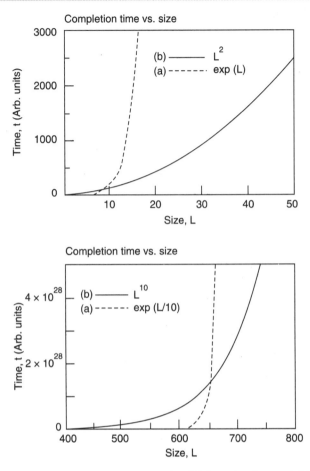

Fig. 13.17 Asymptotic behaviour of computation time for (a) factorization and (b) multiplication problem

13.4.2 Thermodynamics of Computation

Reversibility and entropy are two important thermodynamic concepts which have been extensively used in engineering. Is there any fundamental limit set by laws of physics on the power of computing machines? There are many precedents for this kind of analysis. For example, the efficiency of a heat engine is found to be maximum if all the thermodynamic processes are reversible. Shannon, the pioneer of information science found that there are limits to the amount of information that can be transmitted through a noisy channel. This limit applies no matter how the message is encoded into a signal. It is now believed that the laws of physics do not place any absolute bounds on the speed, reliability and memory capacity of computing machines. *However, if a computer has to run arbitrarily fast, it must operate reversibly i.e.,* it must be constructed from components whose inputs can be deduced from their outputs. This is because irreversible computations involve loss of information. The relation between information and entropy is discussed in section 13.9. Loss of information is related to increase in entropy and entropy changes are related to changes in energy. Hence thermodynamics sets limits on the amount of energy that a system can dissipate as heat. The conventional logic gates are irreversible. For example the inputs of an AND gate cannot be inferred from its output (Fig. 13.18). If the output is 0, the inputs could have been 0/0, 0/1 or 1/0. The most familiar reversible logic gate is the NOT gate. If the output is 0 the input has to be 1 and vice versa. In

1976, Charles Bennett described a universal model for classical computation using reversible gates. He also showed that the computer programs are not slowed down with reversible gates. Later Tomasso Toffoli devised reversible logic gates. This gate is known as Toffoli gate. Toffoli gate has three inputs a, b and c. a and b are known as the first and second control bits while c is the target bit. This gate leaves both control bits unchanged, flips the target bit if both the control bits are set to 1. Otherwise it leaves the target bit unchanged. The truth table of Toffoli gate is shown in Table 13.1.

Name	Graphic symbol	Algebraic function	Truth table
AND		$F = xy$	x y | F 0 0 | 0 0 1 | 0 1 0 | 0 1 1 | 1
OR	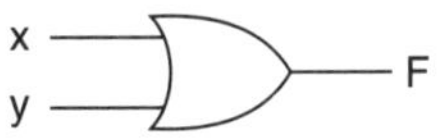	$F = x + y$	x y | F 0 0 | 0 0 1 | 1 1 0 | 1 1 1 | 1
Inverter	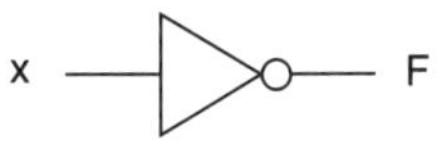	$F = x'$	x | F 0 | 1 1 | 0
Buffer	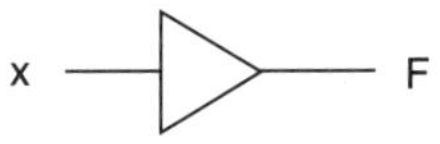	$F = x$	x | F 0 | 0 1 | 1
NAND		$F = (xy)'$	x y | F 0 0 | 1 0 1 | 1 1 0 | 1 1 1 | 0
NOR	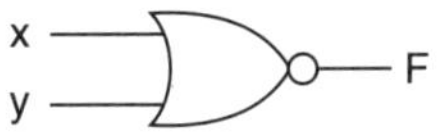	$F = (x + y)'$	x y | F 0 0 | 1 0 1 | 0 1 0 | 0 1 1 | 0
Exclusive-OR (XOR)	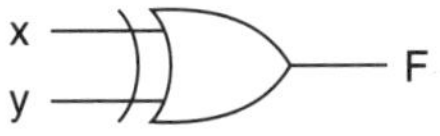	$F = (xy' + x'y)$ $= x \oplus y$	x y | F 0 0 | 0 0 1 | 1 1 0 | 1 1 1 | 0
Exclusive-NOR or equivalence	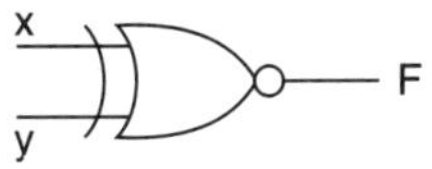	$F = (xy + x'y')$ $= x \odot y$	x y | F 0 0 | 1 0 1 | 0 1 0 | 0 1 1 | 1

Fig. 13.18 Digital logic gates

Table 13.1

Truth table of Toffoli gate					
Input			Output		
a	b	c	a'	b'	c'
0	0	0	0	0	0
0	0	1	0	0	1
0	1	0	0	1	0
0	1	1	0	1	1
1	0	0	1	0	0
1	0	1	1	0	1
1	1	0	1	1	1
1	1	1	1	1	0

Energy required to perform a computation is vanishingly small if computation is reversible. Contrast this with the efficiency of heat engines. The efficiency of a heat engine is maximum when all the thermodynamic processes are reversible. Traditionally in computer science the time and memory requirements are given more importance than the energy requirements. If a computer erases a simple bit of information, the amount of energy dissipated into environment is $k_B T \ln 2$, where T is the temperature of the environment. Present day computers typically dissipate $500 k_B T \ln 2$ energy for each elementary operation. In addition energy is lost due to Joule heating when the current flows in the interconnects.

13.5 CLASSICAL COMPUTATION

A bock diagram of a classical computer is shown in Fig. 13.16. The memory unit stores program as well as input/output and intermediate data. The processor unit performs arithmetic and other data processing tasks as specified by the program. The control unit supervises the flow of information between the various units. The control unit retrieves the instructions one by one from the program which is stored in memory. For each instruction, the control unit informs the processor to execute the operation specified by the instruction. Both program and data are stored in memory. The control unit activates the processor to execute the program instructions and the processor manipulates the data as suggested by the program. The program and the data are transferred into the memory by means of an input device such as teletypewriter or a key board monitor. An output device, such as printer or monitor receives the result of the computations and displays the results. A processor when combined with the control unit forms a CPU.

Digital circuits are invariably constructed with ICs. Digital IC gates are classified not only by their logical operation, but also by the specific logic circuit family to which they belong. Each logic family has its own basic electronic circuit from which more complex digital circuits and functions are developed. The basic circuit in each family is either a NAND gate or a NOR gate. These are universal gates which can perform any Boolean function. The electronic components employed in the construction of the basic circuit are usually used to name the

QUANTUM INFORMATION AND QUANTUM COMPUTATION

logic family. Many different logic families of digital ICs have been introduced commercially. The ones that have achieved widespread popularity are listed below:

TTL–Transistor-Transistor Logic

ECL–Emitter Coupled Logic

MOS–Metal-oxide semiconductor

CMOS–Complimentary Metal-Oxide Semiconductor

I^{2}L–Integrated Injection Logic

More details can be had in the text books on digital electronics and computer design. Classical computers make use of Boolean algebra. Logic gates refer to digital circuits used to implement Boolean algebraic equations. Chips containing more than 100 components are referred to as large scale integrated systems (LSI). Ones with over 10,000 components are known as VLSI systems. Component density, speed and power consumptions are three important design considerations in VLSI system.

A classical computer can be viewed as a collection of bits and gates which act on a certain input *i.e.*, initial state of bits to produce an output state. In a classical computer the processing of bits is done using logic gates. Logic gates are devices that perform elementary operations on bits of information. The Irish logician George Boole showed in the 19th century that any complex logical or arithmetic operation can be accomplished using combination of three simple logic gates NOT, COPY and AND. The various gates used in computers are shown in Fig. 13.18. XOR and NAND gates are essentially irreversible or noninvertible. *When the gates are invertible there is loss of information.* Electronic circuits are made from linear elements such as wires, resistors and capacitors as well as nonlinear elements such as diodes and transistors. These manipulate the bits in different ways. Linear elements alter input signals individually. Nonlinear devices make the input signals passing through them interact. Circuits perform computations by repeating few simple linear and nonlinear tasks over and over again at great speed. NOT and COPY gates are linear while AND gate in nonlinear. The information is stored in magnetic tapes, hard discs. In a classical computer, it is possible in principle to inquire and measure at any time (and without disturbing the computer) the state of any bit in the memory. In a quantum computer such a task is not possible. Also in a classical computer the state of a bit can be copied from one register to another. No-cloning theorem prevents this operation in a quantum computer.

13.6 QUANTUM COMPUTATION

13.6.1 Quantum Gates

Depending on the design of the quantum computer there are different ways of reading or writing a qubit. The method will be explained by considering the simplest case of a hydrogen atom (Fig. 13.19a). Consider a hydrogen atom in its ground state E_0. To write 0 bit on this atom nothing is done. To write 1 the atom is excited to a higher energy E_1. If the photon has energy equal to $E_1 - E_0$, and is applied for a right duration of time, the atom will get excited to energy state E_1. If the atom is already in E_1, then the same photon will bring it to the ground state. Note that the energy of the photon must match the difference between the two energy levels. *If the laser light is applied for half the time it takes for the atom to flip from E_0 to E_1, the atom will be in a superposition state.* If the laser light interacts with an atom which is already in the superposition state, then the atom can make a transition to state E_1 or E_0. The event is probabilistic and the probability can be related to the coefficients of the superposition state as given in eqn. (13.2). Typical quantum gates that can be constructed are the following:

NOT gate: NOT involves bit flipping. This is achieved easily by exciting the atom from the energy state E_0 to E_1 using a laser pulse of energy $(E_1 - E_0)$. The atom can be de-excited from the energy state E_1 to E_0 by using another laser pulse of the same energy. By controlling the length of the laser pulse atom can be prepared in a superposition state. This is shown in Fig. 13.19a

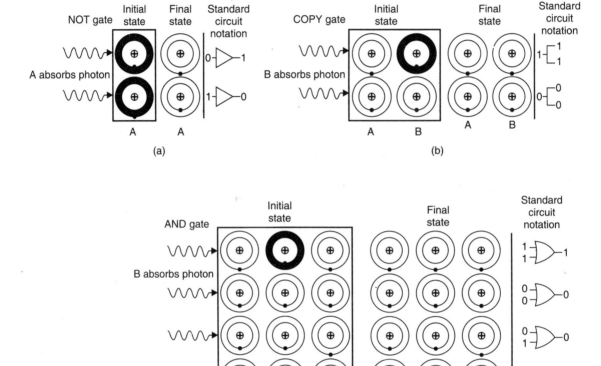

Fig. 13.19 (a) NOT (b) COPY (c) AND quantum gates

COPY gate: COPY gate involves the interaction between two different atoms. Let atom A which can be either in 0 or 1, be in the proximity of another atom B, which is in the ground state. The difference in energy between the states of B will be ΔE_0 if A is in 0 and ΔE_1 if A is in 1. This is because of the interaction between A and B. Now, a pulse of light whose photons have energy equal to ΔE_1 will excite atom B only if the atom A is in 1 state. Hence as shown in the Fig. 13.19b, if A is 1, B becomes 1, and if A is 0, B remains 0.

AND gate: AND gate also depends on atomic interactions. Imagine three atoms A, B and A in close proximity. Now the difference in energy between the ground and excited states of B is a function of the states of the two neighbouring A atoms. Now a laser pulse whose energy equals the difference between the two states of B will excite B only when the atom's neighbouring A's are both 1. Otherwise it will leave atom B unaffected. Clearly this results in the logic operation of AND gate as shown in Fig. 13.19c

Qubit obeys the time dependent Schrödinger equation:

$$i\hbar \frac{\partial \psi}{\partial t} = H\psi$$

where H is the Hamiltonian. In fact all the quantum gates have to mimic the action of the Hamiltonian.

Many of the operations performed by gates are better represented by matrices. Most of the transformations can be visualized by considering the effect of these matrices on a vector in the Bloch sphere. All these matrices have to be unitary.

i.e., $$UU^+ = 1$$

where U^+ is the adjoint of U (obtained by transposing and then taking the complex conjugate of U). Also that the quantum gates are invertible or reversible.

In quantum computers the qubits are processed using special gates. Different types of gates used described below. Following are the action of some of the quantum gates on single qubits.

Z gate: Its action is to leave $|0\rangle$ unchanged and flip the sign of $|1\rangle$ to give $-|1\rangle$. It is represented by the matrix:

$$Z = \begin{bmatrix} 1 & 0 \\ 0 & -1 \end{bmatrix} \begin{bmatrix} |0\rangle \\ |1\rangle \end{bmatrix} = \begin{bmatrix} |0\rangle \\ -|1\rangle \end{bmatrix}$$

Hadmard gate: Its action is to create a superposition state out of the pure state. Geometrically it can be thought of as rotation of the Bloch sphere about y-axis by 90°, followed by a reflection in the x-y mirror plane (Fig. 13.20). It is represented by the matrix operation;

$$H = \frac{1}{\sqrt{2}} \begin{bmatrix} 1 & 1 \\ 1 & -1 \end{bmatrix} \begin{bmatrix} |0\rangle \\ |1\rangle \end{bmatrix} = \frac{1}{\sqrt{2}} \begin{bmatrix} |0\rangle + |1\rangle \\ |0\rangle - |1\rangle \end{bmatrix}$$

Thus $|0\rangle$ is changed to $|0\rangle + |1\rangle$ and $|1\rangle$ is changed to $|0\rangle - |1\rangle$ states respectively.

NOT gate: This changes $|0\rangle$ to $|1\rangle$ and $|1\rangle$ to $|0\rangle$. This is represented by the matrix operation:

$$X = \begin{bmatrix} 0 & 1 \\ 1 & 0 \end{bmatrix} \begin{bmatrix} |0\rangle \\ |1\rangle \end{bmatrix} = \begin{bmatrix} |1\rangle \\ |0\rangle \end{bmatrix}$$

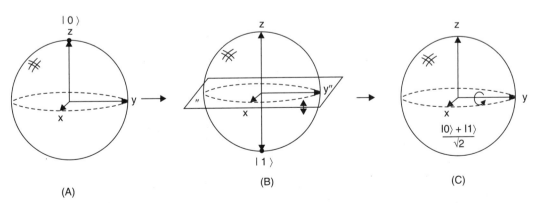

Fig. 13.20 Geometric representation of Hadmard gate operation on the block sphere

The other important gates are:

$$Y = \begin{bmatrix} 0 & -i \\ i & 0 \end{bmatrix}; \text{Phase gate} = \begin{bmatrix} 1 & 0 \\ 0 & i \end{bmatrix}; T = \begin{bmatrix} 1 & 0 \\ 0 & \exp\left(\frac{i\pi}{4}\right) \end{bmatrix} = \exp\left(\frac{i\pi}{8}\right) \begin{bmatrix} \exp\left(\frac{-i\pi}{8}\right) & 0 \\ 0 & \exp\left(\frac{i\pi}{8}\right) \end{bmatrix}$$

Multiple qubit gates:

CNOT gate: The gate has two input qubits known as the control bit and the target bit respectively.

[Control qubit, Target qubit]

If the control qubit is set to 0, then the target bit is left alone. If the control qubit is 1, then the target bit is flipped. The gate operation can be represented as:

$|00\rangle \to |00\rangle;\ |01\rangle \to |01\rangle;\ |10\rangle \to |11\rangle;\ |11\rangle \to |10\rangle$

The matrix representation for the same is given by:

$$CNOT = \begin{bmatrix} 1 & 0 & 0 & 0 \\ 0 & 1 & 0 & 0 \\ 0 & 0 & 0 & 1 \\ 0 & 0 & 1 & 0 \end{bmatrix} \begin{bmatrix} |00\rangle \\ |01\rangle \\ |10\rangle \\ |11\rangle \end{bmatrix} = \begin{bmatrix} |00\rangle \\ |01\rangle \\ |11\rangle \\ |10\rangle \end{bmatrix}$$

The Toffoli gate can be understood as the controlled-controlled NOT. Toffoli gate is a reversible gate and ensures that quantum computers are capable of performing any classical computation as well. It is the general belief that all classical logical circuits can be ultimately explained using quantum mechanics. Quantum gates are reversible. Many classical gates such as NAND are inherently irreversible. Toffoli gates can be used to simulate classical gates. It has three input bits and three output bits as shown in table 13.1. Two of the bits are control bits that are unaffected by the action of Toffoli gate. A third gate is a target bit that is flipped if both control bits are set to 1. Otherwise it is left alone. Mathematically, the effect can be represented as

$(a, b, c) \to (a, b, c \oplus ab)$ $\oplus$ denotes modulo 2 addition

Toffoli gate is the inverse of itself.

It has been proved that all one qubit gates and one two-qubit CNOT gate are sufficient to implement all logical operations on a quantum computer.

13.6.2 Quantum Circuits

A quantum circuit is composed of elementary logic gates connected together by *wires*. The only purpose of wire is to transfer a quantum state from the output of a gate to input of another one and eventually to a measurement device. The *nature of wires* depends on the technological realization of the qubit. For instance wires could be trajectories of flying spins. Two spins may have their trajectory deflected or enter a zone in which they interact (a two qubit gate). A conceptual diagram of a simple circuit involving a number of quantum gates is shown in Fig. 13.21.

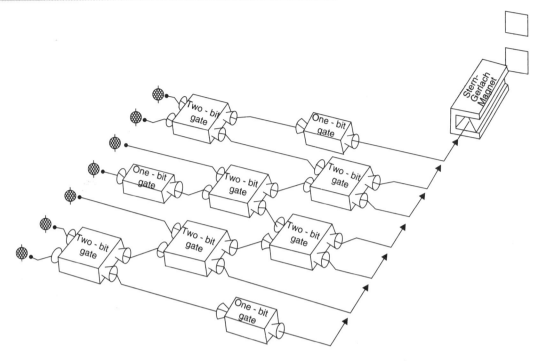

Fig. 13.21 An idealized picture of quantum network

13.6.3 Quantum Algorithms

The input data for quantum computation is fed into a qubit register consisting of a number of qubits. The basic idea in a quantum algorithm is to execute each computational step using one or two qubit gate operations. After every step, the superposition state of the quantum register will change. Each of the superposed state has a measurement probability associated with it that has a wavelike behavior as given by eqn. (13.9). It can interfere with the probabilities of other superposition states constructively or destructively. *Getting the desired answer to a calculation means processing the information in such a way that the undesired solutions interfere destructively, leaving only the desired state or few desired states, at the end.* Needless to say to write such an algorithm is a challenge to physicists, mathematicians and computer scientists. A single final measurement or a series of measurements gives the probability distribution from which a desired answer can be calculated.

The advantage of quantum computation lies in *parallel processing*. In a qubit register at every computational step the initial superposition evolves into different superposition. Each coefficient in the mathematical expression is affected during such an evolution. Hence it is a case of performing massive parallel computation. A quantum computer performs the same mathematical operation on 2^L input numbers at the same time. To accomplish the same task in any classical computer, one needs 2^L steps or 2^L different registers working in parallel. Thus in certain types of computation, it is possible to economize memory and time. Quantum computers are found to be efficient in computational problems related to discrete Fourier transforms. It is this which has made quantum computers ideal for solving some problems like factorizing a number, finding the discrete logarithm or carry out searches in a data base.

Quantum algorithms are superior to classical algorithms at least in three classes of problems. Grover's algorithm for carrying and searches in a database and Shor's algorithm for factorizing a number has been a land mark in quantum computation. Quantum algorithms have proved to be useful in solving following problems:

(a) Given two numbers a and b, to find s such that $b = a^s$.

(b) Given $f(x) = a^x$, to find r such that $f(x+r) = f(x)$

(c) Given N, to find r such that $x^r = 1 (mod\ N)$

(d) *Search algorithms in a data base*

Quantum search algorithms have the potential to be dramatically faster than the conventional algorithms. A classical search of a random list of N items to find a particular one requires an examination of at least $N/2$ of them, to have 50% success probability. A quantum algorithm for the same involves $\sqrt{N}$ steps. A good example is the search in a database. An example will illustrate this point. Let us say the problem is to find a person's name in a telephone directory, given his or her phone number. If the directory contains N entries, then on an average, one has to search through $N/2$ entries before one finds the name of the person. Grover's quantum algorithm does better. It finds the name after searching through $\sqrt{N}$ entries, on an average. So for a directory of 10,000 names the task would require 100 steps instead of 5000 steps on a classical computer. The algorithm works by first creating a superposition of all 10,000 entries in which each entry has the same likelihood of appearing in response to a measurement made on the system. Then to increase the probability of a measurement producing the required entry the superposition is subjected to a series of quantum operations that recognize the required entry and increase its chances of appearing. Note that the recognition is possible because one has the phone number but not the name.

(e) *Factorization*

Factorization of a large composite number is essentially intractable on any classical computer. The largest number that has been factorized has 129 digits. No one can even conceive of how one might factorize a thousand digit number. The computational time needed is much more than the estimated time of the universe. In cryptography this inability of the classical computer to factorize a number is utilized to have a secure key. In quantum computation the factorizing of a number is done using Peter Shor's algorithm. On account of this the security offered in conventional cryptology involving factorization of a number is threatened.

(f) *Simulation algorithms*

A quantum computer is an excellent basic research tool for understanding the principles of quantum mechanics by simulation. Feynman pointed out as far back as in 1982 that simulation of quantum systems on classical computers was not feasible. In simulation experiments one solves the Schrödinger equation. The key challenge in simulating quantum systems is the exponential number of differential equations. For a single qubit evolving according to Schrödinger equation, there are two differential equations. Hence for a system of n qubits there are 2^n differential equations. Solving so many differential equations is a hard problem. Simulating quantum systems by Monte Carlo techniques on a classical computer is a difficult task that takes a large computer time.

13.7 DESIGN OF QUANTUM COMPUTERS

13.7.1 Introduction

Designing of quantum computers and quantum communication systems has proved to be challenging. Many different types of technology have been tried to design a quantum computer. Small quantum computers capable of doing dozens of operations on a few qubits represent the present state of art. The physical devices include trapped ions, photons, and nuclear spins. Experimental demonstration of quantum cryptography has been carried out with success. However, experts are divided in their opinion about the use of quantum computers on a large scale and the viability of quantum networks. Whether the present day classical computers will become obsolete is a question for which there is no certain answer.

The design of quantum computers ought to take into consideration the following:

(a) the qubits ought to be stored for a long time, long enough to complete the computation.

(b) The qubits must be well isolated from environment to minimize errors due to decoherence and noise. This is explained in section 13.8.

(c) Qubits ought to be measured efficiently and reliably

(d) The quantum states of individual qubits ought to be manipulated using suitable gates. Controlled interaction between qubits has to be induced for nonlinear gate action.

(e) The gate action ought to be performed with high precision if the device is to perform reliably.

In the following section some of the physical systems which have been considered for quantum computers are described. The details of the quantum gate operations have been left out.

13.7.2 Ion Trap Quantum Computer

Ion Traps Using Electric Field

A collection of ions stored in a linear trap has been utilized to design a quantum computer. A string of ions can be trapped using a combination of oscillating and static electric fields with the electrode geometry as shown in Fig.13.22. This is also known as linear radio frequency Paul trap. The device sets up a high frequency radio frequency field superimposed on a static electric field. The ions arrange themselves like beads on a string on account of the existing electrostatic repulsion between them. The adjacent ions are separated by a few wavelengths of light. This separation is large enough to ensure that ions can be addressed individually by the laser. The ions also are trapped for long times (~minutes). Although the ions are trapped along the axis of the linear trap they are not at rest, but oscillate around their equilibrium positions. After having trapped them, the next step is to cool these using methods of laser cooling. It is fairly standard today to cool ions to temperatures of the order of millikelvins.

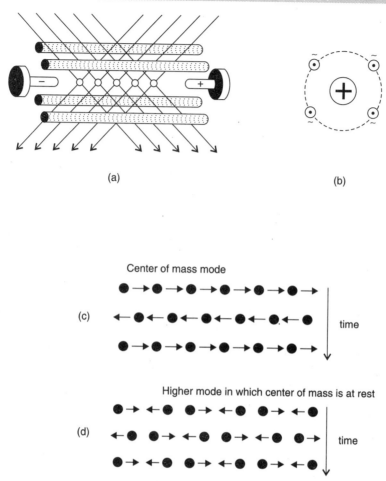

Fig. 13.22 (a) Ion trap computer (b) Electrode configuration-side view (c, d) Vibrational modes of the trapped atoms

Laser Cooling

The phenomenon of laser cooling has been used in atomic physics to limit the motion of thermal motion of atoms. The 1997 Nobel Prize in physics was awarded to Steven Chou, Claude Cohen-Tannoudji and William Philips for this technique. Atoms emitted from an oven will have a spread of velocities around some average value. If this atomic beam encounters a laser beam in the opposite direction, the atoms get slowed down. Atoms have characteristic energy levels that allow them to absorb and emit radiation of specific frequencies. Atoms moving with respect to the laser beam (of speed c) will 'see' the laser frequency shifted because of the Doppler effect. For example, atoms moving toward the laser beam will encounter a laser with high frequency and atoms moving away from the laser beam will encounter a laser beam with low frequency. Even atoms moving in the same direction within the beam of atoms will see slightly different frequencies depending on the velocities of the various atoms. Now if the frequency of the laser beam is tuned to the precise frequency seen by the faster atoms so that those atoms can be excited by absorbing the radiation, those faster atoms will be slowed down by absorbing the momentum of the laser radiation. Essentially the phenomenon can be viewed as scattering of a photon by an atom. Every atom or an ion has a momentum given by the de Broglie relation $p = h/\lambda$. The atoms slow down because photons strike them in opposite direction (Fig. 13.23).

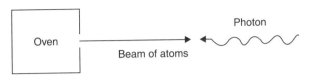

Fig. 13.23 Laser cooling of atoms

The slower atoms will *see* a laser beam that has been Doppler shifted to a lower frequency than is needed to absorb the radiation and these atoms are not as likely to absorb the laser radiation. The net effect is that the atoms as a whole are slowed down and their velocity spread is reduced.

As the atoms slow down, they see that that the Doppler shifted frequencies of the laser change, and the atoms no longer absorb the laser radiation. They continue with the same lower velocity and velocity spread. There are two methods to make the deceleration process continue. In one, the frequency of the laser beam is increased to keep the radiation consistent with the Doppler shifted frequency needed to excite the atoms. In the other method, the laser frequency is kept constant, but sophisticated magnetic fields are used to vary the frequencies needed to excite the atoms by changing the excited atomic energy levels. In the presence of a magnetic field, the energy levels of an atom get affected. By using six intersecting laser beams coming in at different angles atoms are essentially isolated and the average velocity is zero. With this technique atoms have been cooled to temperatures below 0.2 μK.

In the ion trap quantum computer, the ions are cooled till they attain the ground state. On supplying energy the ions get excited to higher vibrational levels. But being quantum particles, the ions can exist in a superposition of the ground state and the vibrational state. So the vibrational states can be used to store a qubit. Since all the ions take part in vibration, this qubit is shared among them. *It is as if the collective motion is a kind of data bus, allowing all the ions to temporarily share the information and become entangled.* If we excite the center of mass mode all the ions will oscillate in phase (Fig.13.22b). If the first ion is excited by a laser and the absorbed photon excites the center of mass mode, then even an ion at the end of the chain will feel this. This provides the necessary interaction for the construction of nonlinear gates. Single qubit gates do not require center of mass mode and they are manipulated by using a suitable laser pulse. The ion trap quantum computer is attractive for the following reasons.

(a) The ions store qubits reliably and for long times.

(b) It provides a means of performing universal qubit quantum gates.

(c) It offers a physical mechanism to perform reliable measurements of qubits.

Ion Traps Using an Optical Cavity

This is similar to ion trap computer. The qubits are represented by ions or atoms and are kept fixed in space in an optical cavity (a region of space bounded by reflecting mirrors). If the wavelength of light is of the order of the dimensions of the cavity, the light can be used to control the quantum states of the atom. The optical cavity has standing electromagnetic waves. A strong coupling can be achieved between single atom or an ion and a single mode of the electromagnetic field. The experimental set up is shown in Fig. 13.24. The coupling can be used to apply quantum gates between the field mode and the ion, thus making it possible to transfer quantum information between separated ion traps via optical cavities and fibers.

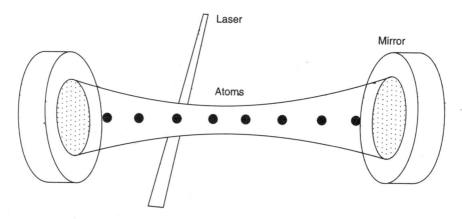

Fig. 13.24 Quantum computer using atoms confined in a radiation field

13.7.3 Nuclear Magnetic Resonance Quantum Computer

Basic Principle

Nuclear Magnetic Resonance (NMR) is based upon the measurement of absorption of electromagnetic radiation in the radio-frequency region of ~ 4 to 600MHz by the nuclei of atoms. The energy levels involved in the absorption of radiation arise because the specimen is placed in a magnetic field. Nucleus is composed of protons and neutrons. Nucleus also carries a spin just like an electron. The nuclear spin I for any nucleus may be zero, a half integer or an integer. Its value is determined by the mass number A and the atomic number Z of the nucleus. The following rules are useful in determining the value of I.

1. If A is odd and Z is even or odd, I is a half integer.
2. If A and Z are both even, I is zero.
3. If A is even and Z is odd, I is an integer.

The spin angular momentum is given by $Ih/2\pi$.

The magnetic moment associated with the nucleus is proportional to the spin angular momentum and is given by:

$$\mu = \gamma\sqrt{I(I+1)}\hbar$$

Here γ is the gyromagnetic ratio. In the presence of a magnetic field (B) the nuclear dipole acquires an additional potential energy (E) given by:

$$E = -\mu B\cos\theta$$

where θ is the angle between the axis of the dipole and the field direction. Classically, the nuclear dipole can have any orientation with respect to the magnetic field. However, owing to space quantization its orientation is limited to few directions. A nucleus having spin I is found to take $(2I + 1)$ discrete orientations. The component of angular momentum for these states in any direction will have values $I, (I-1), (I-2)....-(I-1), -I$. The $(2I+1)$ orientations correspond to discrete energy states. Thus in the presence of the magnetic field a single energy state of the nucleus is split into $(2I+1)$ states. The transitions between these states give rise to emission or absorption of radiation in the radio-frequency region.

QUANTUM INFORMATION AND QUANTUM COMPUTATION

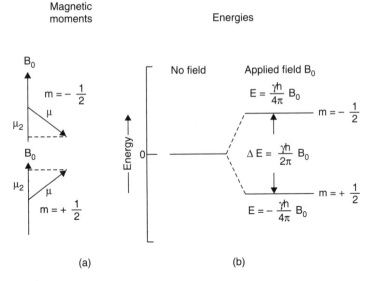

Fig. 13.25 Energy levels of spin ½ nucleus with and without the applied magnetic field

The four nuclei that have been of greatest interest have spin quantum number of ½. These include H^1, C^{13}, F^{19} and P^{31}. For these nuclei, two spin states exist, corresponding to $I = ½$ and $I = -½$. Hence, the nucleus can have only two orientations in the magnetic field. The energies corresponding to these states is given by (Fig. 13.25).

$$E_{+1/2} = -\frac{\gamma \hbar}{2} B$$

$$E_{-1/2} = +\frac{\gamma \hbar}{2} B$$

Thus, the difference in energy ΔE between the two is given by:

$$\Delta E = \frac{\gamma \hbar}{2} B - \left(-\frac{\gamma \hbar}{2}\right) B = \gamma \hbar B \tag{13.13}$$

The transition between these two energy states can be brought about by absorption or emission of electromagnetic radiation of a frequency ν that corresponds to energy ΔE. Thus, by substituting $h\nu = \Delta E$ in eqn (13.13) we obtain the frequency of radiation required to bring about the transition

$$\nu = \frac{\gamma B}{2\pi}$$

NMR experiments are performed by detecting the amount of energy absorbed. The basic features of an NMR spectrometer are (1) a source of radio-frequency radiation, (2) a receiver coil, (3) a dc magnetic field and (4) a recorder or an oscilloscope. Figure 13.26 shows the schematic diagram of an NMR spectrometer. The sample tube is placed between the poles of a powerful magnet. The source of energy, that is, the radio-frequency field, is generated in the coil connected to the *rf* oscillator. The detector coil is placed at right angles to both the direction of the magnetic field and the transmitter coil. The magnet is provided with the sweep coils which are used to vary the field over a range of a few gauss. In a typical experiment, the frequency of the *rf* field is fixed and the magnetic field is varied until the resonance condition is reached. The nuclear magnetic moment transition induces an *emf* in the detector coil which is then amplified and displayed on the recorder or oscilloscope.

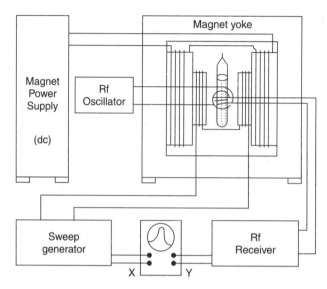

Fig. 13.26 Block diagram of the NMR apparatus

Nowadays what is more popular in the Fourier Transform NMR. Figure 13.27 shows a simplified block diagram of a typical FTNMR spectrometer. The central component of the instrument is a highly stable magnet in which the sample is placed. The sample is surrounded by a transmitter/receiver coil. The excitation pulses are produced by a piezoelectric crystal controlled continuous oscillator having an output frequency v_c. This signal passes into a pulser switch and a power amplifier which creates an intense and reproducible pulse of radio frequency radiation, which passes into the transmitter coil. The pulse shape, pulse length and pulse shape can be adjusted. In the figure the pulse length is shown to be 5μs. The resulting free induction decay (FID) which results due to the changes in the orientation of the nuclear magnetic dipoles is picked up by the same coil which now serves as the receiver. The signal is amplified and is transmitted to a detector. The phase sensitive detector computes the difference between the nuclear signal v_n and the crystal oscillator output v_c. This leads to the low frequency, time domain signal as shown in the right of the figure. This signal is digitized and stored in a computer for frequency analysis by Fourier transform program. This gives the frequency domain spectrum for which the resonance absorption takes place. The resulting frequency domain output is similar to the spectrum produced by a scanning continuous wave experiment.

In NMR quantum computers, the two spin states of an atomic nucleus in a magnetic field are used as the two states of a qubit. Different atoms in a molecule can be distinguished and so a molecule can be used as a quantum computer, with each nucleus providing a single qubit. Simple logic gates which only affect a single qubit are easily implemented using radio frequency fields. These interact strongly with nuclear spins, allowing them to be controlled with great precision. In order to perform interesting computations, however, more complex gates are needed, which allow the state of one qubit to affect other qubits in the computer. This requires some interaction between nuclear spins, so that one spin can sense the state of other spins in the molecule. This is easily achieved, as there is a naturally occurring spin-spin coupling interaction between the various nuclei.

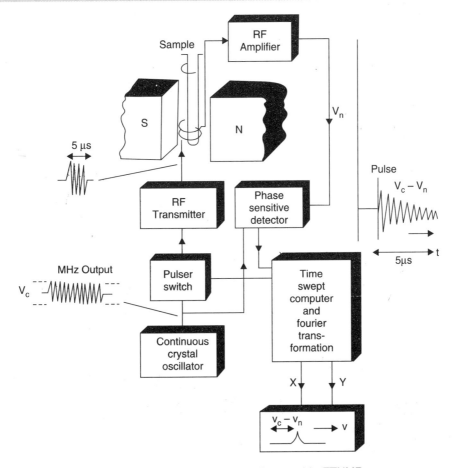

Fig. 13.27 A radiofrequency pulse used in FTNMR

NMR computers are different from other quantum computers in one important respect. NMR signal from a single molecule is far too weak to be detected. Hence, it is necessary to use a large number of identical copies to amplify the signal. This is not difficult as even a few milligrams of a chemical compound will contain the required number of molecules. It is, however, impossible to ensure that all the copies start the calculations in the same initial state, and so different copies will in effect perform different calculations, making it extremely difficult to extract the desired result. This inability has also been overcome.

13.7.4 Quantum Computers with Semiconductor Technology

Quantum computers based on semiconductor technology is an attractive proposition on account of the maturity of semiconductor technology. Solid state realizations of quantum gates and quantum computation principles are presently being studied. One possibility is to use the technology developed for quantum dots. These are regions of semiconductor on the nanometer scale contained within the bulk semiconductor. The motion of electrons is confined to very small regions and hence electrons can exist only in a few well-defined quantum states. Working with semiconductors is now a mature technology and this approach offers the prospect of a new source of single-photon pulses for quantum communications. Another possibility that has been suggested is to use phosphorous nucleus in silicon. With the developments with nano-technology it is quite possible to precisely place single phosphorous atom (nuclear spin – 1/2) within a crystalline wafer of Si (nuclear spin – 0), positioned beneath lithographically patterned electrostatic gates. These gates allow manipulation of the electron cloud surrounding the P^{31} nuclei to perform single qubit operations via modulation of the magnetic field seen by the P^{31} nuclei. Additional gates located above the region separating the P^{31} dopants can be used to artificially create electron distributions connecting adjacent P^{31}, much like a chemical bond, thus allowing two qubit operations to be performed. The fabrication constraints of such a scheme are extremely challenging. The gates should be separated by 100A or less, and the P^{31} dopants must be positioned precisely and in an ordered array. A conceptual picture of the same is shown in Fig.13.28. It is like a conventional metal oxide semiconductor structure, with the exception that what the gate will manipulate is not an electron current, but the position of the electron wave function. The electron surrounds the phosphorous donor as a kind of extended Bohr atom, and using an electric field one can move the electron around. This enables to control the interactions of nuclear spins. *A gates* control the resonance frequency of the nuclear spin qubits while *J-gates* control the electron-mediated coupling between adjacent spins. The ledge over which the gates cross localizes the gate electric field in the vicinity of the donors.

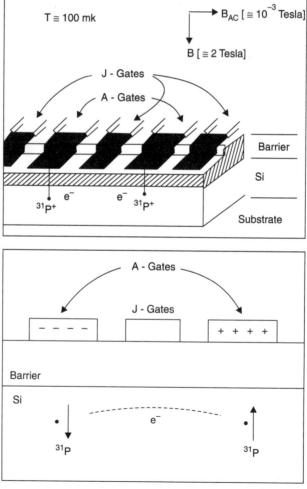

Fig. 13.28 Conceptual diagram of a quantum computer using semiconductor technology

13.8 DISADVANTAGES OF QUANTUM COMPUTERS

13.8.1 Introduction

In a classical computer, one bit of information is represented by the presence or absence of a large number of electrons ($\sim 10^8$) and that is a great advantage. Therefore, a small fluctuation in the number of electrons arising due to noise does not disturb the computer at all. On the contrary, in a quantum computer, the qubit is stored in the state of a single atom or a single photon. Consequently, its operation is more prone to errors. Even worse, quantum computer critically depends upon the survival of the quantum superposition states, which are very sensitive to *decoherence* arising out of interaction with the environment. The number of qubits whose coherence has to be maintained is the most important factor in computation.

Interactions with the environment are the fundamental source of noise in both classical and quantum systems. Modern classical computers with solid state technology are very reliable. Failure rates of the order of 10^{-17} or even less is common in electronic components. Extra efforts needed to protect against noise may not be worthwhile. Quantum computers are very fragile and require substantial protection against noise. As a result, quantum computers are far more susceptible to make errors than conventional digital computers. Thus, the most formidable problems associated with quantum computers are those of *decoherence* and *noise*. A device that works effectively even when its elementary components are imperfect is said to be *fault tolerant*. From a practical point of view, what is important is to devise ways of *fault tolerant computation*.

13.8.2 Decoherence

Apart from the technical difficulty of working with a single atom or ion or photon, one of the biggest obstacles in the operation of quantum computer is quantum decoherence. If the system is not perfectly isolated from its environment, the quantum dynamics of the surrounding apparatus will also be relevant to the operation of quantum computer. Its effect will be to make the evolution of qubit states non-unitary, because it is not a closed system any more. The slightest outside disruption (heat or light for example) can destroy the balance of the quantum states that store the information and this would make the computation impossible. Even the very process of the measuring the state of a qubit can destroy the coherence. Decoherence leads to an exponential increase of the error rate with the input size. Therefore, a given quantum algorithm cannot be considered as efficient any more, regardless of how weak the interaction with environment may be.

All quantum systems suffer from decoherence. But it has very specific implications for quantum computers. Decoherence causes a quantum computer to lose two of its key properties: entanglement between two qubits and interference phenomena. Two simple examples will illustrate these effects. As discussed in section 13.2 an electron passing through two slits in Young's double experiment gives rise to an interference pattern. It is assumed here that electron does not interact with any other particle. If there is an electron near one of the slits, the interference pattern is destroyed. (Fig.13.29) *i.e.*, interaction with the environment has the effect of destroying the superposition state. Any measurement involves interaction with the environment. The interaction with the environment also destroys the entanglement. Consider the EPR pair, where the quantum state of the system is given by $|H\rangle|V\rangle - |V\rangle|H\rangle$. This implies that the two photons in the pair must always have opposite polarizations, but we do not know which one has which polarization. Entanglement is evident here, since the two-photon system cannot be expressed as a product of single-photon states. If a measurement is performed on each of the photons to determine their individual polarizations, the system will experience strong decoherence and the entangle state will become a mixture of classical states, $|H\rangle|V\rangle$ and $|V\rangle|H\rangle$.

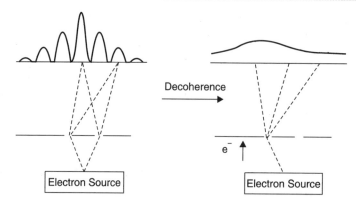

Fig. 13.29 Quantum decoherence in a single particle Young's double slit experiment

A classical computation follows a single definite pathway in time from the beginning to the end. In a quantum computer, the computation can be split up into several pathways that, by the principle of superposition, evolve in time parallel. These pathways recombine at the end because each of them caries a definite phase. They can interfere constructively. Loss of phase coherence will destroy the interference. Therefore, the coherence time τ_c needs to be longer than the running time of the computation. Continuous advances in the experimental techniques aim at increasing τ_c. The minimum time required to execute one quantum gate is denoted by τ_G and is called the clock cycle. It is equal to the time needed to flip a qubit and it is estimated as

$\tau_G = h/2\pi(\Delta E)$, where ΔE is the typical energy splitting in the two-level system. The ratio $M = \tau_c/\tau_G$ is equal to the largest number of steps permitted in a quantum computer using these qubits. In the table below are given the characteristic times in seconds for the various two-level system that be used as qubits.

Table 13.2

System	$\tau_G(s)$	$\tau_c(s)$	M
Nuclear spin	10^{-2}–10^{-8}	10^{3}–10^{6}	10^{5}–10^{14}
Electron spin	10^{-7}	10^{-3}	10^{4}
Trapped ions	10^{-14}	10^{-1}	10^{13}
Electron quantum dot	10^{-6}	10^{-3}	10^{3}

13.8.3 Noise

In classical computers also *noise* has to be taken into account. *Noise* is a term to denote an undesirable signal that is superimposed upon the normal operating signal. The ability of circuits to operate reliably in a noise environment is important in many applications. One defines *noise margin*. This is the maximum noise voltage added to the input signal of a digital circuit, that does not cause an undesirable change in the circuit output. There are two types of noise to be considered viz., *dc noise* and *ac noise*. *DC noise* is caused by the *drift* in the voltage levels of a signal. *AC noise* is a random pulse that may be created by other switching signals. In quantum computers *noise* plays a much more destructive role than in classical computers. Since the information is stored in a *single atom or photon* and gate operations are performed on *single atom or photon*, any fluctuation in the state of a single atom or photon, due to the usual thermal noise or interaction with the environment will have disastrous consequences. Computer architecture may change the effects of noise. If many operations are performed in parallel, the noise is minimal than if the operations are performed serially. Architectural design is a well-developed field of study in classical computers.

13.8.4 Errors in Quantum Computation

The following type of errors might occur in quantum computers due to *decoherence* and *noise*.

Phase Errors: Flipping of the bit $|0\rangle \to |1\rangle$; $|1\rangle \to |0\rangle$; and phase changes of the type $|1\rangle \to -|1\rangle$; $|0\rangle \to -|0\rangle$ can occur. A phase error is serious because changes in state $|\psi^+\rangle \to |\psi^-\rangle$ and $|\phi^+\rangle \to |\phi^-\rangle$ can occur which are orthogonal to each other.

Small errors: Let us say that the qubit is intended to be in the state $|\psi\rangle = a|0\rangle + b|1\rangle$. An error in gate operation might change a and b by an amount of $\sim \varepsilon$. These errors can accumulate over a period of time.

Measurement errors: During the process of measurement an irreversible change in the state of a qubit occurs. This has to be kept in mind while devising the quantum algorithm.

13.8.5 Quantum Error Correction

The presence of decoherence and *noise* has led to the theory of quantum error correcting codes and fault tolerant quantum computation. The effects of *decoherence* and *noise* can be limited by using error correction procedures. In quantum error correction schemes, a single qubit is encoded by a string of qubits. The encoding is carried out such that an error in a single qubit will still allow the original qubit to be inferred from the code string and thus retrieved. Error correction is an important component in classical information theory as well. Following assumptions are made in applying the error correction:

(a) Decoherence occurs independently on each of the qubit of the system

(b) The performance of gate operations on some qubits do not cause decoherence in other qubits of the system

(c) Reliable quantum measurements can be made so that error detection can take place

(d) Systematic errors in the unitary operations associated with quantum gates be made very small.

With these assumptions quantum computation is *possible* even with faulty and decohering components and *noise*.

13.9 QUANTUM INFORMATION THEORY

13.9.1 Information and Entropy

The classical information theory was developed by Shannon. Shannon developed a method to quantify or measure information. This method brings about the relation of information with entropy. Information is sent by a source in the form of symbols. An information source consists of a discrete set of symbols. (A Morse key consists of dashes and dots; A string of 0's and 1's is used in digital communication.) In general, any message emitted by the source consists of a string or sequence of symbols. A channel is any physical medium such as wire, a cable, a radio or a television link. Suppose a source sends one of M possible symbols $s_1, s_2, s_3, \ldots, s_M$ in a statistical independent sequence. It is assumed that the probability of occurrence of a particular symbol sent during a time interval does not depend on the symbols sent by the source. Let $p_1, p_2, p_3, \ldots, p_M$ be the probabilities of the occurrence of symbols. Let the message contain N symbols. Then the symbol s_1 will occur $p_1 N$ times, the symbol s_2 will occur $p_2 N$ times and so on. Treating the messages of unit length, the information content of the ith symbol is defined to be

$\log_2(1/p_i)$. Note that if $p_i=1$, then the information content of the symbol is zero. The information content of the message is given by:

$$I = \sum_{i=1}^{M} Np_i \log_2 (1/p_i) \text{ bits}$$

The average information per symbol is given by:

$$H = \frac{I}{N} = \sum_{i=1}^{M} p_i \log_2 (1/p_i) \text{ bits/symbol}$$

The logarithmic form is taken to ensure that the information content of any complex message is the sum of its individual components. It is to be observed that this definition of information has an earlier precedent in statistical physics or mechanics. In statistical mechanics one determines the macroscopic properties of an ensemble by considering the properties of the constituent atoms or molecules. Both information theory and statistical mechanics are statistical in nature. If we consider a gas, then the macroscopic properties of the gas are defined by temperature, pressure etc. A microstate corresponds to that of an energy state of a molecule. A macrostate corresponds to a large number of microstates. If W is the number of microstates available then entropy(S) is defined as

$$S = k_B \log W$$

Note the W is a measure of *order* in an ensemble. Entropy is thus a direct measure of *disorder*. Greater the disorder greater is the entropy. Assume a gas where all the molecules have the same energy. For such a gas the number of microstate is just one. Hence, entropy is zero. On the other hand, when the state of motion of gas molecules is completely chaotic, then the entropy is large. The entropy of ice is much less than that of water. While the entropy of water is less than that of water vapour. Entropy of a substance in the crystalline state is less when compared with its entropy in the glassy state.

Note that information and entropy are two sides of the same coin. Greater the internal order, greater the knowledge or the information of the system's internal make-up. Thus, *Entropy is the reverse of information*. It is conventional to define information as negative entropy.

13.9.2 Information Communication Channels

Communication channels carry information. They are divided into two categories *viz.*, analog and discrete. An analog channel accepts a continuous amplitude, continuous time electrical waveform as its input and it produces at its output a noisy smeared version of the input wave form. The capacity of an analog channel is it bandwidth. A discrete channel accepts a sequence of symbols as its input and produces an output sequence which is a replica of the input sequence except for occasional errors. The channel capacity is an important parameter of data communication system. It denotes the rate at which data can be transferred over the channel with an arbitrarily small probability of error. It is given by Shannon's theorem:

$$C = B \log_2 \left[1 + \frac{S}{N} \right]$$

where B is the bandwidth, S is the signal power and N is the noise power.

The fundamental results for quantum channels are contained in two theorems. Noiseless channel coding theorem and noisy channel coding theorem. The first one tells how many bits are required to store information by the source. The second theorem indicates how reliably information is transmitted in a noisy communication channel. In a noisy channel, the reliable information transmission is achieved through error correcting codes. Entangled pairs are additional resource in quantum communication theory. The subject of quantum information theory is in its infancy. One is interested to know how much information can be transmitted using a quantum noisy channel and how to protect the information from getting corrupted by noise. These theoretical questions are similar to the ones posed by Shannon, the pioneer of information science. Error correcting codes are routinely used in communication. There are a variety of codes which are used for example in compact disc players, computer modems and satellite communication.

The differences between classical information theory and quantum information theory are summarized here.

No-Cloning: One of the ways of preserving information is to copy it. The *no cloning* theorem forbids it. Observation in quantum mechanics generally destroys the quantum state and makes recovery impossible. In processing of information using a classical computer, information can be copied and replicas can be prepared. This is the case with a recording device or Xeroxing. The no-cloning theorem in quantum information theory prevents such a possibility. This is a property of nature. This follows from the basic idea that quantum systems are perturbed during the act of measurement according to the principle of uncertainty. Eavesdroppers are thus unable to use cloning to try and beat quantum cryptosystem by copying each qubit during transmission. If cloning is possible then it would be possible to send signals faster than light. This would violate the basic postulate of special theory of relativity.

If $|\psi\rangle$ and $|\phi\rangle$ are two non-orthogonal states, it is *impossible* to build a device which will use these states as inputs to give copies of the input states. If $|\psi\rangle$ and $|\phi\rangle$ are orthogonal, it does not prohibit their cloning. Hence, different states of classical information can be thought of as orthogonal quantum states.

13.9.3 Quantum Superdense Coding

Classical communication channel can transmit a minimum of one of bit information *i.e.*, 0 or 1. If the information carrier is a quantum system, then it is no longer true. According to quantum information theory, the minimum information that can be sent using quantum ways is *two qubits* instead of one qubit as in the case of classical information theory. The ability to send more than one qubit of information relies on entanglement. Entangled states permit a completely new way of encoding information.

Consider the Bell states. These four states are each maximally entangled and together they form an orthonormal basis-called the Bell basis-for states of two qubits. The main point here is that each of these states can be prepared from the state $|\psi^-\rangle$ by Alice alone performing purely local operations on her particle. Indeed consider the four 1-qubit unitary operations written in the $\{|0\rangle, |1\rangle\}$ basis:

$$U_{00} = \begin{bmatrix} 1 & 0 \\ 0 & 1 \end{bmatrix}; U_{01} = \begin{bmatrix} 1 & 0 \\ 0 & -1 \end{bmatrix}; U_{10} = \begin{bmatrix} 0 & -1 \\ -1 & 0 \end{bmatrix}; U_{11} = \begin{bmatrix} 0 & 1 \\ 1 & 0 \end{bmatrix}$$

which respectively act on the qubit basis by transforming them as follows

$U_{00}[0, 1] = [0, 1]$ identity operator

$U_{01}[0, 1] = [0, -1]$ phase shift operator

$U_{10}[0, 1] = [-1, 0]$ combined bit flip/phase shift operator

$U_{11}[0, 1] = [1, 0]$ bit flip operator

All the operations listed above are unitary and they do not change the total probability of finding the system in the states $|0\rangle$ and $|1\rangle$. In working with Bell states, all these four operations are relatively easy to perform in experiments with photons, atoms and ions.

The effect of these operations on Bell states (qubit A) is given by:

$$U_{01} |\psi^+\rangle = |0\rangle_A |1\rangle_B - |1\rangle_A |0\rangle_B = |\psi^-\rangle$$
$$U_{01} |\psi^-\rangle = |0\rangle_A |1\rangle_B + |1\rangle_A |0\rangle_B = |\psi^+\rangle$$
$$U_{01} |\phi^+\rangle = |0\rangle_A |0\rangle_B - |1\rangle_A |1\rangle_B = |\phi^-\rangle$$
$$U_{01} |\phi^-\rangle = |0\rangle_A |0\rangle_B + |1\rangle_A |1\rangle_B = |\phi^+\rangle$$

$$U_{10} |\psi^+\rangle = -|1\rangle_A |1\rangle_B + |0\rangle_A |0\rangle_B = |\phi^-\rangle$$
$$U_{10} |\psi^-\rangle = -|1\rangle_A |1\rangle_B - |0\rangle_A |0\rangle_B = -|\phi^+\rangle$$
$$U_{10} |\phi^+\rangle = -|1\rangle_A |0\rangle_B + |0\rangle_A |1\rangle_B = |\psi^-\rangle$$
$$U_{10} |\phi^-\rangle = -|1\rangle_A |0\rangle_B - |0\rangle_A |1\rangle_B = -|\psi^+\rangle$$

$$U_{11} |\psi^+\rangle = |1\rangle_A |1\rangle_B + |0\rangle_A |0\rangle_B = |\phi^+\rangle$$
$$U_{11} |\psi^-\rangle = |1\rangle_A |1\rangle_B - |0\rangle_A |0\rangle_B = |\phi^-\rangle$$
$$U_{11} |\phi^+\rangle = |1\rangle_A |0\rangle_B + |0\rangle_A |1\rangle_B = |\psi^+\rangle$$
$$U_{11} |\phi^-\rangle = |1\rangle_A |0\rangle_B - |0\rangle_A |1\rangle_B = -|\psi^-\rangle$$

Imagine Alice wants to send some information to Bob. Entanglement makes it possible for Alice to send two bits of information to Bob using just one photon, provided Bob has access to both qubits and is able to determine which of the four Bell states they are in. This has been demonstrated using the experimental set up shown in Fig. 13.30. The experiment relies on the process of spontaneous parametric down-conversion in a crystal to produce entangled states of very high quality and intensity. The non-linear properties of the crystal convert a single ultraviolet photon into a pair of infrared photons with entangled polarizations. The experiment uses quarter and half-wave polarizations plates which shift the phase between the two polarization states of a photon by $\lambda/4$ and $\lambda/2$, respectively to make the unitary transformations between the Bell states. After performing these four operations U_{ij} Alice sends this single qubit to Bob. On receiving the qubit from Alice, Bob performs a Bell measurement (distinguishing the four Bell states) on the joint state of the two qubits and reliably reads out the values of ij (table 13.3). Since there are four possibilities, essentially *two* bits of information have been sent to Bob by Alice.

Table 13.3 Experimental detection of Bell States

Alice settings of half wave and quarter wave plates		State sent	Bob's recording of events
$\lambda/2$	$\lambda/4$		
0°	0°	$|\psi^+\rangle$	Coincidence between D_{1H} and D_{1V} or Coincidence between D_{2H} and D_{2V}
0°	90°	$|\psi^-\rangle$	Coincidence between D_{1H} and D_{2V} or Coincidence bwtween D_{2H} and D_{1V}
45°	0°	$|\phi^+\rangle$	Two photons in $D_{1H}/D_{1V}/D_{2H}/D_{2V}$
45°	90°	$|\phi^-\rangle$	Two photons in $D_{1H}/D_{1V}/D_{2H}/D_{2V}$

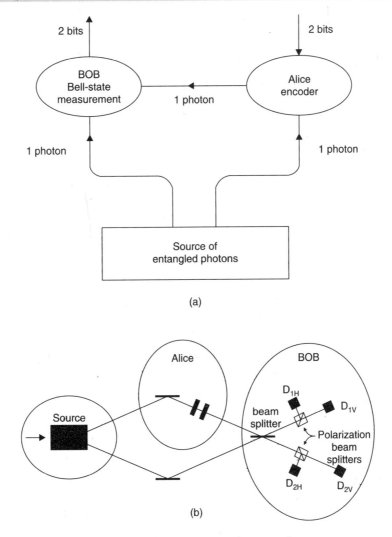

Fig. 13.30 Quantum superdense coding

13.9.4 Quantum Teleportation

Entire quantum particles can be *sent* from one place to another over any distance. The process starts with a sender and a receiver, Alice and Bob. The pair are spatially separated but are in possession of photons A and B respectively, which are entangled. Alice also holds photon C, which is in a state that she wants to teleport to Bob. Entangled particles have the property that a measurement on one immediately determines the state of the other. If Alice performs a procedure that entangles photons A and C, photon B, held by Bob is forced to adopt the original state, a particular polarization, say of Photon C. Bob can only measure this state if Alice sends him details of the type of experiment he must do to get the message. And this can only be done at or below the speed of light. Although only the quantum state of photon C is teleported, when photon B adopts this state, it cannot be distinguished from photon C. To all intents and purposes, it has become photon C. This is what physicists mean when they say photon C has been teleported from Alice to Bob.

Teleportation was first demonstrated using the experimental set up shown in Fig.13.31. Pairs of entangled photons, with polarization orthogonal to each other, are generated by splitting an ultraviolet laser pulse using a crystal called parametric down-converter. One of the pair (photon A) is sent to Alice while the other (photon B) is sent to Bob. Meanwhile, a message photon (C) is prepared in a state that is to be teleported to Bob—in this case 45° polarization. This is sent to Alice and arrives coincidently with photon A at a beam splitter. If the photons leave the splitter and strike both detectors they have become entangled and Alice sends notice of the entanglement to Bob. Bob can then carry out a measurement on photon B to confirm that it is in the 45° polarization state that the message photon C started off in. Quantum teleportation over a distance of ~10km has been demonstrated.

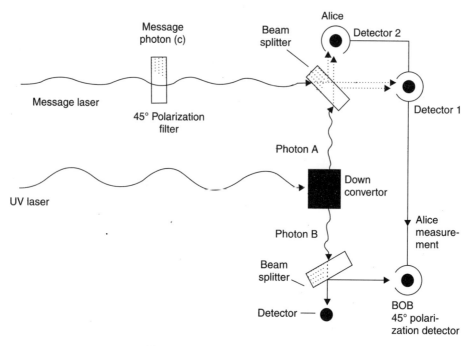

Fig. 13.31 Quantum teleportation

13.9.5 Quantum Cryptography

Cryptology is a mathematical science of secret communication. It has a long and distinguished history of military and diplomatic uses dating back to ancient Greeks. Today the ability to ensure the secrecy of military or diplomatic communications is as vital as ever. But cryptology is also becoming more and more important in everyday life. With the growth of computer networks, internet and electronic mail for business and financial transactions, communication of confidential information is becoming an inevitable necessity. The principles of quantum physics can be effectively used to achieve the desired results.

The two main goals of cryptography are:

(a) sender and the intended recipient must be able to communicate in a form that is *unintelligible to third parties. i.e.,* the message must have secrecy.

(b) Any tampering and altering of the message by the third party ought to be evident to the sender and the receiver. In other words the message must have integrity and authenticity.

These goals can be accomplished with provable security if sender and recipient are in possession of shared secret *key* material. Thus, key material, which is a truly random sequence of digits, is a very valuable commodity even though it conveys no useful information itself. One of the principal problems of cryptography is therefore the so-called *key distribution problem*. How do the sender and intended recipient come into possession of secret key material while being sure that third parties (eavesdroppers) cannot acquire even partial information about it? It is provably impossible to establish a secret key with conventional communication and so key distribution has relied on the establishment of a physically secure channel (trusted couriers) or the conditional security of *difficult* mathematical problems in public key cryptography. However, provably secure key distribution becomes possible with quantum communications. It is this procedure of key distribution that is accomplished by quantum cryptography and *not the transmission of an encrypted message itself*. Hence, a more accurate name is quantum key distribution (QKD).

QKB is based on the fundamental postulate of quantum physics that *every measurement perturbs a system*. Imagine sending a message carried by a single quantum state, such as linearly polarized photons oriented at various angles. If the bits are not altered during transmission, then one can be sure that no eavesdropper has measured the values of those bits. Thus, quantum cryptography turns an apparent limitation—namely that a measurement perturbs the system-into a potentially useful process, in which the perturbation uncovers the presence of an eavesdropper.

Cryptography is the art of hiding information in a string of bits which is meaningless to any unauthorized party. To achieve this goal, an algorithm is used to combine a message with some additional information known as the *key* to produce a cryptogram. This technique is known as *encryption* (Fig.13.32). The person who encrypts and transmits the message is traditionally known as Alice, while, the person who receives it is called Bob. Eve is unauthorized, malevolent eavesdropper. For a crypto-system to be secure, it should be impossible to unlock the cryptogram without Bob's key. In practice, this demand is often softened so that the system is just extremely difficult to crack. The idea is that the message should remain protected as long as the information it contains is valuable.

Cryptosystems come in two main classes. This depends on whether the key is shared in secret or in public. The *one time pad* system, which was proposed by Vernam in 1935, involves sharing a secret key and is the only cryptosystem that provides proven, perfect secrecy. In this scheme, Alice encrypts a message using a randomly generated key and then simply adds each bit of the message to the corresponding bit of the key. The scrambled text is then sent to Bob, who decrypts the message by subtracting the same key. Because the bits of the scrambled text are as random as those of the key, they do not contain any information. Although perfectly secure, the problem with this system is that it is essential for Alice and Bob to share a common *secret key*, which must be at least as long as the message itself. They can only use the key for a single encryption—hence, the name *one-time pad*. If they used the key more than once, Eve could record all of the scrambled messages and start to build up a picture of the key. Furthermore, the key has to be transmitted by some trusted means, such as a courier, or through a personal meeting between Alice and Bob. This procedure can be complex and expensive and may even amount to a loophole in the system. It is interesting to note that if Eve wanted to crack the one-time pad by trying out all possible keys one by one, she would obtain message for each key and would then have to search through all of them. But she would have absolutely no way of knowing which was the right one!

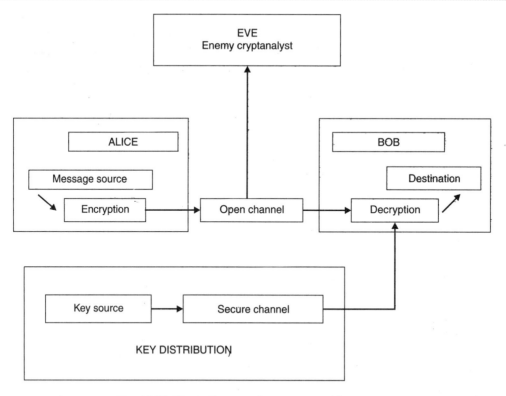

Fig. 13.32 Block diagram of a cryptographic system

The other class of crypto-systems shares a public key. The first *public key crypto-systems* were proposed in 1976. For Alice to transmit a message with a public-key crypto-system, Bob first chooses a private key. He uses this key to compute a public key, which he discloses publicly. Alice then uses this public key, to encrypt her message. She transmits the encrypted message to Bob, who decrypts it with his private key. The encryption-decryption process can be described mathematically as a one-way function with a trapdoor-namely, the private key to obtain the original message. One therefore only need to know this key to obtain the original message. In other words, if Bob knows what the trapdoor is, he can do the reverse calculation and reveal the message from the encrypted text.

These systems are based on so called one-way functions, in which it is easy to compose the function $f(x)$ given the variable x, but difficult to go in the opposite direction and compute x from $f(x)$. In this context, the word "difficult" means that the time to do a task grows exponentially with the number of bits in the input. (*i.e.*, it is a hard problem on the computer). Factoring large integers is a candidate for such a one-way function. For example, it only takes a few seconds to work out that 107×53 is 5671. But it takes much longer to find the prime factors of 5671. However, some of these one-way functions have a *trapdoor* which means that there is in fact an easy of doing the computation in the difficult direction, provided that you have some additional information. For example, if you were told that 107 was one of the prime factors of 5671, the calculation would be relatively easy.

Public-key crypto-systems are convenient and they have become very popular over the last 20 years. The security of Internet, for example, is partially based on such systems. The most common example is the RSA crypto-system which is named after the inventors Ronald Rivest, Adi Shamir and Leonard Adleman. Its secrecy is actually based on the fact that the time needed to calculate the prime factors of an integer increases exponentially with the number of input bits. However, this system suffers from two potential major flaws. First nobody knows for sure if factorization is actually as difficult as we currently think. One could easily improve the safety of the RSA by choosing a longer key, but if an algorithm were found that could factorize numbers quickly, it would immediately annihilate the security. With the development of Shor's algorithm on the quantum computer it is a distinct possibility. Meanwhile, other public-key crypto-systems also rely on unproven assumptions for their security, which could be weakened or suppressed by theoretical or practical advances. One would then have no choice but to turn to secret key crypto-systems.

The principles of cryptography that we have so far described have all been entirely general. Vernam's system, however, requires, Bob and Alice to share a secret key, and it is here that quantum physics enters the scene. Quantum cryptography allows two physically separated parties to create a random secret key without resorting to the services of a courier. It also allows them to verify that the key has not been intercepted. Quantum key distribution is therefore really a better name for quantum cryptography. When used with Vernam's one-time pad scheme, the key allows the message to be transmitted with proven and absolute security. Quantum cryptography is not therefore a totally new crypto-system. But it does allow a key to be securely distributed and is consequently a natural complement to Vernam's cipher.

Consider the BB84 communication protocol named after Charles Bennet and Gilles Brassard in 1984. Alice and Bob are connected by a quantum channel and a classical public channel. If single photons are being used to carry the information, the quantum channel is usually an optical fiber. The public channel, however, can be any communication link, such as a phone line or an Internet connection. In practice, the public link is usually also an optical fiber, with both channels differing only in the intensity of light pulses that code the bits; one photon per bit for the quantum channel, hundreds of photons per bit for the classical public channel.

For sending a polarized photon a calcite crystal is used. Calcite crystal exhibits *birefringence*. A light ray is split into *ordinary* and *extraordinary* rays. Ordinary and the Extraordinary rays are plane polarized and their planes of polarization are mutually perpendicular to each other. The behavior of a single photon entering a calcite crystal is shown in Fig. 13.11. Horizontally polarized photons pass through while the vertically polarized photons are deflected. The diagonally polarized photons are unpolarized at random in either the vertical or horizontal direction and are shifted accordingly.

First, Alice has four polarizers, which can transmit single photons that are linearly polarized either vertically, horizontally; at +45° or at −45°. She sends a series of photons down the quantum channel, having chosen at random one of the polarization states for each photon. She also records her choice.

Second, Bob has two analyzers. One analyzer allows him to distinguish between horizontally and vertically polarized photons. The other allows him to distinguish between photons polarized at +45° and −45°. Bob selects one analyzer at random, and uses it to record each photon. He writes down which analyzer he used and what is recorded. Note that every time Bob uses an analyzer that is not compatible with Alice's choice of polarization, he will not be able to get any information about the state of the photon. For example, if Alice sent a vertically polarized photon and Bob chose the analyzer designed to detect photons at + or − 45°, there is a 50% chance that he will find the photon in either +45° channel or the − 45° channel. And even if he finds out later that he chose the wrong analyzer, he will have no way of finding out which polarization states Alice sent.

Third, after exchanging enough photons, Bob announces on the public channel the sequence of analyzers he used and not the results that he obtained.

Fourth, Alice compares this sequence with the list of bits that she originally sent and tells Bob on the public channel on which occasions his analyzer was compatible with photon's polarization. She does not however tell him which polarization states she sent. If Bob used an analyzer that was not compatible with Alice's photon, the bit is simply discarded. For the bits that remain, Alice and Bob know that they have the same values, provided that an eaves dropper did not perturb the transmission. They can now use these bits to generate a key and send encrypted messages to one another.

The above sequence of operations is shown in Fig. 13.33. To assess the secrecy of their communication, Alice and Bob select a random part of their key and compare it over the public

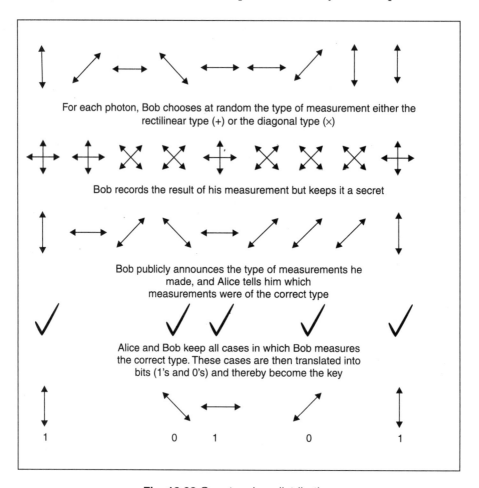

Fig. 13.33 Quantum key distribution

channel. Obviously, the disclosed bits cannot then be used for encryption any more. If their key had been intercepted by an eavesdropper, the correlation between the values of their bits will be reduced. For example, if Eve has the same equipment as Bob and cuts the fiber and measures the signal, she will always get a random bit whenever she chooses the wrong analyzer. *i.e.*, 50% of cases. But having intercepted the signal, Eve still has to send a photon to Bob, to cover her tracks. Therefore, in half of the cases in which Alice's and Bob's analyzers match, Eve would have sent a photon that is incorrectly polarized. However, in half of these cases, the photon would accidentally leave Bob's analyzer through the correct channel—in which case, Eve's presence goes undetected. The point is that if Eve had been listening in, one in four of Alice's and Bob's bit values would disagree. In other words, her eavesdropping strategy could be easily detected. There are other eavesdropping strategies that produce a lower disagreement rate. But since all measurements perturb either the vertical-horizontal polarization states or the diagonal states, or all four states, all eavesdropping strategies perturb the system to some extent. Hence, if Alice and Bob do not notice any discrepancies in the subset of their keys, they can be sure that their key has not been intercepted by Eve. They can then use their key with total confidence to encrypt the message.

In practice there will always be some errors in the transmission. Uncorrelated bits may originate from several experimental imperfections. For example, Alice has to ensure that she creates photons that are in exactly the states she chose. If, for instance, a vertical photon is incorrectly polarized at an angle of 84°, there is a 1% possibility that Bob will find it in the channel for horizontally polarized photons. A similar problem arises for Bob, if his polarizer cannot distinguish perfectly between two-orthogonal states, he will detect photons in the wrong channel from time to time. Another difficulty is ensuring that the encoded bits are maintained during transmission. A vertically polarized photon, for example, should still be vertically polarized by the time it reaches Bob. But due to the birefringence of the fiber, the polarization states received by Bob will, in general, be different from those sent by Alice. Even worse, changes due to the mechanical or thermal environment can produce fluctuations on a timescale of seconds or minutes, which means that the alignment of the two analyzers has to be continuously monitored. This is possible in principle but not very convenient. In fact, the number of transmission errors and hence the quantum-bit error rate is dominated by the noise of the detector. In other words, most errors are not due to photons that have been incorrectly detected. The errors arise when a photon fails to reach a detector as expected and the wrong detector registers a dark count instead. Unfortunately, at the wavelengths where the fiber losses are low (*i.e.*, 130nm) relatively noisy, low-efficiency home made signal photon detectors have to be used. To overcome these problems, Alice and Bob have to apply a classical error-correction algorithm to their data so that they can reduce the errors below an error rate of 10^{-9} – the industry standard for digital telecommunications. And since they cannot be sure if the presence of uncorrelated bits was due to the poor performance of their set-up or to an eavesdropper, they have to assume the worst-case scenario-namely that all of the errors were caused by Eve. To reduce the amount of information the Eve may have obtained, Alice and Bob therefore use a procedure known as "privacy amplification" in which several bits are combined to one. This procedure ensures that the combined bits only correlate if Alice and Bob's initial bits are the same. But Eve ends up with a totally different series of bits, because she only knows a fraction of the initial bits. The problem with privacy amplification is that it shortens the key length a lot and it is only possible up to certain error, which means that Alice and Bob have to be careful to introduce as few errors as possible—when they initially send their quantum bits. Quantum key distribution has been experimentally demonstrated.

13.9.6 Other Applications

Distributed Computation: Assume that there are N quantum computers sharing N entangled states of qubits. If each computer does a particular computation locally then all the computers communicate their results to each other instantaneously. Needless to say, this leads to enhanced power of computation. In distributed computation a number of computers are networked to perform a computation. *Net worked quantum computers require exponentially less communication to solve certain problems when compared with their classical computers.*

Communication through nil capacity channels

According to classical information theory if A and B are connected by a zero capacity channel no information transfer can take place between them. Adding one more zero capacity channel will not make any difference. However, according to quantum information theory, it is possible to send information from A and B if the second zero capacity channel is configured to send information from B to A as shown in Fig.13.34.

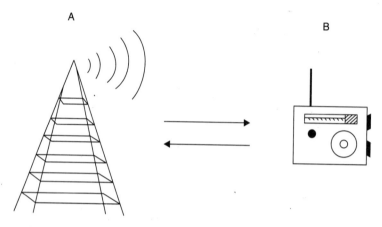

Fig. 13.34 Quantum communication using zero capacity channels

REFERENCES

1. M.A. Nielsen and I.L. Chuang, *Quantum Computation and Quantum Information*, Cambridge University Press, New Delhi, 2002.
2. T. Hey and P. Walters, *The New Quantum Universe*, Cambridge University Press, Cambridge, 2003.
3. S. Adams, *Frontiers–Twentieth Century Physics*, Taylor and Francis, London, 2000.
4. R.L. Liboff, *Introductory Quantum Mechanics*, Pearson Education, New Delhi, 2003.

QUESTIONS

1. What is a qubit?
2. Explain the distinction between a bit and a qubit.
3. Explain single particle interference with suitable examples.
4. Describe Stern Gerlach experiment and discuss its significance.
5. Explain how Stern Gerlach experiment confirms the behaviour of electron spin as a qubit.
6. What is a qubit register? Describe the state of a two and three qubit registers.
7. Explain quantum entanglement.
8. Explain Bell's inequality and its relevance to the postulates of quantum mechanics.
9. Explain the postulates of locality and reality.
10. Explain entanglement swapping.
11. Write a short note on Turing machine.
12. Explain Computational complexity.

13. Give examples of reversible and non-reversible gates.
14. Give examples of quantum gates and write down their matrix representation.
15. Describe the action of a Toffoli gate.
16. What are the advantages of quantum algorithms over classical algorithms?
17. What are the basic differences between classical and quantum computers?
18. Describe an ion trap quantum computer.
19. Describe a NMR quantum computer.
20. Explain decoherence.
21. How do errors in quantum computation arise and how are they overcome?
22. Discuss the relation between information and entropy.
23. Explain quantum superdense coding.
24. Explain quantum teleportation.
25. What are the important characteristics of cryptographic systems?
26. Explain quantum cryptography.
26. How is quantum cryptography superior to classical cryptography?
27. Explain Shannon's theorem.
28. Distinguish between classical and quantum information channels.

14
Basics of Nanoscience and Nanotechnology

14.1 INTRODUCTION

Nanoscience is the study of atoms and molecular structures whose physical size, at least in one dimension is between 1–100nm. Nanotechnology is the design and fabrication of devices using such nanostructures. Miniaturisation has been the dominant trend in the technology of the 20th century. This is amply demonstrated in the fields of electronics, communication and computers. Miniaturisation leads to devices which are faster, more versatile, more efficient, less power consuming and cheaper. Nanotechnology is the logical extension of MEMS (Microelectromechanical systems). However, its scope is immense and many opine that it is the onset of the *second industrial revolution*. It is going to affect the human society in a way more radical than the earlier innovations in science and technology. The present chapter is intended to provide the student a glimpse of the revolution that is underway.

14.2 SCALING LAWS IN MINIATURISATION

In nanotechnology the devices have dimensions of the order of nanometers. A study of the scaling laws helps to get an idea of the magnitude of physical quantities involved in such devices. Scaling is strictly dependent on the size of the physical objects similar to scaling in geometry. It is assumed that the physical properties themselves do not change with the decrease in size which is strictly not correct. With the reduction in size the laws of quantum mechanics begin to operate and might give rise to a behaviour not predicted by scaling laws.

Scaling Laws in Rigid Dynamics

While designing nano or microsystems, the first thing is to decide what forces are to be used for actuation. A matrix formulation is used to describe the scaling results. This helps to show a number of different cases and scale sizes in a simple format. The force scaling vector is defined as:

$$F = [l^F] = \begin{bmatrix} l^1 \\ l^2 \\ l^3 \\ l^4 \end{bmatrix} \tag{14.1}$$

Other quantities are derived based on the above vector.

From paticle kinematics we have

$$s = ut + \frac{1}{2}at^2$$

BASICS OF NANOSCIENCE AND NANOTECHNOLOGY

Assuming $u = 0$,

$$a = \frac{2s}{t^2}$$

The dynamic force F can be expressed as

$$F = ma = \frac{2sM}{t^2} \propto (l)(l^3)(t^{-2}) \qquad (\because M \propto V \propto l^3)$$

Acceleration: The scaling vector for acceleration is obtained as follows:

$$a = \frac{F}{M} = \frac{[l^F]}{[l^3]} = [l^F][l^{-3}] = \begin{bmatrix} l^1 \\ l^2 \\ l^3 \\ l^4 \end{bmatrix}[l^{-3}] = \begin{bmatrix} l^{-2} \\ l^{-1} \\ l^0 \\ l^1 \end{bmatrix} \qquad (14.2)$$

Time: The scaling vector for time is derived as follows:

$$t = \sqrt{\frac{2s}{a}} = \sqrt{\frac{2sM}{F}} \propto \{[l^1][l^3][l^{-F}]\}^{1/2} = [l^2][l^F]^{-1/2}$$

$$= [l^2]\begin{bmatrix} l^1 \\ l^2 \\ l^3 \\ l^4 \end{bmatrix}^{-1/2} = \begin{bmatrix} l^{1.5} \\ l^1 \\ l^{0.5} \\ l^0 \end{bmatrix} \qquad (14.3)$$

Power density: Power consumed per unit volume is an important parameter in the design of any system.

$$\text{Power density} = \frac{P}{V} = \frac{Fs}{tV} = \frac{[l^F][l^1]}{[l^2][l^F]^{-1/2}[l^3]} = \begin{bmatrix} l^{-2.5} \\ l^{-1} \\ l^{0.5} \\ l^2 \end{bmatrix} \qquad (14.4)$$

By combining equations (14.1), (14.2), (14.3) and (14.4) the scaling laws for the rigid body can be summarised as indicated in Table 14.1. This is useful in the design process.

Table 14.1 Scaling laws for rigid dynamics

Order	Force scale	Acceleration	Time	Power density
1	l^1	l^{-2}	$l^{1.5}$	$l^{-2.5}$
2	l^2	l^{-1}	l^1	l^{-1}
3	l^3	l^0	$l^{0.5}$	$l^{0.5}$
4	l^4	l^1	l^0	l^1

Scaling laws in electrostatic forces: The electrostatic energy stored in a parallel plate capacitor plate is given by

$$U = \frac{1}{2}CV^2 = \frac{1}{2}\frac{\varepsilon_0 \varepsilon_r A}{d}V^2$$

Here A is the area of plates and d is the distance between them. V is the potential applied and is equal to $E \times d$ where E is the electric field intensity. Thus keeping the electric field constant, the potential $V \propto d$. Hence scaling for U is of the following form

$$U \propto \frac{(l^0)(l^0)(l^2)(l^1)^2}{(l^1)} = (l^3)$$

This suggests that a factor of 10 decrease in linear dimensions of the plate reduces the energy stored 10^3 times. Since the force is a derivative of energy, it follows that the force between the plates decreases by a factor of 10^2. Note that since capacitance is the ratio of area to the distance between the plates, it scales as l^1.

Scaling in Electricity

The resistance of a wire is given by

$$R = \frac{\rho l}{A} \propto (l^{-1})$$

The resistive power loss is given by:

$$P = \frac{V^2}{R} \propto (l^1) \text{ assuming that the voltage is maintained constant.}$$

Thus the power loss is linear in scaling. The power supply needed for the device is proportional to volume. Hence what is of interest is the ratio of power loss to available power. This ratio is expressed as

$$\frac{P_{loss}}{P_{supply}} \propto \frac{(l^1)}{(l^3)} = (l^{-2})$$

This shows that a decrease in size of the device 10 times actually increases the power dissipation by a factor of 100.

Scaling of Electromagnetic Forces

The inductance of the solenoid is given by

$$L = \mu_0 n^2 l A$$

where n is the total number of turns and l is the length of the solenoid and A is the area enclosed by the coil of the solenoid. Assuming that the area enclosed by the coil is constant

$$L \propto (l^1)$$

Thus both capacitance and inductance scale as l^1. The natural frequency of an LCR circuit would scale as

$$\omega = \frac{1}{\sqrt{LC}} \propto \frac{1}{\sqrt{(l^1)(l^1)}} \propto l^{-1}$$

Thus a reduction in size by a factor of 10 would increase the frequency of an LC circuit by 10. Similarly the Q-factor of LCR circuit would scale as:

$$Q \propto \frac{L}{R} \times \omega \propto \frac{(l^1)}{(l^{-1})} \times (l^{-1}) \propto (l^1)$$

Thus a reduction in size by a factor of 10 would decrease the Q-factor by an equal factor.

BASICS OF NANOSCIENCE AND NANOTECHNOLOGY

The capacitive time constant is independent of scaling. This is because
$$\tau = R \times C \propto (l^{-1})(l^1) \propto (l^0) \propto \text{constant}$$

The inductive time constant scales as follows:
$$\tau = \frac{L}{R} \propto \frac{(l^1)}{(l^{-1})} \propto (l^2)$$

Thus on decreasing the size by a factor of 10, the inductive time constant would decrease by a factor of 100.

The energy stored in an inductor is given by
$$U = \frac{1}{2} L I^2$$

If the radius of the wire forming the inductor is increased the current will increase because the current is proportional to the area of cross-section of the conductor. Thus
$$[U] \propto [l][l^2]^2 \propto [l^5]$$

Hence the force which is proportional to the derivative of magnetic stored energy $\propto l^4$.

Thus a 10 times reduction in size would lead to 10^4 times reduction in the electromagnetic force. This is in sharp contrast to electrostatic force which reduces as l^2. Thus it is not favourable to use electromagnetic force for the purposes of actuation.

Scaling Laws for Thermal Systems

The heat capacity is proportional to mass. Hence

Heat capacity $\propto$ volume $\propto (l^3)$.

The thermal conductance is given by
$$\text{Thermal conductance} \propto \frac{\text{area}}{\text{length}} \propto (l^1)$$

The thermal time constant would scale as follows:
$$\text{Thermal time constant} \propto \frac{\text{heat capacity}}{\text{thermal conductance}} \propto \frac{(l^3)}{(l^1)} \propto (l^2)$$

Scaling Laws for Some Classical Systems

A cantilever is a common arrangement in MEMS and NEMS. The bending stiffness of a cantilever scales as follows:
$$\text{Bending stiffness} \propto \frac{(\text{radius})^4}{(\text{length})^3} \propto \frac{(l^4)}{(l^3)} \propto l^1$$

The natural frequency of the cantilever scales as follows:
$$Frequency \propto \sqrt{\frac{\text{stiffness}}{\text{mass}}} \propto \sqrt{\frac{(l^1)}{(l^3)}} \propto (l^{-1})$$

Thus a reduction in the dimensions of a cantilever by a factor 10 would increase its natural frequency by a factor of 10.

14.3 QUANTUM NATURE OF THE NANOWORLD

The physical and chemical properties of materials at the nanoscale are determined by quantum mechanics. The properties are also affected by dimensionality. A bulk solid has three dimensions. If only one dimension is reduced to the nanorange while the other two dimensions remain unchanged, the nanostructure is known as a one *dimensional quantum well*. If two dimensions are reduced to nanorange, the structure is termed a *quantum wire or a two dimensional quantum well*. A *quantum dot or a three dimensional quantum well* refers to the case when all the three dimensions are in the nanoscale range. The physical as well as the chemical properties of materials are determined by the energy level scheme and the way the electrons are distributed among these energy levels. This is obtained by solving the Schrödinger's equation which forms the basis for quantum mechanics. Electrons obey Fermi-Dirac distribution.

The density of states $D(E)$ refers to the number of energy states available between energy E and $E + dE$ per unit volume and per unit energy interval. This function multiplied by the Fermi-Dirac distribution gives dN, the number of electrons occupying the available energy states between energy E and $E + dE$. These functions for a quantum dot, quantum wire, quantum well and the bulk solid are shown in Fig. 14.1. This is the basic reason for the different behaviour of nanoparticles as compared with the bulk.

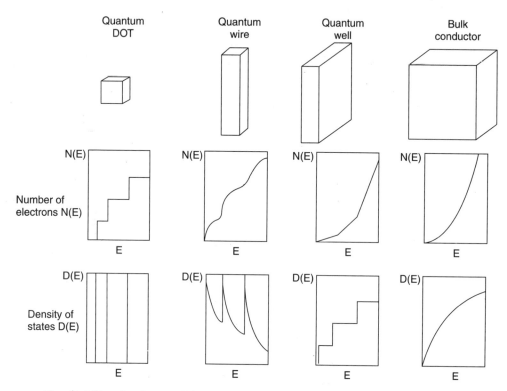

Fig. 14.1 Density of states D(E) and Number of electrons N(E) for a quantum dot, quantum wire, quantum well and bulk conductor N(E) is the integral of dN

In classical physics a particle in a potential well will stay for ever and cannot come out of the well unless energy greater than the potential energy is supplied to it. However according to quantum mechanics, atoms exhibit *tunneling i.e.,* the atoms can come out of the potential even without being supplied the requisite energy to overcome the potential barrier (Fig. 14.2).

This basic idea is made use of in scanning tunneling microscope which is a basic tool of nanotechnology. Tunneling makes nanofabrication difficult and challenging.

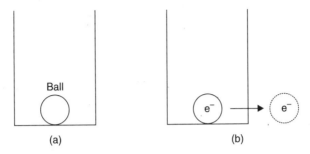

Fig.14.2 (a) A macroscopic ball in a potential well (b) An electron in a potential well exhibits tunneling

14.4 SIZE DEPENDENT PROPERTIES

Many material properties show dependence on the size of the particles constituting them. Typical examples are given here. In a bulk solid the number of atoms at the surface is a very small fraction of the total number of atoms constituting the solid. However, in a nanomaterial the number of atoms at the surface is comparable to the total number of atoms in the material. This is given in Table 14.2. The nanoparticle is assumed to have a diamond lattice structure in the shape of a cube, n unit cells on a side, having a width na, where a is the unit cell dimension. Column 2 gives the size for GaAs which has $a = 0.565$nm. The physical and chemical properties change because of the significant contribution of surface atoms.

Table 14.2 Distribution of surface atoms in the cluster

n	Size (nm)	Total number of atoms	Number of surface atoms	% of atoms on surface
2	1.13	94	48	51.1
3	1.70	279	108	38.7
4	2.26	620	192	31.0
5	2.83	1165	300	25.8
6	3.39	1962	432	22.0
10	5.65	8630	1200	13.9
15	8.48	2.84×10^4	2700	9.5
25	14.1	1.29×10^5	7500	5.8
50	28.3	1.02×10^6	3.0×10^6	2.9
100	56.5	8.06×10^6	1.2×10^7	1.5

The nearest neighbour distance in copper metal as a function of particle size is shown in Fig. 14.3. There is a decrease in the nearest neighbour distance with the decrease in particle size.

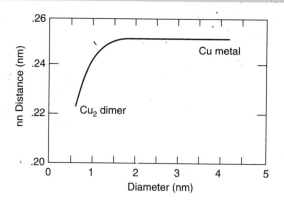

Fig. 14.3 Nearest neighbour distance in copper metal as a function of cluster size

The melting point of gold shows decrease with the decrease of particle size (Fig. 14.4).

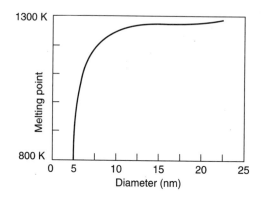

Fig. 14.4 Melting point of gold particles as a function of particle size

Both the Young's modulus and the shear modulus show a decrease with the decrease in particle size (Fig. 14.5). The yield strength (σ) of a metal is found to be function of grain size (d) and is given by the Hall Petch equation

$$\sigma = \sigma_0 + \frac{k}{\sqrt{d}}$$

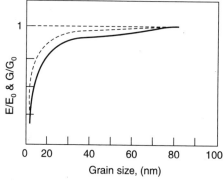

Fig. 14.5 Young's modulus and shear modulus of iron as a function of particle size

The coercive field and the remnant magnetization show a strong dependence on particle size (Figs. 14.6 and 14.7). The magnetic moment of a cluster atoms shows a complex behaviour (Fig. 14.8)

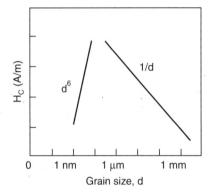

Fig. 14.6 Coercive field as a function of grain size for an alloy of Fe-Nb-Si-B

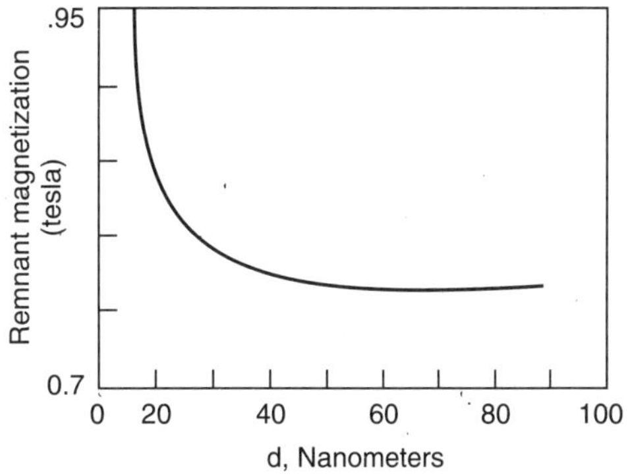

Fig. 14.7 Remnant magnetization as a function of grain size for an alloy of Nd-B-Fe

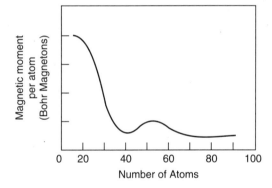

Fig. 14.8 Magnetic moment per atom as a function of number of atoms for rhenium atom clusters

Metals and semiconductors show large changes in optical properties as a function of particle size. The colloidal solution of gold has a deep red colour which becomes yellow with the increase in the size of the gold particle. The optical absorption spectrum of gold particles as a function of particle size is shown in Fig. 14.9. CdSe exhibits photoluminescence. They show large variation in their colour with differing particle size. Figure 14.10 shows the luminescence spectra of CdSe particles as a function of particle size.

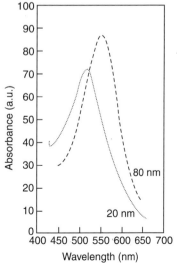

Fig. 14.9 Absorbance of gold particles

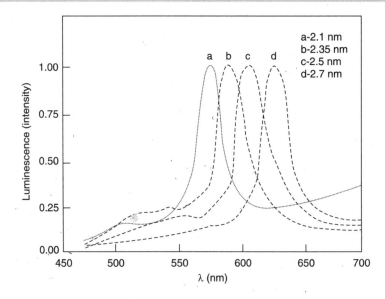

Fig. 14.10 Luminescence intensity as a function of particle size for CdSe

Particle size has drastic effect on phase transitions. Usually there is a shift in the transition temperature. In some ferroelectrics, ferroelectricity disappears below a particular particle size.

The chemical properties of materials depend on the valence electrons and their binding energies. The ionization potential (the energy required to remove an electron from the atom) is found to depend on the cluster size. This radically alters the chemical behaviour of atomic clusters in contrast to the bulk material.

14.5 FABRICATION PROCESSES—TOP DOWN APPROACH

There are two approaches to the fabrication of nanostructures, viz., *top-down approach* and *bottom-up approach*. In the former, atoms, molecules and even nanoparticles are used as building blocks for the creation of complex structures. In the latter, the starting material is in a bulk form. Nanostructures are obtained through careful and controlled removal or division of the bulk material.

14.5.1 Milling

High energy ball milling is of major industrial importance. Coarse grained materials usually metals, but also many ceramics and polymers in the form of powders are crushed mechanically in rotating drums by hard steel and tungsten carbide balls (Fig. 14.11). During the milling processes heat is generated and there is a possibility of unwanted chemical reactions or oxidation taking place. This is usually prevented by using controlled atmospheric conditions. The severe plastic deformation that occurs during the milling process results in nanostructure of the material.

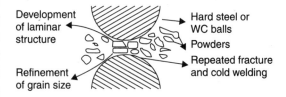

Fig. 14.11 Ball milling technique

14.5.2 Lithography

Lithographic processes using light, electron beam or ion beam is the standard way of fabrication in MEMS (microelectromechanical systems) and microelectronics. This can be extended to NEMS (Nanoelectromechanical systems) as well. In *Lithography* (*in Greek litho means stone and graphien means to write*), an optical image of the desired microstructure pattern is cast on the substrate material coated with a photoresist (photosensitive film). The substrate material can be Si, SiN or SiO_2. The photoresist is exposed to light through a transparent mask which has the desired microstructure pattern (Fig. 14.12 (a)-(c)). Masks are made of quartz. Patterns are photographically reduced from macro or meso sizes to the desired micro sizes. Photoresist materials change their solubility when they are exposed to light. Photoresists that become more soluble on exposure to light are classified as *positive photoresists*. *Negative photoresists* become less soluble on exposure to light. The exposed substrate after development with solvents is shown in Fig. 14.12 (d–f). The retained photoresist materials create the imprinted patterns after development. The portion of the substrate under the shadow of the photoresist is protected from subsequent etching. A permanent pattern is thus created in the substrate after the removal of the photoresist.

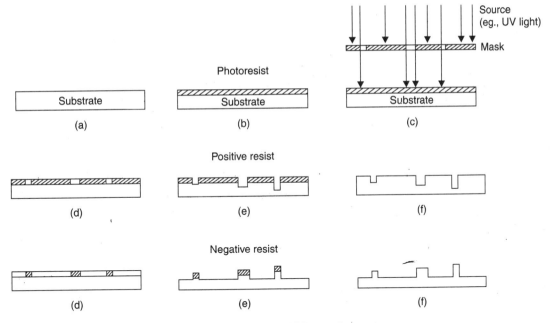

Fig. 14.12 Photolithography

The resolution of the line patterns which are drawn on the mask are determined by the wavelength of light employed. The resolution is roughly equal to half the wavelength of light. With the visible light a resolution of about 200nm is possible. To draw nanostructural patterns obviously the resolution ought to be lower than a nm. This is achieved using electrons, protons or ion beams. The de Broglie wavelengths of these particles when accelerated can be made smaller than a nanometer (Table 14.3). Possible resolution with these beams is roughly half the de Broglie wavelength.

Photoresists are sensitive to light with wavelengths 300–500nm. Photolithography with ultraviolet light ranges from 140–500nm. X-ray lithorgraphy ranges from 0.4–5nm.

Table 14.3 deBroglie wavelengths and velocities of particles

Particle	Kinetic energy	Velocity	$\lambda = h/p$
Electron	1eV	~600km/s	1.2nm
	10eV	~1800km/s	0.4nm
	1keV	~18000km/s	40pm
Proton	1eV	~14km/s	28pm
	10eV	~44km/s	9pm
	1keV	~440km/s	0.9pm
Argon ion	1eV	~2km/s	5fm
	1keV	~30km/s	0.16fm
	1MeV	~2000km/s	0.005fm

14.6 FABRICATION PROCESS—BOTTOM UP APPROACH

Bottom up processes encompass chemical synthesis and highly controlled deposition and growth of materials. Chemical synthesis may be carried out in either solid, liquid or gaseous state. Vapour phase deposition techniques can be used to fabricate thin films, multi-layers, nanotubes, nanofilaments and nanosized particles. The general technique can be classified as either Physical Vapour Deposition (PVD) or Chemcial Vapour deposition (CVD).

14.6.1 Physical Vapour Deposition

The basic steps involved in the process are (a) generation of vapour by boiling or subliming a source material (b) transportation of the vapour from the source to the substrate and (c) condensation of vapour on the cool substrate. Figure 14.13 shows the basic design of the apparatus. In the simplest case, the apparatus consists of an evacuation chamber (~10^{-4} Pa). The chamber is evacuated by an oil diffusion pump equipped with liquid nitrogen trap (to remove water vapour and hydrocarbon contaminants). Starting material usually in the form of a solid powder is placed in a boat made usually from a refractory metal (e.g., Mo or Ta). The boat is resistively heated in the case of solid charges having melting point < 1000°C. The ensuing vapour resulting from the evaporation of the molten solid strikes the substrate positioned above the boat. Vapours condense on the cool substrate and form a thin crystalline film. Typically deposition rates are 0.1–1 μm/s. In order to achieve compositional homogeneity and thickness uniformity, the substrate is often rotated about an axis normal to the plane of the substrate. Figure 14.14 shows the arrangement employed for evaporating multicomponent films. Multifilm deposition can be done sequentially or simultaneously. This is readily accomplished in a multiple source system by opening and closing shutters. After the deposition is complete, an alloy can be formed by elevating the sample temperature and allowing the components to diffuse into each other. When simultaneous deposition of two or more materials is done care must be taken to monitor the composition.

BASICS OF NANOSCIENCE AND NANOTECHNOLOGY

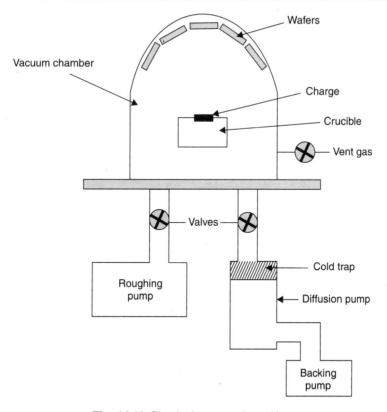

Fig. 14.13 Physical vapour deposition

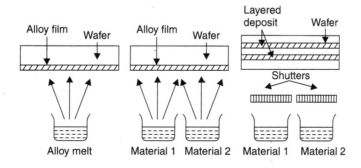

Fig. 14.14 Physical vapour deposition of multicomponent films

14.6.2 Sputtering

This method is used for obtaining thin films of metallic crystals. Sputtering is the process whereby the material in a solid target is ablated by bombardment with energetic ions from an electrical, low pressure plasma struck in a gas. Ejected material from the target in the form of ionized atoms or clusters of atoms pass on to the substrate where the material is deposited. In sputtering the plasma chamber must be arranged such that the high density of ions strike the target containing the material to be deposited. When an energetic ion strikes the surface of the material four things might happen (Fig. 14.15).

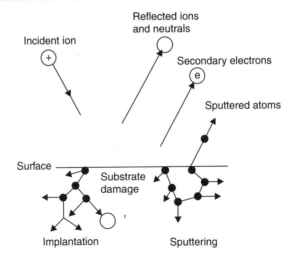

Fig. 14.15 Different processes that occur during sputtering

(a) Ions with low energy may bounce off

(b) Ions having energy < 10 eV may get adsorbed. The energy is transferred to the target atoms increasing their vibrational energy.

(c) Ions having energy > 10 keV, penetrate many layers. They lose energy in the process and the released energy might affect changes in the physical structure. Substrate atoms and clusters of atoms will be ejected from the surface.

(d) For very high energies (i.e., much greater than the energies corresponding to chemical bonds) like those used in implantation, chemical bonding processes can be largely ignored. The target can be considered simply as a collection of atoms. Secondary electrons might be knocked off from the target material.

Figure 14.16 shows the sputtering chamber. A base pressure of ~10^{-4} Pa is maintained by an oil diffusion pump and the sputtering gas (inert gas like argon) is introduced into the chamber at a pressure of ~ 0.1 – 1 Pa. The target material is bonded to one electrode and the substrate to another in a parallel plate capacitor configuration. Usually the target is at the bottom in the sputter mode and loose powders of the material are employed. Sputtering is induced by applying a high d.c. voltage to the target. This attracts the positively charged ions from the plasma struck in the gas which sputter away the target surface. However d.c. sputtering is only feasible for target materials that are metallic.

There are various types of sputtering such as reactive sputtering, magnetron sputtering, and ion beam sputtering. In

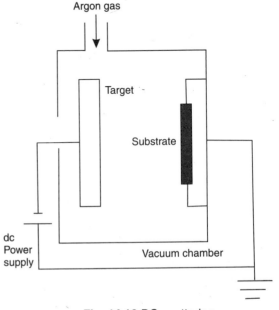

Fig. 14.16 DC sputtering

BASICS OF NANOSCIENCE AND NANOTECHNOLOGY

reactive sputtering, the sputtered atoms react with the substrate yielding the desired material. In magnetron sputtering the sputtered atoms move in the presence of a magnetic field. The magnetic field is used to deflect the electrons from striking the substrate. This helps to prevent the unnecessary heating of the substrate. In ion beam sputtering, an ion beam source is used for sputtering (Fig.14.17).

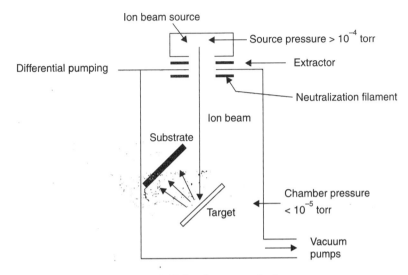

Fig. 14.17 Ion beam sputtering

14.6.3 Laser Ablation

In this technique the solid target material is converted to vapour by focusing laser pulses on the target. The ablated material is deposited on the substrate.

14.6.4 Molecular Beam Epitaxy

A common problem with the simple vapour deposition techniques is the control of composition when the starting material has complex composition as in the case of compound semiconductors. The problem is also severe in case of materials which melt incongruently *i.e.*, composition of the melt and that of the crystal is not the same at the melting temperature. Another cause of compositional variation between the evaporated film and the starting material is that the vapour might contain molecular species that do not preserve the composition of the melt. For example, the stoichiometric compound As_2S_3 for which the vapour is in thermal equilibrium with the melt consists of molecular species As_4S_4 and S_2.

Molecular beam method is a technique that allows precise compositional control. The source emits atoms and molecules (molecular beams) which propagate in straight lines and reach the substrate. The temperature of the source is chosen to obtain the required beam intensity. The beam emerges from very small orifices of an electrically heated Knudsen cell (Fig. 14.18). Knudsen cells are heated enclosures containing the elements required for a molecular beam. The elevated temperature ensures the desired high vapour pressure. Shutters are placed in front of cells to direct the beam at the desired position on the substrate. Each cell is equipped with a computer controlled shutter that allows the composition and the thickness of the growing film to be monitored precisely. Typical MBE thin film growth rates are ~mm per minute or 0.3nm per second. This is equivalent to about one monolayer of atoms per second. Very complicated layer structures such as $Ga_{1-x}Al_x As$ are fabricated. The advantages of MBE include lower growth temperature, better thickness control and interface width.

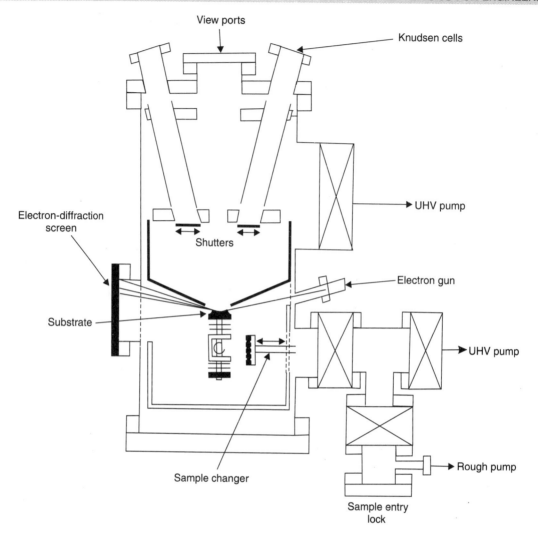

Fig. 14.18 Molecular beam epitaxy

MBE consists of a growth chamber, an auxillary chamber and a load lock (Fig. 14.19). Each chamber has an associated pumping system. The load lock facilitates the installation and removal of samples or substrates without significantly influencing the growth chamber vacuum. The auxillary chamber may contain supplementary surface analytical tools not contained in the growth chamber. These include SIMS (Secondary Ion Mass Spectrometry- to determine the composition of the film; RHEED- Reflection High Energy Electron Diffraction- to determine the crystal structure; XPS-X-ray Photoelectron Spectroscopy to determine the electron structure, ESCA-Electron Spectroscopy for Chemical Analysis for determining the composition of the thin crystal).

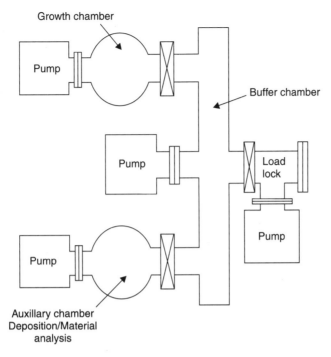

Fig. 14.19 Functional schematic of the basic molecular beam epitaxy

14.6.5 Vapour Phase Expansion

In this method high pressure vapour phase of the material produced in an oven is passed into a low pressure ambient background to produce supersaturation of the vapour. Flow rates can approach supersonic speeds. This process leads to nucleation of atomic clusters. The mass of these clusters can be measured using a mass spectrometer which is linked to the growth apparatus (Fig. 14.20).

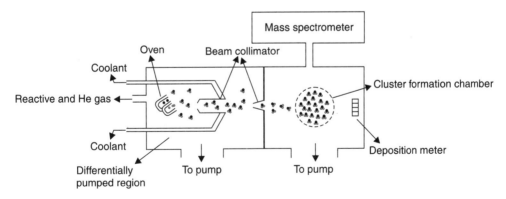

Fig. 14.20 Vapour phase expansion

14.6.6 Vapour Phase Condensation

In this method, the material is evaporated from a temperature controlled crucible into a low pressure, inert gas environment (Fig. 14.21). The metal vapour cools through collisions with inert gas species. It becomes supersaturated and then nucleates homogeneously. The particle size is usually in the range of 1–100nm and can be controlled by varying the inert gas pressure.

Particles can be collected on a cold finger cooled by liquid nitrogen. These particles are scraped, collected in a piston anvil and compacted to produce a dense nanomaterial pellet.

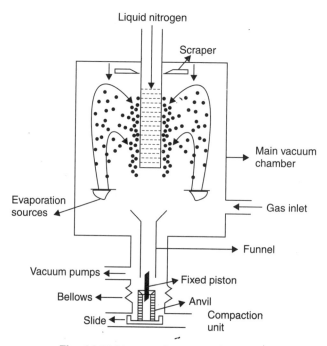

Fig. 14.21 Vapour phase condensation

14.6.7 Chemical Vapour Deposition

Chemical Vapour Deposition (CVD) is the process whereby reaction precursor vapour phase molecular species react, either homogeneously in the gas phase or heterogeneously at the solid-gas interface at the substrate. This produces a composition different from that of the starting materials. The precursor molecules can be made to decompose by means of heat (pyrolysis), absorption of light (photolysis) or in electric plasma formed in the gas. Vapour deposition is sometimes also enhanced by laser. Thermal CVD is the most commonly used method. Some of the typical reactions to grow semiconducting crystals are given below:

$$SiH_4 \text{ (gas)} \longrightarrow Si \text{ (solid)} + H_2 \text{ (gas)}$$

$$Ga(CH_3)_3 \text{(gas)} + AsH_3 \text{(gas)} \longrightarrow GaAs \text{ (solid)} + 3CH_4$$

For silicon dioxide, deposition can be achieved via the reaction of silence at 200–500°C using an N_2 carrier gas *i.e.*,

$$SiH_4 + 2O_2 \longrightarrow SiO_2 + 2H_2O$$

For silicon nitride at 750–850°C, the reaction is

$$3SiH_4 + 4NH_3 \longrightarrow Si_3N_4 + 12H_2$$

For polysilicon a reaction takes place at 500–800°C

$$SiH_4 \longrightarrow Si + 2H_2$$

Figures 14.22 and 14.23 show the apparatus used for CVD with horizontally and vertically mounted substrates. In some cases laser may be used to assist the decomposition of the carrier gas. This is known as Laser Assisted Chemical Vapour Deposition (LACVD). This method has been used to produce nanowires.

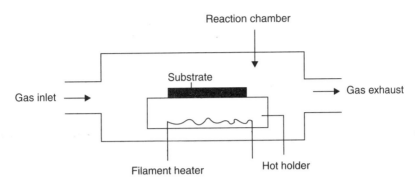

Fig. 14.22 Chemical vapour deposition with substrates mounted horizontal

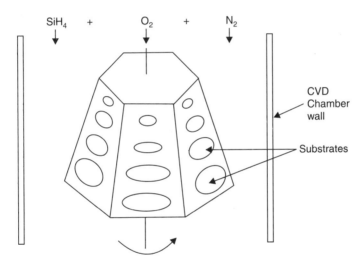

Fig. 14.23 Chemical vapour deposition with substrates mounted vertical

14.6.8 Plasma Enhanced Chemical Vapour Deposition (PECVD)

Energy sources other than heat can be used to initiate molecular deposition in CVD. One obvious alternative source is light photons (typically in the UV region of the electromagnetic spectrum) with energy sufficient to break (or severely weaken) intramolecular chemical bonds. It can cause direct dissociation or facilitate dissociation in association with gas phase collisions. For example, photolysis of disilane (Si_2H_6), leads to high quality films of amorphous hydrogenated silicon, a-Si-H.

Another energy source that can be used to promote CVD is that associated with electrical plasma. In many cases the rates of reaction are affected drastically in the presence of plasma which is source of energetic electrons. Figure 14.24 shows a typical arrangement employed for the purpose. CVD is carried out in the presence of plasma. Plasma is a partially ionized gas. Plasma is characterised by the number density of electrons and their average energy. The glow discharge plasma have usually have energy ~ 1–100 eV and its typical number densities

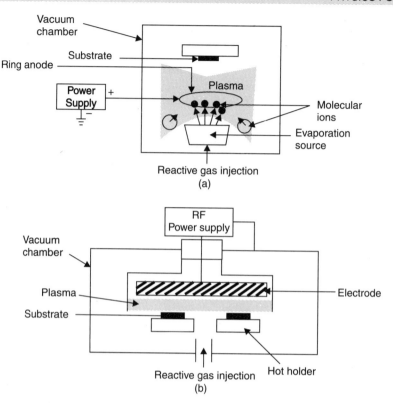

Fig. 14.24 Typical geometries used for plasma enhanced used for plasma enhanced chemical vapour deposition

are $10^{14} - 10^{18}/m^3$. Assuming that the inlet flow of gas contains molecules AB made of atoms A and B, the types of processes that occur in the plasma can be characterized as

Dissociation:	$e^* + AB \rightleftharpoons A + B + e$
Atomic Ionisation:	$e^* + A \rightleftharpoons A^+ + e + e$
Molecular Ionisation:	$e^* + AB \rightleftharpoons AB^+ + e + e$
Atomic Excitation:	$e^* + A \rightleftharpoons A^* + e$
Molecular Excitation:	$e^* + AB \rightleftharpoons AB^* + e$

Here the subscript (*) refers to species whose energy is much larger than the ground state. Dissociated atoms or molecular fragments are called radicals. Radicals have an incomplete bonding and are extremely reactive.

The efficiency of PECVD can be enhanced by the application of magnetic field. Magnetic field can be used to increase the number density of electrons by preventing the ion-electron recombination. This also brings down the effective temperature of the substrate.

14.7 CHEMICAL SYNTHESIS OF NANOPARTICLES

Chemical synthesis provides an economical method for producing large quantities of nanoparticles. In order to avoid coalescence of nanoparticles they need to be isolated from each other using some inert medium. For example nanoparticles can be immobilized in glass, zeolites and polymers etc. which have pores within them (Fig. 14.25).

Fig. 14.25 Atomic cluster assemblies in zeolite $(Na_2, Ca)(Al_2Si_4)O_{12} \cdot 8H_2O$

Nanoparticles can be prevented form coalescing when they are synthesised based on the principles of colloidal chemistry. Colloidal methods are relatively simple and inexpensive. They have been extensively used for the production of metal and semiconductor nanocrystals using optimised reactions and reaction conditions. One problem inherent in many colloidal methods is that colloidal solutions can often age *i.e.,* the particles can grow bigger in size with the passage of time.

14.7.1 Basic Properties of Colloidal Solutions

In a true solution such as one of sugar or salt in water, the solute particles consist of individual molecules or ions. At the other extreme there are *suspensions* in which the particles contain more than one molecule. This agglomeration of molecules can be seen with the naked eye or with the aid of a microscope. Between these extremes are to be found *colloidal dispersions*. In colloidal dispersions particles may contain more than one molecule but are not large enough to be seen in a microscope. Colloidal dispersions pass through most filter papers but can be detected by light scattering and osmosis. By convention, the size of the colloidal dispersion is taken to be 5–200nm.

The unique properties of colloidal systems are due to the fact that the ratio of area to volume is very large. A true solution is a one phase system but a colloidal dispersion behaves like a two phase system. In a colloidal dispersion, for each particle there is a definite surface of separation between the particle and the medium in which the particles are dispersed. The term *dispersed phase* is used to refer to the particles that are present while the medium in which the particles are dispersed is called *dispersion medium*. Both the dispersed phase and the dispersion medium can be in solid, liquid or gas phase. Different types of colloidal systems are shown in Table.14.4.

Table 14.4 Types of colloidal systems

Dispersion medium	*Dispersed phase*	*Name of system*	*Examples*
Gas	Liquid	Aerosol	Fog, mist, clouds
Gas	Solid	Aerosol	Smoke
Liquid	Gas	Foam	Whipped cream
Liquid	Liquid	Emulsion	Milk, mayonnaise
Liquid	Solid	Sol	Gold in water
Solid	Liquid	Gel	Ruby glass, gold in glass
Solid	Gas	Solid foam	Pumice, Styrofoam

14.7.2 Classification of Colloids

Colloids

On the basis of the nature of dispersed phase, colloids have also been classified as *multimolecular colloids*, *macromolecular colloids* and *associated* or *colloidal electrolytes*. Multimolecular colloids are colloidal solutions in which the dispersed phase consists of aggregates of atoms and molecules, each having a diameter less than 1nm. In these aggregates the atoms or molecules are held together by weak van der Waals forces of attraction. Typical examples include gold sol in water and sulphur sol in water which consists of S_8 molecules. *Macromolecular colloids* have macromolecules as the dispersed phase. Starch, cellulose, proteins, synthetic polymers etc. are typical examples. *Associated colloids or colloidal electrolytes* are colloidal solutions in which the colloidal particles behave as normal strong electrolytes at low concentrations. At higher concentrations, they form aggregated particles of colloidal dimensions, called micelles, which exhibit colloidal properties. Soaps and synthetic detergents are examples of this kind.

A further classification of colloids is as *lyophilic* or *solvent-attracting* and lyophobic or *solvent repelling*. If the solvent is water, the term *hydrophilic* and *hydrophobic* is used. An emulsion consists of droplets of one liquid dispersed in another liquid. The droplets are usually from 0.1 to 1 μm in diameter. Thus they are larger than sol particles. Emulsions are generally unstable unless a third substance known as an emulsifying agent or a stabilizing agent is present. Soaps or detergents are effective *emulsifying agents* or *stabilizing agents*, particularly for oil-water emulsions. They consist of long chain of hydrocarbon molecules each having at one end a polar group such as carboxylic acid or sulfonic acid group. These are known as amphiphilic monomers. These molecules are readily adsorbed at the oil-water interface. The polar end prefers to be in water while nonpolar end prefers to be in oil (Fig. 14.26a). Thus amphiphilic monomers when dissolved in water try to stay at water-air interface with head immersed in water and tail standing in air (Fig. 14.26b). On increasing the concentration of monomer of amphiphilic molecules they tend to aggregate forming what are known as micelles (Fig.14.26c). Micelles are spherical aggregate, typically formed by 20–100 monomers. In reverse micelle water is entrapped by amphiphilic molecules (Fig. 14.26d). By choosing appropriate concentrations and conditions, it is not only possible to form micelles, reverse micelles and mixed micelles but also control their size and shape.

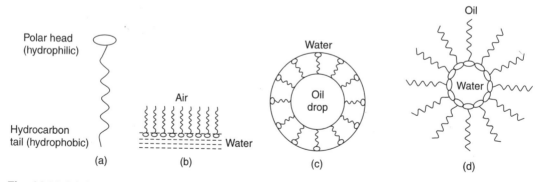

Fig. 14.26 (a) Amphiphilic monomer (b) Monomer is air-water surface (c) Micelle (d) Reverse micelle

Using microemulsions it is possible to entrap the desired nanoparticles in a nanocavity produced by microemulsions. Advantage of using microemulsion lies not only in synthesizing nanoparticles of desired size but also in obtaining biocompatible, biodegradable materials often useful in drug delivery materials to avoid capillary blocking inside the bodies.

A finely dispersed system is in a high free energy state. This is because work has been done to break up the solid. The excess free energy is the energy required to produce the increased surface area. Hence the natural tendency for the colloidal material is to aggregate due to van der Waals forces and lower its energy. However, this process does not take place if there is an energy barrier (Fig. 14.27). The energy of the colloidal solution which is available for aggregation arises from Brownian motion and is typically of the order of k_BT. Agglomeration (the formation of strong compact aggregates of nanoparticles) and also flocculation (the formation of a loose network of particles) can be prevented by increasing the repulsive energy term, which is normally short range. This can be achieved by electrostatic or steric stabilization both of which lead to a repulsive contribution to the potential energy.

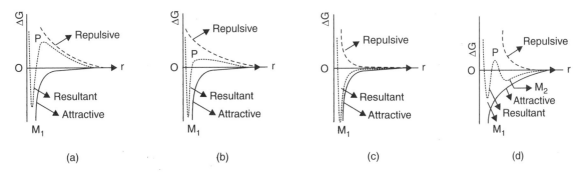

Fig. 14.27 Possible energy balances between attractive and repulsive interparticulate forces as a function of interparticle separation (a) large energy barrier (b) small energy barrier (c) absence of energy barrier (d) barrier permitting weak flocculation of particles

Surface charges on colloidal particles can be easily produced by the ionization of basic or acidic groups when the pH is varied. Stabilizing steric effects can be produced by the attachment of a capping layer to the particle surface. Additional chemicals are added to the colloidal solution which bind to the cluster surface and block vacant coordination sites, thus preventing further growth. These additives can be polymeric surfactants or stabilizers that attach electrostatically to the surface or anionic capping agents which covalently bind to the cluster.

14.7.3 Preparation of Colloidal Solutions

Colloidal solution may be prepared either by coalescing or combining a large number of particles of molecular size into bigger particles of colloidal size (condensation method) or by breaking up the big coarse particles of suspension into fine particles of colloidal solution (disintegration method).

Condensation Methods

By exchange of solvent: Colloidal solution of a substance can be obtained by dissolving the substance in a solvent and then pouring the solution thus obtained into another solvent, in which the substance is insoluble or slightly soluble. Thus when a saturated alcoholic solution of sulphur is poured into water, colloidal solution of sulphur is obtained. Alcohol is then removed by dialysis. P sol may be prepared similarly. Such sols are however unstable.

By change in physical state: Mercury and sulphur sols are produced by allowing their vapours to pass through cold water. Then stabilizing agent (ammonium citrate) is added to the solution.

Controlled condensation: Sols of certain insoluble substances are produced by precipitating them in the presence of some protective colloidal system (*e.g.,* starch, gelatin, glucose, glycerol, etc.). For example, prussian blue sol is obtained by precipitating it in the presence of starch

$$3K_4Fe(CN)_6 + 4FeCl_2 \longrightarrow Fe_4[Fe(CN)_6]_3 \text{ (Prussian blue)} + 12KCl$$

Chemical Methods

By double decomposition: This is the usual way of preparing sols of insoluble inorganic salts. This is based on the fact that when dilute solutions containing the component ions of an insoluble substance are mixed, a precipitate of the substance is obtained. If, however, the precipitated substance has a very low solubility, it is in colloidal state.

Colloidal solution of arsenious sulphide is obtained by passing slowly H_2S gas through a cold dilute solution of As_2O_3 in water (of strength about 2g/litre)

$$As_2O_3 + 3H_2S \longrightarrow As_2S_3 + 3H_2O$$

The excess of H_2S is then removed by passing hydrogen gas through the solution.

Silver halide (*e.g.*, AgCl) sols can be prepared by mixing dilute solutions of silver salts (*e.g.*, $AgNO_3$) and alkali halides (*e.g.*, NaCl) in equivalent amounts.

$$AgNO_3 + NaCl \longrightarrow AgCl \text{ (sol)} + NaNO_3$$

By hydrolysis: Colloidal solutions of many oxides and hydroxides of iron, aluminum, tin, thorium, etc., can be suitably obtained by this method. For example, a deep red colloidal sol of ferric hydroxide is obtained by adding freshly prepared saturated solution of ferric chloride dropwise to vigorously boiling distilled water.

$$FeCl_3 + 3H_2O \longrightarrow Fe(OH)_3 \text{ (sol)} + 3HCl$$

In order to get stable sol, the product is dialyzed rapidly in a parchment bag with running warm water (to free from excess HCl and $FeCl_3$)

By reduction: Colloidal solutions of metals in water are obtained by reducing their salt solutions in water with suitable non-electrolyte reactants (like glucose, thermaldehyde, hydrazine, etc.,) For example ruby red gold sol may be obtained by boiling 3–4 ml of 0.2N K_2CO_3 solution and 1 ml of 1% $AuCl_3$ solution in about 150–200 ml. of distilled water, followed by gradual addition of 1–3ml of 0.3% formaldehyde solution, with continued stirring. K_2CO_3 is added to neutralize the acid formed in order to get stable sol.

$$2AuCl_3 + 3HCHO + 3H_2O \longrightarrow 2Au \text{ (sol)} + 3HCOOH + 6HCl$$

By oxidation: Sols of some non-metals are obtained by this method. For example sulphur sol is best prepared by passing H_2S through a solution of SO_2 in water till the smell of SO_2 nearly removed.

$$2H_2S + SO_2 \longrightarrow 3S \text{ (sol)} + 2H_2O$$

The solution is then boiled with a saturated solution of NaCl (to precipitate sulphur) and then filtered. The precipitate on filter paper is washed with distilled water, till all the chloride ions are removed. On further addition of water, sulphur runs down the filter paper forming a fairly stable colloidal solution of sulphur.

Dispersion or Disintegration Methods

In these methods, large lumps or particles are broken down to a colloidal size in the dispersing phase, by using some appropriate mechanical means. The resulting unstable sols are then stabilized by adding suitable *stabilizers*. The dispersion process may be carried out in any of the following methods

Electrodispersion or Bredig's arc method: Colloidal sols of metals like Au, Ag, Pb, etc., are obtained by the condensation of their vapour. This is accomplished by striking an electric arc between two rods of metal, the ends of which are kept immersed under water in a well cooled vessel containing some stabilizer (Fig. 14.28). The intense local heat of the arc tears off ends of the metal rods to form vapour of metal, which then immediately condense to give particles of colloidal dimensions. The addition of a trace of alkali (KOH) in water stabilizes the colloidal solution so formed.

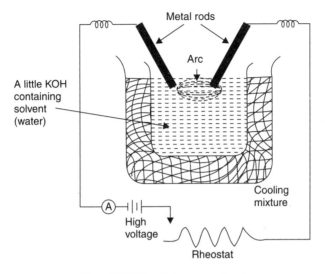

Fig. 14.28 Bredig's arc method

Mechanical dispersion method: Colloidal solutions like black ink, paints, varnishes, ointments, dyestuffs, dental creams etc., are prepared by this method. The substance to be dispersed is first ground as finely as possible by usual methods and then suspended in dispersion medium to get a coarse suspension. The suspension thus obtained is introduced into a colloidal mill, which essentially consists of two metal discs nearly touching each other and rotating in opposite direction at a high speed of about 7000 revolutions per minute (Fig. 14.29). The space between the discs is so-adjusted that coarse suspension gets torn apart to colloidal size. Distance between the discs, actually controls the size of the colloidal particles to be obtained. The colloidal sols are then stabilized by adding a suitable protective colloid like gum-arabic.

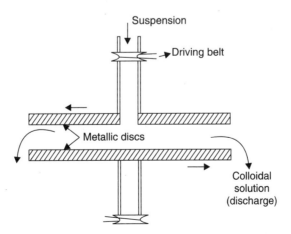

Fig. 14.29 Colloidal mill

Peptization: When to a freshly prepared precipitate, an electrolyte having a common ion (with the precipitate or solid to be dispersed) is added, the former passes into colloidal form. This method of breaking down a precipitate into colloidal sol form by adding an electrolyte solution, is called peptization and the electrolyte used for the purpose is called the peptizing agent.

Freshly precipitated ferric hydroxide can be peptized to a reddish brown sol by the addition of a small quantity of ferric chloride solution (a peptizing agent)

$$Fe(OH)_3(s) + Fe^{3+} \longrightarrow Fe(OH)_3 \cdot Fe^{3+} \text{ (positive sol)}$$

Freshly precipitated AgCl suspension is peptized by shaking it with water containing a little $AgNO_3$.

$$AgCl(s) + Ag^+ \longrightarrow AgCl \cdot Ag^+ \text{ (positive sol)}$$

Although there is no general rule, it is often found that the most effective peptizing agent for a substance is one which contains a common ion.

14.7.4 Sol-Gel Process

This is a good method for obtaining fine nanaoparticles with narrow size distribution and controlled chemical composition at comparatively low temperature. In this process, dispersion of the particles of the metal compound (sol) usually in an aqueous phase is first prepared and then converted into gel. The polymer gel so formed is a three dimensional skeleton surrounding the interconnected pores. Gelation is achieved by means of chemical dehydrating agent with surfactant. 2-ethylhexanol (dehydrating agent) and SPM-80 (surfactant) are commonly used in the process. Typically this involves a hydrolysis reaction followed by condensation or polymerization. In principle, particle size achieved at the precipitation state is maintained during sol and gel formation.

$$Si(OC_2H_5)_4 + C_2H_5OH \longrightarrow (SiO_2)_n \text{ (gel)} + \text{other products}$$

The sol gel process is summarised in Fig. 14.30. Steps 1 to 3 indicate the process upto gelation. While the gelled spheres or collapsed gels (xerogels) can be collected, a better means

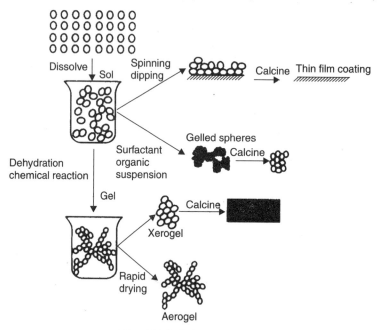

Fig. 14.30 Sol-gel method

of exploiting their surface area is to capture the gel on the surface. This way a greater surface to bulk area ratio is obtained. Another possibility is aerogels. Aerogels are composed of three-dimensional, continuous net works of particles with air (or any other gas) trapped at their intersection. Aerogels are porous and extremely light, yet they can withstand 100 times their weight. Another very clever way of maximizing the surface area is colloidal crystallization on surfaces. In this process, water is very carefully removed so that the sol gel structure is not lost in the precipitate. Nanostructured silica with controlled pore size, shape and ordering can be obtained this way. When surfactants are mixed with water, long range spatially periodic architectures are created in a nanofoam, with lattice parameters in the range of 2 nm to 15 nm.

To obtain nanoparticles in a bulk form the nanoparticles must be consolidated while maintaining the nanosize. The usual goal is to form a high density solid that is free of voids. Sintering is a common technique for consolidating metallic and ceramic materials. The material is first compacted into a low density solid which may contain binders. Next, higher temperatures and sometimes pressures are used to increase the density by the diffusion of vacancies out of the pores. Compacted nanocrystalline metals can have densities which are 96–98% of the bulk.

14.7.5 Pulsed Laser Methods

Pulsed lasers have been used in the synthesis of nanoparticles of silver. Silver nitrate solution and a reducing agent are flowed through a blender line device. In the blender there is a solid disc, which rotates in the solution. The solid disc is subjected to pulses from a laser beam creating hot spots on the surface of the disc. The apparatus is shown in Fig. 14.31. Silver nitrate and the reducing agent react at these hot spots resulting in the formation of small silver particles, which can be separated from the solution using a centrifuge. The size of the particles is controlled by the energy of the laser and the rotation speed of the disc.

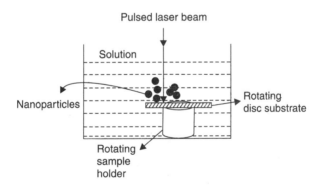

Fig. 14.31 Production of nanoparticles by pulsed laser method

14.7.6 RF Plasma

Figure 14.32 illustrates a method of nanoparticle synthesis that utilizes a plasma generated by RF heating coils. The starting metal is contained in a pestle in an evacuated chamber. The metal is heated above its evaporation point using high voltage RF coils wrapped around the evacuated system in the vicinity of the pestle. Helium gas is then allowed to enter the system forming a high temperature plasma in the region of the coils. The metal vapour nucleates on the He gas atoms and diffuses up to a colder collector and there nanoparticles are formed. The particles are generously passivated by the introduction of some gas such as oxygen. In the case of aluminum nanoparticles the oxygen forms a layer of aluminum oxide about the particle.

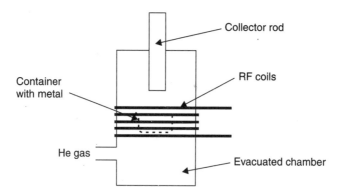

Fig. 14.32 Production of nanoparticles using RF plasma

14.7.7 Thermolysis

Nanoparticles can be made by decomposing solids at high temperature having metal cations and molecular anions or metal organic compounds. The process is called thermolysis. For example, small lithium particles can be made by decomposing lithium azide, LiN_3. The material is placed in an evacuated quartz tube and heated to 400°C in the apparatus shown in Fig. 14.33. At about 370°C the LiN_3 decomposes, releasing N_2 gas, which is observed by an increase in the pressure on the vacuum gauge. In a few minutes the pressure drops back to its original value, indicating that all the N_2 gas has been removed. The remaining lithium atoms coalesce to form small colloidal metal particles. Particles less than 5 nm have been made by this method. Passivation can be achieved by introducing the appropriate gas.

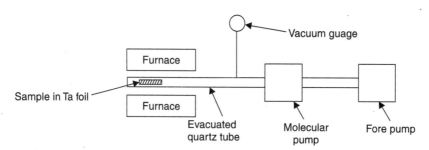

Fig. 14.33 Production of nanoparticles by thermolysis

Nanostructural materials can be made by rapid solidification. This is shown in Fig. 14.34. This is called chill block melt spinning. RF coils are used to melt a metal which is then forced through a nozzle to form liquid stream. This stream is continuously sprayed over the surface of a rotating metal drum under an inert gas atmosphere. The process produces strips or ribbons ranging in thickness from 10 to 100 μm. In another method, called gas atomization, a high velocity inert gas beam impacts a molten metal. The apparatus is shown in Fig. 14.35. A fine dispersion of metal droplets is formed when the metal is impacted by the gas which transfers kinetic energy to the molten metal. This method can be used to produce large quantities of nanostructured powders which are then subjected to hot consolidation to form bulk samples.

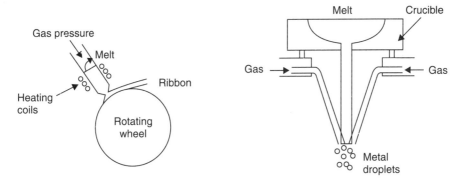

Fig. 14.34 Production of nanometallic films by chill block melt spinning

Fig. 14.35 Production of nanometallic particles by gas atomization

14.7.8 Electrodeposition

Electrodeposition has been used for a long time to make electroplated materials. By carefully controlling the number of electrons transferred, the weight of the material transferred can be determined in accordance with Faraday's laws of electrolysis. According to this law the number of moles of product formed by an electric current is directly proportional to the number of moles of electrons supplied. Thus it is possible to deposit monolayers of atoms by carefully controlling the current and the time of deposition (Fig. 14.36). Using a suitable mask desirable patterns can also be obtained.

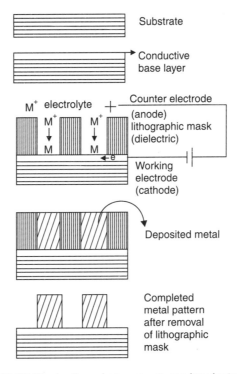

Fig. 14.36 Production of nanostructures by electrolysis

14.7.9 Self Assembly

Self assembly is based on the concept that molecules seek the lowest energy state available to them. The forces involved in self assembly are generally weaker than the bonding forces that hold molecules together. They correspond to weaker aspects of columbic interactions and are found in many places throughout the nature. For example, O—H---O hydrogen bonds in water prevent water molecules from becoming vapour at room temperature. In self assembly, atoms or molecules are deposited on a surface and then molecules align themselves into particular configurations to minimize their total energy. One of the advantages of self assembly is that larger nanostructures can be prepared without manipulating individual nanomolecules or atoms. Some of the examples of self assembled structures are given below:

(a) Nearly perfect monodispersed nanocrystals of CdSe can be formed into three dimensional semiconductor superlattices of quantum dots. Quantum dots are nanosized particles in which the confinement of the electrons and holes shifts the quantised energy levels and thus modifies the optical properties.

(b) Fullerene clusters of C_{60} are found to self assemble into clusters containing specific numbers of C_{60} molecules. These numbers are known as magic numbers.

(c) Stress due to lattice mismatch drives islands to form when germanium is deposited on the (001) surface of silicon. These islands can have shapes which include square based pyramids and domes.

(d) Self assembled monolayers and multilayers of antibodies, enzymes etc., have been prepared on various metallic and inorganic substrates such as Ag, Au, Cu, Ge, Pt, Si, GaAs, SiO_2 and other materials.

14.8 CHARACTERIZATION OF NANOMATERIALS

14.8.1 Introduction

Nanomaterials on account of their atomic and molecular dimensions need special experimental techniques for characterization and measurement of their physical properties. All these techniques are based on years of experience and expertise developed in the field of experimental solid state physics for measurement of properties of materials and their characterization. Microscopy techniques such as scanning electron microscopy (SEM), transmission electron microscopy (TEM), field ion microscopy (FIM), scanning tunneling microscopy (STM) and atomic force microscopy (AFM) are used to measure the surface topography of nanomaterials. Atomic force microscopy is used for nanofabrication as well *i.e.*, to move and place an atom or a molecule on a surface at the desired position. The measurement of mechanical, electrical, optical, thermal and magnetic properties needs sophisticated instrumentation. These topics are beyond the scope of this book. Only the basic experimental tools will be discussed here.

14.8.2 X-ray Scattering

X-ray scattering is used to determine the particle size distribution of nanomaterials. The particle size can be determined from the width of the diffraction peaks using Scherer equation

$$D = \frac{\kappa \lambda}{\beta \cos \theta}$$

where D is the average crystallite size perpendicular to the reflecting planes, λ is the X-ray wavelength, θ is the Bragg angle, β is the width of the peak in radians due to finite size of the crystal and κ is a constant close to unity. The mechanical strain in the lattice also contributes to the width of the Bragg peak. Hence more complicated analysis is needed to determine the particle size.

BASICS OF NANOSCIENCE AND NANOTECHNOLOGY

Particle size distribution can also be determined by studying the angular distribution of elastically scattered intensity of the polarized laser light.

14.8.3 Mass Spectrometry

Nanoparticles could be formed due to clusters of atoms or molecules. The mass of such clusters is determined by mass spectrometry. A typical gas mass spectrometer is shown in Fig.14.37. Nanoparticles are ionized to form positive ions by impact from electrons emitted by the heated filament in an ionization chamber. The newly formed ions are accelerated through a voltage V, then focused by lenses and collimated by slits during their transit to mass analyzer. The magnetic field B of the mass analyzer, orientated normal to the page exerts a force $F = qvB$, which bends the ion beam through an angle of 90° at the radius r, after which the ion beam is detected at the ion collector. The ratio of mass (m) to charge (q) is given by

$$\frac{m}{q} = \frac{B^2 r^2}{2V}$$

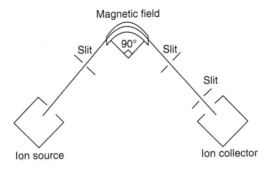

Fig. 14.37 Mass spectrometer

The bending radius r is ordinarily fixed in particular instrument, so either the magnetic field B or the accelerating voltage V can be scanned to focus the ions of various masses at the detector. The charge q of the nanosized ion is ordinarily known and hence mass m can be determined. Some clusters are formed with great probability. The number of atoms corresponding to these clusters is known as magic number. A typical plot for atom clusters is shown in Fig. 14.38.

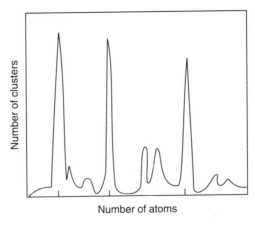

Fig. 14.38 Mass spectrograph showing magic numbers

14.8.4 Field Ion Microscopy

The field ion microscope has resolution of the order of nanometer. This is based on the principle of projecting an image of the object on to the screen as shown in Fig. 14.39a. In a field ion microscope a wire with a fine tip located in a high vacuum chamber is held at a positive potential (between 5 and 20 keV) with respect to a negatively charged conductive fluorescent screen of radius R in a high vacuum and is maintained at cryogenic temperatures. A low pressure inert gas, such as hydrogen or helium is introduced into the chamber. On account of the high electric field, the atoms get ionized. The electrons are attracted by the tip while the positive ions are attracted to the fluorescent screen. On the screen they create luminous spots. In principle each fluorescent point on the screen corresponds to an atom in the tip. Thus the fluorescent image on the screen represents a projected image of the distribution of atoms on the tip.

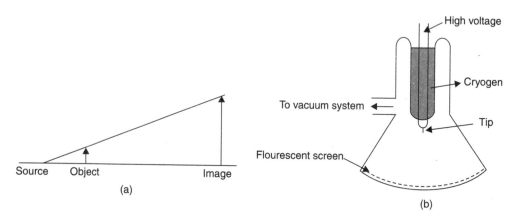

Fig. 14.39 (a) Projection of an image (b) Field ion microscope

14.8.5 Electron Microscopy

Normal eye can resolve fine details ~0.1 – 0.2 mm at the near point vision of 25 cm. The general purpose of microscopy is to resolve fine details that cannot be done with an unaided eye. This is done with the help of an optical microscope. The resolving power of the optical microscope is determined by the wavelength of light. It is roughly equal to half the wavelength of light. Hence the possible resolution is ~200 nm. Electron microscope exploits the wave nature of electron. It is similar to optical microscope. The role of lenses is taken up by electric and magnetic fields which suitable focus the electron beam. There are two types of electron microscope viz., Scanning Electron Microscope (SEM) and Transmission Electron Microscope (TEM). Figure 14.40 shows the construction of an electron microscope. Electrons from a source such as an electron gun enter the sample. They are scattered as they pass through the sample. They are focused by an objective lens and are amplified by the projector lens. According to de Broglie hypothesis, the wavelength of the electron is given by

$$\lambda = \frac{h}{p} = \frac{h}{\sqrt{2m\,e\,V}} = \frac{1.226}{\sqrt{V}}$$ nm where V is the accelerating voltage.

The resolution obtainable with an electron microscope is ~0.2nm.

In the scanning electron microscope, the electron beam scans the surface of the sample. When the electron beam hits the surface secondary electrons are emitted which are collected and amplified. The image is formed on a cathode ray tube. The advantage of SEM is that even thick specimens can be examined.

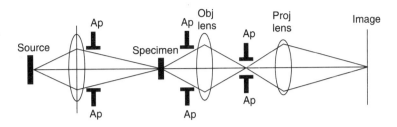

Fig. 14.40 Transmission electron microscope

14.8.6 Scanning Tunneling Microscopy

Scanning tunneling microscope (STM) is based on the quantum mechanical phenomenon of tunneling. It is possible to do real space imaging of surfaces with atomic dimension employing no illumination and no lenses. The experimental arrangement is shown in Fig. 14.41. A sharp conducting tip (often tungsten) acting as the anode is brought close to the surface of the specimen (the cathode). A bias voltage ranging from 1 mV to 1 V is applied between the tip and the sample. Above the specimen surface there is an electron cloud due to surface atoms. When the tip is brought within about 1nm of the sample surface, electrons tunnel across the gap. This causes a current to flow in the circuit. The direction of electron tunneling across the gap depends on the sign of the bias voltage. It is this current that is used to generate an STM image. The tunneling current falls off exponentially with the distance between the tip and the surface. The tip of the sample is scanned laterally using piezoelectric drivers. The STM image reflects the variation in the sample surface topography. Needless to say that the STM ought to be mounted on a good vibration-isolation table. The tunneling current also depends on the atomic species present on the surface and their local chemical environment. STM is designed to operate in constant current mode or constant height mode. In constant height mode the tip travels in a horizontal plane above the sample and the tunneling current varies as a function of the surface topography and the local surface electronic states of the sample. Thus tunneling current constitutes the data set from which the surface topography is reconstructed. In constant current mode, the STM uses a feedback system to keep the tunneling current constant by adjusting the height of the scanner at each measurement point. Thus, for example, when there is an increase in the tunneling current, the tip is moved up away from the sample by adjusting the voltage input to the piezoelectric scanner. Thus the motion of the scanner constitutes the data from which the topography of the surface is reconstructed. STM cannot image insulating materials and that is its disadvantage.

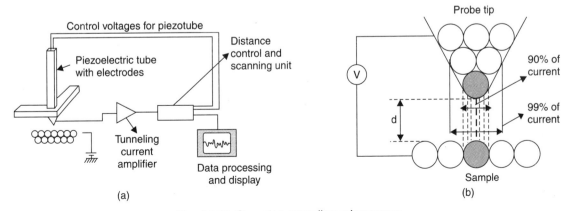

Fig. 14.41 Scanning tunneling microscope

14.8.7 Atomic Force Microscopy

Figure 14.42a shows the construction of an atomic force microscope. Unlike STM, it can be used to probe the surfaces of any material, insulating or otherwise. This also uses a sharp tip about 2μm long and down to a minimum of 20 nm in diameter. The tip is scanned closely over the specimen surface. In AFM the magnitudes of atomic forces rather than tunneling currents are monitored as a function of the probe position on the sample surface. The AFM tip is located at the free end of a cantilever that is 100–200 μm long and forces between the tip and the sample surface cause the cantilever to bend or deflect. A detector is used to measure the cantilever deflection (usually just the displacement in the z-direction although lateral force microscopy is also possible) as the tip is scanned over the sample or alternatively as the sample is scanned under the tip. The measured cantilever deflections allow a computer to generate a map of the surface topography. The position of the cantilever is most commonly detected optically using focused laser beam reflected from the back of the cantilever onto a position sensitive photodetector. Any bending of the cantilever results in a shift in the position of the focused

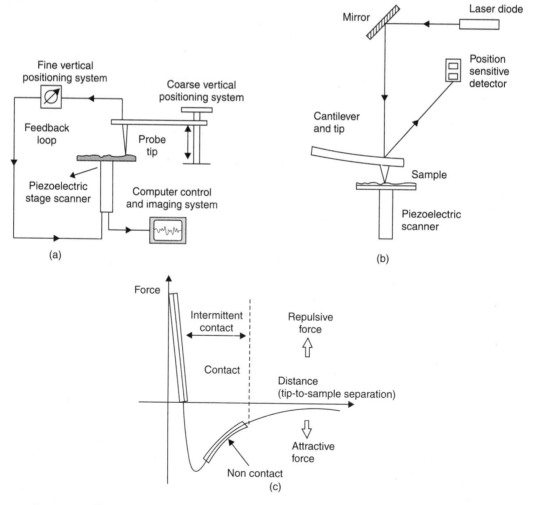

Fig. 14.42 (a) Atomic force microscope (b) Cantilever positioning (c) Atomic force as a function of the distance between the tip and the sample

laser spot on the detector and the PSD can measure such displacements to an accuracy of < 1 nm. AFM is also designed to operate in either constant force mode or constant height mode. In the constant height mode the scanner height is fixed during the scan and the spatial variation of the cantilever deflection is recorded. This forms the data set from which surface topography is constructed. In constant force mode, the deflection of the cantilever is used as an input to a feed back circuit that moves the scanner up and down in the z-direction. Constant force mode data sets are therefore generated form the scanner's motion in the z-direction. The dependence of the atomic force on the distance between the tip and the sample is shown in Fig. 14.42b. When the tip is less than a few angstroms from the sample surface, the interatomic force between the cantilever and the sample is predominantly repulsive, owing to the overlap of electron clouds associated with atoms in the tip with those at the sample surface. This is known as the contact regime. In the non-contact regime, the tip is somewhere between ten to a hundred angstroms from the sample surface and here the interatomic force between the tip and the sample is attractive. This is on account of long range attractive force of attraction on account of van der Waals forces.

AFM is also used for nanofabrication. AFM tips could be used as nano-pens to write nanostructures on a surface. This technique is known as dip pen nanolithography (DPN). The technique is shown in Fig. 14.43. In DPN a reservoir of ink (atoms or molecules) is stored on the top of the scanning probe tip. The tip is moved across the surface leaving molecules behind on the substrate. By suitably moving the tip the desired pattern of molecules can be formed on the surface.

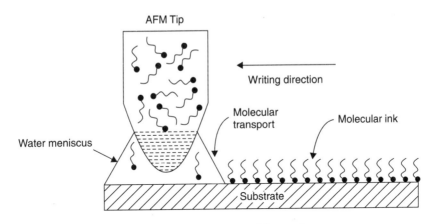

Fig. 14.43 Dip pen nanolithography

14.9 CARBON NANOTUBES

14.9.1 Forms of Carbon

Carbon exists in a large number of allotropic forms. These include diamond, graphite, a variety of discrete molecules such as C_{60}, C_{70} etc., which are collectively known as fullerenes. Diamond structure arises on account of sp^3 hybridisation and is the hardest substance known (Fig. 14.44). Graphite structure arises out of sp^2 hybridisation. It is composed of two dimensional sheets of carbon atoms (Fig. 14.45). Each sheet is a hexagonal net of C atoms. The structure of fullerenes is more complicated (Fig. 14.46).

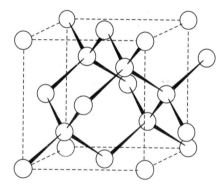

Fig. 14.44 Structure of diamond

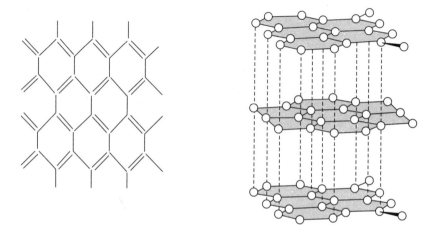

Fig. 14.45 Structure of graphite

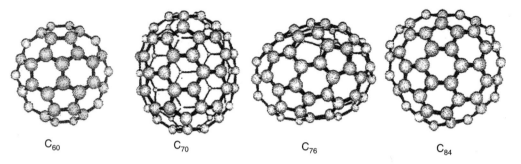

C_{60} C_{70} C_{76} C_{84}

Fig. 14.46 Structure of some fullerenes

14.9.2 Structure

Carbon nanotubes are obtained by rolling up the graphene sheets which occur in graphite. Depending on how the sheet is rolled, they are classified as zigzag, armchair and chiral tubes respectively (Fig. 14.47a–d). The carbon naonotube consisting of one cylindrical graphite sheet is called single walled nanotube (SWNT). Otherwise they are known as multiwalled nanotubes MWNT (Fig. 14.47e).

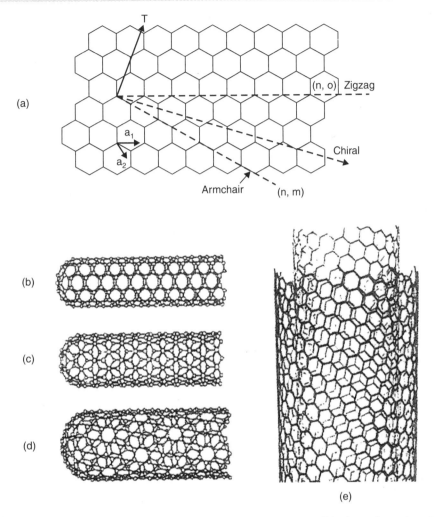

Fig. 14.47 (a) Basis vectors of the two dimensional graphene sheet T is the axis vector about which the sheet is rolled (b) Armchair (c) Zigzag (d) Chiral (e) Multiwall nanotube

14.9.3 Preparation

Carbon nanotubes can be made by chemical vapour deposition, carbon arc methods and laser evaporation. Figure 14.48 shows the chemical vapour deposition method. A hydrocarbon gas

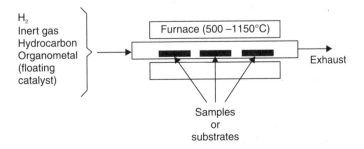

Fig. 14.48 Preparation of CNT by chemical vapour deposition

such as methane (CH_4) is decomposed at 1100°C. As the gas decomposes carbon atoms are produced which condense on a cooler substrate containing various catalysts such as iron. This method produces tubes with open ends which do not occur with other methods. The method is ideally suited for industrial production. Figure 14.49 shows the method for producing carbon nanotubes by carbon arc discharge. A potential of 20–25 V is applied across carbon electrodes 5–20 μm diameter and separated by 1 mm at 500 torr pressure of flowing helium. Carbon atoms are ejected from the positive electrode and form nanotubes on the negative electrode. As the tubes form, the length of the positive electrode decreases, and a carbon deposit is formed on the negative electrode. To produce a single walled nanotube, a small amount of cobalt, nickel or iron is incorporated as a catalyst in the central region of the positive electrode. If no catalysts are used, the tubes are nested or multiwalled types. This method can produce SWNT of diameters 1–5 nm with a length of 1 μm. Figure 14.50 shows the technique of laser ablation to produce carbon nanotubes. The laser beam is used to vaporise a target of a mixture of graphite

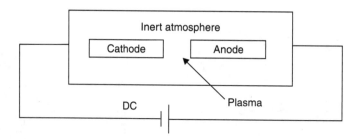

Fig. 14.49 Preparation of CNT by carbon arc discharge

and metal catalyst such as Co or Ni at temperature of approximately 1200°C in a flow of controlled inert gas and pressure. The argon gas sweeps the carbon atoms from the high temperature zone to the colder copper collector on which they condense into nanotubes. Tubes 10–20 nm in diameter and 100 μm long can be made by this method. The mechanism of nanotube growth is not understood since metal catalyst is necessary for the growth of SWNT.

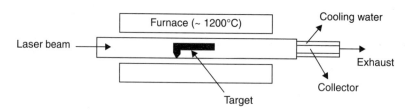

Fig. 14.50 Preparation of CNT by laser ablation

Carbon nanotubes are found in the midst of other carbon materials, such as soot, other forms of amorphous carbon, carbon nanoparticles etc., during production. Basic purification methods that have been used include chromatography, filtration, centrifugation and intercalation techniques.

14.9.4 Properties

Carbon nanotubes are insoluble in water on account of the non-polar nature of bonds in them. They are subject to rules of carbon chemistry. They are not especially reactive. SWNTs exhibit excellent thermal stability in inert atmospheres. They are routinely vacuum annealed at temperatures up to 1200°C. Like graphite they burn when heat treated in air or similar oxidizing environment. Small diameter tubes burn at lower temperatures than large diameter tubes due to difference in strain energy.

Electrical Properties

Carbon nanotubes can be metallic or semiconducting depending on the diameter and chirality of the tube. Chirality refers to how the tubes are rolled with respect to the direction of the T-vector in the graphite plane. A plot of energy gap versus diameter is shown in Fig. 14.51. In the metallic state the conductivity of the nanotubes is very high. It is estimated that they can carry a billion amperes per square centimeter. Copper wire fails at one million amperes per square centimeter because resistive heating melts the wire. The high conductivity arises because there are very few defects to scatter electrons. High currents do not heat the tube.

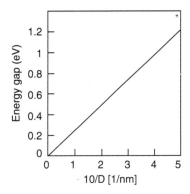

Fig. 14.51 Energy gap as a function of tube diameter

Thermal Properties

Carbon nanotubes have also a very high thermal conductivity, almost a factor of 2 more than that of diamond. Sound propagates through a material primarily via longitudinal acoustic phonons. The thermal conductivity is given by $Cv_s l$ where C is the specific heat, v_s is the sound velocity and l is the mean free path. For SWNT, v_s is ~20 km/s. This anomalously high value stems from the stiffness of the carbn-carbon covalent bond and is directly responsible for the remarkable thermal conductivity of carbon based materials.

Mechanical Properties

Elastic constants and the tensile strength describe the mechanical properties. Mechanical strength is not the same as stiffness. The larger the value of the Young's modulus the less flexible is the material. The longitudinal sound velocity is given by $\left(\dfrac{E}{\rho}\right)^{1/2}$ where E is the Young's modulus and ρ is the density of material. Large value of sound velocity suggests a high value of Young's modulus. The elastic modulus of SWNT is ~1000 GPa. For comparison the moduli of alumina, carbon steel, Kevlar and titanium are approximately 350, 210, 130 and 110 GPa respectively. When carbon tubes are bent, they are very resilient. They buckle like straws but do not break and can be straightened back without any damage. The tensile strength of carbon tubes is about 45 billion pascals. High strength steel alloys break at about 2 billion pascals. Thus nanotubes are about 20 times stronger than steel.

Filled Carbon Nanotubes

A carbon nanotube has one characteristic that virtually no other molecule possesses. It has an interior channel, separated from the exterior environment by an essentially impervious graphene shell. The core of a nanotube filled with some other atom, ion or molecule is expected to generate a new class of one dimensional solids with interesting properties.

Important physical properties of carbon nanotubes are summarised in Table 14.5.

Table 14.5 Physical properties of carbon nanotubes

Properties	*Numerical values*
Typical diameter	1—2 nm
Typical length	100—1000 nm
Intrinsic band gap (metallic/semiconductor)	0 eV—0.5 eV
Work function	~ 5 eV
Resistivity at 300 K	10^{-4}—10^{-3} Ω cm (metallic) 10Ω cm (semiconductor)
Current density	10^7—10^8 A cm^{-2}
Typical field emission current density	10—1000 mA cm^{-2}
Longitudinal sound velocity	~ 20 kms^{-1}
Thermal conductivity at 300 K	20—300 Wm^{-1} K^{-1}
Thermoelectric power at 300 K (bulk sample)	200 mVK^{-1}
Elastic modulus	1000—3000 GPa, harder than steel

14.9.5 Applications

The unusual properties of carbon nanotubes make possible many applications. Some of them are discussed here.

Field Emission

When a small electric field is applied parallel to the axis of a nanotube, electrons are emitted at a very high rate from the ends of the tube. This is called *field emission*. The process of electron emission is governed by the Fowler-Nordheim equation

$$J \propto F^2 \exp\left(-\frac{\phi^{3/2}}{\beta F}\right)$$

where J is the emission current density, F is the applied electric field, β provides a measure of the local field enhancing effect and ϕ is the work function of the material. The work function of the nanotube is ~5 eV, which is not low. However, the molecular sharpness of the nanotube creates a strong local electric field enhancement, promoting field emission. Nanotubes as field emitters are being explored for use in flat panel displays. Nanotube emitters are also being explored to be used as electron sources in portable X-ray generators.

Electronics Applications

A carbon nanotube based field effect transistor and its characteristics are shown in Fig. 14.52. When a small voltage is applied to the gate, the silicon substrate, current flows through the nanotube between the source and the drain. The device is switched on when the current is flowing and off when it is not. A small voltage applied to the gate is found to change the conductivity of the nanotube by a factor $> 1 \times 10^6$, which is comparable to silicon field effect transistors. Carbon nanotube based switches are also being contemplated. Figure 14.53 illustrates the concept. Parallel carbon nanotubes are placed on silicon substrate. The top array and the bottom array are perpendicular to each other and are not in contact. The crossing points represent switches. When the tubes are not touching at the cross points the switch is *off*, otherwise it is on. The on and off configurations can be controlled by passing current through the tubes.

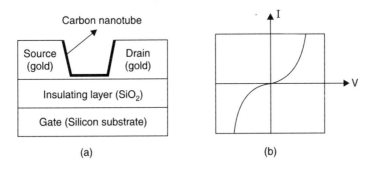

Fig. 14.52 (a) CNT based field effect transistor and its (b) characteristics

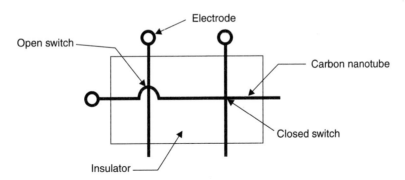

Fig. 14.53 CNT switch arrays on a silicon substrate

Sensor Applications

The field effect transistor based on carbon nanotube is found to be a sensitive detector of various gases such as carbon monoxide, ammonia etc. Nanotubes undergo a large change in electrical conductivity when they are exposed to certain types of gases. This property allows the nanotube to be used as a sensitive, low power device for detecting the presence and concentration of various gases.

Structural Applications

Nanotubes have the ideal combination of mechanical strength and aspect ratio (the ratio of length to diameter) and can be used as fiber reinforcing elements in structural composites. Polymer based composites with carbon nanotubes have been demonstrated to show superior mechanical properties.

Fuel Cells

Carbon nanotubes could be used for storing lithium in lithium based batteries. It is estimated that one lithium atom can be stored for every six carbons of the tube. Nanotubes are also being considered for hydrogen storage applications. A fuel cell consists of two electrodes separated by a special electrolyte that allows hydrogen ions, but not electrons to pass through it. Hydrogen is sent to the anode, where it is ionized. The hydrogen ions diffuse through the electrolyte to the cathode, where electrons, hydrogen, and oxygen combine to form water. The system needs a source of hydrogen. One possibility is to store the hydrogen inside carbon nanotubes. It is estimated that to be useful in this application, the tubes need to hold 6.5% hydrogen by weight. An elegant method to put hydrogen into carbon tube employs the electrochemical cell shown in Fig. 14.54.

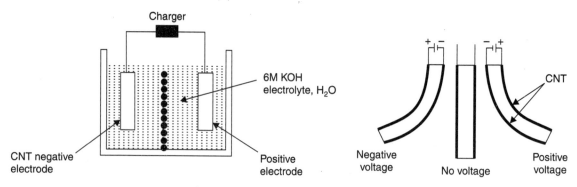

Fig. 14.54 Hydrogen storage in CNT using an electrochemical cell

Fig. 14.55 CNT based actuator

Actuators

Actuators are devices that convert electrical energy into mechanical energy or vice versa. SWNTs deform when they are electrically charged. An actuator based on this property has been demonstrated using single walled carbon nanotube paper. The actuator consists of 3 × 20 mm strips of nanopaper 25–50 µm thick. The two strips are bonded to each other in the manner shown in Fig. 14.55 by double stick scotch tape. An insulating plastic clamp at the upper end supports the paper and holds the electrical contacts in place. The sheets were placed in one molar NaCl solution. Application of a few volts produced deflection of up to a centimeter and could be reversed by changing the polarity of the voltage as shown in Fig. 14.55. Application of an AC voltage produced an oscillation of the cantilever. Such a device has immense potential in NEMS. MWNTs have the potential of being used as structures for nanogears.

Catalysis

Carbon nanotubes are found to be useful as catalytic agents in several chemical reactions.

Thermal Management Systems

Carbon nanotubes are excellent one dimensional thermal conductors. This suggests the use of carbon nanotubes as materials for heat dissipation in many electronic and thermoelectric devices.

REFERENCES

1. M. Ratner and D. Ratner, *Nanotechnology*, Pearson Education, New Delhi, 2003.
2. C.P. Poole Jr and F.J.Owens, *Introduction to Nanotechnology*, Wiley Interscience, 2006.
3. N. P. Mahalik, *Micromanufacturing and Nanotechnology*, Springer, 2006.
4. M. Ventra, S. Evoy and J.R. Heflin, *Introduction to Nanoscale Science and Technology*, Kluwer Academic Publishers, New York, 2004.

QUESTIONS

1. Write a short note on scaling laws.
2. Discuss scaling laws in mechanical, thermal, electrical and electromagnetic systems with suitable examples.
3. Discuss the quantum nature of the nanoworld.
4. Discuss the size dependent physical properties of materials with examples.
5. Describe the various fabrication processes for nanomaterials.
6. Write an essay on lithographic techniques of fabrication.
7. Describe various techniques of physical vapour deposition techniques.
8. Describe various techniques of chemical vapour deposition techniques.
9. Discuss colloidal methods of preparing nanoparticles.
10. Write a short note on self assembly.
11. Explain the working of a scanning electron microscope.
12. Explain the working of a transmission electron microscope.
13. Explain the working of a field ion microscope.
14. Explain the working of a scanning tunneling microscope.
15. Explain the working of an atomic force microscope.
16. Explain the working of a mass spectrometer.
17. Explain how X-ray and light scattering techniques can be used to characterise nanoparticles.
18. Discuss the physical properties of carbon nanotubes in relation to their structure.
19. Discuss the applications of carbon nanotubes.

Index

360° Hologram 320

A

A-scan 132
Absorption 247, 250
    –coefficients 64, 506
    –of sound 51
    –spontaneous emission 250
Acoustic absorption in fluids 103
    –design 69, 70
    –emission 91, 139, 141
    –emission set-up 140
    –hemostatis 146
    –holography 91, 149, 151, 324
    –impedance 49, 95, 96
    –microscope 91, 148
    –surgery 145
Acoustical defects 75
Acoustic emission testing (AET) 139
Acoustics of buildings 44
Acousto-optic devices 91
    –effect 98, 99
    –modulator 151
Active medium 254
Acupuncture 297
Add/drop multiplexer 392
Adjustable reverberation time 83
Advantages of optical fibers 374
Alexandrite laser 260
All-fiber-Mach Zhender interferometric sensor 402
Alloying 292
Ammonia maser 601, 602
Amplitude and phase holograms 314
    –modulation 357
    –resonance 18
    –shift keying (ASK) 393
Analog and digital data transmission 356
Analogue 622

AND gate 652
Angiography 516
Angular offset 350
    –width of the principal maxima 196
Anti-stokes Raman scattering 450
Antiferromagnetic crystal 457
APD 379
Application layer 373
Applications 173
    –of holography 323
    –of Heisenberg's uncertainty principle 472
    –of industrial radiography 514
    –of lasers 288
    –of Michelson's interferometer 173
    –of tunneling 592
Asynchronous transmission 364
Atmospheric pollutant detection 302
    –pollutant detection system 303
Atomic force microscopy 712
Attenuation 347
    –due to absorption 347
    –due to bending losses 349
    –due to coupling losses 350
    –due to scattering 348
    –of ultrasonic waves 100
Average energy photodiode 379, 380, 427
Average energy of the oscillator 421

B

B-scan 133
Bandwidth 20, 21
Band diagram 265
    –of intrinsic semiconductor 264
    –of p-n junction in equilibrium 267
Band edges 612
    –theory 614
Barrier of arbitrary shape 592
    –penetration 580

Basic principles of communication 352
Basics of X-ray scattering 530
Beat length 346
Bell measurements 641
    –state 641
Bell's inequalities 642
Binary semiconductors 540
Biphase-m or bifrequency coding 361
Birefringence 220, 345, 346, 399
Birefringent 219
Birefringent based fiber-optic sensor 404
Black body 414
Black body radiation 412, 413, 414
Bloch sphere 633
Block sphere 653
Body centered cubic 525
Bohr's atomic model 580
Boltzmann constant 425
Bottom-up approach 688, 690
Bow type fibers 347
Bragg condition 451
    –diffraction 98, 99
    –reflection 613
    –reflector 378
    –law 452, 453, 527, 528
Bredig's arc method 703
Bremsstrahlung 502, 503
Brewster's law 218, 219
Bridge 371
Brillouin zone 612, 617
Bus topology 366

C

C-scan 133
Calcite 219, 220
Calculation of diameter of Newton's ring 168
    –of the energy density 421
Calibration 91
Carbon dioxide laser 279
Carbon nanotubes 713, 714, 715, 716
Cardiology 297
Cavitation 136
CCD 436, 438
    –array 438

Characteristic spectrum 501
    –X-ray spectrum 504
    –of the laser light 240
Characterization of nanomaterials 708
Chemical lasers 283
    –synthesis of nanoparticles 698
    –vapour deposition 696, 697
Chromatic and waveguide dispersion 345
    –and waveguide dispersion 345
Circuit switching 369
Circular and elliptic polarization 223
    –polariscope 226
Circulators 374, 385
Cladding 292
Clarity 71
Classical computation 650
Classification of colloids 700
    –of holograms 313
    –of pyrometers 431
CNT based actuator 720
    –field effect transistor 719
    –switch arrays 719
Coaxial cable 353, 354
    –and transmission line 354
Coherence 160, 244
    –length 160
    –time 160
Coherent 392
    –fiber bundle 407
    –fiber optic bundles 407
    –optical communication systems 346
    –or heterodyne optical communication 392
Collision broadening 242, 244
    –damping 16
Colloidal dispersions 699
    –systems 699
Colour centre lasers 255
Colours of thin films 167
Communication protocol 371
Commutation relations 563
Comparison of Bragg and Laue equations 530
Components of a laser 254
Compton effect 446
    –scattering 507
    –shift 447

INDEX

Computation and its complexity 645
Computational complexity 647
Computerised axial tomography 517, 518
Contact EMAT 117
    –method 127
    –Potential 443
Continuous dopplers 145
    –lasers 256
    –spectrum 501
Contrast media 516
Copper and gold vapour lasers 282
COPY gate 652
Copying holograms 320
Coulomb 6
Coummarin dyes 284
Couplers 373, 383
    –and switches 382
Creep 78
Critical angle 329
    –damping constant 8
Critically damped system 9
Cryptographic system 674
Cryptography 628, 673
Crystal structure determination using X-rays 519
Current sensor 403
Cutting 294
Cyclic boundary condition 614
    –redundancy check 360

D

Damped oscillations in a series LCR circuit 26
    –oscillator 15
    –vibration 1
    –vibrations 6
Data communication components 352
    –communication measurement 353
    –link layer 372
    –transmission mode 352
DC sputtering 692
de Broglie wavelengths of electrons 456
    –hypothesis-matter waves 450
Decoherence 665
Definition of hole 618
Delay line 146, 147
Dense wavelength division multiplexing DWDM 363
Density of states 567, 568, 570, 573
Dentistry 297
Dermatology 296
Design of quantum computers 657
Detection of X-rays 508
Determination of Planck's constant 425
Diamond 621
Diffraction 98, 186
    –at a single slit 196
    –grating 199
    –grating-oblique incidence 201
    –of electrons at a single slit 468
    –of light by sound waves 98
    –of sound 49, 50
    –of sound waves 50
    –a qualitative description 196
Diffractometer 519
Diffuse field 47
Diffusion 72
    –or uniformity 72
Digital logic gates 649
    –signals 358
Dip pen nanolithography 713
Direct and indirect band gap semiconductors 265
    –early and reverberation sound 54
    –modulation 381
Directional coupler 383
Directionality 240
Disadvantages of quantum computers 665
Dislocation damping 105
Dispersion 340
    –flattened fibers 351
Displacement sensor 398
Distributed 378
    –bragg reflector 377, 378
    –computation 677
    –feed back laser 377, 378
    –feed back laser and distributed bragg 273
Distribution 673
Doppler broadening 242, 243
Doppler effect 97
    –effect in ultrasound 144
Double heterojunction 270
    –leaf partition 82
Dry friction damping 6

DWDM 392
Dye laser 284, 285

E

Early sound 54
Echelon effect 76
Echocardiography 91
Echoencephalograph 143
Echoencephalography 91, 142
Edge tones 107
Effective mass of electrons 616
Eigen value equation 562
Einstein coefficients 250
Einstein's quantum theory 442
Elastic constants 124
Electrical conduction 614
Electro-optic effect 257, 382
 —materials 382
 —modulator 381
Electrodeposition 707
Electrodispersion 702
Electromagnetic acoustic transducers 116
 —damping 15
Electron diffraction 453
 —gun 210
 —microscope 205
 —microscopy 455, 710
Electron-phonon interactions 106
Electron's existence inside the nucleus 475
Electrons in a periodic potental of a crystal—band theory of crystalline solids 605
Elliptic core fibers 347
EMAT 116, 118
Emissive power 416
Encoding systems 360
Energy bands in solids 619
Energy density 416, 421
 —and intensity of sound 55
 —eigen values 578
 —levels of dye molecule 285
 —of a weakly damped oscillator 10
Entangle photons 644
Entanglement swapping 644, 645
Equation 560
Erbium doped fiber amplifier 262, 386, 387
Erbium–glass laser amplifier 261

Error detection 359
Errors in quantum computation 667
ESCA 445
Evanescent wave 329, 586
 —wave based refractive index sensors 404
 —wave sensor 405
Excimer 282
 —lasers 256, 282
Exclusion principle 413
Expectation values of observations 563
Extrinsic 396
 —absorption 348
 —semiconductors 264
Eyring's formula 58, 61

F

Fabry-Perot 377
Face centered cubic 526
Failure of the classical theory 442
Faraday effect 384, 403
 —rotator 385
FBG 390, 391, 392
Fermi energy for a metal 574
Fermi-dirac 263
 —distribution 264
Ferroelectric 109
Fiber amplifiers and repeaters 385
 —Bragg grating 374, 390, 391
 —brillouin amplifier 389, 390
 —cables 335
 —construction 330
 —dimensions 330
 —fabrication 331
 —laser amplifier 256
 —optic cable 336
 —optic communication system 373
 —optic coupler 586
 —optic gyroscope 402
 —optic radiation thermometer 400, 401, 434
 —optic sensor 346, 586
 —optics 328
 —optics in medicine and industry 407
 —raman amplifier 387
 —sensors 396
 —soliton communication systems 394

INDEX

Fiberscope 407
Field emission 718
    –devices 592
    –ion microscopy 710
Filled carbon nanotubes 718
First critical angle 96
Flexible fiberscope 328, 408
Flexural waves 94
Fluoroscopy 513
Flutter echoes 75
Forced oscillations in a series LCR circuit 27
    –vibrations 1
Fourier 466
Fourier analysis 461
    –hologram 313
    –spectra 461
    –synthesis 461, 462
    –transform 461, 519
Fowler-Nordheim formula 593
FRA 388
Fraunhofer diffraction 187, 191
    –at a circular aperture 202
    –at a single slit 187
    –by a single slit 190
    –due to n parallel slits 193
Free electron lasers 256, 287
Free field 47
    –oscillation in an LC circuit 26
    –particle 563
    –vibration with viscous damping 7
    –vibrations 1, 4
Frequency control 91, 147
    –and calibration 91
    –division multiplexing 362
    –shift keying (FSK) 393
Fresnel diffraction 186
    –number 51
    –zone plate 321, 322
Fresnel's biprism 161
Fringes with white light using the biprism 162
FTNMR 662, 663
Fuel cells 720
Full-duplex 352, 353
Fullerenes 714

G

Gabor zone plate 321, 322
Galton's whistle 107
    –discharge 275
    –dynamic laser 280
    –filled detectors 508
Gas lasers 256, 274
Gateway 371
Gaussian line shape 242
    –wave packet 467
Geiger Muller counters 510
    –tubes 508
Generation and detection of ultrasonic waves 107
Geometrical birefringence 345
    –representation of a qubit 633
GHZ states 642, 643
Graded index fibers 338, 339
Grating spectrum 200
Ground wave propagation 354
Group refractive index 343
    –velocity dispersion 394
    –velocity of waves 460
Grover's algorithm 656
Guided media 353
Gynecology 297

H

Half power points 20, 21
    –value layer 506
Half-duplex 352
Hall Petcht equation 686
Harmonic analysis 461
Hartmann generator 108
Heisenberg's uncertainty principle 466, 471, 474, 557
Helium neon laser 277
Helium-cadmium laser 281
Helmholtz resonator 65, 68, 69
Heterodyne optical communication 392
Hole 619
    –drilling 293
Holes 618
Holey fibers 351

Holographic data storage 326
    –diffraction gratings 326
    –interferometry 323, 324
    –microscopy 324
    –optical elements 326
    –techniques for pattern recognition 325
Holography 150, 310
    –principle 311
Homogeneous broadening 244
Homojunctions and heterojunction lasers 269
Hub 371
Hydrogen storage in CNT 720
Hysteretic damping 6

I

Iintrinsic fiber optic sensor 396
Image intensifiers 513
IME independent Schrödinger 560
Immersion method 127, 128
Impedance matching 96
    –tube method 62
In-line or gabor hologram 313
Indexing 535
Information and entropy 667
    –carrying capacity 361
    –communication channels 668
Infrared camera 436
    –imaging in industry 439
    –imaging in medicine 439
    –imaging in military 439
Infrared imaging systems–IR thermography 434
Infrared pyrometry 429
Inhomogeneous broadening 244
Intensifying screens 514
Intensity modulated sensors 397
Interference 155
    –in newton's rings setup 168
    –in plane parallel films due to reflect 164
    –in plane parallel films due to transmits 165
    –in thin films 163
    –in wedge shaped film 165
Internal friction 52
Interplanar distance 524, 525
    –distances in various crystal systems 527
Intimacy 73

Intrinsic absorption 348
    –semiconductors 263
Introduction 708
Ion beam sputtering 693
    –trap quantum computer 657
    –traps using an optical cavity 659
Ionization chambers, proportional counters 508, 509
Ionospheric propagation 355
Isochromatics 232, 234
    –and isoclinics 231
Isoclinic 232, 234
Isoclinics and isochromatics 233
Isothetic 237

J

Josephson junction 599

K

Kerr effect 381
Kronig-Penney model 605, 606, 622
Kundt's tube 121

L

Lamb waves 92, 94, 96
Laser 240
    –ablation 120, 693
    –amplifier 253
    –amplifier and laser oscillator 252
    –applications in medicine 295
    –assisted chemical vapour deposition (LACVD) 697
    –bar code scanner 301
    –beam focusing 241
    –cooling 658
    –diode 376
    –doppler velocimetry 291
    –fusion 303
    –isotope separation 304
    –isotope separation 305
    –modes 254
    –modes 255
    –optical disc (compact disc) 298
    –oscillator 253

INDEX

–printer 302
Lasers 255
Lateral coherence 244
    –offset 350
Laue equations 528, 529
    –method 533
    –spots 451, 529
Lead zirconate titanate 109
LED 376
Leith-Upatnieks hologram 313
Lensless Fourier hologram 314
LIDAR 289
Light detection using photoelectric effect 443
    –emitting diode 375
    –propagation in fibers 336
Limitations of industrial radiography 514
Line of sight propagation 354
Linear attenuation 506
Linear/non-linear vibration 2
Lintrinsic fiber optic sensor 396
Liquid level sensor 397
    –or dye lasers 255
    –scintillators 512
Lithium drifted silicon detectors 510
Lithography 689
Lithotripsy 91
Liveness 73
Local area network(LAN) 366
    –area networks (LAN) 365
    –area networks (LAN) 365
Locality 641
Logarithmic decrement 12
    –decrement, relaxation time, specific D 11
Long delayed echoes 75
Longitudinal 45
    –coherence 244
    –offset 350
    –or temporal coherence 245
    –waves 92, 93
    –waves, transverse waves 92
Lorentz force 116
Lorentzian and Gaussian line shape function 242
Lorentzian line shape function 242
Low attenuation 350
    –dispersion fibers 350
LPG 390

M

M-scan or motion scan 142
Mach Zhender interferometer 401
Magnetostrictive effect 113
    –materials 114
    –transducers 113, 115
Manchester coding 361
Marking 294
Mass attenuation coefficient for water 507
    –attenuation coefficient of various tissues 515
    –spectrometry 709
Material 6
    –dispersion 340, 344, 345
    –dispersion coefficient 344
    –or chromatic dispersion 342
Materials processing with laser beams 291
Mathematical analysis 233
    –theory of hologram 311
Matter waves 413
Maxwell Boltzmann distribution 248
Maxwellian distribution of velocities 418
MCVD 333
Measurement of absorption coefficient 61
    –of reverberation Time 59, 60
    –of the state of a qubit 634
    –of ultrasonic velocity and attenuation 120
Membrane and panel absorbers 65
    –or panel absorbers 67
Meridional rays 340
Mesh topology 368
Message switching 370
Metal vapour lasers 256
Metropolitan area network (MAN) 366
Michelson interferometer 173, 290
Michelson's interferometer 160, 169 170, 171, 172, 173
Microwaves 354
Miller coding 361
    –indices 519, 520
    –indices for directions in a cubic system 521
    –indices of the planes 522
Milling 688
Minima and secondary maxima 194
Modal dispersion, 340
Mode conversion 96

Models for the black body radiation 416
Modern X-ray diffractometer 534, 535
Modified chemical vapour deposition (MCVD) method 332
Modulation techniques 357
Moire fringes 235, 237
    –patterns 236, 237
    –patterns for 237
Molecular beam epitaxy 693, 694, 695
Monochromaticity 241
Moore's law 629
Moseley's law 504
    –plot 504
Multi quantum well laser diodes 271, 378
Multiple qubit gates 654

N

Nanoscience 680
Nanotechnology 680
Nath diffraction 99
Natural broadening 242
NDT 129
Neodymium laser 259
Network layer 372
    –topologies 366
    –transmission 364, 365
Neutron diffraction 456
Newton's rings 167
Nil capacity channels 678
Nitrogen laser 281
NMR 662
NMR quantum computers 662
No cloning theorem 635
Noise 78, 666
Non-contact EMAT 116, 119, 148
    –thermography 435
Non-radiative transition 541
Non-resonance fluorescence 542
Non-destructive testing (NDT) 126
Non-linear crystal 644
NOT gate 652
NRZ (non return to zero) 361
    –coding 361
Nuclear magnetic resonance quantum computer 660

Number of models in the black body cavity 419
Numerical aperture 336, 337

O

Off-axis 313
    –or Leith-Upatnieks hologram 313
Ohmic contacts 594
Oncology 296
Open wire 353
Operator formalism of Schrödinger equation 562
Operators in Schrödinger representation 562
Ophthalmology 296
Optical amplifiers 373
    –cavity 254
    –communication 306
    –fiber 353, 354, 373
    –fiber gratings 390
    –fibers and cables 330
    –isolators 374, 384
    –modulators 380
    –pyrometry 432
    –receiver 379
    –transmitter 373, 375
Optoisolator 385
Origin of band gaps 612
    –of laue spots 529
    –of X-ray spectrum 502, 503
OSI communication model 371
Otal internal reflection 328
Outside vapour deposition (OVD) method 331
Outside vapour deposition (OVD) technique 332

P

p-i-n photodiode 379, 380, 427
p-n junction 267
Packet switching 370
Pair production 507
PANDA 347
PANDA fibers 347
Panel absorbers 68
Parallax 310
Parallel processing 655
Particle confinement in potential wells 564
    –in a three-dimensional cubical box 571

–in a two-dimensional box of infinite height 568
–nature of X-rays 413
PECVD 698
Peptization 704
Permissible noise levels 78
Phase and group velocity of light waves 465
  –grating 123
  –modulated sensors 401
  –refractive index 343
  –shift keying (PSK) 393
  –velocity 460
  –and group velocity of de broglie waves 464
Phosphors 511
Photoacoustic microscope 152
  –spectroscopy 91, 152
Photoconductive 510
Photoelastic effect 399
Photoelasticity 216, 225, 227
Photoelectric effect 413, 440
Photoelectric emission 507
Photoelectron spectrometer 445, 446
  –spectroscopy 445
Photographic density 316
Photolithography 689
Photoluminescence 541, 687
Photomultiplier tube 444
Photonic band gap 351
  –fibers 351
  –crystal fibers 351
Photovoltaic 510
Physical layer 372
  –properties of carbon nanotubes 718
  –realisation of a qubit 633
  –vapour deposition 690, 691
Piezoelectric element 112
  –transducers 109
Piezoelectricity 109
Piezomagnetism 113
PIN 379
Planck's formula 426
  –law 434
Plane and circular polariscope 225
  –polariscope 226
  –state of stress 230, 231

Plasma activated chemical vapour deposition (PCVD) 332, 333
Plasma enhanced chemical vapour deposition (PECVD) 697
Plate waves 94
PMT 443
Pockels's effect 381, 382
Point and distributed sensors 406
Point to multipoint transmission 364, 365
Point to point transmission 364, 365
Polariscope 232
Polarization by double refraction 217, 219
  –by reflection 217
  –by scattering 217, 222
  –by selective absorption 221
  –dispersion 340, 345
  –maintaining fibers 346
  –of light 216
Polaroid films 222
Polymethane dyes 284
Population inversion 248
Porous absorbers 65, 66
  –materials 65
Potential well of finite height 575
Powder method 533
Power absorbed by a driven oscillator 18
Preparation of CNT by carbon arc discharge 716
  –chemical vapour deposition 715
Preparation of CNT by laser ablation 716
  –of colloidal solutions 701
Presentation layer 373
Pressure, intensity and power in an ultrasonic wave 100
Pressure sensor 398, 399
Principal maxima 194
Principle of complementarity 460
Principles of noise reduction 79
Probabilistic interpretation of wave function 413
Production of entangled states 643
  –of polarized light 217
  –of X-rays 499
Propagation of sound 48
  –of sound reflection and refraction of S 48
Properties of colloidal solutions 699
  –of sound 44

—of ultrasonic waves 92
—of wave functions 560
Proportional counters 509
Public key crypto-systems 674
Pulse amplitude modulation 358
    —code modulation (PCM) 358, 359
    —echo method 122
    —echo system, transmission system 126
    —position modulation 358
    —position modulation (PPM) 358
    —width modulation 358
    —width modulation (PWM) 358
Pulsed doppler 145
    —laser methods 705
    —lasers 256
Pump 254
Pumping 249
Pyroelectric material 429
    —detector 429, 430
    —vidicon 438
Pyrometers 426
Pyrometry 425
PZT 112

Q

Q-factor of a parallel circuit 35
Q-switching 256, 257, 258
QED 412
QKB 673
Quantum algorithms 655
    —circuits 654
    —computation 628, 651
    —computer using semiconductor technology 664
    —computers 628, 630, 665
    —computers with semiconductor technology 663
    —cryptography 672
    —dot 684, 708
Quantum entanglement 640, 641
    —error correction 667
    —gates 651, 652
    —information 628
    —information theory 667
    —key distribution 673, 676
    —model of the atom 413

—network 655
—superdense coding 669, 671
—superposition 630
—teleportation 671, 672
—theory of black body radiation: Planck's Model 422
—turing machine 646
—well 568
—wire 564, 684
Quantum wire 564
Quartz 109, 111, 219, 220
    —transducers, 113
Qubit 633, 636
    —registers 639

R

Radar communication 472
Radiation damping 16
Radiative transition 541
Radio wave propagation 354
Radioactive decay 603
Radioactivity 413
Rainbow hologram 319
Raman scattering 450
Raman-Nath diffraction 98, 123
    —diffraction 99
Range-gated pulsed doppler: 145
Rayleigh 96
    —scattering 348, 388, 450
    —waves 92, 93, 94
    —jeans formula 423
    —model 419
    —criterion for resolving power 203
Real time radiography 514
Reality 641
Receiver 373
Reflection 94
    —and refraction of sound 48
    —and refraction of ultrasonic waves 94
    —and transmission coefficient 96
    —coefficient 585
Reflection hologram 318
Refraction 94
Regenerator 386
Relativistic wave equation for the electron 413

INDEX

Relaxation time 13, 11
Repeater 371, 386
Representation of polarized and unpolarized light 216
Resistive damping 15
Resolving power 455
    –power of a grating 201
    –power of a microscope 203, 204
Resonance curve and Q-factor 30
    –fluorescence 542
    –in a parallel LCR circuit 34
    –in a series LCR circuit—phasor analysis 28
    –quality factor and bandwidth 20
    –system 126, 127
Resonant tunneling 597
    –tunneling diodes 598
Reverberation chamber 61, 83
    –sound 54, 72
    –time 54, 57
    –Sabine's formula 55
RF plasma 705
Riemannian geometry 630
Ring topology 367
Room acoustics 44, 53
    –resonances 75
    –shapes 70
Rotating mirror method 257
Rotational Raman scattering 388
Router 371
Ruby laser 258
RZ 361

S

Sabine's formula 55, 61
Satellite transmission 355
SAW 91
    –devices 148
Scaling in electricity 682
    –laws 680
    –laws for rigid dynamics 681
    –laws for some classical systems 683
    –laws for thermal systems 683
    –laws in electrostatic forces 681
    –laws in rigid dynamics 680
    –of electromagnetic forces 682

Scanning acoustic microscope 149, 150
    –electron microscope (SEM) 710
    –pulse-echo acoustic microscope 150
    –tunneling microscope 594
    –tunneling microscopy 711
Scattering powers of atoms 537
SC, BCC and FCC lattices 525
Schrödinger wave equation 558
Scintillation setectors 511
    –detectors 508
Scintillator dyes 284
Second critical angle 96
    –harmonic generation 260
Selective absorption 77
Self assembly 708
    –phase modulation 394
Semiconductor based detectors 508
    –detector 511
    –detectors 510
    –lasers 262
    –or injection lasers 255
Sensors 328
Session layer 373
Shammou's theorem 472
Sharpness of resonance 22
Shear 94
    –waves 94
Shielded twisted pair 354
SHM 3
Shockwave lithotripsy 145
Shor's algorithm 656
Signal distortion 340
Significance of uncertainty principle 476
Simple cubic 525
Simple harmonic motion 2
Simple heterojunction 269
Simplex 352
    –half-duplex 352
Simulation algorithms 656
Single 82
    –particle interference 459, 630, 632
    –particle quantum interference 631
Sinusoidal zone plate 321
Siren 108, 109
Size dependent properties 685
    –of an atom 474

Skew rays 340
    –and meridional rays 340
Snell's law 219
Sodium iodide 511
Sol-gel process 704
Soldering and welding 293
Solid 6
    –state lasers 258
    –state 255
Sonar 91
Sonoluminescence 91, 136
Sound absorbers 65
    –absorption mechanisms 52
    –concentration 75
    –distortion 77
    –focusing 76
    –reinforcement systems 82
    –shadows 76
Spatial coherence 160, 244, 246
    –coherence and Young's double slit experiment 247
Special X-ray techniques in medicine 516
Specific damping capacity, loss coefficient 13, 11
Spectral linewidth 475
    –wavelength sensitive sensors (fiber Bragg grating sensors 405
Spectrum analysis 461
    –of synchrotron light source 506
    –of the black body 415
Specular reflection 94
Spontaneous emission 247, 250
    –Raman scattering 387
Sputtering 691, 692
Square barrier 587
    –potential 587
Squid 599, 600
SRS 388
Standing waves 612
    –in the impedance tube 63
Star coupler 384
    –topology 367
Statistical interpretation of matter waves 451
Steady state response of an oscillator 14
Stefan's law 416, 424
Step-index fibers 338
    –barrier 581
    –potential 581

Stereograph 310
Stern Gerlach experiment 636, 637, 638
Stimulated emission 247, 250
    –Raman scattering 388
Stokes Raman scattering 450
    –Raman scattering 388
    –shift 543
Strain at a point 227
Stress 227, 229
Stress and strain at a point 227
    –induced birefringence 345
    –optic law for three dimensions 234
    –optic relations–two dimensional case 230
Structural relaxation losses 104
Structure 714
Superposition of waves 155
    –of constant phase difference 156
Superposition of waves of different frequencies 157
    –of equal phase and frequency 155
    –of random phase differences 157
Supersonics 91
Surface acoustic waves 91
    –hardening 293
    –wave techniques 129
Surface waves 93, 94
Switched transmission 364, 365, 369
Switches 383
Switching 368
Symmetric and anti-symmetric wave functions 577
Synchronous transmission 364
Synchrotron light source 505
    –radiation 505
System 126
Systematic absences 537

T

t 328
T-coupler 383, 384
Targeted drug delivery 146
Temperature sensor 400
Temporal coherence 160, 244
Temporal coherence and Young's double slit experiment 246
TGS 438
The argon ion laser 278

INDEX

The positions of maxima and minima 189
The Q-factor of a series LCR circuit and
    selectivity 32
Their remedies 75
Thermal conduction loss 103
    –losses 53, 102
Thermal detectors 428
    –diffuse scattering 539
    –imaging 435
    –relaxation losses 102, 104
    –relaxation losses due to various
        molecular 102
Thermistors 428
Thermodynamics of black body radiation–Wien's
    law 417
    –of computation 648
Thermoelastic effect 105
Thermograms 435
Thermolysis 706
Thermopile 428
Thick and thin holograms 317
Thickness of a plate using Fresnel's biprism 163
    –of a thin sheet of transparent material 162
Thomson scattering 450
Three dimensional grating 519
Time dependent Schrödinger's equation 560, 563
Time division multiplexing (TDM) 362, 363
Time independent Schrödinger wave equation 559
Titanium-Sapphire laser 261
Toffoli gate 649, 650, 654
Tomography 517, 518
Top-down approach 688
Total internal reflection 328, 329
Total radiation pyrometer 432
    –pyrometry 431
Tourmaline 221
Townsend discharge 275
Transient vibrations 2
Transition probabilities 248
Transmission and reflection coefficients 583, 591
    –coefficient 582, 589
    –electron microscope (TEM) 710
    –holograms 317
    –media 353
    –system 129
Transport layer 372

Transverse 94
    –electric atmospheric (TEA) CO_2 laser 280
    –excited atmospheric pressure (TEA)
        configuration 279
    –waves 45, 92
Tree topology 368
Tunable lasers 256
Tunnel diode or Esaki diode 596
Tunneling 580, 684, 685
    –of photons 329, 586
Turing machine 645, 646, 647
Twisted pair 354
    –coaxial cable 353
Two-colour optical pyrometry 433
Two-slit diffraction 191
Types of communication systems 364
    –interference 160
    –optical fibers 337
    –ultrasonic waves 92
Typical semiconductor laser structures 269

U

Ultrasonic attenuation in solids 105
    –mechanisms 101
    –cavitation 91, 135
Ultrasonic cleaning 91, 135, 137
    –diathermy 91, 145
    –drilling 134
    –emulsification 91
    –emulsifier 138
    –emulsifier or homogeniser 137
    –imaging 112, 129, 130
    –soldering 91
    –testing methods 127
    –testing systems 126
    –transducer 111
    –transducer arrays 129
    –welding 91, 134
Ultrasonics 91
    –in electronics 146
    –in optics 148
Ultrasound 91
    –in medicine 141
    –in ophthalmology 143
    –production by laser pulsing 119

Uncertainty principle 413
  —as a consequence of observation 469
Undamped oscillator 14
  —vibration 1
  —system 8
Underwater sonar 124
Unguided media 354
Unshielded twisted pair 354
UPS 445
Urology 297
Use of grids 516

V

Vander Lugt filter 326
Vapour phase condensation 695, 696
  —phase expansion 695
  —axial deposition (VAD) 333
Various accelerating voltages 502
Velocity 616
  —resonance 18
Vertical cavity surface emitting lasers 273
Vibration absorbers 24
  —analysis and control 25
  —isolation 22
Vibrational Raman scattering 388
Vibrations 1
Viscous damping 6
Viscous losses 52, 102
Visualization of X-rays 513

W

Warmth 74
Wave guide dispersion 344
  —packet 461, 463, 464
  —packets in time domain 467
  —properties of electrons 413
Waveguide dispersion 340, 344
Wavelength division multiplexing 363
WDM 390
Wide area network (WAN) 366
Wien's displacement law 415, 424
Wien's law 423

X

X-ray diagnostics with still picture 515
  —diffraction 451, 519, 538
  —diffraction from cubic crystals 535
  —films 514
  —fluorescence 541
  —fluorescence spectrometer 543
  —fluoroscopy 513
  —lasers 256, 286
  —radiography 512, 519
  —radiography in medicine 515
  —scattering 708
  —scattering by a single electron 530
  —scattering by an atom 532
  —scattering from a crystal 533
  —spectrometers 509
  —spectrum 501, 502
  —techniques in materials science 538
  —therapy 519
  —tomography 517
  —topograph 540
  —tube 500
  —tube with a rotating anode 500
X-rays 413, 499
Xanthene dyes 284
XPS 445
XRF 541, 543
XRF spectrometry 543

Y

Young's double experiment 631
  —double experime vith neutrons 457
  —double slit experiment 158, 159, 160, 161
  —experiment with neutrons 459

Z

Zener diode 596
Zero dispersion fibers 351
  —dispersion wavelength 345
  —point energy of the harmonic oscillator 474
Zone plate model of transmission holograms 321